Table of Values

Quantity	Symbol	Value	Units SI	Units CGS
Velocity of light	c	2.997925	10^8 m s^{-1}	10^{10} cm s^{-1}
Proton charge	e	1.60219	10^{-19} C	—
		4.80325	—	10^{-10} esu
Planck's constant	h	6.62620	10^{-34} J s	10^{-27} erg s
	$\hbar = h/2\pi$	1.05459	10^{-34} J s	10^{-27} erg s
Avogadro's number	N	6.02217×10^{23} mole^{-1}	—	—
Atomic mass unit	amu	1.66053	10^{-27} kg	10^{-24} g
Electron rest mass	m	9.10956	10^{-31} kg	10^{-28} g
Proton rest mass	M_p	1.67261	10^{-27} kg	10^{-24} g
Proton mass/electron mass	M_p/m	1836.11	—	—
Reciprocal fine structure constant	a^{-1}	137.036	—	—
Classical electron radius	r_0	2.81794	10^{-15} m	10^{-13} cm
Electron Compton wavelength	λ_e	3.86159	10^{-13} m	10^{-11} cm
Bohr radius	a_H	5.29177	10^{-11} m	10^{-9} cm
Bohr magneton	μ_B	9.27410	10^{-24} J T^{-1}	10^{-21} erg G^{-1}
1 electron volt	eV	1.60219	10^{-19} J	10^{-12} erg
		2.41797×10^{14} Hz	—	—
		8.06546	10^5 m^{-1}	10^3 cm^{-1}
		1.16048×10^4 °K	—	—
Rydberg constant	R_∞	2.17991	10^{-18} J	10^{-11} erg
		13.6058 eV		
Boltzmann constant	k_B	1.38062	10^{-23} J deg^{-1}	10^{-16} erg deg^{-1}
Permittivity of free space	ϵ_0	—	$10^7/4\pi c^2$	1
Permeability of free space	μ_0	—	$4\pi \times 10^{-7}$	1

Source: B. N. Taylor, W. H. Parker, and D. N. Langenberg, Rev. Mod. Phys. **41**, 375 (1969).

INTRODUCTION TO SOLID STATE PHYSICS

CHARLES KITTEL

Introduction
to
Solid State
Physics

FOURTH EDITION

John Wiley & Sons, Inc., New York, London, Sydney, Toronto

Preface to the Fourth Edition

This volume gives an elementary account of central aspects of the physics of solids. The volume was written as a textbook in solid state physics and materials science for senior undergraduate and beginning graduate students of science and engineering. The necessary background is a course in modern atomic physics.

Solid state physics is largely concerned with the remarkable properties exhibited by atoms and molecules because of their association and regular periodic arrangements in crystals. These properties may be understood in terms of simple models of solids. Real solids and amorphous solids are more complicated, but the power and utility of the simple models can hardly be overestimated.

About one-half of the third edition was rewritten to produce the fourth edition, and one hundred forty new illustrations were added. The major changes are:

1. The International System of Units (SI) is introduced in parallel with the CGS-Gaussian System of Units, and the text proper, originally written in the CGS-Gaussian system, is largely bilingual in both systems. The limitations are discussed below. The International System is essentially the same as the MKSA System.

2. New topics, some brief or in problems, include solid state lasers, Josephson junctions, flux quantization, the Mott transition, the Fermi liquid, Zener tunneling, the Kondo effect, helicons, and applications of magnetic resonance. The dielectric function is introduced as a unifying theme in the treatment of electromagnetic propagation, optical modes, plasmons, screening, and polaritons.

3. The chapters on crystal diffraction, energy bands, superconductivity, and magnetic resonance were largely rewritten. A continuous effort was made to produce a clear, intelligible, and well-illustrated text directed to students' needs. I tried to act on any difficulty that any student brought to me.

4. The tables of values of solid state data were considerably expanded and revised. Forty tables of wide application are listed separately in the contents.

The great advances in energy band studies, superconductivity, magnetic resonance, and neutron scattering methods are reflected in the text, as they were in the third edition. Emphasis is given to elementary excitations: phonons, plasmons, polarons, magnons, and excitons.

Nearly every important equation is repeated in SI and in CGS-Gaussian units, wherever these differ. Exceptions to this rule are the figure legends, the Advanced Topics, the chapter summaries, and any long section of text where a single indicated substitution (as of 1 for c or $1/4\pi\epsilon_0$ for 1) will suffice to translate from CGS to SI. Tables are given in conventional units. The contents pages of several chapters include remarks on conventions adopted to make parallel usage simple and natural. Problems should be solved in whichever system is desired by the reader or the teacher.

I have added a few more elements of history, because solid state physics offers the most direct and successful applications of quantum theory to the natural world around us. But aware of the fragmentary record, I thought often of the lines of Jorge Luis Borges: "What is good no longer belongs to anyone . . . but to the language or to tradition."

The sequence of chapters makes it easy to select material for a one semester course. This might include much of the material in Chapters 1 to 11, with additional topics from the later chapters or elsewhere. The selection of subjects for the later chapters should not be seen as an attempt to weigh the importance of the areas of current activity. A single textbook cannot represent the range of current creative activity. The articles in the excellent Seitz-Turnbull-Ehrenreich series should be consulted for subjects not treated in this book and for detailed bibliographies. There are in the literature perhaps ten thousand articles of high quality that usefully could be cited. I have tried to give a helpful, but small, sample of the ones most accessible in English. The translations of earlier editions of this book in French, German, Spanish, Japanese, Russian, Polish, Hungarian, and Arabic often give further references in these languages.

Problems of appreciable length or difficulty are marked by an asterisk. The symbol e denotes the charge on the proton: $e = +4.80 \times 10^{-10}$ esu $= 1.60 \times 10^{-19}$ C. The notation (18) refers to equation number 18 of the current chapter, but (3.18) refers to equation 18 of Chapter 3; figures are referred to in the same way. A caret or "hat" over a vector, as in $\hat{\mathbf{k}}$, denotes a unit vector.

The preparation of this edition was made possible by the cooperation of many colleagues and friends. I would like to mention some of the contributors; there is not space to list all. My new debts include those to T. Nagamiya, V. Heine, D. F. Holcomb, C. H. Townes, A. R. Verma, U. Essmann, H. Träuble, P. W. Montgomery, K. Gschneidner, Jr., H. P. R. Frederikse, W. Känzig, F. A. E. Engel, C. Quate, C. Y. Fong, P. G. Angott, G. Thomas, R. W. De Blois, H. E. Stanley, S. Geller, R. Cahn, R. Gray, P. Richards, and L. Falicov. Parts of the manuscript were reviewed by W. L. McLean and Margaret Geller, and the whole manuscript was reviewed by Robert Kleinberg. I am not certain if Mrs. Madeline Moore could have written this edition without me, but I could not have written it without her. Robert Goff is responsible for the effective and attractive design of the book.

The preface to the third edition included the following statement of acknowledgments.

"The entire manuscript was strongly influenced by detailed criticism by Marvin L. Cohen and Michael Millman; my debt to them is indeed great. Successive drafts were kindly read by Ching Yao Fong and Joseph Ryus, and the problems were checked by Leonard Sander. Individual chapters were reviewed by Adolf Pabst, Charles S. Smith, David Templeton, Raymond Bowers, Sidney Abrahams, Earl Parker, G. Thomas, and M. Tinkham. Walter Marshall very kindly prepared an extensive selection of neutron diffraction results. The preparation of the historical introduction was assisted by Adolf Pabst, P. P. Ewald, Elizabeth Huff, Muriel Kittel, Georgianne Titus, and the physics librarian of the Ecole Normale Supérieure. For their experienced advice in the selection of experimental data for tables of values I am grateful to Leo Brewer, R. M. Bozorth, Norman Phillips, Bernd Matthias, Vera Compton, M. Tinkham, Charles S. Smith, E. Burstein, F. P. Jona and S. Strässler. The illustrations were developed in their final form by Felix Cooper, with early help by Ellis Myers. Credits are given with the individual photographs and figures; exceptional help in their collection was given by Robert van Nordstrand, T. Geballe, W. Parrish, Betsy Burleson, I. M. Templeton, and G. Thomas; also by H. McSkimin, H. J. Williams, R. W. De Blois, E. L. Hahn, A. von Hippel, B. N. Brockhouse, R. C. Miller, R. C. Le Craw, E. W. Müller, P. R. Swann, G. E. Bacon, G. M. Gordon, and Alan Holden."

I am grateful to Yvonne Tsang, Thomas Bergstesser, and Philip Allen for their permission to use the photograph that appears on the jacket.

Berkeley, California   *C. Kittel*

Note to the Reader

Chapters 1 and 2 on crystal structure analysis are fundamental. Every concept developed in Chapter 2 is exploited heavily in the chapters on energy bands and semiconductors; this is particularly true of the reciprocal lattice and of Brillouin zones. The general method developed in Advanced Topic A for x-ray diffraction is repeated in Chapter 9 as the basis of the theory of electron energy bands. Chapter 4 may be omitted on a first reading. Chapters 5 and 6 are concerned with the velocity, quantization, and interaction of elastic waves in crystals; among the topics used later are the enumeration of modes in a Brillouin zone and the number of modes per unit frequency range.

Chapters 7, 8, 9, and 10 are devoted to electrons in metals. Chapters 9 and 10 on energy bands are the most important chapters in the book; the line of development is somewhat new in a textbook, but reflects the attitude of current research in the field. The proof of the Bloch theorem is central for the understanding of the chapters. The properties of holes are discussed with particular care as preparation for Chapter 11 on semiconductors.

Chapter 12 on superconductivity gives the essential experimental facts as viewed in the light of the BCS theory, but at this level it is not possible to give a meaningful derivation of the theory; the author's *Quantum theory of solids* or the book by Ziman may be consulted. Chapters 13 to 17 are devoted to the dielectric and magnetic properties of solids. Chapter 18 is on excitons and optical properties; it contains a discussion of solid state lasers. The final two chapters (19, 20) are concerned largely with imperfections in solids and may be read at any convenient stage.

Contents

Advanced Topics

Guide to Major Tables

Some General References

Atomic physics background

Max Born, *Atomic physics,* Hafner, New York, 7th ed., 1962.

R. L. Sproull, *Modern physics,* Wiley, 2nd ed., 1963.

Statistical physics background

C. Kittel, *Thermal physics,* Wiley, 1969. Cited as *TP.*

Crystallography

F. C. Phillips, *An introduction to crystallography,* Wiley, 3rd ed., 1963.

J. F. Nye, *Physical properties of crystals: their representation by tensors and matrices,* Oxford, 1957.

Problem collection

H. J. Goldsmid, ed., *Problems in solid state physics,* Academic Press, 1968.

Advanced series

F. Seitz, D. Turnbull and H. Ehrenreich, *Solid state physics, advances in research and applications,* Academic Press. Cited as *Solid state physics.*

Advanced texts

R. E. Peierls, *Quantum theory of solids,* Oxford, 1955.

C. Kittel, *Quantum theory of solids,* Wiley, 1963. Cited as *QTS.*

J. M. Ziman, *Principles of the theory of solids,* Cambridge, 1964.

Table of values

American Institute of Physics Handbook, McGraw-Hill, 3rd ed., 1971.

Guides to the literature

The original papers of the early workers in the field are most conveniently identified in the great scientific bibliographic series known as *Poggendorf;* this covers the last 100 years and more. The most valuable modern bibliographic aid is the *Scientific citation index.* For references to data on specific materials, consult the formula indices to *Chemical abstracts* and the subject indices to *Physics abstracts* and *Solid state abstracts.* Good bibliographies often accompany the review articles in *Reports on progress in physics, Critical reviews in solid state sciences, Solid state physics* (see above), *Springer tracts in modern physics,* Reviews of Modern Physics, Soviet Physics (Uspekhi), Comments on Solid State Physics, and Advances in Physics.

INTRODUCTION TO SOLID STATE PHYSICS

1

Crystal Structure

NOTATION: 1 Å = 1 angstrom = 10^{-8} cm = 10^{-10} m. Crystal lattice parameters are sometimes reported in kx-units, with 1.00208 kxu = 1 Å.

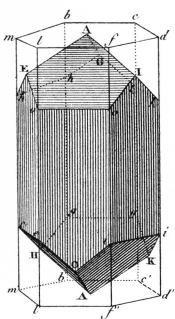

Figure 1 Sketch of a crystal, selected at random from an early mineralogy treatise. (Haüy.)

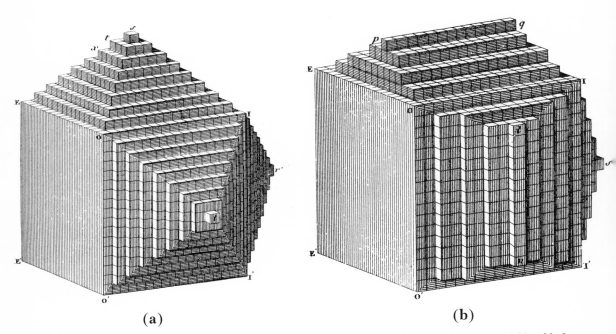

(a) **(b)**

Figure 2 Relation of the external form of crystals to the form of the elementary building blocks. The building blocks are identical in (a) and (b), but different crystal faces are developed. (Haüy, from the atlas to the 1822 edition of his *Traité de cristallographie*.)

The study of the physical properties of the solid state, viewed as a branch of atomic physics, began in the early years of this century. Solid state physics is largely devoted to the study of crystals and of electrons in crystals. A century ago the study of crystals was concerned only with their external form and with symmetry relationships among the various coefficients that describe the physical properties. After 1910 physicists became deeply concerned with atomic models of crystals, following the discovery of x-ray diffraction and the publication of a series of simple and reasonably successful calculations and predictions of crystalline properties.

Many crystalline minerals and gems have been known and described for several thousand years. One of the earliest drawings of a crystal appears in a Chinese pharmacopeia of the eleventh century A.D. Quartz crystals from crowns have been preserved since 768 A.D. in the Shōsōin, the storehouse of the emperors of Japan in Nara. The word crystal referred first only to ice and then to quartz until the late middle ages when the word acquired a more general meaning.

The regularity of the appearance and of the external form of the crystals found in nature (Fig. 1) or grown in the laboratory disposed observers since the seventeenth century to the belief that crystals are formed by a regular repetition of identical building blocks (Fig. 2). When a crystal grows in a constant environment the shape remains unchanged during growth, as if identical elementary building blocks are added continuously to the crystal. We now know the elementary building blocks are atoms or groups of atoms: crystals are a three-dimensional periodic array of atoms.

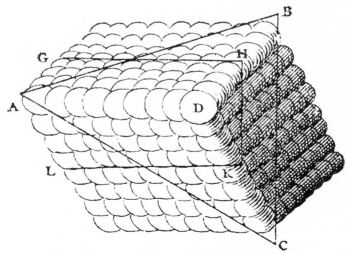

Figure 3 Model of calcite ($CaCO_3$) from C. Huyghens, *Traité de la lumière*, 1690.

In the eighteenth century mineralogists made the important discovery that the index numbers (on a certain scheme of indexing to be described later) of the directions of all faces of a crystal are exact integers. Haüy[1] showed that the arrangement of identical particles in a three-dimensional periodic array could account for this law of rational indices. A. L. Seeber[2] of Freiburg suggested in 1824 that the elementary building blocks of crystals were small spheres, and he proposed an empirical law of interatomic force with both attractive and repulsive regions as needed to cause a lattice array to be the stable equilibrium state of a system of identical atoms.

Probably the most important date in the history of the physics of solids is June 8, 1912, when a paper[3] entitled "Interference effects with Röntgen rays" was laid before the Bavarian Academy of Sciences in Munich. In the first part of the paper Laue develops an elementary theory of diffraction of x-rays by a periodic array of atoms. In the second part Friedrich and Knipping report the first experimental observations of x-ray diffraction by crystals.

The work demonstrated that x-rays are waves, because they can be diffracted. The work also proved decisively that crystals are composed of a periodic array of atoms. This experimental proof marks the beginning of the field of solid state physics as we know it today. With an established atomic model of a crystal, physicists could think or calculate further. Much important pioneer work in physics of solids was done in the years immediately following 1912. The first determinations of crystal structures by x-ray diffraction analysis were reported by W. L. Bragg in 1913: the structures of KCl, NaCl, KBr and KI are given by him in Proc. Roy. Soc. (London) **A89,** 248 (1913).

PERIODIC ARRAYS OF ATOMS

A symbolic language has been built up to describe crystal structures. A person who has learned the language of crystallography can reconstruct a crystal structure from a few printed symbols. We give here several elemen-

[1] R. J. Haüy, *Essai d'une théorie sur la structure des cristaux,* Paris, 1784; *Traité de cristallographie,* Paris, 1801.

[2] A. L. Seeber, "Versuch einer Erklärung des inneren Baues der festen Körper," Annalen der Physik (Gilbert) **76,** 229–248, 349–372 (1824).

[3] "Interferenz-Erscheinungen bei Roentgenstrahlen," W. Friedrich, P. Knipping, and M. Laue, Sitzungsberichte der Bayerischen Akademie de Wissenschaften, Math.-phys. Klasse, pp. 303–322, 1912. This paper, together with several others from the group around Laue, is reprinted with valuable annotations in Vol. 204 of Ostwald's *Klassiker der exakten Wissenschaften,* Akad. Verlag, Leipzig, 1923; the historical excerpts from Laue's Nobel prize lecture are of special interest. For personal accounts of the early years of x-ray diffraction studies of crystals, see P. P. Ewald, ed., *Fifty years of x-ray diffraction,* A. Oosthoek's Uitgeversmij., Utrecht, 1962. Shrewd guesses about the structures of a number of crystals had been made much earlier by W. Barlow, Nature **29,** 186, 205, 404 (1883); he argued from considerations of symmetry and packing (the filling of space).

tary ideas about the language, sufficient to describe the geometry of simple crystal structures.

An ideal crystal is constructed by the infinite regular repetition in space of identical structural units. In the simplest crystals such as copper, silver, gold, and the alkali metals the structural unit contains a single atom. More generally the structural unit contains several atoms or molecules, up to perhaps 100 in inorganic crystals[4] and 10^4 in protein crystals. A crystal may be composed of more than one chemical element (as in NaCl) or it may contain associated groups of identical atoms (as in H_2). We describe the structure of all crystals in terms of a single periodic lattice, but with a group of atoms attached to each lattice point or situated in each elementary parallelepiped. This group of atoms is called the **basis**; the basis is repeated in space to form the crystal. We now make these definitions more precise.

Crystal Translation Vectors and Lattices

An ideal crystal is composed of atoms arranged on a lattice defined by three **fundamental translation vectors a, b, c** such that the atomic arrangement looks the same in every respect when viewed from any point **r** as when viewed from the point

$$\mathbf{r}' = \mathbf{r} + n_1\mathbf{a} + n_2\mathbf{b} + n_3\mathbf{c} , \tag{1}$$

where n_1, n_2, n_3 are arbitrary integers (Fig. 4). Crystallographers may use $\mathbf{a}_1$, $\mathbf{a}_2$, $\mathbf{a}_3$ to denote the fundamental translation vectors.

The set of points $\mathbf{r}'$ specified by (1) for all values of the integers n_1, n_2, n_3 defines a **lattice**. A lattice[5] is a regular periodic arrangement of points in space. A lattice is a mathematical abstraction: the crystal structure is formed only when a basis of atoms is attached identically to each lattice point. The logical relation is

lattice + basis = crystal structure.

The lattice and the translation vectors **a, b, c** are said to be **primitive** if *any* two points **r, r′** from which the atomic arrangement looks the same[6] always satisfy (1) with a suitable choice of the integers n_1, n_2, n_3.

----

[4] The intermetallic compound $NaCd_2$ has a cubic cell of 1192 atoms as its smallest structural unit; see S. Samson, Nature **195**, 259 (1962).

[5] We use the words *lattice* and *lattice points* interchangeably. Bravais noted that the lattice points are the roots of the equation

$$\sin^2(\pi\xi/a) + \sin^2(\pi\eta/b) + \sin^2(\pi\zeta/c) = 0,$$

where ξ, η, ζ are spatial coordinates referred to a system of three coordinate axes, in general oblique. The unit vectors of the coordinate system are denoted by $\hat{\mathbf{a}}$, $\hat{\mathbf{b}}$, $\hat{\mathbf{c}}$.

[6] This definition may sound clumsy, but it guarantees that there is no cell of smaller volume which could serve as a building block for the structure.

An octahedron of fluorite, CaF_2. (Courtesy of Optovac.)

Cleaving a crystal of rocksalt, NaCl. (Courtesy of Optovac.)

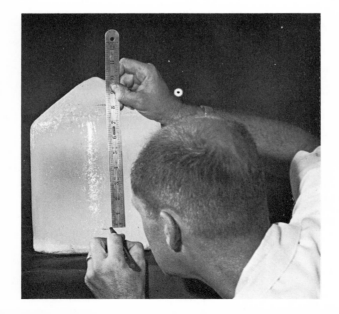

Large single crystal of sodium iodide, with thallium impurity added for use as a detector of gamma radiation. (Courtesy of Crystal Division, Isotopes, Inc.)

Cesium iodide single crystal being raised from the growing furnace. The 60-pound crystal is hanging from a small seed crystal. Support frame over furnace is automatically controlled so that the boule revolves continuously to assure even growth. The boule is slowly raised from melt at rate of about 1 inch per day to assure even vertical growth. Arrow points to boule. (Courtesy of Crystal Division, Isotopes, Inc.)

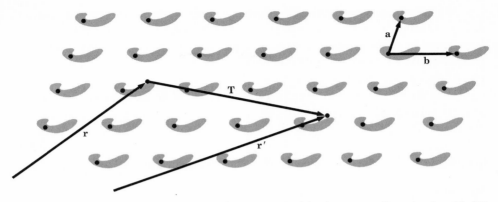

Figure 4 Portion of a crystal of an imaginary protein molecule, in a two-dimensional world. (We picked a protein molecule because it is likely to have no special symmetry of its own.) The atomic arrangement in the crystal looks exactly the same to an observer at $\mathbf{r}'$ as to an observer at $\mathbf{r}$, provided that the vector $\mathbf{T}$ which connects $\mathbf{r}'$ and $\mathbf{r}$ may be expressed as an integral multiple of the vectors $\mathbf{a}$ and $\mathbf{b}$. In this illustration $\mathbf{T} = -\mathbf{a} + 3\mathbf{b}$. The vectors $\mathbf{a}$ and $\mathbf{b}$ are primitive translation vectors of the two-dimensional lattice.

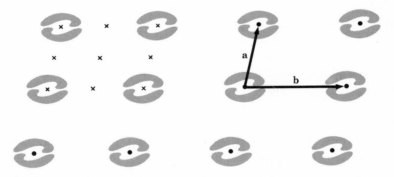

Figure 5 Similar to Fig. 4, but with protein molecules associated in pairs. The crystal translation vectors are $\mathbf{a}$ and $\mathbf{b}$. A rotation of π radians about any point marked × will carry the crystal into itself.

We often use the primitive translation vectors to define the **crystal axes** **a**, **b**, **c** although nonprimitive crystal axes may be used when they are more convenient or simpler. The crystal axes **a**, **b**, **c** form three adjacent edges of a parallelepiped. If there are lattice points only at the corners of the parallelepiped, then it is a primitive parallelepiped.

A **lattice translation operation** or **crystal translation operation** is defined as the displacement of a crystal parallel to itself by a crystal translation vector

$$\mathbf{T} = n_1\mathbf{a} + n_2\mathbf{b} + n_3\mathbf{c} = n_1\mathbf{a}_1 + n_2\mathbf{a}_2 + n_3\mathbf{a}_3 \ , \tag{2}$$

in crystallographic notation. A vector **T** connects any two lattice points.

Symmetry Operations

In describing a crystal structure there are four important questions to answer: What is the lattice?[7] What crystal axes **a**, **b**, **c** do we wish to use to describe the lattice? What is the basis? What are the symmetry operations which carry the crystal structure into itself?

Among the symmetry operations there are the crystal translation operations specified by (2). There are rotation and reflection operations, called **point operations.** About various lattice points or special points within an elementary parallelepiped it may be possible to apply rotation and reflection operations that "carry the crystal structure into itself"; these point operations are in addition to the translation operations. There may be still other symmetry operations, called compound operations, made up of combined translations and point operations. The language of crystallography is largely devoted to the concise description of symmetry operations.

The crystal structure of Fig. 4 has been drawn in such a way that it has only translation symmetry operations, apart from the trivial point operation of rotation by 360°. The crystal structure of Fig. 5 has both translational and point symmetry operations. The primitive translation vectors are **a** and **b**. If we rotate the crystal by π radians (180°) about any point marked $\times$, the crystal structure is carried into itself. This occurs also for equivalent points in other cells, but we have marked the points $\times$ only within one cell.

The Basis and the Crystal Structure

We attach to every lattice point a basis of atoms; every basis is identical in composition, arrangement, and orientation. The formation of a crystal structure by the addition of a basis to every point of a lattice is shown in Fig. 6. In the structures of Figs. 4 and 5 the lattice was indicated by dots; in

---

[7] More than one lattice is always possible for a given structure and more than one set of crystal axes is always possible for a given lattice. We cannot pick the basis until we have selected the lattice and the axes we wish to use. Everything (such as the x-ray diffraction pattern) works out the same in the end as long as the basis is chosen properly for whatever crystal axes we use.

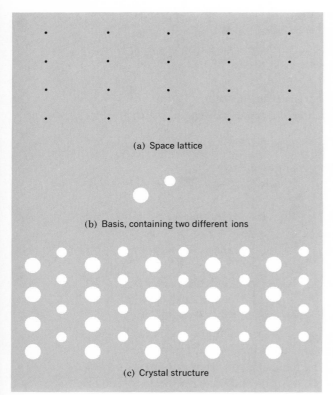

(a) Space lattice

(b) Basis, containing two different ions

(c) Crystal structure

Figure 6 The crystal structure is formed by the addition of the basis (b) to every lattice point of the lattice (a). By looking at (c), you can recognize the basis and then you can abstract the space lattice. It does not matter where the basis is put in relation to a lattice point.

Figure 7a Lattice points of a space lattice in two dimensions. All pairs of vectors **a**, **b** are translation vectors of the lattice. But a_4, b_4, are not *primitive* translation vectors because we cannot form the lattice translation **T** from integral combinations of a_4 and b_4. All other pairs shown of **a** and **b** may be taken as the primitive translation vectors of the lattice. The parallelograms 1, 2, 3 are equal in area and any of them could be taken as the primitive cell. The parallelogram 4 has twice the area of a primitive cell.

Figure 7b Primitive cell of a space lattice in three dimensions.

Figure 7c What are the primitive axes of this "lattice"? The question is meaningless because this is not a lattice in the sense of our definition: the points shown cannot all be reached by vectors of the form $n_1a + n_2b$ for all values of the integers n_1, n_2. But suppose these points were identical atoms: sketch in on the figure a set of lattice points, a choice of primitive axes, a primitive cell, and the basis of atoms associated with a lattice point.

Figure 7d Three linear crystal structures. Lattice points are denoted by $\times$: For structure I we have arbitrarily taken the lattice points to be halfway between two atoms; we might have taken them at any other position, while preserving the length and direction of **a**. In all three structures the lattice points are connected by the translation vector **a**, where **a** is a primitive translation vector for each of the three lattices. The primitive (smallest) basis of structure I consists of one atom at $\frac{1}{2}a$. The primitive basis of structure II consists of two identical atoms, one at u_1a and the other at u_2a. The primitive basis of structure III consists of two different atoms, one at u_1a and the other at u_2a. If we wish to describe structure I in terms of the nonprimitive lattice translation vector a' ($=2a$), then the basis associated with a' consists of two identical atoms, one at $\frac{1}{4}a'$ and the other at $\frac{3}{4}a'$. If we had taken the origin of a' at one of the atoms, the basis would consist of one atom at 0 and a second atom of $\frac{1}{2}a'$. The lattice associated with a' has half the number of lattice points as the lattice associated with **a**.

Fig. 6c the dots are omitted. The number of atoms in the basis may be as low as one, as for many metals and the inert gases, but both inorganic and biochemical structures are known for which the number of atoms in the basis exceeds one thousand. The description of a crystal structure in terms of a lattice plus a basis is very well suited for analysis by x-ray or neutron diffraction, as we show in Chapter 2.

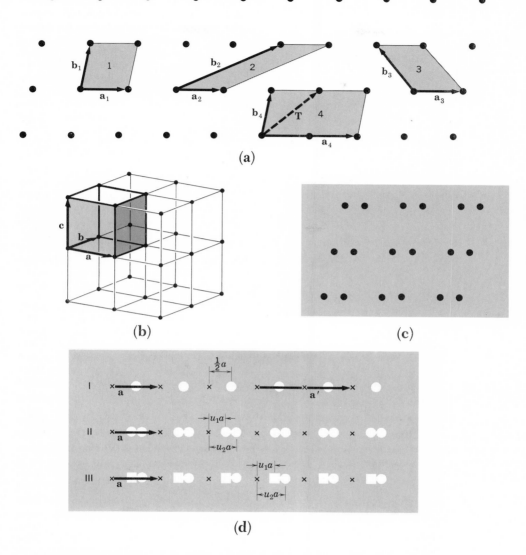

(a)

(b)   (c)

(d)

A basis of N atoms or ions is specified by the set of N vectors

$$\mathbf{r}_j = x_j\mathbf{a} + y_j\mathbf{b} + z_j\mathbf{c}$$

which give the positions of the centers of the atoms of the basis, taken relative to a lattice point. We usually arrange matters so that $0 \leq x_j, y_j, z_j \leq 1$.

Primitive Lattice Cell

The parallelepiped (Fig. 7b) defined by primitive axes $\mathbf{a}$, $\mathbf{b}$, $\mathbf{c}$ is called a **primitive cell**. A primitive cell is a type of **cell**, often called **unit cell**. A cell **will fill all space under the action of suitable crystal translation operations;**

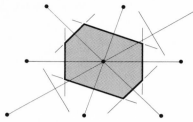

Figure 8 A primitive cell may also be chosen following this procedure: (1) draw lines to connect a given lattice point to all nearby lattice points; (2) at the midpoint and normal to these lines, draw new lines or planes. The smallest volume enclosed in this way is the Wigner-Seitz primitive cell. All space may be filled by these cells, just as by the cells of Fig. 7.

a primitive cell is a minimum-volume cell.[8] There is a density of one lattice point per primitive cell.[9] There are lattice points at the eight corners of the parallelepiped, but each corner point is shared among the eight cells which touch there. The volume V_c of a primitive cell defined by primitive axes **a, b, c** is

$$V_c = |\mathbf{a} \times \mathbf{b} \cdot \mathbf{c}| \ ,\tag{3}$$

by elementary vector analysis. The basis associated with a lattice point of a primitive cell may be called a primitive basis. No basis contains fewer atoms than a primitive basis contains.

Another way of choosing a cell of equal volume V_c is shown in Fig. 8. The cell formed in this way is known to physicists as a **Wigner-Seitz primitive cell.**

FUNDAMENTAL TYPES OF LATTICES[10]

Crystal lattices can be carried into themselves not only by the lattice translations **T** of Eq. (2), but also by various point symmetry operations. A typical symmetry operation is that of rotation about an axis which passes through a lattice point. Lattices can be found such that one-, two-, three-, four-, and six-fold rotation axes are permissible, corresponding to rotations by 2π, $2\pi/2$, $2\pi/3$, $2\pi/4$, and $2\pi/6$ radians and by integral multiples of these rotations. The **rotation axes** are denoted by the symbols 1, 2, 3, 4, and 6.

We cannot find a lattice that goes into itself under other rotations, such as by $2\pi/7$ radians or $2\pi/5$ radians. A single molecule can have any degree of rotational symmetry, but an infinite periodic lattice cannot. We can make a crystal from molecules which individually have a five-fold rotation axis, but we should not expect the lattice to have a five-fold rotation axis. In Fig. 9a we show what happens if we try to construct a periodic lattice having five-fold symmetry: the pentagons do not fit together neatly. We see that we can-

---

[8] There are many ways of choosing the primitive axes and primitive cell for a given lattice (Fig. 7a).

[9] The number of atoms in a primitive cell is the number of atoms in the basis.

[10] The original 1848 study by A. Bravais is reprinted in his *Etudes crystallographiques,* Gautier-Villars, Paris, 1866; a German translation appears in Ostwald's *Klassiker der exakten Wissenschaften* **90,** (1897). For a full discussion of crystal symmetry, see F. Seitz, Z. Krist, **88,** 433 (1934); **90,** 289 (1935); **91,** 336 (1935); **94,** 100 (1936); and Vol. 1 of *International tables for x-ray crystallography,* Kynoch Press, Birmingham, 1952. A particularly readable discussion of space groups is given by F. C. Phillips, *An introduction to crystallography,* Wiley, 1963, 3rd ed.

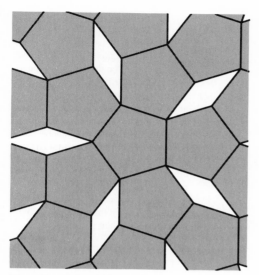

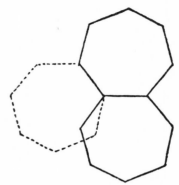

Figure 9a   A five-fold axis of symmetry cannot exist in a lattice because it is not possible to fill all space with a connected array of pentagons. In Fig. 32 we give an example of regular pentagonal packing which does not have the translational invariance of a lattice.

Figure 9b   Kepler's demonstration (*Harmonice mundi,* 1619) that a seven-fold axis of symmetry cannot exist in a lattice. (*Gesammelte Werke,* Vol. 6, Beck, Munich, 1940.)

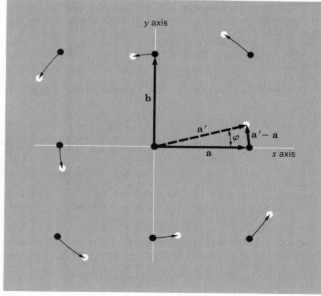

Figure 10   The crystal lattice is rotated by an angle φ about a lattice point. The vector **a** is carried into **a'** by the rotation. For special values of φ the rotated lattice coincides with the original lattice. For a square lattice these special values are $\varphi = \frac{1}{2}\pi$ and multiples thereof, so that the point group of the square lattice includes a four-fold rotation. Wherever the rotated lattice coincides with the original lattice, the vector $\mathbf{a'} - \mathbf{a}$ will be a lattice vector. Such a lattice vector will never be shorter than **a**, because there is no lattice vector shorter than **a**, except zero. Similar requirements determine the special values of φ for all possible lattices.

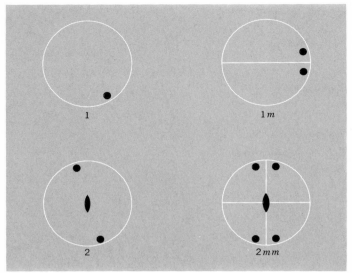

Figure 11 Diagrams illustrating four point groups. The dots denote equivalent points. The point group 1 has no symmetry elements, so here a point has no other point equivalent to it. For 1*m* there is a mirror plane: one dot on reflection across the mirror plane becomes the second dot. With a two-fold axis 2, a rotation of π takes one dot into the other. With a two-fold axis and one mirror plane there is automatically a second mirror plane normal to the first, and we have the point group 2*mm* with four equivalent points.

not combine five-fold point symmetry with the required translational symmetry. The argument against seven-fold symmetry is given in Fig. 9b. The lines of an algebraic argument are indicated in Fig. 10.

A **lattice point group** is defined as the collection of the symmetry operations which, when applied *about a lattice point*, leave the lattice invariant. The possible rotations have been given earlier. We can also have **mirror reflections** *m* about a plane through a lattice point. The **inversion operation** is made up of a rotation of π followed by reflection in a plane normal to the rotation axis; the total effect is to replace **r** by $-\mathbf{r}$.

Locations of equivalent points are shown for four point groups in Fig. 11. The symmetry operations of the group carry one point into all the equivalent positions. The points themselves must not be thought of as possessing symmetry elements; they may be scalene triangles or molecules with no symmetry elements. The symmetry axes and planes of a cube are shown in Fig. 12.

Two-Dimensional Lattice Types

There is an unlimited number of possible lattices because there is no natural restriction on the lengths a, b of the lattice translation vectors or on the angle φ between them. The lattice in Fig. 7a was drawn for arbitrary **a** and **b**. A general lattice such as this is known as an **oblique lattice** and is invariant only under rotation of π and 2π about any lattice point.

But special lattices of the oblique type can be invariant under rotation of $2\pi/3$, $2\pi/4$, or $2\pi/6$, or under mirror reflection. We must impose restrictive conditions on **a** and **b** if we want to construct a lattice which will be invariant under one or more of these new operations. Such restrictions are found below:

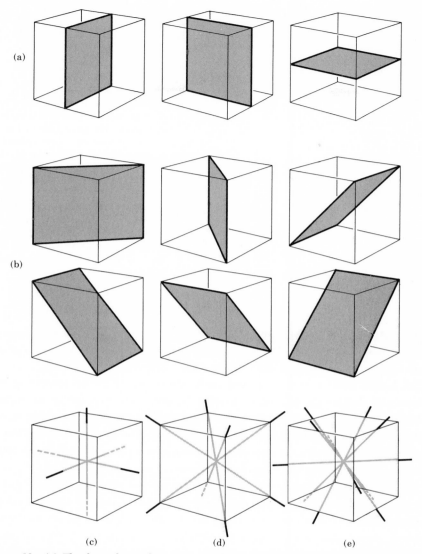

Figure 12 (a) The three planes of symmetry parallel to the faces of a cube. (b) The six diagonal planes of symmetry in a cube. (c) The three tetrad axes of a cube. (d) The four triad axes of a cube. (e) The six diad axes of a cube. The inversion center is not shown.

there are four distinct types of restriction, and each leads to what we may call a **special lattice type.**

Thus there are five distinct lattice types in two dimensions, the oblique lattice and the four special lattices. **Bravais lattice** is the common phrase[11] for a distinct lattice type; we say that there are five Bravais lattices in two dimensions.

[11] We have not succeeded in finding or constructing a definition which starts out "A Bravais lattice is . . ."; the sources we have looked at say "That was a Bravais lattice." The phrase "fundamental type of lattice" is more suggestive.

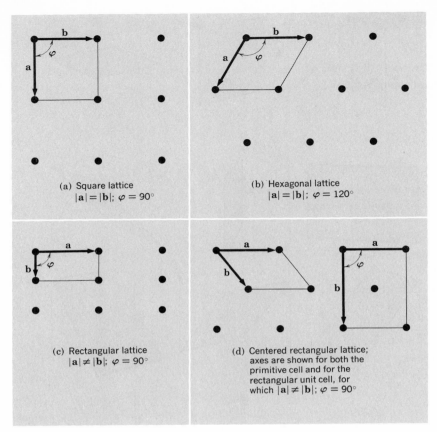

(a) Square lattice
$|\mathbf{a}| = |\mathbf{b}|$; $\varphi = 90°$

(b) Hexagonal lattice
$|\mathbf{a}| = |\mathbf{b}|$; $\varphi = 120°$

(c) Rectangular lattice
$|\mathbf{a}| \neq |\mathbf{b}|$; $\varphi = 90°$

(d) Centered rectangular lattice; axes are shown for both the primitive cell and for the rectangular unit cell, for which $|\mathbf{a}| \neq |\mathbf{b}|$; $\varphi = 90°$

Figure 13

The point operation 4 requires a square lattice (Fig. 13a). The point operations 3 and 6 require a hexagonal lattice (Fig. 13b). This lattice is invariant under a rotation $2\pi/6$ about an axis through a lattice point and normal to the plane.

There are important consequences if the mirror reflection m is present. We write the primitive translation vectors $\mathbf{a}$, $\mathbf{b}$ in terms of the unit vectors $\hat{x}$, $\hat{y}$ along the Cartesian x, y axes:

$$\mathbf{a} = a_x\hat{x} + a_y\hat{y} \; ; \qquad \mathbf{b} = b_x\hat{x} + b_y\hat{y} \; . \tag{4}$$

If the primitive vectors are mirrored in the x axis, then $\mathbf{a}$, $\mathbf{b}$ are transformed by the reflection operation into new vectors $\mathbf{a}'$, $\mathbf{b}'$ given by

$$\mathbf{a}' = a_x\hat{x} - a_y\hat{y} \; ; \qquad \mathbf{b}' = b_x\hat{x} - b_y\hat{y} \; . \tag{5}$$

If the lattice is invariant under the reflection, then $\mathbf{a}'$, $\mathbf{b}'$ must be lattice vectors; that is, they must be of the form $n_1\mathbf{a} + n_2\mathbf{b}$, where n_1 and n_2 are integers. If we take

$$\mathbf{a} = a\hat{x} \; ; \qquad \mathbf{b} = b\hat{y} \; ; \tag{6}$$

then $\mathbf{a}' = \mathbf{a}$ and $\mathbf{b}' = -\mathbf{b}$, so that the lattice is carried into itself. The lattice defined by (6) is rectangular (Fig. 13c).

A second distinct possibility for **a** and **b** gives another type of lattice invariant under reflection. Note that **b'** will be a lattice vector if

$$\mathbf{b'} = \mathbf{a} - \mathbf{b} \; ; \tag{7}$$

then using (5) we have

$$b_x' = a_x - b_x = b_x \; ; \qquad b_y' = a_y - b_y = -b_y \; . \tag{8}$$

These equations have a solution if $a_y = 0$; $b_x = \frac{1}{2}a_x$; thus a possible choice of primitive translation vectors for a lattice with mirror symmetry is

$$\mathbf{a} = a\hat{\mathbf{x}} \; ; \qquad \mathbf{b} = \frac{1}{2}a\hat{\mathbf{x}} + b_y\hat{\mathbf{y}} \; . \tag{9}$$

This choice gives a centered rectangular lattice (Fig. 13d).

We have now exhausted the two-dimensional Bravais lattices which are consistent with the point-group operations applied to lattice points. The five possibilities in two dimensions are summarized in Table 1. The point symmetry given is that of the *lattice;* an actual crystal structure may have lower symmetry than its lattice. Thus it is possible for a crystal with a square lattice to have the operation 4 without having all of the operations *4mm*.

Table 1   The five two-dimensional lattice types

(The notation *mm* means that two mirror lines are present)

Lattice	Conventional cell	Axes of coventional cell	Point-group symmetry of lattice about lattice points
Oblique	Parallelogram	$a \neq b$, $\varphi \neq 90°$	2
Square	Square	$a = b$, $\varphi = 90°$	*4mm*
Hexagonal	60° rhombus	$a = b$, $\varphi = 120°$	*6mm*
Primitive rectangular	Rectangle	$a \neq b$, $\varphi = 90°$	*2mm*
Centered rectangular	Rectangle	$a \neq b$, $\varphi = 90°$	*2mm*

Table 2 **The fourteen lattice types in three dimensions**

System	Number of lattices in system	Lattice symbols	Restrictions on conventional cell axes and angles
Triclinic	1	P	$a \neq b \neq c$ $\alpha \neq \beta \neq \gamma$
Monoclinic	2	P, C	$a \neq b \neq c$ $\alpha = \gamma = 90° \neq \beta$
Orthorhombic	4	P, C, I, F	$a \neq b \neq c$ $\alpha = \beta = \gamma = 90°$
Tetragonal	2	P, I	$a = b \neq c$ $\alpha = \beta = \gamma = 90°$
Cubic	3	P or sc I or bcc F or fcc	$a = b = c$ $\alpha = \beta = \gamma = 90°$
Trigonal	1	R	$a = b = c$ $\alpha = \beta = \gamma < 120°, \neq 90°$
Hexagonal	1	P	$a = b \neq c$ $\alpha = \beta = 90°$ $\gamma = 120°$

Three-Dimensional Lattice Types

In two dimensions the point groups are associated with five different types of lattices. In three dimensions the point symmetry groups require the fourteen different (one general and thirteen special) lattice types shown in Fig. 14 and listed in Table 2. The general lattice type is the triclinic lattice.

The fourteen lattice types are conveniently grouped into seven **systems** according to the seven types of conventional unit cells: triclinic, monoclinic, orthorhombic, tetragonal, cubic, trigonal, and hexagonal. The division into systems is summarized conveniently in terms of the special axial relations for the conventional unit cells. The axes **a, b, c** and angles α, β, γ are defined in Fig. 15.

The unit cells shown in Fig. 14 are the conventional cells, and they are not always primitive cells. Sometimes a nonprimitive cell has a more obvious connection with the point symmetry elements than has a primitive cell. We now discuss the various lattices by their classification in systems.

1. In the **triclinic** system the single lattice type has a primitive (P) unit cell, with three axes of unequal lengths and unequal angles.

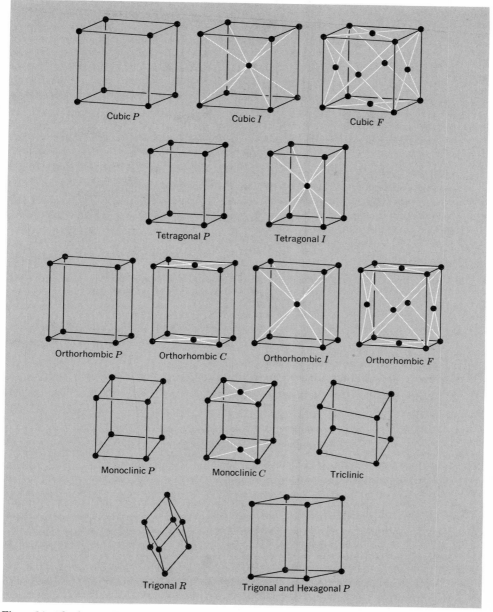

Cubic *P* Cubic *I* Cubic *F*

Tetragonal *P* Tetragonal *I*

Orthorhombic *P* Orthorhombic *C* Orthorhombic *I* Orthorhombic *F*

Monoclinic *P* Monoclinic *C* Triclinic

Trigonal *R* Trigonal and Hexagonal *P*

Figure 14 The fourteen Bravais or space lattices. The cells shown are the conventional cells, which are not always the primitive cells.

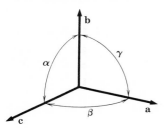

Figure 15 Crystal axes **a**, **b**, **c**. The angle α is included between **b** and **c**.

Table 3 Characteristics of cubic lattices

(Tables of numbers of neighbors and distances in sc, bcc, fcc, hcp, and diamond structures are given on pp. 1037–1039 of J. Hirschfelder, C. F. Curtis, and R. B. Bird, *Molecular theory of gases and liquids,* Wiley, 1964.)

	Simple	Body-centered	Face-centered
Volume, conventional cell	a^3	a^3	a^3
Lattice points per cell	1	2	4
Volume, primitive cell	a^3	$\frac{1}{2}a^3$	$\frac{1}{4}a^3$
Lattice points per unit volume	$1/a^3$	$2/a^3$	$4/a^3$
Number of nearest neighbors[°]	6	8	12
Nearest-neighbor distance	a	$3^{1/2}a/2 = 0.866a$	$a/2^{1/2} = 0.707a$
Number of second neighbors	12	6	6
Second neighbor distance	$2^{1/2}a$	a	a

[°] Nearest neighbors are the nearest lattice points to any given lattice point.

2. In the **monoclinic** system there are two lattice types, one with a primitive unit cell and the other with a nonprimitive conventional cell which may be base-centered (C) with lattice points at the centers of the rectangular cell faces in the *ab* plane.

3. In the **orthorhombic** system there are four lattice types: one lattice has a primitive cell; one lattice is base-centered; one is body-centered (I = German *Innenzentrierte*); and one is face-centered (F).

4. In the **tetragonal** system the simplest unit is a right square prism; this is a primitive cell and is associated with a tetragonal space lattice. A second tetragonal lattice type is body-centered.

5. In the **cubic** system there are three lattices: the simple cubic (sc) lattice, the body-centered cubic (bcc) lattice, and the face-centered cubic (fcc) lattice. The characteristics of the three cubic lattices are summarized in Table 3.

A primitive cell of the body-centered cubic lattice is shown in Fig. 16; the primitive translation vectors are shown in Fig. 17. The primitive translation vectors of the face-centered cubic lattice are shown in Fig. 18. The primitive cells contain only one lattice point, but the conventional cubic cells contain two lattice points (bcc) or four lattice points (fcc).

6. In the **trigonal** system a rhombohedron is usually chosen as the primitive cell. The lattice is primitive.

7. In the **hexagonal** system the conventional cell chosen is a right prism based on a rhombus with an angle of $60°$. The lattice is primitive. The relation of the rhombic cell with a hexagonal prism is shown in Fig. 19.

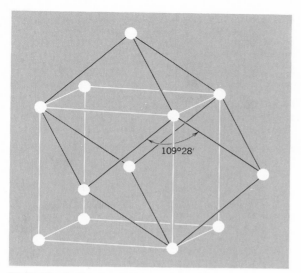

Figure 16 Body-centered cubic lattice, showing a primitive cell. The primitive cell shown is a rhombohedron of edge $\frac{1}{2}\sqrt{3}\,a$, and the angle between adjacent edges is 109°28′.

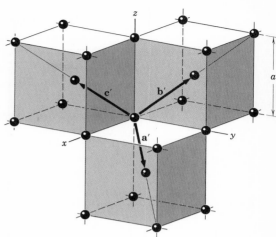

Figure 17 Primitive translation vectors of the body-centered cubic lattice; these vectors connect the lattice point at the origin to lattice points at the body centers. The primitive cell is obtained on completing the rhombohedron. In terms of the cube edge a the primitive translation vectors are

$$\mathbf{a}' = \frac{a}{2}(\hat{x} + \hat{y} - \hat{z}); \quad \mathbf{b}' = \frac{a}{2}(-\hat{x} + \hat{y} + \hat{z});$$

$$\mathbf{c}' = \frac{a}{2}(\hat{x} - \hat{y} + \hat{z}).$$

These primitive axes make angles of 109°28′ with each other.

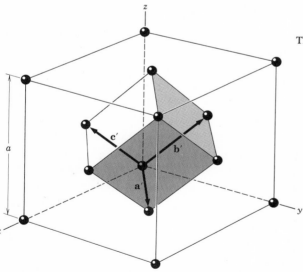

Figure 18 The rhombohedral primitive cell of the face-centered cubic crystal. The primitive translation vectors $\mathbf{a}'$, $\mathbf{b}'$, $\mathbf{c}'$ connect the lattice point at the origin with lattice points at the face centers. As drawn, the primitive vectors are:

$$\mathbf{a}' = \frac{a}{2}(\hat{x} + \hat{y}); \quad \mathbf{b}' = \frac{a}{2}(\hat{y} + \hat{z}); \quad \mathbf{c}' = \frac{a}{2}(\hat{z} + \hat{x}).$$

The angles between the axes are 60°.

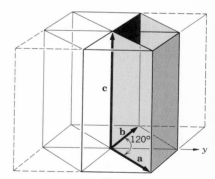

Figure 19 Relation of the primitive cell in the hexagonal system (heavy lines) to a prism of hexagonal symmetry. Here $a = b \neq c$.

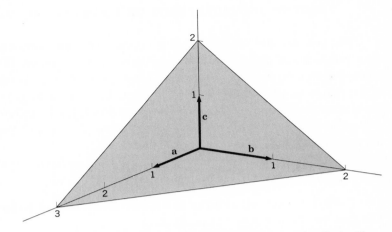

Figure 20 This plane intercepts the **a, b, c** axes at 3**a**, 2**b**, 2**c**. The reciprocals of these numbers are $\frac{1}{3}, \frac{1}{2}, \frac{1}{2}$. The smallest three integers having the same ratio are 2, 3, 3, and thus the Miller indices of the plane are (233).

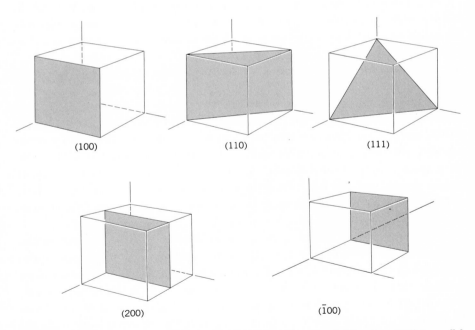

Figure 21 Miller indices of some important planes in a cubic crystal. The plane (200) is parallel to (100).

POSITION AND ORIENTATION OF PLANES IN CRYSTALS

The position and orientation of a crystal plane are determined by any three points in the plane, provided the points are not collinear. If each of the points lies on a crystal axis, the plane may be specified by giving the positions of the points along the axes in terms of the lattice constants. If, for example, the atoms determining the plane have coordinates $(4, 0, 0)$, $(0, 1, 0)$, $(0, 0, 2)$ relative to the axis vectors from some origin, the plane may be specified by the three numbers 4, 1, 2.

But it is more useful for structure analysis to specify the orientation of a plane by **Miller indices**,[12] determined as in Fig. 20:

1. Find the intercepts on the axes **a, b, c** in terms of the lattice constants. The axes may be primitive or nonprimitive.

2. Take the reciprocals of these numbers and then reduce to three integers having the same ratio, usually the smallest three integers. The result is enclosed in parentheses: (hkl).

For the plane whose intercepts are 4, 1, 2 the reciprocals are $\frac{1}{4}$, 1, and $\frac{1}{2}$ and the Miller indices are (142). If an intercept is at infinity, the corresponding index is zero. The Miller indices of some important planes in a cubic crystal are illustrated by Fig. 21.

The indices (hkl) may denote a single plane or a set of parallel planes.[13] If a plane cuts an axis on the negative side of the origin, the corresponding index is negative and is indicated by placing a minus sign above the index: $(h\bar{k}l)$. The cube faces of a cubic crystal are (100), (010), (001), ($\bar{1}$00), (0$\bar{1}$0), and (00$\bar{1}$). Planes equivalent by symmetry may be denoted by curly brackets (braces) around Miller indices; the set of cube faces is {100}. We often speak simply of the 100 faces. If we speak of the (200) plane we mean a plane parallel to (100) but cutting the **a** axis at $\frac{1}{2}$**a**. The formation of the (110), (111), and (322) planes of an fcc crystal structure, starting from (100) planes of atoms, is shown in Figs. 21a, b, c on p. 24.

The indices of a direction in a crystal are expressed as the set of the smallest integers which have the ratio of the components of a vector in the desired direction referred to the axes. The integers are written between square brackets $[hkl]$. In a cubic crystal the x axis is the [100] direction; the $-y$ axis is the [0$\bar{1}$0] direction. Often we speak of the $[hkl]$ and equivalent directions, or simply of the $[hkl]$ directions. In cubic crystals the direction $[hkl]$ is always perpendicular to a plane (hkl) having the same indices (Problem 3), but this is not generally true in other crystal systems.

---

[12] At first sight the usefulness of the Miller indices seems improbable, but Chapter 2 makes clear their convenience and elegance.

[13] Other symbols are frequently used by crystallographers: they may use u, v, w in place of our h, k, l.

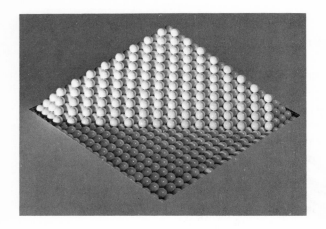

Figure 21a   A (110) plane of an fcc crystal structure, as built up from (100) layers. (This and the accompanying photographs are by J. F. Nicholas, *Atlas of models of crystal surfaces*, Gordon and Breach, 1965).

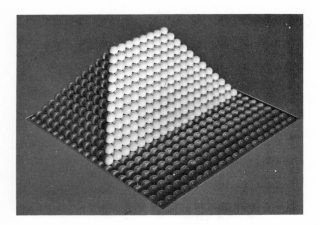

Figure 21b   A (111) plane of an fcc crystal structure, based on (100) layers.

Figure 21c   A (322) plane of an fcc crystal structure, based on (100) layers. The concentration of atoms tends to be lower in planes of high indices than in planes of low indices.

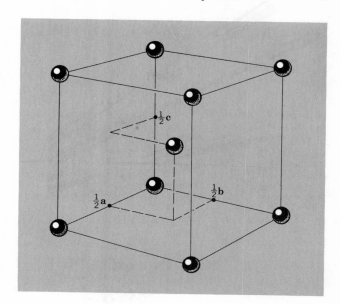

Figure 22   The coordinates of the central point of a cell are $\frac{1}{2}\frac{1}{2}\frac{1}{2}$, in terms of the length of the axes.

POSITION IN THE CELL

The positions of a point in a cell is specified in terms of atomic coordinates x, y, z in which each coordinate is a fraction of the axial length, a, b, or c, in the direction of the coordinate, with the origin taken at a corner of a cell. Thus (Fig. 22) the coordinates of the central point of a cell are $\frac{1}{2}\frac{1}{2}\frac{1}{2}$. The face-centered positions include $\frac{1}{2}\frac{1}{2}0$; $0\frac{1}{2}\frac{1}{2}$; $\frac{1}{2}0\frac{1}{2}$. **The coordinates of atoms in fcc and bcc lattices are usually given in terms of the conventional cubic cell.** Tables of crystal structures usually specify the type and size of the cell, and then they give the values of the atomic coordinates x_i, y_i, z_i for each of the atoms in the cell.

Figure 23 Model of sodium chloride. The sodium ions are smaller than the chlorine ions. (Courtesy of A. N. Holden and P. Singer, from *Crystals and crystal growing.*)

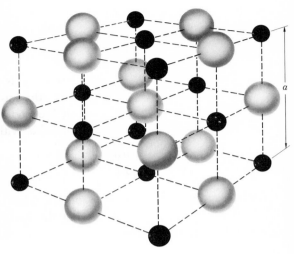

Figure 24 The sodium chloride crystal structure. The space lattice is fcc, and the basis has one Na^+ ion at 0 0 0 and one Cl^- ion at $\frac{1}{2} \frac{1}{2} \frac{1}{2}$.

Figure 25 Natural crystals of lead sulfide, PbS, which has the NaCl crystal structure. (Photograph by Betsy Burleson.)

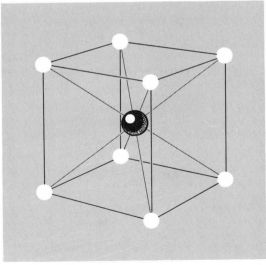

Figure 26 The cesium chloride crystal structure. The space lattice is simple cubic, and the basis has one Cs^+ ion at 0 0 0 and one Cl^- ion at $\frac{1}{2} \frac{1}{2} \frac{1}{2}$.

SIMPLE CRYSTAL STRUCTURES

We discuss briefly a small number of simple crystal structures of general interest, including the sodium chloride, cesium chloride, hexagonal close-packed, diamond, and cubic zinc sulfide structures.

Sodium Chloride Structure

The sodium chloride, NaCl, structure is shown in Figs. 23 and 24. The Bravais lattice is face-centered cubic; the basis consists of one Na atom and one Cl atom separated by one-half the body diagonal of a unit cube. There are four units of NaCl in each unit cube, with atoms in the positions

$$\text{Na:} \quad 0\,0\,0\;; \quad \tfrac{1}{2}\tfrac{1}{2}0\;; \quad \tfrac{1}{2}0\tfrac{1}{2}\;; \quad 0\tfrac{1}{2}\tfrac{1}{2}\,.$$
$$\text{Cl:} \quad \tfrac{1}{2}\tfrac{1}{2}\tfrac{1}{2}\;; \quad 0\,0\tfrac{1}{2}\;; \quad 0\tfrac{1}{2}0\;; \quad \tfrac{1}{2}0\,0\,.$$

Each atom has as nearest neighbors six atoms of the opposite kind.

Representative crystals having the NaCl arrangement include those in the following table. The cube edge a is given in angstroms; $1 \text{ Å} \equiv 10^{-8} \text{ cm} \equiv 10^{10} \text{ m}$.

Crystal	a	Crystal	a
LiH	4.08 Å	AgBr	5.77 Å
NaCl	5.63	MgO	4.20
KCl	6.29	MnO	4.43
PbS	5.92	KBr	6.59

Figure 25 is a photograph of crystals of galena (PbS) from Joplin, Missouri. The Joplin specimens form in beautiful cubes.

Cesium Chloride Structure

The cesium chloride, CsCl, structure is shown in Fig. 26. There is one molecule per primitive cell, with atoms in the body-centered positions:

$$\text{Cs:} \quad 0\,0\,0 \qquad \text{and} \qquad \text{Cl:} \quad \tfrac{1}{2}\tfrac{1}{2}\tfrac{1}{2}\,.$$

The space lattice is simple cubic.[14] Each atom is at the center of a cube of atoms of the opposite kind, so that the coordination number is eight. Representative crystals having the CsCl arrangement include those in the following table:

Crystal	a	Crystal	a
CsCl	4.11 Å	CuZn (β-brass)	2.94 Å
TlBr	3.97	AgMg	3.28
TlI	4.20	LiHg	3.29
NH$_4$Cl	3.87	AlNi	2.88
RbCl (190°C)	3.74	BeCu	2.70

[14] Crystal structures with simple cubic Bravais lattices are not rare, but no common *element* crystallizes in a sc structure, which is favored neither by density or packing nor by directed bonds.

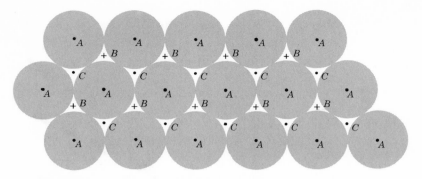

Figure 27a A close-packed layer of spheres is shown, with centers at points marked *A*. A second and identical layer of spheres can be placed over this, with centers over the points marked *B* (or, equivalently, over the points marked *C*). If the second layer goes in over *B*, there are two nonequivalent choices for a third layer. It can go in over *A* or over *C*. If it goes in over *A* the sequence is *ABABAB* . . . and the structure is hexagonal close-packed. If the third layer goes in over *C* the sequence is *ABCABCABC* . . . and the structure is face-centered cubic; the plane is a (111) plane, as in Fig. 27b.

Hexagonal Close-packed Structure (hcp)

There are two ways (Fig. 27a) of arranging equivalent spheres in a regular array to minimize the interstitial volume. One way leads to a structure with cubic symmetry and is the face-centered cubic (cubic close-packed) structure; the other has hexagonal symmetry and is called the hexagonal close-packed structure (Fig. 27c). The fraction of the total volume filled by the spheres is 0.74 for both the fcc and hcp structures.

Spheres may be arranged in a single closest-packed layer by placing each sphere in contact with six others. Such a layer can be either the basal plane of a hcp structure or the (111) plane of the fcc structure. A second similar layer is packed on top of this by placing each sphere in contact with three spheres of the bottom layer, as in Fig. 27a. Next a third layer can be added in two ways: in the fcc structure the spheres in the third layer are placed over the holes in the first layer not occupied by the second layer; in the hexagonal structure the spheres in the third layer are placed directly over the spheres in the first layer. We say that the packing in the fcc structure is *ABCABC* . . . , whereas in the hcp structure the packing is *ABABAB*. . . . The unit cell of the hcp structure is the hexagonal primitive cell; the basis contains two atoms, as shown in Fig. 27d. The fcc primitive cell in Fig. 18 contains one atom.

The c/a ratio for hexagonal closest-packing of spheres is $(\frac{8}{3})^{\frac{1}{2}} = 1.633$. By convention we refer to crystals as hcp even if the actual c/a artio departs somewhat from the theoretical value. Thus zinc with $c/a = 1.86$ ($a = 2.66$ Å; $c = 4.94$ Å) is referred to commonly as hcp, although the interatomic bond angles are quite different from the ideal hcp structure. Magnesium with $c/a = 1.623$ is close to ideal hcp. Many metals transform easily at appropriate temperatures between fcc and hcp structures. We note that the **coordination number,** defined as the number of nearest-neighbor atoms, is the same for the

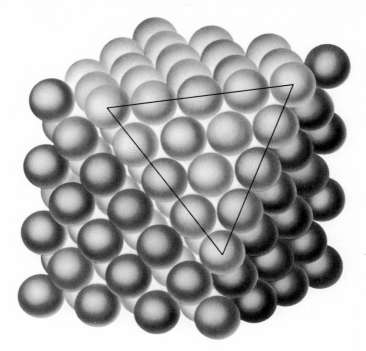

Figure 27b   The fcc structure with one corner sliced off to expose a (111) plane. The (111) planes are close-packed layers of spheres. (After W. G. Moffatt, G. W. Pearsall, and J. Wulff, *Structure*, Vol. 1 of *Structure and properties of materials,* Wiley, 1964.)

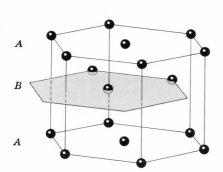

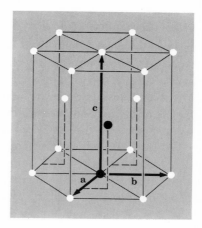

Figure 27c   The hexagonal close-packed structure. The atom positions in this structure do not constitute a space lattice. The space lattice is simple hexagonal with a basis of two identical atoms associated with each lattice point.

Figure 27d   The primitive cell has $a = b$, with an included angle of $120°$. The c axis is normal to the plane of a and b. In ideal hcp we have $c = 1.633 \, a$. The two atoms of one basis are shown as solid in the figure. One atom of the basis is at the origin, 000; the other atom is at $\frac{2}{3} \frac{1}{3} \frac{1}{2}$, which means at $\mathbf{r} = \frac{2}{3}\mathbf{a} + \frac{1}{3}\mathbf{b} + \frac{1}{2}\mathbf{c}$.

two structures. If the binding energy depended only on the number of nearest-neighbor bonds, then there would be no difference in energy between the fcc and hcp structures.

Examples of hexagonal close-packed structures are

Crystal	c/a	Crystal	c/a	Crystal	c/a
He	1.633	Zn	1.861	Zr	1.594
Be	1.581	Cd	1.886	Gd	1.592
Mg	1.623	Co	1.622	Lu	1.586
Ti	1.586	Y	1.570		

Diamond Structure

The space lattice of diamond is face-centered cubic. A primitive basis of two identical atoms at $0\ 0\ 0$; $\frac{1}{4}\frac{1}{4}\frac{1}{4}$ is associated with each lattice point,[15] as shown in Fig. 28. The tetrahedral bonding of the diamond structure is represented in Fig. 29. Each atom has four nearest neighbors and twelve next nearest neighbors. There are eight atoms in a unit cube. The diamond lattice is relatively empty; the maximum proportion of the available volume which may be filled by hard spheres is only 0.34, or about 46 percent of the filling factor for a closest-packed structure. Carbon, silicon, germanium, and gray tin crystallize in the diamond structure, with lattice constants 3.56, 5.43, 5.65, and 6.46 Å, respectively. The diamond structure is the result of covalent bonding, as discussed in Chapter 3.

Cubic Zinc Sulfide Structure

We have seen that the diamond structure is composed of two fcc lattices displaced from each other by one-quarter of a body diagonal. The cubic zinc sulfide or zinc blende structure results from the diamond structure when Zn atoms are placed on one fcc lattice and S atoms on the other fcc lattice, as in Fig. 30. The conventional cell is a cube. The coordinates of the Zn atoms are $0\ 0\ 0$; $0\frac{1}{2}\frac{1}{2}$; $\frac{1}{2}0\frac{1}{2}$; $\frac{1}{2}\frac{1}{2}0$; the coordinates of the S atoms are $\frac{1}{4}\frac{1}{4}\frac{1}{4}$; $\frac{1}{4}\frac{3}{4}\frac{3}{4}$; $\frac{3}{4}\frac{1}{4}\frac{3}{4}$; $\frac{3}{4}\frac{3}{4}\frac{1}{4}$. The space lattice is fcc. There are four molecules of ZnS per conventional cell. About each atom there are four equally distant atoms of the opposite kind arranged at the corners of a regular tetrahedron.

The diamond structure possesses a center of inversion symmetry at the midpoint of each line connecting nearest-neighbor atoms; the ZnS structure does not have inversion symmetry.[16] This is particularly evident if we look at

---

[15] The conventional unit cube contains eight atoms; if we describe the structure in terms of the lattice points at the corners of the unit cube, then the basis will contain eight atoms. There is no way of choosing a primitive cell such that the basis of diamond contains only one atom.

[16] In the inversion operation we carry each point **r** into the point $-\mathbf{r}$. Note that a tetrahedron does not have inversion symmetry about its center.

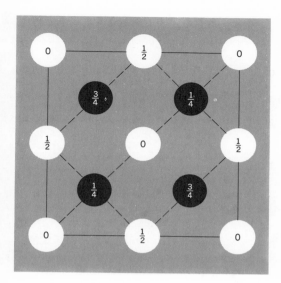

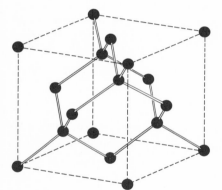

Figure 28 Atomic positions in the cubic cell of the diamond structure projected on a cube face; fractions denote height above the base in units of a cube edge. The points at 0 and $\frac{1}{2}$ are on the fcc lattice; those at $\frac{1}{4}$ and $\frac{3}{4}$ are on a similar lattice displaced among the body diagonal by one-fourth of its length. With a fcc space lattice, the basis consists of two identical atoms at $0\,0\,0; \frac{1}{4}\frac{1}{4}\frac{1}{4}$.

Figure 29 Crystal structure of diamond, showing the tetrahedral bond arrangement.

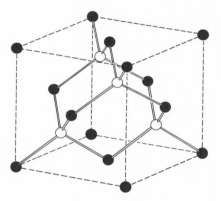

Figure 30 Crystal structure of cubic zinc sulfide.

the arrangement of atoms along a body diagonal. In diamond (Figs. 28 and 29) the order is CC··CC··CC, where the dots represent vacancies. In ZnS (Fig. 30) the order is ZnS··ZnS··ZnS; this order is not invariant under inversion about any point.

Examples of the cubic zinc sulfide structure are

Crystal	a	Crystal	a
CuF	4.26 Å	CdS	5.82 Å
CuCl	5.41	InAs	6.04
AgI	6.47	InSb	6.46
ZnS	5.41	SiC	4.35
ZnSe	5.65	AlP	5.42

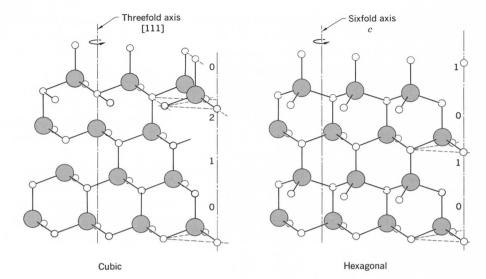

Figure 31 The stacking of tetrahedral layers in cubic and hexagonal ZnS. The large atoms are S; the small atoms are Zn. The vertical axis of hexagonal ZnS is a six-fold screw axis involving a translation of one-half c for each 60 degrees of rotation. The stacking sequences 012012 . . . and 0101 . . . are similar to the sequences of the fcc and hcp structures as shown in Fig. 27a. (After Berry and Mason.)

Hexagonal Zinc Sulfide Structure

A hexagonal crystalline form of diamond was discovered in meteorites and has been synthesized in the laboratory.[17] The hexagonal diamond structure has tetrahedral covalent bonds just as does the cubic diamond structure, and the densities of the two forms are equal. How are the two structures related?

The relation is the same as that of the cubic form of ZnS to the hexagonal[18] form of ZnS. The two forms of diamond are simulated by these structures on replacing all atoms by carbon. Figure 31 gives a view of cubic ZnS with a set of tetrahedral bonds in the upright position, and the figure also gives a view of hexagonal ZnS with a set of tetrahedral bonds in the upright position. The cubic form of zinc sulfide changes to the hexagonal form when heated above 1300°K.

Both cubic and hexagonal structures in this orientation are made up of pairs of layers, one plane in each pair being Zn and the other plane in the pair being S. The layer-pairs in both structures are stacked one above the other,

---

[17] R. E. Hanneman, H. M. Strong, and F. P. Bundy, Science **155**, 955 (1967).

[18] An excellent description of the two structures is given on pp. 308–313 of L. G. Berry and B. Mason, *Mineralogy*, Freeman, 1959. Hexagonal ZnS is usually called **wurtzite**; Wyckoff names the structure **zincite**, after ZnO.

and successive layer-pairs are displaced horizontally. A displacement is necessary to preserve the tetrahedral bond scheme: without a displacement a sequence S-Zn-S would lie in a straight line, contrary to the tetrahedral scheme which requires an angle of 109°28′ between these bonds.

In cubic ZnS the layer-pairs are displaced so that layer-pairs labeled 0, 3, 6 . . . lie exactly above each other, without sidewise displacement. Thus the stacking sequence in cubic ZnS is written as 012012012 . . . . The stacking scheme in hexagonal ZnS is such that layer-pairs 0, 2, 4, . . . lie exactly above each other, and the stacking sequence is written as 010101 . . . . All nearest-neighbor bonds are tetrahedral in both schemes. If we look only at the nearest neighbor atoms to a given atom, we cannot tell whether we are in a cubic or hexagonal crystal, but if we look beyond the nearest-neighbor atoms to the second-nearest-neighbor atoms we can distinguish the cubic and hexagonal structures. The nearest-neighbor bond lengths Zn-S are closely equal in the two structures.

Examples of the hexagonal zinc sulfide structures are

Crystal	a	c	Crystal	a	c
ZnO	3.25 Å	5.12 Å	SiC	3.25 Å	5.21 Å
ZnS	3.81	6.23	Hex. diamond	2.52	4.12
ZnSe	3.98	6.53	CdS	4.13	6.75
ZnTe	4.27	6.99	CdSe	4.30	7.02

OCCURRENCE OF NONIDEAL CRYSTAL STRUCTURES

The ideal crystal of classical crystallographers is formed by the periodic repetition in space of identical units. But no general proof[19] exists to the effect that the ideal crystal is, in fact, the state of minimum energy of the atoms at absolute zero. Many structures occur in nature that are regular without being entirely periodic. The crystallographers' ideal is not necessarily a law of nature. Some of the aperiodic structures may only be metastable, but with very long lifetimes.

Pentagonal Growth

Atoms can aggregate following a perfectly regular rule to produce a structure that is not an ideal crystal. One example, Fig. 32, is the dense packing of hard spheres with a single five-fold symmetry axis, as discussed by Bagley.[20]

[19] According to Courant, the theorem has been proved that the closest-packing arrangement of circles in a plane does form a lattice, but the corresponding theorem for spheres in three dimensions has not been proved.

[20] B. G. Bagley, Nature **208**, 674 (1965). References are given by Bagley to possible experimental realizations of five-fold symmetry in synthetic diamonds, copper dendrites, whiskers of Ni, Fe, and Pt; and cobalt crystals produced by hydrogen reduction of $CoBr_2$.

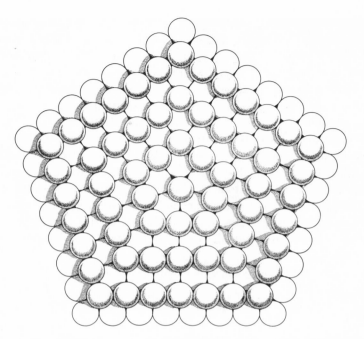

Figure 32 Two planes of hard spheres based on concentric pentagons. The pentagons of the lower plane have an odd number of spheres on each side, and the pentagons on the upper plane have an even number of spheres on each side. The spheres at the corners are counted as one-half with one side, and one-half with the other side. (After B. G. Bagley.)

A plane of spheres is constructed such that the spheres form concentric pentagons with an odd number of balls per pentagon side. We count each ball at a corner as belonging one-half to each adjacent side. The spheres along the sides of a pentagon are in contact, but parallel sides are not in contact. A second plane is constructed as in Fig. 32 so that the spheres form concentric pentagons with an even number of spheres per pentagon side. This plane is placed in contact with the first. The two planes form a layer-pair; identical layer-pairs can be stacked on each other to infinity with their five-fold axes coincident. There is a unique five-fold axis, and not a regular lattice of five-fold axes. The packing fraction is 0.72357, a little less than the packing fraction 0.74048 of fcc and hcp structures (Problem 4).

The pentagonal structure can grow in regular fashion in three dimensions from a suitable nucleus. Although rare, this is a perfectly reasonable way for a structure to grow. The scope of classical crystallography must be extended to include such structures.

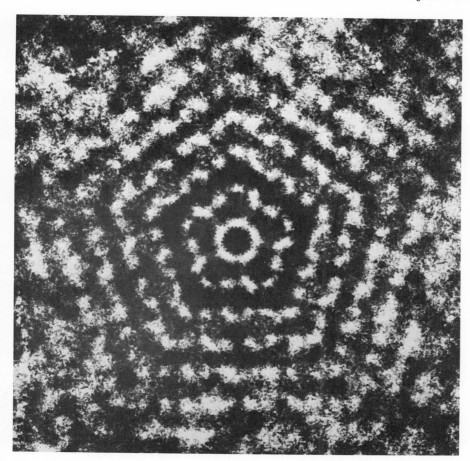

Figure 33   Packing of virus particles (probably R17 virus) about a unique five-fold axis, as observed with an electron microscope at a magnification of 430,000. The symmetry has been emphasized by the photographic superposition of five successive views obtained from rotating the original photograph 72° about the center of the pentagonal group. (After Milman, Uzman, Mitchell, and Langridge.)

Pentagonal packing of virus particles has been observed by G. Milman, B. G. Uzman, A. Mitchell, and R. Langridge, Science **152**, 1381 (1966), as shown in Fig. 33. The individual virus particles probably have icosahedral

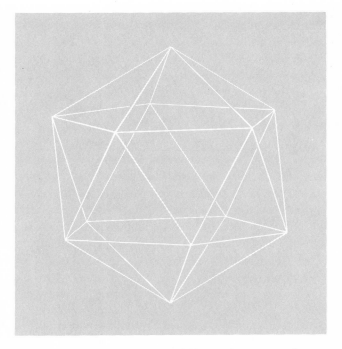

Figure 34 A regular icosahedron has 5-, 3-, and 2-fold axes of symmetry. There are twenty faces. Many virus particles have this form; each particle is constructed from many identical subunits. The symmetry requires at least 60 subunits: see "Structure of small viruses," by F. H. C. Crick and J. D. Watson, Nature **177**, 473–475 (1956); "Physical principles in the construction of regular viruses," D. L. D. Caspar and A. Klug, Cold Spring Harbor Sym. Quant. Biol. **27**, 1 (1962).

symmetry; as in Fig. 34, a five-fold symmetry axis in the particle favors the Bagley growth pattern. Paleontologists use the occurrence of a five-fold axes in fossils as evidence of their biological origin as distinguished from geological artifacts.

Random Stacking and Polytypism

We have seen that the fcc and hcp structures both consist of close-packed planes of atoms (Fig. 27a) and differ only in the stacking sequence of the planes, fcc having the sequence *ABCABC* ... and hcp having the sequence *ABABAB*

· · · . Structures are also known in which the stacking sequence of close-packed layers is random,[21] as in $ACBCABAC$ · · · . This is called **random stacking.** It occurs for example in small particles of cobalt. Cobalt has two common crystal forms, hcp stable under 400°C and fcc stable above. The difference in energy of the two forms at room temperature is small. In some respects a random stacking structure may be thought of as crystalline in two dimensions and glasslike in the third. The structure of a glass has random elements in all three dimensions.

Polytypism[22] is characterized by a stacking sequence with a long repeat unit. The classic example is SiC, which not only occurs in the sequences 012012 · · · as for cubic ZnS and 010101 · · · as for hexagonal ZnS, but also in 45 other known stacking sequences of the hexagonal layers. For example, the polytype of SiC known as 393R has a primitive cell with $a = 3.079$ Å and $c = 989.6$ Å. The longest primitive cell observed for SiC has a repeat distance of 594 layers. The mechanism that induces such long-range crystallographic order is not long-range forces, but the presence of spiral steps due to dislocations in the growth nucleus (Chapter 20). The stacking sequences in polytypism are not random, since a given sequence is repeated many times within a single crystal. Polytypism is encompassed in classical crystallography, but it still surprises us to find it.

Magnetic order. A given magnetic crystal may have different primitive cells according to whether we look at the periodicity of the charge density, which gives the chemical repeat distance, or at the periodicity of the electronic magnetic moment density, which gives the magnetic repeat distance. We may look at the charge distribution by x-ray diffraction (Chapter 2) and at the magnetic moment distribution by neutron diffraction (Chapter 16). The two periodicities need not always be related as integral multiples: if the spin arrangement is helical, the repeat distance along the axis need not be an integral multiple of the repeat distance of the chemical structure.

Collections of Crystal Structure Data

The reader who wishes to look up the crystal structure of a substance may profitably consult the excellent compilation by Wyckoff listed in the references. *Strukturbericht, Structure Reports,* and the journals Acta Crystallographica and Zeitschrift für Kristallographie are valuable aids.

In Table 4 we list for convenience the most common crystal structures and lattice constants of the elements. Hume-Rothery[23] has given a useful series

---

[21] Notice that sequences such as $ACCB$ . . . are excluded in practice: a close-packed plane cannot be put directly on top of another without a large increase in the energy.

[22] See A. R. Verma and P. Krishna, *Polymorphism and polytypism in crystals,* Wiley, 1966. A summary of polytypism in SiC is given by Wyckoff, Vol. I, pp. 113–120. The connection between polytypism and dislocations (Chapter 20) is treated by A. R. Verma in *Silicon carbide,* J. R. O'Connor and J. Smiltens, eds., Pergamon, 1960.

[23] W. Hume-Rothery, *Structure of metals and alloys,* Institute of Metals, London, 4th ed., 1962.

Table 4 Crystal structures of the elements

The data given are at room temperature for the most common form, or at the stated temperature in deg K. For further descriptions of the elements see Wyckoff, Vol. 1, Chap. 2, and Table A-6 of Barrett and Massalski. Structures labeled complex are described there. The notation ABAC refers to the sequence of close-packed planes.

Legend:
- Crystal structure.
- a lattice parameter, in Å.
- c lattice parameter, in Å.

1	2	3	4	5	6	7	8	9	10	11	12	13	14	15	16	17	18
H¹ 4K hcp 3.75 6.12																	**He⁴** 2K hcp 3.57 5.83
Li 78K bcc 3.491	**Be** hcp 2.27 3.59											**B** rhomb.	**C** diamond 3.567	**N** 20K cubic 5.66 (N₂)	**O** complex (O₂)	**F** complex	**Ne** fcc 4.46
Na 5K bcc 4.225	**Mg** hcp 3.21 5.21											**Al** fcc 4.05	**Si** diamond 5.430	**P** complex	**S** complex	**Cl** complex (Cl₂)	**Ar** 4K fcc 5.31
K 5K bcc 5.225	**Ca** fcc 5.58	**Sc** hcp 3.31 5.27	**Ti** hcp 2.95 4.68	**V** bcc 3.03	**Cr** bcc 2.88	**Mn** cubic complex	**Fe** bcc 2.87	**Co** hcp 2.51 4.07	**Ni** fcc 3.52	**Cu** fcc 3.61	**Zn** hcp 2.66 4.95	**Ga** complex	**Ge** diamond 5.658	**As** rhomb.	**Se** hex. chains	**Br** complex (Br₂)	**Kr** 4K fcc 5.64
Rb 5K bcc 5.585	**Sr** fcc 6.08	**Y** hcp 3.65 5.73	**Zr** hcp 3.23 5.15	**Nb** bcc 3.30	**Mo** bcc 3.15	**Tc** hcp 2.74 4.40	**Ru** hcp 2.71 4.28	**Rh** fcc 3.80	**Pd** fcc 3.89	**Ag** fcc 4.09	**Cd** hcp 2.98 5.62	**In** tetr. 3.25 4.95	**Sn** (α) diamond 6.49	**Sb** rhomb.	**Te** hex. chains	**I** complex (I₂)	**Xe** 4K fcc 6.13
Cs 5K bcc 6.045	**Ba** bcc 5.02	**La** hex. 3.77 ABAC	**Hf** hcp 3.19 5.05	**Ta** bcc 3.30	**W** bcc 3.16	**Re** hcp 2.76 4.46	**Os** hcp 2.74 4.32	**Ir** fcc 3.84	**Pt** fcc 3.92	**Au** fcc 4.08	**Hg** rhomb.	**Tl** hcp 3.46 5.52	**Pb** fcc 4.95	**Bi** rhomb.	**Po** sc 3.34	**At** —	**Rn** —
Fr —	**Ra** —	**Ac** fcc 5.31															

Ce fcc 5.16	**Pr** hex. 3.67 ABAC	**Nd** hex. 3.66	**Pm** —	**Sm** complex	**Eu** bcc 4.58	**Gd** hcp 3.63 5.78	**Tb** hcp 3.60 5.70	**Dy** hcp 3.59 5.65	**Ho** hcp 3.58 5.62	**Er** hcp 3.56 5.59	**Tm** hcp 3.54 5.56	**Yb** fcc 5.48	**Lu** hcp 3.50 5.55
Th fcc 5.08	**Pa** tetr. 3.92 3.24	**U** complex	**Np** complex	**Pu** complex	**Am** hex. 3.64 ABAC	**Cm** —	**Bk** —	**Cf** —	**Es** —	**Fm** —	**Md** —	**No** —	**Lw** —

Table 5 Density and atomic concentration of the elements

The data are given at atmospheric pressure and room temperature, or at the stated temperature in deg K. (Crystal modifications as for Table 4.)

Legend (per cell):
→ Density in g cm⁻³ (10³ kg m⁻³)
→ Concentration in 10²² cm⁻³ (10²⁸ m⁻³)
→ Nearest-neighbor distance, in Å (10⁻¹⁰ m)

1	2	3	4	5	6	7	8	9	10	11	12	13	14	15	16	17	18
H 4K 0.088																	**He** 2K 0.205 (at 37 atm)
Li 78K 0.542 4.700 3.023	**Be** 1.82 12.1 2.22											**B** 2.47 13.0	**C** 3.516 17.6 1.54	**N** 20K 1.03	**O**	**F** 1.44	**Ne** 4K 1.51 4.36 3.16
Na 5K 1.013 2.652 3.659	**Mg** 1.74 4.30 3.20											**Al** 2.70 6.02 2.86	**Si** 2.33 5.00 2.35	**P**	**S**	**Cl** 93K 2.03 2.02	**Ar** 4K 1.77 2.66 3.76
K 5K 0.910 1.402 4.525	**Ca** 1.53 2.30 3.95	**Sc** 2.99 4.27 3.25	**Ti** 4.51 5.66 2.89	**V** 6.09 7.22 2.62	**Cr** 7.19 8.33 2.50	**Mn** 7.47 8.18 2.24	**Fe** 7.87 8.50 2.48	**Co** 8.9 8.97 2.50	**Ni** 8.91 9.14 2.49	**Cu** 8.93 8.45 2.56	**Zn** 7.13 6.55 2.66	**Ga** 5.91 5.10 2.44	**Ge** 5.32 4.42 2.45	**As** 5.77 4.65 3.16	**Se** 4.81 3.67 2.32	**Br** 123K 4.05 2.36 3.54	**Kr** 4K 3.09 2.17 4.00
Rb 5K 1.629 1.148 4.837	**Sr** 2.58 1.78 4.30	**Y** 4.48 3.02 3.55	**Zr** 6.51 4.29 3.17	**Nb** 8.58 5.56 2.86	**Mo** 10.22 6.42 2.72	**Tc** 11.50 7.04 2.71	**Ru** 12.36 7.36 2.65	**Rh** 12.42 7.26 2.69	**Pd** 12.00 6.80 2.75	**Ag** 10.50 5.85 2.89	**Cd** 8.65 4.64 2.98	**In** 7.29 3.83 3.25	**Sn** 5.76 3.62 2.81	**Sb** 6.69 3.31 2.91	**Te** 6.25 2.94 2.86	**I** 4.95 2.36 3.54	**Xe** 4K 3.78 1.64 4.34
Cs 5K 1.997 0.905 5.235	**Ba** 3.59 1.60 4.35	**La** 6.17 2.70 3.73	**Hf** 13.20 4.52 3.13	**Ta** 16.66 5.55 2.86	**W** 19.25 6.30 2.74	**Re** 21.03 6.80 2.74	**Os** 22.58 7.14 2.68	**Ir** 22.55 7.06 2.71	**Pt** 21.47 6.62 2.77	**Au** 19.28 5.90 2.88	**Hg** 227K 14.26 4.26 3.01	**Tl** 11.87 3.50 3.46	**Pb** 11.34 3.30 3.50	**Bi** 9.80 2.82 3.07	**Po** 9.31 2.67 3.34	**At** —	**Rn** —
Fr —	**Ra** —	**Ac** 10.07 2.66 3.76															

Lanthanides:

Ce	Pr	Nd	Pm	Sm	Eu	Gd	Tb	Dy	Ho	Er	Tm	Yb	Lu
Ce 6.77 2.91 3.65	**Pr** 6.78 2.92 3.63	**Nd** 7.00 2.93 3.66	**Pm** —	**Sm** 7.54 3.03 3.59	**Eu** 5.25 2.04 3.96	**Gd** 7.89 3.02 3.58	**Tb** 8.27 3.22 3.52	**Dy** 8.53 3.17 3.51	**Ho** 8.80 3.22 3.49	**Er** 9.04 3.26 3.47	**Tm** 9.32 3.32 3.54	**Yb** 6.97 3.02 3.88	**Lu** 9.84 3.39 3.43

Actinides:

Th	Pa	U	Np	Pu	Am	Cm	Bk	Cf	Es	Fm	Md	No	Lw
Th 11.72 3.04 3.60	**Pa** 15.37 4.01 3.21	**U** 19.05 4.80 2.75	**Np** 20.45 5.20 2.62	**Pu** 19.81 4.26 3.1	**Am** 11.87 2.96 3.61	**Cm** —	**Bk** —	**Cf** —	**Es** —	**Fm** —	**Md** —	**No** —	**Lw** —

of tables of crystal structures of elements arranged according to the groups in the periodic table. Values of the density and atomic concentration are given in Table 5.

Many elements occur in several crystal structures and transform from one to the other as the temperature is varied. Other transformations take place under pressure. Sometimes two structures will coexist at the same temperature, although one may be slightly more stable. It may be inferred that the differences in the energies or free energies of certain structures may be very small, often beyond our ability to calculate theoretically. Several examples of transformations follow:

(a) Sodium is bcc at room temperature. It partially transforms[24] to hcp on cooling below 36°K or on deformation below 51°K.

(b) Lithium is bcc at room temperature. At 78°K both bcc and hcp structures coexist. The hcp phase is converted to fcc by cold working at low temperatures.

(c) Cobalt: the stable form of cobalt at room temperature is hcp, although fine powders may be fcc. Above 400°K the stable form is fcc.

(d) Carbon occurs in the diamond, graphite, hexagonal diamond, and amorphous forms, all essentially stable at room temperature.

(e) Iron is bcc up to 910°C; fcc between 910°C and 1400°C; and bcc above 1400°C.

SUMMARY

1. A lattice is an array of points related by the lattice translation operator $\mathbf{T} = n_1\mathbf{a} + n_2\mathbf{b} + n_3\mathbf{c}$, where n_1, n_2, n_3 are integers and $\mathbf{a}$, $\mathbf{b}$, $\mathbf{c}$ are called the crystal axes.

2. To form a crystal we attach to every lattice point an identical basis of N atoms at the positions $\mathbf{r}_j = x_j\mathbf{a} + y_j\mathbf{b} + z_j\mathbf{c}$, with $j = 1, 2, \ldots, N$. Here x, y, z may be selected to have values between 0 and 1.

3. The axes $\mathbf{a}$, $\mathbf{b}$, $\mathbf{c}$ are primitive for the minimum cell volume $|\mathbf{a} \cdot \mathbf{b} \times \mathbf{c}|$ for which the crystal structure can be constructed from a lattice translation operator $\mathbf{T}$ and a basis at every lattice point.

4. It may be convenient (particularly for cubic crystals) to describe a crystal in terms of a conventional cell with a volume an integral multiple of the volume of a primitive cell.

5. A plane in the crystal is denoted by the Miller indices (hkl) and a direction is denoted by $[hkl]$.

6. Important simple structures include the bcc, fcc, hcp, diamond, NaCl, CsCl, cubic ZnS, and hexagonal ZnS structures.

[24] C. S. Barrett, Acta Cryst. **9**, 671 (1955).

Problems

1. *Diamond structure.* (a) How many atoms are there in the primitive cell of diamond? (b) What is the length in angstroms of a primitive translation vector? (c) Show that the angle between the tetrahedral bonds of diamond is $109°28'$. (*Hint:* As in Fig. 16, this is the angle between the body diagonals of a cube.) (d) How many atoms are there in the conventional cubic unit cell?

2. *Five-fold axis.* A lattice cannot have five-fold rotational symmetry. Give an algebraic proof of this statement by considering a vector $\mathbf{a}$ taken to be the shortest nonvanishing translation of the lattice; show that the vector $\mathbf{a}'' + \mathbf{a}'$ is shorter than $\mathbf{a}$, where $\mathbf{a}'$, $\mathbf{a}''$ are vectors obtained from $\mathbf{a}$ by rotations of $\pm 2\pi/5$.

3. *Perpendicular to plane.* Prove that in a cubic crystal a direction $[hkl]$ is perpendicular to a plane (hkl) having the same indices.

4. *Packing fraction.* Show that the maximum proportion of the available volume which may be filled by hard spheres arranged on various lattices is: simple cubic, 0.52; body-centered cubic, 0.68; face-centered cubic, 0.74.

5. *Hcp structure.* (a) Show that the c/a ratio for an ideal hexagonal close-packed structure is $(\frac{8}{3})^{1/2} = 1.633$. If c/a is significantly larger than this value, the crystal structure may be thought of as composed of planes of closely-packed atoms, the planes being loosely stacked. (b) Show that no hcp space lattice exists (with a basis of *one* atom per lattice point). *Hint:* Show that no lattice vectors $\mathbf{a}$, $\mathbf{b}$, $\mathbf{c}$ can be found such that the set of translations $\mathbf{T}$ of Eq. (2) forms a hcp lattice. We can choose $\mathbf{a}$, $\mathbf{b}$ to form a hexagonal net in a basal plane, but the problem is with $\mathbf{c}$.

6. *Crystal structures.* Describe the crystal structures of hexagonal ZnO and NiAs. Specify the lattice and the basis which you use.

7. *Sublattices.* Show that a bcc lattice may be decomposed into two sc lattices A, B with the property that none of the nearest-neighbor lattice points to a lattice point on A lie on A, and similarly for the B lattice. Show that to obtain the same property a sc lattice is decomposed into two fcc lattices, and a fcc lattice into four sc lattices. These considerations are of interest for antiferromagnetism (Chapter 16).

8. *Closest-packing of fibers.* Find the closest-packing arrangement of identical infinite straight fibers of circular cross-section. What is the packing fraction for this arrangement?

9. *Three-fold axes and cubic crystals.* Show that a crystal which has four three-fold point symmetry axes making tetrahedral angles (Problem 1) with each other is consistent with a cubic lattice. (This is the minimum symmetry requirement for a cubic crystal.)

10. *Cesium chloride structure.* Consider a Cs^+ ion at the position 000 in the simple cubic primitive cell. (a) Give the number of first, second, and third nearest neighbor ions. The first nearest neighbors are Cl^-, the second nearest are Cs^+, and the third nearest are Cs^+. *Hint:* The sequence of the neighbors can be seen most easily by considering a (110) plane. This plane contains the three principal cubic directions: cube edge, face diagonal, and body diagonal. (b) Give the atomic coordinates x, y, z of all the first, second, and third nearest neighbors to the Cs^+ ion at 000.

References

ELEMENTARY

A. Holden, *Nature of solids*, Columbia University Press, 1965.

A. Holden and P. Singer, *Crystals and crystal growing* (Anchor S7) Doubleday, 1960.

L. G. Berry and B. Mason, *Mineralogy*, Freeman, 1959.

B. Chalmers, J. G. Holland, K. A. Jackson, and R. B. Williamson, *Crystallography: A programmed course in three dimensions*, Books 1 through 6. Appleton, 1965.

CRYSTALLOGRAPHY

M. J. Buerger, *Elementary crystallography*, Wiley, 1963.

F. C. Phillips, *An introduction to crystallography*, Wiley, 3rd ed., 1963.

GEOMETRY

D. Hilbert and S. Cohn-Vossen, *Geometry and the imagination*, Chelsea, New York, 1952. See especially Chapter 2.

H. S. M. Coxeter, *Regular polytopes*, Methuen, London, 1948. A treatment of the geometrical properties of polyhedra.

CRYSTAL GROWTH

W. D. Lawson and S. Nielsen, *Preparation of single crystals*, Butterworths, 1958.

A. Smakula, *Einkrystalle*, Springer, 1962.

J. J. Gilman, *The art and science of growing crystals*, Wiley, 1963.

W. G. Pfann, *Zone melting*, Wiley, 1966. The process of zone melting makes possible the production of materials of very high purity. See also the article by Pfann in Scientific American, Dec. 1967.

Journal of Crystal Growth.

Kristall und Technik. (A journal.)

CLASSICAL TABLES AND HANDBOOKS

P. H. Groth, *Chemische Krystallographie*, 5 volumes, W. Englemann, Leipzig, 1906.

International tables for x-ray crystallography, 3 volumes, Kynoch Press, Birmingham, 1952–1962.

C. Palache, H. Berman, and C. Frondel, *Dana's system of mineralogy*, Vols. I, II, and III, Wiley, 7th ed., 1944, 1951, 1962.

R. W. G. Wyckoff, *Crystal structures*, Interscience, 2nd ed., 1963.

W. B. Pearson, *Handbook of lattice spacings and structures of metals and alloys*, Pergamon, 1958, 1967 (2 vols.).

J. D. H. Donnay and G. Donnay, *Crystal data, determinative tables*, Amer. Cryst. Assoc., 1963, 2nd ed.

C. S. Barrett and T. B. Massalski, *Structure of metals: crystallographic methods, principles, data*, McGraw-Hill, 3rd ed. 1966.

2

Crystal Diffraction and the Reciprocal Lattice

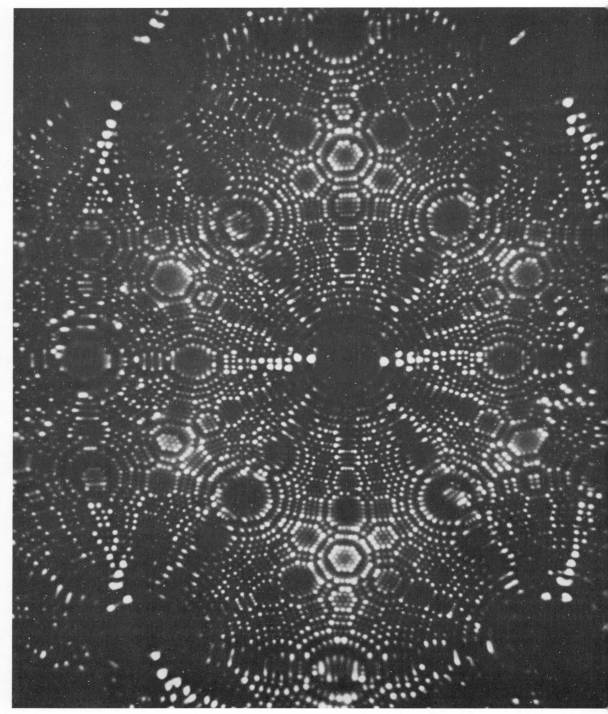

Figure A Field ion microscope image of a clean tungsten tip of radius approximately 450 Å. (Courtesy of E. W. Müller.)

Dear Nikolai Aleksandrovich,   2 October 1912

You conclude your letter by saying that the human eye shall never see atoms. You wrote this at approximately the same time people saw atoms with their own eyes; if not the atoms themselves, then the photographic images caused by them . . . . For us crystallographers this discovery is of prime importance . . . .

E. Fedorov

Excerpt from a letter to N. A. Morozov, as reproduced in *Fifty years of x-ray crystallography*, ed., P. P. Ewald.

In some situations it is possible to form a direct microscope image of the structure of a crystal. An electron microscope with a resolving power of 2 Å can resolve the prominent planes of laminar crystals such as graphite, but the resolution of the electron microscope does not at present permit the accurate direct determination of unknown crystal structures. A direct image[1] of the atoms on the surface of a tip of tungsten is shown in Fig. A; a direct image of atomic planes in a crystal is shown in Fig. B, taken with an electron microscope. But, in general, to explore the structure of crystals we study the diffraction patterns of waves that interact with atoms and that have a wavelength comparable with the interatomic spacing (10^{-8} cm) in crystals. Radiation of longer wavelength cannot resolve the details of structure on an atomic scale, and radiation of much shorter wavelength is diffracted through inconveniently small angles. The citation above from a letter by Fedorov refers to the first observations of the diffraction of x-rays by crystals.

Our object in this chapter is to show how to use the diffraction of waves by the crystal to determine the size of the cell, the position of the nuclei, and the distribution of electrons within the cell.

---

[1] A review of the method of field emission microscopy is given by R. H. Good and E. W. Müller, *Encyclo. of physics* **21**, 176–231 (1956); see also E. W. Müller, ASTM Special Tech. Publ. No. 340, pp. 80–98 (1962). Single atoms of a heavy element resting on a carbon substrate have been seen with a high-resolution scanning electron microscope by A. V. Crewe, J. Wall, and J. Langmore, Science **168**, 1338 (1970); their electron-optical system produces a focused spot of electrons about 5 Å in diameter.

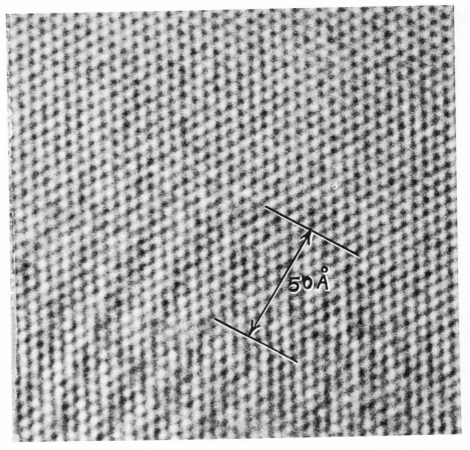

Figure B Electron microscope photograph of a crystal plane of $Al_2O_3 \cdot 4SiO_2 \cdot H_2O$ at a magnification of 3,250,000. (Courtesy of Perkin-Elmer-Hitachi.)

The Incident Beam

We study crystal structure through the diffraction of photons, neutrons, and, less often, electrons. The angle through which a wave is diffracted by a crystal depends chiefly on the crystal structure and on the wavelength of the radiation.

X-rays. The energy of an x-ray photon is related to its wavelength λ by $\epsilon = h\nu = hc/\lambda$, where $h = 6.62 \times 10^{-27}$ erg sec $= 6.62 \times 10^{-34}$ J sec is Planck's constant. In laboratory units,

$$\lambda(\text{Å}) = \frac{12.4}{\epsilon(\text{keV})} , \tag{1}$$

where λ is the wavelength in angstrom units ($1 \text{ Å} = 10^{-8}$ cm $= 10^{-10}$ m) and ϵ is the energy in kilo-electron-volts ($1 \text{ eV} = 1.60 \times 10^{-12}$ erg $= 1.60 \times 10^{-19}$ J).

Figure 1 Wavelength versus particle energy, for photons, neutrons, and electrons.

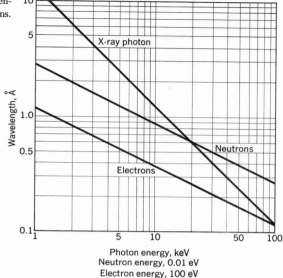

Photon energy, keV
Neutron energy, 0.01 eV
Electron energy, 100 eV

We see from Fig. 1 that for crystal studies we require photon energies in the 10 to 50 keV range. X-rays are generated both by the deceleration of electrons in metal targets and by the inelastic excitation of the core electrons in the atoms of the target. The first process gives a broad continuous spectrum; the second gives sharp lines. The radiation from a copper target bombarded by electrons shows a strong line (the $K\alpha_1$ line) at 1.541 Å; a molybdenum target has its $K\alpha_1$ line at 0.709 Å.

When an atom is exposed to electromagnetic radiation, the atomic electrons may scatter part or all of the radiation elastically, at the frequency of the incident radiation. At optical wavelengths such as 5000 Å the superposition of the waves scattered elastically by the individual atoms in a crystal results in ordinary optical refraction. But when the wavelength of the radiation is comparable with or smaller than the lattice constant, we may find one or more diffracted beams in directions quite different from the incident direction.

Neutrons. The energy of a neutron is related to its de Broglie wavelength λ by $\epsilon = h^2/2M_n\lambda^2$, where $M_n = 1.675 \times 10^{-24}$ g is the mass of the neutron. (We recall that $\epsilon = p^2/2M_n$, and the wavelength λ is related to the momentum p by $\lambda = h/p$.) In laboratory units

$$\lambda(\text{Å}) \simeq \frac{0.28}{[\epsilon(\text{eV})]^{\frac{1}{2}}} \, , \tag{2}$$

where ϵ is the neutron energy in eV. From Fig. 1 we see that $\lambda = 1$ Å for $\epsilon \approx 0.08$ eV.

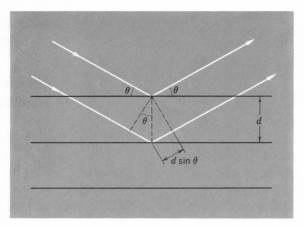

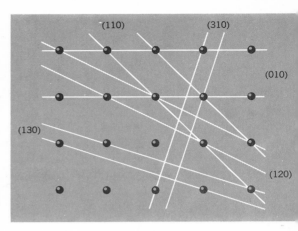

Figure 2 Derivation of the Bragg equation $2d \sin \theta = n\lambda$; here d is the spacing of parallel atomic planes and $2n\pi$ is the difference in phase between reflections from successive planes. What do we mean by a set of parallel reflecting planes? Any set of parallel planes will do, provided each plane passes through at least three non-colinear lattice points! See Fig. 3 for several examples. *The reflecting planes have nothing to do with the surface planes bounding the particular specimen,* because the x-rays or neutrons see all!

Figure 3 Several types of reflecting planes in a simple cubic crystal lattice. The planes shown are labeled by their Miller indices. We have shown in each case a set of two parallel planes. The closest distance between parallel planes tends to decrease as the indices increase; thus high index reflections require shorter wavelengths. In principle the number of different types of reflecting planes is unlimited if the crystal is infinite.

Because of their magnetic moment, neutrons interact with the magnetic electrons of a solid, and neutron methods are exceedingly valuable in structural studies of magnetic crystals. In nonmagnetic materials the neutron interacts only with the nuclei of the constituent atoms.

Electrons. The energy of an electron is related to its de Broglie wavelength λ by $\epsilon = h^2/2m\lambda^2$, where $m = 0.911 \times 10^{-27}$ g is the mass of the electron. In laboratory units

$$\lambda(\text{Å}) \cong \frac{12}{[\epsilon(\text{eV})]^{\frac{1}{2}}} , \tag{3}$$

Electrons are charged and interact strongly with matter; they penetrate a relatively short distance into a crystal. Structural studies by electron diffraction are important for surfaces, films, very thin crystals, and gases.

Bragg Law

W. L. Bragg[2] presented a simple explanation of the observed angles of the diffracted beams from a crystal. Suppose that the incident waves are reflected specularly[3] from parallel planes of atoms in the crystal, with each

[2] W. L. Bragg, Proc. Cambridge Phil. Soc. **17**, 43 (1913). The Bragg derivation is simple but is convincing only because it reproduces the results of Laue, Eq. (22). Diffraction from a single plane of atoms is the subject of Problem 6.

[3] In specular (mirrorlike) reflection the angle of incidence is equal to the angle of reflection.

plane reflecting only a very small fraction of the radiation, as with a very lightly silvered mirror. The diffracted beams are found only when the reflections from parallel planes of atoms interfere constructively, as in Fig. 2. We consider elastic scattering, so that the wavelength of the photon or neutron is not changed on reflection. Inelastic scattering (scattering accompanied by the excitation of elastic waves in the crystal) is treated in Chapter 5.

Consider a series of parallel lattice planes spaced equal distances d apart. The radiation is incident in the plane of the paper. The path difference for rays reflected from adjacent planes is $2d \sin \theta$, where θ is measured from the plane. Constructive interference of the radiation reflected from successive planes occurs whenever the path difference is an integral number n of wavelengths λ. Thus the condition for constructive reflection of the incident radiation is

$$2d \sin \theta = n\lambda \ . \tag{4}$$

This is the **Bragg law.** Observe that although the reflection from each plane is assumed to be specular, only for certain values of θ will the reflections from all parallel planes add up in phase to give a strong reflected (diffracted) beam. Of course, if each plane were perfectly reflecting, then only the first plane of a parallel set would see the radiation and any wavelength would be reflected.

The Bragg law is a consequence of the periodicity of the space lattice. The law does not refer to the arrangement or basis of atoms associated with each lattice point. The composition of the basis determines the relative intensity of the various orders n of diffraction from a given set of parallel planes.

Bragg reflection can occur only for wavelengths $\lambda \leq 2d$. This is why we cannot use visible light! Consider 1.54 Å radiation incident on a cubic crystal with a lattice constant 4.00 Å. In the first-order ($n = 1$) reflection from parallel (100) planes of the crystal we have

$$\theta = \sin^{-1}(\lambda/2d) = \sin^{-1}(1.54/8.00) = 11° \ .$$

As the wavelength is decreased, the angle is decreased. For gamma-rays glancing angles must be used.

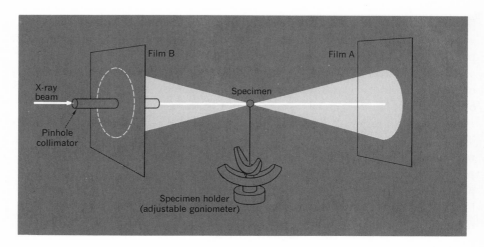

Figure 4 (*above*) A flat plate camera. With a continuous spectrum x-ray beam and a single crystal specimen, the camera produces Laue patterns. The adjustable mount is convenient for the orientation of single crystals, as is often needed in other solid state experiments. The film B is used for back-reflection Laue patterns. (Courtesy of Philips Electronic Instruments.)

Figure 5 (*right*) Laue pattern of a silicon crystal in approximately the [100] orientation. Note that the pattern is nearly invariant under a rotation of $2\pi/4$. The invariance follows from the four-fold symmetry of silicon about a [100] axis. The black center is a cut-out in the film. (Courtesy of J. Washburn.)

EXPERIMENTAL DIFFRACTION METHODS

The Bragg law (4) requires that θ and λ be matched: *x-rays of wavelength λ striking a three-dimensional crystal at an arbitrary angle of incidence will in general not be reflected.* To satisfy the Bragg law it is necessary to scan in either wavelength or angle. We do this experimentally by providing for a continuous range of values of either λ or θ, usually of θ. The standard methods of diffraction used in crystal structure analysis are designed expressly to accomplish this. Three methods are employed, sometimes with elaborate modifications, in current research.

Laue Method

In the **Laue method** a single crystal is held stationary in a beam of x-ray or neutron radiation of continuous wavelength. The crystal selects out and diffracts the discrete values of λ for which planes exist of spacing d and incidence angle θ satisfying the Bragg law. The Laue method is convenient for the rapid determination of crystal orientation and symmetry. It is also used to study the extent of crystalline imperfection under mechanical and thermal treatment.

A Laue x-ray camera is illustrated schematically in Fig. 4. A source is used which produces a beam of x-rays over a wide range of wavelengths, perhaps from 0.2 Å to 2 Å. A pinhole arrangement produces a well-collimated beam. The dimensions of the single-crystal specimen need not be greater than 1 mm. Flat film is placed to receive the diffracted beams. The diffraction pattern consists of a series of spots, shown for a silicon crystal in Fig. 5.

Each reflecting plane in the crystal selects from the incident beam a wavelength satisfying the Bragg equation $2d \sin \theta = n\lambda$. The pattern will show the symmetry of the crystal: if a crystal with four-fold axis of symmetry is oriented with the axis parallel to the beam, then the Laue pattern will show the four-fold symmetry, as in Fig. 5. The Laue pattern is widely used to orient crystals for solid state experiments.

The Laue method is practically never used for crystal structure determination. Because of the wide range of wavelengths, it is possible for several wavelengths to reflect in different orders from a single plane, so that different orders of reflection may superpose on a single spot. This makes difficult the determination of reflected intensity, and thus the determination of the basis.

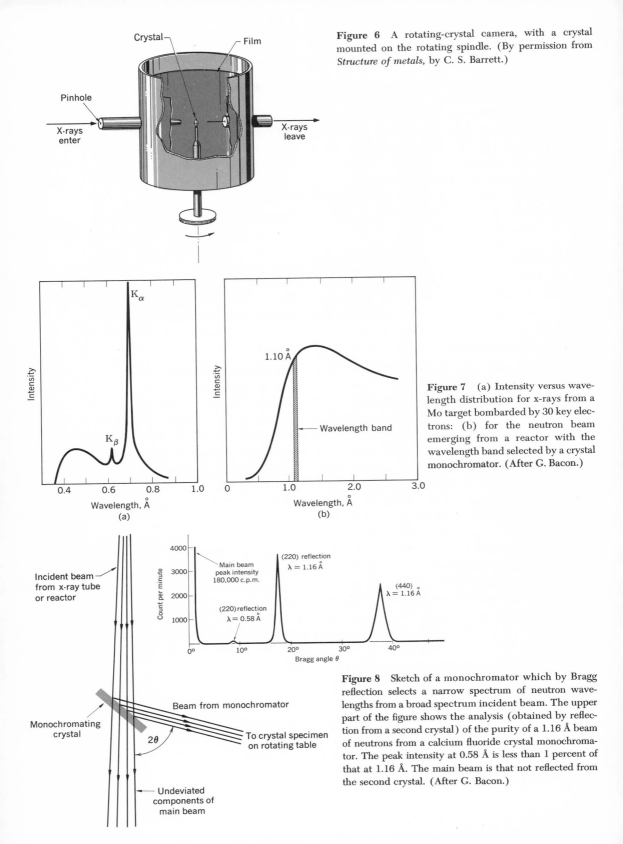

Figure 6 A rotating-crystal camera, with a crystal mounted on the rotating spindle. (By permission from *Structure of metals*, by C. S. Barrett.)

Crystal

Film

Pinhole

X-rays enter

X-rays leave

K_α

K_β

Intensity

Wavelength, Å

(a)

1.10 Å

Intensity

Wavelength band

Wavelength, Å

(b)

Figure 7 (a) Intensity versus wavelength distribution for x-rays from a Mo target bombarded by 30 key electrons: (b) for the neutron beam emerging from a reactor with the wavelength band selected by a crystal monochromator. (After G. Bacon.)

Incident beam from x-ray tube or reactor

Main beam peak intensity 180,000 c.p.m.

(220) reflection λ = 1.16 Å

(440) λ = 1.16 Å

(220) reflection λ = 0.58 Å

Count per minute

Bragg angle θ

Beam from monochromator

Monochromating crystal

2θ

To crystal specimen on rotating table

Undeviated components of main beam

Figure 8 Sketch of a monochromator which by Bragg reflection selects a narrow spectrum of neutron wavelengths from a broad spectrum incident beam. The upper part of the figure shows the analysis (obtained by reflection from a second crystal) of the purity of a 1.16 Å beam of neutrons from a calcium fluoride crystal monochromator. The peak intensity at 0.58 Å is less than 1 percent of that at 1.16 Å. The main beam is that not reflected from the second crystal. (After G. Bacon.)

Figure 9 A small rotating-crystal spectrometer at Harwell. The large can contains a neutron counter surrounded by shielding material. Most counters used in neutron diffraction studies are filled with boron trifluoride enriched in B^{10}. (Courtesy of G. Bacon.)

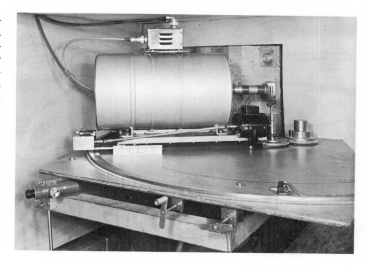

Rotating-Crystal Method

In the **rotating-crystal method** a single crystal is rotated about a fixed axis in a beam of monoenergetic x-rays or neutrons. The variation in the angle θ brings different atomic planes into position for reflection. A simple rotating-crystal x-ray camera is shown in Fig. 6. The film is mounted in a cylindrical holder concentric with a rotating spindle on which the single crystal specimen is mounted. The dimensions of the crystal usually need not be greater than 1 mm. The incident x-ray beam is made nearly monochromatic by a filter or by reflection from an earlier crystal. The beam is diffracted from a given crystal plane whenever in the course of rotation the value of θ satisfies the Bragg equation. Beams from all planes parallel to the vertical rotation axis will lie in the horizontal plane. Planes with other orientations will reflect in layers above and below the horizontal plane.

The intensity distribution of the radiation from a 30 keV x-ray tube with a molybdenum target is shown by Fig. 7a. The distribution of neutrons emerging from a nuclear reactor is shown by Fig. 7b. If we reflect the x-ray or neutron beam from a monochromating crystal, as in Fig. 8, we get the crosshatched distribution of Fig. 7b. A simple rotating-crystal neutron spectrometer is shown in Fig. 9.

Several variations of the rotating-crystal method are in common use. In *oscillating-crystal* photographs the crystal is oscillated through a limited angular range, instead of being rotated through 360°. The limited range reduces the possibility of overlapping reflections. The *Weissenberg goniometer* and also the *precession cameras* shift the film in synchronism with the oscillation of the crystal. Modern methods use diffractometers in which scintillation counters or proportional counter tubes are used to detect the diffracted radiation. These

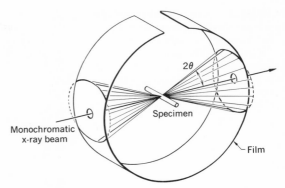

Figure 10 X-ray powder diffraction camera. The specimen is a polycrystalline powder. (Courtesy of Philips Electronic Instruments.)

methods allow automatic collection of data: complex structures may exhibit 10,000 diffracted rays.

Nearly all crystals with simple structures were solved by x-ray analysis a long time ago. One present center of interest in x-ray structure analysis is in the determination of the configuration of enzymes with a molecular weight between 10,000 and 100,000. The crystallization of an enzyme and the subsequent x-ray analysis of the structure of the crystal is the most effective method for the determination of the shape of the molecule. The coordinates of 500 to 5000 atoms in a cell are wanted, so at least this number of x-ray reflection lines is required. Computer programs have enormously simplified the problem of structure determination.

Powder Method

In the **powder method** (Fig. 10) the incident monochromatic radiation strikes a finely powdered specimen or a fine-grained polycrystalline specimen contained in a thin-walled capillary tube. The distribution of crystallite orientations will be nearly continuous. The powder method is convenient because single crystals are not required. Diffracted rays go out from individual crystallites which happen to be oriented with planes making an incident angle θ with the beam satisfying the Bragg equation. An early neutron powder spectrometer as used at Oak Ridge is shown in Fig. 11. Figures 12 and 13 are examples of powder-pattern results. Diffracted rays leave the specimen along the generators of cones concentric with the original beam. The generators make an angle 2θ with the direction of the original beam, where θ is the Bragg angle. The cones intercept the film in a series of concentric rings.

Figure 11 An early neutron spectrometer for powder studies, after E. O. Wollan and C. G. Shull, Phys. Rev. **73**, 830 (1948). The sketch shows the monochromating crystal (detailed in left center) collimating slits, shielding, second spectrometer with location of powder specimen and counter.

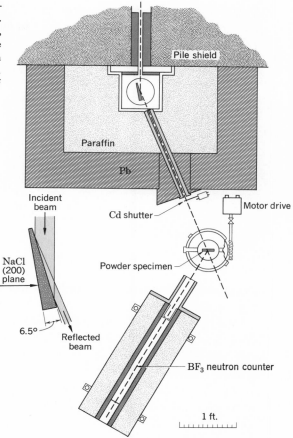

Pile shield

Paraffin

Pb

Incident
beam

NaCl
(200)
plane

Cd shutter

Motor drive

Powder specimen

6.5°

Reflected
beam

BF$_3$ neutron counter

1 ft.

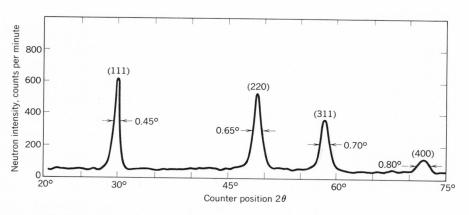

Figure 12 Neutron diffraction pattern for powdered diamond. (After G. Bacon.)

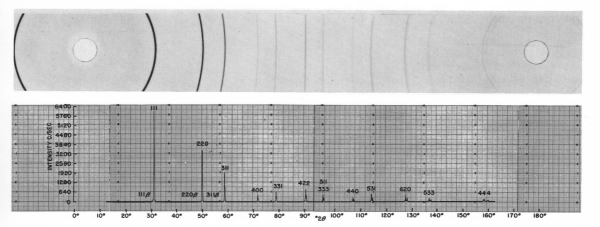

Figure 13 Powder camera and x-ray diffractometer recording of silicon. The upper figure is a film recording and the lower is a counter recording of the diffracted beams. (Courtesy of W. Parrish, Philips Laboratories and Philips Electronic Instruments.)

LAUE DERIVATION OF AMPLITUDE OF SCATTERED WAVE

The Bragg derivation of the diffraction condition gives a neat and clear statement of the condition for the constructive interference of waves scattered by point charges at the lattice points of a space lattice. When we are concerned with the *intensity* of scattering from a spatial distribution of electrons within each cell, we must carry out a deeper analysis. The simplest procedure (due to Laue) is to add up the contributions of the wavelets scattered from each volume element of the crystal. Another method of analysis, given in Advanced Topic A, looks for solutions of the electromagnetic wave equation in a medium with the dielectric constant a periodic function of position within the crystal.

The problem treated by Laue is to find the directions of the waves that leave a crystal, given the direction of the incident wave (Fig. 14). The final result is given in (20) and (22) below. We assume that the response of the crystal is linear, so that the frequency ω' of the diffracted wave generated by the response of the crystal is identical with the frequency ω of the incident wave. The magnitude of the wavevector[4] of a wave in vacuum satisfies $\omega = ck$, so that if $\omega' = \omega$, then $k' = k$, where k' is the magnitude of the wavevector of a diffracted wave in vacuum. To summarize,

$$\omega' = \omega \; ; \qquad k' = k \; . \tag{5}$$

We want to find the direction of a diffracted wave $\mathbf{k}'$ in terms of the incident wave $\mathbf{k}$ and of the primitive vectors $\mathbf{a}$, $\mathbf{b}$, $\mathbf{c}$ of the crystal lattice.

---

[4] The wavevector $\mathbf{k}$ is normal to the planes of equal phase and has the magnitude $2\pi/\lambda$, where λ is the wavelength.

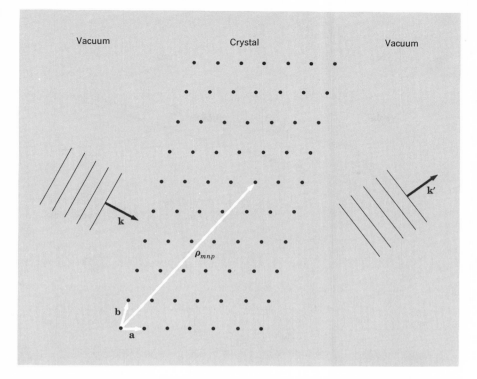

Figure 14 An electromagnetic wave of wavevector **k** is incident upon a crystal. We want to find the wavevectors **k'** of the outgoing waves created by diffraction by the atoms of the crystal. The vector ρ_{mnp} runs from the lattice point at the origin O to the lattice point at

$$\rho_{mnp} = m\mathbf{a} + n\mathbf{b} + p\mathbf{c} ,$$

where m, n, p are integers. If the phase factor of the incident wave at the origin is 1, the phase factor of the incident wave at ρ_{mnp} is $\exp[i\mathbf{k} \cdot \rho_{mnp}]$.

One component of the electric field of the incident wave in free space has the form

$$E(\mathbf{x}) = E_0\, e^{i(\mathbf{k} \cdot \mathbf{x} - \omega t)} .$$ (6)

It is the real part of this expression that has physical meaning, but we can take the real part at the end of the calculation. This wave impinges on a scattering center at position ρ, and the response of the scattering center is to create a scattered wave. The form of the scattered wave is

$$E_{sc} = C E(\rho)\,\frac{e^{ikr}}{r} = C E_0 e^{i\mathbf{k} \cdot \rho}\,\frac{e^{i(kr - \omega t)}}{r} ,$$ (7)

where we have omitted an unimportant angular factor. The amplitude of the scattered wave is proportional to the amplitude (6) of the incident wave[5] at ρ: this gives the factor $E(\rho)$, and C is a constant of proportionality which involves

---

[5] We assume the crystal is small, so that to a first approximation we may neglect the attenuation of the incident wave within the crystal.

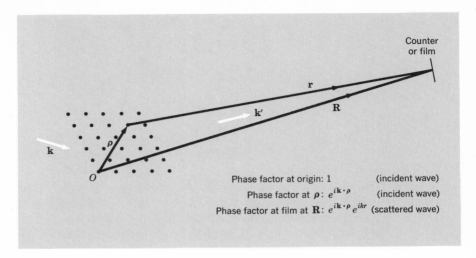

Phase factor at origin: 1   (incident wave)

Phase factor at ρ: $e^{i\mathbf{k}\cdot\rho}$ (incident wave)

Phase factor at film at $\mathbf{R}$: $e^{i\mathbf{k}\cdot\rho}\,e^{ikr}$ (scattered wave)

Figure 15 The vector **r** satisfies $\rho + \mathbf{r} = \mathbf{R}$, or $\mathbf{r} = \mathbf{R} - \rho$. We square both sides to obtain

$$r^2 = (\mathbf{R} - \rho)^2 = R^2 - 2\rho R \cos(\rho,\mathbf{R}) + \rho^2 \; .$$

We take the square root:

$$r = R\left[1 - \frac{2\rho}{R}\cos(\rho,\mathbf{R}) + \left(\frac{\rho}{R}\right)^2\right]^{\frac{1}{2}} \cong R\left(1 - \frac{\rho}{R}\cos(\rho,\mathbf{R}) + \cdots\right) ,$$

where in the development of the square root we have neglected terms of order $(\rho/R)^2$ and higher.

the details of the scattering center. The factor $1/r$ conserves the flow of scattered energy, and the entire expression is a solution of the radial wave equation

$$\left(\frac{\partial^2}{\partial r^2} + \frac{2}{r}\frac{\partial}{\partial r} - \frac{1}{c^2}\frac{\partial^2}{\partial t^2}\right)E_{sc}(r) = 0 \; , \tag{8}$$

a form of the classical electromagnetic wave equation in vacuum:

$$\nabla^2 E = \frac{1}{c^2}\frac{\partial^2 E}{\partial t^2} \; . \tag{9}$$

Our program is to sum the wavelets scattered by all centers in the crystal to give the total scattered wave amplitude at a photon counter located at a point **R** relative to the origin O in the crystal. We see from the legend of Fig. 15 that the distance between the scattering center and **R** is

$$r \cong R - \rho\cos(\rho,\mathbf{R}) \; , \tag{10}$$

if the film is very far from the crystal in comparison with the extent of the crystal.

The spatial phase factors of the scattered wave in (7) may be combined to give

$$e^{i(\mathbf{k}\cdot\rho + kr)} = e^{ikR}e^{i[\mathbf{k}\cdot\rho - k\rho\cos(\rho,\mathbf{R})]} \; , \tag{11}$$

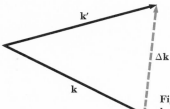

Figure 16 Definition of the scattering vector $\Delta\mathbf{k}$ such that $\mathbf{k} + \Delta\mathbf{k} = \mathbf{k}'$. In elastic scattering the magnitudes satisfy $k' = k$.

by use of (10). But the scattered wavevector $k' = k$ by (5), and the direction of $\mathbf{k}'$ may be defined as the direction of $\mathbf{R}$. Thus

$$k\rho \cos(\rho,\mathbf{R}) = k'\rho \cos(\rho,\mathbf{k}') = \mathbf{k}' \cdot \rho \ , \tag{12}$$

by definition of the scalar product. It follows that the phase factor (11) may be written as

$$e^{i(\mathbf{k} \cdot \rho + kr)} = e^{ikR} e^{i(\mathbf{k} \cdot \rho - \mathbf{k}' \cdot \rho)} = e^{ikR} e^{-i\rho \cdot \Delta\mathbf{k}} \ , \tag{13}$$

where $\Delta\mathbf{k}$ is defined as the change of the wavevector on scattering, as in Fig. 16:

$$\Delta\mathbf{k} = \mathbf{k}' - \mathbf{k} \ ; \qquad \mathbf{k}' = \mathbf{k} + \Delta\mathbf{k} \ . \tag{14}$$

The **scattering vector** $\Delta\mathbf{k}$ plays an important part in the theory.

We may now write (7) for the wave scattered from a center at ρ_{mnp} in the form

$$E_{sc}(r) = \left(\frac{CE_0 \, e^{ikR} \, e^{-i\omega t}}{R} \right) \exp\left(-i\rho_{mnp} \cdot \Delta\mathbf{k} \right) \ , \tag{15}$$

where as a reasonable approximation r has been replaced by R in the denominator. The total scattering in a given direction from a lattice of point atoms is determined by sum of E_{sc} in (15) over all lattice points. The interesting quantity is the sum of the phase factors:

$$\mathfrak{a} \equiv \sum_{mnp} \exp\left(-i\rho_{mnp} \cdot \Delta\mathbf{k} \right) \ , \tag{16}$$

for this sum determines the allowed scattering directions, as shown below. More generally the scattering is from an electron distribution throughout the crystal. If the scattering from a volume element dV of the crystal is proportional to the local electron concentration $n(\rho)$, then the scattering amplitude is proportional to the integral

$$\int dV \, n(\rho) e^{-i\rho \cdot \Delta\mathbf{k}} \ . \tag{17}$$

Scattering from Lattice of Point Atoms

We consider a finite crystal in the form of a parallelepiped. There are identical point scattering centers at every lattice point

$$\rho_{mnp} = m\mathbf{a} + n\mathbf{b} + p\mathbf{c} \ , \tag{18}$$

where m, n, p are integers which run from 0 to $M - 1$. The crystal has M^3 primitive cells. As seen in (16), the total scattered radiation is proportional to

$$\mathcal{a} \equiv \sum_{mnp} \exp[-i(m\mathbf{a} + n\mathbf{b} + p\mathbf{c}) \cdot \Delta\mathbf{k}] \ . \tag{19}$$

We call $\mathcal{a}$ the **scattering amplitude.**

The sum over the lattice points is a maximum when

$$\rho_{mnp} \cdot \Delta\mathbf{k} = (m\mathbf{a} + n\mathbf{b} + p\mathbf{c}) \cdot \Delta\mathbf{k} = 2\pi(\text{integer}) \tag{20}$$

for all lattice points ρ_{mnp}, because now each term of the form $\exp(-i\rho_{mnp} \cdot \Delta\mathbf{k})$ is equal to unity. When $\Delta\mathbf{k}$ satisfies (20), the sum for the scattering amplitude over a crystal with M^3 lattice points gives

$$\mathcal{a}_{\max} = M^3 \ . \tag{21}$$

A departure of $\Delta\mathbf{k}$ from a value that satisfies (20) will drastically reduce the value of the sum in (19). It is left to Problem 5 to show that the scattering amplitude vanishes as the number of lattice points increases, unless $\Delta\mathbf{k}$ exactly satisfies the diffraction condition (20).

Diffraction Conditions

The quantity $\Delta\mathbf{k} = \mathbf{k}' - \mathbf{k}$ satisfies the diffraction condition (20) if the following three equations are satisfied simultaneously for integral values of h,k,l:

$$\boxed{\mathbf{a} \cdot \Delta\mathbf{k} = 2\pi h; \qquad \mathbf{b} \cdot \Delta\mathbf{k} = 2\pi k; \qquad \mathbf{c} \cdot \Delta\mathbf{k} = 2\pi l \ .} \tag{22}$$

These equations are known as the **Laue equations** for the diffraction maxima. They may be solved for the vector $\Delta\mathbf{k}$. If $\Delta\mathbf{k}$ satisfies (22), then the scattering amplitude (19) becomes

$$\mathcal{a} = \sum_{mnp} \exp[-2\pi i(mh + nk + pl)] \ , \tag{23}$$

where $mh + nk + pl$ assumes only integral values because h, k, l, m, n, p are all integers. For a crystal in the form of a parallelepiped of edges Ma, Mb, Mc

we then have the result

$$\mathcal{a} = \sum_{m=0}^{M-1} \sum_{n=0}^{M-1} \sum_{p=0}^{M-1} (1) = M^3 \tag{24}$$

when Δk satisfies the Laue equations for a diffraction maximum.[6]

The solution of the Laue equations is particularly simple if the crystal axes $\mathbf{a}, \mathbf{b}, \mathbf{c}$ are mutually orthogonal, for then

$$\Delta \mathbf{k} = 2\pi \left(\frac{h}{a} \hat{\mathbf{a}} + \frac{k}{b} \hat{\mathbf{b}} + \frac{l}{c} \hat{\mathbf{c}} \right) \tag{25}$$

is the solution, where $\hat{\mathbf{a}}, \hat{\mathbf{b}}, \hat{\mathbf{c}}$ are unit vectors along the crystal axes and h, k, l are integers. If the crystal axes are not orthogonal, the result (25) is not a solution of the Laue equations, for now quantities such as $\mathbf{a} \cdot \hat{\mathbf{b}}$ are not all zero. For this problem we need to develop the concept of vectors in the reciprocal lattice. This concept is so useful, elegant, and general that we shall use it throughout the text for the expression of all the results of wave problems in crystals, including the theory of energy bands. The concept is due to J. Willard Gibbs.

RECIPROCAL LATTICE

Consider the expression

$$\Delta \mathbf{k} = h\mathbf{A} + k\mathbf{B} + l\mathbf{C} , \tag{26}$$

where h, k, l are the integers of the Laue equations (22) and $\mathbf{A}, \mathbf{B}, \mathbf{C}$ are vectors to be determined. On direct substitution we see that (26) will be a solution of (22) if all these relations are satisfied:

$$
\begin{array}{lll}
\mathbf{A} \cdot \mathbf{a} = 2\pi , & \mathbf{B} \cdot \mathbf{a} = 0 , & \mathbf{C} \cdot \mathbf{a} = 0 ; \\
\mathbf{A} \cdot \mathbf{b} = 0 , & \mathbf{B} \cdot \mathbf{b} = 2\pi , & \mathbf{C} \cdot \mathbf{b} = 0 ; \\
\mathbf{A} \cdot \mathbf{c} = 0 , & \mathbf{B} \cdot \mathbf{c} = 0 , & \mathbf{C} \cdot \mathbf{c} = 2\pi .
\end{array}
\tag{27}
$$

We see from the first column of (27) that $\mathbf{A}$ must be perpendicular to $\mathbf{b}$ and to $\mathbf{c}$. A vector perpendicular to $\mathbf{b}$ and $\mathbf{c}$ is given by the vector product $\mathbf{b} \times \mathbf{c}$; to form $\mathbf{A}$ there remains only the question of normalizing $\mathbf{b} \times \mathbf{c}$ to satisfy the equation $\mathbf{A} \cdot \mathbf{a} = 2\pi$. We can satisfy all the equations (27) by choosing

$$\boxed{\mathbf{A} = 2\pi \frac{\mathbf{b} \times \mathbf{c}}{\mathbf{a} \cdot \mathbf{b} \times \mathbf{c}} ; \qquad \mathbf{B} = 2\pi \frac{\mathbf{c} \times \mathbf{a}}{\mathbf{a} \cdot \mathbf{b} \times \mathbf{c}} ; \qquad \mathbf{C} = 2\pi \frac{\mathbf{a} \times \mathbf{b}}{\mathbf{a} \cdot \mathbf{b} \times \mathbf{c}} .} \tag{28}$$

---

[6] Like the Bragg law, the Laue equations are necessary conditions for a diffracted beam. If the crystal cell contains more than one atom the equations are not sufficient conditions, for the scattered amplitude will vanish if the structure factor (discussed below) is zero.

These[7] are the **fundamental vectors of the reciprocal lattice.** They are orthogonal only if **a**, **b**, **c** are orthogonal. We have written all the denominators as **a · b × c**, because by elementary vector algebra

$$\mathbf{b \cdot c \times a = c \cdot a \times b = a \cdot b \times c} \ ,$$

whose magnitude is the volume of the cell of the crystal lattice. The reciprocal lattice should be chosen to be consistent with the crystal lattice, with either primitive **a**, **b**, **c**, or conventional **a**, **b**, **c**.

It is as important to be able to visualize the reciprocal lattice as to visualize the real crystal lattice. Every crystal structure has two important lattices associated with it, the crystal lattice and the reciprocal lattice. A diffraction pattern of a crystal is a map of the reciprocal lattice of the crystal, in contrast to a microscope image which is a map of the real crystal structure. The two lattices are related by the definitions (28). When we rotate a crystal, we rotate both the direct lattice and the reciprocal lattice. Vectors in the crystal lattice have the dimensions of [length]; vectors in the reciprocal lattice have the dimensions of [length]$^{-1}$. The crystal lattice is a lattice in real or ordinary space; the reciprocal lattice is a lattice in the associated Fourier space. The term **Fourier space** is motivated below. Wavevectors are always drawn in Fourier space.

The lattice points ρ_{mnp} of the crystal lattice are given by

$$\rho_{mnp} = m\mathbf{a} + n\mathbf{b} + p\mathbf{c} \ , \qquad (m, n, p = \text{integers}) \ . \qquad (29)$$

Similarly, we define the **reciprocal lattice points** or **reciprocal lattice vectors G** in Fourier space as

$$\boxed{\mathbf{G} = h\mathbf{A} + k\mathbf{B} + l\mathbf{C} \ , \qquad (h, k, l = \text{integers}) \ .} \qquad (30)$$

Every position in Fourier space has a meaning, but there is a special importance to the reciprocal lattice points defined by (30). These points are of the form (26), so that if $\Delta \mathbf{k}$ is equal to any reciprocal lattice vector **G**, then the Laue equations for a diffraction maximum are satisfied.

To see the significance of the **G**'s, form the scalar product:

$$\begin{aligned} \mathbf{G} \cdot \rho_{mnp} &= (h\mathbf{A} + k\mathbf{B} + l\mathbf{C}) \cdot (m\mathbf{a} + n\mathbf{b} + p\mathbf{c}) \\ &= 2\pi(hm + kn + lp) = 2\pi(\text{integer}). \end{aligned}$$

It follows that

$$\exp\left(i\mathbf{G} \cdot \rho_{mnp}\right) = 1 \ .$$

[7] Crystallographers usually omit the factor of 2π from the definitions and then denote the reciprocal lattice vectors by $\mathbf{b}_1$, $\mathbf{b}_2$, $\mathbf{b}_3$; most solid state physicists include the factor 2π. Both groups have vigorous feelings, and one of the most distinguished crystallographers has written that "Willard Gibbs will rotate at increasing speed in his grave as long as this blur is perpetrated by solid state physicists in connection with his beautiful invention of reciprocal lattice vectors". (Private communication from P. P. Ewald.)

EXAMPLE: *Fourier Analysis of Periodic Distributions.* The electron concentration $n(\rho)$ in a crystal may be expressed as a Fourier series[8]

$$n(\rho) = \sum_{\mathbf{K}} n_{\mathbf{K}}\, e^{i\mathbf{K}\cdot\rho} \ , \qquad (31)$$

where ρ is any point in the crystal. We now prove an important theorem: for any function that has the translational periodicity of the lattice, the only values of $\mathbf{K}$ that appear in the corresponding Fourier series are the reciprocal lattice vectors $\mathbf{G}$ as defined by (30).

We form $n(\rho + \rho_{mnp})$, where $\rho_{mnp} = m\mathbf{a} + n\mathbf{b} + p\mathbf{c}$ is a crystal lattice translation:

$$n(\rho + \rho_{mnp}) = \sum_{\mathbf{K}} n_{\mathbf{K}} e^{i\mathbf{K}\cdot\rho} e^{i\mathbf{K}\cdot\rho_{mnp}} \ . \qquad (32)$$

This will have the desired translational periodicity and be equal to $n(\rho)$ only if

$$\mathbf{K}\cdot(m\mathbf{a} + n\mathbf{b} + p\mathbf{c}) = 2\pi(\text{integer}) \ . \qquad (33)$$

But this is precisely the form of the condition $\mathbf{G}\cdot\rho_{mnp} = 2\pi(\text{integer})$ that we encountered above. Thus (31) reduces to

$$\boxed{n(\rho) = \sum_{\mathbf{G}} n_{\mathbf{G}}\, e^{i\mathbf{G}\cdot\rho} \ ,} \qquad (34)$$

where $\mathbf{G} = h\mathbf{A} + k\mathbf{B} + l\mathbf{C}$ is any reciprocal lattice vector. The quantities h, k, l are integers, by definition of the $\mathbf{G}$'s.

The set of quantities $n_{\mathbf{G}}$ describe completely the distribution of electrons in the entire crystal, and they also describe the diffraction of x-rays. To show this connection we combine (17) with (34):

$$\int dV\, n(\rho)e^{-i\rho\cdot\Delta\mathbf{k}} = \sum_{\mathbf{G}} n_{\mathbf{G}} \int dV\, e^{i\mathbf{G}\cdot\rho}e^{-i\Delta\mathbf{k}\cdot\rho} \ . \qquad (35)$$

The integral is equal to the volume V if $\Delta\mathbf{k}$ is equal to a reciprocal lattice vector $\mathbf{G}$, and the integral may be shown to be equal to zero if $\Delta\mathbf{k}$ is not equal to a reciprocal lattice vector. Eq. (35) tells us that $n_{\mathbf{G}}$ is a measure of the amplitude of the diffracted beam.

---

[8] A Fourier series expansion in one dimension of a real periodic function $n(x)$ is usually written as

$$n(x) = C_0 + \sum_{p>0} (C_p \cos px + S_p \sin px) \ , \qquad (31a)$$

where all coefficients C and S are real. The series

$$\begin{aligned} n(x) &= n_0 + \sum_{p\neq 0} n_p e^{ipx} \\ &= n_0 + \sum_{p>0} [n_p(\cos px + i\sin px) + n_{-p}(\cos px - i\sin px)] \end{aligned} \qquad (31b)$$

can be made identically equal to (31a) by the suitable choice of the real and imaginary parts of the complex coefficients n_p and n_{-p}:

$$C_p = \Re\{n_p + n_{-p}\} \ ; \qquad S_p = in_p - in_{-p} = -\Im\{n_p - n_{-p}\} \ ; \qquad n_p^* = n_{-p} \ . \quad (31c)$$

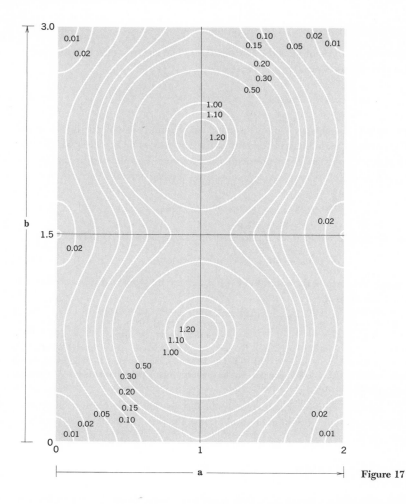

Figure 17

In Figs. 17 and 18 we give an example of the Fourier analysis of a periodic distribution in a crystal. Figure 17 shows a single rectangular primitive cell with axes **a, b** and a basis of two identical atoms. Contour lines of constant electron concentration are drawn. The shaded area in Fig. 18 is a cell of the reciprocal lattice. Reciprocal lattice points are shown together with the values of n_G for the Fourier coefficients of the charge distribution of Fig. 17.

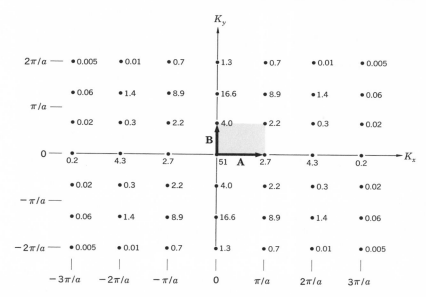

Figure 18 The probability density may be expressed as a Fourier series $n(\mathbf{r}) = \sum_{G} n_G e^{i\mathbf{G}\cdot\mathbf{r}}$, where the **G**'s are vectors in the reciprocal lattice. Here we give values of $10^3 \times |n_G|$ at reciprocal lattice points near the origin in Fourier space, for the charge distribution of Fig. 17. There are actually an infinite number of reciprocal lattice points. Notice how fast the values of the Fourier coefficients drop off as we move out from $\mathbf{G} = 0$. This characteristic decrease makes it difficult to detect diffracted beams of x-rays at high values of G. The charge distribution·was taken as Gaussian to simplify the numerical work. The distribution about a hydrogen atom is not Gaussian, but varies as $e^{-\lambda r}$, where λ is a constant. (Courtesy of Y. Tsang.)

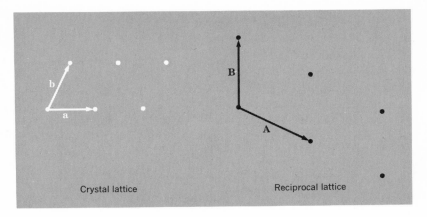

Figure 19 Reciprocal lattice in two dimensions: **A** and **B** are perpendicular to sets of planes (lines) in the crystal lattice, namely, to the lines parallel to **b** and **a**, respectively. Any vector between reciprocal lattice points is perpendicular to some lattice plane in the crystal lattice.

EXAMPLE: *Reciprocal Lattice in Two Dimensions.* A particular two-dimensional lattice (Fig. 19) has the basis vectors $\mathbf{a} = 2\hat{\mathbf{x}}$; $\mathbf{b} = \hat{\mathbf{x}} + 2\hat{\mathbf{y}}$. Find the basis vectors of the reciprocal lattice.

We can use our three-dimensional definitions in this two-dimensional problem if we suppose that **c** is parallel to the z axis, for then the plane of the reciprocal lattice vectors **A** and **B** will be in the plane of **a** and **b**. Let us take $\mathbf{c} = \hat{\mathbf{z}}$. Now

$$\mathbf{c} \times \mathbf{a} = \hat{\mathbf{z}} \times (2\hat{\mathbf{x}}) = 2\hat{\mathbf{y}} \ ;$$

$$\mathbf{b} \times \mathbf{c} = \hat{\mathbf{x}} \times \hat{\mathbf{z}} + 2\hat{\mathbf{y}} \times \hat{\mathbf{z}} = -\hat{\mathbf{y}} + 2\hat{\mathbf{x}} \ ; \qquad \mathbf{a} \cdot \mathbf{b} \times \mathbf{c} = 4 \ .$$

These results substituted in (28) give

$$\mathbf{A} = \pi\hat{\mathbf{x}} - \tfrac{1}{2}\pi\hat{\mathbf{y}} \ ; \qquad \mathbf{B} = \pi\hat{\mathbf{y}} \ ,$$

as sketched in Fig. 19.

Ewald Construction

There are two conditions for a diffracted beam, the first on the frequency and the second on the wavevector. The two conditions combined lead to a most helpful geometrical construction for the diffraction equation. We let **k** denote the wavevector and ω the frequency of the incident beam; **k**′ and ω′ refer to the scattered beam. Then:

(a) The scattering is elastic, so that the energy of the x-ray quantum is conserved:

$$\hbar\omega' = \hbar\omega \ . \tag{36}$$

Now by the dispersion relation for electromagnetic waves in free space we have $\omega' = ck'$ and $\omega = ck$, whence

$$k' = k \ . \tag{37}$$

(b) The scattering condition on the wavevector is that $\Delta \mathbf{k} = \mathbf{G}$, or

$$\mathbf{k}' = \mathbf{k} + \mathbf{G} \ , \tag{38}$$

by the definition (14) of $\Delta \mathbf{k}$ and the solution (30) of the Laue equations. In effect, Eq. (38) is a selection rule[9] for the scattered wavevector $\mathbf{k}'$.

We take the square of both sides of (38) to obtain

$$k'^2 = (\mathbf{k} + \mathbf{G})^2 = k^2 + 2\mathbf{k}\cdot\mathbf{G} + G^2 \ . \tag{39}$$

Now by (37) we have $k'^2 = k^2$, so that

G = recipr. lattice. vector

$$\boxed{2\mathbf{k}\cdot\mathbf{G} + G^2 = 0 \ .} \tag{40}$$

We shall encounter this equation again and again in problems of wave propagation in periodic lattices. It can be shown to be equivalent to the Bragg equation $2d \sin\theta = n\lambda$, and it is convenient to refer to (40) as the Bragg condition. We do not prove the equivalence because we shall always use the diffraction condition in the form (40). This expression is used below in the construction of Brillouin zones.

---

[9] The selection rule $\mathbf{k}' = \mathbf{k} + \mathbf{G}$ may be viewed as an expression of the conservation of momentum in a crystal. In free space the momentum of a photon of energy $\hbar\omega$ is $\hbar\omega/c$, which is equal to $\hbar k$. Now a plane wave of form $e^{i\mathbf{k}\cdot\mathbf{r}}$ in free space is modulated after entering the crystal by the periodic distribution of electronic charge or local refractive index, so that in the crystal

$$e^{i\mathbf{k}\cdot\mathbf{r}} \rightarrow e^{i\mathbf{k}\cdot\mathbf{r}} \sum_{\mathbf{G}} C_{\mathbf{G}} e^{i\mathbf{G}\cdot\mathbf{r}} \ ,$$

where the $C_{\mathbf{G}}$ are constants which depend on the electron distribution. We see that a plane wave in free space will contain in the crystal additional wavevector components $\mathbf{k} + \mathbf{G}$. These components form the diffracted waves. An individual diffracted wave then looks as if it has the photon momentum $\hbar(\mathbf{k} + \mathbf{G})$, for this is clearly the momentum in the beam when the photon leaves the crystal to reenter free space.

Conservation of the total momentum of the system is assured if the crystal then recoils with momentum $-\hbar\mathbf{G}$, as in Fig. 20a. The crystal recoil is too small to detect, although similar recoil processes have been detected for atoms and for nuclei. The recoil velocity of a crystal of mass 1 g for a reflection with reciprocal lattice vector of length 1×10^{-8} cm is

$$v = \frac{\hbar G}{M} \approx 10^{-19} \frac{\text{cm}}{\text{sec}} \ ,$$

too small to be detectable.

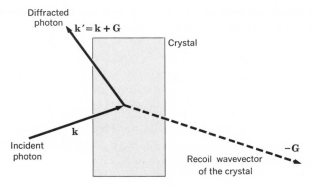

Figure 20a In a reflection described by $\mathbf{k}' = \mathbf{k} + \mathbf{G}$ the crystal specimen recoils with momentum $-\hbar\mathbf{G}$, which assures that the momentum $\hbar\mathbf{k}' - \hbar\mathbf{G}$ of the total system, specimen plus photon, after reflection is equal to the total momentum $\hbar\mathbf{k}$ of the system in the initial condition with the crystal at rest. The momentum of the incident photon is $\hbar\mathbf{k}$, where $k = \omega/c$.

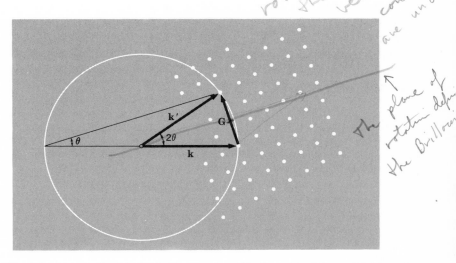

Figure 20b The points on the right-hand side are reciprocal lattice points of the crystal. The vector $\mathbf{k}$ is drawn in the direction of the incident x-ray beam and it terminates at any reciprocal lattice point. We draw a sphere of radius $k = 2\pi/\lambda$ about the origin of $\mathbf{k}$. A diffracted beam will be formed if this sphere intersects any other point in the reciprocal lattice. The sphere as drawn intercepts a point connected with the end of $\mathbf{k}$ by a reciprocal lattice vector $\mathbf{G}$. The diffracted x-ray beam is in the direction $\mathbf{k}' = \mathbf{k} + \mathbf{G}$. This construction is due to P. P. Ewald.

In Fourier space the selection rules (37) and (38) have the geometrical significance shown in Fig. 20b. Notice that the length of the vector $\mathbf{k}'$ will be equal to the length of $\mathbf{k}$ if $\mathbf{k}'$ terminates somewhere on the spherical surface of radius k. Furthermore, if both $\mathbf{k}$ and $\mathbf{k}'$ terminate on reciprocal lattice points they must be connected by a reciprocal lattice vector, whence it follows that $\mathbf{k}' = \mathbf{k} + \mathbf{G}$. The construction in the figure is known as the **Ewald construction** and is used particularly in the analysis of x-ray and neutron diffraction. The following section describes the Brillouin construction which is often used for the description of electron states in solids and, although rarely used for x-rays, gives a clear picture of the diffraction conditions.

BRILLOUIN ZONES

A Brillouin zone is defined as a Wigner-Seitz cell in the reciprocal lattice. (The Wigner-Seitz cell of the direct lattice was described by Fig. 1.8). The definition of the Brillouin zone gives a vivid geometrical interpretation of the diffraction condition $2\mathbf{k} \cdot \mathbf{G} + G^2 = 0$. First it is convenient to substitute $-\mathbf{G}$ for $\mathbf{G}$ to obtain the diffraction condition in the form

$$2\mathbf{k} \cdot \mathbf{G} = G^2 . \tag{41}$$

The indicated replacement does not change the meaning of the diffraction condition because if $\mathbf{G}$ is a vector in the reciprocal lattice, then $-\mathbf{G}$ is also a vector in the reciprocal lattice. Let us rewrite (41) as

$$\mathbf{k} \cdot (\tfrac{1}{2}\mathbf{G}) = (\tfrac{1}{2}G)^2 . \tag{42}$$

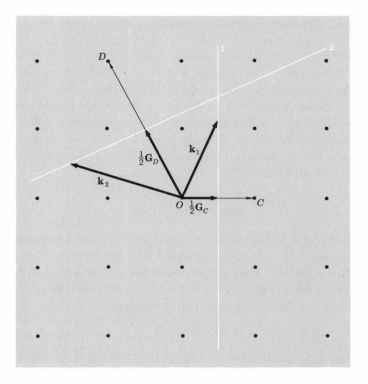

Figure 21 Reciprocal lattice points near the point O at the origin of the reciprocal lattice. The reciprocal lattice vector $\mathbf{G}_C$ connects points OC; and $\mathbf{G}_D$ connects OD. Two planes 1 and 2 are drawn which are the perpendicular bisectors of $\mathbf{G}_C$ and $\mathbf{G}_D$, respectively. Any vector from the origin to the plane 1, such as $\mathbf{k}_1$, will satisfy the diffraction condition $\mathbf{k}_1 \cdot (\tfrac{1}{2}\mathbf{G}_C) = (\tfrac{1}{2}\mathbf{G}_C)^2$. Any vector from the origin to the plane 2, such as $\mathbf{k}_2$, will satisfy the diffraction condition $\mathbf{k}_2 \cdot (\tfrac{1}{2}\mathbf{G}_D) = (\tfrac{1}{2}\mathbf{G}_D)^2$.

We construct a plane normal to the vector $\mathbf{G}$ at the midpoint; then (Fig. 21) **any vector k from the origin to the plane will satisfy the diffraction condition.** The plane thus described forms a part of the boundary of a Brillouin zone.

A reciprocal lattice vector has a definite length and a definite orientation relative to the crystal axes **a, b, c** of the crystal specimen under study. An x-ray beam incident on the crystal will be diffracted if its wavevector has the magnitude and direction required by (42), and the diffracted beam will be in the direction of the vector $\mathbf{k} + \mathbf{G}$.

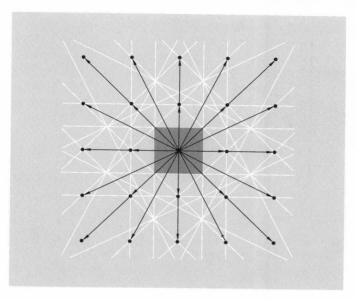

Figure 22 Square reciprocal lattice with reciprocal lattice vectors shown as fine black lines. The lines shown in white are perpendicular bisectors of the reciprocal lattice vectors. The central square is the smallest volume about the origin which is bounded entirely by white lines. The square is the Wigner-Seitz primitive cell of the reciprocal lattice.

The set of planes that are the perpendicular bisectors of the various reciprocal lattice vectors are of particular importance in the theory of wave propagation in crystals, because a wave whose wavevector drawn from the origin terminates on any of these planes will satisfy the conditions for diffraction. These planes divide up the Fourier space of the crystal into odd bits and pieces, as shown in two dimensions in Fig. 22. The central square of Fig. 22 is a primitive cell of the reciprocal lattice, and we recognize that the square has been drawn according to the prescription in Chapter 1 for the primitive Wigner-Seitz cell, except that there the cell was in real space and here it is in Fourier space.

The central cell in the reciprocal lattice is of special importance in the theory of solids, and we call it the first Brillouin zone. **The first Brillouin zone is the smallest volume entirely enclosed by planes that are the perpendicular bisectors of the reciprocal lattice vectors drawn from the origin.** The first

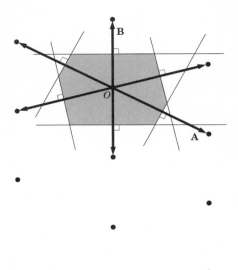

Figure 23 (***left above***) Construction of the first Brillouin zone for an oblique lattice in two dimensions. The crystal lattice is similar to that of Fig. 19. We first draw a number of vectors from O to nearby points in the reciprocal lattice. Next we construct lines perpendicular to these vectors at their midpoints. The smallest enclosed area is the first Brillouin zone.

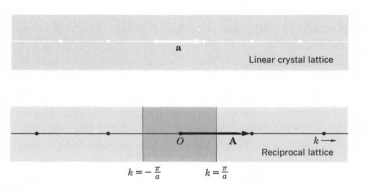

Figure 24 Crystal and reciprocal lattices in one dimension. The basis vector in the reciprocal lattice is **A**, of length equal to $2\pi/a$. The shortest reciprocal lattice vectors from the origin are **A** and $-\mathbf{A}$. The perpendicular bisectors of these vectors form the boundaries of the first Brillouin zone. The boundaries are at $k = \pm\pi/a$.

Brillouin zone of an oblique lattice in two dimensions is constructed in Fig. 23 and of a linear lattice in one dimension in Fig. 24. The zone boundaries of the linear lattice are at $k = \pm\pi/a$, where a is the primitive of the crystal lattice. Historically, Brillouin zones are not part of the folklore of x-ray diffraction analysis of crystal structures, but the zones are an essential part of the analysis

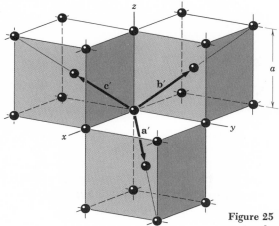

Figure 25 Primitive basis vectors of the body-centered cubic lattice.

of the electronic energy-band structure of crystals, as discussed in Chapters 9 and 10. The special importance of the first zone is developed in Chapter 10. **The Brillouin construction exhibits all the incident wavevectors k which can be Bragg-reflected by the crystal.**

Reciprocal Lattice to sc Lattice. The primitive translation vectors of a simple cubic lattice may be taken as the set

$$\mathbf{a} = a\hat{\mathbf{x}} \; ; \qquad \mathbf{b} = a\hat{\mathbf{y}} \; ; \qquad \mathbf{c} = a\hat{\mathbf{z}} \; . \tag{43}$$

The volume of the cell is $\mathbf{a} \cdot \mathbf{b} \times \mathbf{c} = a^3$. The primitive translation vectors of the reciprocal lattice are found from the standard prescription (28):

$$\mathbf{A} = 2\pi \frac{\mathbf{b} \times \mathbf{c}}{\mathbf{a} \cdot \mathbf{b} \times \mathbf{c}} = \frac{2\pi}{a}\hat{\mathbf{x}} \; ; \qquad \mathbf{B} = \frac{2\pi}{a}\hat{\mathbf{y}} \; ; \qquad \mathbf{C} = \frac{2\pi}{a}\hat{\mathbf{z}} \; . \tag{44}$$

Thus the reciprocal lattice is itself a simple cubic lattice, now of lattice constant $2\pi/a$.

The boundaries of the first Brillouin zones are the planes normal to the six vectors

$$\pm\tfrac{1}{2}\mathbf{A} = \pm\frac{\pi}{a}\hat{\mathbf{x}}; \qquad \pm\tfrac{1}{2}\mathbf{B} = \pm\frac{\pi}{a}\hat{\mathbf{y}}; \qquad \pm\tfrac{1}{2}\mathbf{C} = \pm\frac{\pi}{a}\hat{\mathbf{z}} \; . \tag{45}$$

The six planes bound a cube of edge $2\pi/a$ and of volume $(2\pi/a)^3$; this cube is the first Brillouin zone of the sc crystal lattice.

Reciprocal Lattice to bcc Lattice. The primitive translation vectors of the bcc lattice as shown in Fig. 25 are

$$\begin{aligned} \mathbf{a}' &= \tfrac{1}{2}a(\hat{\mathbf{x}} + \hat{\mathbf{y}} - \hat{\mathbf{z}}) \; ; \\ \mathbf{b}' &= \tfrac{1}{2}a(-\hat{\mathbf{x}} + \hat{\mathbf{y}} + \hat{\mathbf{z}}) \; ; \\ \mathbf{c}' &= \tfrac{1}{2}a(\hat{\mathbf{x}} - \hat{\mathbf{y}} + \hat{\mathbf{z}}) \; , \end{aligned} \tag{46}$$

where a is the side of the conventional unit cube and $\hat{\mathbf{x}}, \hat{\mathbf{y}}, \hat{\mathbf{z}}$ are orthogonal unit vectors parallel to the cube edges. The volume of the primitive cell is

$$V = |\mathbf{a}' \cdot \mathbf{b}' \times \mathbf{c}'| = \tfrac{1}{2}a^3 \; . \tag{47}$$

The primitive translations **A**, **B**, **C** of the reciprocal lattice are defined by (28). We have, using (46) and (47),

$$\mathbf{A} = \frac{2\pi}{a}(\hat{\mathbf{x}} + \hat{\mathbf{y}}) \; ; \qquad \mathbf{B} = \frac{2\pi}{a}(\hat{\mathbf{y}} + \hat{\mathbf{z}}) \; ; \qquad \mathbf{C} = \frac{2\pi}{a}(\hat{\mathbf{x}} + \hat{\mathbf{z}}) \; . \tag{48}$$

Note by comparison with Fig. 1.18 that these are just the primitive vectors of an fcc lattice. Thus *the fcc lattice is the reciprocal lattice of the bcc lattice.*

If h, k, l are integers, the general reciprocal lattice vector is

$$\mathbf{G} = h\mathbf{A} + k\mathbf{B} + l\mathbf{C} = \frac{2\pi}{a}\left[(h + l)\hat{\mathbf{x}} + (h + k)\hat{\mathbf{y}} + (k + l)\hat{\mathbf{z}}\right] \; . \tag{49}$$

The shortest nonzero **G**'s are the following twelve vectors, where all choices of sign are independent:

$$\frac{2\pi}{a}\,(\pm\hat{\mathbf{x}} \pm \hat{\mathbf{y}}) \; ; \qquad \frac{2\pi}{a}\,(\pm\hat{\mathbf{y}} \pm \hat{\mathbf{z}}) \; ; \qquad \frac{2\pi}{a}\,(\pm\hat{\mathbf{x}} \pm \hat{\mathbf{z}}) \; . \tag{50}$$

The primitive cell of the reciprocal lattice may be taken to be the parallelepiped described by **A**, **B**, and **C** defined by (48). The volume of the primitive cell of the reciprocal lattice is $|\mathbf{A} \cdot \mathbf{B} \times \mathbf{C}| = 2(2\pi/a)^3$. The parallelepiped contains one reciprocal lattice point, because each of the eight corner points is shared among eight parallelepipeds: one parallelepiped contains one-eighth of each of eight corner points. But it is customary in solid state physics to take the primitive cell of the reciprocal lattice as the smallest volume bounded by planes normal to each of the (shorter) **G**'s at its midpoint. Each of the new cells contains one lattice point; the point is at the center of the cell. The cell is the Wigner-Seitz cell of the reciprocal lattice, so that it is the first Brillouin zone of the bcc lattice.

This zone is formed from the planes normal to the twelve vectors of Eq. (50) at their midpoints. The zone is a regular twelve-faced solid, a rhombic dodecahedron, as shown in Fig. 26. The vectors from the origin to the center of each face are one-half of the vectors (50), or

$$\frac{\pi}{a}\,(\pm\hat{\mathbf{x}} \pm \hat{\mathbf{y}}) \; ; \qquad \frac{\pi}{a}\,(\pm\hat{\mathbf{y}} \pm \hat{\mathbf{z}}) \; ; \qquad \frac{\pi}{a}\,(\pm\hat{\mathbf{x}} \pm \hat{\mathbf{z}}) \; . \tag{51}$$

All choices of sign are independent, giving twelve vectors.

Reciprocal Lattice to fcc Lattice. The primitive translation vectors of the fcc lattice as shown in Fig. 27 are:

$$\mathbf{a}' = \tfrac{1}{2}a(\hat{\mathbf{x}} + \hat{\mathbf{y}}) \; ; \qquad \mathbf{b}' = \tfrac{1}{2}a(\hat{\mathbf{y}} + \hat{\mathbf{z}}) \; ; \qquad \mathbf{c}' = \tfrac{1}{2}a(\hat{\mathbf{z}} + \hat{\mathbf{x}}) \; , \tag{52}$$

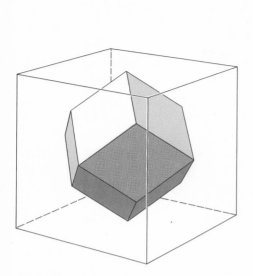

Figure 26 (*above*)   First Brillouin zone of the body-centered cubic lattice. The figure is a regular rhombic dodecahedron.

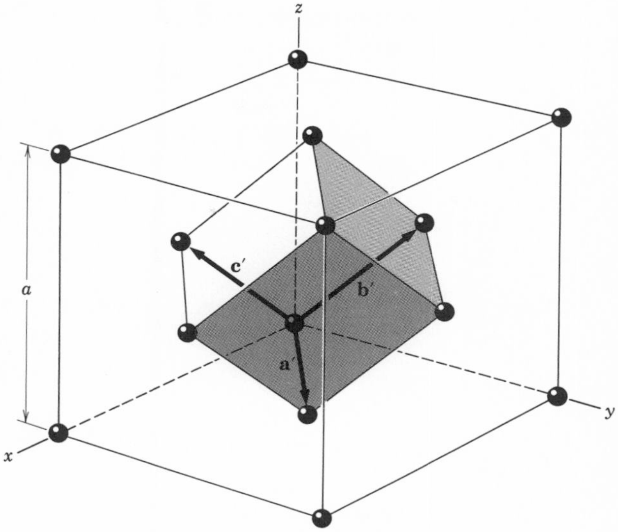

Figure 27   Primitive basis vectors of the face-centered cubic lattice.

which are parallel to the vectors of (48). The volume of the primitive cell is

$$V = |\mathbf{a}' \cdot \mathbf{b}' \times \mathbf{c}'| = \tfrac{1}{4}a^3 \ .$$

From the definitions (28) the primitive translation vectors $\mathbf{A}$, $\mathbf{B}$, $\mathbf{C}$ of the reciprocal lattice to the fcc lattice are

$$\mathbf{A} = \frac{2\pi}{a}\,(\hat{\mathbf{x}} + \hat{\mathbf{y}} - \hat{\mathbf{z}}) \ ;$$

$$\mathbf{B} = \frac{2\pi}{a}\,(-\hat{\mathbf{x}} + \hat{\mathbf{y}} + \hat{\mathbf{z}}) \ ; \qquad (53)$$

$$\mathbf{C} = \frac{2\pi}{a}\,(\hat{\mathbf{x}} - \hat{\mathbf{y}} + \hat{\mathbf{z}}) \ .$$

These are primitive translation vectors of a bcc lattice, so that the bcc lattice is the reciprocal lattice of the fcc lattice. The volume of the primitive cell in the reciprocal lattice is $|\mathbf{A} \cdot \mathbf{B} \times \mathbf{C}| = 4(2\pi/a)^3$.

We have now for the reciprocal lattice vectors

$$\mathbf{G} = \frac{2\pi}{a}\,[(h - k + l)\hat{\mathbf{x}} + (h + k - l)\hat{\mathbf{y}} + (-h + k + l)\hat{\mathbf{z}}] \ , \quad (54)$$

where h, k, l are any integers. The shortest nonzero $\mathbf{G}$'s are the eight vectors:

$$\frac{2\pi}{a}\,(\pm\hat{\mathbf{x}} \pm \hat{\mathbf{y}} \pm \hat{\mathbf{z}}) \ . \qquad (55)$$

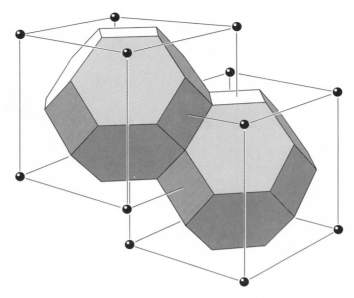

Figure 28 Brillouin zones of the face-centered cubic lattice. The cells are in reciprocal space, and the reciprocal lattice is body-centered, as drawn.

The boundaries of a primitive cell in the reciprocal lattice may be determined for the most part by the eight planes normal to these vectors at their midpoints. But the corners of the octahedron thus formed are cut by the planes that are the perpendicular bisectors of the six other reciprocal lattice vectors:[10]

$$\frac{2\pi}{a}(\pm 2\hat{x}) \; ; \qquad \frac{2\pi}{a}(\pm 2\hat{y}) \; ; \qquad \frac{2\pi}{a}(\pm 2\hat{z}) \; . \qquad (56)$$

The primitive cell is the smallest bounded volume about the origin, so that it is the truncated octahedron shown in Fig. 28. This is the first Brillouin zone of the fcc crystal lattice.

STRUCTURE FACTOR OF THE BASIS

Equation (22) determines all the reflections $\Delta\mathbf{k}$ possible for a given crystal lattice. The possible reflections are described by the reciprocal lattice points $\mathbf{G}(hkl) = h\mathbf{A} + k\mathbf{B} + l\mathbf{C}$ and may be labeled as an (hkl) reflection. The intensities of the various reflections depend on the contents of the cell: on the number, position, and electron distribution of the atoms in the cell. We now develop the connection between the contents of the cell and the intensities of the reflections.

[10] Note that $(2\pi/a)(2\hat{x})$ is a reciprocal lattice vector because it is equal to $\mathbf{A} + \mathbf{C}$.

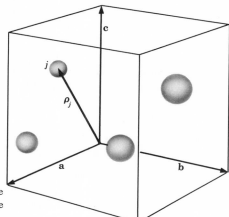

Figure 29 Position of the jth atom within the unit cell is specified by $\boldsymbol{\rho}_j = x_j\mathbf{a} + y_j\mathbf{b} + z_j\mathbf{c}$, where x_j, y_j, z_j are constants.

Let there be s atoms within each cell and let the nucleus of the jth atom be at the position (Fig. 29) specified by the vector

$$\boldsymbol{\rho}_j = x_j\mathbf{a} + y_j\mathbf{b} + z_j\mathbf{c} \tag{57a}$$

relative to the lattice point

$$\boldsymbol{\rho}_{mnp} = m\mathbf{a} + n\mathbf{b} + p\mathbf{c}$$

which is the origin of $\boldsymbol{\rho}_j$. This lattice point is associated with the cell under consideration, so that we may label the cell by mnp. But let the master origin be at the lattice point

$$\boldsymbol{\rho}_{000} = 0 \ ;$$

then relative to this lattice point the position of the jth atom in the cell mnp is

$$\boldsymbol{\rho}_j + \boldsymbol{\rho}_{mnp} \ .$$

The electrons are not concentrated at the nucleus of each atom, but are spread out in the vicinity. Their distribution in the crystal may be described as the superposition of electron concentration functions c_j associated with each atom. The function

$$c_j(\boldsymbol{\rho} - \boldsymbol{\rho}_j - \boldsymbol{\rho}_{mnp}) \tag{57b}$$

defines the electron concentration at the position $\boldsymbol{\rho}$ near the atom j of the cell mnp. Thus the total electron concentration $n(\boldsymbol{\rho})$ in the crystal may be written as the sum

$$n(\boldsymbol{\rho}) = \sum_{mnp} \sum_{j=1}^{s} c_j(\boldsymbol{\rho} - \boldsymbol{\rho}_j - \boldsymbol{\rho}_{mnp}) \ . \tag{57c}$$

The sum over j is over the s atoms of one basis, and the sum over mnp is over all lattice points, taken earlier as M^3 in number. The decomposition of $n(\boldsymbol{\rho})$

given by (57c) is not unique if the charge distributions of different ions overlap, for we cannot always say how much charge is associated with each atom, but this is not an important difficulty.

As suggested by (17), the total scattering amplitude in the crystal for the scattering vector $\Delta \mathbf{k}$ may be written as

$$\mathcal{A}_{\Delta \mathbf{k}} = \int dV\, n(\boldsymbol{\rho}) e^{-i\boldsymbol{\rho} \cdot \Delta \mathbf{k}} = \sum_{mnp} \sum_{j} \int dV\, c_j(\boldsymbol{\rho} - \boldsymbol{\rho}_j - \boldsymbol{\rho}_{mnp}) e^{-i\boldsymbol{\rho} \cdot \Delta \mathbf{k}} \ . \quad (58a)$$

The contribution to $\mathcal{A}$ of the single term $c_j(\boldsymbol{\rho} - \boldsymbol{\rho}_j - \boldsymbol{\rho}_{mnp})$ in (58a) is

$$\begin{aligned}
\int dV\, c_j(\boldsymbol{\rho} - \boldsymbol{\rho}_j - \boldsymbol{\rho}_{mnp}) e^{-i\boldsymbol{\rho} \cdot \Delta \mathbf{k}} \\
= \int dV\, c_j(\boldsymbol{\rho}') \, e^{-i\boldsymbol{\rho}' \cdot \Delta \mathbf{k}} \exp\left[-i(\boldsymbol{\rho}_j + \boldsymbol{\rho}_{mnp}) \cdot \Delta \mathbf{k}\right] \\
= f_j \exp\left[-i(\boldsymbol{\rho}_j + \boldsymbol{\rho}_{mnp}) \cdot \Delta \mathbf{k}\right] \ .
\end{aligned} \quad (58b)$$

Here we have made the substitution

$$\boldsymbol{\rho}' \equiv \boldsymbol{\rho} - \boldsymbol{\rho}_j - \boldsymbol{\rho}_{mnp}$$

and have introduced

$$\boxed{f_j = \int dV\, c_j(\boldsymbol{\rho}') e^{-i\boldsymbol{\rho}' \cdot \Delta \mathbf{k}} \ ,} \quad (59a)$$

called the **atomic form factor** (see the following section).

The scattering amplitude is now

$$\begin{aligned}
\mathcal{A}_{\Delta \mathbf{k}} &= \sum_{m} \sum_{j} f_j \exp\left[-i(\boldsymbol{\rho}_j + \boldsymbol{\rho}_{mnp}) \cdot \Delta \mathbf{k}\right] \\
&= \left(\sum_{mnp} \exp\left(-i\boldsymbol{\rho}_{mnp} \cdot \Delta \mathbf{k}\right)\right)\left(\sum_{j} f_j \, \exp\left(-i\boldsymbol{\rho}_j \cdot \Delta \mathbf{k}\right)\right),
\end{aligned}$$

or

$$\mathcal{A}_{\mathbf{G}} = M^3 \mathcal{S}_{\mathbf{G}} \ , \quad (59b)$$

where we have used the earlier result (21) that

$$\sum_{mnp} \exp\left(-i\boldsymbol{\rho}_{mnp} \cdot \Delta \mathbf{k}\right)$$

is nonzero only when $\Delta \mathbf{k}$ is equal to a reciprocal lattice vector. The sum

$$\boxed{\mathcal{S}_{\mathbf{G}} = \sum_{j} f_j \exp\left(-i\boldsymbol{\rho}_j \cdot \mathbf{G}\right)} \quad (59c)$$

is called the **structure factor** of the basis.

We speak of a reflection (hkl) when the reciprocal lattice vector is $\mathbf{G} = 2\pi(h\mathbf{A} + k\mathbf{B} + l\mathbf{C})$. For this reflection we have, with $\boldsymbol{\rho}_j$ given by (57a),

$$\boldsymbol{\rho}_j \cdot \mathbf{G} = (x_j \mathbf{a} + y_j \mathbf{b} + z_j \mathbf{c}) \cdot (h\mathbf{A} + k\mathbf{B} + l\mathbf{C}) = 2\pi(x_j h + y_j k + z_j l) \ , \quad (60)$$

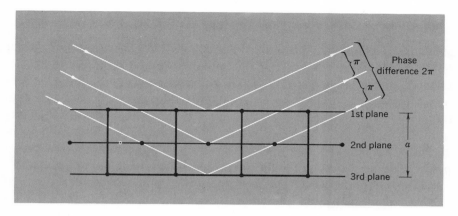

Figure 30 Explanation of the absence of a (100) reflection from a body-centered cubic lattice. The phase difference between successive planes is π, so that the reflected amplitude from two adjacent planes is $1 + e^{-i\pi} = 1 - 1 = 0$.

so that the structure factor for the reflection may be written as

$$S(hkl) = \sum_j f_j \exp\left[-i2\pi(x_j h + y_j k + z_j l)\right] . \qquad (61)$$

The structure factor need not be real; the scattered intensity involves $S^\circ S$, where S° is the complex conjugate of S. Our first interest is in the zeros of S, for at a zero there is no intensity in a reflection $\mathbf{G}$ permitted by the space lattice. The structure factor can cancel some reflections allowed by the space lattice, and the missing reflections help us in the determination of the structure.

Structure Factor of the bcc Lattice. The basis of the bcc structure referred to the usual cubic unit cell has identical atoms at $0\ 0\ 0$ and $\frac{1}{2}\frac{1}{2}\frac{1}{2}$. For one atom $x_1 = y_1 = z_1 = 0$ and for the other atom $x_2 = y_2 = z_2 = \frac{1}{2}$. Thus (61) becomes

$$S(hkl) = f\{1 + \exp\left[-i\pi(h + k + l)\right]\} , \qquad (62)$$

where f is the scattering power of an atom. The value of S is zero whenever the exponential has the value -1. Now the exponential has the value -1 whenever its argument is $-i\pi$ times an odd integer. We have

$$S = 0 \qquad \text{when } h + k + l = \text{odd integer;}$$
$$S = 2f \qquad \text{when } h + k + l = \text{even integer.}$$

Metallic sodium has a bcc structure. The diffraction spectrum does not contain lines such as (100), (300), (111), or (221), but lines such as (200), (110), and (222) will be present; here the indices (hkl) are referred to a cubic cell.

What is the physical interpretation of the result that the (100) reflection vanishes for the bcc lattice? The (100) reflection normally occurs when the reflections from the first and third planes in Fig. 30 differ in phase by 2π. These planes bound the unit cube. In the bcc lattice there is an intervening plane of atoms, labeled the second plane in the figure, which is equal in scattering power to the other planes. Situated midway between them, it gives a reflection retarded in phase by π with respect to the first plane, thereby canceling the contribution from that plane. The cancellation of the (100) reflection occurs in the bcc lattice because the planes are identical in composition. In the CsCl structure (Fig. 1.26) this cancellation does not occur; the planes of Cs and Cl ions alternate, but the scattering power of Cs is much greater than the scattering power of Cl, because Cs^+ has 54 electrons and Cl^- has only 18 electrons.

Structure Factor of the fcc Lattice. The basis of the fcc structure referred to the usual cubic unit cell has identical atoms at $0\ 0\ 0$; $0\ \frac{1}{2}\ \frac{1}{2}$; $\frac{1}{2}\ 0\ \frac{1}{2}$; $\frac{1}{2}\ \frac{1}{2}\ 0$. Thus (61) becomes

$$S(hkl) = f\{1 + \exp[-i\pi(k+l)] + \exp[-i\pi(h+l)] + \exp[-i\pi(h+k)]\}\ . \quad (63)$$

If all indices are even integers, $S = 4f$; similarly if all indices are odd integers. But if only one of the integers is even, two of the exponents will be odd multiples of $-i\pi$ and S will vanish. If only one of the integers is odd, the same argument applies and S will also vanish. Thus in the fcc lattice no reflections can occur for which the indices are partly even and partly odd. The point is beautifully illustrated by Fig. 31: both KCl and KBr have an fcc lattice, but KCl simulates an sc lattice because the K^+ and Cl^- ions have equal numbers of electrons. The allowed reflections for aluminum, which has the fcc structure, are indicated in Fig. 32.

ATOMIC SCATTERING FACTOR OR FORM FACTOR

In the expression (61) for the geometrical structure factor there occurs the quantity f_j, which we said was a measure of the scattering power of the jth atom in the unit cell. What determines f_j? For x-rays the scattering power involves only the atomic electrons, because the nuclei are too heavy to respond significantly to an x-ray photon.

The value of f involves the number and distribution of atomic electrons, and it also involves the wavelength and angle of scattering of the radiation. These factors enter because of interference effects arising from the finite extent of the atoms. We now give a classical calculation of the scattering factor.

The scattered radiation from a single atom must take account of interference effects within the atom. We defined in (59a) the function

$$f_G = \int dV\, c(\mathbf{r}) e^{-i\mathbf{r}\cdot\mathbf{G}}\ , \quad (64)$$

with the integral extended over the electron concentration $c(\mathbf{r})$ associated with a single atom. We call f the **atomic scattering factor** or **form factor**. Let $\mathbf{r}$ make an angle α with $\mathbf{G}$; then $\mathbf{r}\cdot\mathbf{G} = rG\cos\alpha$. If the electron distribu-

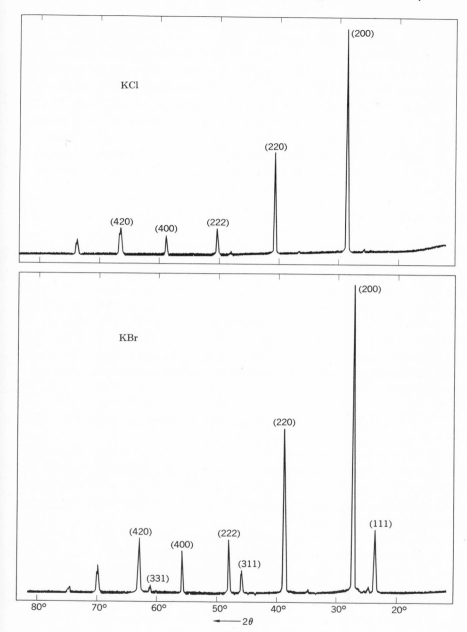

Figure 31 Comparison of x-ray reflections from KCl and KBr powders. In KCl the numbers of electrons of K^+ and Cl^- ions are equal. The scattering amplitudes $f(K^+)$ and $f(Cl^-)$ are almost exactly equal, so that the crystal looks to x-rays as if it were a monatomic simple cubic lattice of lattice of lattice constant $a/2$. Only even integers occur in the reflection indices when these are based on a cubic lattice of lattice constant a. In KBr the form factor of Br^- is quite different than that of K^+, and all reflections of the fcc lattice are present. (Courtesy of Robert van Nordstrand.)

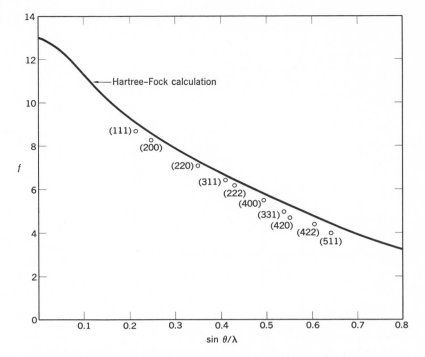

Figure 32 Absolute experimental atomic scattering factors for metallic aluminum, after Batterman, Chipman, and DeMarco. Each observed reflection is labeled. The incident radiation was MoKα with λ = 0.709 Å. Note that no reflections occur for which the indices are partly even and partly odd; this is what we expect for an fcc crystal, according to (63).

tion is spherically symmetric about the origin, then

$$f_G \equiv 2\pi \int r^2 \, dr \, d(\cos \alpha) c(r) e^{-iGr \cos \alpha} = 2\pi \int dr \, r^2 c(r) \cdot \frac{e^{iGr} - e^{-iGr}}{iGr} \; ,$$

where we have integrated over $d(\cos \alpha)$ between -1 and 1. Thus the atomic scattering factor is given by

$$f_G = 4\pi \int dr \, c(r) r^2 \frac{\sin Gr}{Gr} \; . \tag{65}$$

If the same total electron density were concentrated at the origin where $r = 0$, then in the integral in (65) only $Gr = 0$ would contribute to the integrand. In this limit $(\sin Gr)/Gr = 1$, and for all G

$$f_G = 4\pi \int dr \, c(r) r^2 = Z \; , \tag{66}$$

the number of atomic electrons. Therefore f_G *is the ratio of the radiation amplitude scattered by the actual electron distribution in an atom to that scattered by one electron localized at a point.*

Figure 33 Electron density in vicinity of midpoint between two neighboring Si atoms in the [111] direction in the lattice. The solid curve is calculated from the observed f values; the dashed curve is the superposition of calculated free atom charge densities. Part of the difference is due to the expansion of charge distributions in the crystal, and part is due to an extra concentration of charge in the chemical bond. (After Göttlicher et al.)

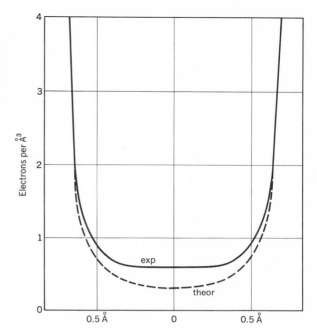

At $\theta = 0$ we see by the same argument that $G = 0$, and f_G reduces to the value Z. We also see from the solution of Problem 3 that at very short wavelengths interference effects reduce the scattered amplitude drastically.[11]

The overall electron distribution in a solid is fairly close to that of the appropriate free atoms. This statement does not mean that the outermost or valence electrons are not redistributed in forming the solid; it means only that the x-ray reflection intensities are represented well by the free atom values of the form factors. As an example, Batterman and co-workers[12] find agreement within 1 percent in a comparison of the x-ray intensities of Bragg reflections of metallic iron, copper, and aluminum with the theoretical free atom values from wavefunction calculations. The results for aluminum are shown in Fig. 32.

There have been many attempts to obtain direct x-ray evidence about the actual electron distribution in a covalent chemical bond, particularly in crystals having the diamond structure.[13] The question lies near the limits of what can be explored by x-ray diffraction methods. There is some indication in silicon that midway between two nearest-neighbor atoms there is an appreciable increase in electron concentration over what is expected from the overlap of the electron densities calculated for two free atoms of Si (see Fig. 33).

---

[11] In a more accurate treatment one finds that the amplitude and phase of scattering are rather different for those inner electrons whose binding energies approach the energy of the x-ray photons. This effect, widely referred to as "anomalous dispersion," introduces complications which can be very useful in structure determinations.

[12] B. W. Batterman, D. R. Chipman, and J. J. DeMarco, Phys. Rev. **122**, 68 (1961); see also L. D. Jennings et al., Phys. Rev. **135**, 1612 (1964).

[13] For diamond, see G. B. Carpenter, J. Chem. Phys. **32**, 525 (1960); for silicon, see S. Göttlicher et al., Z. f. physikalische Chemie **21**, 133 (1959).

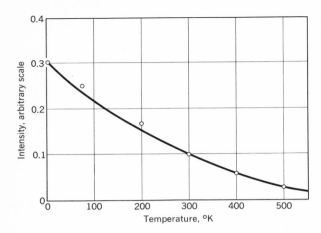

Figure 34 Temperature dependence of the integrated x-ray intensity of MoKα radiation reflected from the (800) planes of copper, after P. A. Flinn, G. M. McManus, and J. A. Rayne, Phys. Rev. **123**, 809 (1961).

Temperature Dependence of the Reflection Lines

"... I came to the conclusion that the sharpness of the interference lines would not suffer but that their intensity should diminish with increasing angle of scattering, the more so the higher the temperature."

(P. Debye)

As the temperature of the crystal is increased, the intensity of the Bragg-reflected beams decreases, but the angular width of the reflected line is unchanged. Experimental intensities for a reflection line of copper are shown in Fig. 34. It is surprising that we can get a sharp x-ray reflection from atoms in large amplitude random thermal motion, with instantaneous nearest-neighbor spacings differing by perhaps 10 percent at room temperature. Before the Laue experiment was done, but when the proposal was being discussed[14] in coffee houses in Munich, the objection was made that the instantaneous positions of the atoms in a crystal at room temperature are very far from a regular periodic array, because of this large thermal fluctuation. Therefore, the argument went, one should not expect a well-defined diffracted beam.

But there is a well-defined diffracted beam. The essential argument was given by Debye in 1912. Consider the expression (19) for the radiation amplitude scattered by a crystal: let the position of the atom nominally at ρ_0 contain a term $\mathbf{u}(t)$ fluctuating in time:

$$\rho(t) = \rho_0 + \mathbf{u}(t) \ . \tag{67}$$

We suppose that each atom fluctuates independently about its own equilibrium position.[15] Then the thermal-average scattered amplitude of (59c) in the

---

[14] P. P. Ewald, private communication.

[15] This is the Einstein model of a solid; it is not a very good model at low temperatures, but it works well at high temperatures. For our present purpose it leads us to a simple result. For a more realistic treatment of scattering by thermal fluctuations, see *QTS*, Chap. 20.

direction of a diffraction maximum can be written

$$\langle a \rangle = a_0 \langle \exp(-i\mathbf{u} \cdot \mathbf{G}) \rangle \ , \tag{68}$$

where $\mathbf{G}$ is equal to the change in wavevector on reflection and $\langle \cdots \rangle$ denotes thermal average. The appearance of a_0 assures us that all diffraction lines will be sharp.

The exponential factor in (68) reduces the intensity. The series expansion of the exponential is

$$\langle \exp(-i\mathbf{u} \cdot \mathbf{G}) \rangle = 1 - i\langle \mathbf{u} \cdot \mathbf{G} \rangle - \tfrac{1}{2}\langle (\mathbf{u} \cdot \mathbf{G})^2 \rangle + \cdots \ . \tag{69}$$

But $\langle \mathbf{u} \cdot \mathbf{G} \rangle = 0$, because $\mathbf{u}$ is a random thermal motion uncorrelated with the direction of $\mathbf{G}$. Further,

$$\langle (\mathbf{u} \cdot \mathbf{G})^2 \rangle = \tfrac{1}{3}\langle u^2 \rangle G^2 \ . \tag{70}$$

The factor $\tfrac{1}{3}$ arises by geometrical averaging in three dimensions, because only the component of $\mathbf{u}$ along $\mathbf{G}$ is involved. We can limit ourselves to (69) to get a feeling for the physics, but it is useful to note that the function

$$\exp[-\tfrac{1}{6}\langle u^2 \rangle G^2] = 1 - \tfrac{1}{6}\langle u^2 \rangle G^2 + \cdots \tag{71}$$

has the same series expansion as (69) for the first two terms shown here. For a harmonic oscillator it is a fact that all terms in the series (69) and (71) can be shown to be identical. Then the scattered intensity, which is the square of the amplitude, is

$$I = I_0 \exp[-\tfrac{1}{3}\langle u^2 \rangle G^2] \ , \tag{72}$$

where I_0 is the scattered intensity from the rigid lattice as treated earlier. The exponential factor is known as the **Debye-Waller factor**.

Here $\langle u^2 \rangle$ is the mean square displacement of an atom. The thermal average potential energy $\langle U \rangle$ of a classical harmonic oscillator in three dimensions is $\tfrac{3}{2}k_B T$, whence

$$\langle U \rangle = \tfrac{1}{2}C\langle u^2 \rangle = \tfrac{1}{2}M\omega^2\langle u^2 \rangle = \tfrac{3}{2}k_B T \ , \tag{73}$$

where C is the force constant, M is the mass of an atom, and ω is the frequency of the oscillator. We have used the result $\omega^2 = C/M$. Thus the scattered intensity is

$$I(hkl) = I_0 \exp\left[-\frac{k_B T G^2}{M\omega^2}\right] , \tag{74}$$

where hkl are the indices of $\mathbf{G} = h\mathbf{A} + k\mathbf{B} + l\mathbf{C}$. This classical result is a good approximation at high temperatures.

For quantum oscillators $\langle u^2 \rangle$ does not vanish even at $T = 0$ because of zero-point motion. We continue with the independent harmonic oscillator model for the motion of an atom: at absolute zero the motion can be described in terms of the zero point energy $\frac{3}{2}\hbar\omega$. This is the energy of a three-dimensional quantum harmonic oscillator in its ground state referred to the classical energy of the same oscillator at rest. Half of the energy of an oscillator is potential energy, so that (73) gives for the average potential energy in the ground state:

$$\langle U \rangle = \tfrac{1}{2}M\omega^2 \langle u^2 \rangle = \tfrac{3}{4}\hbar\omega \ , \tag{75}$$

or

$$\langle u^2 \rangle = \frac{3\hbar}{2M\omega} \ , \tag{76}$$

whence, by (72),

$$I = I_0 \exp\left[-\frac{3\hbar G^2}{2M\omega}\right] \tag{77}$$

at absolute zero. If $G = 10^9$ cm^{-1}, $\omega = 10^{14}$ sec^{-1}, and $M = 10^{-22}$ g, then the argument of the exponential is approximately 0.1, so that $I/I_0 \cong 0.9$. For these numbers 90 percent of the beam is elastically scattered at absolute zero and 10 percent is inelastically scattered. The loss of energy from the x-ray beam in the inelastic component reappears in the energy of the atom which is put into an excited oscillator state.

We see from (74) and from Fig. 34 that the intensity of the diffracted line decreases (but not catastrophically) as the temperature is increased. The reflections of low G are less affected than the reflections of high G. The intensity we have calculated is that of the coherent diffraction or elastic scattering in the well-defined Bragg directions. The part of the intensity lost from these directions as the temperature is increased appears as a somewhat diffuse background. That is, the lost intensity reappears in the inelastic photon scattering, discussed in Chapter 5. In inelastic scattering the x-ray photon causes the excitation or de-excitation of a lattice vibration, and the photon changes direction and energy.

At a given temperature the Debye-Waller factor of a diffraction line decreases with an increase in the magnitude of the reciprocal lattice vector $\mathbf{G}$ associated with the reflection. The larger $|\mathbf{G}|$ is, the weaker will be the reflection at high temperatures. The temperature dependence of the reflected

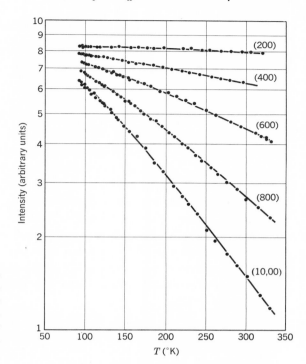

Figure 35 The dependence of intensity on temperature for the $(h00)$ x-ray reflections of aluminum. Reflections $(h00)$ with h odd are forbidden for an fcc structure. [After R. M. Nicklow and R. A. Young, Phys. Rev. **152**, 591 (1966).]

intensity for the $(h00)$ reflections of aluminum is shown in Fig. 35. The theory we have worked out here for x-ray reflection describes equally well the **Mössbauer effect,**[16] the recoilless emission of gamma rays by nuclei bound in crystals. The theory also applies to neutron scattering.

A beam of x-rays also can be absorbed in a crystal by inelastic processes connected with the photo-ionization of electrons and with the Compton scattering. In the photoeffect the x-ray photon is absorbed and an electron is ejected from an atom. In the Compton effect the photon is scattered by an electron: the photon loses energy and the electron is ejected from an atom. The depth of penetration[17] of the x-ray beam depends on the solid and on the photon energy, but 1 cm is typical. A diffracted beam in Bragg reflection usually will be formed in a much shorter distance, perhaps 10^{-5} to 10^{-4} cm in an ideal crystal.

[16] *QTS*, Chap. 20.
[17] See W. Heitler, *Quantum theory of radiation*, Clarendon Press, 1954, 3rd ed., p. 223.

SUMMARY

This has been a long and important chapter. It is good to summarize the steps which relate the crystal structure to the relative intensities of the diffraction pattern produced by the structure. Suppose that we have guessed the structure. We want to predict the diffraction pattern of the assumed structure and test the prediction against the observed diffraction pattern. The observed pattern may be available in the form of a map in reciprocal space giving the values of $\Delta \mathbf{k} \equiv \mathbf{k}' - \mathbf{k}$ for which diffracted beams are found.

First: We choose a set of translation vectors $\mathbf{a}$, $\mathbf{b}$, $\mathbf{c}$ for the assumed structure. The vectors need not necessarily be primitive translation vectors. From $\mathbf{a}$, $\mathbf{b}$, $\mathbf{c}$ we form the vectors $\mathbf{A}$, $\mathbf{B}$, $\mathbf{C}$ which are the fundamental vectors of the reciprocal lattice. We plot the points $\mathbf{G} = h\mathbf{A} + k\mathbf{B} + l\mathbf{C}$ for integral values of h, k, l. Some or all of these points should coincide with the points of the experimental map of $\Delta \mathbf{k}$. If no points coincide, we have most probably chosen incorrect vectors $\mathbf{a}$, $\mathbf{b}$, $\mathbf{c}$. We can adjust the $\mathbf{a}$, $\mathbf{b}$, $\mathbf{c}$ and thus the $\mathbf{A}$, $\mathbf{B}$, $\mathbf{C}$ until some points $\mathbf{G}$ coincide with the observed $\Delta \mathbf{k}$. The corresponding $\mathbf{a}$, $\mathbf{b}$, $\mathbf{c}$ define the crystal lattice.

Second: Every $\Delta \mathbf{k}$ now falls on some $\mathbf{G}$, but some $\mathbf{G}$'s will not fall on an observed $\Delta \mathbf{k}$ if the structure factor $\mathcal{S}$ is zero for these values of $\mathbf{G}$. We calculate $\mathcal{S}$ for our assumed basis associated with $\mathbf{a}$, $\mathbf{b}$, $\mathbf{c}$ and see if the zeros of $\mathcal{S}$ fall at the values of $\mathbf{G}$ for which no diffracted beam was observed. We adjust the positions x_i, y_i, z_i of the atoms in the assumed basis until the zeros of $\mathcal{S}$ coincide with the positions of the "absent" reflections. The corresponding atomic coordinates x_i, y_i, z_i define the basis associated with the crystal lattice.

Third: Values of the atomic form factor f for the atoms and $\mathbf{G}$'s of interest may be found in the *International tables for x-ray crystallography*.[18] The exact indices of the zeros of $\mathcal{S}$ usually may be found without specific use of the f's. We should compare at least qualitatively the predicted and observed values of the relative intensities of the diffracted beams. To make this comparison we use the f's to calculate $|\mathcal{S}(hkl)|^2$. After we multiply this by the Debye-Waller temperature factor (72) we obtain a rough measure of the approximate relative intensity. The lost intensity reappears as long, low wings about the position of the diffraction line.

We recall here the memorable results of the chapter:

1. Various statements of the Bragg condition:

$$2d \sin \theta = n\lambda \; ; \qquad \Delta \mathbf{k} = \mathbf{G} \; ; \qquad 2\mathbf{k} \cdot \mathbf{G} = G^2 \; .$$

2. Laue conditions:

$$\mathbf{a} \cdot \Delta \mathbf{k} = 2\pi h \; ; \qquad \mathbf{b} \cdot \Delta \mathbf{k} = 2\pi k \; ; \qquad \mathbf{c} \cdot \Delta \mathbf{k} = 2\pi l \; .$$

[18] Vol. III, pp. 201–227.

3. The primitive translation vectors of the reciprocal lattice are

$$\mathbf{A} = 2\pi \frac{\mathbf{b} \times \mathbf{c}}{\mathbf{a} \cdot \mathbf{b} \times \mathbf{c}} \; ; \qquad \mathbf{B} = 2\pi \frac{\mathbf{c} \times \mathbf{a}}{\mathbf{a} \cdot \mathbf{b} \times \mathbf{c}} \; ; \qquad \mathbf{C} = 2\pi \frac{\mathbf{a} \times \mathbf{b}}{\mathbf{a} \cdot \mathbf{b} \times \mathbf{c}} \; .$$

Here $\mathbf{a}$, $\mathbf{b}$, $\mathbf{c}$ are the primitive translation vectors of the crystal lattice.

4. A reciprocal lattice vector has the form

$$\mathbf{G} = h\mathbf{A} + k\mathbf{B} + l\mathbf{C}$$

where h, k, l are integers or zero.

5. The scattered amplitude in the direction $\Delta \mathbf{k} = \mathbf{k}' - \mathbf{k} = \mathbf{G}$ is proportional to

$$a = \mathcal{S}_\mathbf{G} \times \left(\sum_{mnp} \exp\left(-i\boldsymbol{\rho}_{mnp} \cdot \mathbf{G} \right) \right) .$$

6. Geometrical structure factor:

$$\mathcal{S}_\mathbf{G} \equiv \sum_{j=1}^{N} f_j \exp\left(-i\boldsymbol{\rho}_j \cdot \mathbf{G} \right) = \sum f_j \exp\left[-i2\pi(x_j h + y_j k + z_j l) \right] \; ,$$

where j runs over the N atoms of the basis, and f_j is the atomic form factor (65) of the jth atom of the basis. The expression on the right-hand side is written for a reflection (hkl), for which $\mathbf{G} = h\mathbf{A} + k\mathbf{B} + l\mathbf{C}$.

7. Any function which is invariant under a lattice translation may be expanded in a Fourier series of the form

$$n(\boldsymbol{\rho}) = \sum_\mathbf{G} n_\mathbf{G} e^{i\mathbf{G} \cdot \boldsymbol{\rho}} \; .$$

8. The first Brillouin zone is the Wigner-Seitz primitive cell of the reciprocal lattice. Any wave whose wavevector $\mathbf{k}$ from the origin terminates on a surface of the Brillouin zone will be diffracted by the crystal.

9. *Crystal lattice*   *First Brillouin zone*

    Simple cubic   cube

    Body-centered cubic   rhombic dodecahedron (Fig. 26)

    Face-centered cubic   truncated octahedron (Fig. 28)

10. Thermal motion does not broaden a diffraction line, but only reduces the intensity. The lost intensity reappears as long, low wings about the position of the diffraction line.

Problems

1. *Brillouin zone of orthorhombic lattice.* An orthorhombic lattice has the three primitive axis vectors:

$$\mathbf{a} = 5.0\hat{\mathbf{x}} \; ; \qquad \mathbf{b} = 2.0\hat{\mathbf{y}} \; ; \qquad \mathbf{c} = 1.0\hat{\mathbf{z}} \; ,$$

all in Å. Give the dimensions and form of the first Brillouin zone.

2. *Hexagonal space lattice.* The primitive translation vectors of the hexagonal space lattice may be taken as

$$\mathbf{a} = (3^{\frac{1}{2}}a/2)\hat{\mathbf{x}} + (a/2)\hat{\mathbf{y}} \; ;$$
$$\mathbf{b} = -(3^{\frac{1}{2}}a/2)\hat{\mathbf{x}} + (a/2)\hat{\mathbf{y}} \; ;$$
$$\mathbf{c} = c\hat{\mathbf{z}} \; .$$

(a) Show from $V = |\mathbf{a} \cdot \mathbf{b} \times \mathbf{c}|$ that the volume of the primitive cell is $(3^{\frac{1}{2}}/2)a^2c$.

(b) Show that the primitive translations of the reciprocal lattice are

$$\mathbf{A} = (2\pi/3^{\frac{1}{2}}a)\hat{\mathbf{x}} + (2\pi/a)\hat{\mathbf{y}} \; ;$$
$$\mathbf{B} = -(2\pi/3^{\frac{1}{2}}a)\hat{\mathbf{x}} + (2\pi/a)\hat{\mathbf{y}} \; ;$$
$$\mathbf{C} = (2\pi/c)\hat{\mathbf{z}} \; ,$$

so that the lattice is its own reciprocal, but with a rotation of axes.

(c) Describe and sketch the first Brillouin zone of the hexagonal space lattice.

3. *Form factor of uniform sphere.* Find the atomic form factor f for a uniform distribution of Z electrons inside a sphere of radius R. *Hint:* In (65) set $c(r) = c$; then

$$f = (4\pi c/G^3)\int_0^{GR} x \sin x \, dx;$$ now evaluate the integral. For $GR \gg 1$ the form factor

is proportional to

$$f \propto G^{-2} \cos GR \; ,$$

so that the scattering amplitude decreases as G increases.

4. *Structure factor of diamond.* The crystal structure of diamond is described in Chapter 1. If the unit cell is taken as the conventional cube the basis consists of eight atoms. (a) Find the structure factor of this basis. (b) Find the zeros of $\mathcal{S}$ and show that the allowed reflections of the diamond structure satisfy $h + k + l = 4n$ where all indices are even and n is any integer; or else all indices are odd. However, the forbidden reflection (222) will be observed if there is an extra concentration of electrons midway between two nearest neighbor carbon atoms (see Fig. 36).

5. *Width of diffraction maximum.* We suppose that in a linear crystal there are identical point scattering centers at every lattice point $\rho_m = m\mathbf{a}$, where m is an integer. By analogy with (19) the total scattered radiation amplitude seen at $\mathbf{R}$ will be proportional to $a = \Sigma \exp[-im\mathbf{a} \cdot \Delta\mathbf{k}]$. The sum over M lattice points has the value

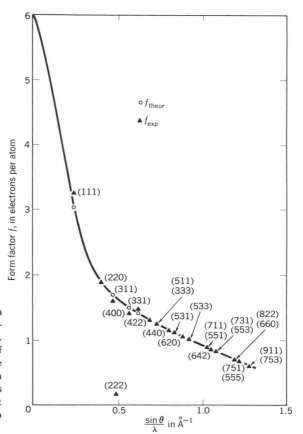

Figure 36 Experimental and theoretical atomic form factors from x-ray reflections of diamond, after S. Göttlicher and E. Wölfel, Z. Elektrochemie **63**, 891 (1959). The theoretical form factors are for a lattice of atoms of spherically symmetric charge density, as calculated by the Hartree method. The presence of the (222) forbidden reflection indicates an extra concentration of electrons (perhaps 0.4 electron) in the bond between nearest neighbors, above what one finds from the simple overlap of two spherical charge distributions.

$$\mathcal{a} = \frac{1 - \exp\left[-iM(\mathbf{a}\cdot\Delta\mathbf{k})\right]}{1 - \exp\left[-i(\mathbf{a}\cdot\Delta\mathbf{k})\right]} ,$$

by use of the series

$$\sum_{m=0}^{M-1} x^m = \frac{1 - x^M}{1 - x} .$$

(a) The scattered intensity is proportional to $|\mathcal{a}|^2$. Show that

$$|\mathcal{a}|^2 \equiv \mathcal{a}^{\circ}\,\mathcal{a} = \frac{\sin^2 \frac{1}{2}M(\mathbf{a}\cdot\Delta\mathbf{k})}{\sin^2 \frac{1}{2}(\mathbf{a}\cdot\Delta\mathbf{k})} .$$

(b) We know that a diffraction maximum appears when $\mathbf{a}\cdot\Delta\mathbf{k} = 2\pi h$, where h is an integer. We change $\Delta\mathbf{k}$ slightly and define ϵ in $\mathbf{a}\cdot\Delta\mathbf{k} = 2\pi h + \epsilon$ such that ϵ gives the position of the first zero in $\sin \frac{1}{2}M(\mathbf{a}\cdot\Delta\mathbf{k})$. Show that $\epsilon = 2\pi/M$, so that the width of the diffraction maximum is proportional to $1/M$ and can be extremely narrow for macroscopic values of M. The same result holds true for a three-dimensional crystal.

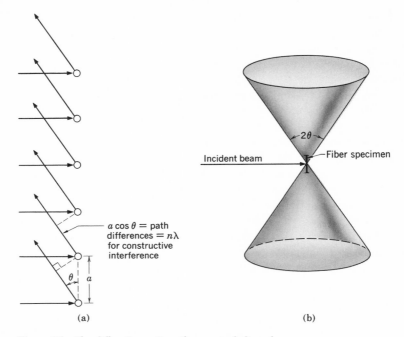

(a)   (b)

Figure 37 The diffraction pattern from a single line of lattice constant a in a monochromatic x-ray beam perpendicular to the line. (a) The condition for constructive interference is $a \cos \theta = n\lambda$, where n is an integer. (b) For given n the diffracted rays lie on the surface of a cone.

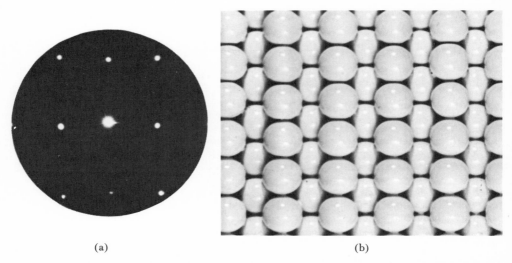

(a)   (b)

Figure 38 (a) Backward scattering pattern of 76 eV electrons incident normally on the (110) face of a nickel crystal; a model of the surface is shown in (b). (Courtesy of A. U. MacRae.)

6. *Diffraction from a linear array and a square array.* The diffraction pattern of a linear structure of lattice constant a is explained[19] in Fig. 37. Somewhat similar structures are important in molecular biology: DNA and many proteins are linear helices[20]. (a) A cylindrical film is exposed to the diffraction pattern of Fig. 37b; the axis of the cylinder is coincident with the axis of the linear structure or fiber. Describe the appearance of the diffraction pattern on the film. (b) A flat photographic plate is placed behind the fiber and normal to the incident beam. Sketch roughly the appearance of the diffraction pattern on the plate. (c) A single plane of atoms forms a square lattice of lattice constant a. The plane is normal to the incident x-ray beam. Sketch roughly the appearance of the diffraction pattern[21] on the photographic plate. *Hint:* The diffraction from a plane of atoms can be inferred from the patterns for two perpendicular lines of atoms. (d) Figure 38 shows the electron diffraction pattern in the backward direction, from the nickel atoms on the (110) surface of a nickel crystal. Explain the orientation of the diffraction pattern in relation to the atomic positions of the surface atoms shown in the model. Assume that only the surface atoms are effective in the reflection of low-energy electrons.

7. *Diatomic line.* Consider a line of atoms $ABAB \ldots AB$, with an A—B bond length of $\frac{1}{2}a$. The form factors are f_A, f_B for atoms A, B, respectively. The incident beam of x-rays is perpendicular to the line of atoms. (a) Show that the interference condition is $n\lambda = a \cos \theta$, where θ is the angle between the diffracted beam and the line of atoms. (b) Show that the intensity of the diffracted beam is proportional to $|f_A - f_B|^2$ for n odd, and to $|f_A + f_B|^2$ for n even. (c) Explain what happens if $f_A = f_B$.

8. *Reciprocal lattice and allowed reflections.* (a) Why are there fewer reciprocal lattice points **G** in a given volume of Fourier space if the unit cell of the crystal lattice is primitive than if the unit cell is nonprimitive? (b) In view of (a), how can the allowed reflections from a given structure be independent of the choice of the unit cell of the crystal lattice?

[19] Another viewpoint is useful: for a linear lattice the diffraction pattern is described by the single Laue equation $\mathbf{a} \cdot \Delta\mathbf{k} = 2\pi q$, where q is an integer. The lattice sums which led to the other Laue equations do not occur for a linear lattice. Now $\mathbf{a} \cdot \Delta\mathbf{k} = $ const. is the equation of a plane; thus the reciprocal lattice becomes a set of parallel planes normal to the line of atoms. What happens to the reciprocal lattice for a single plane of atoms? (Now there are two Laue equations.) Care must be taken with the concept of reciprocal lattice points when discussing one- or two-dimensional structures, because the reciprocal lattice point becomes a reciprocal lattice line for a planar structure and a reciprocal lattice plane for a linear structure.

[20] The diffraction pattern of a helix is treated by W. Cochran, F. H. C. Crick, and V. Vand, Acta Cryst. **5**, 581 (1952); A. Klug, F. H. C. Crick, and H. W. Wyckoff, Acta Cryst. **11**, 199 (1958). An elementary account is given by C. Kittel, Amer. Jour. Physics **36**, 610 (1968).

[21] See L. H. Germer, "Structure of crystal surfaces," Sci. American, March 1965, p. 32.

References

X-RAY DIFFRACTION

C. S. Barrett and T. B. Massalski, *Structure of metals: crystallographic methods, principles, data,* McGraw-Hill, 1966, 3rd ed. (An excellent guide to the practical use of diffraction methods.)

W. H. Bragg and W. L. Bragg, "Structure of the diamond," Proc. Roy. Soc. (London) **A89,** 277 (1913).

W. L. Bragg, "Structure of some crystals as indicated by their diffraction of x-rays," Proc. Roy. Soc. (London) **A89,** 248 (1913). (A careful treatment and analysis; the first correct structure determinations are given here.)

M. J. Buerger, *Crystal structure analysis,* Wiley, 1960.

A. Guinier, *X-ray diffraction in crystals, imperfect crystals and amorphous bodies,* Freeman, 1963.

W. C. Hamilton," The revolution in crystallography," Science **169,** 133 (1970).

K. C. Holmes and D. M. Blow, *Use of x-ray diffraction in the study of protein and nuclei acid structure,* Interscience, 1966.

R. W. James, *The optical principles of the diffraction of x-rays,* G. Bell and Sons, Ltd., London, new ed., 1950.

R. W. James, "Dynamical theory of x-ray diffraction," *Solid state physics* **15,** 53-220 (1963).

H. Lipson and W. Cochran, *The determination of crystal structures,* G. Bell and Sons, Ltd., London, 3rd revised and enlarged ed., 1966.

L. Muldawer, "Resource letter XR-1 on x-rays," Amer. J. Physics **37,** 132 (1969). (Annotated bibliography).

W. H. Zachariasen, *Theory of x-ray diffraction in crystals,* Wiley, 1945.

NEUTRON DIFFRACTION

G. E. Bacon, *Neutron diffraction,* Oxford, 2nd ed., 1962.

ELECTRON DIFFRACTION

R. Gevers, in Strumane et al., ed., *Interaction of radiation in solids,* North-Holland, Amsterdam (1964).

3

Crystal Binding

Advanced Topic Relevant to this Chapter:

B. Derivation of the van der Waals interaction

NOTATION: The choice of units is involved in this chapter only in the form of the Coulomb interaction, $\pm q^2/r$ in CGS and $\pm q^2/4\pi\epsilon_0 r$ in SI, where ϵ_0 is the permittivity of free space. Where this makes a difference we write only the CGS form, so labeled, because the conversion to SI is trivial here.

Table 1 Cohesive Energies* of the elements

Energy required to form separated neutral atoms from the solid at 0°K; the values in parentheses are at 298.15°K or at the melting point, whichever temperature is lower. To obtain the energy in J mol⁻¹, multiply the energy in kcal mol⁻¹ by 4.184 = 10³. To obtain the energy in ergs per atom, multiply the energy in eV per atom by 1.6019 × 10⁻¹².

Each cell lists: symbol / eV per atom / kcal per mole.

1	2	3	4	5	6	7	8	9	10	11	12	13	14	15	16	17	18
H 4.48 / 103																	He
Li 1.65 / 38.0	Be 3.33 / 76.9											B 5.81 / 134.	C 7.36 / 170.	N —/(114)	O —/(60)	F —/(20)	Ne 0.02 / 0.45
Na 1.13 / 26.0	Mg 1.53 / 35.3											Al 3.34 / 76.9	Si 4.64 / 107	P —/(79.2)	S 2.86 / 66.1	Cl —/(32.2)	Ar 0.080 / 1.85
K 0.941 / 21.7	Ca 1.825 / 42.1	Sc 3.93 / 90.6	Ti 4.855 / 112.0	V 5.30 / 122.	Cr 4.10 / 94.5	Mn 2.98 / 68.7	Fe 4.29 / 98.9	Co 4.387 / 101.2	Ni 4.435 / 102.3	Cu 3.50 / 80.8	Zn 1.35 / 31.1	Ga 2.78 / 64.2	Ge 3.87 / 89.3	As 3.0 / 69.	Se 2.13 / 49.2	Br 1.22 / (28.2)	Kr 0.116 / 2.67
Rb 0.858 / 19.8	Sr —/(39.1)	Y 4.387 / 101.2	Zr 6.316 / 145.7	Nb 7.47 / 172.	Mo 6.810 / 157.1	Tc	Ru 6.615 / 152.6	Rh 5.752 / 132.7	Pd 3.936 / 90.8	Ag 2.96 / 68.3	Cd 1.160 / 26.76	In 2.6 / 59	Sn 3.12 / 71.9	Sb 2.7 / 62.	Te 2.0 / 46	I —/(25.6)	Xe —/(3.57)
Cs 0.827 / 19.1	Ba 1.86 / (42.8)	La 4.491 / 103.6	Hf 6.35 / 146.	Ta 8.089 / 186.6	W 8.66 / 200.	Re 8.10 / 187.	Os —/(187)	Ir 6.93 / 160.	Pt 5.852 / 135.0	Au 3.78 / 87.3	Hg (0.694) / (16.0)	Tl 1.87 / 43.2	Pb 2.04 / 47.0	Bi 2.15 / 49.6	Po —/(34.5)	At	Rn
Fr	Ra	Ac															

Ce 4.77 / 110	Pr 3.9 / 89	Nd 3.35 / 77.2	Pm	Sm 2.11 / 48.6	Eu 1.80 / 41.5	Gd 4.14 / 95.4	Tb 4.1 / 94	Dy 3.1 / 71	Ho 3.0 / 70	Er 3.3 / 77	Tm 2.6 / 59	Yb 1.6 / 36	Lu (4.4) / (102)
Th 5.926 / 136.7	Pa 5.46 / 126	U 5.405 / 124.7	Np 4.55 / 105	Pu 4.0 / 92	Am 2.6 / 60	Cm	Bk	Cf	Es	Fm	Md	No	Lw

*Data furnished by Leo Brewer and reduced by S. Strässler.

The world about us contains condensed matter in very many forms. There are biological materials, DNA and enzymes, and geological materials, granite and mica. There are thousands of metallic alloys and millions of organic compounds. All these are built up from atoms of one hundred chemical elements. Solid state physics has been largely concerned only with single crystals of the elements and of simple compounds; of these we have gained a deep understanding and insight. Measurements on single crystals are always much more significant and informative than are measurements on polycrystalline specimens, but amorphous materials are of enormous and increasing practical importance.

The observed differences between the forms of condensed matter are caused by differences in the distribution of the electrons and nuclei, or particularly of the outermost electrons and the ion cores. In studying a crystal we first ask ourselves "Where are the nuclei and the electrons?" The determination of the structure of the solid often can be handled by the diffraction methods of Chapter 2.

In this chapter we are concerned with the question: What holds a crystal together? *The attractive electrostatic interaction between the negative charges of the electrons and the positive charges of the nuclei is entirely responsible for the cohesion of solids.* Magnetic forces have only a weak effect on cohesion, and gravitational forces are negligible. Given a knowledge of the spatial distribution and velocity distribution of electrons and nuclei in a crystal (as determined in principle by quantum mechanics), we can calculate the binding energy of the crystal. Specialized terms are employed only to categorize distinctive situations: exchange energy, van der Waals forces, resonance stabilization energy, and covalent bonds.

To bind atoms into solids by the electrostatic attraction between the valence electrons and the ion cores we can do four things, which are not all compatible:

1. The positive ion cores should be kept apart, in order to minimize the Coulomb repulsion of like charges.

2. The valence electrons should also be kept apart.

3. The valence electrons should be kept close to the positive ions, to maximize the Coulomb attraction of unlike charges.

4. These three suggestions may lower the potential energy of the system, but they must be carried out in such a way that the kinetic energy of the system is not much increased. By quantum theory any localization of electrons tends to increase the kinetic energy.[1]

---

[1] Suppose we localize an electron in one dimension in a region of extent Δx; the Heisenberg uncertainty principle tells us that the related spread in momentum $\Delta(mv) \geq h/2\pi \, \Delta x$, where h is Planck's constant. Thus the value of the kinetic energy $\frac{1}{2}mv^2$ is at least $h^2/2m(2\pi \, \Delta x)^2$. If $\Delta x \approx 10^{-8}$ cm, then the kinetic energy is $\approx 5 \times 10^{-12}$ erg ≈ 3 eV.

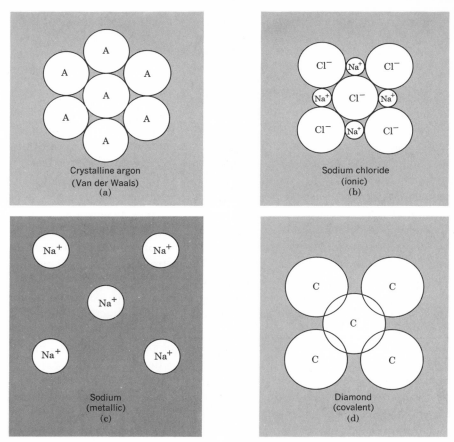

Figure 1 The principal types of crystalline binding forces. In (a) neutral atoms are bound together weakly by the van der Waals forces associated with fluctuations in the charge distributions. In (b) electrons are transferred from the alkali atoms to the halogen atoms, and the resulting ions are held together by attractive electrostatic forces between the positive and negative ions. In (c) the valence electrons are taken away from each alkali atom to form a community electron sea in which the positive ions are dispersed. In (d) the neutral atoms appear to be bound together by the overlapping parts of their electron distributions.

To discuss cohesion we compare the total energy of the solid, kinetic plus potential, with the energy of the same number of free neutral atoms at infinite separation. A crystal can only be stable if its total energy is lower than the total energy of the atoms or molecules when free. The difference (free atom energy) − (crystal energy) is defined as the **cohesive energy.**

Values of the cohesive energy of the crystalline elements, referred to separated neutral atoms, are given in Table 1. The values are usually obtained from a combination of thermodynamic and spectroscopic data. Notice the wide variation in cohesive energy between different columns of the table. The inert gas crystals (at the right) are weakly bound, with cohesive energies less than a few percent of the elements in the C, Si, Ge . . . column. The alkali metal crystals (at the left) have intermediate values of the cohesive energy. The transition element metals (in the middle columns) are quite strongly bound. Ionic compounds (not shown) are also strongly bound.

In Fig. 1 we picture the principal types of crystalline binding forces. In this chapter we shall try to understand at least qualitatively the reasons for some of the differences in the binding forces.

CRYSTALS OF INERT GASES

In many respects the inert gases form the simplest crystals known to us. The properties of inert gas crystals at absolute zero are summarized in Table 2. The crystals are transparent insulators and are weakly bound, with low melting temperatures. The crystals are composed of atoms which have very high ionization energies (see Table 3). The outermost electronic shells of the atoms are completely filled, and the distribution of electronic charge in the free atom is spherically symmetric. In the crystal there is every reason for the inert gas atoms to pack together as closely as possible: the crystal structures (Fig. 2) actually are all cubic close-packed (fcc), except[2] for He[3] and He[4].

Table 2 Properties of inert gas crystals

(Extrapolated to 0°K and zero pressure)

	Nearest-neighbor distance, in Å	Experimental cohesive energy[°]		Melting point, deg K	Ionization[†] potential of free atom, eV	Parameters in Lennard-Jones potential,[‡] Eq. 4	
		kJ/mole	eV/atom			ϵ, in 10^{-16} erg	σ, in Å
He		(liquid at zero pressure)			24.58	14	2.56
Ne	3.13	1.88	0.02	24	21.56	50	2.74
Ar	3.76	7.74	0.080	84	15.76	167	3.40
Kr	4.01	11.2	0.116	117	14.00	225	3.65
Xe	4.35	16.0	0.17	161	12.13	320	3.98

[°] See E. R. Dobbs and G. O. Jones, Rpts. Prog. Phys. **20**, 516 (1957); the values for Ar and Kr were furnished by L. Brewer.

[†] See C. E. Moore, *Atomic energy levels,* Circular of the National Bureau of Standards 467 Vol. I, p. XL. These volumes are the canonical source for data about the electronic energy states of free atoms.

[‡] N. Bernardes, Phys. Rev. **112**, 1534 (1958).

[2] Zero-point motion (kinetic energy at absolute zero) is a quantum effect which plays a dominant role in He[3] and He[4]. They do not solidify at zero pressure even at absolute zero. The average fluctuation at 0°K of a He atom from its equilibrium position is of the order of 30–40 percent of the nearest-neighbor distance. See L. H. Nosanow and G. L. Shaw, Phys. Rev. **128**, 546 (1962). The heavier the atom, the less important are the zero-point effects, as we shall see. If we omit zero-point motion, we use the range parameter σ and Eq. (15) to calculate a molar volume of 9 cm³ mol⁻¹ for solid helium, as compared with the observed values of 27.5 and 36.8 cm³ mol⁻¹ for liquid He[4] and liquid He[3], respectively. To understand the ground state of helium we must take account of the zero-point motion of the atoms.

Figure 2 Cubic close-packed (fcc) crystal structure of the inert gases Ne, Ar, Kr, and Xe. The lattice parameters of the cubic cells are 4.46, 5.31, 5.64, and 6.13 Å, respectively, at 4°K.

What holds an inert gas crystal together? We believe that the electron distribution in the crystal cannot be significantly distorted from the electron distribution around the free atoms, because the cohesive energy of an atom in the crystal is only one percent or less of the ionization energy of an atomic electron, from Table 2. Thus not much energy is available to distort the free atom charge distributions. Part of this distortion is the van der Waals interaction.

Van der Waals-London Interaction

Consider two identical inert gas atoms at a separation R large in comparison with any reasonable measure of the radii of the atoms. What interactions exist between the two neutral atoms?

If the charge distributions on the atoms were rigid, the interaction between atoms would be zero, because the electrostatic potential of a spherical distribution of electronic charge is canceled outside a neutral atom by the electrostatic potential of the charge on the nucleus. Then the inert gas atoms could show no cohesion and could not condense, contrary to experiment. It is true that the time-average electric moments are all zero. But the electrons are in

Table 3 Ionization energies of the elements

The total energy required to remove the first two electrons is the sum of the first and second ionization potentials. *Source:* National Bureau of Standards Circular 467.

⟶ Energy to remove one electron, in eV.
⟶ Energy to remove two electrons, in eV.

1	2	3	4	5	6	7	8	9	10	11	12	13	14	15	16	17	18
H 13.595																	**He** 24.58 / 78.98
Li 5.39 / 81.01	**Be** 9.32 / 27.53											**B** 8.30 / 33.45	**C** 11.26 / 35.64	**N** 14.54 / 44.14	**O** 13.61 / 48.76	**F** 17.42 / 52.40	**Ne** 21.56 / 62.63
Na 5.14 / 52.43	**Mg** 7.64 / 22.67											**Al** 5.98 / 24.80	**Si** 8.15 / 24.49	**P** 10.55 / 30.20	**S** 10.36 / 34.0	**Cl** 13.01 / 36.81	**Ar** 15.76 / 43.38
K 4.34 / 36.15	**Ca** 6.11 / 17.98	**Sc** 6.56 / 19.45	**Ti** 6.83 / 20.46	**V** 6.74 / 21.39	**Cr** 6.76 / 23.25	**Mn** 7.43 / 23.07	**Fe** 7.90 / 24.08	**Co** 7.86 / 24.91	**Ni** 7.63 / 25.78	**Cu** 7.72 / 27.93	**Zn** 9.39 / 27.35	**Ga** 6.00 / 26.51	**Ge** 7.88 / 23.81	**As** 9.81 / 30.0	**Se** 9.75 / 31.2	**Br** 11.84 / 33.4	**Kr** 14.00 / 38.56
Rb 4.18 / 31.7	**Sr** 5.69 / 16.72	**Y** 6.5 / 18.9	**Zr** 6.95 / 20.98	**Nb** 6.77 / 21.22	**Mo** 7.18 / 23.25	**Tc** 7.28 / 22.54	**Ru** 7.36 / 24.12	**Rh** 7.46 / 25.53	**Pd** 8.33 / 27.75	**Ag** 7.57 / 29.05	**Cd** 8.99 / 25.89	**In** 5.78 / 24.64	**Sn** 7.34 / 21.97	**Sb** 8.64 / 25.1	**Te** 9.01 / 27.6	**I** 10.45 / 29.54	**Xe** 12.13 / 33.3
Cs 3.89 / 29.0	**Ba** 5.21 / 15.21	**La** 5.61 / 17.04	**Hf** 7. / 22.	**Ta** 7.88 / 24.1	**W** 7.98 / 25.7	**Re** 7.87 / 24.5	**Os** 8.7 / 26.	**Ir** 9.	**Pt** 8.96 / 27.52	**Au** 9.22 / 29.7	**Hg** 10.43 / 29.18	**Tl** 6.11 / 26.53	**Pb** 7.41 / 22.44	**Bi** 7.29 / 23.97	**Po** 8.43	**At**	**Rn** 10.74
Fr	**Ra** 5.28 / 15.42	**Ac** 6.9 / 19.0															

Ce 6.91	**Pr** 5.76	**Nd** 6.31	**Pm**	**Sm** 5.6	**Eu** 5.67	**Gd** 6.16	**Tb** 6.74	**Dy** 6.82	**Ho**	**Er**	**Tm**	**Yb** 6.2	**Lu** 5.0
Th	**Pa**	**U** 4	**Np**	**Pu**	**Am**	**Cm**	**Bk**	**Cf**	**Es**	**Fm**	**Md**	**No**	**Lw**

motion around the nucleus even in the lowest electronic state, and at any instant of time there is likely to be a nonvanishing electric dipole moment from this motion.[3] An instantaneous dipole moment of magnitude p_1 on one atom (Fig. 3) produces an electric field E of magnitude $2p_1/R^3$ at the center of the second atom distant R from the first atom. This field will induce an instantaneous dipole moment $p_2 = \alpha E = 2\alpha p_1/R^3$ on the second atom; here α is the electronic polarizability, defined in Eq. (13.31) as the dipole moment per unit electric field.

The standard result of electrostatics for the energy of interaction of two dipoles of moment $\mathbf{p}_1$ and $\mathbf{p}_2$ separated by $\mathbf{R}$ is

$$\text{(CGS)} \qquad U(R) = \frac{\mathbf{p}_1 \cdot \mathbf{p}_2}{R^3} - \frac{3(\mathbf{p}_1 \cdot \mathbf{R})(\mathbf{p}_2 \cdot \mathbf{R})}{R^5} \ . \tag{1a}$$

Because the induced moments in the van der Waals interaction are parallel, we have for the potential energy of the dipole moments

$$\text{(CGS)} \qquad U(R) \approx -\frac{2p_1 p_2}{R^3} = -\frac{4\alpha p_1^2}{R^6} \ . \tag{1b}$$

The interaction is attractive. To obtain (1) in SI, multiply the right-hand side by $1/4\pi\epsilon_0$.

The coefficient αp_1^2 may be estimated as follows: Electronic polarizabilities have dimensions [length]^3 as we see from the relation $p_2 = 2\alpha p_1/R^3$ above; and the relevant length is an atomic radius, denoted by r_0. Dipole moments have dimensions [charge] $\times$ [length] and have magnitudes of the order of er_0. Thus

$$\text{(CGS)} \quad U(R) \approx -\frac{4e^2 r_0^5}{R^6}$$

$$\approx -\frac{4(5 \times 10^{-10})^2 (1 \times 10^{-8})^5}{R^6} \approx -\frac{10^{-58}}{R^6} \ , \tag{2}$$

in ergs for R in cm. We have taken $r_0 \approx 10^{-8}$ cm.

We write the interaction as

$$U(R) = -\frac{C}{R^6} \ . \tag{3}$$

---

[3] The semiclassical model leads to the correct result, but the language is not to be taken entirely literally. A simple quantum-mechanical model (two harmonic oscillators) is developed in Advanced Topic B.

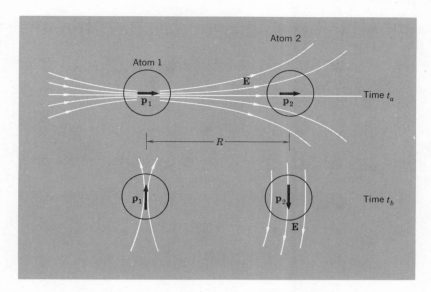

Figure 3 Origin of the van der Waals interaction, according to a classical argument. At one instant of time there is a dipole moment p_1 on atom 1. This produces an electric field E at atom 2, which acquires an induced dipole moment p_2. Diagrams are shown for two times, t_a and t_b. The interaction is always attractive: the closer the atoms, the tighter the binding.

This is known variously as the **van der Waals interaction,**[4] the London interaction, or the induced dipole-dipole interaction. It is the principal attractive interaction in crystals of inert gases and also in crystals of many organic molecules. Taking $C \approx 10^{-58}$ erg-cm^6, the interaction[5] at a separation $R = 4$ Å as for krypton has the value $U \approx 2 \times 10^{-14}$ erg, whence in temperature units $U/k_B \approx 100°$K, of the order of magnitude of the melting temperature of inert gas crystals.

Because of the R^{-6} dependence the interaction increases rapidly at shorter distances. For example, at the interatomic separation of metallic copper, 2.55 Å, the van der Waals interaction between the Cu$^+$ ion cores is $\approx 2 \times 10^{-13}$ ergs for the above value of C. A better value of C may be as much as ten times greater, which would account for a substantial part of the high cohesive energy of copper (Table 1).

[4] The quantum-mechanical theory is discussed by H. Margenau, Revs. Mod. Phys. **11**, 1 (1939); see also *Advances in chemical physics* **12**: *Intermolecular forces*, J. O. Hirschfelder, ed., Interscience, 1967, and S. Doniach, Phil. Mag. **8**, 129 (1963).

[5] It is easier to grasp the magnitude of an interaction energy U if it is expressed in terms of an effective temperature which we define by the relation $k_B T = U$, where k_B is the Boltzmann constant.

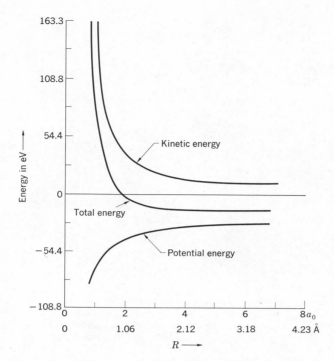

Figure 4 Kinetic, potential, and total energy of a hydrogen atom in a rigid spherical container of radius R. The total energy of the atom is increased as the radius of the container is decreased. [From S. R. de Groot and C. A. ten Seldam, *Physica* **12**, 669 (1946).]

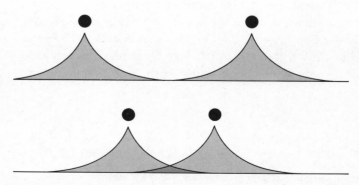

Figure 5 Electronic charge distributions overlap as atoms approach. The solid circles denote the nuclei.

Repulsive Interaction

We imagine that we squeeze the electron distribution of an atom by confinement of the atom within a rigid spherical container. The electronic kinetic energy of the atom will be increased by the confinement, as shown in Fig. 4 for a hydrogen atom. The increase in energy acts as a repulsive force on the container, a force which resists compression. The confinement effect is one contribution to the repulsive energy of atoms in a crystal. A more important contribution to the repulsion may arise from the overlap of the electron distributions of two atoms close to each other.

As the two atoms are brought together their charge distributions gradually overlap (Fig. 5), thereby changing the electrostatic energy of the system. At sufficiently close separations the overlap energy is repulsive. For atoms with filled electronic shells the overlap energy may be repulsive[6] at all distances of interest (say 0.5 Å to 5 Å), in large part because of the Pauli exclusion principle.

The elementary statement of the **Pauli exclusion principle** is that two electrons cannot have all their quantum numbers equal. The Pauli principle prevents two electrons from occupying the same quantum state: when the charge distributions of two atoms overlap there is a tendency for electrons from atom B to occupy in part states of atom A already occupied by electrons of atom A, and vice versa. The Pauli principle prevents multiple occupancy, and electron distributions of atoms with closed shells can overlap only if accompanied by the partial promotion of electrons to unoccupied higher energy states of the atoms. Thus the electron overlap increases the total energy of the system and gives a repulsive contribution to the interaction. An extreme example in which the overlap is complete is shown in Fig. 6.

We make no attempt here to evaluate the repulsive interaction from first principles. Experimental data on the inert gases can be fitted well by an empirical repulsive potential of the form B/R^{12}, where B is a positive constant, when used together with a long-range attractive potential of the form of (3). The constants B and C are empirical parameters determined from independent measurements made in the gas phase; the data used include the virial coefficients and the viscosity. It is usual to write the total potential energy of two atoms at separation R as

$$U(R) = 4\epsilon \left[\left(\frac{\sigma}{R} \right)^{12} - \left(\frac{\sigma}{R} \right)^{6} \right] , \qquad (4)$$

where ϵ and σ are the new parameters, with $4\epsilon\sigma^6 \equiv C$ and $4\epsilon\sigma^{12} \equiv B$. The po-

---

[6] The overlap energy naturally depends on the radial distribution of charge about each atom. The mathematical calculation is always complicated even if the charge distribution is known. For a clear discussion for two hydrogen atoms, see Chapter 12 of L. Pauling and E. B. Wilson, *Introduction to quantum mechanics*, McGraw-Hill, 1935.

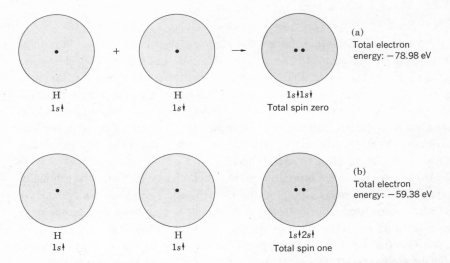

Figure 6 The effect of Pauli principle on the repulsive energy: in an extreme example, two hydrogen atoms are pushed together until the protons are almost in contact. The energy of the electron system alone can be taken from observations on atomic He, which has two electrons. In (a) the electrons have antiparallel spins and the Pauli principle has no effect: the electrons are bound by -78.98 eV. In (b) the spins are parallel: the Pauli principle forces the promotion of an electron from a $1s\uparrow$ orbital of H to a $2s\uparrow$ orbital of He. The electrons now are bound by -59.38 eV, less than (a) by 19.60 eV. This is the amount by which the Pauli principle has increased the repulsion. We have omitted the repulsive Coulomb energy of the two protons, which is the same in both (a) and (b).

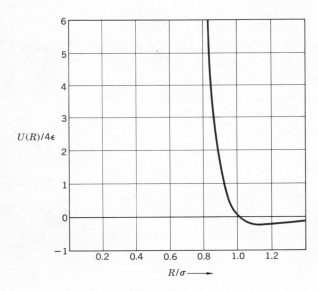

Figure 7 Form of the Lennard-Jones potential (4) which describes the interaction of two inert gas atoms. The minimum occurs at $R/\sigma = 2^{1/6} \cong$ 1.12. Notice how steep the curve is inside the minimum, and how flat it is outside the minimum. The value of U at the minimum is $-\epsilon$; and $U = 0$ at $R = \sigma$. The minimum of U is at $R \cong 1.122\sigma$.

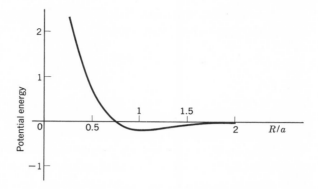

Figure 8 Coulomb energy of two spheres of radius a, as a function of the distance R between their centers. Each sphere carries a charge $+q$ at its center, and has a charge $-q$ uniformly distributed throughout the volume of the sphere. The charges are taken as rigid, so that van der Waals and polarization effects are not included in this model. The model takes no account of the exclusion principle; in this respect the calculation is analogous to that for curve N in Fig. 15 below. (Courtesy of C. Y. Fong.)

tential (4) is known as the Lennard-Jones potential, and is plotted in Fig. 7. The force between the two atoms is given by $-dU/dR$. Values of ϵ and σ as recommended by Bernardes are given in Table 2; the values can be obtained from gas-phase data, so that calculations on properties of the solid do not involve any disposable parameters.

Other empirical forms for the repulsive interaction are widely used,[7] in particular the exponential form $\lambda \exp(-R/\rho)$, where ρ is a measure of the range of the interaction. This is generally as easy to handle analytically as the inverse power law form. The classical Coulomb energy of two neutral atoms with static electrons distributed uniformly inside spheres is plotted in Fig. 8. The interaction is repulsive at small distances because of the electrostatic repulsion of the two protons.

Equilibrium Lattice Constants

If we neglect the kinetic energy of the inert gas atoms, the cohesive energy of an inert gas crystal is given by summing the Lennard-Jones potential (4) over all pairs of atoms in the crystal. If there are N atoms in the crystal, the total potential energy is

$$U_{\text{tot}} = \tfrac{1}{2}N(4\epsilon)\left[\sum_{j}{}'\left(\frac{\sigma}{p_{ij}R}\right)^{12} - \sum_{j}{}'\left(\frac{\sigma}{p_{ij}R}\right)^{6}\right], \qquad (5)$$

where $p_{ij}R$ is the distance between reference i atom and any other atom j, expressed in terms of the nearest neighbor distance R. The factor $\tfrac{1}{2}$ occurs with the

[7] For discussions of repulsive terms, see M. P. Tosi, *Solid state physics* **16**, 1 (1964); Hirschfelder, Curtis, and Bird, *Molecular theory of gases and liquids*, Wiley, 1954; F. G. Fumi and M. P. Tosi, J. Phys. Chem. Solids **25**, 31, 45 (1964).

N to avoid counting twice each pair of atoms. The summations in (5) have been evaluated,[8] and for the fcc structure

$$\sideset{}{'}\sum_{j} p_{ij}^{-12} = 12.13188 \;\; ; \qquad \sideset{}{'}\sum_{j} p_{ij}^{-6} = 14.45392 \; . \tag{6}$$

There are twelve nearest-neighbor sites in the fcc structure; we see that the series are rapidly converging and have values not far from twelve. Thus the nearest neighbors contribute most of the interaction energy of inert gas crystals. The corresponding sums for the hcp structure are 12.13229 and 14.45489.

If we take U_{tot} in (5) as the total energy of the crystal, the equilibrium value R_0 is given by requiring that U_{tot} be a minimum with respect to variations in the nearest neighbor distance R:

$$\frac{dU_{tot}}{dR} = 0 = -2N\epsilon \left[(12)(12.13)\frac{\sigma^{12}}{R^{13}} - (6)(14.45)\frac{\sigma^{6}}{R^{7}} \right] , \tag{7}$$

whence

$$R_0/\sigma = 1.09 \; , \tag{8}$$

the same for all elements with an fcc structure. The observed values of R_0/σ, using the independently determined values of σ given in Table 2, are:

	Ne	Ar	Kr	Xe
R_0/σ	1.14	1.11	1.10	1.09

The agreement with (8) is remarkable. The slight departure of R_0/σ from the universal value 1.09 predicted for inert gases can be explained by quantum effects.[9] From measurements on the gas phase we have predicted the lattice constant of the crystal.

We have made the implicit assumption that the fcc structure is the structure of minimum energy for the Lennard-Jones interaction (5). Calculation[10] suggests that the hcp structure will have a lower energy at absolute zero, by 0.01 percent. But by experiment the fcc structure is the stable structure except for helium.

[8] J. E. Lennard-Jones and A. E. Ingham, Proc. Roy. Soc. (London) **A107**, 636 (1925). Summations for the exponents -4 to -30 are given on p. 1040 of the book by Hirschfelder, Curtis, and Bird.

[9] N. Bernardes, Phys. Rev. **112**, 1534 (1958); see also L. H. Nosanow and G. L. Shaw, Phys. Rev. **128**, 546 (1962); T. R. Koehler, Phys. Rev. Letters **17**, 89 (1966).

[10] T. A. K. Barron and C. Domb, Proc. Roy. Soc. (London) **A227**, 447 (1955); see also R. Bullough, H. R. Glyde, and J. A. Venables, Phys. Rev. Letters **17**, 249 (1966).

Cohesive Energy

The cohesive energy of inert gas crystals at absolute zero and at zero pressure is obtained by substituting (6) and (8) in (5):

$$U_{\text{tot}}(R) = 2N\epsilon \left[(12.13)\left(\frac{\sigma}{R}\right)^{12} - (14.45)\left(\frac{\sigma}{R}\right)^{6} \right] , \qquad (9)$$

and, at $R = R_0$,

$$U_{\text{tot}}(R_0) = -(2.15)(4N\epsilon) , \qquad (10)$$

the same for all inert gases. This is the calculated cohesive energy if the kinetic energy is zero. Bernardes has calculated quantum-mechanical corrections including kinetic energy contributions; the corrections act to reduce the binding and amount to 28, 10, 6, and 4 percent of Eq. (10) for Ne, Ar, Kr, and Xe, respectively.

The heavier the atom, the smaller the quantum correction. We can understand the origin of the quantum correction by consideration of a simple model in which an atom is confined by fixed boundaries. If the particle has the quantum wavelength λ, where λ is determined by the boundaries, then the particle has kinetic energy $p^2/2M = (h/\lambda)^2/2M$ with the de Broglie relation $p = h/\lambda$ for the connection between the momentum and the wavelength of a particle. On this model the quantum zero-point correction to the energy is inversely proportional to the mass, in quite good agreement with the ratio of the corrections cited above for neon (atomic weight 20.2) and xenon (atomic weight 130). The final calculated cohesive energies agree with the experimental values of Table 2 within 1 to 7 percent.

One consequence of the quantum kinetic energy is that a crystal of the isotope Ne^{20} is observed to have a larger value of the equilibrium lattice constant than a crystal of Ne^{22}. The higher quantum kinetic energy of the lighter isotope expands the lattice, because the kinetic energy is reduced by expansion. The observed[11] lattice constants (extrapolated to absolute zero from 2.5°K) are

$$\begin{array}{ll} Ne^{20} & 4.4644 \text{ Å } ; \\ Ne^{22} & 4.4559 \text{ Å } . \end{array}$$

Compressibility and Bulk Modulus

An independent test of the theory is provided by the **bulk modulus,** defined as

$$B = -V\frac{dp}{dV} , \qquad (11)$$

where V is the volume and p the pressure. The compressibility is defined as the reciprocal of the bulk modulus. At absolute zero the entropy is constant, so that by the thermodynamic identity $dU = -p\,dV$ is the change in energy accom-

[11] D. N. Batchelder, D. L. Losee, and R. O. Simmons, Phys. Rev. **173**, 873 (1968).

panying a change dV in volume. Thus $dp/dV = -d^2U/dV^2$, and

$$B = V \frac{d^2U}{dV^2} \ . \tag{12}$$

The bulk modulus is a measure of the stiffness of the crystal or of the energy required to produce a given deformation. The higher the bulk modulus, the stiffer is the crystal.

The volume occupied by N atoms in an fcc lattice of lattice constant a is $V = \frac{1}{4}Na^3$, because $\frac{1}{4}a^3$ is the volume per atom (Chapter 1). In terms of the nearest-neighbor distance $R = a/\sqrt{2}$, we have $V = NR^3/\sqrt{2}$. The potential energy (9) may be written as

$$U_{\text{tot}}(V) = \frac{b_{12}}{V^4} - \frac{b_6}{V^2} \ , \tag{13}$$

where the parameters b_{12} and b_6 follow from (5) and (6):

$$b_{12} \equiv \tfrac{1}{2}(12.13)N^5\epsilon\sigma^{12} \ ; \qquad b_6 \equiv (14.45)N^3\epsilon\sigma^6 \ .$$

At equilibrium under zero pressure

$$\frac{dU_{\text{tot}}}{dV} = 0 = -\frac{4b_{12}}{V^5} + \frac{2b_6}{V^3} \ , \tag{14}$$

whence the equilibrium volume is given by

$$V_0 = \left(\frac{2b_{12}}{b_6}\right)^{\frac{1}{2}} = \left(\frac{12.13}{14.45}\right)^{\frac{1}{2}}N\sigma^3 \ . \tag{15}$$

The bulk modulus is

$$B = \left(V \frac{d^2U}{dV^2}\right)_{V_0} = \frac{20b_{12}}{V_0^5} - \frac{6b_6}{V_0^3} = \sqrt{2}\,\frac{b_6^{\frac{5}{2}}}{b_{12}^{\frac{3}{2}}} \ , \tag{16}$$

of the order of magnitude ϵ/σ^3.

With the empirical two-atom potential (4) inferred from gas-phase data we can give an excellent account of the observed properties of the inert gas crystals Ne, Ar, Kr, and Xe. Quantum corrections improve the agreement further.

Figure 9 Electron density distribution in the base plane of NaCl, after x-ray studies by G. Schoknecht, Z. Naturforschg. **12a**, 983 (1957).

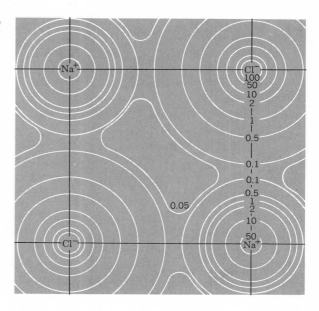

IONIC CRYSTALS

Ionic crystals are made up of positive and negative ions. The ions arrange themselves with the Coulomb attraction between ions of opposite sign stronger than the Coulomb repulsion between ions of the same sign. The **ionic bond** is the bond resulting from the electrostatic interaction of oppositely charged ions. Two common crystal structures found for ionic crystals, the sodium chloride and the cesium chloride structures, were shown in Figs. 23 to 26 of Chapter 1.

The electronic configurations of all ions of a simple ionic crystal correspond to closed electronic shells, as in the inert gas atoms. In lithium fluoride the configuration of the neutral atoms are, according to the periodic table in the front end papers of this book, Li: $1s^2 2s$, F: $1s^2 2s^2 2p^5$. The singly charged ions have the configurations Li$^+$: $1s^2$, F$^-$: $1s^2 2s^2 2p^6$, as for helium and neon, respectively. Inert gas atoms have closed shells, and the charge distributions are spherically symmetric. We expect that the charge distributions on each ion in an ionic crystal will have approximately spherical symmetry, with some distortion near the region of contact with neighboring atoms. This picture is confirmed by x-ray studies of electron distributions (Fig. 9).

A quick estimate suggests that we are not misguided in looking to electrostatic interactions for a large part of the binding energy of an ionic crystal. The distance between a positive ion and the nearest negative ion in crystalline sodium chloride is 2.81×10^{-8} cm, and the attractive Coulomb part of the potential energy of the two ions by themselves is 5.1 eV. This value may be compared (Fig. 10) with the known value from Table 5 of 7.9 eV per molecule unit for the cohesive energy of crystalline NaCl with respect to separated Na$^+$ and Cl$^-$ ions. The order of magnitude agreement between the values is quite sug-

112

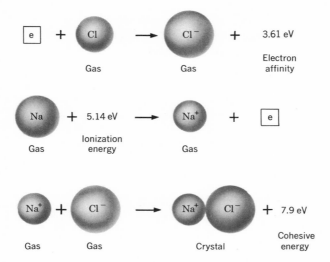

Figure 10 The energy per molecule unit of a crystal of sodium chloride is $(7.9 - 5.1 + 3.6) = 6.4$ eV lower than the energy of separated neutral atoms. The cohesive energy with respect to separated *ions* is 7.9 eV per molecule unit. All values on the figure are experimental. Values of the ionization energy are given in Table 3, and values of the electron affinity are given in Table 4.

gestive. We now calculate the lattice energy more closely. The estimate in Fig. 10 uses the experimental value given in Table 4 of the electron affinity of the Cl^- ion.

Electrostatic or Madelung Energy

The long range interaction between ions with charge $\pm q$ is the electrostatic interaction $\pm q^2/r$ attractive between ions of opposite charge and repulsive between ions of the same charge. The ions arrange themselves in whatever crystal structure gives the strongest attractive interaction compatible with the repulsive interaction which sets in at short distances between ion cores. The repulsive interactions between ions with inert gas configurations are similar to those between inert gas atoms. The van der Waals part of the attractive interaction in ionic crystals makes a relatively small contribution to the cohesive energy, of the order of 1 or 2 percent. The main contribution to the binding energy of ionic crystals is electrostatic and is called the **Madelung energy**.

If U_{ij} is the interaction energy between ions i and j, we may define a sum U_i which includes all interactions involving the ion i:

$$U_i = \sum_j{}' U_{ij} \; , \qquad (17)$$

where the summation includes all ions except $j = i$. We suppose that U_{ij} may be written as the sum of a central field repulsive potential of the form $\lambda \exp(-r/\rho)$, where λ and ρ are empirical parameters, and a Coulomb potential

Table 4   Electron affinities of negative ions

[Largely from B. L. Moiseiwitsch, *Advances in atomic and molecular physics* **1**, 61 (1965).]

Note: The electron affinity is positive for a stable negative ion. The static electric field of a neutral atom is by itself insufficient to bind an additional electron, but the electron induces electric dipole and higher multipole moments in the atom, causing an attractive potential proportional to $-1/r^4$ at large distances. In many instances this polarization potential is strong enough to bind the extra electron to the free atom.

Atom	Electron affinity energy, in eV	
	Theory	Experiment
H	0.7542	0.77 ± 0.02
Li	0.58	—
C	1.17	1.25 ± 0.03
N	-0.27	—
O	1.22	1.465 ± 0.005
F	3.37	3.448 ± 0.005
Na	0.78	—
Al	0.49	—
Si	1.39	—
P	0.78	—
S	2.12	2.07 ± 0.07
Cl	3.56	3.613 ± 0.003
Br	—	3.363 ± 0.003
I	—	3.063 ± 0.003
W	—	0.50 ± 0.3
Re	—	0.15 ± 0.1

$\pm q^2/r$. Thus

(CGS)
$$U_{ij} = \lambda \exp\left(-r_{ij}/\rho\right) \pm \frac{q^2}{r_{ij}}, \tag{18}$$

where the $+$ sign is taken for the like charges and the $-$ sign for unlike charges. The repulsive term describes the fact that filled electronic shells act as if they are fairly hard, and each ion resists overlapping with the electron distributions of neighboring ions. We have seen that this is in large measure an effect of the Pauli exclusion principle. We regard the strength λ and range ρ as constants to be determined from observed values[12] of the lattice constant and compressibility; we have used the exponential form of the empirical repulsive potential rather than the R^{-12} form used for the inert gases. The change of form is made partly for variety in the discussion and partly because it may give a better representation of the repulsive interaction.

[12] For the ions we do not have gas-phase data available to permit the independent determination of λ and ρ. We note that ρ is a measure of the range of the repulsive interaction: when $r = \rho$, the repulsive interaction is reduced to e^{-1} of the value at $r = 0$. In SI units the Coulomb interaction is $\pm q^2/4\pi\epsilon_0 r$; we write this section in CGS units in which the Coulomb interaction is $\pm q^2/r$.

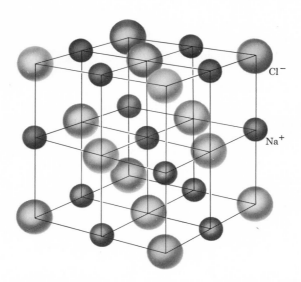

Figure 11 We may construct the sodium chloride crystal structure by arranging Na$^+$ and Cl$^-$ ions alternately at the lattice points of a simple cubic lattice. In the crystal each ion is surrounded by six nearest neighbors of the opposite charge and twelve next-nearest neighbors of the same charge as the reference ion. The Na$^+$ ion carries a single positive charge, so that the electronic configuration is identical with neon, and the Cl$^-$ ion carries a single negative charge and is in the argon configuration. The space lattice is fcc, as discussed in Chap. 1.

In the NaCl structure (Fig. 11) the value of U_i does not depend on whether the reference ion i is a positive or a negative ion. The sum (17) can be arranged to converge rapidly, so that its value will not depend on the particular location of the reference ion in the crystal, as long as it is not near the surface. Neglecting surface effects, we may write the total lattice energy U_{tot} of a crystal composed of N molecules or $2N$ ions as, with U_i defined in (17),

$$U_{tot} = NU_i \ . \tag{19}$$

Here N, rather than $2N$, occurs because in taking the total lattice energy we must count each *pair* of interactions only once or each bond only once. The total lattice energy (19) is the energy required to separate the crystal into individual ions at an infinite distance apart.

It is convenient again to introduce quantities p_{ij} such that $r_{ij} \equiv p_{ij}R$, where R is the nearest-neighbor separation in the crystal. If we include the repulsive interaction only among nearest neighbors, we have

$$\text{(CGS)} \qquad U_{ij} = \begin{cases} \lambda \exp\left(-R/\rho\right) - \dfrac{q^2}{R} & \text{(nearest neighbors)} \\[2mm] \pm \dfrac{1}{p_{ij}} \dfrac{q^2}{R} & \text{(otherwise).} \end{cases} \tag{20}$$

Thus

(CGS)
$$U_{\text{tot}} = NU_i = N\left(z\lambda e^{-R/\rho} - \frac{\alpha q^2}{R}\right) , \tag{21}$$

where z is the number of nearest neighbors of any ion and

$$\boxed{\alpha \equiv \sum_j \frac{(\pm)}{p_{ij}} \equiv \text{Madelung constant.}} \tag{22}$$

The sum should include the nearest-neighbor contribution, which is just z. The $(\pm)$ sign is discussed just before (26). The value of the Madelung constant is of central importance in the theory of an ionic crystal. Methods for its calculation are discussed below.

At the equilibrium separation $dU_{\text{tot}}/dR = 0$, so that

(CGS)
$$N\frac{dU_i}{dR} = -\frac{Nz\lambda}{\rho}\exp\left(-R/\rho\right) + \frac{N\alpha q^2}{R^2} = 0 , \tag{23}$$

or

(CGS)
$$R_0{}^2 \exp\left(-R_0/\rho\right) = \frac{\rho\alpha q^2}{z\lambda} . \tag{24}$$

This determines the equilibrium separation R_0 if the parameters ρ, λ of the repulsive interaction are known. To convert to SI, replace q^2 by $q^2/4\pi\epsilon_0$.

The total lattice energy of the crystal of $2N$ ions at their equilibrium separation R_0 may be written, using (21) and (24), as

(CGS)
$$U_{\text{tot}} = -\frac{N\alpha q^2}{R_0}\left(1 - \frac{\rho}{R_0}\right) . \tag{25}$$

The term $-N\alpha q^2/R_0$ is the Madelung energy. We shall find in (32) that ρ is of the order of $0.1R_0$, so that the cohesive energy is dominated by the Madelung contribution. The low value of ρ/R_0 means that the repulsive interaction is steep and has a very short range.

Evaluation of the Madelung Constant[13]

The first calculation of the Coulomb energy constant α was made by Madelung.[14] A powerful general method for lattice sum calculations was developed by Ewald,[15] and Evjen and Frank[16] have given simple methods which arrange the counting in rapidly convergent ways.

The definition (22) of the Madelung constant α is

$$\alpha = \sum_j{}' \frac{(\pm)}{p_{ij}} . \qquad \left[\text{ot} \quad \sum_j \frac{\pm\, 8.92}{p_{ij}} \; ? \right]$$

[13] A detailed review and bibliography is given by M. P. Tosi, *Solid state physics* **16**, 1 (1964).

[14] E. Madelung, Physik. Z. **19**, 524 (1918).

[15] P. P. Ewald, Ann. Physik **64**, 253 (1921); see also Appendix A to the second edition of this book.

[16] H. M. Evjen, Phys. Rev. **39**, 675 (1932); F.C. Frank, Phil. Mag. **41**, 1287 (1950).

i.c. is $\alpha_{NaCl} = \frac{1}{4} \alpha_{MgO}$?

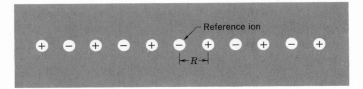

Figure 12 Line of ions of alternating signs, with distance R between ions.

If we take the reference ion as a negative charge the plus sign will be used for positive ions and the minus sign for negative ions. An equivalent definition is

$$\frac{\alpha}{R} = \sum_{j}{}' \frac{(\pm)}{r_j} \ , \tag{26}$$

where r_j is the distance of the jth ion from the reference ion and R is the nearest-neighbor distance. It must be emphasized that the value given for α will depend on whether it is defined in terms of the nearest-neighbor distance R or in terms of the lattice parameter a or in terms of some other relevant length. *Beware!*

As an example, we compute the value of the Madelung constant for the infinite line of ions of alternating sign in Fig. 12. We pick a negative ion as reference ion, and let R denote the distance between adjacent ions. Then

$$\frac{\alpha}{R} = 2\left[\frac{1}{R} - \frac{1}{2R} + \frac{1}{3R} - \frac{1}{4R} + \cdots \right] \ ,$$

or

$$\alpha = 2\left[1 - \frac{1}{2} + \frac{1}{3} - \frac{1}{4} + \cdots \right] \ ;$$

the factor 2 occurs because there are two ions, one to the right and one to the left, at equal distances r_j. We sum this series by the expansion

$$\log (1 + x) = x - \frac{x^2}{2} + \frac{x^3}{3} - \frac{x^4}{4} + \cdots \ .$$

Thus for the one-dimensional chain the Madelung constant is

$$\alpha = 2 \log 2 \ . \tag{27}$$

In three dimensions the series presents greater difficulty. It is not possible to write down the successive terms by a casual inspection. More important, the series will not converge unless the successive terms in the series are arranged so that the contributions from the positive and negative terms nearly cancel.

In the sodium chloride structure (Fig. 11) the nearest neighbors to the negative reference ion are six positive ions at $p = 1$, giving a positive contribution to α of $\frac{6}{1}$; there are twelve negative ions at $p = 2^{\frac{1}{2}}$, giving $-12/2^{\frac{1}{2}}$;

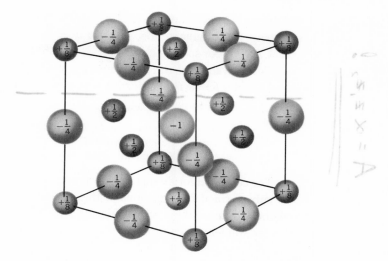

Figure 13 Fractional charge assignments for the NaCl structure, arranged according to the Evjen method for calculating the Madelung constant. Face atoms carry charge $+\frac{1}{2}$; edge atoms $-\frac{1}{4}$; corner atoms $+\frac{1}{8}$.

eight positive ions at $p = 3^{\frac{1}{2}}$, giving $8/3^{\frac{1}{2}}$; six negative ions at $p = 2$, giving $-\frac{6}{2}$; etc. Thus

$$\alpha = \frac{6}{1} - \frac{12}{2^{\frac{1}{2}}} + \frac{8}{3^{\frac{1}{2}}} - \frac{6}{2} + \cdots$$

$$= 6.000 - 8.485 + 4.620 - 3.000 + \cdots .$$

The convergence is obviously poor.

We may improve the convergence by arranging to work with neutral, or nearly neutral, groups of ions, if necessary dividing an ion among different groups and using fractional charges. The physical motivation for working with a neutral group is that its potential falls off faster[17] at a distance than if the group has an excess of charge.

In the sodium chloride structure we obtain nearly neutral groups by considering the charges on cubes, counting charges on cube faces as shared between two cells, on edges as shared between four cells, and on corners as shared between eight cells. The first cube (Fig. 13) surrounding the negative reference ion intercepts six positive charges on cube faces, twelve negative charges on cube edges, and eight positive charges at cube corners. The contribution to α from the first cube is

$$\frac{\frac{6}{2}}{1} - \frac{\frac{12}{4}}{2^{\frac{1}{2}}} + \frac{\frac{8}{8}}{3^{\frac{1}{2}}} = 1.46 .$$

[17] The potential of a charge goes as $1/r$; of a dipole as $1/r^2$; of a quadrupole as $1/r^3$, etc.

On taking into account in similar fashion the ions in the next larger cube[18] enclosing the original cube, we get $\alpha = 1.75$, close to the accurate value $\alpha = 1.747565$ for sodium chloride.

Typical values of the Madelung constant are listed below, based on unit charges and *referred to the nearest-neighbor distance:*

Structure	α
Sodium chloride, NaCl	1.747565
Cesium chloride, CsCl	1.762675
Zinc blende, cubic ZnS	1.6381

The cesium chloride structure is shown in Fig. 1.26. Each ion is at the center of a cube formed by eight ions of the opposite charge. For the same nearest-neighbor distance the cesium chloride structure has a slightly (~ 1 percent) stronger Madelung contribution to the cohesive energy than in the sodium chloride structure, as the value of the Madelung constant α is higher for cesium chloride. But there are more nearest neighbors in CsCl, so that the repulsive energy is higher. Each ion has eight nearest neighbors contributing to the repulsive energy, whereas there are only six in sodium chloride. The repulsive energy is about 10 percent of the total energy of sodium chloride; roughly, we might expect the repulsive energy to be perhaps $\frac{8}{6} \times 10$ percent ≈ 13 percent of the total energy in cesium chloride. This increase outweighs the Madelung energy difference and may favor the sodium chloride structure by a small amount. Problem 3 involves a comparison of the NaCl and cubic ZnS structures.

Many more ionic crystals are known with the sodium chloride structure than with the cesium chloride structure, but the differences in binding energy are small. We can decide which structure will be stable for a particular salt only by a consideration of second-order contributions to the energy. A detailed discussion of the stability of the two lattices is given in the review article by Tosi.

Bulk Modulus

We found in (12) that the bulk modulus at absolute zero is given by $B = V \, d^2U/dV^2$, where V is the volume. For the NaCl structure the volume occupied by N molecules is $V = 2NR^3$, where R is the nearest-neighbor distance. This follows because the volume per molecule is $\frac{1}{4}a^3$ and $a = 2R$, as in Fig. 11.

Now

$$\frac{dU}{dV} = \frac{dU}{dR}\frac{dR}{dV}; \qquad \frac{dR}{dV} = \frac{1}{dV/dR} = \frac{1}{6NR^2}; \qquad (28)$$

$$\frac{d^2U}{dV^2} = \frac{d^2U}{dR^2}\left(\frac{dR}{dV}\right)^2 + \frac{dU}{dR}\frac{d^2R}{dV^2}. \qquad (29)$$

[18] The remainder of fractional charges not counted in the first cube is to be counted as part of the second cube.

At the equilibrium separation $R = R_0$ and $dU/dR = 0$, whence

$$B = V \frac{d^2U}{dR^2}\left(\frac{1}{6NR^2}\right)^2 = \boxed{\frac{1}{18NR_0}\frac{d^2U}{dR^2}} \cdot \qquad (30)$$

(handwritten, left:) $B = V \dfrac{d^2 U}{d V^2}$

(handwritten, right:) ok

(handwritten, far right:) $\left(\dfrac{d V}{d R}\right)^2 = \left(\dfrac{1}{6NR^2}\right)^2$

By (21) and (24),

(CGS) $$\left(\frac{d^2U}{dR^2}\right)_{R_0} = \frac{Nz\lambda}{\rho^2}e^{-R_0/\rho} - \frac{2N\alpha q^2}{R_0^3} = \frac{N\alpha q^2}{R_0^3}\left(\frac{R_0}{\rho} - 2\right) ,$$

whence

(CGS) $$B = \frac{\alpha q^2}{18R_0^4}\left(\frac{R_0}{\rho} - 2\right) . \qquad (31)$$

(handwritten margin:) $\dfrac{z \lambda}{\rho} e^{-r_0/\rho = R_0}$ $\dfrac{z e^2}{\rho^2}$

(handwritten right margin:)
$V = 2NR^3$
$\dfrac{\partial V}{\partial R} = 6NR^2$
$\dfrac{\partial R}{\partial V} = \left(\dfrac{1}{6NR^2}\right)^2$

We may solve (31) for ρ using observed values of R_0 and the bulk modulus. We may then calculate the cohesive energy from (25) and compare with experimental results. Let us carry out the calculation for KCl. The experimental value of the bulk modulus of KCl extrapolated to absolute zero temperature is $B = 1.97 \times 10^{11}$ dynes/cm^2 or 1.97×10^{10} N/m^2, using Eq. (4.29) and Table 4.1. The nearest-neighbor separation R_0 is 3.14×10^{-8} cm; α is 1.75; and thus (31) gives

(CGS) $$\frac{R_0}{\rho} = \frac{18R_0^4B}{\alpha q^2} + 2 \cong 10.4 , \qquad (32)$$

(handwritten:) α not scaled by $z_1 z_2$

so that the range of the repulsive interaction is $\rho \cong 0.30 \times 10^{-8}$ cm.

With this value of R_0/ρ, the calculated cohesive energy (25) becomes

(CGS) $$\frac{U_{\text{tot}}}{N} = -\frac{\alpha q^2}{R_0}\left(1 - \frac{\rho}{R_0}\right) \cong -7.26 \text{ eV} , \qquad (33)$$

in excellent agreement with the observed value -7.397 eV for KCl near $0°$K (Table 5).

The repulsive energy parameter λ may be found from (24):

(CGS) $$z\lambda = \frac{\rho\alpha q^2}{R_0^2}e^{R_0/\rho} \cong 3.8 \times 10^{-8} \text{ erg} , \qquad (34)$$

with $z = 6$ for the number of nearest neighbors. The Madelung and repulsive

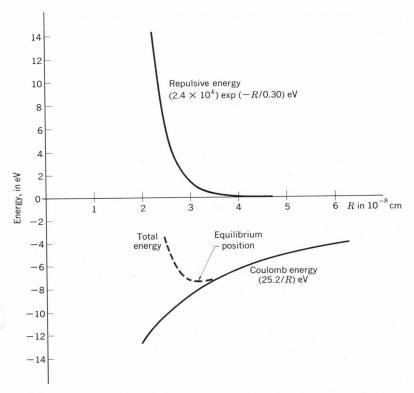

Figure 14 Energy per molecule of KCl crystal, showing Madelung and repulsive contributions.

contributions to the binding of a KCl crystal are shown in Fig. 14.

Properties of alkali halide crystals having the sodium chloride structure are given in Table 5. The values of the repulsive parameters λ and ρ are very sensitive to the values of the input data B and R_0, but the cohesive energy turns out to be relatively insensitive[19] to the value adopted for ρ. The calculated values of the cohesive energy are in exceedingly good agreement with the observed values.

[19] Notice that in our example (33) to (35) we used a value of the bulk modulus of KCl at $0°K$ which differs by 10 percent from that given in Table 5 for room temperature. A 10 percent change in B changes λ by perhaps a factor of 3. The cohesive energy is not very sensitive to the value of the bulk modulus.

Table 5 Properties of alkali halide crystals with the NaCl structure.

[Data from various tables by M. P. Tosi, *Solid state physics* **16**, 1 (1964).]

All values (except those in brackets) at room temperature and atmospheric pressure, with no correction for changes in R_0 and U from absolute zero. Values in brackets at absolute zero temperature and zero pressure, from private communication by L. Brewer.

	Nearest-neighbor separation R_0, in Å	Bulk modulus B, in 10^{11} dynes/cm^2	Repulsive energy parameter° $z\lambda$, in 10^{-8} erg	Repulsive range parameter† ρ, in Å	Cohesive energy compared to free ions, in kcal/mole	
					Experimental	Calculated‡
LiF	2.014	6.71	0.296	0.291	−242.3 [−246.8]	−242.2
LiCl	2.570	2.98	0.490	0.330	−198.9 [−201.8]	−192.9
LiBr	2.751	2.38	0.591	0.340	−189.8	−181.0
LiI	3.000	(1.71)	0.599	0.366	−177.7	−166.1
NaF	2.317	4.65	0.641	0.290	−214.4 [−217.9]	−215.2
NaCl	2.820	2.40	1.05	0.321	−182.6 [−185.3]	−178.6
NaBr	2.989	1.99	1.33	0.328	−173.6 [−174.3]	−169.2
NaI	3.237	1.51	1.58	0.345	−163.2 [−162.3]	−156.6
KF	2.674	3.05	1.31	0.298	−189.8 [−194.5]	−189.1
KCl	3.147	1.74	2.05	0.326	−165.8 [−169.5]	−161.6
KBr	3.298	1.48	2.30	0.336	−158.5 [−159.3]	−154.5
KI	3.533	1.17	2.85	0.348	−149.9 [−151.1]	−144.5
RbF	2.815	2.62	1.78	0.301	−181.4	−180.4
RbCl	3.291	1.56	3.19	0.323	−159.3	−155.4
RbBr	3.445	1.30	3.03	0.338	−152.6	−148.3
RbI	3.671	1.06	3.99	0.348	−144.9	−139.6

° Calculated from B and R_0, using Eqs. (24) and (31); for the NaCl structure $z = 6$.

† Calculated from B and R_0, using Eq. (31).

‡ Calculated from B and R_0, using Eq. (33).

COVALENT CRYSTALS

The covalent bond is the classical electron pair or homopolar bond of chemistry, particularly of organic chemistry. It is a strong bond: the bond between two carbon atoms in diamond has a cohesive energy of 7.3 eV with respect to separated neutral atoms. This is comparable with the bond strength in ionic crystals, in spite of the fact that the covalent bond acts between neutral atoms. The covalent bond has strong directional properties. Thus carbon, silicon, and germanium have the diamond structure,[20] with atoms joined to four nearest neighbors at tetrahedral angles, even though this arrangement gives a low filling of space. For spheres the diamond structure fills 0.34 of the available space, compared with 0.74 for a close-packed structure (Problem 1.4). The tetrahedral bond allows only four nearest neighbors, whereas a close-packed structure has twelve.

The covalent bond is usually formed from two electrons, one from each atom participating in the bond. The electrons forming the bond tend to be partly localized in the region between the two atoms joined by the bond. The spins of the two electrons in the bond are antiparallel.

The Pauli principle gives a repulsive interaction between atoms with filled shells. If the shells are not filled, electron overlap can be accommodated without excitation of electrons to high energy states. Compare the bond length (2 Å) of Cl_2 with the interatomic distance (3.76 Å) of Ar in solid Ar; also compare the cohesive energies given in Table 1. The difference between Cl_2 and Ar_2 is that the Cl atom has five electrons in the 3p shell and the Ar atom has six, filling the shell, so that the repulsive interaction is stronger in Ar than in Cl.

The elements C, Si, and Ge lack four electrons with respect to filled shells, and thus these elements (for example) can have an attractive interaction associated with charge overlap. The electron configuration of carbon is $1s^2 2s^2 2p^2$. Further details can be found in works on quantum chemistry, where it is shown that to form a tetrahedral system of covalent bonds the carbon atom must first be promoted to the electronic configuration $1s^2 2s 2p^3$. This promotion from the ground state requires only 4 eV, an amount more than regained when the bonds are formed.

There is a continuous range of crystals between the ionic and the covalent limits. It is often important to estimate the extent a given bond is ionic or covalent. A semi-empirical theory of the fractional ionic or covalent character of a bond in a dielectric crystal has been developed with considerable success by J. C. Phillips; some of his conclusions are given in Table 6. We think of

[20] We should not overemphasize the similarity of the bonding of carbon and silicon. Carbon gives biology, but silicon gives geology.

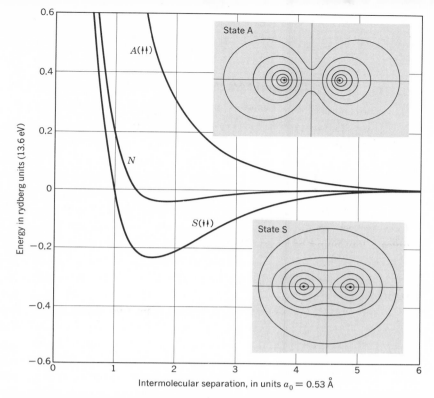

Figure 15 Energy of molecular hydrogen (H_2) referred to separated neutral atoms. A negative energy corresponds to binding. The curve N refers to a classical calculation with free atom charge densities; A is the result for parallel electron spins, taking the Pauli exclusion principle into account, and S (the stable state) for antiparallel spins. The density of charge is represented by contour lines for the states A and S.

NaCl as an ionic crystal and of SiC and GaAs[21] as largely covalent. Atoms with nearly filled shells (Na, Cl) tend to be ionic, whereas atoms in columns III, IV, and V of the periodic table tend to be covalent (In, C, Ge, Si, As). The strength of the covalent bond is indicated in Table 7.

The binding of molecular hydrogen (H_2) is a simple example of a covalent bond. The problem is discussed in detail by Pauling and Wilson.[22] The strongest binding (Fig. 15) occurs when the spins of the two electrons are antiparallel. The binding depends on the relative spin orientation not because there are strong magnetic dipole forces between the spins, but because the Pauli principle modifies the distribution of charge according to the spin orientation. This spin-dependent Coulomb energy is called the **exchange interaction.**

[21] Gallium arsenide has the cubic ZnS crystal structure (Chapter 1). The x-ray structure factor of the (200) reflection depends on the difference of the atomic scattering factors of the Ga and As atoms. If the atoms were present as the ions Ga⁻ and As⁺, the number of electrons would be equal and the (200) structure factor would show only the different distributions of the electrons. If the atoms were present as Ga and As, there would be a difference of two in the number of atomic electrons, so that the structure factor would be larger than for the ionic case; see the measurements by J. J. Demarco and R. J. Weiss [Physics Letters **13**, 209 (1964)].

[22] L. Pauling and E. B. Wilson, *Introduction to quantum mechanics*, McGraw-Hill, 1935.

Table 6   Fractional ionic character of bonds in binary crystals

[After J. C. Phillips, Phys. Rev. Letters **22,** 705 (1969);
Physics Today **23,** Feb. 1970, p. 23]

Crystal	Fractional ionic character	Crystal	Fractional ionic character
Si	0.00		
SiC	0.18	CuCl	0.75
Ge	0.00	CuBr	0.74
ZnO	0.62	AgCl	0.86
ZnS	0.62	AgBr	0.85
ZnSe	0.63	AgI	0.77
ZnTe	0.61		
		MgO	0.84
CdO	0.79	MgS	0.79
CdS	0.69	MgSe	0.79
CdSe	0.70		
CdTe	0.67	LiF	0.92
		NaCl	0.94
InP	0.44	RbF	0.96
InAs	0.35		
InSb	0.32		
GaAs	0.32		
GaSb	0.26		

Table 7   Energy values for single covalent bonds

(After Pauling)

Bond	Bond energy		Bond	Bond energy	
	eV	kcal/mole		eV	kcal/mole
H—H	4.5	104	P—P	2.2	51
C—C	3.6	83	O—O	1.4	33
Si—Si	1.8	42	Te—Te	1.4	33
Ge—Ge	1.6	38	Cl—Cl	2.5	58

METAL CRYSTALS

Metals are characterized by high electrical conductivity, and a large number of the electrons in a metal must be free to move about, usually one or two per atom. The electrons available to move about are called conduction electrons. In some metals the interaction of the ion cores with the conduction electrons always makes a large contribution to the binding energy but, as compared with the free atom, the characteristic feature of metallic binding is the reduction in the kinetic energy of the valence electron in the metal. This point is developed in Chapter 10.

As in Fig. 1, we may think of an alkali metal crystal as an array of positive charges embedded in a more-or-less uniform sea of negative charge. In the transition metals there may be additional binding from inner electron shells. Transition metals and the metals immediately following them in the periodic table have large d-electron shells and are characterized by high binding energy (Table 1). This may be caused in part by covalent bonding and in part by the van der Waals interaction of the cores. In iron and tungsten, for example, the d-electrons make a substantial contribution to the binding energy.

The binding energy of an alkali metal crystal is very considerably less than that of an alkali halide crystal: the bond formed by a quasi-free conduction electron is not very strong. The interatomic distances are relatively large in the alkali metals because the kinetic energy of the conduction electrons is lower at large interatomic distances. This leads to weak binding.

In general, metals tend to crystallize in relatively close-packed structures: hcp, fcc, bcc, and some other closely-related structures.

We close this section with a quotation from Wigner and Seitz:[23]

"If one had a great calculating machine, one might apply it to the problem of solving the Schrödinger equation for each metal and obtain thereby the interesting physical quantities, such as the cohesive energy, the lattice constant, and similar parameters. It is not clear, however, that a great deal would be gained by this. Presumably the results would agree with the experimentally determined quantities and nothing vastly new would be learned from the calculation. It would be preferable instead to have a vivid picture of the behavior of the wave functions, a simple description of the essence of the factors which determine cohesion and an understanding of the origins of variation in properties from metal to metal."

[23] E. P. Wigner and F. Seitz, *Solid state physics* **1**, 97 (1955).

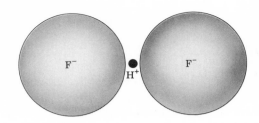

Figure 16

(a) The hydrogen difluoride ion HF_2^- is stabilized by a hydrogen bond. The sketch is of an extreme model of the bond, extreme in the sense that the proton is shown bare of electrons.

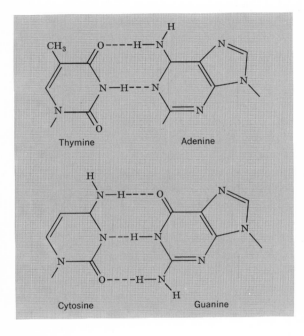

Thymine   Adenine

Cytosine   Guanine

(b) Hydrogen bonding between organic bases as in DNA, after F. H. C. Crick and J. D. Watson.

HYDROGEN-BONDED CRYSTALS

Because neutral hydrogen has only one electron, it should form a covalent bond with only one other atom. It is known, however, that under certain conditions an atom of hydrogen is attracted by rather strong forces to two atoms, thus forming what is called a **hydrogen bond**[24] between them, with a bond energy of the order of 0.1 eV. It is believed that the hydrogen bond is largely ionic in character, being formed only between the most electronegative atoms, particularly F, O, and N. In the extreme ionic form of the hydrogen bond the hydrogen atom loses its electron to another atom in the molecule; the bare proton forms the hydrogen bond. The small size of the proton permits only two nearest-neighbor atoms, because the atoms adjacent to the proton are so close that more than two of them would get in each other's way; thus the hydrogen bond connects only two atoms (Fig. 16a).

[24] G. Pimentel and A. McClellan, *Hydrogen bond*, Freeman, 1960.

The hydrogen bond is an important part of the interaction between H_2O molecules and is responsible together with the electrostatic attraction of the electric dipole moments for the striking physical properties of water and ice.[25] The hydrogen bond restrains protein molecules to their normal geometrical arrangements. It is also responsible for the polymerization of hydrogen fluoride and formic acid, for example. It is important in certain ferroelectric crystals, such as potassium dihydrogen phosphate. It is important also in molecular genetics,[26] by controlling in part the pairing possible between the two strands of a DNA molecule (Fig. 16b).

ATOMIC RADII

Distances between atoms in crystals can be measured very accurately by x-ray diffraction methods, often to 1 part in 10^5. Can we say that the observed distance between atoms or ions may be assigned, so much to atom A, and so much to atom B? Can a definite meaning be assigned to the radius of an atom or ion, irrespective of the nature and composition of the crystal?

Strictly, the answer is *no*. The charge distribution around an atom is not limited by a rigid spherical boundary. The size of a sodium atom depends on whether it is free, or in a metal, or in an ionic crystal. The radius of a sodium atom in metallic sodium might be taken as 1.86 Å, which is one-half the nearest-neighbor distance of 3.72 Å. Electron diffraction study of gaseous F_2 gives a F-F internuclear distance of 1.44 Å, one-half of which is 0.72 Å. Adding 1.86 Å and 0.72 Å gives 2.58 Å as an estimate of the Na-F bond length. The actual Na-F distance in crystalline sodium fluoride is rather smaller, 2.32 Å, so that taking the mean of the atomic radii is not very accurate here. The values of the ionic radii of Na^+ and F^- given in Table 8 are 0.98 Å and 1.33 Å, respectively. The sum of the ionic radii is 2.31 Å. The close agreement with the observed value of 2.32 Å for the crystal is not surprising: values of ionic radii in tables are usually selected so that their sums represent on the average the internuclear distances in crystals at room temperature.

---

[25] See L. Pauling, *General chemistry*, Freeman, 1956, pp. 327–331.

[26] F. H. C. Crick and J. D. Watson, Proc. Roy. Soc. (London) **A223**, 80 (1954); see also F. W. Stahl, *Mechanics of inheritance*, Prentice-Hall, 1964; G. Stent, *Molecular biology of bacterial viruses*, Freeman, 1963.

But used with care and in the proper context, the concept of atomic radius can be useful and fruitful. The interatomic distance between C atoms in diamond is 1.54 Å; one-half of this is 0.77 Å. In silicon, which has the same crystal structure, one-half the interatomic distance is 1.17 Å. Now SiC crystallizes in two forms, in both of which each atom is surrounded by four atoms of the opposite kind. If we add the C and Si radii just given, we predict 1.94 Å for the length of the C-Si bond, in fair agreement with the 1.89 Å observed for the bond length. This is the kind of agreement (a few percent) which we shall find in using tables of atomic radii.[27]

Tetrahedral Covalent Radii

Pauling has proposed the set of empirical tetrahedral covalent atomic radii included in Table 8 for atoms in crystals with coordination number four, such as the diamond, cubic ZnS, and hexagonal ZnS structures. A large number of observed interatomic distances in appropriate compounds agree closely with the sums of the tetrahedral radii.

---

[27] For references on atomic and ionic radii, see L. Pauling, *The nature of the chemical bond,* Cornell University Press, N.Y., 1960, 3rd ed., Chapters 7, 11, 13; Landolt-Börnstein, *Tabellen* 1:4 (1950), p. 521 et seq; J. C. Slater, J. Chem. Phys. **41**, 3199 (1964); B. J. Austin and V. Heine, J. Chem. Physics **45**, 928 (1966); R. G. Parsons and V. F. Weisskopf, Zeits. f. Physik **202**, 492 (1967); S. Geller, Z. Kristallographie **125**, 1 (1967); see Table 2 of H. G. F. Winkler, *Struktur und Eigenschaften der Kristalle*, Springer, 1955, 2nd ed. A detailed analysis of ionic radii in oxides and fluorides has been made by R. D. Shannon and C. T. Prewitt, Acta Cryst. **B25**, 925 (1969). A critical revision of the ionic radii in alkali halides has been proposed by M. Tosi, ref. 13.

Table 8 Atomic and ionic radii

(Values approximate only: see Table 9 for use of the standard radii of ions.
Units are 1 Å = 10^{-10} m.)

Legend:

→ Standard radii for ions in inert gas (filled shell) configuration

→ Radii of atoms when in tetrahedral covalent bonds

→ Radii of ions in valence state indicated in superscript

1	2	3	4	5	6	7	8	9	10	11	12	13	14	15	16	17	18
H																	He
Li 0.68	Be 0.30 / 1.06											B 0.16 / 0.88	C 0.77	N 0.70	O 1.46 / 0.66	F 1.33 / 0.64	Ne 1.58
Na 0.98	Mg 0.65 / 1.40											Al 0.45 / 1.26	Si 0.38 / 1.17	P 1.10	S 1.90 / 1.04	Cl 1.81 / 0.99	Ar 1.88
K 1.33	Ca 0.94	Sc 0.68	Ti 0.60 / 2+0.90	V 2+0.88	Cr 2+0.84	Mn 2+0.80	Fe 2+0.76	Co 2+0.74	Ni 2+0.72	Cu 1.35 / +0.96	Zn 1.31 / 2+0.83	Ga 1.26 / 3+0.62	Ge 1.22 / 4+0.44	As 1.18 / 3+0.69	Se 1.14	Br 1.95 / 1.11	Kr 2.00
Rb 1.48	Sr 1.10	Y 0.88	Zr 0.77	Nb 0.67	Mo	Tc	Ru	Rh	Pd 2+0.86	Ag 1.52 / +1.13	Cd 1.48 / 2+1.03	In 1.44 / 3+0.92	Sn 1.40 / 4+0.74	Sb 1.36 / 3+0.90	Te 1.32	I 2.16 / 1.28	Xe 2.17
Cs 1.67	Ba 1.29	La 1.04	Hf	Ta	W	Re	Os	Ir	Pt	Au +1.37	Hg 1.48 / 2+1.12	Tl 3+1.05	Pb 4+0.84	Bi	Po	At	Rn
Fr 1.75	Ra 1.37	Ac 1.11															

Ce	Pr	Nd	Pm	Sm	Eu	Gd	Tb	Dy	Ho	Er	Tm	Yb	Lu
0.92 / 3+1.11		3+1.08		3+1.04		3+1.02		3+0.99		3+0.96		3+0.94	

Th	Pa	U	Np	Pu	Am	Cm	Bk	Cf	Es	Fm	Md	No	Lw
0.99	0.90	0.83 / 4+1.05											

Ionic Crystal Radii

In Table 8 we include a set of ionic crystal radii in inert gas configurations, after Zachariasen. The ionic radii are to be used in conjunction with Table 9. Let us consider $BaTiO_3$ (Fig. 14.2), with a measured average lattice constant of 4.004 Å at room temperature. Each Ba^{++} ion has twelve nearest O^{--} ions, so that the coordination number is twelve and the correction Δ_{12} of Table 9 applies. If we suppose that the structure is determined by the Ba-O contacts, we have, from Table 8, $D_{12} = 1.29 + 1.46 + 0.19 = 2.94$ Å, or $a = 4.16$ Å; if the Ti-O contact determines the structure, we have $D_6 = 0.60 + 1.46 = 2.06$, or $a = 4.12$ Å. The actual lattice constant is somewhat smaller than the estimates and may perhaps suggest that the bonding is not purely ionic, but is partly covalent. For sodium chloride, probably principally ionic, we have $D_6 = 0.98 + 1.81 = 2.79$, or $a = 5.58$ Å, whereas 5.63 Å is the observed value at room temperature.

Bond lengths and bond energies of the single, double, and triple carbon bonds are given in Table 10.

Table 9   Use of the standard radii of ions given in Table 8

(After Zachariasen, unpublished)

The interionic distance D is represented by $D_N = R_C + R_A + \Delta_N$, for ionic crystals, where N is the coordination number of the cation (positive ion), R_C and R_A are the standard radii of the cation and anion, and Δ_N is a correction for coordination number. Room temperature.

N	Δ_N(Å)	N	Δ_N(Å)	N	Δ_N(Å)
1	-0.50	5	-0.05	9	$+0.11$
2	-0.31	6	0	10	$+0.14$
3	-0.19	7	$+0.04$	11	$+0.17$
4	-0.11	8	$+0.08$	12	$+0.19$

Table 10   Carbon-carbon bonds

Type	Bond length in Å	Bond energy in eV
C—C	1.54	3.60
C=C	1.33	6.37
C≡C	1.20	8.42

SUMMARY

1. Crystals of inert gas atoms are bound by the van der Waals interaction (induced dipole-dipole interaction), and this varies with distance as $1/R^6$.

2. The repulsive interaction between atoms arises generally from three mechanisms: (a) the increase of the kinetic energy of the outermost electrons, which is a consequence of their confinement; (b) the electrostatic repulsion of overlapping charge distributions; and (c) the Pauli principle, which compels overlapping electrons of parallel spin to enter orbitals of higher energy.

3. Ionic crystals are bound by the electrostatic attraction of charged ions of opposite sign. The electrostatic energy of a structure of $2N$ ions of charge $\pm q$ is

(CGS)
$$ U = -N\alpha \frac{q^2}{R} = -N\sum \frac{(\pm)q^2}{r_{ij}} \ , $$

where α is the Madelung constant and R is the distance between nearest neighbors.

4. Metals are bound in large measure by the reduction in the kinetic energy of the valence electrons in the metal as compared with the free atom.

5. A covalent bond is characterized by the overlap of charge distributions of antiparallel electron spin. The Pauli contribution to the repulsion is reduced for antiparallel spins, and this makes possible a greater degree of overlap. The overlapping electrons bind their associated ion cores by electrostatic attraction.

Problems

1. *Ionic radii.* (a) Check the Zachariasen radii (Table 8) against observed lattice constants for CsCl, NaCl, and KBr. (b) Check the tetrahedral radii against CuF, ZnS, and InSb. *Note:* The lattice constants are given in Chapter 1.

2. *Linear ionic crystal.* Consider a line of $2N$ ions of alternating charge $\pm q$ with a repulsive potential energy A/R^n between nearest neighbors. (a) Show that at the equilibrium separation

(CGS)
$$ U(R_0) = -\frac{2Nq^2 \log 2}{R_0}\left(1 - \frac{1}{n}\right) \ . $$

(b) Let the crystal be compressed so that $R_0 \rightarrow R_0(1 - \delta)$. Show that the work done in compressing a unit length of the crystal has the leading term $\frac{1}{2}C\delta^2$, where

(CGS)
$$ C = \frac{(n-1)q^2 \log 2}{R_0^2} \ . $$

To obtain the results in SI, replace q^2 by $q^2/4\pi\epsilon_0$. *Note:* We should not expect to obtain this result from the expression for $U(R_0)$, but we must use the complete expression for $U(R)$.

3. **Cubic ZnS structure.** Using λ and ρ from Table 5 and the Madelung constants given in the text, calculate the cohesive energy of KCl in the cubic ZnS structure described in Chapter 1. Compare with the value calculated for KCl in the NaCl structure.

4. **Bulk modulus of LiF.** From the experimental values of the cohesive energy and the nearest-neighbor distance, calculate the bulk modulus of LiF. Compare with the observed value.

5. **Hydrogen bonds in ice.** What is the evidence for hydrogen-bonding in ice? (Consult the book by Pauling and also the excellent article "Ice" by L. K. Runnels, Sci. American, Dec. 1966, pp. 118–126 and p. 156.)

References

M. Born and K. Huang, *Dynamical theory of crystal lattices,* Oxford, 1954. The classic work on ionic crystals.

G. Leibfried, in *Encyclo. of physics* **7**/1, 1955. Superb review of lattice properties.

M. P. Tosi, "Cohesion of ionic solids in the Born model," *Solid state physics* **16**, 1 (1964).

Advances in chemical physics **12**: *Intermolecular forces,* J. O. Hirschfelder, ed., Interscience, 1967.

L. Pauling, *Nature of the chemical bond,* Cornell University Press, 1960, 3rd ed.; see Chap. 13, "Sizes of ions and the structure of ionic crystals." Particularly valuable for the discussion of the principles determining the structure of complex ionic crystals, pp. 543–562.

J. Hirschfelder, C. F. Curtis, and R. B. Bird, *Molecular theory of gases and liquids,* Wiley, 1964.

G. K. Horton, "Ideal rare gas crystals," American Journal of Physics **36**, 93 (1968). A good review of the lattice properties of crystals of the inert gases.

4

Elastic Constants
and
*Elastic Waves**

° Note to the reader: This chapter is included primarily for reference and need not necessarily be studied in a first reading of the book. The material is not widely available in a simple form.

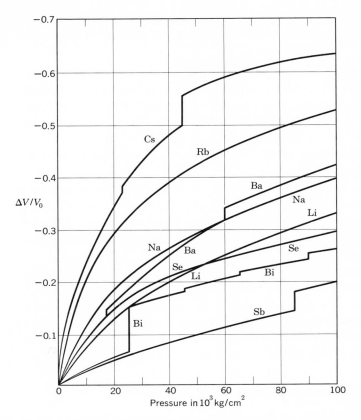

Figure 1 The change of relative volume of a number of substances shown as a function of pressure up to 100,000 kg/cm² (metric atmospheres). The bulk modulus satisfies Hooke's law over a region of pressure in which a part of one of these curves may be represented closely by a straight line, which may mean over perhaps 10^4 atmospheres, or a dilation of -0.05. The vertical breaks mark transformations to another crystal structure and occasionally to another electronic configuration for the ion cores. [After P. W. Bridgman, Endeavour **10**, 68 (1951).]

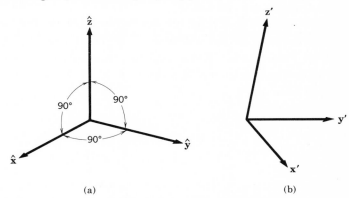

Figure 2 Coordinate axes for the description of the state of strain; the orthogonal unit axes in the unstrained state (a) are deformed in the strained state (b).

This chapter deals with the elastic properties of a crystal viewed as a homogeneous continuous medium rather than as a periodic array of atoms. The continuum approximation is usually valid for elastic waves of wavelengths λ longer than 10^{-6} cm, which means for frequencies below 10^{11} or 10^{12} Hz. Higher frequencies are not easily obtainable at present by electronic means; to study higher frequency elastic waves we use the inelastic scattering methods discussed in Chapter 5.

The frequency region for which the continuum approximation is valid is of great interest in solid state physics. Ultrasonic waves are used to measure elastic constants and to study lattice defects, the electronic structure of metals, and superconductivity. There are also numerous technological applications of elastic waves in solids.

Some of the material below looks complicated because of the unavoidable multiplicity of subscripts on the symbols. The basic physical ideas are simple: we use Hooke's law and Newton's second law. **Hooke's law** states that in an elastic solid the strain is directly proportional to the stress. The law applies to small strains only (Fig. 1). We say that we are in the **nonlinear region** when the strains are so large that Hooke's law is no longer satisfied.

ANALYSIS OF ELASTIC STRAINS

We specify the strain in terms of the components e_{xx}, e_{yy}, e_{zz}, e_{xy}, e_{yz}, e_{zx} which are defined below. We treat infinitesimal strains only. We shall not distinguish in our notation between isothermal (constant temperature) and adiabatic (constant entropy) deformations. The small differences between the isothermal and adiabatic elastic constants are not often of importance at room temperature and below.

We imagine that three orthogonal vectors $\hat{x}$, $\hat{y}$, $\hat{z}$ of unit length are embedded securely in the unstrained solid, as shown in Fig. 2. After a small uniform deformation[1] of the solid has taken place the axes are distorted in orientation and in length. The new axes x', y', z' may be written in terms of the old axes:

$$x' = (1 + \epsilon_{xx})\hat{x} + \epsilon_{xy}\hat{y} + \epsilon_{xz}\hat{z} \; ;$$
$$y' = \epsilon_{yx}\hat{x} + (1 + \epsilon_{yy})\hat{y} + \epsilon_{yz}\hat{z} \; ; \tag{1}$$
$$z' = \epsilon_{zx}\hat{x} + \epsilon_{zy}\hat{y} + (1 + \epsilon_{zz})\hat{z} \; .$$

The coefficients $\epsilon_{\alpha\beta}$ define the deformation; they are dimensionless and have values $\ll 1$ if the strain is small. The original axes were of unit length, but the

---

[1] In a uniform deformation each primitive cell of the crystal is deformed in the same way.

Uniform strain

Nonuniform strain

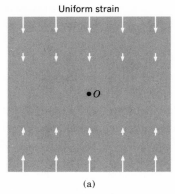

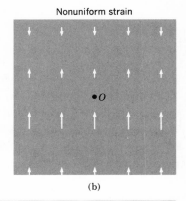

(a)   (b)

Figure 3   (a) Displacement vectors **R** of Eq. (4) in a uniform strain and (b) in a nonuniform strain. The origin is at O. (c) $\mathbf{A} \cdot \mathbf{B} \times \mathbf{C}$ is equal to the volume of the parallelepiped having edges **A, B, C**. Recall that $\mathbf{B} \times \mathbf{C}$ is a vector perpendicular to the plane of **B** and **C**, and of magnitude equal to the area of the parallelogram having **B, C** as sides.

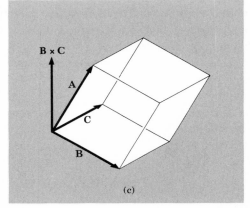

(c)

new axes will not necessarily be of unit length. For example,

$$\mathbf{x}' \cdot \mathbf{x}' = 1 + 2\epsilon_{xx} + \epsilon_{xx}{}^2 + \epsilon_{xy}{}^2 + \epsilon_{xz}{}^2 \; ,$$

whence $x' \cong 1 + \epsilon_{xx} + \cdots$. The fractional changes of length of the $\hat{\mathbf{x}}$, $\hat{\mathbf{y}}$, and $\hat{\mathbf{z}}$ axes are ϵ_{xx}, ϵ_{yy}, ϵ_{zz}, respectively, to the first order.

What is the effect of the deformation (1) on a point (or atom) originally at $\mathbf{r} = x\hat{\mathbf{x}} + y\hat{\mathbf{y}} + z\hat{\mathbf{z}}$? The origin is taken at some other atom. If the deformation is uniform (Fig. 3a), then after the deformation the point will be at the position[2] $\mathbf{r}' = x\mathbf{x}' + y\mathbf{y}' + z\mathbf{z}'$. The **displacement R** of the deformation is defined by

$$\mathbf{R} \equiv \mathbf{r}' - \mathbf{r} = x(\mathbf{x}' - \hat{\mathbf{x}}) + y(\mathbf{y}' - \hat{\mathbf{y}}) + z(\mathbf{z}' - \hat{\mathbf{z}}) \; , \qquad (2)$$

or, from (1),

$$\mathbf{R}(\mathbf{r}) \equiv (x\epsilon_{xx} + y\epsilon_{yx} + z\epsilon_{zx})\hat{\mathbf{x}} + (x\epsilon_{xy} + y\epsilon_{yy} + z\epsilon_{zy})\hat{\mathbf{y}}$$
$$+ (x\epsilon_{xz} + y\epsilon_{yz} + z\epsilon_{zz})\hat{\mathbf{z}} \; . \qquad (3)$$

This may be written in a more general form by introducing u, v, w such that

---

[2] This is obviously correct if we choose the $\hat{\mathbf{x}}$ axis such that $\mathbf{r} = x\hat{\mathbf{x}}$; then $\mathbf{r}' = x\mathbf{x}'$ by definition of $\mathbf{x}'$.

the displacement is given by

$$\mathbf{R}(\mathbf{r}) = u(\mathbf{r})\hat{\mathbf{x}} + v(\mathbf{r})\hat{\mathbf{y}} + w(\mathbf{r})\hat{\mathbf{z}} \ . \tag{4}$$

If the deformation is nonuniform as in Fig. 3b we must relate u, v, w to the local strains. We take the origin of $\mathbf{r}$ close to the region of interest; then comparison of (3) and (4) gives, by Taylor series expansion of $\mathbf{R}$ using $\mathbf{R}(0) = 0$,

$$x\epsilon_{xx} \cong x\frac{\partial u}{\partial x} \ ; \qquad y\epsilon_{yx} = y\frac{\partial u}{\partial y} \ ; \qquad \text{etc.} \tag{5}$$

The derivatives are independent of the origin chosen for $\mathbf{R}$.

It is usual to work with coefficients $e_{\alpha\beta}$ rather than $\epsilon_{\alpha\beta}$. We define the **strain components** e_{xx}, e_{yy}, e_{zz} by the relations

$$e_{xx} \equiv \epsilon_{xx} = \frac{\partial u}{\partial x} \ ; \qquad e_{yy} \equiv \epsilon_{yy} = \frac{\partial v}{\partial y} \ ; \qquad e_{zz} \equiv \epsilon_{zz} = \frac{\partial w}{\partial z} \ , \tag{6}$$

using (5). The other strain components e_{xy}, e_{yz}, e_{zx} are defined in terms of the changes in angle between the axes: using (1) we may define

$$
\begin{aligned}
e_{xy} &\equiv \mathbf{x}' \cdot \mathbf{y}' \cong \epsilon_{yx} + \epsilon_{xy} = \frac{\partial u}{\partial y} + \frac{\partial v}{\partial x} \ ; \\[2mm]
e_{yz} &\equiv \mathbf{y}' \cdot \mathbf{z}' \cong \epsilon_{zy} + \epsilon_{yz} = \frac{\partial v}{\partial z} + \frac{\partial w}{\partial y} \ ; \\[2mm]
e_{zx} &\equiv \mathbf{z}' \cdot \mathbf{x}' \cong \epsilon_{zx} + \epsilon_{xz} = \frac{\partial u}{\partial z} + \frac{\partial w}{\partial x} \ .
\end{aligned}
\tag{7}
$$

We may replace the $\cong$ signs by $=$ signs if we neglect terms of order ϵ^2. The six coefficients $e_{\alpha\beta}(=e_{\beta\alpha})$ completely define the strain. The strains as defined are dimensionless.

Dilation

The fractional *increase* of volume associated with a deformation is called the dilation. The dilation is negative for hydrostatic pressure. The unit cube of edges $\hat{\mathbf{x}}$, $\hat{\mathbf{y}}$, $\hat{\mathbf{z}}$ has a volume after deformation of

$$V' = \mathbf{x}' \cdot \mathbf{y}' \times \mathbf{z}' \ , \tag{8}$$

by virtue of a well-known result for the volume of a parallelepiped having edges $\mathbf{x}'$, $\mathbf{y}'$, $\mathbf{z}'$ (Fig. 3c). From (1) we have

$$\mathbf{x}' \cdot \mathbf{y}' \times \mathbf{z}' = \begin{vmatrix} 1 + \epsilon_{xx} & \epsilon_{xy} & \epsilon_{xz} \\ \epsilon_{yx} & 1 + \epsilon_{yy} & \epsilon_{yz} \\ \epsilon_{zx} & \epsilon_{zy} & 1 + \epsilon_{zz} \end{vmatrix} \cong 1 + e_{xx} + e_{yy} + e_{zz} \ . \tag{9}$$

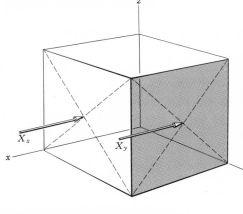

Figure 4 Stress component X_x is a force applied in the x direction to a unit area of a plane whose normal lies in the x direction; X_y is applied in the x direction to a unit area of a plane whose normal lies in the y direction.

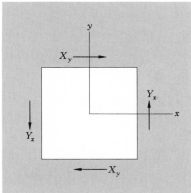

Figure 5 Demonstration that for a body in static equilibrium $Y_x = X_y$. The sum of the forces in the x direction is zero. The sum of the forces in the y direction is also zero. The total force vanishes. The total torque about the origin is also zero if $Y_x = X_y$.

Products of two strain components have been neglected. The dilation δ is then given by

$$\delta \equiv \frac{V' - V}{V} \cong e_{xx} + e_{yy} + e_{zz} \ . \tag{10}$$

Stress Components

The force acting on a unit area in the solid is defined as the stress. There are nine stress components: X_x, X_y, X_z, Y_x, Y_y, Y_z, Z_x, Z_y, Z_z. The capital letter indicates the direction of the force, and the subscript indicates the normal to the plane to which the force is applied. In Fig. 4 the stress component X_x represents a force applied in the x direction to a unit area of a plane whose normal lies in the x direction; the stress component X_y represents a force applied in the x direction to a unit area of a plane whose normal lies in the y direction. The number of independent stress components is reduced from nine to six by applying to an elementary cube (as in Fig. 5) the condition that the angular acceleration vanish,[3] and hence that the total torque must be zero. It follows that

$$Y_z = Z_y \; ; \qquad Z_x = X_z \; ; \qquad X_y = Y_x \; . \qquad (11)$$

The six independent stress components may be taken as $X_x, Y_y, Z_z, Y_z, Z_x, X_y$.

Stress components have the dimensions of force per unit area or energy per unit volume. The strain components are ratios of lengths and are dimensionless.

ELASTIC COMPLIANCE AND STIFFNESS CONSTANTS

Hooke's law states that for sufficiently small deformations the strain is directly proportional to the stress, so that the strain components are linear functions of the stress components:

$$
\begin{aligned}
e_{xx} &= S_{11}X_x + S_{12}Y_y + S_{13}Z_z + S_{14}Y_z + S_{15}Z_x + S_{16}X_y \; ; \\
e_{yy} &= S_{21}X_x + S_{22}Y_y + S_{23}Z_z + S_{24}Y_z + S_{25}Z_x + S_{26}X_y \; ; \\
e_{zz} &= S_{31}X_x + S_{32}Y_y + S_{33}Z_z + S_{34}Y_z + S_{35}Z_x + S_{36}X_y \; ; \\
e_{yz} &= S_{41}X_x + S_{42}Y_y + S_{43}Z_z + S_{44}Y_z + S_{45}Z_x + S_{46}X_y \; ; \\
e_{zx} &= S_{51}X_x + S_{52}Y_y + S_{53}Z_z + S_{54}Y_z + S_{55}Z_x + S_{56}X_y \; ; \\
e_{xy} &= S_{61}X_x + S_{62}Y_y + S_{63}Z_z + S_{64}Y_z + S_{65}Z_x + S_{66}X_y \; .
\end{aligned}
\qquad (12)
$$

Conversely, the stress components are linear functions of the strain components:

$$
\begin{aligned}
X_x &= C_{11}e_{xx} + C_{12}e_{yy} + C_{13}e_{zz} + C_{14}e_{yz} + C_{15}e_{zx} + C_{16}e_{xy} \; ; \\
Y_y &= C_{21}e_{xx} + C_{22}e_{yy} + C_{23}e_{zz} + C_{24}e_{yz} + C_{25}e_{zx} + C_{26}e_{xy} \; ; \\
Z_z &= C_{31}e_{xx} + C_{32}e_{yy} + C_{33}e_{zz} + C_{34}e_{yz} + C_{35}e_{zx} + C_{36}e_{xy} \; ; \\
Y_z &= C_{41}e_{xx} + C_{42}e_{yy} + C_{43}e_{zz} + C_{44}e_{yz} + C_{45}e_{zx} + C_{46}e_{xy} \; ; \\
Z_x &= C_{51}e_{xx} + C_{52}e_{yy} + C_{53}e_{zz} + C_{54}e_{yz} + C_{55}e_{zx} + C_{56}e_{xy} \; ; \\
X_y &= C_{61}e_{xx} + C_{62}e_{yy} + C_{63}e_{zz} + C_{64}e_{yz} + C_{65}e_{zx} + C_{66}e_{xy} \; .
\end{aligned}
\qquad (13)
$$

The quantities $S_{11}, S_{12} \ldots$ are called **elastic compliance constants or elastic constants**; the quantities $C_{11}, C_{12}, \ldots$ are called the **elastic stiffness constants** or moduli of elasticity. Other names are also current. The S's have the dimensions of [area]/[force] or [volume]/[energy]. The C's have the dimensions of [force]/[area] or [energy]/[volume].

Elastic Energy Density

The 36 constants in (12) or in (13) may be reduced in number by several considerations. The elastic energy density U is a quadratic function of the strains, in the approximation of Hooke's law (recall the expression for the energy of a

---

[3] This does not mean we cannot treat problems in which there is an angular acceleration; it just means that we can use the static situation to define the elastic constants.

stretched spring). Thus we may write

$$U = \frac{1}{2} \sum_{\lambda=1}^{6} \sum_{\mu=1}^{6} \tilde{C}_{\lambda\mu} e_\lambda e_\mu \ , \qquad (14)$$

where the indices 1 through 6 are defined as:

$$1 \equiv xx \ ; \quad 2 \equiv yy \ ; \quad 3 \equiv zz \ ; \quad 4 \equiv yz \ ; \quad 5 \equiv zx \ ; \quad 6 \equiv xy \ . \quad (15)$$

The $\tilde{C}$'s are related to the C's of (13), as we see in (17) below.

The stress components are found from the derivative of U with respect to the associated strain component. This result follows from the definition of potential energy. Consider the stress X_x applied to one face of a unit cube, the opposite face being held at rest:

$$X_x = \frac{\partial U}{\partial e_{xx}} \equiv \frac{\partial U}{\partial e_1} = \tilde{C}_{11} e_1 + \frac{1}{2} \sum_{\beta=2}^{6} (\tilde{C}_{1\beta} + \tilde{C}_{\beta 1}) e_\beta \ . \qquad (16)$$

Note that only the combination $\frac{1}{2}(\tilde{C}_{\alpha\beta} + \tilde{C}_{\beta\alpha})$ enters the stress-strain relations. It follows that the elastic stiffness constants are symmetrical:

$$C_{\alpha\beta} = \frac{1}{2}(\tilde{C}_{\alpha\beta} + \tilde{C}_{\beta\alpha}) = C_{\beta\alpha} \ . \qquad (17)$$

Thus the thirty-six elastic stiffness constants are reduced to twenty-one.

Elastic Stiffness Constants of Cubic Crystals

The number of independent elastic stiffness constants is reduced further if the crystal possesses symmetry elements. We now show that in cubic crystals there are only three independent stiffness constants.

We assert that the elastic energy density of a cubic crystal is

$$U = \frac{1}{2}C_{11}(e_{xx}^2 + e_{yy}^2 + e_{zz}^2) + \frac{1}{2}C_{44}(e_{yz}^2 + e_{zx}^2 + e_{xy}^2)$$
$$+ C_{12}(e_{yy}e_{zz} + e_{zz}e_{xx} + e_{xx}e_{yy}) \ , \quad (18)$$

and that no other quadratic terms occur; that is

$$(e_{xx}e_{xy} + \cdots) \ ; \qquad (e_{yz}e_{zx} + \cdots) \ ; \qquad (e_{xx}e_{yz} + \cdots) \quad (19)$$

do not occur.

The minimum[4] symmetry requirement for a cubic structure is the existence of four three-fold rotation axes. The axes are in the [111] and equivalent

---

[4] See the standard stereographic drawings of the symmetry elements of the cubic point group as given in texts on crystallography; the cubic group with the smallest number of symmetry axes is labeled 23.

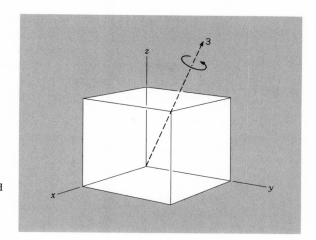

Figure 6 Rotation by $2\pi/3$ about the axis marked 3 changes $x \to y$; $y \to z$; and $z \to x$.

directions (Fig. 6). The effect of a rotation of $2\pi/3$ about these four axes is to interchange the x, y, z axes according to the schemes

$$x \to y \to z \to x \; ; \qquad -x \to z \to -y \to -x \; ;$$
$$x \to z \to -y \to x \; ; \qquad -x \to y \to z \to -x \; , \qquad (20)$$

according to the axis chosen. Under the first of these schemes, for example,

$$e_{xx}^2 + e_{yy}^2 + e_{zz}^2 \to e_{yy}^2 + e_{zz}^2 + e_{xx}^2 \; ,$$

and similarly for the other terms in parentheses in (18). Thus (18) is invariant under the operations considered. But each of the terms exhibited in (19) is odd in one or more indices. A rotation in the set (20) can be found which will change the sign of the term, because $e_{xy} = -e_{x(-y)}$, for example. Thus the terms (19) are not invariant under the required operations.

It remains to verify that the numerical factors in (18) are correct. By (16)

$$\partial U/\partial e_{xx} = X_x = C_{11}e_{xx} + C_{12}(e_{yy} + e_{zz}) \; . \qquad (21)$$

The appearance of $C_{11}e_{xx}$ agrees with (13). On further comparison, we see that

$$C_{12} = C_{13} \; ; \qquad C_{14} = C_{15} = C_{16} = 0 \; . \qquad (22)$$

Further, from (18),

$$\partial U/\partial e_{xy} = X_y = C_{44}e_{xy} \; ; \qquad (23)$$

on comparison with (13) we have

$$C_{61} = C_{62} = C_{63} = C_{64} = C_{65} = 0 \; ; \qquad C_{66} = C_{44} \; . \qquad (24)$$

Thus from (18) we find that the array of values of the elastic stiffness constants is reduced for a cubic crystal to the matrix

	e_{xx}	e_{yy}	e_{zz}	e_{yz}	e_{zx}	e_{xy}
X_x	C_{11}	C_{12}	C_{12}	0	0	0
Y_y	C_{12}	C_{11}	C_{12}	0	0	0
Z_z	C_{12}	C_{12}	C_{11}	0	0	0
Y_z	0	0	0	C_{44}	0	0
Z_x	0	0	0	0	C_{44}	0
X_y	0	0	0	0	0	C_{44}

(25)

For cubic crystals the stiffness and compliance constants are related by

$$C_{44} = 1/S_{44} \; ; \qquad C_{11} - C_{12} = (S_{11} - S_{12})^{-1} \; ;$$
$$C_{11} + 2C_{12} = (S_{11} + 2S_{12})^{-1} \; . \quad (26)$$

These relations follow on evaluating the inverse matrix to (25).

Bulk Modulus and Compressibility

Consider the uniform dilation $e_{xx} = e_{yy} = e_{zz} = \frac{1}{3}\delta$. For this deformation the energy density (18) of a cubic crystal is

$$U = \tfrac{1}{6}(C_{11} + 2C_{12})\delta^2 \; . \qquad (27)$$

We may define the **bulk modulus** B by the relation

$$U = \tfrac{1}{2}B\delta^2 \; , \qquad (28)$$

which is equivalent to the definition $-V \, dp/dV$ used in Chapter 3. For a cubic crystal

$$B = \tfrac{1}{3}(C_{11} + 2C_{12}) \; . \qquad (29)$$

The **compressibility** K is defined as $K \equiv 1/B$. Values of B and K are given in Table 1.

Table 1 Isothermal bulk modulii and compressibilities of the elements at room temperature

After K. Gschneidner, Jr., *Solid state physics* **16**, 275–426 (1964); several data are from F. Birch, in *Handbook of physical constants,* Geological Society of America Memoir **97**, 107–173 (1966.) Original references should be consulted when values are needed for research purposes. Values in parentheses are estimates. Letters in parentheses refer to the crystal form. Letters in brackets refer to the temperature:

[a] = 77°K; [b] = 273°K; [c] = 1°K; [d] = 4°K; [e] = 81°K.

Bulk modulus in units 10^{12} dyne/cm^2 or 10^{11} N/m^2.
Compressibility in units 10^{-12} cm^2/dyne or 10^{-11} m^2/N.

H [d] 0.002 500																	**He** [d] 0.00 1168
Li 0.116 8.62	**Be** 1.003 0.997											**B** 1.78 0.562	**C** (d) 5.45 0.183	**N** [e] 0.012 80	**O**	**F**	**Ne** [d] 0.010 100
Na 0.068 14.7	**Mg** 0.354 2.82											**Al** 0.722 1.385	**Si** 0.988 1.012	**P** (b) 0.304 3.29	**S** (r) 0.178 5.62	**Cl**	**Ar** [a] 0.016 93.8
K 0.032 31.	**Ca** 0.152 6.58	**Sc** 0.435 2.30	**Ti** 1.051 0.951	**V** 1.619 0.618	**Cr** 1.901 0.526	**Mn** 0.596 1.68	**Fe** 1.683 0.594	**Co** 1.914 0.522	**Ni** 1.86 0.538	**Cu** 1.37 0.73	**Zn** 0.598 1.67	**Ga** [b] 0.569 1.76	**Ge** 0.772 1.29	**As** 0.394 2.54	**Se** 0.091 11.0	**Br**	**Kr** [a] 0.018 56
Rb 0.031 32.	**Sr** 0.116 8.62	**Y** 0.366 2.73	**Zr** 0.833 1.20	**Nb** 1.702 0.587	**Mo** 2.725 0.366	**Tc** (2.97) (0.34)	**Ru** 3.208 0.311	**Rh** 2.704 0.369	**Pd** 1.808 0.553	**Ag** 1.007 0.993	**Cd** 0.467 2.14	**In** 0.411 2.43	**Sn** (g) 1.11 0.901	**Sb** 0.383 2.61	**Te** 0.230 4.35	**I**	**Xe**
Cs 0.020 50.	**Ba** 0.103 9.97	**La** 0.243 4.12	**Hf** 1.09 0.92	**Ta** 2.00 0.50	**W** 3.232 0.309	**Re** 3.72 0.269	**Os** (4.18) (0.24)	**Ir** 3.55 0.282	**Pt** 2.783 0.359	**Au** 1.732 0.577	**Hg** [c] 0.382 2.60	**Tl** 0.359 2.79	**Pb** 0.430 2.33	**Bi** 0.315 3.17	**Po** (0.26) (3.8)	**At**	**Rn**
Fr (0.020) (50.)	**Ra** (0.132) (7.6)	**Ac** (0.25) (4.)															

Lanthanides

Ce (γ) 0.239 4.18	**Pr** 0.306 3.27	**Nd** 0.327 3.06	**Pm** (0.35) (2.85)	**Sm** 0.294 3.40	**Eu** 0.147 6.80	**Gd** 0.383 2.61	**Tb** 0.399 2.51	**Dy** 0.384 2.60	**Ho** 0.397 2.52	**Er** 0.411 2.43	**Tm** 0.397 2.52	**Yb** 0.133 7.52	**Lu** 0.411 2.43
Th 0.543 1.84	**Pa** (0.76) (1.3)	**U** 0.987 1.01	**Np** (0.68) (1.5)	**Pu** 0.54 1.9	**Am**	**Cm**	**Bk**	**Cf**	**Es**	**Fm**	**Md**	**No**	**Lw**

144

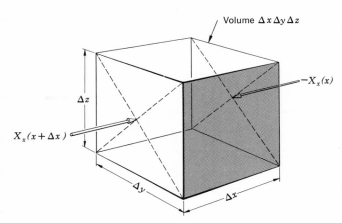

Volume $\Delta x\,\Delta y\,\Delta z$

$-X_x(x)$

Δz

$X_x(x+\Delta x)$

Δy

Δx

Figure 7 Cube of volume $\Delta x\,\Delta y\,\Delta z$ acted on by a stress $-X_x(x)$ on the face at x, and $X_x(x+\Delta x) \cong X_x(x) + \dfrac{\partial X_x}{\partial x}\Delta x$ on the parallel face at $x + \Delta x$. The net force is $\left(\dfrac{\partial X_x}{\partial x}\Delta x\right)\Delta y\,\Delta z$. Other forces in the x direction arise from the variation across the cube of the stresses X_y and X_z, which are not shown. The net x component of the force on the cube is

$$F_x = \left(\frac{\partial X_x}{\partial x} + \frac{\partial X_y}{\partial y} + \frac{\partial X_z}{\partial z}\right)\Delta x\,\Delta y\,\Delta z.$$

The force equals the mass of the cube times the component of the acceleration in the x direction. The mass is $\rho\,\Delta x\,\Delta y\,\Delta z$ and the acceleration is $\partial^2 u/\partial t^2$.

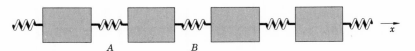

A   B

Figure 8 If springs A and B are stretched equally, the block between them experiences no net force. This illustrates the fact that a uniform stress X_x in a solid does not give a net force on a volume element. If the spring at B is stretched more than the spring at A, the block between them will be accelerated by the force $X_x(B) - X_x(A)$.

ELASTIC WAVES IN CUBIC CRYSTALS

By considering as in Fig. 7 the forces acting on an element of volume in the crystal we obtain the equation of motion in the x direction

$$\rho\frac{\partial^2 u}{\partial t^2} = \frac{\partial X_x}{\partial x} + \frac{\partial X_y}{\partial y} + \frac{\partial X_z}{\partial z}\;; \tag{30}$$

here ρ is the density and u is the displacement in the x direction. There are similar equations for the y and z directions. From (13) and (25) it follows that for a cubic crystal

$$\rho\frac{\partial^2 u}{\partial t^2} = C_{11}\frac{\partial e_{xx}}{\partial x} + C_{12}\left(\frac{\partial e_{yy}}{\partial x} + \frac{\partial e_{zz}}{\partial x}\right) + C_{44}\left(\frac{\partial e_{xy}}{\partial y} + \frac{\partial e_{zx}}{\partial z}\right)\;; \tag{31}$$

here the x, y, z directions are parallel to the cube edges. Using the definitions (6) and (7) of the strain components we have

$$\rho\frac{\partial^2 u}{\partial t^2} = C_{11}\frac{\partial^2 u}{\partial x^2} + C_{44}\left(\frac{\partial^2 u}{\partial y^2} + \frac{\partial^2 u}{\partial z^2}\right) + (C_{12} + C_{44})\left(\frac{\partial^2 v}{\partial x\,\partial y} + \frac{\partial^2 w}{\partial x\,\partial z}\right) \tag{32a}$$

where u, v, w are the components of the displacement $\mathbf{R}$ as defined by (4).

The corresponding equations of motion for $\partial^2 v/\partial t^2$ and $\partial^2 w/\partial t^2$ are found directly from (32a) by symmetry:

$$\rho\,\frac{\partial^2 v}{\partial t^2} = C_{11}\frac{\partial^2 v}{\partial y^2} + C_{44}\left(\frac{\partial^2 v}{\partial x^2} + \frac{\partial^2 v}{\partial z^2}\right) + (C_{12} + C_{44})\left(\frac{\partial^2 u}{\partial x\,\partial y} + \frac{\partial^2 w}{\partial y\,\partial z}\right)\;;$$
(32b)

$$\rho\,\frac{\partial^2 w}{\partial t^2} = C_{11}\frac{\partial^2 w}{\partial z^2} + C_{44}\left(\frac{\partial^2 w}{\partial x^2} + \frac{\partial^2 w}{\partial y^2}\right) + (C_{12} + C_{44})\left(\frac{\partial^2 u}{\partial x\,\partial z} + \frac{\partial^2 v}{\partial y\,\partial z}\right)\;.$$
(32c)

We now look for simple special solutions of these equations.

Waves in the [100] Direction

One solution of (32a) is given by a longitudinal wave

$$u = u_0 \exp\left[i(Kx - \omega t)\right]\;,$$
(33)

where u is the x component of the particle displacement. Both the wavevector and the particle motion are along the x cube edge. Here $K = 2\pi/\lambda$ is the wavevector and $\omega = 2\pi\nu$ is the angular frequency. If we substitute (33) into (32a) we find

$$\omega^2\rho = C_{11}K^2\;;$$
(34)

thus the velocity ω/K of a longitudinal wave in the [100] direction is

$$v_s = \nu\lambda = \omega/K = (C_{11}/\rho)^{\frac{1}{2}}\;.$$
(35)

Consider a transverse or shear wave with the wavevector along the x cube edge and with the particle displacement v in the y direction:

$$v = v_0 \exp\left[i(Kx - \omega t)\right]\;.$$
(36)

On substitution in (32b) this gives the dispersion relation

$$\omega^2\rho = C_{44}K^2\;;$$
(37)

thus the velocity ω/K of a transverse wave in the [100] direction is

$$v_s = (C_{44}/\rho)^{\frac{1}{2}}\;.$$
(38)

The identical velocity is obtained if the particle displacement is in the z direction. Thus for **K** parallel to [100] the two independent shear waves have equal velocities. This is not true for **K** in a general direction in the crystal.

Waves in the [110] Direction

There is a special interest in waves that propagate in a face diagonal direction of a cubic crystal, because the three elastic constants can be found simply from the three propagation velocities in this direction.

Consider a shear wave that propagates in the xy plane with particle displacement w in the z direction

$$w = w_0 \exp\left[i(K_x x + K_y y - \omega t)\right] , \tag{39}$$

whence (32c) gives

$$\omega^2 \rho = C_{44}(K_x{}^2 + K_y{}^2) = C_{44}K^2 , \tag{40}$$

independent of propagation direction in the plane.

Consider other waves that propagate in the xy plane with particle motion in the xy plane: let

$$u = u_0 \exp\left[i(K_x x + K_y y - \omega t)\right] ; \qquad v = v_0 \exp\left[i(K_x x + K_y y - \omega t)\right] . \tag{41}$$

From (32a) and (32b),

$$\omega^2 \rho u = (C_{11}K_x{}^2 + C_{44}K_y{}^2)u + (C_{12} + C_{44})K_x K_y v ; \\ \omega^2 \rho v = (C_{11}K_y{}^2 + C_{44}K_x{}^2)v - (C_{12} + C_{44})K_x K_y u . \tag{42}$$

This pair of equations has a particularly simple solution for a wave in the [110] direction, for which $K_x = K_y = K/\sqrt{2}$. The condition for a solution is that the determinant of the coefficients of u and v in (42) should equal zero:

$$\begin{vmatrix} -\omega^2\rho + \tfrac{1}{2}(C_{11} + C_{44})K^2 & \tfrac{1}{2}(C_{12} + C_{44})K^2 \\ \tfrac{1}{2}(C_{12} + C_{44})K^2 & -\omega^2\rho + \tfrac{1}{2}(C_{11} + C_{44})K^2 \end{vmatrix} = 0 . \tag{43}$$

This equation has the roots

$$\omega^2\rho = \tfrac{1}{2}(C_{11} + C_{12} + 2C_{44})K^2 ; \qquad \omega^2\rho = \tfrac{1}{2}(C_{11} - C_{12})K^2 . \tag{44}$$

The first root describes a longitudinal wave; the second root describes a shear wave. How do we determine the direction of particle displacement? The

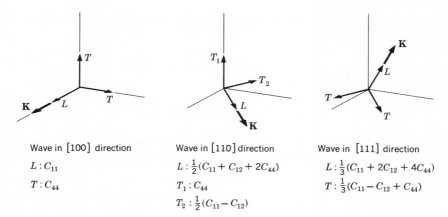

Wave in [100] direction

$L : C_{11}$

$T : C_{44}$

Wave in [110] direction

$L : \frac{1}{2}(C_{11} + C_{12} + 2C_{44})$

$T_1 : C_{44}$

$T_2 : \frac{1}{2}(C_{11} - C_{12})$

Wave in [111] direction

$L : \frac{1}{3}(C_{11} + 2C_{12} + 4C_{44})$

$T : \frac{1}{3}(C_{11} - C_{12} + C_{44})$

Figure 9 Effective elastic constants for the three modes of elastic waves in the principal propagation directions in cubic crystals. The two transverse modes are degenerate for propagation in the [100] and [111] directions.

first root when substituted into the upper equation of (42) gives

$$\tfrac{1}{2}(C_{11} + C_{12} + 2C_{44})K^2u = \tfrac{1}{2}(C_{11} + C_{44})K^2u + \tfrac{1}{2}(C_{12} + C_{44})K^2v , \quad (45)$$

whence the displacement components satisfy $u = v$. Thus the particle displacement is along [110] and parallel to the **K** vector (Fig. 9). The second root of (44) when substituted into the upper equation of (42) gives

$$\tfrac{1}{2}(C_{11} - C_{12})K^2u = \tfrac{1}{2}(C_{11} + C_{44})K^2u + \tfrac{1}{2}(C_{12} + C_{44})K^2v , \quad (46)$$

whence $u = -v$. The particle displacement is along [1$\bar{1}$0] and perpendicular to the **K** vector.

Selected values of the adiabatic elastic stiffness constants of cubic crystals at low temperatures and at room temperature are given in Table 2. Notice the general tendency for the elastic constants to decrease as the temperature is increased. Results for silver are plotted in Fig. 10 and for BaF$_2$ in Fig. 11. Fur-

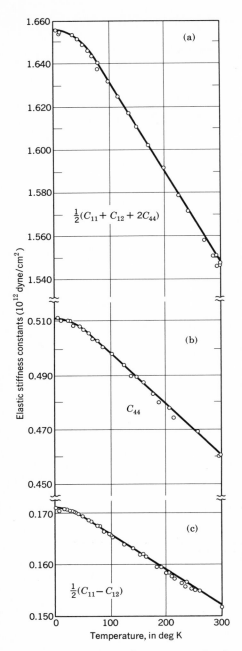

Figure 10 Elastic stiffness constants for silver as a function of temperature, after J. R. Neighbours and G. A. Alers, Phys. Rev. **111**, 707 (1958). The constants plotted are (a): $\frac{1}{2}(C_{11} + C_{12} + 2C_{44})$; (b): C_{44}; (c): $\frac{1}{2}(C_{11} - C_{12})$. These combinations correspond to the three waves in the [110] direction.

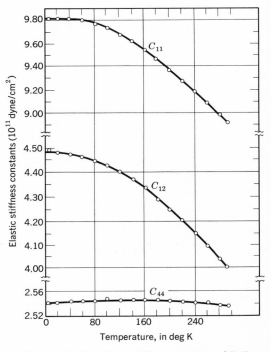

Figure 11 The elastic stiffness constants of BaF$_2$ as a function of temperature, after D. Gerlich, Phys. Rev. **135**, A1331 (1964).

**Table 2   Adiabatic elastic stiffness constants of cubic crystals
at low temperature and at room temperature**

The values given at $0°K$ were obtained by extrapolation of measurements carried out down to $4°K$. The table was compiled with the assistance of Professor Charles S. Smith.*

Crystal	Stiffness constants, in 10^{12} dyne/cm^2 (10^{11} N/m^2)			Temperature, °K	Density, g/cm^3
	C_{11}	C_{12}	C_{44}		
W	5.326	2.049	1.631	0	19.317
	5.233	2.045	1.607	300	—
Ta	2.663	1.582	0.874	0	16.696
	2.609	1.574	0.818	300	—
Cu	1.762	1.249	0.818	0	9.018
	1.684	1.214	0.754	300	—
Ag	1.315	0.973	0.511	0	10.635
	1.240	0.937	0.461	300	—
Au	2.016	1.697	0.454	0	19.488
	1.923	1.631	0.420	300	—
Al	1.143	0.619	0.316	0	2.733
	1.068	0.607	0.282	300	—
K	0.0416	0.0341	0.0286	4	
	0.0370	0.0314	0.0188	295	
Pb	0.555	0.454	0.194	0	11.599
	0.495	0.423	0.149	300	—
Ni	2.612	1.508	1.317	0	8.968
	2.508	1.500	1.235	300	—
Pd	2.341	1.761	0.712	0	12.132
	2.271	1.761	0.717	300	—
V	2.324	1.194	0.460	0	6.051
	2.280	1.187	0.426		—
LiF	1.246	0.424	0.649	0	2.646
	1.112	0.420	0.628	300	—
KCl	0.483	0.054	0.066	4	2.038
	0.403	0.066	0.063	300	—
BaF$_2$	0.981	0.448	0.254	0	—
	0.891	0.400	0.254	300	4.886

* References to the original papers are given in the article by C. Kittel in *Phonons*, R. W. H. Stevenson, ed., Oliver & Boyd, 1966.

**Table 3   Adiabatic elastic stiffness constants
of several cubic crystals
at room temperature or 300°K.**

Most of these data are taken from the compilation by H. B. Huntington; see also D. I. Bolef and M. Menes, J. Appl. Physics **31**, 1010 (1960).

	Stiffness constants, in 10^{12} dyne/cm^2 or 10^{11} N/m^2		
	C_{11}	C_{12}	C_{44}
Diamond	10.76	1.25	5.76
Na	0.073	0.062	0.042
Li	0.135	0.114	0.088
Ge	1.285	0.483	0.680
Si	1.66	0.639	0.796
GaSb	0.885	0.404	0.433
InSb	0.672	0.367	0.302
MgO	2.86	0.87	1.48
NaCl	0.487	0.124	0.126
RbBr	0.317	0.042	0.039
RbI	0.256	0.031	0.029
CsBr	0.300	0.078	0.076
CsI	0.246	0.067	0.062

ther values at room temperature alone are given in Table 3.

Useful published tables of elastic constants include the following:

H. B. Huntington, *Solid state physics* **7**, 213 (1958);

R. F. S. Hearmon, Advances in Physics **5**, 323 (1956);

K. S. Aleksandrov and T. V. Ryzhova, Soviet Physics (Crystallography) **6**, 228, (1961).

There are three normal modes of wave motion in a crystal for a given magnitude and direction of the wavevector **K**. In general, the polarizations (directions of particle displacement) of these modes are not exactly parallel or perpendicular to **K**. In the special propagation directions [100], [111], and [110] of a cubic crystal two of the three modes for a given **K** are such that the particle motion is exactly transverse to **K** and in the third mode the motion

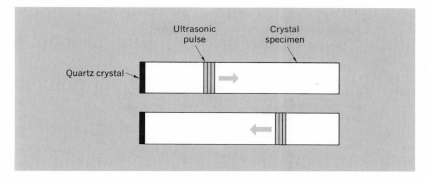

Figure 12   In the ultrasonic pulse method for the determination of elastic wave velocities a pulse of sound is generated by a piezoelectric transducer. The pulse makes successive reflections and is detected each time it reaches the transducer.

is exactly longitudinal (parallel to **K**). The analysis is much simpler in these special directions than in general directions.[5]

EXPERIMENTAL DETERMINATION OF ELASTIC CONSTANTS

The classic methods for the measurement of the elastic constants of crystals are described by Hearmon.[6] Since his review the use of the ultrasonic pulse method has become widespread[7] because of its adaptability to a wide range of experimental conditions, although many other methods are valuable and are used. In this method an ultrasonic pulse generated by a quartz[8] transducer is transmitted through the test crystal and reflected from the rear surface of the crystal back to the transducer (Fig. 12). The elapsed time between initiation and receipt of the pulse is measured by standard electronic methods (Figs. 13 to 15). The velocity is obtained by dividing the round-trip distance by the elapsed time. In a representative arrangement the experimental frequency may be 15 MHz, and the pulse length 1 μsec. The wavelength is of the order 3×10^{-2}

---

[5] See W. P. Mason, *Physical acoustics and the properties of solids*, Van Nostrand, 1958.

[6] R. F. S. Hearmon, Revs. Modern Phys. **18**, 409 (1946).

[7] See, for example, H. Huntington, Phys. Rev. **72**, 321 (1947); J. K. Galt, Phys. Rev. **73**, 1460 (1948).

[8] A slab may be cut from a quartz crystal in such a way that either a longitudinal or a transverse wave is excited in the slab by the application of an r-f electric field across electrodes evaporated on opposite surfaces. The theory of piezoelectric excitation is discussed briefly in Chapter 13.

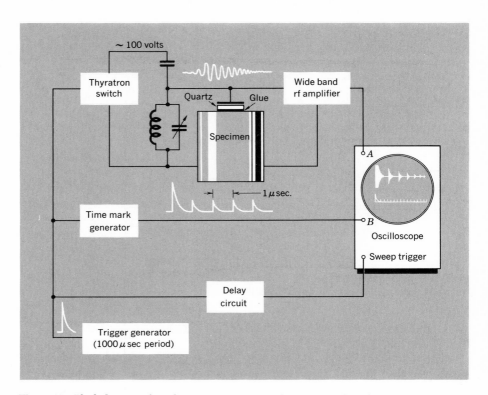

Figure 13 Block diagram of an electronic apparatus used to measure the velocity of sound by the ultrasonic pulse-echo technique. The trigger generator starts the process by closing the thyratron switch; this discharges a condenser through the LC network and imposes the indicated voltage waveform on the quartz transducer glued to the specimen. The waveform becomes a sound pulse which bounces back and forth within the specimen to produce the train of echoes shown on the oscilloscope face. The time mark generator and sweep delay circuit allow the time of arrival of some later echo to be measured accurately. [After G. A. Alers and J. R. Neighbours, J. Phys. Chem. Solids **7**, 58, (1958).]

Figure 14   Photograph of a single crystal of aluminum mounted in a spring-loaded holder for making velocity of sound measurements by the ultrasonic pulse-echo technique. The top face of the crystal is a (110) crystal plane on which is glued a quartz transducer with a metal electrode and rf lead in place. (Courtesy of the Scientific Laboratory, Ford Motor Company.)

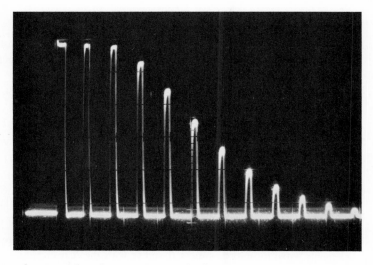

Figure 15   Successive ultrasonic pulse echoes across a crystal. The time interval between successive pulses is measured on the oscilloscope. This interval is the round-trip travel time of the ultrasonic pulse. The attenuation of the wave may be found from the decrease in pulse height of successive pulses, with allowance for losses on reflection at the ends. The voltage across the transducer is directly proportional to the stress in the wave. (Courtesy of H. J. McSkimin.)

cm. The crystal specimen may be of the order of 1 cm in length. References to ultrasonic techniques at microwave frequencies are given by E. H. Jacobsen in the book by Bak; see also H. E. Bömmel and K. Dransfeld, Phys. Rev. **117**, 1245 (1960); T. O. Woodruff and H. Ehrenreich, Phys. Rev. **123**, 1553 (1961).

The elastic stiffness constants C_{11}, C_{12}, C_{44} of a cubic crystal may be determined from the velocities of three waves, as we have noted. The same crystal orientation can be used for all three waves, but it is necessary to alter the cut and mounting of the quartz crystal transducer in order to excite the desired direction of particle motion in the crystal specimen.

Third-Order Elastic Constants

In the region of Hooke's law the elastic energy density is quadratic in the strain components, as in (14). Beyond this region higher order products of strains are needed. The third-order elastic stiffness constants relate the energy to products of three strain components. They are the lowest order constants to enter the description of nonlinear effects (Chapter 6), such as the interaction of phonons and thermal expansion. The third-order constants can be determined from velocity measurements on small amplitude sound waves in statically stressed media. A good account of the definitions of the third-order constants and the theory of their measurement is given by K. Brugger, Phys. Rev. **133**, A1611 (1964); R. N. Thurston and K. Brugger, Phys. Rev. **133**, A1604 (1964).

Problems

1. **Young's modulus and Poisson's ratio.** A cubic crystal is subject to tension in the [100] direction. Find expressions in terms of the elastic stiffnesses for Young's modulus and Poisson's ratio as defined in Fig. 16.

2. **Longitudinal wave velocity.** Show that the velocity of a longitudinal wave in the [111] direction of a cubic crystal is given by $v_s = [\frac{1}{3}(C_{11} + 2C_{12} + 4C_{44})/\rho]^{\frac{1}{2}}$. *Hint:* For such a wave $u = v = w$. Let $u = u_0 e^{iK(x+y+z)/\sqrt{3}} e^{-i\omega t}$, and use Eq. (32a).

3. **Transverse wave velocity.** Show that the velocity of transverse waves in the [111] direction of a cubic crystal is given by $v_s = [\frac{1}{3}(C_{11} - C_{12} + C_{44})/\rho]^{\frac{1}{2}}$. *Hint:* See Problem 2.

4. **Effective shear constant.** Show that the shear constant $\frac{1}{2}(C_{11} - C_{12})$ in a cubic crystal is defined by setting $e_{xx} = -e_{yy} = \frac{1}{2}e$ and all other strains equal to zero, as in Fig. 17. *Hint:* Consider the energy density (18); look for a C' such that $U = \frac{1}{2}C'e^2$.

5. **Determinantal approach.** It is known[9] that an R-dimensional square matrix with all elements equal to unity has roots R and 0, with the R occurring once and the zero occurring $R - 1$ times. If all elements have the value p, then the roots are Rp and 0. (a) Show that if the diagonal elements are q and all other elements are p, then there

---

[9] The elements of the determinantal equation are $a_{ij} = 1 - \lambda\delta_{ij}$, where λ is a root; see Advanced Topic L. Here δ_{ij} is the Kronecker delta function.

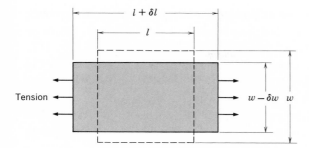

Figure 16 Young's modulus is defined as stress/strain for a tensile stress acting in one direction, with the sides of specimen left free. Poisson's ratio is defined as $(\delta w/w)/(\delta l/l)$ for this situation.

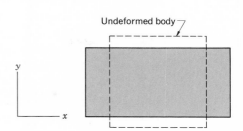

Figure 17 This deformation is compounded from the two shears $e_{xx} = -e_{yy}$.

is one root equal to $(R-1)p + q$ and $R-1$ roots equal to $q - p$. (b) Show from the elastic equation (32) for a wave in the [111] direction of a cubic crystal that the determinantal equation which gives ω^2 as a function of K is

$$\begin{vmatrix} q - \omega^2\rho & p & p \\ p & q - \omega^2\rho & p \\ p & p & q - \omega^2\rho \end{vmatrix} = 0 \;,$$

where $q \equiv \frac{1}{3}K^2(C_{11} + 2C_{44})$ and $p \equiv \frac{1}{3}K^2(C_{12} + C_{44})$. This expresses the condition that three linear homogeneous algebraic equations for the three displacement components u, v, w have a solution. Use the result of part (a) to find the three roots of ω^2; check with the results given for Problems 2 and 3.

6. *General propagation direction.* (a) By substitution in (32) find the determinantal equation which expresses the condition that the displacement

$$\mathbf{R}(\mathbf{r}) = [u_0\hat{x} + v_0\hat{y} + w_0\hat{z}] \exp [i(\mathbf{K} \cdot \mathbf{r} - \omega t)]$$

be a solution of the elastic wave equations in a cubic crystal. (b) The sum of the roots of a determinantal equation is equal to the sum of the diagonal elements a_{ii}. Show from part (a) that the sum of the squares of the three elastic wave velocities in any direction in a cubic crystal is equal to $(C_{11} + 2C_{44})/\rho$. Recall that $v_s^2 = \omega^2/K^2$.

°7. *Stability criteria.* The criterion that a cubic crystal with one atom in the primitive cell be stable against small homogeneous deformations is that the energy density (18) be positive for all combinations of strain components. What restrictions are thereby imposed on the elastic stiffness constants? (In mathematical language the problem is to find the conditions that a real symmetric quadratic form should be positive definite. The solution is given in books on algebra; see also Korn and Korn, *Mathematical handbook*, McGraw-Hill, 1961, Sec. 13.5–6.) *Ans.* $C_{44} > 0$, $C_{11} > 0$, $C_{11}^2 - C_{12}^2 > 0$, and $C_{11} + 2C_{12} > 0$. For an example of the instability which results when $C_{11} \cong C_{12}$ see L. R. Testardi et al., Phys. Rev. Letters **15**, 250 (1965).

References

J. F. Nye, *Physical properties of crystals: their representation by tensors and matrices*, Oxford, 1957.

H. B. Huntington, "Elastic constants of crystals," *Solid state physics* **7**, 213 (1958).

A. E. H. Love, *A treatise on the mathematical theory of elasticity*, Dover paperback, 1944.

W. P. Mason, *Physical acoustics and the properties of solids*, Van Nostrand, 1958.

W. P. Mason, ed., *Physical acoustics*, in several volumes, Academic Press, Vol. I, 1964.

C. Zener, *Elasticity and anelasticity of metals*, University of Chicago Press, 1948.

T. A. Bak, ed., *Phonons and phonon interactions*, Benjamin, 1964.

5

Phonons
and
Lattice Vibrations

NOTATION: In the CGS-Gaussian system we have $D = \epsilon E$, where ϵ is called the dielectric constant or relative dielectric constant (relative to vacuum); for the Fourier components we have $D(\omega, \mathbf{K}) = \epsilon(\omega, \mathbf{K})E(\omega, \mathbf{K})$, where $\epsilon(\omega, \mathbf{K})$ is called the dielectric function. In SI we write $D = \epsilon\epsilon_0 E$, where ϵ is the dielectric constant and ϵ_0 is the permittivity of free space; also, $D(\omega, \mathbf{K}) = \epsilon(\omega, \mathbf{K})\epsilon_0 E(\omega, \mathbf{K})$. The values of ϵ and of $\epsilon(\omega, \mathbf{K})$ are identical in the two systems. See also the contents page of Chapter 13.

Sign	Name	Field
 ———→	Electron	——
 ∿∿∿→	Photon	Electromagnetic wave
 ⸺W⸺→	Phonon	Elastic wave
 ⸺‖⊢→	Plasmon	Collective electron wave
 ⸺℮℮℮→	Magnon	Magnetization wave
–	Polaron	Electron + elastic deformation
–	Exciton	Polarization wave

Figure 1 Some of the important elementary excitations in solids. The signs shown are used in this text. The origins of the names of the excitations are discussed by C. T. Walker and G. A. Slack, Am. J. Phys. **38**, 1380 (1970).

QUANTIZATION OF LATTICE VIBRATIONS

The energy in a lattice vibration or elastic wave is quantized. The quantum of energy in an elastic wave is called a **phonon,** in analogy with the photon, which is the quantum of energy in an electromagnetic wave (Fig. 1). We first review the story of the photon. Almost all of the concepts, such as the wave-particle duality, which apply to photons apply equally well to phonons. Sound waves in crystals are composed of phonons. Thermal vibrations in crystals are thermally excited phonons, analogous to the thermally excited photons of black-body electromagnetic radiation in a cavity.

Quantum theory began in 1900 when Max Planck showed that quantization of energy would explain the observed distribution in frequency of the electromagnetic energy radiated by a black body in thermal equilibrium. Planck assumed that the energy of each mode of oscillation of the electromagnetic field in a cavity is equal to an integral multiple of $h\nu$. The energy of one photon is $\epsilon = h\nu$; the energy of n photons in a mode of frequency ν is

$$\epsilon = nh\nu , \tag{1}$$

where n is a positive integer or zero, and the constant h (known now as Planck's constant) has the value 6.6262×10^{-27} erg sec. We have for convenience omitted the usual zero-point term $\frac{1}{2}h\nu$ from (1); it has no direct effect on the matters we now consider. It is more common to write (1) as $\epsilon = n\hbar\omega$, where $\omega \equiv 2\pi\nu$ is the angular frequency and $\hbar \equiv h/2\pi \cong 1.0546 \times 10^{-27}$ erg sec.

We know from innumerable diffraction experiments that the electromagnetic field has many aspects of a wave. What we learn from the Planck distribution law is that the energy in the electromagnetic field is quantized. The same considerations apply to elastic waves.

What is the experimental evidence that the energy of an elastic wave is quantized? The earliest evidence was the observation that the lattice contribution to the heat capacity of solids (Chapter 6) always approaches zero as the temperature approaches zero; this can be explained only if the lattice vibrations are quantized.

X-rays and neutrons are scattered inelastically by crystals, with energy and momentum changes corresponding to the creation or the absorption of one or more phonons. By measuring the recoil of the scattered x-ray or neutron we determine the properties of individual phonons, such as the dispersion relations of frequency versus wavevector.

PHONON MOMENTUM

A phonon of wavevector $\mathbf{K}$ interacts with other particles and fields as if[1] it had a momentum $\hbar\mathbf{K}$. A phonon on a lattice does not really have momentum; we show in Problem 5 that only the $\mathbf{K} = 0$ phonon carries physical momentum, for this mode corresponds to a uniform translation of the system. But for most practical purposes a phonon acts as if its momentum were $\hbar\mathbf{K}$. Sometimes $\hbar\mathbf{K}$ is called the **crystal momentum**.

There exist in crystals wavevector selection rules for allowed transitions between quantum states: these selection rules involve $\mathbf{K}$. In Chapter 2 we saw that the elastic scattering (Bragg diffraction) of an x-ray photon by a crystal is governed by the wavevector selection rule

$$\mathbf{k}' = \mathbf{k} + \mathbf{G} , \tag{2}$$

where $\mathbf{G}$ is a vector in the reciprocal lattice; $\mathbf{k}$ is the wavevector of the incident photon and $\mathbf{k}'$ is the wavevector of the scattered photon. In the reflection process (2) the crystal as a whole will recoil with momentum $-\hbar\mathbf{G}$, but this is rarely considered explicitly. The total wavevector of interacting waves is conserved in a periodic lattice, but only with the possible addition of a reciprocal lattice vector. The true momentum of the whole system is rigorously conserved.

If the scattering of the photon is inelastic, with the creation of a phonon of wavevector $\mathbf{K}$, then the wavevector selection rule becomes[2]

$$\mathbf{k}' + \mathbf{K} = \mathbf{k} + \mathbf{G} . \tag{3}$$

---

[1] For a discussion of this delicate point, see G. Leibfried in *Encyclo. of physics* **VII**/1, 104 (1955); G. Süssman, Z. Naturf. **11a**, 1 (1956); G. Beck, Anals. Acad. Brasil. Ci. **26**, 64 (1954). The reason that phonons on a lattice do not carry momentum is that a phonon coordinate (except for $K = 0$) involves *relative* coordinates of the atoms. Thus in an H_2 molecule the internuclear vibrational coordinate $\mathbf{r}_1 - \mathbf{r}_2$ is a relative coordinate and does not carry linear momentum; the center of mass coordinate $\frac{1}{2}(\mathbf{r}_1 + \mathbf{r}_2)$ corresponds to the uniform mode and can carry linear momentum.

[2] We can exhibit by an example the mathematics involved in the different selection rules for a lattice and for a continuum. Suppose two phonons $\mathbf{K}_1$, $\mathbf{K}_2$ interact through third-order anharmonic terms in the elastic energy as discussed in Chapter 6 to create a third phonon $\mathbf{K}_3$. The probability of the collision will involve the product of the three phonon wave amplitudes, summed over all lattice sites:

$$(\text{phonon } \mathbf{K}_1 \text{ in})(\text{phonon } \mathbf{K}_2 \text{ in})(\text{phonon } \mathbf{K}_3 \text{ out}) \propto \sum_n e^{-i\mathbf{K}_1 \cdot \mathbf{r}_n} e^{-i\mathbf{K}_2 \cdot \mathbf{r}_n} e^{i\mathbf{K}_3 \cdot \mathbf{r}_n}$$
$$= \sum_n \exp\left[i(\mathbf{K}_3 - \mathbf{K}_1 - \mathbf{K}_2) \cdot \mathbf{r}_n\right] .$$

This sum in the limit of a large number of lattice sites approaches zero unless $\mathbf{K}_3 = \mathbf{K}_1 + \mathbf{K}_2$ or unless $\mathbf{K}_3 = \mathbf{K}_1 + \mathbf{K}_2 + \mathbf{G}$. If either of these conditions is satisfied, of which the first is merely a special case of the second, the sum is equal to the number of lattice sites N. A similar sum was considered in Problem 2.5.

In a continuum the matrix element in the same problem involves

$$\int d^3x \exp\left[i(\mathbf{K}_3 - \mathbf{K}_1 - \mathbf{K}_2) \cdot \mathbf{r}\right] = (2\pi)^3 \delta(\mathbf{K}_3 - \mathbf{K}_1 - \mathbf{K}_2) ,$$

where δ is the Dirac delta function. In the continuum no reciprocal lattice vectors are defined; if they were defined as a limit, the shortest nonzero vector would have infinite length.

If a phonon $\mathbf{K}$ is absorbed in the process, we have instead the relation

$$\mathbf{k'} = \mathbf{k} + \mathbf{K} + \mathbf{G} \ . \tag{4}$$

Relations (4) and (3) are natural extensions of (2).

INELASTIC SCATTERING OF PHOTONS BY LONG WAVELENGTH PHONONS

Consider a photon of frequency $\nu = \omega/2\pi$ which propagates in a crystal. If the crystal is viewed as a continuum of refractive index n, then the wavevector of the photon is determined by the relation

$$\omega = \frac{ck}{n} \qquad \text{or} \qquad \lambda\nu = \frac{c}{n} \ , \tag{5}$$

where c is the velocity of light. The momentum of the photon is

$$\mathbf{p} = \hbar\mathbf{k} \ . \tag{6}$$

Let a photon in this beam interact with a phonon beam or sound wave in the crystal. The photon can be scattered by the sound wave. The interaction may occur because the elastic strain field of the sound wave changes the local concentration of atoms and therefore the refractive index of the crystal. Thus the sound wave modulates the optical properties of the medium. Reciprocally, the electric field of the light wave modulates the elastic properties of the medium.

In a crystal a photon can create or absorb a phonon. The photon will be scattered in the process; its wavevector will change from $\mathbf{k}$ to $\mathbf{k'}$ and its frequency from ω to ω'. Suppose that a phonon is created with a wavevector $\mathbf{K}$ and angular frequency Ω. The kinematics of the collision event (Fig. 2) are simple. By conservation of energy

$$\hbar\omega = \hbar\omega' + \hbar\Omega \ . \tag{7}$$

By the wavevector selection rule

$$\mathbf{k} = \mathbf{k'} + \mathbf{K} \ , \tag{8}$$

where for simplicity we do not include in (8) the possibility (3) that the scattering may be combined with a Bragg diffraction involving a reciprocal lattice vector of the crystal lattice. If the velocity of sound v_s is constant, we have $\Omega = v_s K$, because $\lambda\Omega/2\pi = v_s$.

Now a phonon can carry off only a small part of the energy of the incident photon: the velocity of sound v_s is very much less than the velocity of light c. For a phonon wavevector K comparable in magnitude to the photon wavevector k, it follows that $ck \gg v_s K$. Now $\omega = ck$ and $\Omega = v_s K$; thus $\omega \gg \Omega$. It follows from (7) that $\omega' \cong \omega$ and $k' \cong k$.

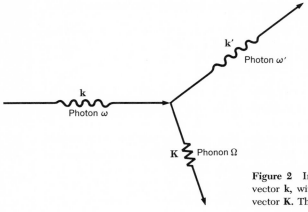

Figure 2 Inelastic scattering of a photon of wavevector **k**, with the production of a phonon of wavevector **K**. The scattered photon has wavevector **k**′.

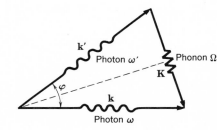

Figure 3 Selection rule diagram for the process of Fig. 2. If $k = k'$, the triangle is isosceles. The base of the triangle is $K = 2k \sin \frac{1}{2}\varphi$.

If $k' \cong k$, we see from Fig. 3 that

$$K \cong 2k \sin \tfrac{1}{2}\varphi \; , \tag{9}$$

or, using $k = \omega n/c$ from (5),

$$v_s K \cong \frac{2v_s \omega n}{c} \sin \tfrac{1}{2}\varphi \; . \tag{10}$$

Because $\Omega = v_s K$, the phonons produced when photons are scattered inelastically at an angle φ from the incident direction will be of frequency

$$\Omega \cong (2v_s \omega n/c) \sin \tfrac{1}{2}\varphi \; , \tag{11}$$

where n is the refractive index of the crystal.

EXAMPLE: *Generation of Phonons.* The maximum phonon frequency generated by scattering of visible light of vacuum wavelength $\lambda = 4000$ Å is, for $v_s \approx 5 \times 10^5$ cm/sec and $n \approx 1.5$,

$$\Omega \approx 2(5 \times 10^5)(2\pi)(1.5)/(4 \times 10^{-5}) \approx 2 \times 10^{11} \text{ rad s}^{-1} , \qquad (12)$$

using (11) with $\sin \frac{1}{2}\varphi = 1$. At this frequency $K = \Omega/v_s \approx 4 \times 10^5$ cm^{-1}. The fractional change in the frequency of the light on scattering is 5×10^{-5}.

The scattering of visible light from an intense laser source has been used[3] to generate phonons in the microwave frequency range in quartz and in sapphire. The photon frequency shifts observed are in close agreement with the shifts calculated from Eq. (11), using values of the velocity of sound measured at lower frequencies by ultrasonic methods.

The scattering of light by phonons in solids and liquids is known as **Brillouin scattering.**[4] The spectrum after scattering of monochromatic light incident on water is shown in Fig. 4. The excitation of microwave phonons in crystals has been detected by the diffraction of light.[5]

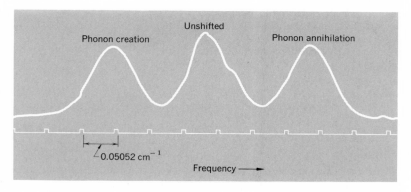

Figure 4 Spectrum of light at 6328 Å scattered at right angles in water at room temperature. The unshifted central peak at the laser frequency is due largely to Tyndall scattering from tiny suspended particles in the water. The line width is due to the slit width of the spectrograph. The spectrum was swept out on the recorder in 5 minutes. The phonon frequency was determined from this trace to be $(4.33 \pm 0.02) \times 10^9$ Hz. The velocity is calculated from Eq. (11) to be $(1.457 \pm 0.010) \times 10^5$ cm/sec. [After G. B. Benedek et al., J. Opt. Soc. Amer. **54**, 1284 (1964).]

[3] See R. Y. Chiao, C. H. Townes, and B. P. Stoicheff, Phys. Rev. Letters **12**, 592 (1964).

[4] L. Brillouin, Ann. phy. **17**, 88 (1922); a review of the Raman effect in crystals is given by R. Loudon, Adv. in Physics **13**, 424–482 (1964).

[5] K. N. Baranskii, Soviet Phys. Doklady **2**, 237 (1957); see also Fig. 1 of H. E. Bömmel and K. Dransfeld, Phys. Rev. Letters **1**, 234 (1958).

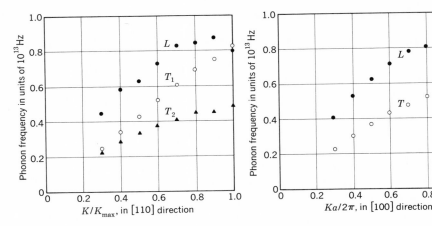

Figure 5 Dispersion curves determined by the inelastic scattering of x-rays for phonons propagating along the [110] axis in aluminum. The longitudinal wave is shown by solid circles; the transverse wave T_1 for which the particle motion is parallel to the [001] axis is shown by open circles; and the transverse wave T_2 parallel to the [110] axis is shown by solid triangles. [After C. B. Walker, Phys. Rev. **103**, 547 (1956).]

Figure 6 Dispersion curves for elastic waves propagating along the [100] axis in aluminum, as observed by inelastic scattering of x-rays. The longitudinal and transverse waves are shown, respectively, by the solid and open circles. (After C. B. Walker.)

INELASTIC SCATTERING OF X-RAYS BY PHONONS

One method of studying the phonon spectrum of solids is by the inelastic scattering of x-rays. The principles described above can be applied to the inelastic or diffuse scattering of x-ray photons in processes in which one phonon is created or absorbed. Results obtained by Walker for aluminum are plotted in Figs. 5 and 6.

In such experiments we want to find the phonon frequency as a function of the phonon wavevector **K**. The wavevector is determined by application of the general condition (10) or (11) for conservation of wavevector. Unfortunately, it is difficult to determine directly the small frequency shift of the scattered x-ray beam. Neutron scattering experiments have the advantage that the energy shift can usually be measured directly.

INELASTIC SCATTERING OF NEUTRONS BY PHONONS

A neutron sees the crystal lattice chiefly by interaction with the nuclei of the atoms. The kinematics of the scattering of a neutron beam by a crystal lattice are described by the general wavevector selection rule (3) or (4):

$$\mathbf{k} = \mathbf{k'} + \mathbf{G} \pm \mathbf{K} \;, \tag{13}$$

and by the requirement of conservation of energy. Here $\mathbf{K}$ is the wavevector of the phonon created $(+)$ or absorbed $(-)$ in the process, and $\mathbf{G}$ is any reciprocal lattice vector.

The kinetic energy of the incident neutron is $p^2/2M_n$, where M_n is the mass of the neutron. The momentum $\mathbf{p}$ is given by $\hbar\mathbf{k}$, where $\mathbf{k}$ is the wavevector of the neutron. Thus $\hbar^2 k^2/2M_n$ is the kinetic energy of the incident neutron. If $\mathbf{k'}$ is the wavevector of the scattered neutron, the energy of the scattered neutron is $\hbar^2 k'^2/2M_n$. The statement of conservation of energy is

$$\frac{\hbar^2 k^2}{2M_n} = \frac{\hbar^2 k'^2}{2M_n} \pm \hbar\omega_{\mathbf{K}} \;, \tag{14}$$

where $\hbar\omega_{\mathbf{K}}$ is the energy of the phonon created $(+)$ or absorbed $(-)$ in the process.

To determine the dispersion relation[6] using (13) and (14) it is necessary in the experiment to find the energy gain or loss of the scattered neutrons as a function of the scattering direction $\mathbf{k} - \mathbf{k'}$.

An example of an accurate determination of phonon spectra in a metal is shown by Fig. 7a for sodium. In Fig. 7b phonon spectra are given for an ionic crystal, potassium bromide.

Under favorable conditions neutron scattering is the ideal method for the determination of phonon spectra. The method is not applicable when the absorption of neutrons by the nuclei of the crystal is high. In some circumstances it is possible to obtain important data on phonon lifetimes from the angular width of the scattered neutron beam. The spectrometer used at Chalk River, one of the centers of research in this field, is shown in Figs. 8 and 9. Another major facility is at Brookhaven on Long Island.

---

[6] A geometrical construction to illustrate the kinematics of inelastic neutron scattering is given on p. 377 of *QTS*. An excellent review of inelastic neutron scattering is given by B. N. Brockhouse, S. Hautecler, and H. Stiller, in Strumane et al., ed., *Interaction of radiation with solids*, North-Holland, 1963.

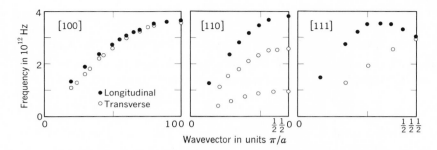

Figure 7a The dispersion curves of sodium for phonons propagating in the [001], [110], and [111] directions at 90°K, as determined by inelastic scattering of neutrons. [Woods, Brockhouse, March and Bowers, Proc. Phys. Soc. London **79**, pt. 2, 440 (1962).]

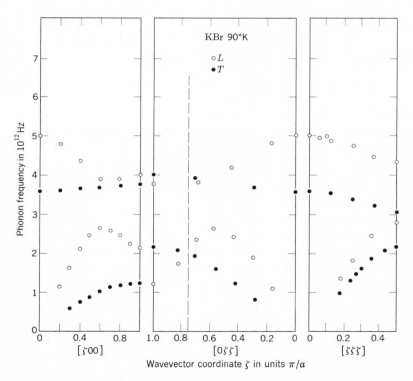

Figure 7b Dispersion curves for the optical and acoustic branches of the phonon spectrum of potassium bromide at 90°K, after A. D. B. Woods, B. N. Brockhouse, R. A. Cowley, and W. Cochran, Phys. Rev. **131**, 1025 (1963). The data can be fitted very well by a simple model called the shell model, discussed by A. D. B. Woods, W. Cochran, and B. N. Brockhouse, Phys. Rev. **119**, 980 (1960); B. J. Dick and A. W. Overhauser, Phys. Rev. **112**, 90 (1958); J. E. Hanlon and A. W. Lawson, Phys. Rev. **113**, 472 (1959).

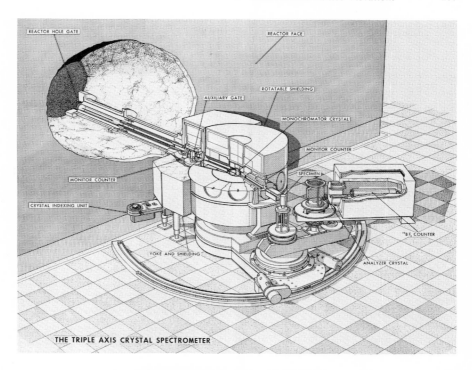

THE TRIPLE AXIS CRYSTAL SPECTROMETER

Figure 8 (*above*) Cutaway drawing of the Brockhouse triple-axis crystal spectrometer, as in Fig. 9.

Figure 9 (*right*)   Triple-axis crystal spectrometer at the high flux reactor at the Chalk River Nuclear Laboratories of Atomic Energy of Canada Limited. The spectrometer was developed by Dr. B. N. Brockhouse and colleagues to carry out studies of phonons in crystals using the inelastic scattering of neutrons.

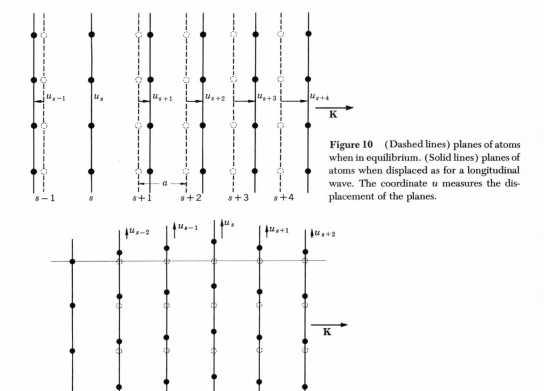

Figure 10 (Dashed lines) planes of atoms when in equilibrium. (Solid lines) planes of atoms when displaced as for a longitudinal wave. The coordinate u measures the displacement of the planes.

Figure 11 Planes of atoms as displaced during passage of a transverse wave.

VIBRATION OF MONATOMIC LATTICES

We now extend the discussion of the elastic vibrations of crystals to the short wavelength range, where the wavelength of the lattice wave is comparable with the lattice constant of the crystal. The periodicity of the crystal structure has important consequences for elastic waves just as for x-rays.

To simplify the problem we consider elastic waves propagating in directions such that the wave polarizations are purely transverse or purely longitudinal.[7] (We consider only crystals for which the primitive basis contains only one atom.) In a cubic crystal these are the 100, 111, and 110 directions. When a wave propagates along one of these directions, entire planes of atoms in the crystal move in phase. The motion is parallel to the propagation direction if the wave is longitudinal and perpendicular to the propagation direction if the wave is transverse. We can learn a great deal about the forces which connect

[7] In general directions of propagation the polarizations are not purely longitudinal or transverse.

different planes of atoms from the relation between frequency ω and wave-vector $\mathbf{K}$ for these special modes of propagation for which the mathematics is simple.

If planes of atoms are displaced as a whole parallel or perpendicular to the wavevector $\mathbf{K}$ during the passage of a wave, we can describe by a single co-ordinate u_s the displacement of the plane s from equilibrium. The problem is then one dimensional. We suppose for the present that all planes of atoms are identical. Figure 10 shows the particle displacement for a longitudinal wave, Fig. 11 for a transverse wave.

We assume that the force on the plane labeled s as caused by the displacement of the plane labeled $s + p$ is proportional to the difference $u_{s+p} - u_s$ of their displacements, so that the total force on plane s is given by

$$F_s = \sum_p C_p(u_{s+p} - u_s) \ . \tag{15}$$

This expression is linear in the displacements and is of the form of Hooke's law. The constant C_p is the force constant between planes removed from each other by p. The C_p's will differ for longitudinal and transverse waves. It is convenient to regard C_p as defined for one atom of the plane, so that F_s is the force on one atom in the plane s.

What is the connection between C_p and the potential energy of two atoms? Let $U(R_0)$ be the potential energy between two atoms at their equilibrium separation R_0. If the separation is increased by ΔR, the new potential energy is

$$U(R) = U(R_0) + \left(\frac{dU}{dR}\right)_{R_0} \Delta R + \tfrac{1}{2}\left(\frac{d^2U}{dR^2}\right)_{R_0} (\Delta R)^2 + \cdots \ . \tag{15a}$$

The contribution of this bond to the force between two planes is

$$F = -\frac{dU}{d\,\Delta R} = -\left(\frac{dU}{dR}\right)_{R_0} - \left(\frac{d^2U}{dR^2}\right)_{R_0} \Delta R + \cdots \ , \tag{15b}$$

but we are not concerned with the term $-(dU/dR)_{R_0}$ because it is independent of ΔR and also because, when summed over all planes that interact with a given plane, the total force must be zero at the equilibrium separation. Now a force constant C is defined by $F = -C\,\Delta R$, so that

$$C = \left(\frac{d^2U}{dR^2}\right)_{R_0} \tag{15c}$$

gives the contribution of this pair of atoms to the force constant. A plot of C

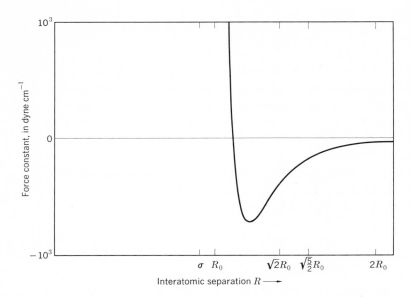

Figure 12 Plot of the force constant $C = \partial^2 U/\partial R^2$ between two atoms of argon, according to the Lennard-Jones potential. The separations at the 1st, 2nd, 3rd, and 4th nearest neighbor distances in the crystal are marked on the horizontal axis. The force constant between a pair of atoms at $R_0 = 1.11\sigma$, the equilibrium separation in the crystal, is $+6.6 \times 10^3$ dyne cm^{-1} or 6.6 N m^{-1}. (Courtesy of R. Gray.)

versus R for the Lennard-Jones potential for argon is given in Fig. 12. To obtain C_p itself we must sum contributions from all pairs of atoms in the two planes, and then divide by the number of atoms in one plane.

The equation of motion of the plane s is

$$M \frac{d^2 u_s}{dt^2} = \sum_p C_p (u_{s+p} - u_s) \; , \qquad (16)$$

where M is the mass of an atom. In the summation p runs over all positive and negative integers.

We look for a solution to (16) of the form of a traveling wave:

$$u_{s+p} = u\, e^{i(s+p)Ka} e^{-i\omega t} \qquad (17)$$

where a is the spacing[8] between planes and K is the wavevector. Then (16) reduces to

$$-\omega^2 M u e^{isKa} e^{-i\omega t} = \sum_p C_p (e^{i(s+p)Ka} - e^{isKa}) u e^{-i\omega t} \; . \qquad (18)$$

[8] The value of the interplanar spacing for given lattice will depend on the direction of **K**.

We may cancel $ue^{isKa}e^{-i\omega t}$ from both sides, so that

$$\omega^2 M = -\sum_p C_p(e^{ipKa} - 1) \ . \tag{19}$$

As the primitive basis contains only one atom it follows from the translational symmetry that $C_p = C_{-p}$, and we may regroup (19) as

$$\omega^2 M = -\sum_{p>0} C_p(e^{ipKa} + e^{-ipKa} - 2) \ . \tag{20}$$

Using the identity $2\cos pKa \equiv e^{ipKa} + e^{-ipKa}$ we have the dispersion relation

$$\boxed{\ \omega^2 = \frac{2}{M} \sum_{p>0} C_p(1 - \cos pKa) \ . \ } \tag{21}$$

We notice that the slope of ω versus K is always zero at $K = \pm\pi/a$. We have

$$\frac{d\omega^2}{dK} = \frac{2}{M} \sum_{p>0} paC_p \sin pKa = 0 \tag{22}$$

at $K = \pm\pi/a$, for here $\sin pKa = \sin(\pm p\pi) = 0$. This result suggests that there is a special significance to phonon wavevectors that lie on the Brillouin zone boundaries, just as for the photon wavevectors discussed in Chapter 2.

If there are interactions only among nearest-neighbor planes, then (21) reduces to

$$\omega^2 = (2C_1/M)(1 - \cos Ka) \ . \tag{23a}$$

By a trigonometric identity this may be written as

$$\omega^2 = (4C_1/M)\sin^2 \tfrac{1}{2}Ka \ ; \qquad \omega = (4C_1/M)^{\frac{1}{2}}|\sin \tfrac{1}{2}Ka| \ . \tag{23b}$$

We arrange the sign of the square root so that the frequency ω is always positive for a stable lattice. We have plotted ω^2 versus Ka in Fig. 13a and ω versus Ka in Fig. 13b. Both curves are periodic functions of K, with period $2\pi/a$.

First Brillouin Zone

What range of K is physically significant for phonons? From (17) the ratio of the displacements of two successive planes is given by

$$\frac{u_{s+1}}{u_s} = \frac{ue^{i(s+1)Ka}e^{-i\omega t}}{ue^{isKa}e^{-i\omega t}} = e^{iKa} \ . \tag{24}$$

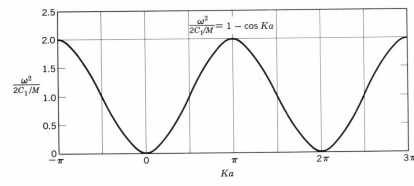

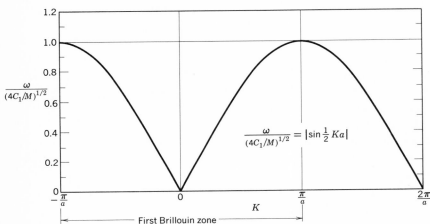

Figure 13a Plot of ω^2 versus K for lattice with interactions only between nearest neighbor planes. The interplanar force constant is C_1 and the interplanar spacing is a.

Figure 13b Plot of ω versus K for model of Fig. 12. The region of $K \ll 1/a$ or $\lambda \gg a$ corresponds to the continuum approximation; here ω is directly proportional to K.

The range $-\pi$ to $+\pi$ for the phase Ka of the exponential e^{iKa} covers all independent values of the exponential. There is absolutely no point in saying that two adjacent atoms are out of phase by more than $+\pi$: thus a relative phase of 1.2π, for example, is physically identical with a relative phase of -0.8π, and a relative phase of 4.2π is identical with 0.2π. We want both positive and negative values of K because waves can propagate to the right or to the left. Thus the range of independent values of K can be specified by[9]

$$-\pi \le Ka \le \pi , \qquad \text{or} \qquad -\frac{\pi}{a} \le K \le \frac{\pi}{a} . \tag{25}$$

This range of values of K is referred to as the **first Brillouin zone** of the linear lattice, as defined in Chapter 2. The extreme values of K in this zone are

$$K_{\max} = \pm \frac{\pi}{a}, \tag{26}$$

where $K_{\max}$ may be of the order of 10^8 cm^{-1}.

[9] Note that this represents a real difference from the behavior of an elastic continuum. In the continuum limit $a \to 0$ and $K_{\max} \to \pm\infty$.

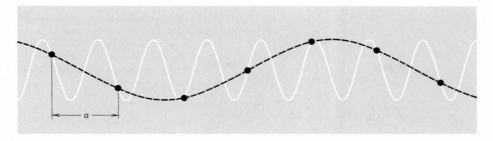

Figure 13c   The wave represented by the solid curve conveys no information not given by the dashed curve. Only wavelengths longer than $2a$ are needed to represent the motion. (Courtesy of P. Hansma.)

Suppose we use in (17) values of K outside of the first Brillouin zone (Fig. 13c). Such values merely reproduce lattice motions already described by values of K within the limits $\pm\pi/a$. We show that we may treat a value of K outside of these limits by subtracting the appropriate integral multiple of $2\pi/a$ that will give a wavevector inside these limits.

Suppose K lies outside the first zone, but a related wavevector K' defined by $K' \equiv K - 2\pi n/a$ lies within the first zone, where n is an integer. Then the displacement ratio (24) becomes

$$\frac{u_{s+p+1}}{u_{s+p}} = e^{iKa} \equiv e^{2\pi ni}e^{i(Ka-2\pi n)} \equiv e^{iK'a} \; , \qquad (27)$$

because $e^{i2\pi n} = 1$. Thus the displacement can always be described by a wavevector value lying within the first zone. We note that $2\pi n/a$ is a reciprocal lattice vector because $2\pi/a$ is a reciprocal lattice vector. By the subtraction of an appropriate reciprocal lattice vector from K, we always obtain an equivalent wavevector in the first zone.

At the boundaries $K_{\max} = \pm\pi/a$ of the Brillouin zone the solution

$$u_s = ue^{isKa}e^{-i\omega t}$$

does not represent a traveling wave, but a standing wave.[10] At the zone boundaries $sK_{\max}a = \pm s\pi$, whence

$$u_s = ue^{\pm is\pi}e^{-i\omega t} = u(-1)^s e^{-i\omega t} \; . \qquad (28)$$

This is a standing wave. We show below that the group velocity is zero. For this wave alternate atoms move in opposite phases, because $\cos s\pi = \pm1$ according to whether s is an even or an odd integer. The wave moves neither to the right nor to the left.

---

[10] We shall find this property in Chapter 9 for conduction electron wavefunctions at the zone boundaries.

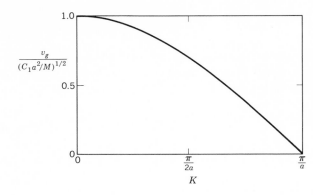

$$\frac{v_g}{(C_1 a^2/M)^{1/2}}$$

K

Figure 14 Group velocity v versus K, for model of Fig. 13. At the zone boundary the group velocity is zero. The range of K of laboratory-produced ultrasonic waves is too limited at present to be visible at the left of the graph.

This situation is equivalent to Bragg reflection of x-rays: when the Bragg condition is satisfied a traveling wave cannot propagate in a lattice, but through successive reflections back and forth a standing wave is set up. The critical value $K_{\max} = \pm\pi/a$ found here satisfies the Bragg condition $2d \sin\theta = n\lambda$: we have $\theta = \frac{1}{2}\pi$, $d = a$, $K = 2\pi/\lambda$, $n = 1$, so that $\lambda = 2a$. With x-rays it is possible to have n equal to other integers besides unity because the amplitude of the wave has a meaning in the space between atoms, but the displacement amplitude of an elastic wave has a meaning only at the atoms themselves.

Group Velocity

The velocity of a wave packet is the group velocity, given from physical optics as

$$v_g = \frac{d\omega}{dK}\;, \qquad \text{or} \qquad \mathbf{v}_g = \text{grad}_{\mathbf{K}}\,\omega(\mathbf{K}) \tag{29a}$$

in two or three dimensions, with $\text{grad}_{\mathbf{K}}$ as the gradient with respect to $\mathbf{K}$. The group velocity is the velocity of energy transmission in the medium. For the dispersion relation (23) the group velocity (Fig. 14) is

$$v_g = (C_1 a^2/M)^{\frac{1}{2}} \cos \tfrac{1}{2}Ka\;. \tag{29b}$$

Our general result (22) shows that the group velocity is zero at the edge of the zone. This is what we expect from a standing wave!

Long Wavelength or Continuum Limit

For $pKa \ll 1$ we have $\cos pKa \cong 1 - \frac{1}{2}(pKa)^2$, and the dispersion relation (21) becomes

$$\omega^2 = K^2 \left(\frac{a^2}{M}\right) \sum_{p>0} p^2 C_p\;. \tag{30}$$

But from Chapter 4 we know that $\omega^2 = K^2 \times$ (elastic stiffness/density). The p^2 term in the summation in (30) will tend to make long-range force components play an important role in determining the macroscopic elastic constants.

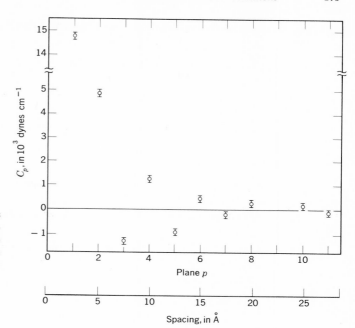

Figure 15   The interplanar force constants for longitudinal waves in the [100] direction in lead at 100°K, after Brockhouse et al., Phys. Rev. **128**, 1099 (1962). The horizontal axis gives the distance of the plane from the reference plane. (1×10^3 dyne cm^{-1} = 1 N m^{-1}.)

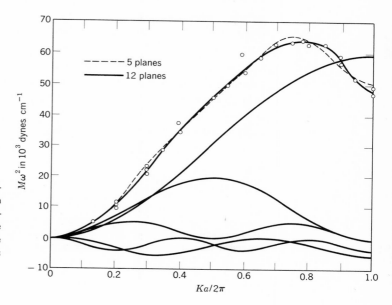

Figure 16   Values of $M\omega^2$ for the longitudinal branch in the [100] direction in Pb, plotted against the reduced wavevector. The fitted curves with twelve planes (good fit) and with five planes are shown. The first five Fourier components are also plotted.

Derivation of Force Constants from Experimental Dispersion Relation

In many metals the effective forces may be of quite long range. Effects have been found which connect planes of atoms separated by as many as twenty planes.[11] It is quite simple to make a statement about the range of the forces if we know the dispersion relation for ω.

We solve for the C_p by multiplying both sides of (21) by $\cos rKa$, where r is an integer, and integrating over the range of independent values of K:

$$M \int_{-\pi/a}^{\pi/a} dK \, \omega_K^2 \cos rKa = 2 \sum_{p>0} C_p \int_{-\pi/a}^{\pi/a} dK \, (1 - \cos pKa) \cos rKa$$

$$= -2\pi C_r/a \ . \quad (31a)$$

The integral vanishes except for $p = r$. Thus

$$C_p = -\frac{Ma}{2\pi} \int_{-\pi/a}^{\pi/a} dK \, \omega_K^2 \cos pKa \ . \quad (31b)$$

This important result[12] expresses the force constant from the pth plane of atoms in terms of the Fourier cosine transform of ω^2 as a function of K. It holds only for monatomic lattices.

LATTICE WITH TWO ATOMS PER PRIMITIVE CELL

With crystals having more than one atom per primitive cell the vibrational spectrum shows new features. We consider two atoms per primitive cell, as in the NaCl structure or the diamond structure. For each polarization mode in a given propagation direction the dispersion relation ω versus K develops two branches, known as the **acoustical** and **optical branches**, with phonons named to suit. We have longitudinal LA and transverse acoustical TA phonons, and longitudinal LO and transverse optical TO phonons, as in Fig. 17a.

If there are p atoms in the primitive cell, there will be $3p$ branches to the phonon dispersion relation: 3 acoustical branches and $3p - 3$ optical branches. Thus diamond with two carbon atoms in a primitive cell has six phonon branches: one LA, one LO, two TA, and two TO, as in Fig. 17b.

[11] J. M. Rowe, B. N. Brockhouse, and E. C. Svensson, Phys. Rev. Letters **14**, 554 (1965).

[12] The result is due to A. J. E. Foreman and W. M. Lomer, Proc. Phys. Soc. (London) **B70**, 1143 (1957). The application to the observed dispersion relations of lead (Figs. 15 and 16) is considered by Brockhouse et al., Phys. Rev. **128**, 1099 (1962). The origin of the long-range forces observed in several metals is discussed by W. A. Harrison, Phys. Rev. **129**, 2512 (1963) and by S. H. Koenig, Phys. Rev. **135**, A1693 (1964). The first treatment along these lines was given for sodium by T. Toya in 1958.

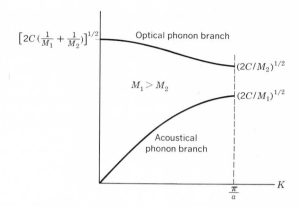

Figure 17a   Optical and acoustical phonon branches of the dispersion relation for a diatomic linear lattice, showing the limiting frequencies at $K = 0$ and $K = K_{max} = \pi/a$. The lattice constant is a.

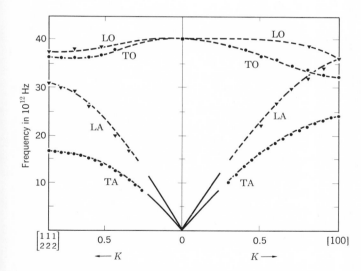

Figure 17b   Experimental dispersion curves ν versus K for diamond in the (100) and (111) directions, where K is the reduced wavevector in units of π/a. Notice the existence of optical and acoustical branches, as appropriate to a crystal with two atoms (even though identical) per primitive cell. The right-hand half of the figure is for phonons in the (100) direction, and the left-hand half is for phonons in the (111) direction. In the propagation directions shown the transverse modes are doubly-degenerate: there are two independent polarization directions for each point on the TA and TO curves. (After J. L. Warren, R. G. Wenzel, and J. L. Yarnell, *Inelastic scattering of neutrons*, IAEA, Vienna, 1965.)

We consider a cubic crystal where atoms of mass M_1 lie on one set of planes and atoms of mass M_2 lie on planes interleaved between those of the first set (Fig. 17c). It is not essential that the masses be different, but either the force constants or the masses will be different if the two atoms of the basis are in fact nonequivalent. Let a denote the repeat distance of the lattice in the direction normal to the lattice planes considered. We consider only waves that propagate in a symmetry direction for which a single plane contains only a single type of ion; such directions are [111] in the NaCl structure and [100] in the CsCl structure.

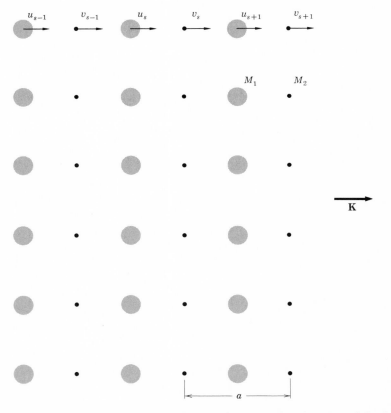

Figure 17c A diatomic crystal structure with masses M_1, M_2 connected by force constant C between adjacent planes. The displacements of atoms M_1 are denoted by u_{s-1}, u_s, u_{s+1}, ..., and of atoms M_2 by v_{s-1}, v_s, v_{s+1}. The repeat distance is a in the direction of the wavevector K. The atoms are shown in their undisplaced positions.

We write the equations of motion under the assumption that each plane interacts only with its nearest-neighbor planes and that the force constants are identical between all pairs of nearest-neighbor planes. We refer to Fig. 17c to obtain

$$M_1 \frac{d^2 u_s}{dt^2} = C(v_s + v_{s-1} - 2u_s) \ ;$$

$$M_2 \frac{d^2 v_s}{dt^2} = C(u_{s+1} + u_s - 2v_s) \ . \tag{32}$$

We look for a solution in the form of a traveling wave, now with different amplitudes u, v on alternate planes:

$$u_s = u \, e^{isKa} e^{-i\omega t} \ ; \qquad v_s = v \, e^{isKa} e^{-i\omega t} \ . \tag{33}$$

On substitution of (33) in (32) we have

$$-\omega^2 M_1 u = Cv(1 + e^{-iKa}) - 2Cu \ ;$$

$$-\omega^2 M_2 v = Cu(e^{iKa} + 1) - 2Cv \ . \tag{34}$$

These homogeneous linear equations in the two unknowns u, v have a nontrivial solution only if the determinant of the coefficients of u and v vanishes:

$$\begin{vmatrix} 2C - M_1\omega^2 & -C(1 + e^{-iKa}) \\ -C(1 + e^{iKa}) & 2C - M_2\omega^2 \end{vmatrix} = 0 \ , \tag{35}$$

or

$$M_1 M_2 \omega^4 - 2C(M_1 + M_2)\omega^2 + 2C^2(1 - \cos Ka) = 0 \ . \tag{36}$$

We can solve this equation exactly for ω^2, but it is simpler to examine the limiting cases $Ka \ll 1$ and $Ka = \pm\pi$ at the zone boundary. For small Ka we have $\cos Ka \cong 1 - \frac{1}{2}K^2 a^2 + \cdots$, and the two roots of (36) are

$$\omega^2 \cong 2C\left(\frac{1}{M_1} + \frac{1}{M_2}\right) \qquad \text{(optical branch)}; \tag{37}$$

$$\omega^2 \cong \frac{\frac{1}{2}C}{M_1 + M_2} K^2 a^2 \qquad \text{(acoustical branch)}. \tag{38}$$

The range of the first Brillouin zone is $-\pi/a \leq K \leq \pi/a$, where a is the repeat distance of the lattice. At $K_{\max} = \pm\pi/a$ the roots are

$$\omega^2 = \frac{2C}{M_1} \ ; \qquad \omega^2 = \frac{2C}{M_2} \ . \tag{39}$$

The dependence of ω on K is shown in Fig. 17a for $M_1 > M_2$.

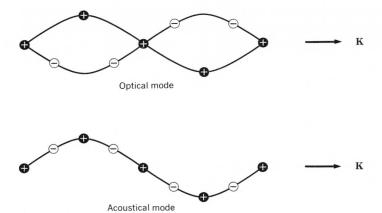

Figure 18 Transverse optical and transverse acoustical waves in a diatomic linear lattice, illustrated by the particle displacements for the two modes at the same wavelength.

The particle displacements in the transverse acoustical (TA) and transverse optical (TO) branches are shown in Fig. 18. For the optical branch at $K = 0$ we find, from (34) and (37),

$$\frac{u}{v} = -\frac{M_2}{M_1} . \tag{40}$$

The atoms vibrate against each other, but their center of mass is fixed. If the two atoms carry opposite charges, we may excite a motion of this type with the electric field of a light wave,[13] so that the branch is called the optical branch.[14] If a photon is absorbed by the crystal with the creation of a single phonon,[15] then by wavevector conservation $\mathbf{k}_{photon} = \mathbf{K}_{phonon}$. The photon wavevectors at the relevant frequencies ($\sim 10^{13}$ Hz) are of the order of 10^3 cm^{-1}, but phonon wavevectors run up to 10^8 cm^{-1}. Thus phonons excited by photons in direct processes have small wavevectors.

Figure 17a shows that wavelike solutions do not exist for certain frequencies, here between $(2C/M_1)^{\frac{1}{2}}$ and $(2C/M_2)^{\frac{1}{2}}$. This is a characteristic feature of elastic waves in polyatomic lattices. There is a **frequency gap** at the boundary $K_{max} = \pm\pi/a$ of the first Brillouin zone. If we look for solutions in the gap with ω real, then the wavevector K will be complex, so that the wave is damped in space. The same effect is demonstrated for photons in Advanced Topic A.

---

[13] Effects of the magnetic field of the light wave are weaker because the magnetic force on a charge involves v/c, where v is the velocity of an ion in the lattice.

[14] Absorption frequencies in the optical branch lie in the infrared portion of the spectrum.

[15] This process is different from that considered earlier in which the photon is scattered by creating or absorbing a phonon.

Another solution besides (40) for the amplitude ratio at small K is $u = v$; the atoms (and their center of mass) move together, as in the long wavelength acoustical vibrations of Eq. (38); hence the term acoustical branch.

OPTICAL PROPERTIES IN THE INFRARED

We consider the response to infrared photons of a diatomic crystal of ions of charge $\pm e$. In the long wavelength or $K = 0$ limit the displacements u_s, v_s are independent of the index s. The equations of motion in a local electric field $Ee^{-i\omega t}$ are, on adding a force term to (34),

$$-\omega^2 M_1 u = 2C(v - u) + eE \;\; ;$$
$$-\omega^2 M_2 v = 2C(u - v) - eE \;\; . \tag{41}$$

We divide the first equation by M_1 and the second by M_2. We subtract the second equation from the first to obtain

$$u - v = \frac{eE/\mu}{\omega_T{}^2 - \omega^2} \;\; , \tag{42}$$

with the definitions

$$\frac{1}{\mu} \equiv \frac{1}{M_1} + \frac{1}{M_2} \;\; ; \qquad \omega_T{}^2 \equiv \frac{2C}{\mu} \;\; . \tag{43}$$

Here μ is the reduced mass of an ion pair, and ω_T is the frequency in the optical branch as $K \to 0$. Equation (42) exhibits a resonance at $\omega = \omega_T$.

The application of the electric field causes the positive and negative ions to be displaced in opposite directions, thereby polarizing the crystal. The dielectric polarization $\mathbf{P}$ is defined as the dipole moment per unit volume. If there are N positive and N negative ions per unit volume, then the contribution to the polarization from the relative displacement of the ions is

$$P(\text{ionic}) = Ne(u - v) = \frac{Ne^2/\mu}{\omega_T{}^2 - \omega^2}E \;\; . \tag{44}$$

Note the resonant behavior at $\omega = \omega_T$. In (41) to (44) the field E is the local electric field[16] and is not necessarily identical with the macroscopic average electric field of Maxwell's equations, as discussed in Chapter 13. However, Eq. (44) always leads to a frequency-dependent dielectric function of the general form

$$\epsilon(\omega) = \epsilon(\infty) + \frac{\omega_T{}^2}{\omega_T{}^2 - \omega^2} \cdot [\epsilon(0) - \epsilon(\infty)] \;\; , \tag{45}$$

---

[16] We show in Chapter 13 that the local field at a cubic site is $\mathbf{E} + 4\pi\mathbf{P}/3$ in CGS or $\mathbf{E} + \mathbf{P}/3\epsilon_0$ in SI.

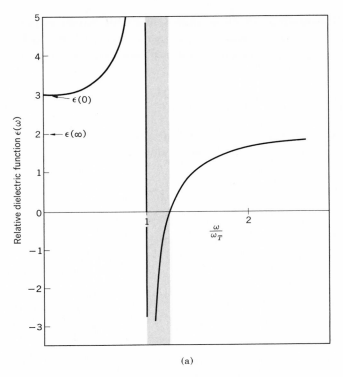

(a)

Figure 19a Plot of $\epsilon(\omega)$ from (45) for $\epsilon(\infty) = 2$ and $\epsilon(0) = 3$. The dielectric constant is negative between $\omega = \omega_T$ and $\omega = \omega_L = (3/2)^{\frac{1}{2}}\omega_T$; that is, between the pole (infinity) of $\epsilon(\omega)$ and the zero of $\epsilon(\omega)$. Incident electromagnetic waves of frequencies such that $\omega_T < \omega < \omega_L$ will not propagate in the medium, but will be reflected at the boundary of the medium. Here all ϵ's are relative dielectric functions.

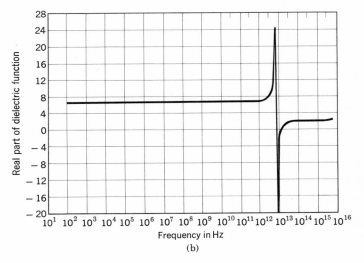

(b)

Figure 19b Relative dielectric function (real part) of SrF_2 as measured over a wide frequency range, exhibiting the decrease of the ionic polarizability at high frequencies. (Courtesy of A. von Hippel.)

as plotted in Fig. 19a. This result applies to a cubic crystal with two atoms per primitive cell. Here $\epsilon(0)$ is the static dielectric constant and $\epsilon(\infty)$ is the high frequency dielectric function, but defined to include the core electron contribution. Experimental results are shown in Fig. 19b.

ZEROS AND POLES OF THE DIELECTRIC FUNCTION

An unexpected consequence of the dielectric function (45) is that electromagnetic waves will not propagate in a forbidden frequency region

$$\omega_T^2 < \omega^2 < \omega_L^2 \equiv \omega_T^2 \cdot \frac{\epsilon(0)}{\epsilon(\infty)} \ . \tag{46}$$

The relation on the right defines the frequency ω_L; we give the significance of ω_L below. With this definition (45) may be written as

$$\epsilon(\omega) = \epsilon(\infty)\left(\frac{\omega_L^2 - \omega^2}{\omega_T^2 - \omega^2}\right) \ . \tag{47}$$

At frequencies between ω_T and ω_L the dielectric function is negative.

But the relation between frequency and wavevector for electromagnetic waves in a dielectric is

$$\text{(CGS)} \quad c^2 K^2 = \epsilon(\omega)\omega^2 \ ; \qquad \boxed{\text{(SI)} \quad c^2 K^2 = \epsilon(\omega)\epsilon_0\omega^2} \ , \tag{48}$$

as follows from the electromagnetic wave equation. If ω is real and $\epsilon(\omega)$ is negative, then K must be imaginary, and the wave assumes the form

$$e^{iKx} \rightarrow e^{-|K|x} \ , \tag{49}$$

damped in space and nonpropagating. Such a wave can be transmitted only through a thin slab of thickness of the order of $1/|K|$. Hence the phrase **forbidden gap**: waves incident on the crystal in the frequency region between ω_T and ω_L will be reflected. The forbidden frequency gap is quite different from the gap involved in Bragg reflection, because the infrared reflection has nothing to do directly with the periodicity of the crystal lattice.

The upper bound of the forbidden band of frequencies falls at ω_L, the zero of (47):

$$\epsilon(\omega_L) = 0 \ . \tag{50}$$

The frequency ω_L defined by (46) or (50) may be identified with the **longitudinal optical phonon frequency** for small K. We may establish this surprising connection quite generally as follows. The Maxwell equation

$$\text{div } \mathbf{D} = 0 \qquad \text{or} \qquad \epsilon(\omega) \text{ div } \mathbf{E} = 0 \tag{51}$$

has two types of roots. For one type of root div $\mathbf{E} = 0$ is always satisfied if the wavevector **K** is perpendicular to **E**, as in a transverse optical mode, Fig. 19c. The other type of root arises when $\epsilon(\omega) = 0$, as in (50): for this root **D** = 0, characteristic of a longitudinal optical mode.[17]

By this identification of the roots the definition (46) of ω_L becomes a meaningful physical result:

$$\frac{\omega_T{}^2}{\omega_L{}^2} = \frac{\epsilon(\infty)}{\epsilon(0)} , \tag{52}$$

where ω_T is the TO frequency and ω_L is the LO frequency, both for small K. The relation (52) is known as the **Lyddane-Sachs-Teller relation.**[18] Notice that $\epsilon(0) \to \infty$ as $\omega_T \to 0$; this has implications for ferroelectricity (Chapter 14).

What is the physical significance of a pole (an infinity) of the dielectric function $\epsilon(\omega)$? We saw that the pole at $\omega = \omega_T$ is the frequency of a transverse optical mode before the introduction of coupling with transverse electromagnetic waves. With coupling the significance of the pole at ω_T remains unchanged, provided that we look at wavevectors high enough to get us out of the crossover region where the phonon and photons are strongly mixed together (see Fig. 20, p. 186).

For a free mechanical transverse vibration the polarization P will be much larger than the electric field E, because in a transverse polarization wave the electric field arises by coupling with the electromagnetic field, and the coupling is large only near the crossover region. Because

$$(\text{CGS}) \quad P = \frac{\epsilon(\omega) - 1}{4\pi} E ; \qquad (\text{SI}) \quad P = [\epsilon(\omega) - 1]\epsilon_0 E , \tag{53}$$

we obtain a "mechanical" mode when $\epsilon(\omega) \gg 1$, essentially a pole of $\epsilon(\omega)$. Thus **the poles of the dielectric function are associated with transverse optical phonons,** for wavevectors above 10^4 cm^{-1}.

---

[17] We offer two arguments. By the geometry of a longitudinal polarization wave there is a depolarization field **E** = -4π**P**, as discussed in Chapter 13; thus **D** $\equiv$ **E** + 4π**P** = 0. In SI units, **D** = ϵ_0**E** + **P** = 0. Another argument is that there is no magnetic component in a longitudinal wave, so that the Maxwell equation

$$(\text{CGS}) \; \text{curl } \mathbf{H} = \frac{1}{c}\frac{\partial \mathbf{D}}{\partial t} = -\frac{i\omega\epsilon(\omega)}{c}\mathbf{E} ; \qquad (\text{SI}) \quad \text{curl } \mathbf{H} = -i\omega\epsilon(\omega)\epsilon_0\mathbf{E} ,$$

is satisfied for **H** = 0 only if $\epsilon(\omega) = 0$. In the SI equation, $\epsilon(\omega)$ denotes the relative dielectric constant or the dielectric function, and has the same value as in the CGS equation.

[18] The extension to complex structures has been treated by several workers; see, especially, A. S. Barker, Jr., "Infrared dielectric behavior of ferroelectric crystals," in *Ferroelectricity*, E. F. Weller, ed., Elsevier, 1967. Barker gives an interesting derivation of the LST relation by a causality argument.

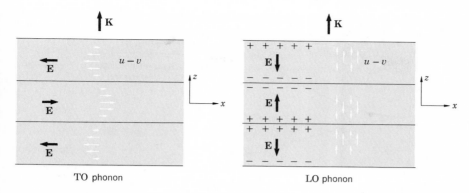

TO phonon   LO phonon

Figure 19c   The sets of arrows represent the relative displacements of the positive and negative ions at one instant of time for a wave in an optical mode traveling along the z axis. The planes of nodes (zero displacement) are shown; for long wavelength phonons the nodal planes are separated by many planes of atoms. In the transverse optical phonon mode pictured on the left the particle displacement is perpendicular to the wavevector $\mathbf{K}$; the macroscopic electric field in an infinite medium will lie only in the $\pm x$ direction for the mode shown, and by the symmetry of the problem $\partial E_x/\partial x = 0$. It follows that div $\mathbf{E} = 0$ for a TO phonon. In the longitudinal optical phonon mode on the right the particle displacements and hence the dielectric polarization $\mathbf{P}$ are parallel to the wavevector. The macroscopic electric field $\mathbf{E}$ satisfies $\mathbf{D} = \mathbf{E} + 4\pi\mathbf{P} = 0$ in CGS or $\epsilon_0\mathbf{E} + \mathbf{P} = 0$ in SI; by symmetry $\mathbf{E}$ and $\mathbf{P}$ are parallel to the z axis, and $\partial E_z/\partial z \neq 0$. Thus div $\mathbf{E} \neq 0$ for an LO phonon, and $\epsilon(\omega)$ div $\mathbf{E}$ is zero only if $\epsilon(\omega) = 0$.

We compare values of ω_L/ω_T obtained by inelastic neutron scattering[19] with experimental values of $[\epsilon(0)/\epsilon(\infty)]^{\frac{1}{2}}$ obtained by dielectric measurements:

	NaI	KBr	GaAs
ω_L/ω_T	1.44 ± 0.05	1.39 ± 0.02	1.07 ± 0.02
$[\epsilon(0)/\epsilon(\infty)]^{\frac{1}{2}}$	1.45 ± 0.03	1.38 ± 0.03	1.08

The agreement with the prediction of (52) is excellent.

[19] A. D. B. Woods et al., Phys. Rev. **131**, 1025 (1963); J. L. T. Waugh and G. Dolling, Phys. Rev. **132**, 2410 (1963).

186

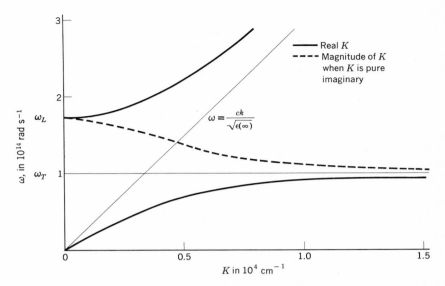

Figure 20 Coupled modes of photons and transverse optical phonons in an ionic crystal. The fine horizontal line represents oscillators of frequency ω_T in the absence of coupling to the electromagnetic field, and the fine line labeled $\omega = cK/\sqrt{\epsilon(\infty)}$ corresponds to electromagnetic waves in the crystal, but uncoupled to the lattice oscillators ω_T. The heavy lines are the dispersion relations in the presence of coupling between the lattice oscillators and the electromagnetic wave. The coupling increases the dielectric function by

$$\frac{\omega_T{}^2}{\omega_T{}^2 - \omega^2}[\epsilon(0) - \epsilon(\infty)] \ ,$$

as in Eq. (45). One effect of the coupling is to create the frequency gap between ω_L and ω_T: within this gap the wavevector is pure imaginary of magnitude given by the broken line in the figure. In the gap the wave attenuates as $e^{-|K|x}$, and we see from the plot that the attenuation is much stronger near ω_T than near ω_L. The range of K in this figure is of the order of 10^4 cm^{-1}, which is to the extreme left of the dispersion relations in Fig. 17a which are for oscillators not coupled to the electromagnetic field. In the present figure no acoustical phonons are shown nor are LO phonons shown, for neither are coupled to the transverse electromagnetic field. The quanta of the coupled photon-phonon modes are called **polaritons**.

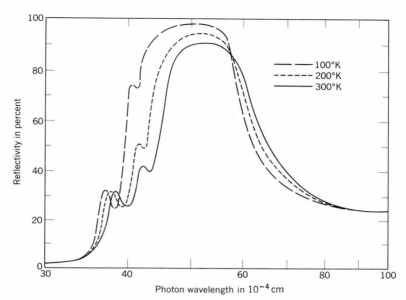

Figure 21 Reflectivity of a thick crystal of NaCl at several temperatures, versus wavelength. The nominal values of ω_L and ω_T at room temperature correspond to wavelengths of 38 and 61×10^{-4} cm, respectively. [After A. Mitsuishi et al., J. Opt. Soc. Am. **52**, 14 (1962).]

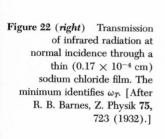

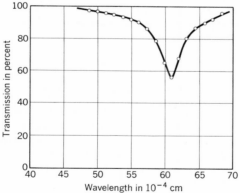

Figure 22 (*right*) Transmission of infrared radiation at normal incidence through a thin (0.17×10^{-4} cm) sodium chloride film. The minimum identifies ω_T. [After R. B. Barnes, Z. Physik **75**, 723 (1932).]

In Fig. 20 we show the solutions of the dispersion relation $\omega^2 \epsilon(\omega) = c^2 K^2$ for electromagnetic waves in matter. There are two branches, with a forbidden frequency gap between ω_T and ω_L. Undamped electromagnetic waves with frequencies within the gap cannot propagate in a thick crystal. The reflectivity of a crystal surface is expected to be high in this frequency region, as in Fig. 21. For films of thickness less than a wavelength the situation is changed: because for frequencies in the gap the wave attenuates as $e^{-|K|x}$ it is possible for the radiation to be transmitted through a film for small values of $|K|$, but for the large values of $|K|$ near ω_T the wave will be reflected, as in Fig. 22. By reflection at nonnormal incidence the frequency ω_L of longitudinal optical phonons can be observed, as in Fig. 23.

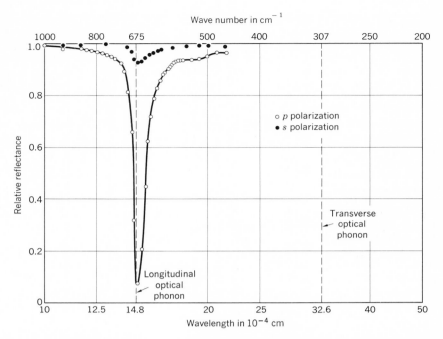

Wave number in cm^{-1}

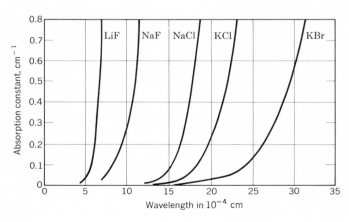

Figure 23 (*above*)   Reflectance versus wavelength of a LiF film backed by silver, for radiation incident near 30°. The longitudinal optical phonon absorbs strongly the radiation polarized (*p*) in the plane normal to the film, but absorbs hardly at all the radiation polarized (*s*) parallel to the film. [After D. W. Berreman, Phys. Rev. **130**, 2193 (1963).]

Figure 24 Infrared absorption of crystals. In the plot the absorption constant is the quantity α in the expression $I(x) = I_0 10^{-\alpha x}$ for the intensity. (After G. Joos.)

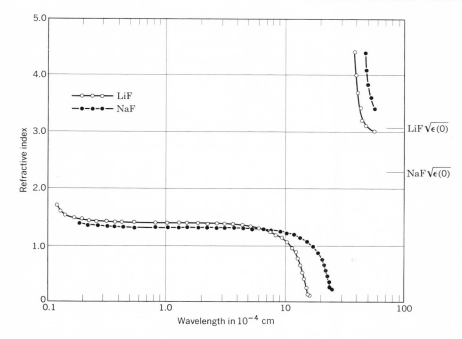

Figure 25 Wavelength dependence of the index of refraction of crystals of LiF and NaF. [After H. W. Hohls, Ann. Physik **29**, 433 (1937).]

The wavelength corresponding to ω_T is known as the residual ray wavelength or, from the German, the *Reststrahl* wavelength. The dispersive aspect of the optical properties of ionic crystals is applied in prisms in infrared spectroscopy. The absorption coefficients and refractive indices of a number of crystals in the infrared are shown in Figs. 24 and 25. Large single crystals are produced commercially for use in prisms, and also for use as windows and lens elements. For some purposes it is desired to have ω_T at as low frequency as possible: then the masses of the constituent atoms should be as heavy as practical. A mixed thallium bromide–thallium iodide crystal known as KRS-5 is widely used because of the heavy masses of the atoms.

Experimental values of $\epsilon(0)$, $\epsilon(\infty)$, and ω_T are given in Table 1, largely after a review by E. Burstein. The values of ω_L are calculated from the other data using the LST relation, Eq. (52). Values of ω_T at a temperature of 4°K have been given by Jones et al.[20]

[20] G. O. Jones et al., Proc. Roy. Soc. (London) **A261**, 10 (1961).

Table 1   Infrared lattice vibration parameters of NaCl and CsCl type crystals.

(At room temperature)

Crystal	Static dielectric constant $\epsilon(0)$	Optical dielectric constant $\epsilon(\infty)$	ω_T in 10^{13} rad s^{-1} experimental	ω_L in 10^{13} rad s^{-1} calculated from three preceding columns by LST relation
LiH	12.9	3.6	11.	21.
LiF	8.9	1.9	5.8	12.
LiCl	12.0	2.7	3.6	7.5
LiBr	13.2	3.2	3.0	6.1
NaF	5.1	1.7	4.5	7.8
NaCl	5.9	2.25	3.1	5.0
NaBr	6.4	2.6	2.5	3.9
KF	5.5	1.5	3.6	6.1
KCl	4.85	2.1	2.7	4.0
KI	5.1	2.7	1.9	2.6
RbF	6.5	1.9	2.9	5.4
RbI	5.5	2.6	1.4	1.9
CsCl	7.2	2.6	1.9	3.1
CsI	5.65	3.0	1.2	1.6
TlCl	31.9	5.1	1.2	3.0
TlBr	29.8	5.4	0.81	1.9
AgCl	12.3	4.0	1.9	3.4
AgBr	13.1	4.6	1.5	2.5
MgO	9.8	2.95	7.5	14.

LOCAL PHONON MODES

The phonon spectrum in a crystal is modified by lattice defects and impurity atoms. Consider the substitution of a light ion for a heavy ion, such as an H$^-$ ion substituted for a Cl$^-$ ion in a KCl crystal (Fig. 26). This impurity is called a U-center. Physical insight suggests that there is a high frequency mode of motion in which the light H$^-$ ion moves back and forth in the heavy cage of K$^+$ ions by which it is surrounded. This mode has an electric dipole moment. The crystal lattice near the H$^-$ ion will be deformed slightly during the motion, but the amplitude of the deformation should decrease rapidly with the distance from the H$^-$ ion. Such a vibration is called a **localized phonon.** The earliest theoretical studies of localized phonons were by Lifshitz.[21]

---

[21] Full references to the theoretical literature are found in Chap. 5 of A. A. Maradudin, E. W. Montroll, and G. H. Weiss, *Theory of lattice dynamics in the harmonic approximation,* Academic Press, 1963.

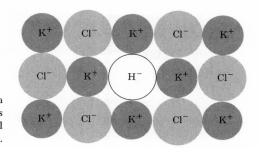

Figure 26 An H⁻ ion substituted for a Cl⁻ ion in a crystal of KCl; such an impurity center is known as a U-center. High frequency local phonon modes are associated with the H⁻ ions.

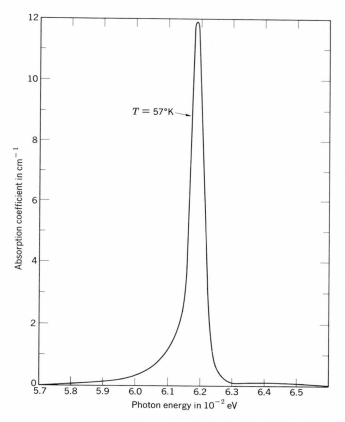

Figure 27 Infrared absorption by H⁻ ions in KCl. The concentration of H⁻ ions is 3×10^{17} cm⁻³. The photon energy at the center of the line corresponds to a wavelength of about 21×10^{-4} cm. [After G. Schaefer, J. Phys. Chem. Solids **12**, 233 (1960).]

Experiments on optical absorption due to localized phonons associated with H⁻ ions in alkali halides have been carried out by Schaefer, whose results for KCl are given in Fig. 27. Local phonons have also been observed by neutron scattering.

The simplest local phonon problem is that of a linear lattice of atoms all of mass M, except for one atom of mass $M' < M$. We shall show that one of the normal modes of the lattice is localized around the light atom, and the corresponding frequency is raised above the upper limit ω_{max} of the unperturbed lattice. We assume only nearest-neighbor interactions, and that these are the same between M' and M as between M and M.

Let the light atom be at the origin, $s = 0$. The equations of motion for the lattice are:

$$M' \frac{d^2 u_0}{dt^2} = C(u_1 + u_{-1} - 2u_0) \; ; \tag{54}$$

$$M \frac{d^2 u_1}{dt^2} = C(u_2 + u_0 - 2u_1) \; ; \quad \text{etc.} \tag{55}$$

We look for a solution which is exponentially damped as we go away from $s = 0$, and which in the limit $M' \to M$ approaches the form of the highest frequency normal mode of the unperturbed lattice. The solution at the zone boundary for the unperturbed lattice is given by (28): $u_s = u(0) \cos s\pi \, e^{-i\omega t} \equiv u(0)(-1)^s e^{-i\omega t}$. For the perturbed lattice let us try

$$u_s = u_0(-1)^s e^{-i\omega t} e^{-|s|\alpha} \; , \tag{56}$$

where α is to be determined. On substitution in (55) we find

$$\omega^2 = (C/M)(2 + e^{-\alpha} + e^{\alpha}) \; , \tag{57}$$

while on substitution in (54) we find

$$\omega^2 = (C/M')(2 + 2e^{-\alpha}) \; . \tag{58}$$

Equations (57) and (58) are consistent if $e^{\alpha} = (2M - M')/M'$, whence

$$\omega^2 = \omega_{max}{}^2 \cdot \frac{M^2}{2MM' - M'^2} \; , \tag{59}$$

where $\omega_{max} = (4C/M)^{\frac{1}{2}}$ is the cutoff frequency [see (23)] of the unperturbed lattice, for which $M = M'$. If $M' \ll M$, then (59) reduces to $\omega^2 = \omega_{max}{}^2(M/2M')$.

Other impurity modes[22] are possible in a diatomic crystal with two different atoms per primitive cell. Figure 28 shows the atomic motions in what are called local, gap, and resonant modes. We have already seen that a **local mode** may be formed when a light atom replaces a heavier atom, as when H^- replaces Cl^- in KCl. The frequency of a local mode is above the maximum phonon frequency of the pure crystal. In a **gap mode** the frequency lies within the gap

[22] Helpful references are given by A. J. Sievers, A. A. Maradudin, and S. S. Jaswal, Phys. Rev. **138**, A272 (1965) and K. F. Renk, Z. Physik **201**, 445 (1967).

between the acoustical and optical branches. A gap mode is observed when I^- replaces Cl^- in KCl.

Certain kinds of impurities, particularly very heavy ones, can give rise to quasilocalized **resonant modes** whose frequencies lie in the range of the allowed phonon frequencies of the perfect host crystal; such modes are characterized by a greatly enhanced amplitude of vibration of the impurity atom. This is observed when Ag^+ replaces K^+ in KI.

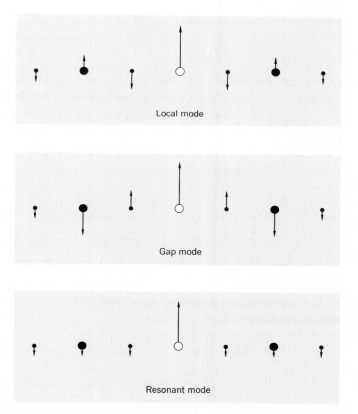

Figure 28 Particle amplitudes associated with local, gap, and resonant modes in a crystal with two atoms per primitive cell. The white circle denotes an impurity atom. Calculations of the displacement amplitudes for localized and gap modes are given by K. F. Renk, Z. Physik **201**, 445 (1967) and for resonant modes by R. Weber, *Thesis*, Freiburg, 1967.

SUMMARY

1. The quantum unit of a lattice vibration is a phonon. If the angular frequency is ω, the energy of the phonon is $\hbar\omega$.

2. When a phonon of wavevector $\mathbf{K}$ is created by the inelastic scattering of a photon or neutron from wavevector $\mathbf{k}$ to $\mathbf{k}'$, the wavevector selection rule which governs the process is

$$\mathbf{k} = \mathbf{k}' + \mathbf{K} + \mathbf{G} \ ,$$

where $\mathbf{G}$ is a reciprocal lattice vector.

3. All lattice waves can be described by wavevectors that lie within the first Brillouin zone in reciprocal space.

4. If there are p atoms in the primitive cell, the phonon dispersion relation will have 3 acoustical phonon branches and $3p - 3$ optical phonon branches.

5. In a diatomic crystal there is a forbidden gap getween ω_T and ω_L for coupled phonon-photon modes. Transverse waves will not propagate in the crystal for frequencies in this gap.

6. The Lyddane-Sachs-Teller relation

$$\frac{\omega_L{}^2}{\omega_T{}^2} = \frac{\epsilon(0)}{\epsilon(\infty)}$$

relates ω_L and ω_T to the dielectric constants at low and high frequencies.

7. The zeros of the dielectric function $\epsilon(\omega)$ or $\epsilon(\omega, K)$ are associated with longitudinal modes, and the poles are associated with transverse modes.

8. Localized phonon modes are associated with point and line defects and with surfaces of the crystal.

Problems

1. *Vibrations of square lattice.* We consider transverse vibrations of a planar square lattice of rows and columns of identical atoms, and let $u_{l,m}$ denote the displacement normal to the plane of the lattice of the atom in the lth column and mth row (Fig. 29). The mass of each atom is M, and C is the force constant for nearest neighbor atoms.

(a) Show that the equation of motion is

$$M(d^2u_{lm}/dt^2) = C[(u_{l+1,m} + u_{l-1,m} - 2u_{lm}) + (u_{l,m+1} + u_{l,m-1} - 2u_{lm})]$$

(b) Assume solutions of the form

$$u_{lm} = u(0)e^{i(lK_xa+mK_ya-\omega t)} \ ,$$

where a is the spacing between nearest-neighbor atoms. Show that the equation of motion is satisfied if

$$\omega^2 M = 2C(2 - \cos K_xa - \cos K_ya) \ .$$

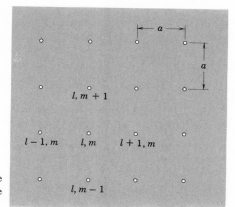

Figure 29 Square array of lattice constant a. The displacements considered are normal to the plane of the lattice.

This is the dispersion relation for the problem.

(c) Show that the region of **K** space for which independent solutions exist may be taken as a square of side $2\pi/a$. This is the first Brillouin zone of the square lattice. Sketch ω versus K for $K = K_x$ with $K_y = 0$, and for $K_x = K_y$.

(d) For $Ka \ll 1$, show that

$$\omega = (Ca^2/M)^{\frac{1}{2}}(K_x{}^2 + K_y{}^2)^{\frac{1}{2}} = (Ca^2/M)^{\frac{1}{2}}K \ .$$

2. *Monatomic linear lattice.* Consider a longitudinal wave

$$u_s = u \cos(\omega t - sKa)$$

which propagates in a monatomic linear lattice of atoms of mass M, spacing a, and nearest-neighbor interaction C.

(a) Show that the total energy of the wave is

$$E = \tfrac{1}{2}M \sum_s (du_s/dt)^2 + \tfrac{1}{2}C \sum_s (u_s - u_{s+1})^2 \ ,$$

where s runs over all atoms.

(b) By substitution of u_s in this expression, show that the time-average total energy per atom is

$$\tfrac{1}{4}M\omega^2 u^2 + \tfrac{1}{2}C(1 - \cos Ka)u^2 = \tfrac{1}{2}M\omega^2 u^2 \ ,$$

where in the last step we have used the dispersion relation (23a) for this problem.

3. *Continuum wave equation.* Show that for long wavelengths the equation of motion (16) reduces to the continuum elastic wave equation

$$\frac{\partial^2 u}{\partial t^2} = v^2 \frac{\partial^2 u}{\partial x^2} \ ,$$

where v is the velocity of sound.

4. *Optical phonons.* For optical phonons with $K = 0$ the relative displacement of adjacent ions on a diatomic lattice is given by $u - v$, in the notation of (33). Estimate the

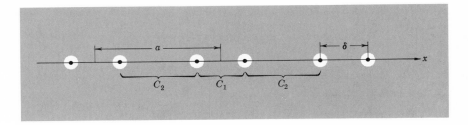

Figure 30 A linear lattice of atoms of equal mass; the primitive basis consists of two atoms separated by δ. The lattice constant is a. Two force constants are shown.

root-mean-square magnitude of $u - v$ in Å for 1 cm^3 of KCl, with 100 phonons excited in this mode.

5. **Momentum of phonon.** (a) Show from (17) that the linear momentum of a crystal in which a wave of wavevector K is excited is

$$\mathbf{p} = -i\omega M N \mathbf{u}\, e^{-i\omega t} \sum_{r=0}^{N-1} e^{irKa} \; ,$$

where the linear crystal consists of N atoms of mass M.
(b) For $K \neq 0$, show that the summation in part (a) is equal to

$$\frac{1 - e^{iNKa}}{1 - e^{iKa}} \; .$$

(c) Let us apply the periodic boundary condition $u_r = u_{r+N}$. Show that this restricts K to values such that $\exp(iNKa) = 1$. Using the result of part (b), we see that $\mathbf{p} = 0$ except for $K = 0$. Thus a phonon really carries zero momentum, except for $K = 0$.

6. **Basis of two identical atoms.** Determine and sketch the longitudinal phonon acoustical and optical spectrum of a linear lattice of primitive lattice constant a, having a basis of two identical atoms of mass M at equilibrium separation $\delta < \frac{1}{2}a$. Both atoms of the basis are on the line. The force constant is C_1 between the atoms of the basis, and C_2 between one atom of the basis and the nearer of the two atoms belonging to the nearest basis (Fig. 30). *Note:* the structure is not entirely unlike a linear lattice of molecular hydrogen.

7. **Basis of two unlike atoms.** For the problem treated by (32) to (39), find the amplitude ratios u/v for the two branches at $K_{\max} = \pi/a$. Show that at this value of K the two lattices act as if decoupled: one lattice remains at rest while the other lattice moves.

8. **Kohn anomaly.** We suppose that the interplanar force constant C_p in Eq. (15) is of the form

$$C_p = A\frac{\sin pk_0 a}{pa} \; ,$$

where A and k_0 are constants and p runs over all integers. Such a form is expected in metals. Use this and Eq. (21) to find an expression for ω^2 and also for $\partial \omega^2/\partial K$. Prove that $\partial \omega^2/\partial K$ is infinite when $K = k_0$. Thus a plot of ω^2 versus K or of ω versus K has a vertical tangent at k_0: there is a kink at k_0 in the phonon dispersion relation $\omega(K)$. A related effect was predicted by W. Kohn, Phys. Rev. Letters **2**, 393 (1959).

References

A. A. Maradudin, E. W. Montroll, and G. H. Weiss, "Theory of lattice dynamics in the harmonic approximation," *Solid state physics,* Supp. **3** (1963).

S. S. Mitra, "Vibrational spectra of solids," *Solid state physics* **13**, 1–80 (1962).

W. Cochran, "Lattice vibrations," Repts. Prog. Phys. **26**, 1 (1963).

M. Born and K. Huang, *Dynamical theory of crystal lattices,* Oxford, 1954.

L. Brillouin. *Wave propagation in periodic structures,* Dover, 1953.

T. A. Bak, ed., *Phonons and phonon interactions,* Benjamin, 1964.

R. F. Wallis, ed., *Lattice dynamics* (Copenhagen conference 1963), Pergamon, 1965.

P. W. Kruse, L. D. McGlauchlin, and R. B. McQuistan, *Elements of infrared technology,* Wiley, 1962.

D. E. McCarthy, "Reflection and transmission of infrared materials: 1, Spectra from 2–50 microns; 2, Bibliography," Applied Optics **2**, 591, 596 (1963).

G. Leibfried, "Gittertheorie der mechanischen und thermischen Eigenschaften der Kristalle," *Encyclo. of physics* **7/1**, 104–324 (1955).

R. W. H. Stevenson, ed., *Phonons* (Aberdeen 1965 summer school), Oliver and Boyd, 1966.

R. H. Enns and R. R. Haering, eds., *Phonons and their interactions,* Gordon and Breach, 1969.

A. Hadni, *L'infrarouge lointain,* Presse Universitaires, 1969. Review of research in the far infrared.

6

Thermal Properties
of Insulators

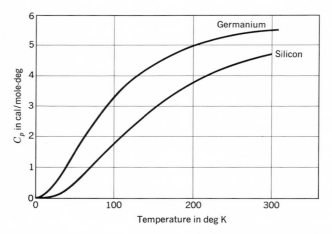

Figure 1　Heat capacity of silicon and germanium. Note the decrease at low temperatures. To convert to J/mole-deg, multiply by 4.186.

We discuss the Einstein and Debye approximations to the heat capacity associated with the lattice vibrations of crystals; the features of more exact calculations are indicated. We then consider effects of anharmonic lattice interactions, including thermal expansion, the Grüneisen relation, and the thermal conductivity of insulators. The thermal properties of metals are considered in Chapter 7, of superconductors in Chapter 12, and the thermal properties particular to magnetic materials are considered in Chapters 15 and 16.

LATTICE HEAT CAPACITY

By heat capacity we shall usually mean the heat capacity at constant volume, which is more fundamental than the heat capacity at constant pressure, which is what the experiments determine.[1] The heat capacity at constant volume is defined as

$$C_V \equiv T \left(\frac{\partial S}{\partial T} \right)_V = \left(\frac{\partial E}{\partial T} \right)_V , \tag{1}$$

where S is the entropy, E the energy, and T the temperature.

The experimental facts about the heat capacity of representative inorganic solids are these:

1. In the room-temperature range the value of the heat capacity of nearly all solids is close to $3Nk_B$, or 25 J mol^{-1} deg^{-1}.

2. At lower temperatures the heat capacity drops markedly (Fig. 1) and approaches zero as T^3 in insulators and as T in metals. If the metal becomes a superconductor, the drop is faster than T.

3. In magnetic solids there is a large contribution to the heat capacity over the temperature range in which the magnetic moments become ordered.[2] Below 0.1°K the ordering of nuclear moments may give very large heat capacities.

---

[1] A simple thermodynamic relation connects C_V and C_p:

$$C_p - C_V = 9\alpha^2 BVT ,$$

where α is the temperature coefficient of linear expansion, V the volume, and B the bulk modulus. The fractional difference between C_p and C_V is usually small and often may be neglected, particularly below room temperature.

[2] A change in the degree of order means a change in the entropy and thus a contribution to the heat capacity.

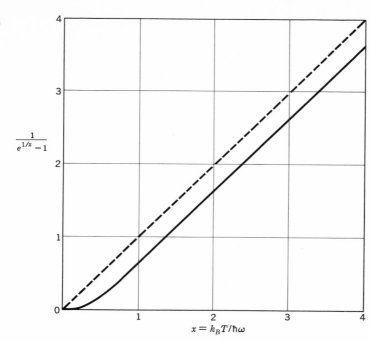

$$\frac{1}{e^{1/x}-1}$$

$$x = k_B T/\hbar\omega$$

Figure 2 Plot of Planck distribution function. Note that at high temperatures the occupancy of a state is approximately linear in the temperature. The function $\langle n \rangle + \frac{1}{2}$, which is not plotted, approaches the dashed line as asymptote at high temperatures. The dashed line is the classical limit.

The normal modes of vibration of a lattice are independent if Hooke's law is applicable. Thus the energy of a lattice mode depends only on its frequency ω and phonon occupancy n, and is independent of the occupancy of the other lattice modes. In thermal equilibrium at temperature T the occupancy is given by the Planck distribution

$$\langle n \rangle = \frac{1}{e^{\hbar\omega/k_B T} - 1} \text{ , or } \quad \langle n \rangle + \frac{1}{2} = \frac{1}{2} \text{ ctnh } \frac{\hbar\omega}{2k_B T} \text{ ,} \qquad (2)$$

where the $\langle \cdots \rangle$ denotes the average in thermal equilibrium and k_B is the Boltzmann constant. A graph of $\langle n \rangle$ is given in Fig. 2.

REVIEW: *Planck Distribution.*[3] Consider a set of identical harmonic oscillators in thermal equilibrium. The ratio of the number of oscillators in their $(n + 1)$th quantum state of excitation to the number in the nth quantum state is

$$N_{n+1}/N_n = e^{\hbar\omega/\tau} \text{ ,} \qquad \tau \equiv k_B T \text{ ,} \qquad (3)$$

according to the Boltzmann factor. Thus the fraction of the total number of oscillators in the nth quantum state is

$$\frac{N_n}{\sum\limits_{s=0}^{\infty} N_s} = \frac{e^{-n\hbar\omega/\tau}}{\sum\limits_{s=0}^{\infty} e^{-s\hbar\omega/\tau}} \qquad (4)$$

[3] See C. Kittel, *Thermal physics*, Wiley, 1969, Chapters 9, 15, 16.

From (4) we see that the average excitation quantum number of an oscillator is

$$\langle n \rangle = \frac{\sum_s s e^{-s\hbar\omega/\tau}}{\sum_s e^{-s\hbar\omega/\tau}} \quad . \tag{5}$$

The summation in the denominator of (5) is of the form,

$$\sum_s x^s = \frac{1}{1-x} \quad , \tag{6}$$

with $x = \exp(-\hbar\omega/\tau)$. The numerator in (5) is of the form

$$\sum_s s x^s = x \frac{d}{dx} \sum_s x^s = \frac{x}{(1-x)^2} \quad . \tag{7}$$

Thus we may rewrite (5) in the form of the Planck distribution:

$$\langle n \rangle = \frac{x}{1-x} = \frac{1}{e^{\hbar\omega/\tau} - 1} \quad . \tag{8}$$

We remarked in Fig. 2 that

$$\langle n \rangle \cong k_B T / \hbar\omega \tag{9}$$

if $\hbar\omega < k_B T$, because

$$e^{\hbar\omega/\tau} \cong 1 + (\hbar\omega/\tau) + \cdots . \tag{10}$$

When (9) is satisfied we say that the occupation is classical in the sense that each oscillator has energy $\langle n \rangle \hbar\omega \simeq k_B T$. This is roughly satisfied even for $\hbar\omega/k_B T \sim 1$. At low temperatures $\hbar\omega/k_B T \gg 1$, and we have instead

$$\langle n \rangle \cong e^{-\hbar\omega/\tau} \quad . \tag{11}$$

Einstein Model

The average energy of an oscillator of frequency ω is $\langle n \rangle \hbar\omega$. For N oscillators in one dimension, all having the same resonance frequency, the energy E is

$$E = N\langle n \rangle \hbar\omega = \frac{N\hbar\omega}{e^{\hbar\omega/\tau} - 1} \quad . \tag{12}$$

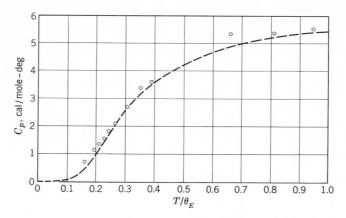

Figure 3 Comparison of experimental values of the heat capacity of diamond with values calculated on the Einstein model, using the characteristic temperature $\theta_E = \hbar\omega/k_B = 1320°$K. [After A. Einstein, Ann. Physik **22**, 180 (1907).] To convert to J/mole-deg, multiply by 4.186.

The heat capacity of the oscillators is

$$C_V = \left(\frac{\partial E}{\partial T}\right)_V = Nk_B \left(\frac{\hbar\omega}{\tau}\right)^2 \frac{e^{\hbar\omega/\tau}}{(e^{\hbar\omega/\tau} - 1)^2} \ , \tag{13}$$

as plotted in Fig. 3.

This is the result of the Einstein model for the contribution of N oscillators of the same resonance frequency to the heat capacity of a solid. In three dimensions N is replaced by $3N$ because each of the N atoms has three degrees of freedom; then the high-temperature limit of (13) becomes $3Nk_B$, the Dulong and Petit value. At low temperatures (13) decreases as in Fig. 3 but gives $C_V \propto e^{-\hbar\omega/\tau}$, whereas the experimental lattice contribution is known to be $C_V \propto T^3$, which can be accounted for by the Debye model discussed later. The limitation of the Einstein model is that all the elastic waves in a solid do not have the same frequency. At the time Einstein wanted to show[4] that mechanical oscillators should be quantized just as Planck had quantized radiation oscillators. Einstein's example was a quick demonstration of why the heat capacity of solids dropped to zero as $T \to 0$. But the Einstein model is often used to approximate part of the phonon spectrum, particularly the contribution of optical phonons.

Enumeration of Normal Modes

The energy in thermal equilibrium of a collection of oscillators of different frequencies ω_K is

$$E = \sum_K n_K \hbar\omega_K \ , \tag{14}$$

where each n_K is related to ω_K by the Planck distribution. It is often convenient to replace the summation in (14) by an integral. Suppose that the crystal has

[4] See the historical note by M. J. Klein, Physics Today, p. 38 (Jan. 1965).

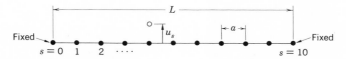

Figure 4   Elastic line of $N + 1$ atoms, with $N = 10$, for boundary conditions that the end atoms $s = 0$ and $s = 10$ are fixed. The particle displacements in the normal modes for either longitudinal or transverse displacements are of the form $u_s \propto \sin sKa$. This form is automatically zero at the atom at the end $s = 0$, and we choose K to make the displacement zero at the other end, $s = 10$.

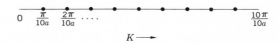

Figure 5   The boundary condition $\sin sKa = 0$ for $s = 10$ can be satisfied by choosing $K = \pi/10a$, $2\pi/10a, \ldots, 9\pi/10a$, where $10a$ is the length L of the line. The present figure is in K space. The dots are not atoms but are the allowed values of K. Of the $N + 1$ particles on the line, only $N - 1$ are allowed to move, and their most general motion can be expressed in terms of the $N - 1$ allowed values of K. This quantization of K has nothing to do with quantum mechanics but follows classically from the boundary conditions that the end atoms be fixed. (There are three types of polarization possible for each K value: two transverse, in which the particles move up and down in the plane of the page or in and out of the plane, and one longitudinal, in which the motion is parallel to the line of atoms.)

$\mathfrak{D}(\omega)\, d\omega$ modes of vibration in the frequency range ω to $\omega + d\omega$. Then the energy is

$$E = \int d\omega\; \mathfrak{D}(\omega)\; n(\omega,T)\; \hbar\omega \; ; \qquad (15)$$

the heat capacity is found from this by differentiation of $n(\omega,T)$ with respect to temperature. The central problem is to find $\mathfrak{D}(\omega)$, the number of modes per unit frequency range. This function is also called the density of modes.

Density of Modes in One Dimension

Let us first consider the elastic problem for a one-dimensional line (Fig. 4) of length L carrying $N + 1$ particles at separation a. We suppose that the particles $s = 0$ and $s = N$ at the ends of the line are held fixed. Each normal mode is a standing wave:

$$u_s = u(0)e^{-i\omega_K t} \sin sKa \; , \qquad (16)$$

where ω_K is related to K by the appropriate dispersion relation, as treated in Chapter 5. Here K is restricted by the fixed-end boundary conditions to the values

$$K = \frac{\pi}{L}\,, \quad \frac{2\pi}{L}\,, \quad \frac{3\pi}{L}\,, \ldots, \quad \frac{(N-1)\pi}{L}\,, \qquad (17)$$

as in Fig. 5.

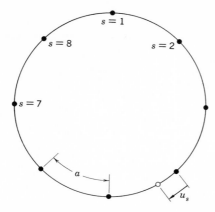

Figure 6 Consider N particles constrained to remain on a circular ring. The particles can oscillate if connected by elastic springs. In a normal mode the displacement u_s of atoms s will be of the form $\sin sKa$ or $\cos sKa$: these are independent modes. By the geometrical periodicity of the ring the boundary condition is that $u_{N+s} = u_s$ for all s, so that the NKa must be an integral multiple of 2π. For $N = 8$ the allowed independent values of K are 0, $2\pi/8a$, $4\pi/8a$, $6\pi/8a$, and $8\pi/8a$. The value $K = 0$ has a meaning only for the cosine form, because $\sin s0a = 0$. The value $8\pi/8a$ also has a meaning only for the cosine form, because $\sin (s8\pi a/8a) = \sin s\pi = 0$. The three other values of K are allowed for both the sine and cosine modes, giving a total of eight allowed modes for the eight particles. Thus the *periodic* boundary condition leads to one allowed mode per particle, exactly as for the fixed-end boundary condition of Fig. 5. If we had taken the modes in the complex form exp $(isKa)$, the periodic boundary condition would lead to the eight modes with $K = 0$, $\pm2\pi/Na$, $\pm4\pi/Na$, $\pm6\pi/Na$, and $8\pi/Na$, as in Eq. (20).

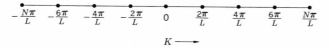

Figure 7 Allowed values of wavevector K for periodic boundary conditions applied to a linear lattice of periodicity $N = 8$ atoms on a line of length L. The $K = 0$ solution is the uniform mode. The special points $\pm N\pi/L$ represent only a *single* solution because $e^{i\pi s}$ is identical to $e^{-i\pi s}$; thus there are eight allowed modes, with displacements of the sth atom described by 1, $e^{\pm i\pi s/4}$, $e^{\pm i\pi s/2}$, $e^{\pm i3\pi s/4}$, $e^{i\pi s}$.

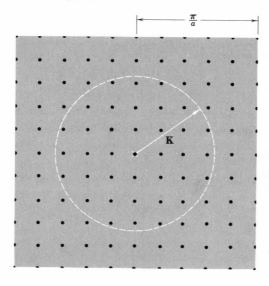

Figure 8 Allowed values in Fourier space of the phonon wavevector $\mathbf{K}$ for a square lattice of lattice constant a, with periodic boundary conditions applied over a square of side $L = 10a$. The uniform mode is marked with a cross. There is one allowed value of $\mathbf{K}$ per area element $(2\pi/10a)^2 = (2\pi/L)^2$, so that within the circle of area πK^2 the smoothed number of allowed points is $\pi K^2 (L/2\pi)^2$.

The solution for $K = \pi/L$ has

$$u_s \propto \sin(s\pi a/L) \qquad (18)$$

and vanishes for $s = 0$ and $s = N$ as required. The solution for $K = N\pi/L = \pi/a = K_{max}$ has

$$u_s \propto \sin s\pi \; ; \qquad (19)$$

this permits no motion of any atom, because $\sin s\pi$ vanishes at each atom. Thus there are $N - 1$ allowed independent values of K in (17). This number is equal to the number of particles allowed to move. Each allowed value of K is associated with a solution of the form (16). For the one-dimensional line of lattice constant a there is one mode for each interval $\Delta K = \pi/L$, so that the number of modes per unit range of K is L/π for $K \leq \pi/a$; and 0 for $K > \pi/a$.

There is another device for enumerating modes that is often used and that is equally valid. We consider the medium as unbounded, but require that the solutions be periodic over a large distance L, so that $u(sa) = u(sa + L)$. The method of **periodic boundary conditions** (Figs. 6 and 7) does not change the physics of the problem in any essential respect for a large system. Then in the running wave solution $u_s = u(0) \exp[i(sKa - \omega_K t)]$ the allowed values of K are

$$K = 0 \; , \quad \pm \frac{2\pi}{L} \; , \quad \pm \frac{4\pi}{L} \; , \quad \pm \frac{6\pi}{L} \; , \quad \ldots \; , \quad \frac{N\pi}{L} \qquad (20)$$

This method of enumeration gives the same number of modes (one per mobile atom) as given by (18), but we have now both plus and minus values of K, with the interval $\Delta K = 2\pi/L$ between successive values of K. For periodic boundary conditions the number of modes per unit range of K is $L/2\pi$ for $-\pi/a \leq K \leq \pi/a$; and 0 otherwise. If for convenience in counting we restrict K to positive values, we regain the earlier result L/π. The situation in a two-dimensional lattice is portrayed in Fig. 8.

We need to know $\mathfrak{D}(\omega)$, the number of modes per unit frequency range. The number of modes $\mathfrak{D}(\omega) \, d\omega$ in $d\omega$ at ω is given in one dimension by

$$\mathfrak{D}(\omega) \, d\omega = \frac{L}{\pi} \frac{dK}{d\omega} \, d\omega = \frac{L}{\pi} \cdot \frac{d\omega}{d\omega/dK} \; . \qquad (21)$$

We can obtain the group velocity $d\omega/dK$ from the dispersion relation ω versus K. There is a singularity in $\mathfrak{D}(\omega)$ whenever the dispersion relation $\omega(K)$ is horizontal; that is, whenever the group velocity is zero.

In the continuum or Debye approximation $\omega = vK$, so that $d\omega/dK = v$, the constant velocity of sound. In one dimension we have from (21) that

$$\mathfrak{D}(\omega) = \frac{L}{\pi v} \tag{22}$$

for $\omega \leq v\pi/a$ and $\mathfrak{D}(\omega) = 0$ otherwise. The spectrum is cut off at $\omega_D = v\pi/a$ in order to give the correct value N for the total number of modes. The result (22) is the density of modes for each polarization type. If there are three modes of different polarizations for each K value, then (22) is to be summed over the three polarizations, using the appropriate values of the velocity of sound v for each polarization.

For N Einstein oscillators at frequency ω_E we have

$$\mathfrak{D}(\omega) = N\,\delta(\omega - \omega_E) \ , \tag{23a}$$

where δ is the Dirac delta function. This function is defined by the property

$$\int_{-\infty}^{\infty} dx\, f(x)\, \delta(x - a) = f(a) \ , \tag{23b}$$

for any function $f(x)$. The delta function may be viewed as the limit of a very sharply peaked function.

The dispersion relation for a monatomic line of atoms with nearest-neighbor interactions is necessarily of the form of Eq. (5.23) for interacting parallel layers:

$$\omega = \omega_m |\sin \tfrac{1}{2} Ka| \ , \tag{24}$$

where a is the spacing and ω_m is the maximum frequency. We may solve this for K as a function of ω:

$$K = \frac{2}{a} \sin^{-1} \frac{\omega}{\omega_m} \ , \tag{25}$$

whence

$$\frac{dK}{d\omega} = \frac{2}{a} \cdot \frac{1}{(\omega_m{}^2 - \omega^2)^{\frac{1}{2}}} \ . \tag{26}$$

From (22) the density of modes is

$$\mathfrak{D}(\omega) = \frac{L}{\pi} \cdot \frac{dK}{d\omega} = \frac{2L}{\pi a} \cdot \frac{1}{(\omega_m{}^2 - \omega^2)^{\frac{1}{2}}} \ , \tag{27}$$

as plotted in Fig. 9. The singularity in the density of modes results from the zero of $d\omega/dK$ at $K = \pi/a$.

The dispersion relation for a diatomic linear lattice is given by Eq. (5.36) under the assumption of nearest-neighbor interactions. The density of modes

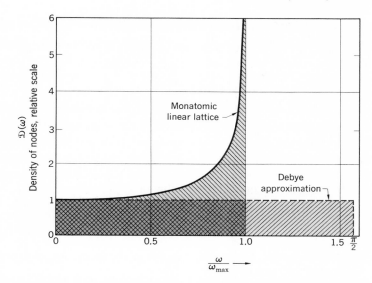

Figure 9 Density of phonon modes $\mathfrak{D}(\omega)$ for monatomic line of atoms with nearest-neighbor interactions, after Eq. (27), as compared with the density of modes in the Debye or continuum approximation calculated from Eq. (22) for the same velocity of sound in the limit of low frequencies. Note that the lattice model has a singularity which is absent in the Debye approximation. The Debye spectrum must be cut off at $\omega_D = \pi\omega_{\max}/2$ if the total number of states is to be equal to the number of atoms.

in the acoustical branch will be somewhat similar to (27); in the optical branch of the linear lattice the density of modes has singularities at the upper and lower limiting frequencies. If the mass of one of the ions is much less than that of the other ion, we see from (5.36) that the frequency in the optical branch is approximately independent of the wavevector. Thus $d\omega/dK$ is nearly zero, and the density of modes in the optical branch may be approximated by a delta function at the appropriate frequency. This is an example of the Einstein model. The thermal energy in the acoustical modes may be approximated by the Debye model, and the thermal energy in the optical modes may be treated by the Einstein model. The energies and, hence, the heat capacities of the two types of modes are additive.

Density of Modes in Three Dimensions

We apply periodic boundary conditions over N^3 atoms within a cube of side L, so that $\mathbf{K}$ is determined by the condition

$$\exp\left[i(K_x x + K_y y + K_z z)\right]$$
$$\equiv \exp\left\{i[K_x(x + L) + K_y(y + L) + K_z(z + L)]\right\} , \quad (28)$$

whence

$$K_x, K_y, K_z = 0 ; \quad \pm\frac{2\pi}{L} ; \quad \pm\frac{4\pi}{L} ; \quad \ldots ; \quad \frac{N\pi}{L} . \quad (29)$$

Therefore there is one allowed value of $\mathbf{K}$ per volume $(2\pi/L)^3$ in $\mathbf{K}$ space, or

$$\left(\frac{L}{2\pi}\right)^3 = \frac{V}{8\pi^3} \quad (30)$$

allowed values of $\mathbf{K}$ per unit volume of $\mathbf{K}$ space, for each polarization. The volume of the specimen is $V = L^3$.

EXAMPLE:[5] *General result for $\mathfrak{D}(\omega)$.* We want to find a general expression for $\mathfrak{D}(\omega)$, the number of modes per unit frequency range, when we are given the phonon dispersion relation $\omega(\mathbf{K})$. The number of allowed values of $\mathbf{K}$ for which the phonon frequency is between ω and $\omega + d\omega$ is

$$\mathfrak{D}(\omega)\, d\omega = \left(\frac{L}{2\pi}\right)^3 \int_{\text{shell}} d^3K , \quad (31)$$

by (30); here the integral is extended over the volume of the shell in $\mathbf{K}$ space bounded by the two surfaces on which the phonon frequency is constant, one surface on which the frequency is ω and the other on which the frequency is $\omega + d\omega$.

The real problem is to evaluate the volume of this shell. We let dS_ω denote an element of area (Fig. 10a) on the surface in $\mathbf{K}$ space of the selected constant frequency ω. The element of volume between the constant frequency surfaces ω and $\omega + d\omega$ is a right cylinder of base dS_ω and altitude $dK_\perp$, so that

$$\int_{\text{shell}} d^3K = \int dS_\omega\, dK_\perp .$$

Here $dK_\perp$ is the perpendicular distance (Fig. 10b) between the surface ω constant and the surface $\omega + d\omega$ constant. The value of $dK_\perp$ will vary from one point to another on the surface.

The gradient of ω, which is $\nabla_\mathbf{K}\omega$, is also normal to the surface ω constant, and the quantity

$$|\nabla_\mathbf{K}\omega|\, dK_\perp = d\omega , \quad (33a)$$

---

[5] This example may be skipped in a first reading.

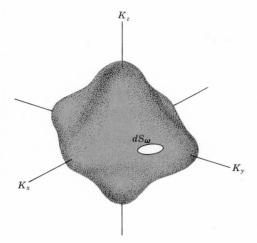

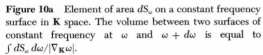

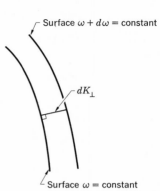

Figure 10a Element of area dS_ω on a constant frequency surface in **K** space. The volume between two surfaces of constant frequency at ω and $\omega + d\omega$ is equal to $\int dS_\omega \, d\omega/|\nabla_\mathbf{K}\omega|$.

Figure 10b The quantity $dK_\perp$ is the perpendicular distance between two constant frequency surfaces in **K** space, one at frequency ω and the other at frequency $\omega + d\omega$.

is the difference in frequency between the two surfaces connected by $dK_\perp$. Thus the element of volume is

$$dS_\omega \, dK_\perp = dS_\omega \frac{d\omega}{|\nabla_\mathbf{K}\omega|} = dS_\omega \frac{d\omega}{v_g} \ , \tag{33b}$$

where $v_g = |\nabla_\mathbf{K}\omega|$ is the magnitude of the group velocity of a phonon. For (31) we have

$$\mathfrak{D}(\omega) \, d\omega = \left(\frac{L}{2\pi}\right)^3 \int \frac{dS_\omega}{v_g} \, d\omega \ . \tag{33c}$$

We divide both sides by $d\omega$ and write $V = L^3$ for the volume of the crystal: the result for the density of modes is

$$\boxed{\mathfrak{D}(\omega) = \frac{V}{(2\pi)^3} \int \frac{dS_\omega}{v_g}} \tag{34}$$

The integral is taken over the area of the surface ω constant, in **K** space. The result refers to a single branch of the dispersion relation. We can use this result also in electron band theory (see Chapters 9 and 10).

There is a special interest in the contribution to $\mathfrak{D}(\omega)$ from points at which the group velocity is zero. Such critical points produce singularities (known as Van Hove[6] singularities) in the distribution function. An elementary discussion is given in Advanced Topic C, where it is shown that saddle points in $\omega(\mathbf{K})$ are of particular importance.

[6] L. Van Hove, Phys. Rev. **89**, 1189 (1953); see also H. P. Rosenstock, Phys. Rev. **97**, 290 (1955); J. C. Phillips, Phys. Rev. **104**, 1263 (1956).

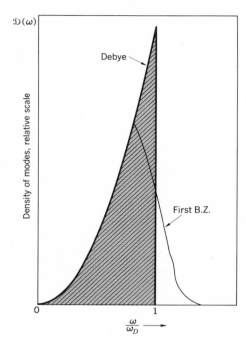

Figure 11 Density of modes $\mathfrak{D}(\omega)$ versus ω, with assumed constant phonon velocity, for integration in **K** space over the Debye sphere (shaded area) and over the cube which forms the first Brillouin zone of a monatomic simple cubic lattice. (Courtesy of C. Y. Fong.)

In the continuum or Debye approximation the velocity of sound is taken as constant: $\omega(\mathbf{K}) = vK$. The total number of modes with wavevector less than K is, by (30), given by $(L/2\pi)^3$ times the volume of a sphere of radius K, so that

$$N = \left(\frac{L}{2\pi}\right)^3 \frac{4\pi}{3} K^3 = \left(\frac{L}{2\pi}\right)^3 \frac{4\pi\omega^3}{3v^3} = \frac{V\omega^3}{6\pi^2 v^3} \tag{35}$$

for each polarization type. The density of modes is

$$\mathfrak{D}(\omega) = \frac{dN}{d\omega} = \frac{V\omega^2}{2\pi^2 v^3} \tag{36}$$

for each polarization type.

If there are N primitive cells in the specimen, the total number of acoustic phonon modes is N, and the cutoff frequency ω_D is determined by (35) as

$$\omega_D{}^3 = \frac{6\pi^2 v^3 N}{V} . \tag{37}$$

To this frequency there corresponds a cutoff wavevector in **K** space:

$$K_D = \frac{\omega_D}{v} = \left(\frac{6\pi^2 N}{V}\right)^{\frac{1}{3}}.$$

(38)

On the Debye model we do not allow modes of wavevector larger than K_D; the number of modes of $K \leq K_D$ exhausts the number of degrees of freedom of a monatomic lattice.

Thus in the Debye approximation we not only replace the actual density of modes by the result (36) that follows from the linear dispersion relation $\omega = vK$, but we also replace the correct region of integration over the Brillouin zone in **K** space by a spherical region. The density of modes given by (36) for the Debye approximation is plotted in Fig. 11, along with the density of modes for the correct region of integration for a simple cubic lattice, still with the assumption of constant velocity of sound. If $v = 5 \times 10^5$ cm/sec and $N/V = 10^{23}$ atoms/cm³, then

$$\omega_D \simeq 1 \times 10^{14} \text{ rad s}^{-1}$$

and

$$K_D \simeq 2 \times 10^8 \text{ cm}^{-1} = 2 \times 10^{10} \text{ m}^{-1}.$$

(39)

The distribution function $\mathfrak{D}(\omega)$ may also be calculated from experimental or from realistic theoretical dispersion relations. The labor involved may be appreciable, and the problem is often handled with the aid of electronic computers. We need to find ω over a very fine mesh grid in **K** space, and we then construct a histogram giving the number of points lying in small equal intervals of ω values. Calculations for aluminum are shown in Figs. 12 and 13. The effect of the Van Hove singularities is evident (*overleaf*).

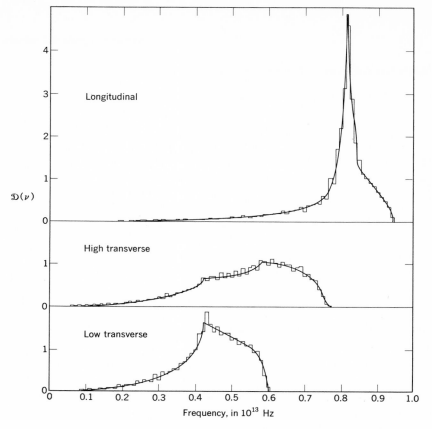

Figure 12 Phonon density of modes $\mathfrak{D}(\nu)$ for the three branches of aluminum; here $\nu = \omega/2\pi$. The histograms are obtained from computed frequencies for 2791 wavevectors. [After C. B. Walker, Phys. Rev. **103**, 547 (1956). For further results on aluminum see R. Stedman, L. Almqvist, and G. Nilsson, Phys. Rev. **162**, 549 (1967).]

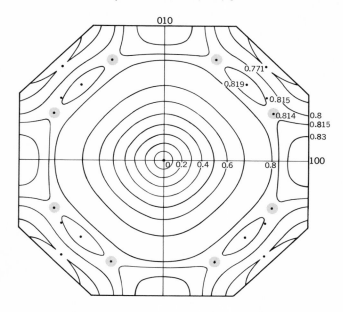

Figure 13 Surfaces of constant frequency in aluminum, for longitudinal phonons. The section shown is the (100) plane of the reciprocal lattice. The gray circles lie on one set of saddle points. (After C. B. Walker.)

Debye Model of the Lattice Heat Capacity

The energy (15) is given by

$$E = \int d\omega \, \mathfrak{D}(\omega) \, n(\omega) \, \hbar\omega = \int_0^{\omega_D} d\omega \left(\frac{V\omega^3}{2\pi^2 v^3}\right)\left(\frac{\hbar\omega}{e^{\hbar\omega/\tau} - 1}\right), \qquad (40)$$

for each polarization type. For brevity we assume that the phonon velocity is independent of the polarization, so that we multiply (40) by the factor 3 to obtain

$$E = \frac{3V\hbar}{2\pi^2 v^3} \int_0^{\omega_D} d\omega \, \frac{\omega^3}{e^{\hbar\omega/\tau} - 1} = \frac{3Vk_B^4 T^4}{2\pi^2 v^3 \hbar^3} \int_0^{x_D} dx \, \frac{x^3}{e^x - 1}, \qquad (41)$$

where $x \equiv \hbar\omega/\tau \equiv \hbar\omega/k_B T$ and

$$x_D \equiv \frac{\hbar\omega_D}{k_B T} \equiv \frac{\theta}{T}. \qquad (42)$$

This defines the **Debye temperature** θ in terms of ω_D defined by (37). We may express θ as

$$\theta = \frac{\hbar v}{k_B} \cdot \left(\frac{6\pi^2 N}{V}\right)^{\frac{1}{3}}, \qquad (43)$$

so that (41) becomes

$$E = 9Nk_B T\left(\frac{T}{\theta}\right)^3 \int_0^{x_D} dx \, \frac{x^3}{e^x - 1}, \qquad (44)$$

where N is the number of atoms in the specimen and $x_D = \theta/T$.

The heat capacity is found most easily by differentiating the middle expression of (41) with respect to temperature. Then

$$C_V = \frac{3V\hbar^2}{2\pi^2 v^3 k_B T^2} \int_0^{\omega_D} d\omega \, \frac{\omega^4 e^{\hbar\omega/\tau}}{(e^{\hbar\omega/\tau} - 1)^2} = 9Nk_B\left(\frac{T}{\theta}\right)^3 \int_0^{x_D} dx \, \frac{x^4 e^x}{(e^x - 1)^2}. \qquad (45)$$

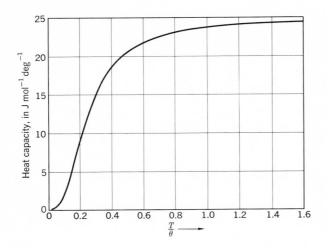

Figure 14 Heat capacity C_V of a solid, according to the Debye approximation. The vertical scale is in J mol^{-1} deg^{-1}. The horizontal scale is the temperature normalized to the Debye temperature θ. The region of the T^3 law is below 0.1θ. The asymptotic value at high values of T/θ is 24.943 J mol^{-1} deg^{-1}.

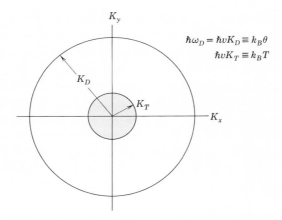

$$\hbar\omega_D = \hbar v K_D \equiv k_B\theta$$
$$\hbar v K_T \equiv k_B T$$

Figure 15 To obtain a qualitative explanation of the Debye T^3 law, we suppose that all phonon modes of wavevector less than K_T have the classical thermal energy k_BT and that modes between K_T and the Debye cutoff K_D are not excited at all. Of the $3N$ possible modes, the fraction excited is $(K_T/K_D)^3 = (T/\theta)^3$, because this is the ratio of the volume of the inner sphere to the outer sphere. The energy is

$$E \approx k_BT \cdot 3N\left(\frac{T}{\theta}\right)^3 \; ,$$

and the heat capacity is

$$C_V = \frac{\partial E}{\partial T} \approx 4Nk_B\left(\frac{T}{\theta}\right)^3 \; .$$

Tables have been calculated on the Debye theory for E, C_V, and other quantities and are given in the Landolt-Börnstein tables and also in the Jahnke-Emde-Lösch tables. The heat capacity is plotted in Fig. 14. At $T \gg \theta$ the heat capacity approaches the classical value of $3Nk_B$.

Debye T^3 Law. At very low temperatures we may approximate (44) by letting the upper limit go to infinity. We have

$$\int_0^\infty dx \, \frac{x^3}{e^x - 1} = \int_0^\infty dx \, x^3 \sum_{s=1}^\infty e^{-sx} = 6 \sum_1^\infty \frac{1}{s^4} = \frac{\pi^4}{15} , \qquad (46)$$

where the sum over s^{-4} is to be found in standard tables. Thus $E \cong 3\pi^4 Nk_B T^4 / 5\theta^3$ for $T \ll \theta$, and

$$C_V \cong \frac{12\pi^4}{5} Nk_B \left(\frac{T}{\theta}\right)^3 = 234 \, Nk_B \left(\frac{T}{\theta}\right)^3 . \qquad (47)$$

This exhibits the Debye T^3 approximation. For sufficiently low temperatures the Debye approximation should be quite good, as here only long wavelength acoustic modes are excited. These are just the modes that may be treated as in an elastic continuum with macroscopic elastic constants. The energy of short wavelength modes is too high to allow them to be populated at low temperatures, according to (11).

We can understand the T^3 region by a simple argument (Fig. 15). Only those lattice modes having $\hbar\omega < k_B T$ will be excited to any appreciable extent at a low temperature T. The excitation of these modes will be approximately classical with an energy close to $k_B T$, according to Eq. (9) and Fig. 2. The volume in **K** space occupied by the excited modes is of the order of $(K_T/K_D)^3$, where K_T is a wavevector defined such that $\hbar v K_T = k_B T$ and K_D is the Debye cutoff wavevector defined by (38), or $(T/\theta)^3$ of the total volume in **K** space. Thus there are of the order of $N(T/\theta)^3$ excited modes, each having energy $k_B T$. The internal energy is $\sim Nk_B T(T/\theta)^3$, and the heat capacity is $\sim 4Nk_B(T/\theta)^3$. The large numerical factor, 234, in (47) may be traced to the inclusion in accordance with convention of the factor $(6\pi^2)^{\frac{1}{3}}$ in the definition of θ in (43).

Methods for the determination of a suitable average sound velocity to be used in calculating θ have been given by Blackman.[7] For actual lattices the temperatures at which the T^3 approximation holds true are quite low. It may be necessary to be below $T = \theta/50$ to get reasonably pure T^3 behavior. The heat capacity is, however, relatively insensitive to changes in the density of modes.

The best practical way to obtain the density of modes is to measure the dispersion relation in selected crystal directions by inelastic neutron scattering and then to make a theoretical analytic fit to give the dispersion relation in a general direction, from which $\mathfrak{D}(\omega)$ may be calculated by an electronic computer.

[7] M. Blackman, Repts. Prog. Phys. **8**, 11 (1941); for a comparison of the Debye θ determined from elastic constants and from calorimetry, see G. A. Alers and J. R. Neighbours, Rev. Mod. Phys. **31**, 675 (1959).

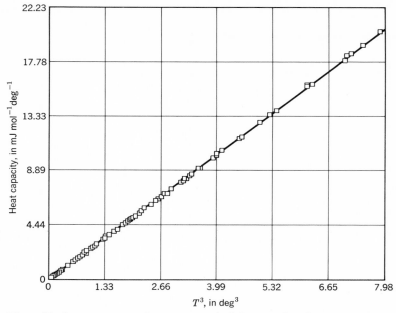

Figure 16 Low temperature heat capacity of solid argon, plotted against T^3. In this temperature region the experimental results are in excellent agreement with the Debye T^3 law with $\theta_0 = 92.0$ deg. (Courtesy of L. Finegold and N. E. Phillips.)

Selected values of θ are given in Table 1. Note for example in the alkali metals that the heavier atoms have the lowest θ's, because the velocity of sound decreases as the density increases. Experimental results for argon, a dielectric crystal with one atom per primitive cell, are given in Fig. 16.

ANHARMONIC CRYSTAL INTERACTIONS

The theory of lattice vibrations we have discussed in this and in the preceding chapters has been limited in the potential energy to terms quadratic in the interatomic displacements. This is the harmonic theory; among its consequences are:

1. There is no thermal expansion.
2. Adiabatic and isothermal elastic constants are equal.
3. The elastic constants are independent of pressure and temperature.
4. The heat capacity becomes constant at high temperatures $T > \theta$.
5. Two lattice waves do not interact; a single wave does not decay or change form with time.

In real crystals none of these consequences is satisfied accurately. The deviations may be attributed to the neglect of anharmonic (higher than quadratic) terms in the interatomic displacements. We shall discuss only some of the simpler aspects of anharmonic effects; for further details consult the references at the end of the chapter.

Table 1 Debye Temperature θ_0 in deg K

(The subscript zero on the θ denotes the low temperature limit of the experimental values.)

1	2	3	4	5	6	7	8	9	10	11	12	13	14	15	16	17	18
Li 344	Be 1440											B	C 2230	N	O	F	Ne 75
Na 158	Mg 400											Al 428	Si 645	P	S	Cl	Ar 92
K 91	Ca 230	Sc 360.	Ti 420	V 380	Cr 630	Mn 410	Fe 470	Co 445	Ni 450	Cu 343	Zn 327	Ga 320	Ge 374	As 282	Se 90	Br	Kr 72
Rb 56	Sr 147	Y 280	Zr 291	Nb 275	Mo 450	Tc	Ru 600	Rh 480	Pd 274	Ag 225	Cd 209	In 108	Sn w 200	Sb 211	Te 153	I	Xe 64
Cs 38	Ba 110	La β 142	Hf 252	Ta 240	W 400	Re 430	Os 500	Ir 420	Pt 240	Au 165	Hg 71.9	Tl 78.5	Pb 105	Bi 119	Po	At	Rn
Fr	Ra	Ac															

Ce	Pr	Nd	Pm	Sm	Eu	Gd 200	Tb	Dy 210	Ho	Er	Tm	Yb 120	Lu 210
Th 163	Pa	U 207	Np	Pu	Am	Cm	Bk	Cf	Es	Fm	Md	No	Lw

Most of the data were supplied by N. Pearlman; references are given in the *A.I.P. Handbook*, 3rd ed.

219

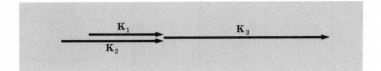

Figure 17 Collinear interaction of two longitudinal phonons to produce a third longitudinal phonon. In the nondispersive region of the phonon spectrum it is possible to satisfy simultaneously with this arrangement the energy and wavevector conservation relations $\omega_1 + \omega_2 = \omega_3$ and $\mathbf{K}_1 + \mathbf{K}_2 = \mathbf{K}_3$. In the Shiren experiment, power conversions up to 70 percent were observed over an interaction path of 2 cm, for input power levels ~0.5 watts/cm^2.

Beautiful demonstrations of anharmonic effects are the experiments on the interaction of two phonons to produce a third phonon at a frequency $\omega_3 = \omega_1 + \omega_2$. Shiren[8] describes an experiment in which a beam of longitudinal phonons of frequency 9.20 GHz interacts in an MgO crystal with a parallel beam of longitudinal phonons at 9.18 GHz. The interaction of the two beams (Fig. 17) produced a third beam of longitudinal phonons at $9.20 + 9.18 = 18.38$ GHz.

Rollins[9] et al. carried out the elegant experiment pictured in Fig. 18. Two narrow ultrasonic beams are allowed to interact near the center of a large circular disk. They verified that a third beam is generated at the volume of interaction only when energy and wavevector are conserved between the three beams (Fig. 19).

The three-phonon processes are caused by third-order terms in the lattice potential energy. A typical term might be

$$U_3 = Ae_{xx}e_{yy}e_{zz} \ , \tag{48}$$

where the e's are strain components and A is a constant. The A's have the same dimensions as the elastic stiffness constants in Chapter 4, but may have values perhaps an order of magnitude larger than the C's. The physics of the phonon interaction can be stated simply: the presence of one phonon causes a periodic elastic strain which (through the anharmonic interaction) modulates in space and time the elastic constant of the crystal. A second phonon perceives the modulation of the elastic constant and is thereupon scattered to produce a third phonon, just as from a moving three-dimensional grating. In the absence of anharmonic terms the acoustic modulation does not occur.

[8] N. S. Shiren, Phys. Rev. Letters **11**, 3 (1963).

[9] F. R. Rollins, Jr., L. H. Taylor, and P. H. Todd, Jr., Phys. Rev. **136**, A597 (1964).

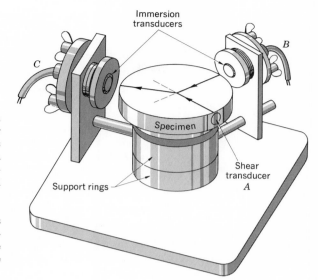

Figure 18 Ultrasonic experiment on three-phonon interactions. In a typical experiment the transducer A generates a 10-MHz shear wave which interacts near the center of the disc-shaped specimen with a 15 MHz longitudinal wave generated by B, to produce by their interaction a 25 MHz longitudinal wave detected by the transducer C: $L(15) + T(10) \rightarrow L(25)$. The wavevectors satisfy $\mathbf{K}_{15} + \mathbf{K}_{10} = \mathbf{K}_{25}$. The angle φ is easily calculated from this equation and the wave velocities. The whole apparatus is immersed in a suitable fluid to provide coupling between the immersion transducers and the specimen. (After Rollins, Taylor, and Todd.)

Figure 19 The displacement amplitude of the generated beam should be proportional to the product of the amplitudes of the two primary beams. This curve (due to Rollins et al.) shows the experimental linear relationship between $(V_1 V_2)$ and V_3 where V_1 and V_2 are the voltages applied to the two primary transducers and V_3 is the transducer voltage produced by the interaction wave.

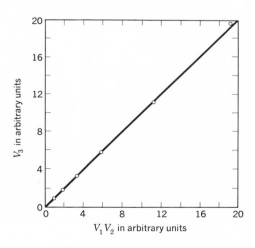

Thermal Expansion

We may understand the origin of thermal expansion by considering for a classical oscillator the effect of anharmonic terms in the potential energy on the mean separation of a pair of atoms at a temperature T. We take the potential energy of the atoms at a displacement x from their equilibrium separation at $0°K$ as

$$U(x) = cx^2 - gx^3 - fx^4 , \tag{49}$$

with c, g, and f all positive. The term in x^3 represents the asymmetry of the mutual repulsion of the atoms and the term in x^4 represents the softening of the vibration at large amplitudes. The minimum of (49) at $x = 0$ is not an absolute minimum, but for small oscillations the form is an adequate representation of an interatomic potential, such as Fig. 3.5.

We calculate the average displacement by using the Boltzmann distribution function, which weights the possible values of x according to their thermodynamic probability:

$$\langle x \rangle = \frac{\displaystyle\int_{-\infty}^{\infty} dx \, x e^{-\beta U(x)}}{\displaystyle\int_{-\infty}^{\infty} dx \, e^{-\beta U(x)}} , \tag{50}$$

with $\beta \equiv 1/k_B T$. For displacements such that the anharmonic terms in the energy are small in comparison with $k_B T$, we may expand the integrands in (50) as

$$\int dx \, x e^{-\beta U} \cong \int dx \, e^{-\beta c x^2}(x + \beta g x^4 + \beta f x^5) = (3\pi^{\frac{1}{2}}/4)(g/c^{\frac{5}{2}})\beta^{-\frac{3}{2}} ; \tag{51}$$

$$\int dx \, e^{-\beta U} \cong \int dx \, e^{-\beta c x^2} = (\pi/\beta c)^{\frac{1}{2}} , \tag{52}$$

whence

$$\langle x \rangle = \frac{3g}{4c^2} k_B T \tag{53}$$

in the classical region.[10] Note that in (51) we have left cx^2 in the exponential, but we have expanded $\exp(\beta g x^3 + \beta f x^4) \cong 1 + \beta g x^3 + \beta f x^4 + \cdots$. Several values of the linear expansion coefficient are given in Table 2. Measurements for solid argon are shown in Fig. 20.

---

[10] Compare with the discussion of the anharmonic oscillator on pp. 227–229 of the *Berkeley physics course*, Vol. I., McGraw-Hill (1965).

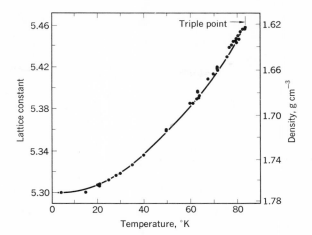

Figure 20 Lattice constant of solid argon as a function of temperature, as measured by various workers [O. G. Peterson, D. N. Batchelder, and R. O. Simmons, Phys. Rev. **150**, 703 (1966)].

Table 2 Linear thermal expansion coefficients near room temperature.[*]

$$\alpha = \frac{1}{\ell} \frac{\partial \ell}{\partial T}$$

Substance	$\alpha \times 10^6$, per deg K	Substance	$\alpha \times 10^6$, per deg K
Li	45	Fe	11.7
Na	71	Ni	12.5
K	83	Cr	7.5
Cs	97	Mo	5.2
Cu	17.0	Ta	6.6
Ag	18.9	W	4.6
Au	13.9	Ir	6.5
Ca	22.5	Pd	11.6
Al	23.6	Pt	8.9
Pb	28.8		

[*] See W. B. Pearson, *A handbook of lattice spacings and structures of metals and alloys*, Pergamon, 1958. For data at low temperature, see K. Andres, Phys. kondens. Materie **2**, 294 (1964).

THERMAL CONDUCTIVITY

The thermal conductivity coefficient K of a solid is most easily defined with respect to the steady-state flow of heat down a long rod with a temperature gradient dT/dx:

$$Q = K\frac{dT}{dx} , \tag{54}$$

where Q is the flux of thermal energy (energy transmitted across unit area per unit time).

The form of the equation (54) which defines the conductivity implies that the process of thermal energy transfer is a random process. The energy does not simply enter one end of the specimen and proceed directly in a straight path to the other end, but rather the energy diffuses through the specimen, suffering frequent collisions. If the energy were propagated directly through the specimen without deflection, then the expression for the thermal flux would not depend on the temperature gradient, but only on the difference in temperature ΔT between the ends of the specimen, regardless of the length of the specimen. The random nature of the conductivity process brings the temperature gradient and a mean free path into the expression for the thermal flux.

From the kinetic theory of gases we find below in a certain approximation the following expression for the thermal conductivity:

$$K = \tfrac{1}{3}Cv\ell , \tag{55}$$

where C is the heat capacity per unit volume, v is the average particle velocity, and ℓ is the mean free path of a particle between collisions. This result was applied first by Debye to describe thermal conductivity in dielectric solids, with C as the heat capacity of the phonons, v the phonon velocity, and ℓ the phonon mean free path. Several representative values of the mean free path are given in Table 3.

We give first the elementary kinetic theory which leads to (55). The flux of particles in the x direction is $\tfrac{1}{2}n\langle|v_x|\rangle$, where n is the concentration of molecules; in equilibrium there is a flux of equal magnitude in the opposite direction. The $\langle\cdots\rangle$ denote average value. If c is the heat capacity of a particle, then in moving from a region at local temperature $T + \Delta T$ to a region at local temperature T a particle will give up energy $c\,\Delta T$. Now ΔT between the ends of a free path of the particle is given by

$$\Delta T = \frac{dT}{dx}\ell = \frac{dT}{dx}v_x\tau , \tag{56}$$

where τ is the average time between collisions.

The net flux of energy (from both senses of the particle flux) is therefore

$$Q = n\langle v_x{}^2\rangle\, c\tau\,\frac{dT}{dx} = \tfrac{1}{3}n\langle v^2\rangle\, c\tau\,\frac{dT}{dx} . \tag{57}$$

Table 3   Phonon mean free paths.

[Calculated from (55), taking $v = 5 \times 10^5$ cm/sec as a representative sound velocity. The ℓ's obtained in this way refer to the umklapp processes defined by (61)].

Crystal	T, °C	C, in J cm^{-3}deg^{-1}	K, watt cm^{-1}deg^{-1}	ℓ, in Å
Quartz°	0	2.00	0.13	40
	−190	0.55	0.50	540
NaCl	0	1.88	0.07	23
	−190	1.00	0.27	100

° Parallel to optic axis.

If, as for phonons, v is constant, we may write (57) as

$$Q = \tfrac{1}{3}Cv\ell\,\frac{dT}{dx}\,, \tag{58}$$

with $\ell \equiv v\tau$ and $C \equiv nc$. Thus $K = \tfrac{1}{3}Cv\ell$.

Lattice Thermal Resistivity

The phonon mean free path ℓ is determined principally by two processes, geometrical scattering and scattering by other phonons. If the forces between atoms were purely harmonic, there would be no mechanism for collisions between different phonons, and the mean free path would be limited solely by collisions of a phonon with the crystal boundary, and by lattice imperfections. There are situations where these effects are dominant. With anharmonic lattice interactions such as (48) there is a coupling between different phonons which limits the value of the mean free path. The exact normal modes of the anharmonic system are no longer like pure phonons. We consider first the thermal resistivity from lattice interactions.

The theory of the effect of anharmonic coupling on thermal resistivity is a complicated problem. An approximate calculation has been given by Debye,[11] and Peierls[12,13] has considered the problem in great detail. They both show that ℓ is proportional to $1/T$ at high temperatures, in agreement with many experiments. We can understand this dependence in terms of the number of

[11] P. Debye, "Zustandsgleichung und Quantenhypothese mit einem Anhang über Wärmeleitung." In *Vorträge über die kinetische Theorie der Materie und der Elektrizität*, von M. Planck et al. (Mathematische Vorlesungen an der Universität Göttingen: VI.) Leipzig, Teubner, 1914, pp. 19–60.

[12] R. Peierls, Ann. Physik **3**, 1055 (1929).

[13] C. Herring, Phys. Rev. **95**, 954 (1954); J. Callaway, Phys. Rev. **113**, 1046 (1959); R. E. Nettleton, Phys. Rev. **132**, 2032 (1963); and the reviews cited at the end of the chapter. The Callaway and Nettleton papers contribute to an understanding of the combined effects of lattice and impurity scattering; see also M. G. Holland, Phys. Rev. **132**, 2461 (1963); P. Erdös, Phys. Rev. **138**, A1200 (1965).

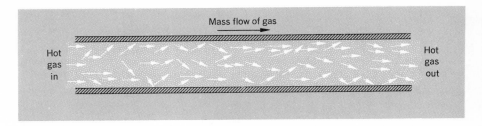

Figure 21a Flow of gas molecules down a long open tube with frictionless walls. Elastic collision processes among the gas molecules do not change the momentum or energy flux of the gas because in each collision the velocity of the center of mass of the colliding particles and their energy remain unchanged. Thus energy is transported from left to right without being driven by a temperature gradient. Therefore the thermal resistivity is zero and the thermal conductivity is infinite.

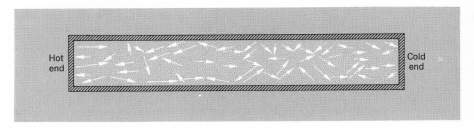

Figure 21b The usual definition of thermal conductivity in a gas refers to a situation where no mass flow is permitted. Here the tube is closed at both ends, preventing the escape or entrance of molecules. With a temperature gradient the colliding pairs with above-average center of mass velocities will tend to be directed to the right, those with below-average velocities will tend to be directed to the left. A slight concentration gradient, high on the right, will be set up to enable the net mass transport to be zero while allowing a net energy transport from the hot to the cold end.

phonons with which a given phonon can interact: at high temperature the total number of excited phonons is proportional to T, according to (9). The collision frequency of a given phonon should be proportional to the number of phonons with which it can collide, whence $\ell \propto 1/T$.

To define a thermal conductivity there must exist mechanisms in the crystal whereby the distribution of phonons may be brought locally into thermal equilibrium. Without such mechanisms we may not speak of the phonons at one end of the crystal as being in thermal equilibrium at a temperature T_2 and those at the other end in equilibrium at T_1. It is not sufficient for thermal conductivity to have only a way of limiting the mean free path, but there must also be a way of establishing a true equilibrium distribution of phonons.

Phonons collisions with a static imperfection or a crystal boundary will not by themselves establish thermal equilibrium, because such collisions do not change the energy of individual phonons: the frequency ω_2 of the scattered phonon is equal to the frequency ω_1 of the incident phonon.

It is rather remarkable also that a three-phonon collision process

$$\mathbf{K}_1 + \mathbf{K}_2 = \mathbf{K}_3 \tag{59}$$

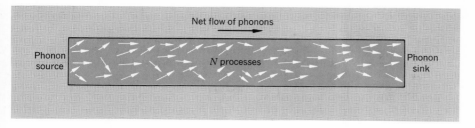

Net flow of phonons

Phonon source

N processes

Phonon sink

Figure 21c In a crystal we may arrange to create phonons chiefly at one end, as by illuminating the left end with a lamp. From that end there will be a net flux of phonons toward the right end of the crystal. If only *N* processes ($\mathbf{K}_1 + \mathbf{K}_2 = \mathbf{K}_3$) occur, the phonon flux is unchanged in momentum on collision and some phonon flux will persist down the length of the crystal. On arrival of phonons at the right end we can arrange in principle to convert most of their energy to radiation, thereby creating a sink for the phonons. Just as in (a) the thermal resistivity is zero.

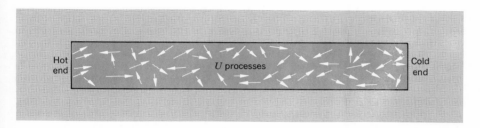

Hot end

U processes

Cold end

Figure 21d In *U* processes ($\mathbf{K}_1 + \mathbf{K}_2 = \mathbf{K}_3 + \mathbf{G}$, as in Fig. 22b) there is a large net change in phonon momentum in each collision event. An initial net phonon flux will rapidly decay as we move to the right. The ends may act as sources and sinks. Net energy transport under a temperature gradient occurs somewhat as in (b).

will not establish equilibrium, but for a subtle reason: the total momentum of the phonon gas is not changed by such a collision. An equilibrium distribution of phonons at a temperature T can move down the crystal with a drift velocity which is not disturbed by three-phonon collisions[14] of the form (59). For in such collisions the phonon momentum

$$\mathbf{J} = \sum_K \hbar \mathbf{K} n_\mathbf{K} \tag{60}$$

is conserved, because on collision the change in $\mathbf{J}$ is $\mathbf{K}_3 - \mathbf{K}_2 - \mathbf{K}_1 = 0$. Here $n_\mathbf{K}$ is the number of phonons having wavevector $\mathbf{K}$. For a distribution with $\mathbf{J} \neq 0$, collisions such as (70) are incapable of establishing complete thermal equilibrium because they leave $\mathbf{J}$ unchanged. If we start a distribution of hot phonons down a rod with $\mathbf{J} \neq 0$, the distribution will propagate down the rod with $\mathbf{J}$ unchanged. Therefore there is no thermal resistance. The problem as illustrated in Fig. 21 is like that of the collisions between molecules of a gas in a straight tube with frictionless walls.

[14] See R. E. Peierls, *Quantum theory of solids*, Oxford, 1955, pp. 41–45. The same result holds for collision processes with any number of phonons.

Umklapp Processes

Peierls pointed out that the important three-phonon processes for thermal conductivity are not of the form $\mathbf{K}_1 + \mathbf{K}_2 + \mathbf{K}_3$, as in (59), but are of the form

$$\mathbf{K}_1 + \mathbf{K}_2 = \mathbf{K}_3 + \mathbf{G} \ , \tag{61}$$

where $\mathbf{G}$ is a reciprocal lattice vector (Fig. 22). We recall that $\mathbf{G}$ may occur in all momentum conservation laws in crystal lattices. We have already encountered in Chapters 2 and 5 examples of wave interaction processes in crystals for which the total wavevector change need not be zero, but may be a reciprocal lattice vector. Such processes are always possible in periodic lattices, whereas in a true continuum $\mathbf{G}$ is always zero.

Processes or collisions in which $\mathbf{G} \neq 0$ are called **umklapp processes,** after the German for "flipping over." This term refers to the circumstance (as in Fig. 22b) that a collision of two phonons both having a positive K_x can by umklapp give after collision a phonon with a negative K_x. Umklapp processes are also called U processes. Collisions in which $\mathbf{G} = 0$ are called **normal processes** or N processes. A typical umklapp process is shown in Fig. 23 for a linear lattice. At high temperatures ($T > \theta$) all phonons are excited because $k_B T > \hbar\omega_{max}$. A substantial proportion of all phonon collisions will then be U processes, with the attendant high momentum change in the collision. We can estimate the thermal resistivity without particular distinction between N and U processes; by the earlier argument about nonlinear effects we expect to find a lattice thermal resistivity $\propto T$ at high temperatures.

The energy of phonons $\mathbf{K}_1$, $\mathbf{K}_2$ suitable for umklapp to occur is of the order of $\frac{1}{2}k_B\theta$, because each of the phonons 1 and 2 must have wavevectors of the order of $\frac{1}{2}G$ in order for the collision (61) to be possible. (If both phonons have low K, and therefore low energy, there is no way to get from their collision a phonon of wavevector comparable to G. The umklapp process must conserve energy, just as for the normal process.) At low temperatures the number of suitable phonons of the high energy $\frac{1}{2}k_B\theta$ required may be expected to vary roughly as $e^{-\theta/2T}$, according to the Boltzmann factor. The exponential form is in good agreement with experiment. In summary, the phonon mean free path which enters (55) is the mean free path for umklapp collisions between phonons and not for all collisions between phonons.

Imperfections

Geometrical effects may also be important in limiting the mean free path. We must consider scattering by crystal boundaries, the distribution of isotopic masses in natural chemical elements, chemical impurities, lattice imperfections, and amorphous structures.

When at low temperatures the mean free path ℓ becomes comparable with the width of the test specimen, the value of ℓ is limited by the width, and the thermal conductivity becomes a function of the dimensions of the specimen.

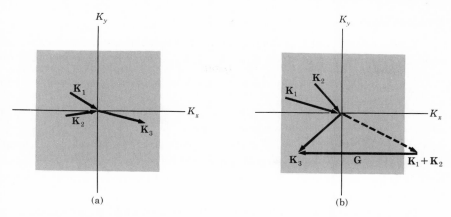

(a)   (b)

Figure 22   (a) Normal $\mathbf{K}_1 + \mathbf{K}_2 = \mathbf{K}_3$ and (b) umklapp $\mathbf{K}_1 + \mathbf{K}_2 = \mathbf{K}_3 + \mathbf{G}$ phonon collision processes in a two-dimensional square lattice. The square in each figure represents the first Brillouin zone in the phonon $\mathbf{K}$ space; this zone contains all the possible independent values of the phonon wavevector. Vectors $\mathbf{K}$ which have arrowheads at the center of the zone represent phonons absorbed in the collision process; those with arrowheads away from the center of the zone represent phonons emitted in the collision. We see in (b) that in the umklapp process the direction of the x-component of the phonon flux has been reversed. The reciprocal lattice vector $\mathbf{G}$ as shown is of length $2\pi/a$, where a is the lattice constant of the crystal lattice, and is parallel to the $\mathbf{K}_x$ axis. For all processes, N or U, energy must be conserved, so that $\omega_1 + \omega_2 = \omega_3$.

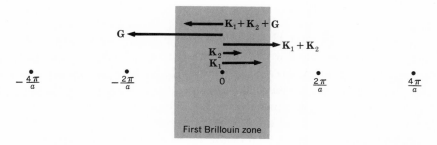

Figure 23   Central reciprocal lattice points for monatomic linear crystal of lattice constant a. A typical umklapp process is shown: here a phonon of wavevector $\mathbf{K}_1$ collides with a phonon of wavevector $\mathbf{K}_2$. The sum $\mathbf{K}_1 + \mathbf{K}_2$ lies outside the first Brillouin zone of the reciprocal lattice, but by the argument of Eq. (5.24) such a $\mathbf{K}$ value is always equivalent to a wavevector $\mathbf{K}_1 + \mathbf{K}_2 + \mathbf{G}$ inside the first zone, where $\mathbf{G}$ is a suitable reciprocal lattice vector. For the process shown $G = -2\pi/a$. For the thermal conductivity there is a difference between processes in which $\mathbf{K}_1 + \mathbf{K}_2$ lies inside the first Brillouin zone and those in which $\mathbf{K}_1 + \mathbf{K}_2$ lies outside and has to be brought back by adding a suitable $\mathbf{G}$.

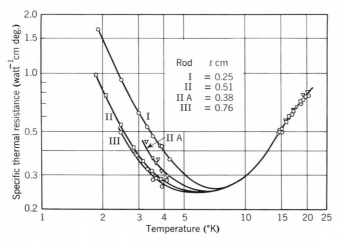

Figure 24 Thermal resistivity of a single crystal of potassium chloride as measured by Biermasz and de Haas. Below 5°K the resistivity is a function of the crystal thickness t, because the phonon mean free path is determined by the crystal dimensions. The increase in thermal resistivity at low temperatures is caused by the decrease in lattice heat capacity; the increase in resistivity above 10°K is caused by the exponential onset of umklapp processes.

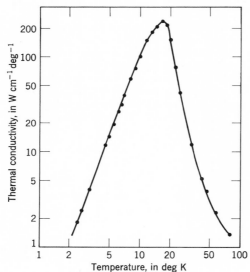

Figure 25 Thermal conductivity of a highly purified crystal of sodium fluoride, after H. E. Jackson, C. T. Walker, and T. F. McNelly, Phys. Rev. Letters 25, 26 (1970). This crystal was used in the study of the propagation of heat pulses and second sound. (The propagation of second sound in crystals is treated in QTS, pp. 28–30.)

This effect was discovered by de Haas and Biermasz,[15] and the explanation was suggested by Peierls and worked out by Casimir;[16] results of measurements on potassium chloride crystals are given in Fig. 24. The abrupt decrease in thermal conductivity of pure crystals at low temperatures is caused by the size effect. At low temperatures the umklapp process becomes ineffective in limiting the thermal conductivity, and the size effect becomes dominant, as shown also in Fig. 25. One would expect then that the phonon mean free path would be constant and of the order of the diameter D of the specimen, so that

$$K \approx CvD \ . \tag{62}$$

[15] W. J. de Haas and T. Biermasz, Physica 2, 673 (1935); 4, 752 (1937); 5, 47, 320, 619 (1938); see also R. Berman, Proc. Roy. Soc. (London) A208, 90 (1951).

[16] H. B. G. Casimir, Physica 5, 495 (1938); R. E. B. Makinson, Proc. Cambridge Phil. Soc. 34, 474 (1938).

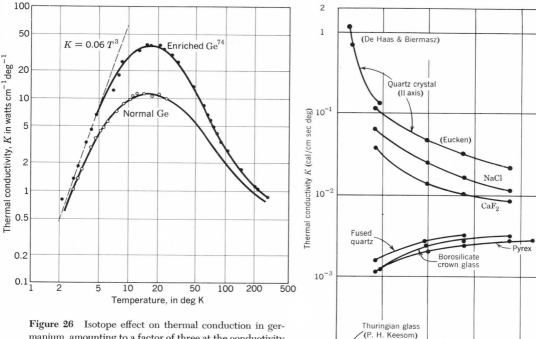

Figure 26 Isotope effect on thermal conduction in germanium, amounting to a factor of three at the conductivity maximum. The enriched specimen is 96 percent Ge^{74}; natural germanium is 20 percent Ge^{70}, 27 percent Ge^{72}, 8 percent Ge^{73}, 37 percent Ge^{74}, and 8 percent Ge^{76}. Below $5°K$ the enriched specimen has $K = 0.060\ T^3$, which agrees well with Casimir's theory for thermal resistance caused by boundary scattering. The conductivity data lead to a boundary-scattering mean free path of 1.80 mm, as compared with 1.57 mm calculated from the area of the cross-section. [After T. H. Geballe and G. W. Hull, Phys. Rev. **110**, 773 (1958).]

Figure 27 Temperature dependence of the thermal conductivity of various crystals and glasses.

The only temperature-dependent term on the right is C, the heat capacity, which varies as T^3 at low temperatures. We may therefore expect the thermal conductivity to vary as T^3 at low temperatures. The size effect enters whenever the phonon mean free path becomes comparable with the diameter of the specimen.

In an otherwise perfect crystal the distribution of isotopes of the chemical elements often provides an important mechanism for phonon scattering. (The random distribution of isotopes disturbs the periodicity of the density as seen by an elastic wave.) In some substances the importance of scattering of phonons by isotopes is comparable to scattering by other phonons, even at room temperature.[17] Results for germanium are shown in Fig. 26.

In glasses the thermal conductivity (Fig. 27) decreases as the temperature is lowered, even at room temperature. The values of the thermal conductivity

[17] See J. M. Ziman, *Electronics and phonons*, Oxford, 1960, Sec. 8.6.

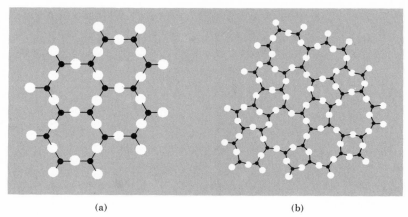

| (a) | (b) |

Figure 28 Schematic two-dimensional analogs, after Zachariasen, illustrating the difference between: (a) the regularly repeating structure of a crystal and (b) the random network of a glass. The solid circles are oxygen atoms.

at room temperature run about an order of magnitude lower for glasses than for crystals. The mean free path in quartz glass at room temperature is 8 Å, which is of the order of magnitude of the dimensions of a silicon dioxide tetrahedron (7 Å). A glass such as fused quartz is made up of a random, but continuous, network[18] (Fig. 28) of silicon-oxygen bonds. The effective crystallite size is only of the order of a single tetrahedron of the structure. We expect that (except at low temperatures, where the phonon wavelengths are so long that the structure looks uniform) the phonon mean free path will be constant, limited by the crystallite size, and the decline in the conductivity as the temperature is lowered may be attributed to the decline in the heat capacity.

Dielectric crystals may have thermal conductivities as high as metals. Synthetic sapphire (Al_2O_3) has one of the highest values of the conductivity: nearly 200 watt cm^{-1} deg^{-1} at 30°K. Glasses have values as low as 5×10^{-4} watt cm^{-1} deg^{-1} at 2°K, and Berman[19] has suggested the conductivity of microcrystalline graphite at 1°K may be 10^{-5} watt cm^{-1} deg^{-1}. The maximum of the thermal conductivity in sapphire is greater than the maximum of 50 watt cm^{-1} deg^{-1} in copper.[20] Metallic gallium, however, has a conductivity[21] of 845 watt cm^{-1} deg^{-1} at 1.8°K. The electronic contribution to the thermal conductivity of metals is treated in Chapter 7.

[18] W. H. Zachariasen, J. Am. Chem. Soc. **54**, 3841 (1932); B. E. Warren, J. Appl. Phys. **8**, 645 (1937); **13**, 602 (1942); E. U. Condon, "Physics of the glassy state," Am. J. Phys. **22**, 43, 132, 224, 310 (1954).

[19] R. Berman, Phys. Rev. **76**, 315 (1949).

[20] R. Berman and D. K. C. MacDonald, Proc. Roy. Soc. (London) **A211**, 122 (1952); the peak value of the conductivity of NaF in the experiment of Fig. 25 is 240 W cm^{-1} deg^{-1}.

[21] R. I. Boughton and M. Yaqub, Phys. Rev. Letters **20**, 108 (1968).

SUMMARY

We give an alternative development[22] of the Debye theory in terms of standing waves in a cube of side L with the condition of zero displacement on all boundaries. The isotropic wave equation is

$$\nabla^2 u = \frac{1}{v^2} \frac{\partial^2 u}{\partial t^2} \; ; \qquad v = \text{velocity of sound.}$$

The solutions for the particle displacement u are of the form

$$u = A e^{-i\omega t} \sin \frac{n_x \pi x}{L} \sin \frac{n_y \pi y}{L} \sin \frac{n_z \pi z}{L} \; ,$$

where A is a constant and

$$\omega^2 = \left(\frac{v\pi}{L}\right)^2 (n_x{}^2 + n_y{}^2 + n_z{}^2) = \left(\frac{v\pi}{L}\right)^2 n^2 \; ,$$

with n_x, n_y, n_z any triplet of positive integers. The maximum value of n is determined by

$$\frac{1}{8} \cdot \frac{4\pi}{3} n_{\max}{}^3 = N \; ,$$

where the left-hand side is the volume of the positive octant of a sphere in the space of the n_x, n_y, n_z. The total number of modes of any single polarization type is N, the number of atoms.

The thermal energy of the phonons is, with $\tau \equiv k_B T$,

$$E = \sum_{n_x n_y n_z} \frac{\hbar \omega_{\mathbf{n}}}{e^{\hbar \omega_{\mathbf{n}}/\tau} - 1} = \frac{3}{8} \cdot 4\pi \int_0^{n_{\max}} n^2 \, dn \, \frac{\hbar \omega_{\mathbf{n}}}{e^{\hbar \omega_{\mathbf{n}}/\tau} - 1} \; .$$

With $\omega_n = v\pi n/L$ and $x = (\pi v \hbar / L \tau) n$, we have

$$E = \frac{3\pi}{2} \tau^4 \left(\frac{L}{\hbar \pi v}\right)^3 \int_0^{x_m} \frac{x^3 \, dx}{e^x - 1} \; .$$

Here

$$x_m \equiv \frac{\pi v \hbar}{L k_B T} n_{\max} = \frac{\hbar v}{k_B T} \cdot \left(\frac{6\pi^2 N}{V}\right)^{\frac{1}{3}} = \frac{\theta}{T} \; .$$

---

[22] We follow here the development of *TP*, Chapters 10 and 16.

Other major topics in this chapter are:

(a) $\qquad \mathfrak{D}(\omega) = \dfrac{V}{(2\pi)^3} \displaystyle\int \dfrac{dS_\omega}{v_g} \; ; \qquad v_g = |\nabla_{\mathbf{K}}\omega| \;\; .$

(b) Van Hove singularities.

(c) Effects of anharmonic crystal interactions.

(d) Thermal conductivity $K = \frac{1}{3}Cv\ell$.

(e) Role of umklapp collisions in the thermal resistivity.

Problems

1. ***Heat capacity of one-dimensional lattice.*** Show that the heat capacity of a monatomic lattice in one dimension in the Debye approximation is proportional to T/θ for low temperatures $T \ll \theta$, where θ is the effective Debye temperature in one dimension defined as $\theta = \hbar\omega_m/k_B = \hbar\pi v_0/k_B a$; here k_B is the Boltzmann constant and a the interatomic separation.

2. ***Energy and partition function.*** Show that the expression for the average energy of a system may be written as

$$\langle E \rangle = k_B T^2 \frac{d \log Z}{dT} \; ,$$

where the partition function Z is defined for a classical one-dimensional system by

$$Z = \iint dp \, dx \, \exp\left[-E(p,x)/k_B T\right] \; ;$$

here p is the momentum.

3. ***Heat capacity of anharmonic oscillator.*** Using the anharmonic potential $U(x) = cx^2 - gx^3 - fx^4$, show that the approximate heat capacity of the classical anharmonic oscillator in one dimension is

$$C \cong k_B \left[1 + \left(\frac{3f}{2c^2} + \frac{15g^2}{8c^3}\right) k_B T\right] \; .$$

Note: $\log(1 + \delta) \cong \delta - \frac{1}{2}\delta^2$ for $\delta \ll 1$; the calculation is shorter if the partition function (Problem 2) is employed. We assume that the oscillations about $x = 0$ are small so that $c\langle x^2 \rangle$ is dominant in U; leave $-\beta cx^2$ in the exponential, but expand the other terms. It is necessary to retain terms of order g^3 in all expansions. Here $\beta = 1/k_B T$.

4. ***Three-phonon interactions.*** Consider a crystal for which $\omega_L = v_L K$ and $\omega_T = v_T K$, where v_L, v_T are independent of $\mathbf{K}$. The subscripts L, T denote longitudinal and transverse. If $v_L > v_T$, show that the normal three-phonon process $T + L \leftrightarrow T$ cannot satisfy conservation of energy and wavevector.

References

A. A. Maradudin, E. W. Montroll, and G. H. Weiss, "Theory of lattice dynamics in the harmonic approximation," *Solid state physics*, Supp. **3**, 1963.

G. Leibfried and W. Ludwig, "Theory of anharmonic effects in crystals," *Solid state physics* **12**, 276–444 (1961).

THERMAL CONDUCTIVITY

P. G. Klemens, "Thermal conductivity and lattice vibration modes," *Solid state physics* **7**, 1–98 (1958); see also *Encyclo. of physics* **14**, 198 (1956).

K. Mendelssohn and H. M. Rosenberg, "Thermal conductivity of metals at low temperatures," *Solid state physics* **12**, 223–274 (1961).

H. M. Rosenberg, *Low temperature solid state physics*, Oxford, 1963, Chap. 3.

J. M. Ziman, *Electrons and phonons*, Oxford, 1960, Chap. 8.

7

Free Electron
Fermi Gas I

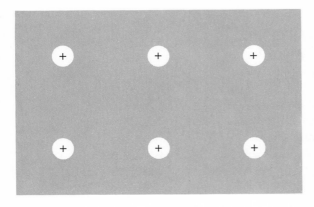

Figure 1 Schematic model of a crystal of sodium metal. The atomic cores are Na^+ ions; they are immersed in a sea of conduction electrons. The conduction electrons are derived from the $3s$ valence electrons of the free atoms. The atomic cores contain ten electrons in the configuration $1s^2\ 2s^2\ 2p^6$. In an alkali metal the atomic cores occupy a relatively small part (~ 15 percent) of the total volume of the crystal, but in a noble metal (Cu, Ag, Au) the atomic cores are relatively larger and may be in contact with each other. The common crystal structure at room temperature is bcc for the alkali metals and fcc for the noble metals.

"In a theory which has given results like these, there must certainly be a great deal of truth."

(H. A. Lorentz)

We can understand a number of important physical properties of metals, particularly the simple metals, in terms of the free electron model. According to this model the most weakly bound electrons of the constituent atoms move about freely through the volume of the metal. The valence electrons of the atoms become the conductors of electricity in the metal and are called conduction electrons. The forces between the conduction electrons and the ion cores are neglected in the free electron approximation: all calculations proceed as if the conduction electrons were free to move everywhere within the specimen. The total energy is all kinetic energy; the potential energy is neglected.

Even in metals for which the free electron model works well, the actual charge distribution of the conduction electrons is known to reflect the strong electrostatic potential of the ion cores. The usefulness of the free electron model is greatest for experiments that depend essentially upon the kinetic properties of the conduction electrons. Chapters 9 and 10 are concerned with the effects of the conduction electron interaction with the lattice.

The simplest metals are the alkali metals (lithium, sodium, potassium, cesium, and rubidium). Conduction electrons in all metals are observed to act very much like free electrons, except those metals where the d shell electrons overlap or fall close in energy to the conduction band. Electrons in d shells tend to be more localized and less mobile than electrons in s and p shells. Other simple metals are Be, Mg, Ca, Sr, Ba, Al, Ga, In, Tl, Zn, Cd, Hg, and Pb. Metals that are not simple are the noble metals (copper, silver, gold), the transition metals, and the lanthanide and actinide metals.

Conduction electrons in a simple metal arise from the valence electrons of the constituent atoms. In a sodium atom the valence electron is in a $3s$ state; in the metal this electron becomes a conduction electron, roving throughout the crystal. A monovalent crystal which contains N atoms will have N conduction electrons and N positive ion cores. The ten electrons of the Na^+ ion core fill the $1s$, the $2s$, and the $2p$ states in the free ion; the distribution of core electrons is essentially the same in the metal as in the free ion.

The ion cores fill only about 15 percent of the volume of a sodium crystal, as in Fig. 1. The radius of the free Na^+ ion (Table 3.7) is 0.98 Å, whereas one-half of the nearest neighbor distance of the metal (Table 1.5) is 1.85 Å.

The interpretation of metallic properties in terms of free electrons was developed long before the invention of quantum mechanics. The classical theory had several conspicuous successes and several remarkable failures. The successes include the derivation of the form of Ohm's law, which connects the electric current with the electric field, and the derivation of the relation between the electrical conductivity and the thermal conductivity. The classical theory completely fails to explain the heat capacity and the paramagnetic susceptibility of the conduction electrons.

There is a further difficulty: using the classical theory we cannot understand the occurrence of long electronic mean free paths. From many types of experiments it is abundantly clear that a conduction electron in a metal can move freely in a straight path over many atomic distances, undeflected by collisions with other conduction electrons or by collisions with the atom cores. In a very pure specimen at low temperatures the mean free path may be as long as 10^8 or 10^9 interatomic spacings (more than 1 cm), vastly longer than we would expect from the known sizes of atoms. We must ask why condensed matter is so transparent to conduction electrons. The conduction electrons act in this respect as a gas of noninteracting particles.

There are two parts to the answer to our question: (a) A conduction electron is not deflected by ion cores arranged on a *periodic* lattice because matter waves propagate freely in a periodic structure. We showed free propagation of x-rays in periodic lattices in Chapter 2; we discuss electron waves in lattices in Chapter 9. (b) A conduction electron is scattered only infrequently by other conduction electrons. This property we shall show is a consequence of the Pauli exclusion principle. By a **free electron Fermi gas** we shall mean a gas of free and noninteracting electrons which are subject to the Pauli principle.

ENERGY LEVELS AND DENSITY OF ORBITALS IN ONE DIMENSION

We first discuss the behavior of a free electron gas in a one-dimensional world, taking account of quantum theory and of the Pauli principle. Consider an electron of mass m confined to a length L by infinite barriers (Fig. 2). The wavefunction $\psi_n(x)$ of the electron is a solution of the Schrödinger equation $\mathcal{H}\psi = \epsilon\psi$; with the neglect of potential energy we have $\mathcal{H} = p^2/2m$, where p is the momentum. In quantum theory p may be represented by $-i\hbar\, d/dx$, so that

$$\mathcal{H}\psi_n = -\frac{\hbar^2}{2m}\frac{d^2\psi_n}{dx^2} = \epsilon_n\psi_n \; , \tag{1}$$

where ϵ_n is the energy of the electron in the orbital.[1] The boundary conditions are

$$\psi_n(0) = 0 \; ; \qquad \psi_n(L) = 0 \; , \tag{2}$$

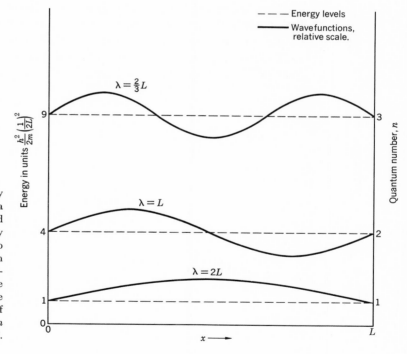

Figure 2   First three energy levels and wavefunctions of a free electron of mass m confined to a line of length L. The energy levels are labeled according to the quantum number n which gives the number of half-wavelengths in the wavefunction. The wavelengths are indicated on the wavefunctions. The energy ϵ_n of the level of quantum number n is equal to $(h^2/2m)(n/2L)^2$.

---

[1] We use the term **orbital** to denote a solution of the wave equation for a system of only one electron. The term allows us to distinguish between an exact quantum state of the wave equation of a system of N electrons and an approximate quantum state which we construct by assigning the N electrons to N different orbitals, where each orbital is a solution of a wave equation for one electron. The orbital model is exact only if there are no interactions between electrons.

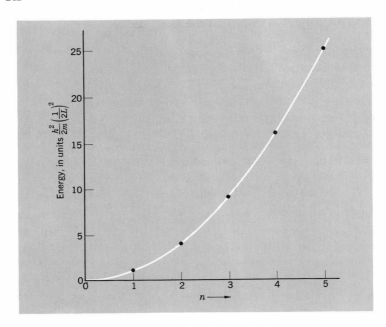

Figure 3 The energy is a quadratic function of the quantum number n, for a free electron confined to a length L in one dimension.

as imposed by the infinite potential energy barrier; they are satisfied if the wavefunction is sinelike with an integral number n of half-wavelengths between 0 and L:

$$\psi_n \propto \sin\left(\frac{2\pi}{\lambda_n}x\right) ; \qquad \tfrac{1}{2}n\lambda_n = L . \tag{3}$$

Thus

$$\psi_n = A \sin\left(\frac{n\pi}{L}x\right) , \tag{4}$$

where A is a constant. We see that (4) is a solution of (1), because

$$\frac{d\psi_n}{dx} = A\left(\frac{n\pi}{L}\right)\cos\left(\frac{n\pi}{L}x\right) ; \qquad \frac{d^2\psi_n}{dx^2} = -A\left(\frac{n\pi}{L}\right)^2\sin\left(\frac{n\pi}{L}x\right) ,$$

whence the energy ϵ_n is given by

$$\epsilon_n = \frac{\hbar^2}{2m}\left(\frac{n\pi}{L}\right)^2 , \tag{5}$$

as shown in Fig. 3.

Suppose we want to accommodate N electrons on the line. The elementary statement of the **Pauli exclusion principle** is that *no two electrons can have all their quantum numbers identical.* That is, each oribtal[2] can be occupied by at most one electron. This applies to electrons in atoms, molecules, or solids. In a solid the quantum numbers of an electron in a conduction electron orbital are n and m_s, where n is any positive integer in (4) and $m_s = \pm\frac{1}{2}$, according to the spin orientation. Each pair of orbitals labeled by the quantum number n can accommodate two electrons, one with its spin up and one with its spin down. If there are eight electrons, then in the ground state of the system the filled orbitals are those given in the table:

n	m_s	Electron occupancy	n	m_s	Electron occupancy
1	↑	1	4	↑	1
1	↓	1	4	↓	1
2	↑	1	5	↑	0
2	↓	1	5	↓	0
3	↑	1			
3	↓	1			

Let n_F denote the topmost filled energy level, where we start filling the levels from the bottom ($n = 1$) and continue filling the higher levels with electrons until all N electrons are accommodated. It is convenient to suppose that N is an even number. The condition $2n_F = N$ determines n_F, the value of n for the uppermost filled level. The **Fermi energy** ϵ_F is defined as the energy of the topmost filled level. By (5) with $n = n_F$ we have

$$\epsilon_F = \frac{\hbar^2}{2m}\left(\frac{n_F\pi}{L}\right)^2 = \frac{\hbar^2}{2m}\left(\frac{N\pi}{2L}\right)^2 , \tag{6}$$

in one dimension.

---

[2] More than one orbital may have the same energy. The number of orbitals with the same energy is called the **degeneracy**.

EFFECT OF TEMPERATURE ON THE FERMI-DIRAC
DISTRIBUTION FUNCTION

The ground state is the state of the system at absolute zero. What happens as the temperature is increased? This is a standard problem in elementary statistical mechanics and the solution (Advanced Topic E) is given by the Fermi-Dirac distribution function. The kinetic energy of the electron gas increases as the temperature is increased: some energy levels are occupied which were vacant at absolute zero, and some levels are vacant which were occupied at absolute zero.

The situation is illustrated by Fig. 4, where the plotted curves are of the function

$$f(\epsilon) = \frac{1}{e^{(\epsilon - \mu)/k_B T} + 1} .$$

(7)

This is the **Fermi-Dirac distribution function:** it gives the probability that an orbital at energy ϵ will be occupied in an ideal electron gas in thermal equilibrium.

The quantity μ is a function of the temperature; μ is to be chosen for the particular problem in such a way that the total number of particles in the system comes out correctly—that is, equal[3] to N. At absolute zero $\mu = \epsilon_F$, because in the limit $T \to 0$ the function $f(\epsilon)$ changes discontinuously from the value 1 (filled) to the value 0 (empty) at $\epsilon = \epsilon_F = \mu$. *At all temperatures $f(\epsilon)$ is equal to $\frac{1}{2}$ when $\epsilon = \mu$,* for then the denominator of (7) has the value 2. The quantity μ is called the **chemical potential,**[4] and we see that at absolute zero the chemical potential is equal to the Fermi energy. At low temperatures μ is close to ϵ_F in value, as in Fig. 5. The Fermi energy was defined as the energy of the topmost filled orbital at absolute zero.

The high energy tail of the distribution is that part for which $\epsilon - \mu \gg k_B T$; here the exponential term is dominant in the denominator of (7), so that $f(\epsilon) \cong e^{(\mu - \epsilon)/k_B T}$. Note that this is essentially the Boltzmann distribution.

---

[3] If the orbital energies are ϵ_i, then we must have $\sum_i f(\epsilon_i) = N$ at all temperatures. In integral form $\int_0^\infty d\epsilon\, f(\epsilon) \mathfrak{D}(\epsilon) = N$, where the density of orbitals $\mathfrak{D}(\epsilon)$ is defined below.

[4] In the presence of an external field ϵ is changed but so also is μ. The new value of μ is usually called the electrochemical potential. The term chemical potential is often reserved for the electrochemical potential minus the potential energy of the particle due to the external field. Such a usage is common in semiconductor physics in connection with junction devices. In *TP*, chemical potential means the same thing as the electrochemical potential.

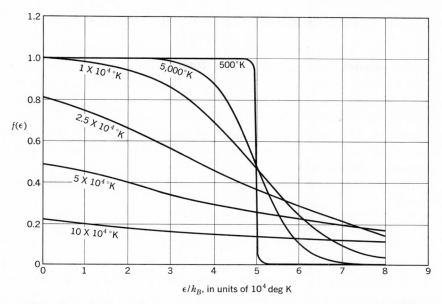

Figure 4   Fermi-Dirac distribution function at various temperatures, for $T_F \equiv \epsilon_F/k_B = 50{,}000$ deg K. The results apply to a gas in three dimensions. The total number of particles is constant, independent of temperature. (Courtesy of B. Feldman.)

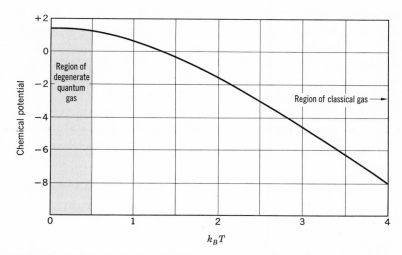

Figure 5   Plot of the chemical potential versus temperature for a gas of noninteracting fermions in three dimensions. For convenience in plotting, the particle concentration has been chosen so that $\mu(0) \equiv \epsilon_F = (\frac{3}{2})^{\frac{2}{3}}$.

FREE ELECTRON GAS IN THREE DIMENSIONS

The free-particle Schrödinger equation in three dimensions is

$$-\frac{\hbar^2}{2m}\left(\frac{\partial^2}{\partial x^2} + \frac{\partial^2}{\partial y^2} + \frac{\partial^2}{\partial z^2}\right)\psi_{\mathbf{k}}(\mathbf{r}) = \epsilon_{\mathbf{k}}\psi_{\mathbf{k}}(\mathbf{r}) \ . \tag{8}$$

If the electrons are confined to a cube of edge L, the analog to the wavefunction (4) is

$$\psi_{\mathbf{n}}(\mathbf{r}) = A \sin(\pi n_x x/L) \sin(\pi n_y y/L) \sin(\pi n_z z/L) \ , \tag{9}$$

where n_x, n_y, n_z are positive integers. This is a standing wave.

It is convenient to introduce wavefunctions which satisfy periodic boundary conditions, as we did for phonons in Chapter 6. We now require the wavefunctions to be periodic in x, y, z with period L. Thus

$$\psi(x + L, y, z) = \psi(x, y, z) \ , \tag{10}$$

and similarly for the y and z coordinates. Wavefunctions satisfying the free-particle Schrödinger equation (8) and the periodicity condition (10) are of the form of a traveling plane wave:

$$\boxed{\psi_{\mathbf{k}}(\mathbf{r}) = e^{i\mathbf{k}\cdot\mathbf{r}} \ ,} \tag{11}$$

provided that the components of the wavevector $\mathbf{k}$ satisfy

$$k_x = 0 \ ; \quad \pm\frac{2\pi}{L} \ ; \quad \pm\frac{4\pi}{L} \ ; \quad \dots , \tag{12}$$

and similarly for k_y and k_z. That is, any component of $\mathbf{k}$ is of the form $2n\pi/L$, where n is a positive or negative integer. The components of $\mathbf{k}$ are the quantum numbers of the problem, along with the quantum number m_s for the spin direction. We confirm that these values of k_x satisfy (10), for

$$\exp[ik_x(x + L)] = \exp[i2n\pi(x + L)/L] =$$
$$\exp(i2n\pi x/L)\exp(i2n\pi) = \exp(i2n\pi x/L) = \exp(ik_x x) \ . \tag{13}$$

On substituting (11) in (8) we have

$$\epsilon_{\mathbf{k}} = \frac{\hbar^2}{2m}k^2 = \frac{\hbar^2}{2m}(k_x{}^2 + k_y{}^2 + k_z{}^2) \tag{14}$$

for the energy $\epsilon_{\mathbf{k}}$ of the orbital with wavevector $\mathbf{k}$. The magnitude of the wavevector is related to the wavelength λ by

$$k = \frac{2\pi}{\lambda} \ . \tag{15}$$

The linear momentum $\mathbf{p}$ may be represented in quantum mechanics by the operator $\mathbf{p} = -i\hbar\nabla$, whence for the orbital (11)

$$\mathbf{p}\psi_{\mathbf{k}}(\mathbf{r}) = -i\hbar\nabla\psi_{\mathbf{k}}(\mathbf{r}) = \hbar\mathbf{k}\psi_{\mathbf{k}}(\mathbf{r}) \ , \tag{16}$$

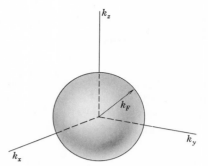

Figure 6   In the ground state of a system of N free electrons the occupied orbitals of the system fill a sphere of radius k_F, where $\epsilon_F = \hbar^2 k_F{}^2/2m$ is the energy of an electron having a wavevector k_F on the surface of a sphere.

so that the plane wave $\psi_{\mathbf{k}}$ is an eigenfunction of the linear momentum with the eigenvalue $\hbar\mathbf{k}$. The particle velocity in the orbital $\mathbf{k}$ is given by

$$\mathbf{v} = \frac{\hbar\mathbf{k}}{m} \ . \tag{17}$$

In the ground state of a system of N free electrons the occupied orbitals may be represented as points inside a sphere in $\mathbf{k}$ space. The energy at the surface of the sphere is the Fermi energy; the wavevectors at the Fermi surface have a magnitude k_F such that (Fig. 6):

$$\epsilon_F = \frac{\hbar^2}{2m} k_F{}^2 \ . \tag{18}$$

From the conditions (12) we see that there is one allowed wavevector—that is, one distinct triplet of quantum numbers k_x, k_y, k_z—for the volume element $(2\pi/L)^3$ of $\mathbf{k}$ space. Thus in the sphere of volume $4\pi k_F{}^3/3$ the total number of orbitals is

$$2 \cdot \frac{4\pi k_F{}^3/3}{(2\pi/L)^3} = \frac{V}{3\pi^2} k_F{}^3 = N \ , \tag{19}$$

where the factor 2 on the left comes from the two allowed values of m_s, the spin quantum number, for each allowed value of $\mathbf{k}$. We have set the number of orbitals equal to N, the number of electrons. Then

$$k_F = \left(\frac{3\pi^2 N}{V}\right)^{\frac{1}{3}} \ ; \tag{20}$$

this depends only on the particle concentration and not on the mass.

Using (18),
$$\boxed{\epsilon_F = \frac{\hbar^2}{2m}\left(\frac{3\pi^2 N}{V}\right)^{\frac{2}{3}} \ .} \tag{21}$$

This relates the Fermi energy to the electron concentration N/V and the mass m. The electron velocity v_F at the Fermi surface is

$$v_F = \left(\frac{\hbar k_F}{m}\right) = \left(\frac{\hbar}{m}\right)\left(\frac{3\pi^2 N}{V}\right)^{\frac{1}{3}} \ . \tag{22}$$

Table 1   Calculated free electron Fermi surface parameters for metals

[All metals at room temperature except for Na, K, Rb, Cs at 5°K and Li at 78°K]

Notes: The electron concentration N/V is found by multiplying the valency by the atomic concentration given in Table 1.5. In units cm^{-1} for k_F, $cm\ s^{-1}$ for v_F, and cm^3 for V, we have the following numerical relations: the Fermi wavevector is $k_F = (3\pi^2 N/V)^{\frac{1}{3}} = (29.609\ N/V)^{\frac{1}{3}}$; and the Fermi velocity is $v_F = \hbar k_F/m = 1.157\ k_F$. The Fermi energy is $\epsilon_F = \frac{1}{2}mv_F^2$, or $\epsilon_F(eV) = 0.284 \times 10^{-15}\ v_F^2$; and $T_F(\deg K) = 1.16 \times 10^4\ \epsilon_F(eV)$.

Valency	Metal	Electron concentration, in cm^{-3}	Radius° parameter r_s	Fermi wavevector, in cm^{-1}	Fermi velocity, in $cm\ s^{-1}$	Fermi energy, in eV	Fermi temperature, $T_F \equiv \epsilon_F/k_B$ in deg K
1	Li	4.70×10^{22}	3.25	1.11×10^8	1.29×10^8	4.72	5.48×10^4
	Na	2.65	3.93	0.92	1.07	3.23	3.75
	K	1.40	4.86	0.75	0.86	2.12	2.46
	Rb	1.15	5.20	0.70	0.81	1.85	2.15
	Cs	0.91	5.63	0.64	0.75	1.58	1.83
	Cu	8.45	2.67	1.36	1.57	7.00	8.12
	Ag	5.85	3.02	1.20	1.39	5.48	6.36
	Au	5.90	3.01	1.20	1.39	5.51	6.39
2	Be	24.2	1.88	1.93	2.23	14.14	16.41
	Mg	8.60	2.65	1.37	1.58	7.13	8.27
	Ca	4.60	3.27	1.11	1.28	4.68	5.43
	Sr	3.56	3.56	1.02	1.18	3.95	4.58
	Ba	3.20	3.69	0.98	1.13	3.65	4.24
	Zn	13.10	2.31	1.57	1.82	9.39	10.90
	Cd	9.28	2.59	1.40	1.62	7.46	8.66
3	Al	18.06	2.07	1.75	2.02	11.63	13.49
	Ga	15.30	2.19	1.65	1.91	10.35	12.01
	In	11.49	2.41	1.50	1.74	8.60	9.98
4	Pb	13.20	2.30	1.57	1.82	9.37	10.87
	Sn(w)	14.48	2.23	1.62	1.88	10.03	11.64

° The dimensionless radius parameter is defined as $r_s = r_0/a_H$, where a_H is the first Bohr radius $(0.529 \times 10^{-8}\ cm)$ and r_0 is the radius of a sphere that contains one electron.

Calculated values of k_F, v_F, and ϵ_F are given in Table 1 for selected metals; also given are values of the quantity T_F which is defined as ϵ_F/k_B. (The quantity T_F has nothing to do with the temperature of the electron gas!)

We now find an expression for the number of orbitals per unit energy range, $\mathfrak{D}(\epsilon)$, often called the density of states. We use (21) to express the total number of orbitals of energy $\leq \epsilon_F$ in terms of ϵ_F:

$$N = \frac{V}{3\pi^2}\left(\frac{2m\epsilon_F}{\hbar^2}\right)^{\frac{3}{2}} , \tag{23}$$

so that the density of orbitals at the Fermi energy is

$$\boxed{\mathfrak{D}(\epsilon_F) \equiv \frac{dN}{d\epsilon_F} = \frac{V}{2\pi^2}\cdot\left(\frac{2m}{\hbar^2}\right)^{\frac{3}{2}}\cdot\epsilon_F^{\frac{1}{2}} .} \tag{24}$$

This result may be obtained and expressed most simply by writing (23) as

$$\log N = \frac{3}{2}\log\epsilon_F + \text{constant} ; \qquad \frac{dN}{N} = \frac{3}{2}\cdot\frac{d\epsilon_F}{\epsilon_F} ,$$

whence

$$\mathfrak{D}(\epsilon_F) \equiv \frac{dN}{d\epsilon_F} = \frac{3N}{2\epsilon_F} . \tag{25}$$

Within a factor of the order of unity, the number of orbitals per unit energy range at the Fermi energy is just the total number of conduction electrons divided by the Fermi energy.

These results apply to free electrons, with ϵ proportional to k^2. We can obtain a result for a general relation $\epsilon(\mathbf{k})$ by direct analogy with the argument of Eq. (6.34):

$$\mathfrak{D}(\epsilon) = \frac{2V}{(2\pi)^3}\int\frac{dS_\epsilon}{|\text{grad}_\mathbf{k}\,\epsilon|} , \tag{26}$$

where the factor of 2 arises from the two spin orientations; V is the volume of the specimen; and dS_ϵ is the element of area in $\mathbf{k}$ space of the surface of constant energy ϵ.

HEAT CAPACITY OF THE ELECTRON GAS

The question which caused the greatest difficulty in the early development of the electron theory of metals concerns the heat capacity of the conduction electrons. Classical statistical mechanics predicts that a free particle should have a heat capacity of $\frac{3}{2}k_B$, where k_B is the Boltzmann constant. If N atoms each give one valence electron to the electron gas, and the electrons are freely mobile, then the electronic contribution to the heat capacity should be $\frac{3}{2}Nk_B$. But the observed electronic contribution at room temperature is usually less

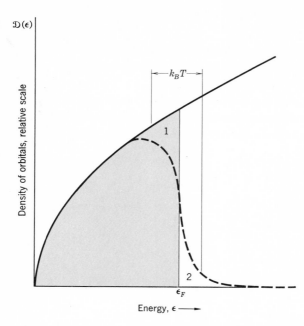

Figure 7 Density of single-particle states as a function of energy, for a free electron gas in three dimensions. The dashed curve represents the density $f(\epsilon, T)\mathfrak{D}(\epsilon)$ of filled orbitals at a finite temperature, but such that $k_B T$ is small in comparison with ϵ_F. The shaded area represents the filled orbitals at absolute zero. The average energy is increased when the temperature is increased from 0 to T, for electrons are thermally excited from region 1 to region 2.

than 0.01 of this value. This discrepancy distracted the early workers, such as Lorentz: how can the electrons participate in electrical conduction processes as if they were mobile, while not contributing to the heat capacity? The question was answered only upon the discovery of the Pauli exclusion principle and the Fermi distribution function. Fermi found the correct equation, and he could then write, "One recognizes that the specific heat vanishes at absolute zero and that at low temperatures it is proportional to the absolute temperature."

When we heat the specimen from absolute zero not every electron gains an energy $\sim k_B T$ as expected classically, but only those electrons in orbitals within an energy range $k_B T$ of the Fermi level are excited thermally; these electrons gain an energy which is itself of the order of $k_B T$, as in Fig. 7. This gives an immediate qualitative solution to the problem of the heat capacity of the conduction electron gas. If N is the total number of electrons, only a fraction of the order of T/T_F can be excited thermally at temperature T, because only these lie within an energy range of the order of $k_B T$ of the top of the energy distribution. Each of these NT/T_F electrons has a thermal energy of the order of $k_B T$, and so the total electronic thermal energy ΔE is of the order of

$$\Delta E \approx \frac{NT}{T_F} k_B T \ .$$

The electronic heat capacity is given by

$$C_{el} = \frac{\partial\,\Delta E}{\partial T} \approx N k_B \cdot \frac{T}{T_F} \tag{27}$$

and is directly proportional to T, in agreement with the experimental results discussed in the following section. At room temperature C_{el} in (27) is smaller than the classical value $\frac{3}{2}Nk_B$ by a factor of the order of 0.01 or less, for a typical value $\epsilon_F/k_B \equiv T_F \sim 5 \times 10^4$ deg.

We now derive a quantitative expression for the electronic heat capacity valid at low temperatures $k_BT \ll \epsilon_F$. The argument is simple, but devious. The increase ΔE in the total energy (Fig. 7) of a system of N electrons when heated from 0 to T is

$$\Delta E = \int_0^\infty d\epsilon \, \epsilon \, \mathfrak{D}(\epsilon) \, f(\epsilon) - \int_0^{\epsilon_F} d\epsilon \, \epsilon \, \mathfrak{D}(\epsilon) \; . \tag{28}$$

Here $f(\epsilon)$ is the Fermi-Dirac function and $\mathfrak{D}(\epsilon)$ is the number of orbitals per unit energy range. We multiply the number of particles

$$N = \int_0^\infty d\epsilon \, f(\epsilon) \, \mathfrak{D}(\epsilon)$$

by ϵ_F to obtain

$$\epsilon_F N = \epsilon_F \int_0^\infty d\epsilon \, f(\epsilon) \, \mathfrak{D}(\epsilon) \; . \tag{29}$$

We now differentiate (28) and (29):

$$C_{el} = \frac{\partial \, \Delta E}{\partial T} = \int_0^\infty d\epsilon \, \epsilon \, \mathfrak{D}(\epsilon) \frac{\partial f}{\partial T} \; ; \tag{30}$$

$$0 = \epsilon_F \frac{\partial N}{\partial T} = \int_0^\infty d\epsilon \, \epsilon_F \, \mathfrak{D}(\epsilon) \frac{\partial f}{\partial T} \; , \tag{31}$$

and subtract the second line from the first to find the electronic heat capacity in the form

$$C_{el} = \frac{\partial \, \Delta E}{\partial T} = \int_0^\infty d\epsilon (\epsilon - \epsilon_F) \frac{\partial f}{\partial T} \mathfrak{D}(\epsilon) \; . \tag{32}$$

At the low temperatures $(k_BT/\epsilon_F < 0.01)$ of interest the derivative $\partial f/\partial T$ is large only at energies near ϵ_F, so that we may evaluate $\mathfrak{D}(\epsilon)$ at ϵ_F and take it outside of the integrand:

$$C_{el} \cong \mathfrak{D}(\epsilon_F) \int_0^\infty d\epsilon \, (\epsilon - \epsilon_F) \frac{\partial f}{\partial T} \; . \tag{33}$$

Examination[5] of Fig. 4 suggests that to the first order in T we may in the expression (7) for f replace the chemical potential μ by the constant Fermi energy ϵ_F as defined by $\epsilon_F \equiv \mu(0)$. Then

$$\frac{\partial f}{\partial T} = \frac{\epsilon - \epsilon_F}{k_B T^2} \cdot \frac{e^{(\epsilon - \epsilon_F)/k_B T}}{[e^{(\epsilon - \epsilon_F)/k_B T} + 1]^2} \; ,$$

---

[5] The present derivation was suggested by J. Twidell, private communication.

and, setting

$$x \equiv (\epsilon - \epsilon_F)/k_B T \ ,$$

it follows from (33) that

$$C_{el} = \mathfrak{D}(\epsilon_F)(k_B{}^2 T) \int_{-\epsilon_{F/k_B T}}^{\infty} dx \, x^2 \frac{e^x}{(e^x + 1)^2} \ . \tag{34}$$

Because the factor e^x in the integrand is negligible at $x = -\epsilon_F/k_B T$, we may safely replace the lower limit by $-\infty$. The integral becomes[6]

$$\int_{-\infty}^{\infty} dx \, x^2 \frac{e^x}{(e^x + 1)^2} = \frac{\pi^2}{3} \ , \tag{35}$$

or

$$\boxed{C_{el} = \tfrac{1}{3}\pi^2 \mathfrak{D}(\epsilon_F) k_B{}^2 T \ .} \tag{36}$$

From (25) we have for a free electron gas:

$$\mathfrak{D}(\epsilon_F) = \frac{3N}{2\epsilon_F} = \frac{3N}{2k_B T_F} \ , \tag{37}$$

with $k_B T_F \equiv \epsilon_F$. Thus

$$C_{el} = \tfrac{1}{2}\pi^2 N k_B \cdot \frac{k_B T}{\epsilon_F} = \tfrac{1}{2}\pi^2 N k_B \cdot \frac{T}{T_F} \ , \tag{38}$$

in agreement with the qualitative result (27).

Experimental Heat Capacity of Metals

At temperatures much below the Debye temperature and very much below the Fermi temperature, the heat capacity of metals at constant volume may be written as the sum of electronic and lattice contributions:

$$C = \gamma T + AT^3 \ ,$$

where γ and A are constants characteristic of the material and are given by (36) and (6.47). The electronic term is linear in T and is dominant at sufficiently low temperatures. It is convenient to exhibit the experimental values of C as a plot of C/T versus T^2:

$$\frac{C}{T} = \gamma + AT^2 \ , \tag{39}$$

for then the points should lie on a straight line with slope A and intercept γ. Such a plot for potassium is shown in Fig. 8. The apparatus used for these measurements is shown in Fig. 9. Observed values of γ are given in Table 2; values of the Debye θ derived from observed values of A were given in Table 6.1.

---

[6] See, for example, integral 313.11b of the Gröbner and Hofreiter tables, Vol. 2; note that the integrand is an even function of x. See also *TP*, p. 234.

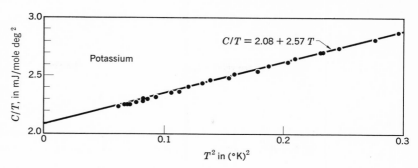

Figure 8 Experimental heat capacity values for potassium, plotted as C/T versus T^2. The solid points were determined with an adiabatic demagnetization cryostat. [After W. H. Lien and N. E. Phillips, Phys. Rev. **133**, A1370 (1964).]

Potassium

$$C/T = 2.08 + 2.57\,T$$

C/T, in mJ/mole deg ^{2}

T^2 in $(°K)^2$

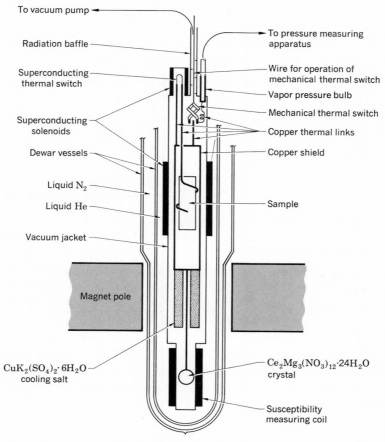

To vacuum pump

Radiation baffle

Superconducting thermal switch

Superconducting solenoids

Dewar vessels

Liquid N_2

Liquid He

Vacuum jacket

Magnet pole

$CuK_2(SO_4)_2 \cdot 6H_2O$ cooling salt

To pressure measuring apparatus

Wire for operation of mechanical thermal switch

Vapor pressure bulb

Mechanical thermal switch

Copper thermal links

Copper shield

Sample

$Ce_2Mg_3(NO_3)_{12} \cdot 24H_2O$ crystal

Susceptibility measuring coil

Figure 9 Apparatus for heat capacity measurements between 0.05 and 1°K. The mechanical thermal switch is used to cool the sample and paramagnetic salts to 1°K by evaporation of liquid He. Temperatures below 1°K are produced by adiabatic demagnetization (Chapter 15) of $CuK_2(SO_4)_2 \cdot 6H_2O$, and determined from the magnetic susceptibility of $Ce_2Mg_3(NO_3)_{12} \cdot 24H_2O$, which follows a Curie law at these temperatures. Thermal contact between the sample and the paramagnetic salts is made through the superconducting thermal switch, a Pb wire that can be made either normal (good heat conductor) or superconducting (poor heat conductor) by adjustment of the current in the surrounding superconducting solenoid. A resistance thermometer, which is calibrated against the magnetic thermometer, and a resistance heater are attached directly to the sample. (Courtesy of N. Phillips.)

254

Table 2 Experimental and free electron values of electronic heat constant γ of metals

(From compilations kindly furnished by N. Phillips and N. Pearlman.)

Observed γ in mJ mol^{-1} deg^{-2}.
Calculated free electron γ in mJ mol^{-1} deg^{-2}.
m^*/m = (observed γ)/(free electron γ).

1	2	3	4	5	6	7	8	9	10	11	12	13	14	15
Li 1.63 0.749 2.18	**Be** 0.17 0.500 0.34											**B**	**C**	**N**
Na 1.38 1.094 1.26	**Mg** 1.3 0.992 1.3											**Al** 1.35 0.912 1.48	**Si**	**P**
K 2.08 1.668 1.25	**Ca** 2.9 1.511 1.9	**Sc** 10.7	**Ti** 3.35	**V** 9.26	**Cr** 1.40	**Mn(γ)** 9.20	**Fe** 4.98	**Co** 4.73	**Ni** 7.02	**Cu** 0.695 0.505 1.38	**Zn** 0.64 0.753 0.85	**Ga** 0.596 1.025 0.58	**Ge**	**As** 0.19
Rb 2.41 1.911 1.26	**Sr** 3.6 1.790 2.0	**Y** 10.2	**Zr** 2.80	**Nb** 7.79	**Mo** 2.0	**Tc** —	**Ru** 3.3	**Rh** 4.9	**Pd** 9.42	**Ag** 0.646 0.645 1.00	**Cd** 0.688 0.948 0.73	**In** 1.69 1.233 1.37	**Sn** (w) 1.78 1.410 1.26	**Sb** 0.11
Cs 3.20 2.238 1.43	**Ba** 2.7 1.937 1.4	**La** 10.	**Hf** 2.16	**Ta** 5.9	**W** 1.3	**Re** 2.3	**Os** 2.4	**Ir** 3.1	**Pt** 6.8	**Au** 0.729 0.642 1.14	**Hg(α)** 1.79 0.952 1.88	**Tl** 1.47 1.29 1.14	**Pb** 2.98 1.509 1.97	**Bi** 0.008

The values given for γ are believed to be reliable to within perhaps 2 percent. The values refer to one mole.

The observed values of the coefficient γ are of the expected magnitude, but often do not agree very closely with the value calculated for free electrons of mass m by use of the relation (38):

$$\gamma(\text{free}) = \tfrac{1}{3}\pi^2 \mathfrak{D}(\epsilon_F) k_B{}^2 = \frac{\pi^2 N_0 k_B{}^2 z}{2\epsilon_F} , \qquad (40)$$

for a mole of material, where N_0 is Avogadro's number and z is the valency of the element. It is common practice to express the ratio of the observed to the free electron values of the electronic heat capacity as a ratio of a **thermal effective mass** m_{th}^* to the electron mass m, where m_{th}^* is defined by the relation

$$\frac{m_{\text{th}}^*}{m} \equiv \frac{\gamma(\text{observed})}{\gamma(\text{free})} . \qquad (41)$$

This form arises in a natural way because ϵ_F in the expression (40) for γ is inversely proportional to the mass of the electron, whence $\gamma \propto m$. Values of the ratio (41) are given in Table 2. The departure of the ratio from unity involves three separate effects:

1. The interaction of the conduction electrons with the periodic potential of the rigid crystal lattice. The effective mass of an electron in this potential is called the band effective mass and is treated in Chapters 9 and 10.

2. The interaction of the conduction electrons with phonons. An electron tends to polarize or distort the lattice in its neighborhood, so that the moving electron tries to drag nearby ions along, thereby increasing the effective mass of the electron.[7] In ionic crystals the effect has a name, the polaron effect (Chapter 11).

3. The interaction of the conduction electrons with themselves. A moving electron causes an inertial reaction in the surrounding electron gas, thereby increasing the effective mass of the electron. The effects of electron-electron interactions are usually described within the framework of what is called the Landau theory of a Fermi liquid.

The Fermi Liquid

A Fermi gas is a system of noninteracting identical particles subject to the Pauli exclusion principle. The same system with interactions is called a **Fermi liquid.** Conduction electrons in a metal form a Fermi liquid, and liquid He^3 is also a Fermi liquid.

---

[7] See, for example, the detailed discussion of Zn and Cd by P. B. Allen, M. L. Cohen, L. M. Falicov, and R. V. Kasowski, Phys. Rev. Letters **21**, 1794 (1968).

The theory of the Fermi liquid is due to Landau.[8] The object of the theory is to give a unified account of the effect of interactions between particles on the properties of a system of fermions. The results of the theory are expressed in terms of macroscopic parameters that sometimes may be calculated from first principles and sometimes may be determined experimentally.[9]

Landau's theory of the Fermi liquid gives a good account of the low-lying single particle excitations of the system of interacting electrons. These single particle excitations are called **quasiparticles**; they have a one-to-one correspondence with the single particle excitations of the free electron gas. A quasiparticle may be thought of as a single particle accompanied by a distortion cloud in the electron gas. One effect of the Coulomb interactions between electrons is to change the effective mass of the electron; in the alkali metals the increase is roughly of the order of 25 percent. Other metals have smaller values of the parameter r_s (see Table 1) than the alkalis, and here the effective mass correction from Coulomb interactions is smaller and may be slightly negative.

The theory predicts two types of collective modes[10] of wave propagation (collective excitations) in a Fermi liquid, one mode called zero sound in which a distortion of the Fermi surface from sphericity drives the oscillation, and another mode like the spin waves of Chapter 16.

---

[8] L. Landau, "Theory of a Fermi liquid," Soviet Physics JETP **3**, 920 (1957); see also D. Pines and P. Nozieres, *Theory of quantum liquids*, Benjamin, 1966, Vol. I.

[9] See, for example, the analysis of Na and K by T. M. Rice, Phys. Rev. **175**, 858 (1968), and also T. M. Rice, Annals Phys. **31**, 100 (1965).

[10] L. Landau, "Oscillations in a Fermi liquid," Soviet Physics JETP **5**, 10 (1957).

ELECTRICAL CONDUCTIVITY AND OHM'S LAW

The momentum of a free electron is related to the wavevector by (17):

$$m\mathbf{v} = \hbar\mathbf{k} \ . \tag{42}$$

In an electric field $\mathbf{E}$ and magnetic field $\mathbf{B}$ the force $\mathbf{F}$ on the electron[11] is $-e[\mathbf{E} + (1/c)\mathbf{v} \times \mathbf{B}]$, so that Newton's second law of motion becomes

$$\text{(CGS)} \qquad \boxed{\mathbf{F} = m\frac{d\mathbf{v}}{dt} = \hbar\frac{d\mathbf{k}}{dt} = -e\left(\mathbf{E} + \frac{1}{c}\mathbf{v} \times \mathbf{B}\right) \ .} \tag{43}$$

In the absence of collisions the Fermi sphere (Fig. 10) in $\mathbf{k}$ space is displaced at a uniform rate by a constant applied electric field. We integrate (43) with $\mathbf{B} = 0$ to obtain

$$\mathbf{k}(t) - \mathbf{k}(0) = -e\mathbf{E}t/\hbar \ . \tag{44}$$

If the field is applied at time $t = 0$ to an electron gas that fills the Fermi sphere centered at the origin of $\mathbf{k}$ space, then at a later time τ the sphere will be displaced to a new center at

$$\delta\mathbf{k} = -e\mathbf{E}\tau/\hbar \ . \tag{45}$$

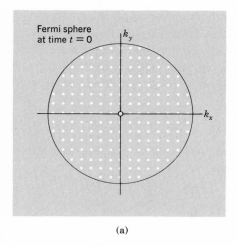

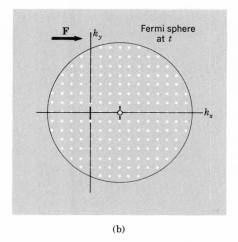

(a)   (b)

Figure 10 (a) The Fermi sphere encloses the occupied electron orbitals in $\mathbf{k}$ space in the ground state of the electron gas. The net momentum is zero, because for every orbital $\mathbf{k}$ there is an occupied orbital at $-\mathbf{k}$. (b) Under the influence of a constant force $\mathbf{F}$ acting for a time interval t every orbital has its $\mathbf{k}$ vector increased by $\delta\mathbf{k} = \mathbf{F}t/\hbar$. This is equivalent to a displacement of the whole Ferm sphere by $\delta\mathbf{k}$. The total momentum is $N\hbar\delta\mathbf{k}$, if there are N electrons present. The application of the force increases the energy of the system by $N(\hbar\delta\mathbf{k})^2/2m$.

[11] The charge on the proton is written as e.

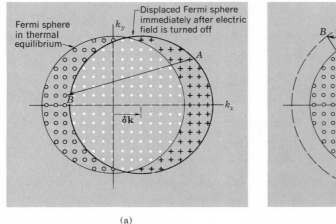
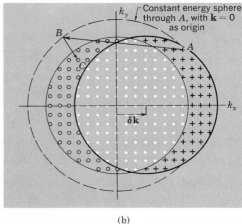

(a)   (b)

Figure 11 (a) When the applied force is turned off, collision processes tend to return the system to the ground state. We need to transfer electrons from those filled orbitals marked with crosses ($+$) to those empty orbitals marked with circles ($\bigcirc$). An electron at A can make a transition to an empty orbital, say B, by the emission of a phonon of suitable wavevector and frequency. (b) Elastic scattering of an electron at A by a static imperfection or impurity can carry the electron to any point such as B which lies on the sphere of constant energy ϵ_A. Elastic scattering will reduce the total momentum to zero by redistributing the occupied orbitals ($+$), but phonon processes such as $B \to C$ are needed to return the distribution to the ground state.

Because of collisions of electrons with impurities, lattice imperfections, and phonons, the displaced sphere may be maintained in a steady state in an electric field. The effects of the collisions on the distribution after the applied field is switched off are shown in Fig. 11. If collision time is τ, the displacement of the Fermi sphere in the steady state is given by (45). The incremental velocity is

$$\delta\mathbf{v} = -\frac{e\mathbf{E}\tau}{m} \; . \tag{46}$$

If in a constant electric field $\mathbf{E}$ there are n electrons of charge $q = -e$ per unit volume, the electric current density is

$$\mathbf{j} = nq \, \delta\mathbf{v} = \frac{ne^2\tau\mathbf{E}}{m} \; , \tag{47}$$

using (46). This is in the form of Ohm's law. The electrical conductivity σ is defined by $\mathbf{j} = \sigma\mathbf{E}$, so that

$$\boxed{\sigma = \frac{ne^2\tau}{m} \; .} \tag{48}$$

The electrical resistivity ρ is defined as the reciprocal of the conductivity, so that

$$\rho = \frac{m}{ne^2\tau} \; . \tag{49}$$

Values of the electrical conductivity and resistivity of the elements are given in Table 3.

It is easy to understand the result (48) for the conductivity. We expect the charge transported to be proportional to the charge density ne; the factor e/m enters because the acceleration in a given electric field is proportional to e and inversely proportional to the mass m; and the time τ describes the free time during which the field acts on the carrier. (We suppose that each collision removes all memory of the drift velocity.)

It is possible to obtain crystals of copper so pure that their conductivity at liquid helium temperatures (4°K) is nearly 10^5 times that at room temperature; for these conditions $\tau \approx 2 \times 10^{-9}$ sec at 4°K. The mean free path ℓ of a conduction electron is defined as

$$\ell = v_F \tau \ , \tag{50}$$

where v_F is the velocity at the Fermi surface. We see from Fig. 11 that all colli-

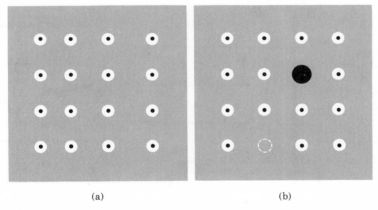

(a)   (b)

Figure 12   Electrical resistivity in most metals arises from collisions of electrons with irregularities in the lattice, as in (a) by phonons and in (b) by impurities and vacant lattice sites.

sions involve only electrons near the Fermi surface. From Table 1 we have $v_F = 1.57 \times 10^8$ cm/sec for Cu; the mean free path is

$$\ell(300°K) \approx 3 \times 10^{-6} \text{ cm} \ ; \qquad \ell(4°K) \approx 0.3 \text{ cm} \ . \tag{51}$$

Mean free paths as long as 10 cm have been observed in very pure metals in the liquid helium temperature range.

Experimental Electrical Resistivity of Metals

The electrical resistivity of most metals is dominated at room temperature (300°K) by collisions of the conduction electrons with lattice phonons and at liquid helium temperature (4°K) by collisions with impurity atoms and mechanical imperfections in the lattice (Fig. 12).

Table 3 Electrical conductivity and resistivity of metals at 295°K

(Resistivity values as given by G. T. Meaden, *Electrical resistance of metals*, Plenum, 1965; residual resistivities have been subtracted.)

Conductivity in units of 10^5 (ohm-cm)$^{-1}$.
Resistivity in units of 10^{-6} ohm-cm.

Li	Be											B	C	N	O	F	Ne
1.07 / 9.32	3.08 / 3.25																
Na	Mg											Al	Si	P	S	Cl	Ar
2.11 / 4.75	2.33 / 4.30											3.65 / 2.74					
K	Ca	Sc	Ti	V	Cr	Mn	Fe	Co	Ni	Cu	Zn	Ga	Ge	As	Se	Br	Kr
1.39 / 7.19	2.78 / 3.6	0.21 / 46.8	0.23 / 43.1	0.50 / 19.9	0.78 / 12.9	0.072 / 139.	1.02 / 9.8	1.72 / 5.8	1.43 / 7.0	5.88 / 1.70	1.69 / 5.92	0.67 / 14.85					
Rb	Sr	Y	Zr	Nb	Mo	Tc	Ru	Rh	Pd	Ag	Cd	In	Sn (w)	Sb	Te	I	Xe
0.80 / 12.5	0.47 / 21.5	0.17 / 58.5	0.24 / 42.4	0.69 / 14.5	1.89 / 5.3	~0.7 / ~14.	1.35 / 7.4	2.08 / 4.8	0.95 / 10.5	6.21 / 1.61	1.38 / 7.27	1.14 / 8.75	0.91 / 11.0	0.24 / 41.3			
Cs	Ba	La	Hf	Ta	W	Re	Os	Ir	Pt	Au	Hg liq.	Tl	Pb	Bi	Po	At	Rn
0.50 / 20.0	0.26 / 39.	0.13 / 79.	0.33 / 30.6	0.76 / 13.1	1.89 / 5.3	0.54 / 18.6	1.10 / 9.1	1.96 / 5.1	0.96 / 10.4	4.55 / 2.20	0.10 / 95.9	0.61 / 16.4	0.48 / 21.0	0.086 / 116.	0.22 / 46.		
Fr	Ra	Ac															

Ce	Pr	Nd	Pm	Sm	Eu	Gd	Tb	Dy	Ho	Er	Tm	Yb	Lu
0.12 / 81.	0.15 / 67.	0.17 / 59.		0.10 / 99.	0.11 / 89.	0.070 / 134.	0.090 / 111.	0.11 / 90.0	0.13 / 77.7	0.12 / 81.	0.16 / 62.	0.38 / 26.4	0.19 / 53.
Th	Pa	U	Np	Pu	Am	Cm	Bk	Cf	Es	Fm	Md	No	Lw
0.66 / 15.2		0.39 / 25.7	0.085 / 118.	0.070 / 143.									

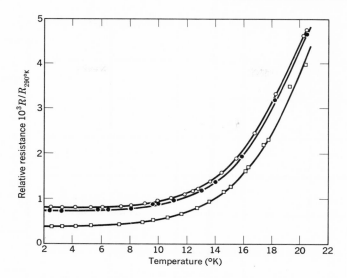

Figure 13 Resistance of sodium below 20°K, as measured on three specimens by MacDonald and Mendelssohn, Proc. Roy. Soc. (London) **A202**, 103 (1950).

The resistivity of a metal containing impurity atoms may usually be written in the form

$$\rho = \rho_L + \rho_i \ , \tag{52}$$

where ρ_L is the resistivity caused by thermal motion of the lattice, and ρ_i is the resistivity caused by scattering of the electron waves by impurity atoms which disturb the periodicity of the lattice. If the concentration of impurity atoms is small, ρ_i is found to be independent of temperature; this statement is known as **Matthiessen's rule**. The **residual resistivity** is the extrapolated resistivity at 0°K and is equivalent to ρ_i, because ρ_L vanishes as $T \to 0$. Measurements on sodium in Fig. 13 show that the residual resistance may vary from specimen to specimen, whereas the resistivity caused by thermal motion is independent of the specimen. The **resistivity ratio** of a specimen is usually defined as the ratio of its resistivity at room temperature to its resistivity at liquid helium temperature. In exceptional specimens the ratio may be as high as 10^5 or even 10^6; in certain alloys the ratio may be as low as 2.

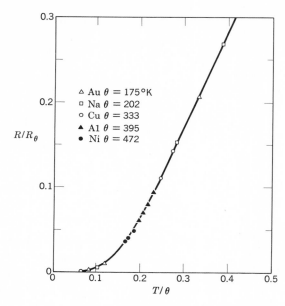

Figure 14 Theoretical (Grüneisen) temperature variation of electrical resistance, and experimental values of various metals. (After Bardeen.)

The lattice or phonon contribution to the electrical resistivity depends on temperature in simple metals essentially as $\rho_L \propto T$, except at low temperatures. For $T \ll \theta$ we have $\rho_L \propto T^5$. Experimental results are shown in Fig. 14. The direct proportionality to T at high temperatures follows because the probability of scattering of an electron is proportional to the number of phonons. (At high temperatures the number of phonons is a measure of the mean square local strain). Reference to detailed theoretical calculations of the conductivity of metals are given by Meaden.[12]

THERMAL CONDUCTIVITY OF METALS

In Chapter 6 we found an expression $K = \frac{1}{3}Cv\ell$ for the thermal conductivity by particles of velocity v, heat capacity C per unit volume, and mean free path ℓ. The thermal conductivity of a Fermi gas follows from (38) for the heat capacity, and with $\epsilon_F = \frac{1}{2}mv_F{}^2$:

$$K_{el} = \frac{\pi^2}{3} \cdot \frac{nk_B{}^2 T}{mv_F{}^2} \cdot v_F \cdot \ell = \frac{\pi^2 nk_B{}^2 T\tau}{3m} \ . \tag{53}$$

Here $\ell = v_F\tau$; the electron concentration is n, and τ is the collision time.

Do the electrons or the phonons carry the greater part of the heat current in a metal? At room temperature normal pure metals tend to have values of the thermal conductivity one or two orders of magnitude higher than for dielectric solids, so that under these conditions the electrons must carry almost all the heat current. In pure metals the electronic contribution is dominant at

[12] G. T. Meaden, *Electrical resistance of metals*, Plenum, 1965; see, in particular, A. Hasegawa, J. Phys. Soc. Japan **19**, 504 (1964).

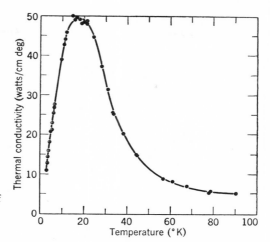

Figure 15 The thermal conductivity of copper, after Berman and MacDonald.

all temperatures. In impure metals or in disordered alloys the phonon contribution may be comparable with the electronic contribution.

Measurements on copper are shown in Fig. 15. Experimental curves for many metals at low temperatures are given by H. M. Rosenberg, Phil. Trans. Roy. Soc. (London) **A247**, 441–497 (1955), and R. L. Powell and W. A. Blanpied, "Thermal conductivities of metals and alloys at low temperatures," National Bureau of Standards Circular 556.

Ratio of Thermal to Electrical Conductivity

The **Wiedemann-Franz law** states that for metals at not too low temperatures the ratio of the thermal conductivity to the electrical conductivity is directly proportional to the temperature, with the value of the constant of proportionality independent of the particular metal. This result was most important in the history of the theory of metals, for it supported the picture of an electron gas. It can be explained by using (48) for σ and (53) for K:

$$\frac{K}{\sigma} = \frac{\pi^2 k_B{}^2 T n \tau / 3m}{n e^2 \tau / m} = \frac{\pi^2}{3} \left(\frac{k_B}{e} \right)^2 T \ . \tag{54}$$

The **Lorenz number** L is defined as

$$L \equiv \frac{K}{\sigma T} \ , \tag{55}$$

and according to (54) should have the value

$$L = \frac{\pi^2}{3} \left(\frac{k_B}{e} \right)^2 = 2.72 \times 10^{-13} \ \text{esu/deg}^2 = 2.45 \times 10^{-8} \ \text{watt-ohm/deg}^2 \ , \tag{56}$$

This remarkable result involves neither n nor m. It does not involve τ if the

Table 4   Experimental Lorenz numbers

Metal	$L \times 10^8$ watt-ohms/deg^2		Metal	$L \times 10^8$ watt-ohms/deg^2	
	0°C	100°C		0°C	100°C
Ag	2.31	2.37	Pb	2.47	2.56
Au	2.35	2.40	Pt	2.51	2.60
Cd	2.42	2.43	Sn	2.52	2.49
Cu	2.23	2.33	W	3.04	3.20
Ir	2.49	2.49	Zn	2.31	2.33
Mo	2.61	2.79			

relaxation times are identical for electrical and thermal processes. Experimental values of L at 0°C and at 100°C as given in Table 4 are in good agreement with (56). On purely classical theory with a Maxwellian distribution of velocities, the result is $L = 3(k_B/e)^2$, very close to (56), and also in fair agreement with experiment. The quotation from Lorentz that stands at the beginning of this chapter refers to this happy agreement.

At low temperatures $(T \ll \theta)$ the value of the Lorenz number tends to decrease; for pure copper near 15°K the observed value is an order of magnitude smaller than (56). The reason is attributed to a difference in the collision averages involved in the thermal and electrical conductivities, so that the thermal and electrical relaxation times are not identical in value.

Problems

1. **Particle in a box.** (a) Using the boundary condition $\psi = 0$ on the surface of a cube of side L, find all the wavefunctions for the first three distinct energy levels. (b) Give an expresssion for the energy of each level. (c) What is the degeneracy of each level? That is, what is the number of independent wavefunctions having the same energy? (Omit the electron spin from the enumeration.)

2. **Kinetic energy of electron gas.** Show that the kinetic energy of a three-dimensional gas of N free electrons at 0°K is

$$E_0 = \tfrac{3}{5} N \epsilon_F .$$

3. **Pressure and bulk modulus of an electron gas.** (a) Derive a relation connecting the pressure and volume of an electron gas at 0°K. *Hint:* Use the result of Problem 2 and the relation between ϵ_F and electron concentration. The result may be written as $P = \tfrac{2}{3}(E_0/V)$. (b) Show that the bulk modulus $B = -V(\partial P/\partial V)$ of an electron gas at 0°K is $B = \tfrac{5}{3}P = 10E_0/9V$. (c) Estimate for potassium, using Table 1, the value of the electron gas contribution to B and compare it with the experimental bulk modulus, using Table 4.2 and Eq. (4.29).

4. *Chemical potential.* Find an exact transcendental equation for the chemical potential $\mu(T)$ of a Fermi gas in two dimensions. *Note:* The density of orbitals of a free electron gas in two dimensions is independent of energy: $\mathfrak{D}(\epsilon) = m/\pi\hbar^2$ per unit area of specimen.

5. *Boundary conditions and wavefunctions.* (a) A crystal has nonorthogonal primitive axes **a**, **b**, **c**. With the periodic boundary condition over the surfaces of the parallelepiped of edges $N_1\mathbf{a}$, $N_2\mathbf{b}$, $N_3\mathbf{c}$, show that the solutions of the wave equation for free electrons are of the form

$$\psi \propto \exp\left[i\left(\frac{l}{N_1}\mathbf{A} + \frac{m}{N_2}\mathbf{B} + \frac{n}{N_3}\mathbf{C}\right)\cdot\mathbf{r}\right],$$

where l, m, n are any positive or negative integers; N_1, N_2, N_3 are also integers; **A**, **B**, **C** are reciprocal lattice vectors. (b) Give the form of the solution of the same problem for the boundary condition $\psi = 0$ on the surfaces of the parallelepiped.

6. *Fermi gases in astrophysics.* (a) Given $M_\odot = 2 \times 10^{33}$ g for the mass of the Sun, estimate the number of electrons in the Sun. In a white dwarf star this number of electrons may be ionized and contained in a sphere of radius 2×10^9 cm; find the Fermi energy of the electrons in eV. (b) The energy of an electron in the relativistic limit $\epsilon \gg mc^2$ is related to the wavevector as $\epsilon \cong pc = \hbar ck$. Show that the Fermi energy in this limit is $\epsilon_F \approx \hbar c(N/V)^{\frac{1}{3}}$, roughly. (c) If the above number of electrons were contained within a pulsar of radius 10 km, show that the Fermi energy would be $\approx 10^8$ eV. This value explains why pulsars are believed to be composed largely of neutrons rather than of protons and electrons, for the energy release in the reaction $n \to p + e^-$ is only 0.8×10^6 eV, which is not large enough to enable many electrons to form a Fermi sea. The neutron decay proceeds only until the electron concentration builds up enough to create a Fermi level of 0.8×10^6 eV, at which point the neutron, proton, and electron concentrations are in equilibrium.

7. *Liquid He³.* The atom He³ has spin $\frac{1}{2}$ and is a fermion. The density of liquid He³ is 0.081 g cm⁻³ near absolute zero. Calculate the Fermi energy ϵ_F and the Fermi temperature T_F. (For a survey of the properties of liquid He³, see J. Wilks, *Properties of liquid and solid helium,* Clarendon Press, 1967.)

References

H. M. Rosenberg, *Low temperature solid state physics,* Oxford, 1963, Chaps. 4 and 5.

J. M. Ziman, *Electrons and phonons,* Oxford, 1960, Chap. 9.

D. N. Langenberg, "Resource letter OEPM-1 on the ordinary electronic properties of metals," American J. Physics **36**, 777 (1968). An excellent bibliography on transport effects, anomalous skin effect, Azbel-Kaner cyclotron resonance, magnetoplasma waves, size effects, conduction electron spin resonance, optical spectra and photoemission, quantum oscillations, magnetic breakdown, ultrasonic effects, and the Kohn effect.

8

Free Electron
Fermi Gas II

In this chapter we complete the treatment of the most important properties of the free electron Fermi gas. Our object is to get a good physical feeling for the behavior of free electrons before we go on in Chapter 9 to treat the modifications introduced by interactions of the conduction electrons with the crystal lattice. We start with a discussion of the response of a free electron gas to an applied electric field. The static response of the electron gas leads to electrostatic screening of the Coulomb interaction. The dynamic response of the electron gas leads to the characteristic metallic reflection of light and to the excitation of plasmons, a collective mode of motion of the electron gas.

The Maxwell equation that involves the dielectric response of a medium is usually written as

(CGS) $$\text{curl } \mathbf{H} = \frac{4\pi}{c}\sigma\mathbf{E} + \frac{1}{c}\frac{\partial \mathbf{D}}{\partial t} \; ; \tag{1a}$$

(SI) $$\text{curl } \mathbf{H} = \sigma\mathbf{E} + \frac{\partial \mathbf{D}}{\partial t} \; ,$$

Here σ is the electrical conductivity, $\mathbf{H}$ the magnetic field, $\mathbf{E}$ the electric field, and $\mathbf{D}$ the displacement. The displacement is defined as $\mathbf{D} \equiv \mathbf{E} + 4\pi\mathbf{P}$ in CGS or $\mathbf{D} \equiv \epsilon_0\mathbf{E} + \mathbf{P}$ in SI. The polarization $\mathbf{P}$ is the dipole moment per unit volume.[1] The terms $\sigma\mathbf{E}$ and $\partial\mathbf{P}/\partial t$ both arise from charge displacements, the first from the motion of free charges and the second from bound charges. This separation is unnecessary, and we may usefully define the polarization $\mathbf{P}$ to include the polarization of both bound and free charges, so that the Maxwell equation will be written as

(CGS) $$\text{curl } \mathbf{H} = \frac{1}{c}\frac{\partial}{\partial t}(\mathbf{E} + 4\pi\mathbf{P}) \; ; \tag{1b}$$

(SI) $$\text{curl } \mathbf{H} = \frac{\partial}{\partial t}(\epsilon_0\mathbf{E} + \mathbf{P}) \; .$$

DIELECTRIC RESPONSE OF AN ELECTRON GAS

In the absence of collisions the equation of motion of a free electron in an electric field is

$$m\frac{d^2x}{dt^2} = -eE \; . \tag{2}$$

---

[1] A detailed discussion of the definitions of dielectric quantities is given in Chapter 13.

If x and E have the time dependence $e^{-i\omega t}$, then (2) becomes

$$-\omega^2 mx = -eE \;\; ; \qquad x = \frac{eE}{m\omega^2} \; . \tag{3}$$

The dipole moment p of the electron is

$$p = -ex = -\frac{e^2 E}{m\omega^2} \; , \tag{4a}$$

and the polarization, defined as the dipole moment per unit volume, is

$$P = -nex = -\frac{ne^2}{m\omega^2} E \; , \tag{4b}$$

where n is the electron concentration.

The **dielectric function** at frequency ω is defined as

(CGS) $$\epsilon(\omega) \equiv \frac{D(\omega)}{E(\omega)} \equiv 1 + 4\pi \frac{P(\omega)}{E(\omega)} \;\; ; \tag{5}$$

(SI) $$\epsilon(\omega) = \frac{D(\omega)}{\epsilon_0 E(\omega)} = 1 + \frac{P(\omega)}{\epsilon_0 E(\omega)} \; ,$$

where the notation emphasizes that E, P, and D have the common frequency ω. We use (4b) and (5) to find for the dielectric function of the free electron gas

(CGS) $$\epsilon(\omega) = 1 - \frac{4\pi ne^2}{m\omega^2} \;\; ; \tag{6}$$

(SI) $$\epsilon(\omega) = 1 - \frac{ne^2}{\epsilon_0 m\omega^2} \; .$$

We define the **plasma frequency**[2] by the relation

(CGS) $$\omega_p{}^2 \equiv \frac{4\pi ne^2}{m} \;\; ; \tag{7}$$

(SI) $$\omega_p{}^2 = \frac{ne^2}{\epsilon_0 m} \; .$$

---

[2] A plasma is a medium with equal concentration of positive and negative charges, of which at least one charge type is mobile. In a solid the negative charges of the conduction electrons are balanced by an equal concentration of positive charge of the ion cores.

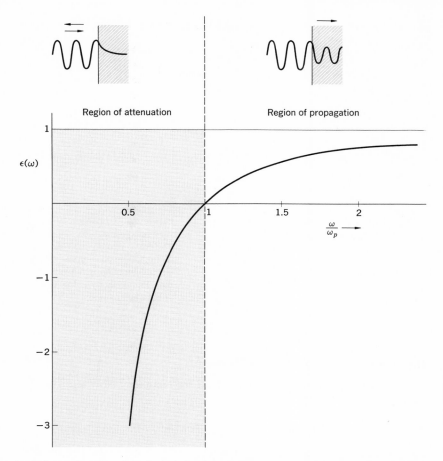

Figure 1 Dielectric function $\epsilon(\omega)$ of a free electron gas versus frequency in units of the plasma frequency ω_p. Electromagnetic waves propagate only when $\epsilon(\omega)$ is positive. Electromagnetic waves are totally reflected from the medium when $\epsilon(\omega)$ is negative.

so that we may write the dielectric function as

$$\epsilon(\omega) = 1 - \frac{\omega_p{}^2}{\omega^2} \ . \tag{8}$$

This function is plotted in Fig. 1.

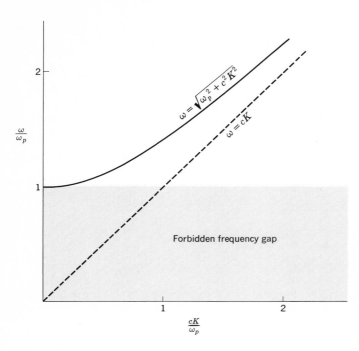

Figure 2a Dispersion relation for transverse electromagnetic waves in a plasma. The group velocity $v_g = d\omega/dK$ is the slope of the dispersion curve. Although the dielectric function is between zero and one, the group velocity is less than the velocity of light in vacuum.

Transverse Optical Modes in a Plasma

We see that $\epsilon(\omega)$ is negative for all frequencies below ω_p. The dispersion relation

$$\omega^2\epsilon(\omega) = c^2K^2 \tag{9}$$

for electromagnetic waves has no wavelike solutions when the dielectric constant is negative. Instead, the solutions are of the form $e^{-|K|x}$ in the frequency region $0 < \omega \leq \omega_p$. Waves incident on the medium in this frequency region will be totally reflected. An electron gas acts as a high-pass filter: the gas is transparent only when $\omega > \omega_p$, for here the dielectric function is positive. If we use (8) for the dielectric function, the dispersion relation may be written as

$$\omega^2 = \omega_p{}^2 + c^2K^2 \; ; \tag{10}$$

this applies to transverse electromagnetic waves in a plasma (Fig. 2a).

Values of the plasma frequency ω_p and of the wavelength $\lambda_p \equiv 2\pi c/\omega_p$ for electron concentrations of interest in solids are given below. A wave will propagate in the medium only if its free space wavelength is less than λ_p; otherwise the wave is reflected.

n in electrons/cm^3	10^{22}	10^{18}	10^{14}	10^{10}
ω_p in rad s^{-1}	5.7×10^{15}	5.7×10^{13}	5.7×10^{11}	5.7×10^9
λ_p in cm	3.3×10^{-5}	3.3×10^{-3}	0.33	33

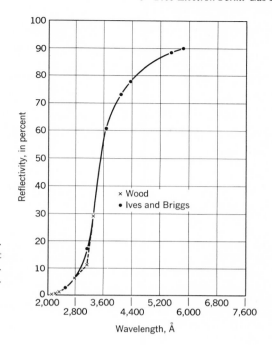

Figure 2b Reflecting power of potassium. Below 3000 Å most of the radiation incident on thin films of potassium is transmitted. [After H. E. Ives and H. B. Briggs, J. Optical Soc. Amer. **26**, 238 (1936).]

Table 1   Ultraviolet transmission limits of alkali metals, in Å

(A metal film is transparent for $\lambda < \lambda_p$)

	Li	Na	K	Rb	Cs
λ_p, calculated, mass m	1550	2090	2870	3220	3620
λ_p, observed	1550	2100	3150	3400	—

Transparency of Alkali Metals in the Ultraviolet

From the preceding discussion of the dielectric function we conclude that simple metals should reflect light in the visible region and be transparent to ultraviolet light. The effect was discovered by Wood[3] and explained by Zener.[4] A comparison of calculated and observed cutoff wavelengths is given in Table 1. The reflection of light from a metal is entirely similar to the reflection of radio waves from the ionosphere, for the free electrons in the ionosphere make the dielectric constant negative at low frequencies. Experimental results for potassium are shown in Fig. 2b.

[3] R. W. Wood, Phys. Rev. **44**, 353 (1933); R. W. Wood and C. Lukens, Phys. Rev. **54**, 332 (1938); H. E. Ives and H. B. Briggs, J. Opt. Soc. Am. **26**, 238 (1936); **27**, 181 (1937). For a review of the optical properties of metals, see M. P. Givens, *Solid state physics* **6**, 313 (1958).

[4] C. Zener, Nature **132**, 968 (1933).

Longitudinal Optical Modes in a Plasma

We saw in Chapter 5 that the zeros of the dielectric function determine the frequencies of the longitudinal optical modes. That is, the condition

$$\epsilon(\omega_L) = 0 \tag{11}$$

determines the longitudinal frequency ω_L. (This condition is also discussed in Advanced Topic D.) For the zero of the dielectric function of an electron gas we have

$$\epsilon(\omega_L) = 1 - \frac{\omega_p{}^2}{\omega_L{}^2} = 0 \;, \tag{12}$$

whence

$$\omega_L = \omega_p \;. \tag{13}$$

Thus there is a free longitudinal oscillation mode (Fig. 3) of an electron gas at the plasma frequency described earlier as the cutoff of transverse electromagnetic waves, Eq. (10).

A longitudinal plasma oscillation is shown in Fig. 4 as a uniform displacement of an electron gas in a thin metallic slab. The electron gas is moved as a whole with respect to the positive ion background. The displacement u creates an electric field $E = 4\pi neu$ that acts as a restoring force. The equation of motion of a unit volume of the electron gas is

(CGS) $$nm\frac{d^2u}{dt^2} = -neE = -4\pi n^2 e^2 u \;, \tag{14}$$

or

(CGS) $$\frac{d^2u}{dt^2} + \omega_p{}^2 u = 0 \;; \qquad \omega_p = \left(\frac{4\pi ne^2}{m}\right)^{\frac{1}{2}} \;. \tag{15}$$

This is the equation of motion of a simple harmonic oscillator of frequency ω_p, the plasma frequency. The expression for ω_p is identical with (7), which arose in quite a different connection. In SI units, the displacement u creates the electric field $E = neu/\epsilon_0$, whence $\omega_p = (ne^2/\epsilon_0 m)^{\frac{1}{2}}$.

A plasma oscillation of small wavevector will have approximately the frequency ω_p. It turns out[5] that the wavevector dependence of the dispersion relation for longitudinal oscillations is given by

$$\omega \cong \omega_p\left(1 + \frac{3k^2 v_F{}^2}{10\omega_p{}^2} + \cdots\right) \;, \tag{16}$$

plotted in Fig. 5. Here v_F is the electron velocity at the Fermi energy.

---

[5] D. Pines, *Elementary excitations in solids,* Benjamin, 1963, Chap. 3. The theoretical discussion of plasmon excitations in metals is due largely to Bohm and Pines.

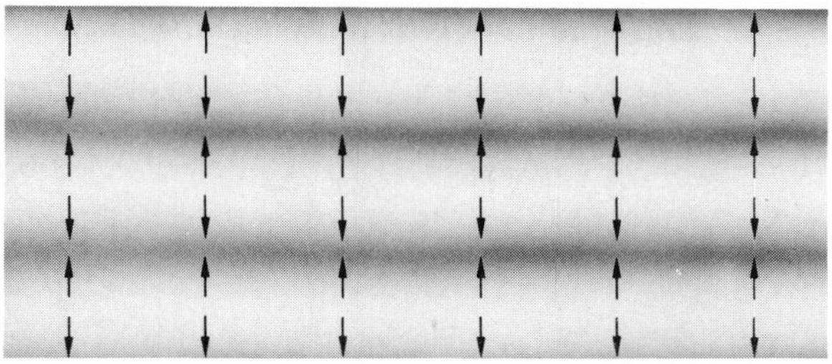

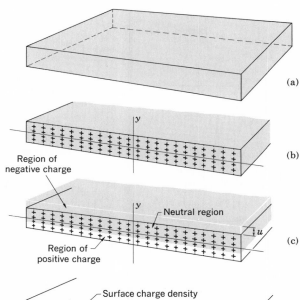

Figure 3 A plasma oscillation of finite wavelength. The arrows indicate the direction of displacement of the electrons.

Figure 4 In (a) is shown a thin slab or film of a metal. A cross section is shown in (b), with the positive ion cores indicated by + signs and the electron sea indicated by the gray background. The slab is electrically neutral. In (c) the negative charge has been displaced upward uniformly by a small distance u, shown exaggerated in the figure. As in (d), this displacement establishes a surface charge density $-neu$ on the upper surface of the slab and $+neu$ on the lower surface, where n is the electron concentration. An electric field $E = 4\pi neu$ is produced inside the slab. This field tends to restore the electron sea to its equilibrium position (b). In SI units, $E = neu/\epsilon_0$.

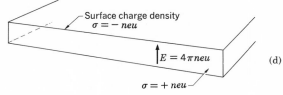

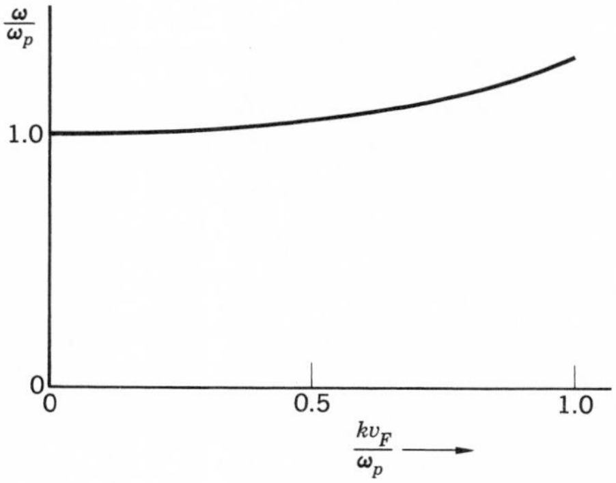

Figure 5 Theoretical dispersion relation for longitudinal modes of plasma. Here ω_p is the plasma frequency for $k \to 0$ and v_F is the electron velocity at the Fermi surface.

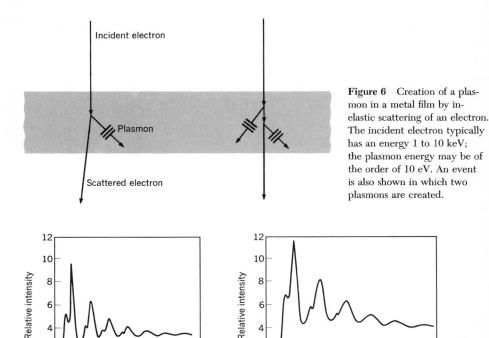

Figure 6 Creation of a plasmon in a metal film by inelastic scattering of an electron. The incident electron typically has an energy 1 to 10 keV; the plasmon energy may be of the order of 10 eV. An event is also shown in which two plasmons are created.

Figure 7 Energy loss spectra for electrons reflected from films of (a) aluminum and (b) magnesium, for primary electron energies of 2020 eV. The twelve loss peaks observed in Al are made up of combinations of 10.3 and 15.3 eV losses, where the 10.3 eV loss is due to surface plasmons (see Prob. 5) and the 15.3 eV loss is due to volume plasmons, as described by Eq. (23). The ten loss peaks observed in Mg are made up of combinations of 7.1 eV surface plasmons and 10.6 eV volume plasmons. [After C. J. Powell and J. B. Swan, Phys. Rev. **115**, 869 (1959); **116**, 81 (1959).]

PLASMONS

A plasma oscillation in a metal is a collective longitudinal excitation of the conduction electron gas. A **plasmon** is a quantized plasma oscillation; we may excite a plasmon by passing an electron through a thin metallic film (Fig. 6) or by reflecting an electron (or a photon) from a film. The charge of the electron couples with the electrostatic field fluctuations of the plasma oscillations. The reflected or transmitted electron will show an energy loss equal to integral multiples of the plasmon energy. Experimental excitation spectra for Al and Mg are shown in Fig. 7. A comparison of observed and calculated values of plasmon energies is given in Table 2; references to further data are given in the review by Raether.

It is equally possible to excite collective plasma oscillations in dielectric films; results for several dielectrics are included[6] in Table 2. The calculated plasma energies of Si, Ge, and InSb are based on four valence electrons per atom. In a dielectric the plasma oscillation is physically the same as in a metal: the entire valence electron sea oscillates back and forth with respect to the ion cores.

Table 2   Volume plasmon energies

| Material | Observed $\hbar\omega_p$, in eV | Calculated $\hbar\omega_p$, in eV | | Reference |
		Free electron, Eq. (7)	Corrected for core polarization[7]	
Metals				
Li	7.12	8.02	7.96	a
Na	5.71	5.95	5.58	a
	5.85			b
K	3.72	4.29	3.86	a
	3.87			c
Mg	10.6	10.9		d
Al	15.3	15.8		e
Dielectrics				
Si	16.4–16.9	16.0		
Ge	16.0–16.4	16.0		
InSb	12.0–13.0	12.0		

a. C. Kunz, Physics Letters **15**, 312 (1965).
b. J. B. Swan, Phys. Rev. **135**, A1467 (1964).
c. J. L. Robins and F. E. Best, Proc. Phys. Soc. (London) **79**, 110 (1962).
d. C. J. Powell and J. B. Swan, Phys. Rev. **116**, 81 (1959).
e. C. J. Powell and J. B. Swan, Phys. Rev. **115**, 869 (1959).

[7] The ion core polarization correction follows from the equivalence of the plasmon frequency to the root of the dielectric function:

$$\text{(CGS)} \qquad \epsilon(\omega) = \epsilon_{\text{core}} - \frac{4\pi n e^2}{m\omega^2} = 0 \qquad (17)$$

at the plasmon frequency. Thus the plasmon frequency as corrected for ion core polarization is

$$\text{(CGS)} \quad \omega_p' = \left(\frac{4\pi n e^2}{\epsilon_{\text{core}} m}\right)^{\frac{1}{2}} ; \qquad \text{(SI)} \quad \omega_p' = \left(\frac{n e^2}{\epsilon_{\text{core}} \epsilon_0 m}\right)^{\frac{1}{2}} . \qquad (18)$$

Ion core polarizabilities are given in Table 13.1, and values of ϵ_{core} may be calculated with the help of Eq. (13.35). For Na and K we calculate $\epsilon_{\text{core}} = 1.14$ and 1.24, respectively.

[6] The data are from the analysis by H. R. Philipp and H. Ehrenreich, Phys. Rev. **129**, 1550 (1963). Under observed values, the left-hand entry is from optical measurements and the right-hand entry is from electron energy loss measurements.

Electrostatic Screening

If we immerse a point test charge q at rest in a metal, the electron concentration near the test charge will be perturbed in such a way that the electric field of the charge is essentially canceled by the induced disturbance in the electron concentration. We say that the test charge is screened by the electron gas. There is a **screening length** within which the screening tends to be ineffective and outside which the screening becomes progressively more complete.

We give now an approximate treatment of the static screening problem. The Poisson equation of electrostatics is

(CGS) $$\nabla^2 \varphi = 4\pi e [n(\mathbf{r}) - n_0] \; ; \tag{19}$$

(SI) $$\nabla^2 \varphi = (e/\epsilon_0)[n(\mathbf{r}) - n_0] \;,$$

where $\varphi(\mathbf{r})$ is the electrostatic potential, and $[n(\mathbf{r}) - n_0]$ is the deviation from uniform electron concentration.

We can find another equation that relates the electrostatic potential to the electron concentration, for the electrochemical potential of the electron gas must be constant in equilibrium,[8] independent of position. In a region of the specimen where there is no electrostatic potential the chemical potential μ is related to the uniform concentration n_0 by

$$\mu = \epsilon_F^{\,0} = \frac{\hbar^2}{2m} (3\pi^2 n_0)^{\frac{2}{3}} \tag{20a}$$

at absolute zero, according to (7.21). In a region where the electrostatic potential is $\varphi(\mathbf{r})$, the electrochemical potential is

$$\mu = \epsilon_F(\mathbf{r}) - e\varphi(\mathbf{r}) \cong \frac{\hbar^2}{2m} [3\pi^2 n(\mathbf{r})]^{\frac{2}{3}} - e\varphi(\mathbf{r}) \;. \tag{20b}$$

It is an approximation (called the Thomas-Fermi approximation) that the electrochemical potential can be written in this way; the expression should be valid for electrostatic potentials that vary slowly in comparison with the wavelength of an electron. If the electrochemical potential is constant as the electrostatic potential varies, we must have (Fig. 8):

$$\frac{\hbar^2}{2m} [3\pi^2 n(\mathbf{r})]^{\frac{2}{3}} - e\varphi(\mathbf{r}) = \frac{\hbar^2}{2m} [3\pi^2 n_0]^{\frac{2}{3}} \;. \tag{21}$$

By a Taylor series expansion (21) may be written as

$$\frac{d\epsilon_F}{dn_0} [n(\mathbf{r}) - n_0] \cong e\varphi(\mathbf{r}) \;. \tag{21a}$$

[8] See Chapter 7.

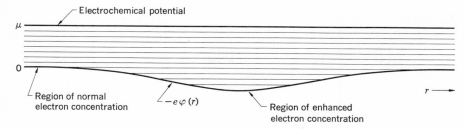

Figure 8   In thermal and diffusive equilibrium the electrochemical potential is constant; to maintain it constant we increase the electron concentration in regions of space where the potential energy is low, and we decrease the concentration where the potential is high.

From (20a) we have $d\epsilon_F/dn_0 = 2\epsilon_F/3n_0$, whence

$$n(\mathbf{r}) - n_0 \cong \frac{3}{2} n_0 \frac{e\varphi(\mathbf{r})}{\epsilon_F} \ . \tag{21b}$$

Thus (19) becomes

(CGS) $$\nabla^2\varphi = \frac{6\pi n_0 e^2}{\epsilon_F}\varphi = \lambda^2\varphi \ ; \tag{22}$$

(SI) $$\nabla^2\varphi = \frac{3n_0 e^2}{2\epsilon_0\epsilon_F}\varphi = \lambda^2\varphi \ ,$$

where

(CGS) $$\lambda^2 \equiv \frac{6\pi n_0 e^2}{\epsilon_F} = \frac{4me^2 n_0^{\frac{1}{3}}}{\hbar^2} \cdot \left(\frac{3}{\pi}\right)^{\frac{1}{3}} \cong \frac{4n_0^{\frac{1}{3}}}{a_0} \ , \tag{23}$$

where a_0 is the Bohr radius.

We look for a potential with spherical symmetry which is a solution of

$$\left(\frac{d^2}{dr^2} + \frac{2}{r}\frac{d}{dr}\right)\varphi(\mathbf{r}) = \lambda^2\varphi(\mathbf{r}) \ , \tag{24}$$

where the differential operator is the radial part of ∇^2 in spherical coordinates. The desired solution is

$$\varphi(\mathbf{r}) = \frac{qe^{-\lambda r}}{r} \ , \tag{25}$$

as we see on substitution in (24), for

$$\frac{d\varphi}{dr} = -\frac{qe^{-\lambda r}}{r}\left(\lambda + \frac{1}{r}\right) \ ; \qquad \frac{d^2\varphi}{dr^2} = \frac{qe^{-\lambda r}}{r}\left(\lambda^2 + \frac{2\lambda}{r} + \frac{2}{r^2}\right) \ . \tag{26}$$

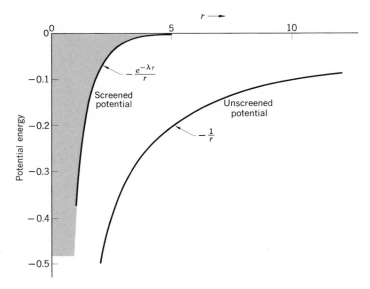

Figure 9 Comparison of screened and unscreened Coulomb potentials for a unit positive charge embedded in a Fermi gas of electrons. The screening length $1/\lambda$ has been taken as equal to unity.

Equation (25) defines the **screened Coulomb potential.** The **screening length** is $1/\lambda$ (Fig. 9). It is plotted in Fig. 10 as a function of electron concentration: for copper with $n_0 = 8.5 \times 10^{22}$ electrons/cm³ the screening length is 0.55 Å. Improved calculations of screening effects are discussed in QTS, Chapter 6.

It is often helpful to discuss the screening effect of an electron gas on an external potential of sinusoidal form. The analysis is carried out in Advanced Topic D, where the results are expressed in terms of the **dielectric function** $\epsilon(\mathbf{K}, \omega)$ which gives the response of the electron gas to an external potential of wavevector $\mathbf{K}$ and frequency ω. Much of solid state theory tends to be expressed in terms of the Fourier transforms $u(\mathbf{K}, \omega)$ of functions of position and time. Thus functions such as the dielectric function which describe the response at $\mathbf{K}$ and ω are of importance in the theory.

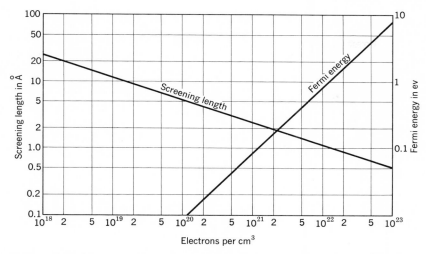

Figure 10   Log-log plot of the screening length $1/\lambda$ on the Thomas-Fermi model as a function of electron concentration, from Eq. (23). The Fermi energy is also plotted from Eq. (7.21).

ELECTRON-ELECTRON COLLISIONS

It is an astonishing property of metals that conduction electrons, although crowded together only 1 to 3 Å apart, travel long distances between collisions with each other. The mean free paths for electron-electron collisions are longer than 10^4 Å at room temperature and longer than 10 cm at 1°K. Two factors are responsible for these long mean free paths, without which the free electron model of metals would have little value: the most powerful factor is the exclusion principle (Fig. 11), and the second factor is the screening of the Coulomb interaction between two electrons.

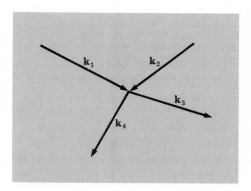

Figure 11   A collision between two electrons of wavevector k_1 and k_2. After the collision the particles have wavevector k_3 and k_4. The Pauli exclusion principle allows collisions only to final states k_3, k_4 which were vacant before the collision.

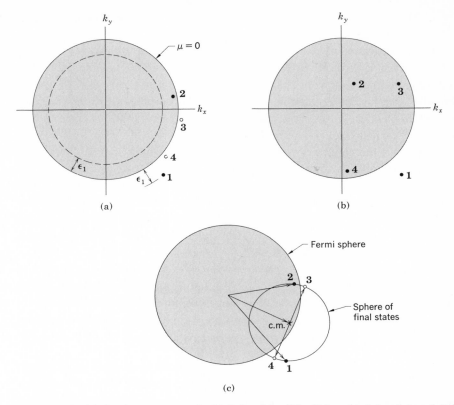

Figure 12 In (a) the electrons in initial orbitals **1** and **2** collide. If the orbitals **3** and **4** are initially vacant, the electrons **1** and **2** can occupy orbitals **3** and **4** after the collision. Energy and momentum are conserved. In (b) the electrons in initial orbitals **1** and **2** have no vacant final orbitals available that allow energy to be conserved in the collision. Orbitals such as **3** and **4** would conserve energy and momentum, but they are already filled with other electrons. In (c) we have denoted with x the wavevector of the center of mass **1** and **2**. All pairs of orbitals **3** and **4** conserve momentum and energy if they lie at opposite ends of a diameter of the small sphere. The small sphere was drawn from the center of mass to pass through **1** and **2**. But not all pairs of points **3**, **4** are allowed by the exclusion principle, for both **3**, **4** must lie outside the Fermi sphere: the fraction allowed is $\approx \epsilon_1/\epsilon_F$.

We now show how the exclusion principle reduces the collision frequency of an electron that has a low excitation energy ϵ_1 outside a filled Fermi sphere (Fig. 12). We estimate the effect of the exclusion principle on the two-body collision

$$1 + 2 \rightarrow 3 + 4$$

between an electron in the excited orbital **1** and an electron in the filled orbital **2** in the Fermi sea. It is convenient to refer all energies to the Fermi level μ taken as the zero of energy; thus ϵ_1 will be positive and ϵ_2 will be negative. Because of the exclusion principle the orbitals **3** and **4** of the electrons after collision must lie outside the Fermi sphere, all orbitals within the sphere being already occupied; thus both energies ϵ_3, ϵ_4 must be positive.

The conservation of energy requires that $|\epsilon_2| < \epsilon_1$, for otherwise $\epsilon_3 + \epsilon_4 = \epsilon_1 + \epsilon_2$ could not be positive. This means that collisions are possible only if the orbital **2** lies within a shell of thickness ϵ_1 within the Fermi surface, as in Fig. 12a. Thus the fraction $\approx \epsilon_1/\epsilon_F$ of the electrons in filled orbitals provides a suitable target[8a] for electron **1**. But even if the target electron **2** is in the suitable energy shell, only a small fraction of the final orbitals compatible with conservation of energy and momentum are allowed by the exclusion principle. This gives a second factor of ϵ_1/ϵ_F. In Fig. 12c we show a small sphere on which all pairs of orbitals **3, 4** on opposite ends of a diameter satisfy the conservation laws, but collisions can occur only if both orbitals **3, 4** lie outside the Fermi sea. The product of the two fractions is $(\epsilon_1/\epsilon_F)^2$. If ϵ_1 corresponds to $1°$K and ϵ_F to $5 \times 10^4 °$K, we have $(\epsilon_1/\epsilon_F)^2 \approx 4 \times 10^{-10}$, the factor by which the exclusion principle reduces the collision rate.

The argument is not changed for a thermal distribution of electrons at a low temperature such that $k_B T \ll \epsilon_F$. We replace ϵ_1 by the thermal energy $\approx k_B T$, and now the rate at which electron-electron collisions take place is reduced below the classical value by $(k_B T/\epsilon_F)^2$, so that the effective collision cross-section is

$$\sigma_{el-el} \approx \left(\frac{k_B T}{\epsilon_F}\right)^2 \sigma_0 \ , \tag{27}$$

where σ_0 is the cross-section for the screened Coulomb interaction. The interaction of an electron with another electron has a range of the order of the screening length $1/\lambda$. From numerical calculations the effective cross-section with screening for collisions between electrons is of the order of 10^{-15} cm^2 or 10 Å^2 in typical metals. The effect of screening in electron-electron collisions is greatly to reduce the scattering cross-section below the value expected from the Rutherford scattering equation for the unscreened Coulomb potential, but the greatest reduction by far is caused by the Pauli principle factor $(k_B T/\epsilon_F)^2$.

At room temperature in a typical metal $k_B T/\epsilon_F$ is $\sim 10^{-2}$, so that $\sigma \sim 10^{-4}\sigma_0 \sim 10^{-19}$ cm^2. The mean free path for electron-electron collisions is therefore of the order

$$\ell_{el-el} \approx \frac{1}{n\sigma} \sim 10^{-4} \text{ cm}$$

at room temperature. This is longer than the mean free path due to electron-phonon collisions by at least a factor of 10, so that at room temperature collisions with phonons are dominant. At liquid helium temperatures a contribution proportional to T^2 has been reported[9] in the resistivity of indium and aluminum,

---

[8a] The calculation is given by P. Morel and P. Nozières, Phys. Rev. **126**, 1909 (1962).

[9] J. C. Garland and R. Bowers, Phys. Rev. Letters **21**, 1007 (1968).

and such a temperature dependence is consistent with the electron-electron scattering cross-section (27). The mean free path in indium at 2°K is of the order of 30 cm, as expected. We see that the Pauli principle explains one of the central problems of the theory of metals: how do the electrons travel long distances without colliding with each other?

MOTION IN MAGNETIC FIELDS

By the argument of (7.45) we have the equation of motion for the displacement $\delta\mathbf{k}$ of a Fermi sphere of particles acted on by a force $\mathbf{F}$:

$$\hbar\left(\frac{d}{dt} + \frac{1}{\tau}\right)\delta\mathbf{k} = \mathbf{F} \ . \tag{28}$$

The free particle acceleration term is $(\hbar d/dt)\,\delta\mathbf{k}$ and the effect of collisions (the friction) is $\hbar\,\delta\mathbf{k}/\tau$, where τ is the collision time. Consider now the motion of the system in a uniform magnetic field $\mathbf{B}$. The Lorentz force on an electron is

(CGS) $$\mathbf{F} = -e\left(\mathbf{E} + \frac{1}{c}\mathbf{v} \times \mathbf{B}\right) \ ; \tag{29}$$

(SI) $$\mathbf{F} = -e(\mathbf{E} + \mathbf{v} \times \mathbf{B}) \ .$$

If $m\,\delta\mathbf{v} = \hbar\,\delta\mathbf{k}$, then the equation of motion is

(CGS) $$m\left(\frac{d}{dt} + \frac{1}{\tau}\right)\delta\mathbf{v} = -e\left(\mathbf{E} + \frac{1}{c}\delta\mathbf{v} \times \mathbf{B}\right) . \tag{30}$$

We have written $\delta\mathbf{v}$ in the force term as the average of $\mathbf{v}$ over the Fermi sphere.

Cyclotron Frequency

Consider first the free motion of the system, for $\mathbf{B}$ parallel to the z axis. For convenience we let $\tau \to \infty$; and we take $\mathbf{E} = 0$; the equations could equally well be solved for finite τ. The condition for a well-defined resonance line[9a] is that $\omega_c\tau > 1$, where ω_c is defined by (33) below. Now (30) becomes

(CGS) $$m\frac{d}{dt}\,\delta v_x = -\frac{eB}{c}\,\delta v_y \ ; \qquad m\frac{d}{dt}\,\delta v_y = \frac{eB}{c}\,\delta v_x \ . \tag{31}$$

[9a] This is similar to the condition for a sharp resonance response in a damped harmonic oscillator with the equation of motion

$$m\left(\frac{d^2x}{dt^2} + \frac{1}{\tau}\frac{dx}{dt} + \omega_0{}^2x\right) = Ee^{-i\omega t} \ .$$

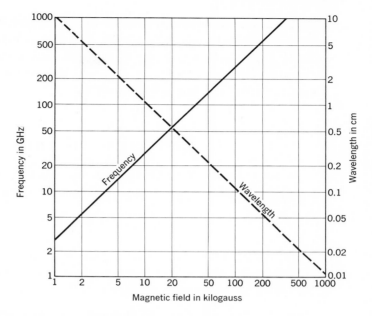

Figure 13 Cyclotron resonance frequency f_c in GHz versus magnetic field in kG, for free electron. Also plotted is the free space electromagnetic wavelength λ_c for cyclotron resonance versus magnetic field.

These equations have the solution

$$\delta v_x = v_0 \cos \omega_c t \; ; \qquad \delta v_y = v_0 \sin \omega_c t \; , \qquad (32)$$

where

$$\text{(CGS)} \quad \omega_c \equiv \frac{eB}{mc} \; ; \qquad\qquad \text{(SI)} \quad \omega_c = \frac{eB}{m} \; . \qquad (33)$$

This is the **cyclotron frequency** for a free electron. Numerically (Fig. 13)

$$f_c(MHz) \cong 2.80 \, B(\text{gauss}) = 2.80 \times 10^{-4} \, B(\text{tesla}) \; , \qquad (34)$$

with $f_c \equiv \omega_c/2\pi$. The constant v_0 in (32) is not the Fermi velocity, but it is the magnitude of whatever initial drift velocity is given to the Fermi sea.

For a free electron in a field of 10 kG we have $\omega_c = 1.76 \times 10^{11}$ rad s^{-1}. If the relaxation times (as for pure copper) are 2×10^{-14} sec and 2×10^{-9} sec at 300°K and 4°K, respectively, then $\omega_c \tau = 3.5 \times 10^{-3}$ and 3.5×10^2. The cyclotron orbit is quite incomplete at room temperature, but many cycles are completed at helium temperatures.

Static Magnetoconductivity

An important situation is the following: let a static magnetic field $\mathbf{B}$ lie along the z axis. Then the equations of motion for electrons are

$$(\text{CGS}) \qquad m\left(\frac{d}{dt} + \frac{1}{\tau}\right)\delta v_x = -e\left(E_x + \frac{B}{c}\,\delta v_y\right) \;;$$

$$m\left(\frac{d}{dt} + \frac{1}{\tau}\right)\delta v_y = -e\left(E_y - \frac{B}{c}\,\delta v_x\right) \;;$$

$$m\left(\frac{d}{dt} + \frac{1}{\tau}\right)\delta v_z = -eE_z \;. \tag{35}$$

The results in SI are obtained by replacing c by 1. In the steady state in a static electric field the time derivatives are zero, so that (35) reduces to

$$\delta v_x = -\frac{e\tau}{m}E_x - \omega_c\tau\,\delta v_y \;; \qquad \delta v_y = -\frac{e\tau}{m}E_y + \omega_c\tau\,\delta v_x \;; \qquad \delta v_z = -\frac{e\tau}{m}E_z \;. \tag{36}$$

On solving for δv_x and δv_y we have

$$\delta v_x = -\frac{e\tau/m}{1 + (\omega_c\tau)^2}\,(E_x - \omega_c\tau E_y) \;;$$

$$\delta v_y = -\frac{e\tau/m}{1 + (\omega_c\tau)^2}\,(E_y + \omega_c\tau E_x) \;. \tag{37}$$

The current density for electrons is given by $\mathbf{j} = n(-e)\,\delta\mathbf{v}$. The components of the electric current density are, with $\sigma_0 \equiv ne^2\tau/m$,

$$j_x = \sigma_{xx}E_x + \sigma_{xy}E_y = \frac{\sigma_0}{1 + (\omega_c\tau)^2}\,(E_x - \omega_c\tau E_y) \;;$$

$$j_y = \sigma_{yx}E_x + \sigma_{yy}E_y = \frac{\sigma_0}{1 + (\omega_c\tau)^2}\,(\omega_c\tau E_x + E_y) \;;$$

$$j_z = \sigma_{zz}E_z = \sigma_0 E_z \;. \tag{38}$$

The z component of the current is not affected by a magnetic field in the z direction. The current density can be written in matrix form as

$$\begin{pmatrix} j_x \\ j_y \\ j_z \end{pmatrix} = \frac{\sigma_0}{1 + (\omega_c\tau)^2} \begin{pmatrix} 1 & -\omega_c\tau & 0 \\ \omega_c\tau & 1 & 0 \\ 0 & 0 & 1 + (\omega_c\tau)^2 \end{pmatrix} \begin{pmatrix} E_x \\ E_y \\ E_z \end{pmatrix} \;. \tag{39}$$

We see from (38) that the diagonal components σ_{xx} and σ_{yy} of the magnetoconductivity tensor decrease monotonically as the magnetic field (or ω_c) is increased. The magnitude of the off-diagonal components σ_{xy} and σ_{yx} at first

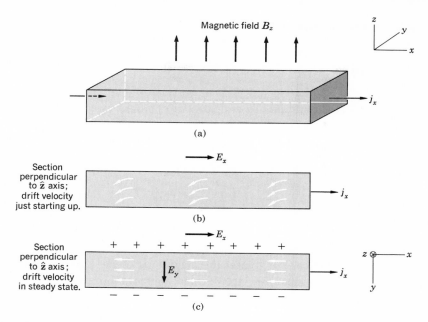

Figure 14 The standard geometry for the Hall effect: a rod-shaped specimen of rectangular cross-section is placed in a magnetic field B_z, as in (a). An electric field E_x applied across the end electrodes causes an electric current density j_x to flow down the rod. The drift velocity of the electrons immediately after the electric field is applied is shown in (b). The deflection in the $-y$ direction is caused by the magnetic field. Electrons accumulate on one face of the rod and a positive ion excess is established on the opposite face until, as in (c), the transverse electric field (Hall field) just cancels the force due to the magnetic field.

increase and later decrease as B is increased. However, to find the electrical resistivity in a magnetic field we must specify the experimental geometry (see Problem 4).

Hall Effect

Consider a rod-shaped specimen in a longitudinal electric field E_x and a transverse magnetic field, as in Fig. 14. If current cannot flow out of the rod in the y direction we must have $j_y = 0$. From (38) this is possible only if there is a transverse electric field

(CGS)
$$E_y = -\omega_c \tau E_x = -\frac{eB\tau}{mc} E_x \; ;$$
(40)

(SI)
$$E_y = -\omega_c \tau E_x = -\frac{eB\tau}{m} E_x \; .$$

We can measure this transverse electric field. It is known as the **Hall field.** The quantity

$$R_H = \frac{E_y}{j_x B}$$

(41)

is called the **Hall constant.** To evaluate (41) on our simple model we substitute (40) in the first line of (38). We have

(CGS) $$R_H = -\frac{eB\tau E_x/mc}{ne^2\tau E_x B/m} = -\frac{1}{nec} \; ;$$ (42)

(SI) $$R_H = -\frac{1}{ne} \; .$$

This is negative for free electrons, for e is positive by definition. The lower the carrier concentration, the greater the magnitude of the Hall constant. Measuring R_H is a way of measuring the carrier concentration.

The simple result (42) follows from the assumption that all relaxation times are equal, independent of the velocity of the electron. A numerical factor of order unity enters if the relaxation time is a function of the velocity. The expression becomes somewhat more complicated if both electrons and holes (see below) contribute to the conductivity, as considered in Problem 11.3. The theory of the Hall effect again becomes simple in high magnetic fields[10] such that $\omega_c\tau \gg 1$, where ω_c is the cyclotron frequency and τ the relaxation time.

Observed values of the Hall constant for several metals are compared in Table 3 with values calculated directly from the concentration of charge carriers. The most accurate measurements are made on pure specimens at low temperatures and strong magnetic fields by the method of helicon resonance, Problem 7. We see that the accurate values of the Hall constant for the monovalent metals sodium and potassium are in excellent agreement with the values calculated for one valence electron per atom, using (42). Notice next the values for the trivalent elements aluminum and indium: the experimental values are in close concord with the values calculated for one positive charge carrier (or hole) per atom and would disagree in sign and magnitude with values calculated for three electrons per atom. The problem of Hall constants with positive signs arises also for Be and As in the table, for a positive Hall constant must be associated with the motion of positive charge carriers. The problem was recognized in the early work on the Hall effect; as Lorentz wrote,

"It seems to prove that we must imagine two kinds of free electrons, the motion of the positive ones predominating in one body and that of the negative ones in the other."

[10] A review of galvanomagnetic effects in high fields is given by E. Fawcett, Advances in Physics **13**, 139 (1964).

But some metals are not filled with positrons while others are filled with electrons; we shall see in the next chapter that energy-band theory accounts for the motion of electrons in some situations as if they were endowed with a positive charge. The orbits of such electrons are called hole orbits. We shall also be able to account for the large magnitude of the Hall constant in semimetals (such as As, Sb, Bi) and in semiconductors.

Table 3   Comparison of observed Hall constants with those calculated on free electron theory

[The experimental values of R_H as obtained by conventional methods are summarized from data at room temperature presented in the Landolt-Börnstein tables, Vol. II.6, p. 161 (1959); the values obtained by the helicon wave method at 4°K are by J. M. Goodman, Phys. Rev. **171**, 641 (1968). The values of the carrier concentration n are from Table 1.5 except for Na, K, Al, In, where Goodman's values are used. To convert the value of R_H in CGS units to the value in volt-cm/amp-gauss, multiply by 9×10^{11}; to convert R_H in CGS to m³/coulomb, multiply by 9×10^{13}.]

Metal	Method	Experimental R_H, in 10^{-24} CGS units	Assumed carriers per atom	Calculated $-1/nec$, in 10^{-24} CGS units
Li	conv.	−1.89	1 electron	−1.48
Na	helicon	−2.619	1 electron	−2.603
	conv.	−2.3		
K	helicon	−4.946	1 electron	−4.944
	conv.	−4.7		
Rb	conv.	−5.6	1 electron	−6.04
Cu	conv.	−0.6	1 electron	−0.82
Ag	conv.	−1.0	1 electron	−1.19
Au	conv.	−0.8	1 electron	−1.18
Be	conv.	+2.7	—	—
Mg	conv.	−0.92	—	—
Al	helicon	+1.136	1 hole	+1.135
	conv.	−0.43		
In	helicon	+1.774	1 hole	+1.780
As	conv.	+50.	—	—
Sb	conv.	−22.	—	—
Bi	conv.	−6000.	—	—

Problems

1. **Thomas-Fermi approximation.** In the Thomas-Fermi approximation the electron concentration $n(\mathbf{r})$ is related to the electrostatic potential $\varphi(\mathbf{r})$ by

$$n(\mathbf{r}) \cong n_0 + \frac{3}{2} n_0 \frac{e\varphi(\mathbf{r})}{\epsilon_F} \;,$$

where n_0 is the electron concentration at a position where $\varphi = 0$; the Fermi energy is ϵ_F. (a) Using this result, show that for $x > 0$ the electrostatic potential is

$$\varphi(x) = -\frac{2\pi\sigma}{k_s} e^{-k_s x} \;,$$

where σ is an external charge density on the plane $x = 0$. (b) Find an expression for k_s in terms of the unperturbed Fermi velocity v_F and the unperturbed plasma frequency ω_p. (c) Estimate k_s for sodium.

2. **Hagen-Rubens relation for infrared reflectivity of metals.** The complex refractive index $n + i\kappa$ of a metal for $\omega\tau \ll 1$ is given by

(CGS) $$\epsilon(\omega) \equiv (n + i\kappa)^2 = \frac{1 + 4\pi i \sigma_0}{\omega} \;,$$

where σ_0 is the conductivity for static fields. Using the relation

$$R = \frac{(n-1)^2 + \kappa^2}{(n+1)^2 + \kappa^2}$$

for the reflection coefficient at normal incidence, show that

(CGS) $$R \cong 1 - \left(\frac{2\omega}{\pi\sigma_0}\right)^{\frac{1}{2}} \;.$$

This is the Hagen-Rubens relation. Assume that $\omega \ll \sigma_0$. *Note:* For a discussion of experiments on Al see H. E. Bennett, M. Silver, and E. J. Ashley, J. Opt. Soc. Am. **53**, 1089 (1963).

3. **Refractive index for x-rays.** Estimate the dielectric constant and refractive index of metallic Na for x-rays of energy 10 kev. Neglect the ionization energy of the electrons in comparison with the photon energy: thus *all* the electrons of Na are to be treated as free electrons in this experiment. Assume that the relaxation time τ is infinite.

4. **Magnetoresistance.** The transverse magnetoresistivity of a solid is defined as E_x/j_x for the standard geometry of Fig. 14. Show that (38) leads to $j_x = \sigma_0 E_x$, because $j_y = 0$ for this geometry. Thus the resistivity is independent of the magnetic field, whereas experiments generally show a resistivity which increases as the magnetic field is increased. This defect of our model is caused in part by the unrealistic assumption that all electrons have the identical relaxation time τ, independent of the electron velocity.

°5. **Surface plasmons.** Consider a semi-infinite plasma on the positive side of the plane $z = 0$. A solution of Laplace's equation $\nabla^2\varphi = 0$ in the plasma is

$$\varphi_i(x,z) = A \cos kx \, e^{-kz} \;,$$

whence $E_{zi} = kA \cos kx\, e^{-kz}$; $E_{xi} = kA \sin kx\, e^{-kz}$. (a) Show that in the vacuum

$$\varphi_0(x,z) = A \cos kx\, e^{kz}$$

for $z < 0$ satisfies the boundary condition that the tangential component of **E** be continuous at the boundary; i.e., find E_{xo}. (b) Note that $\mathbf{D}_i = \epsilon(\omega)\mathbf{E}_i$; $\mathbf{D}_o = \mathbf{E}_o$. Show that the boundary condition that the normal component of **D** be continuous at the boundary requires that $\epsilon(\omega) = -1$, whence from (8) we have

$$\omega_s{}^2 = \tfrac{1}{2}\omega_p{}^2 \qquad (43)$$

for the frequency ω_s of a surface plasma oscillation. Refs.: R. H. Ritchie, Phys. Rev. **106**, 874 (1957); E. A. Stern and R. A. Ferrell, Phys. Rev. **120**, 130 (1960); surface plasma resonance experiments in small spherical silver and gold particles are discussed by V. Kreibig and P. Zacharias, Z. Physik **231**, 128 (1970).

°6. *Interface plasmons.* We consider the plane interface $z = 0$ between metal 1 at $z > 0$ and metal 2 at $z < 0$. Metal 1 has bulk plasmon frequency ω_{p1}; metal 2 has ω_{p2}. The dielectric constants in both metals are those of free electron gases. Show that surface plasmons associated with the interface have the frequency

$$\omega = [\tfrac{1}{2}(\omega_{p1}{}^2 + \omega_{p2}{}^2)]^{\frac{1}{2}} \;. \qquad (44)$$

Such plasmons have been observed at an Al/Mg interface; see C. Kunz, Z. Physik **196**, 311 (1966).

°7. *Helicon waves.* There exists in pure metals at low temperatures an unusual mode of propagation of electromagnetic waves. The helicon modes were first observed by R. Bowers and co-workers, and they offer a means for the measurement of Hall constants. We consider a constant applied magnetic field B_a in the z direction. At frequencies $\omega \ll 1/\tau$ the drift velocity components are given by (36). (a) Show that (36) may be written as

$$E_x = \rho_0 j_x - R_H B j_y \;; \qquad E_y = R_H B j_x + \rho_0 j_y \;, \qquad (45)$$

where the static resistivity $\rho_0 = 1/\sigma_0 = m/ne^2\tau$ and the Hall constant $R_H = -1/nec$ in CGS. In (45) we may approximate B by B_a. (b) With the neglect of the displacement current, show that the Maxwell equations (written for a nonmagnetic medium)

(CGS) $$\operatorname{curl}\mathbf{B} = \frac{4\pi}{c}\mathbf{j}\;; \qquad \operatorname{curl}\mathbf{E} = -\frac{1}{c}\frac{\partial\mathbf{B}}{\partial t}\;, \qquad (46)$$

may be combined to give, with $\operatorname{curl}\operatorname{curl}\mathbf{E} = -\nabla^2\mathbf{E}$,

(CGS) $$\nabla^2\mathbf{E} = \frac{4\pi}{c^2}\frac{\partial\mathbf{j}}{\partial t}\;. \qquad (47)$$

(c) Assume **E**, **j** of the form $e^{i(kz-\omega t)}$ and make the high field approximation that $R_H B_a \gg \rho_0$. With the neglect of ρ_0, show that (47) has a solution if

(CGS) $$\omega = k^2\frac{|R_H|B_a c^2}{4\pi} = k^2\frac{B_a c}{2\pi ne}\;. \qquad (48)$$

This is the dispersion relation for helicon waves, and by it the Hall constant may be determined.

References

D. Pines, "Electron interaction in metals," *Solid state physics* **1**, 367 (1955).

H. Raether, "Solid state excitations by electrons; plasma oscillations and single electron transitions," *Ergebnisse der exacten Naturwissenschaften* (*Springer tracts in modern physics*) **38**, 84 (1965).

M. P. Givens, "Optical properties of metals," *Solid state physics* **6**, 313 (1958).

F. Stern, "Elementary theory of the optical properties of metals," *Solid state physics* **15**, 300 (1963).

A. C. Smith, J. F. Janak, and R. B. Adler, *Electronic conduction in solids,* McGraw-Hill, 1967.

E. A. Kaner and V. G. Skobov, "Electromagnetic waves in metals in a magnetic field," Advances in Physics **17**, 605 (1968).

A. T. Stewart, "Positron annihilation in metals," in *Positron annihilation,* eds. A. T. Stewart and L. O. Roellig, Academic Press, 1966.

HISTORICAL DEVELOPMENT:

H. A. Lorentz, *Collected papers,* vol. 8, Martinus Nijhoff, The Hague, 1935. A good view of the historical development of the prequantum electron theory of metals.

P. Drude, "Zur Elektronentheorie der Metalle," Annalen der Physik **1**, 566 (1900); **3**, 369 (1900). Two pioneer papers in the electron theory of the transport properties of metals; the earlier paper contains the first derivation of the Wiedemann-Franz ratio.

L. L. Campbell, *Galvanomagnetic and thermomagnetic effects,* Longmans, Green (1923). Includes an account of the early history of the Hall effect.

9

Energy Bands I

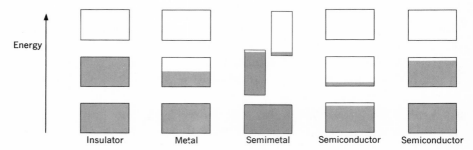

Figure 1 Schematic electron occupancy of allowed energy bands for an insulator, metal, semimetal, and semiconductor. The vertical extent of the boxes indicates the allowed energy regions; the shaded areas indicate the regions filled with electrons.

This and the following chapter are not the easiest in the book,[1] but they are the most important. Here we find all the important new concepts associated with the quantum theory of solids: energy bands, band gaps, Fermi surfaces, effective masses, and holes. An account is given of some of the central experiments used to determine the shape of the **Fermi surface,** defined as the surface of constant energy ϵ_F in **k** space.

The free electron model of metals developed in the preceding chapters gave us considerable insight into several of the electronic properties of metals, yet there are other electronic properties of solids for which the free electron model gives us no help. The model cannot help us understand why some chemical elements crystallize to form good conductors of electricity and others to form insulators; still others form semiconductors, with electrical properties varying markedly with temperature. Yet the distinction between the resistivity values of normal metallic conductors and insulators is striking: the resistivity of a pure metal at low temperatures may be of the order of 10^{-10} ohm-cm, and the resistivity of a good insulator may be as high as 10^{22} ohm-cm. E. M. McMillan has remarked that this observed range of 10^{32} in resistivity may be the widest range of any common physical property of solids.

Every solid contains electrons. The important question for electrical conductivity is how the electrons respond to an applied electric field. We shall see that electrons in crystals are arranged in energy bands (Fig. 1) separated by regions in energy for which no wavelike electron orbitals exist. Such forbidden regions are called **energy gaps** or **band gaps,** and will be shown to result from the interaction of the conduction electron waves with the ion cores of the crystal. The crystal will behave as an insulator if the number of electrons is such that the allowed energy bands are either filled or empty, for then no electrons can move in an electric field. The crystal will act as a metal if one or more bands are partly filled, say 10 to 90 percent filled. The crystal is a semiconductor or a semimetal[2] if all bands are entirely filled, except for one or two bands slightly filled or slightly empty.

To understand the difference between insulators and conductors, we must extend the free electron model to take account of the periodic lattice of the solid. The possibility of a band gap is the most important new property that emerges. We shall encounter other quite remarkable properties of electrons in crystals. For example, they respond to applied electric or magnetic fields as if the electrons were endowed with an effective mass m^*, which may be larger or smaller than the free electron mass, or may even be negative.

---

[1] For semipopular accounts of energy band theory see A. R. Mackintosh, "Fermi surface of metals," Scientific American, July, 1963, and J. Ziman, *Electrons in metals—a short guide to the Fermi surface,* Taylor and Francis, London, 1963.

[2] In a **semimetal** (such as bismuth) one band is almost filled and another band is nearly empty at absolute zero, but a pure **semiconductor** becomes an insulator at absolute zero. See Chapter 11.

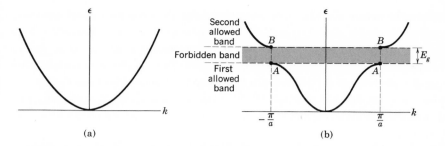

Figure 2 (a) Plot of energy ϵ versus wavevector k for a free electron. (b) Plot of energy versus wavevector for an electron in a monatomic linear lattice of lattice constant a. The energy gap E_g shown is associated with the first Bragg reflection at $k = \pm\pi/a$; other gaps are found at $\pm n\pi/a$, for integral values of n. A similar plot for x-rays is given in Fig. A.1.

On the free electron model the allowed energy values are distributed continuously from zero to infinity. We saw in Chapter 7 that

$$\epsilon_{\mathbf{k}} = \frac{\hbar^2}{2m}(k_x^2 + k_y^2 + k_z^2) \; , \tag{1}$$

where, for periodic boundary conditions over a cube of side L,

$$k_x, k_y, k_z = 0 \; ; \quad \pm\frac{2\pi}{L} \; ; \quad \pm\frac{4\pi}{L} \; ; \quad \ldots \; , \tag{2}$$

as in (7.12). The free electron wavefunctions are of the form

$$\psi_{\mathbf{k}}(\mathbf{r}) = e^{i\mathbf{k}\cdot\mathbf{r}} \; ; \tag{3}$$

they represent running waves and carry momentum $\mathbf{p} = \hbar\mathbf{k}$.

We know that Bragg reflection is a characteristic feature of wave propagation in crystals. Bragg reflection of electron waves in crystals is the cause of energy gaps.[3] There arise substantial regions of energy in which wavelike solutions of the Schrödinger equation do not exist, as in Fig. 2. These **energy gaps** are of decisive significance in determining whether a solid is an insulator or a conductor.

NEARLY FREE ELECTRON MODEL

The band structure of a crystal often can be described by the nearly free electron model for which the band electrons are treated as perturbed only weakly by the periodic potential of the ion cores. Often the gross overall aspects

---

[3] An energy gap for an electron is directly analogous to the frequency gap for x-rays exhibited in Fig. A.1.

of the band structure and the intricate detail of the observed Fermi surfaces can be explained on this model; we shall also find some situations where it is not applicable. But this model answers almost all the qualitative questions about the behavior of electrons in metals. The reasons for the validity of the model are discussed under the heading of pseudopotentials in Chapter 10. For pedagogical purposes another model, the beautiful Kronig-Penney model, is often developed, but it does not correspond closely to reality nor does the solution afford deep insight.

Let us understand physically the origin of forbidden bands, considering first the simple problem of a linear solid of lattice constant a. The low energy portions of the band structure are shown qualitatively in Fig. 2, in (a) for entirely free electrons and in (b) for electrons which are nearly free, but with an energy gap at $k = \pm\pi/a$. The Bragg condition $(\mathbf{k} + \mathbf{G})^2 = k^2$ for diffraction of a wave of wavevector $\mathbf{k}$ becomes in one dimension

$$k = \pm\tfrac{1}{2}G = \pm n\pi/a \ , \tag{4}$$

where $G = \pm 2n\pi/a$ is a reciprocal lattice vector and n is an integer. The first reflections and the first energy gap occur at $k = \pm\pi/a$. Other energy gaps occur for the other values of the integer n in (4).

The reflection at $k = \pm\pi/a$ arises because the wave reflected from one atom in the linear lattice interferes constructively with the wave reflected from a nearest-neighbor atom. The difference in phase between the two reflected waves is just $\pm 2\pi$ for these two values of k. The region in k space between $-\pi/a$ and π/a is called the **first Brillouin zone** of this lattice, as in Chapter 2.

At $k = \pm\pi/a$ the wavefunctions are not the traveling waves $e^{i\pi x/a}$ and $e^{-i\pi x/a}$ of the free electron model. We shall show that the solutions at these particular k values are made up *equally* of waves traveling to the right and to the left: the solutions are standing waves. We first give a plausibility argument. When the Bragg condition is satisfied a wave traveling in one direction is soon Bragg-reflected and then travels in the opposite direction. Each subsequent Bragg reflection reverses the direction of travel again. The only time-independent situation is formed by standing waves. We can form two different standing waves from the traveling waves $e^{i\pi x/a}$ and $e^{-i\pi x/a}$:

$$\psi(+) = e^{i\pi x/a} + e^{-i\pi x/a} = 2\cos(\pi x/a) \ ;$$
$$\psi(-) = e^{i\pi x/a} - e^{-i\pi x/a} = 2i\sin(\pi x/a) \ . \tag{5}$$

The standing waves are made up each of equal portions of right- and left-directed travelling waves. The standing waves are labeled $(-)$ or $(+)$ according to whether they do or do not change sign when $-x$ is substituted for x. We have not normalized the forms (5).

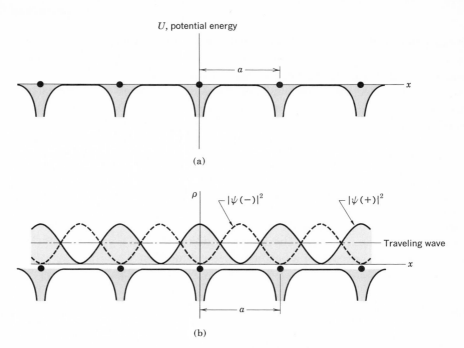

(a)

(b)

Figure 3 (a) Variation of potential energy of a conduction electron in the field of the ion cores of a linear lattice. (b) Distribution of probability density ρ in the lattice for $|\psi(-)|^2 \propto \sin^2 \pi x/a$; $|\psi(+)|^2 \propto \cos^2 \pi x/a$; and for a traveling wave. The wavefunction $\psi(+)$ piles up electronic charge on the cores of the positive ions, thereby lowering the potential energy in comparison with the average potential energy seen by a traveling wave. The wavefunction $\psi(-)$ piles up charge in the region between the ions and removes it from the ion cores; thereby raising the potential energy in comparison with that seen by a traveling wave. This figure is the key to understanding the origin of the energy gap.

Origin of the Energy Gap

The two standing waves $\psi(+)$ and $\psi(-)$ pile up electrons at different regions relative to the ions, and therefore the two waves have different values of the potential energy. This is the origin of the energy gap. In quantum mechanics the probability density ρ of a particle is $|\psi|^2$. For a pure traveling wave e^{ikx} we have $\rho = e^{-ikx}e^{ikx} = 1$, so that the charge density is constant. The charge density is not constant for linear combinations of plane waves. Consider the standing wave $\psi(+)$ in (5); for this we have

$$\rho(+) = |\psi(+)|^2 \propto \cos^2 \frac{\pi x}{a} \ ,$$

which piles up negative charge on or near the positive ions[4] located at $x = 0$, $a, 2a, \ldots$ where the potential energy is lowest. Figure 3a indicates schematically

---

[4] The origin of x is at the center of an ion.

the variation of the electrostatic potential energy of a conduction electron in the field of the positive ion cores of the monatomic linear lattice. The ion cores bear a positive charge, because in the metal each atom has given up one or more valence electrons to form the conduction band. The potential energy of an electron in the field of a positive ion is negative, that is, attractive. In (b) we sketch the distribution of electron density corresponding to the standing waves $\psi(+)$, $\psi(-)$, and to a traveling wave.

For the standing wave $\psi(-)$ we have the probability density

$$\rho(-) = |\psi(-)|^2 \propto \sin^2 \frac{\pi x}{a} \;,$$

which distributes electrons preferentially midway between ion cores, that is, away from the ion cores. On calculating average values of the potential energy over the three charge distributions, we expect to find the potential energy of $\rho(+)$ lower than that of a traveling wave, whereas the potential energy of $\rho(-)$ is higher than that of a traveling wave. If the potential energies of $\rho(-)$ and $\rho(+)$ differ by an amount E_g, we have the energy gap in Fig. 2 of width E_g. The wavefunction below the energy gap at points A in Fig. 2 will be $\psi(+)$, and the wavefunction above the energy gap at points B will be $\psi(-)$.

WAVE EQUATION OF ELECTRON IN A PERIODIC POTENTIAL

We considered above the approximate form we expect the solution of the Schrödinger equation to have if the wavevector is at a zone boundary, such as at $k = \pi/a$. We now treat in detail the wave equation and its solution at general values of k.

Let $U(x)$ denote the potential energy of an electron in a linear lattice of lattice constant a. We know that the potential energy is invariant under a crystal lattice translation:

$$U(x) = U(x + a) \;. \tag{6}$$

In Chapter 2 we showed that a function that is invariant under a crystal lattice translation may be expanded as a Fourier series in the reciprocal lattice vectors **G**. We write the Fourier series for the potential energy as

$$U(x) = \sum_G U_G e^{iGx} \;. \tag{7}$$

The values of the coefficients U_G for actual crystal potentials tend to decrease rapidly with increasing magnitude of G. For a bare Coulomb potential U_G decreases[5] as $1/G^2$.

We want the potential energy $U(x)$ to be a real function, which means

----

[5] *QTS*, p. 5.

that $U(x)$ must be equal to its complex conjugate $U^*(x)$. From the Fourier series we must have

$$\sum_G U_G e^{iGx} = \sum_G U_G^* e^{-iGx} \ , \qquad (8)$$

with use of $(e^{iGx})^* = e^{-iGx}$. We can satisfy (8) if

$$U_{-G} = U_G^* \ . \qquad (9)$$

This is a restriction on the Fourier series for $U(x)$.

We may be able to choose the origin such that $U(x)$ is an even function of x, which means that $U(x) = U(-x)$. This leads to a further restriction on the U_G. We use (7) to form

$$U(-x) = \sum_G U_G e^{-iGx} \ . \qquad (10)$$

On comparison of (10) with (7) we see that the two series will be identical if

$$U_G = U_{-G} \ . \qquad (11)$$

When the requirement (11) is combined with the reality requirement (9) we have $U_G = U_G^*$. We have found that the U_G themselves must be real if $U(x)$ is an even real function of x. With (11) we may write

$$U(x) = \sum_{G>0} U_G(e^{iGx} + e^{-iGx}) = 2\sum_{G>0} U_G \cos Gx \ . \qquad (12)$$

We have set $U_0 = 0$ for convenience.

The wave equation of an electron in the crystal is $\mathcal{H}\psi = \epsilon\psi$, where $\mathcal{H}$ is the hamiltonian and ϵ is the energy eigenvalue. The solutions ψ are called eigenfunctions or orbitals. Explicitly, the wave equation is

$$\left(\frac{1}{2m}p^2 + U(x)\right)\psi(x) = \left(\frac{1}{2m}p^2 + \sum_G U_G e^{iGx}\right)\psi(x) = \epsilon\psi(x) \ , \qquad (13)$$

using the Fourier series (7) for the potential energy. The momentum operator p is represented by $-i\hbar d/dx$, so that $p^2 = -\hbar^2 d^2/dx^2$. Equation (13) is written in the one-electron approximation in which the orbital $\psi(x)$ describes the motion of one electron in the potential of the ion cores and in the average potential of the other conduction electrons.

The wavefunction $\psi(x)$ may be expressed as a Fourier series[6] summed over all values of the wavevector permitted by the boundary conditions, so that

$$\psi(x) = \sum_K C(K)e^{iKx} \ , \qquad (14)$$

where K is real. The set of values of K is of the form $2\pi n/L$, for these values

[6] This analysis parallels that of Advanced Topic A for electromagnetic waves in a crystal. Advanced Topic F gives another very useful approach, the **tight-binding approximation**, to the energy band problem.

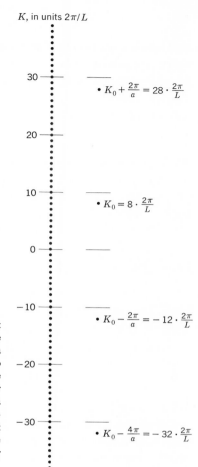

K, in units $2\pi/L$

$\bullet \ K_0 + \dfrac{2\pi}{a} = 28 \cdot \dfrac{2\pi}{L}$

$\bullet \ K_0 = 8 \cdot \dfrac{2\pi}{L}$

$\bullet \ K_0 - \dfrac{2\pi}{a} = -12 \cdot \dfrac{2\pi}{L}$

$\bullet \ K_0 - \dfrac{4\pi}{a} = -32 \cdot \dfrac{2\pi}{L}$

Figure 4 The points in the left-hand column represent the values of wavevector $K = 2\pi n/L$ allowed by the periodic boundary condition on the wavefunction over a ring of circumference L. The allowed values continue to $\pm \infty$. The points in the right-hand column represent the first few wavevectors which may enter into the Fourier expansion (14) of a wavefunction $\psi(x)$, starting from a particular wavevector $K_0 = 8 \cdot (2\pi/L)$. We have supposed in the construction of the right-hand column that the ring is composed of 20 primitive cells; thus the lattice constant $a = \frac{1}{20}L$ and the shortest reciprocal lattice vector has the length $2\pi/a = 20 \cdot (2\pi/L)$.

satisfy the periodic boundary conditions over the length L; here n is any integer, positive or negative. We do not assume that $\psi(x)$ is periodic in the fundamental lattice translation a; the translational properties of $\psi(x)$ will emerge in (34) below.

We shall first prove the very important result that not all wavevectors of the set $2\pi n/L$ enter the Fourier expansion (14) of any particular solution ψ of the periodic potential problem. If one wavevector K_0 is allowed—that is, if K_0 is known to be contained in a particular ψ—then the other wavevectors in the expansion of this ψ will be shown to have the form $K_0 + G$, where G is any reciprocal lattice vector. We can label a wavefunction ψ that contains a Fourier component K_0 as $\psi(K_0)$ or equally well as $\psi(K_0 + G)$, for if K_0 enters the Fourier series then $K_0 + G$ also enters, as we show. The wavevectors $K_0 + G$, running over G, are a very restricted subset of the wavevectors $2\pi n/L$, as illustrated by Fig. 4.

Our problem is to determine the values of the coefficients C in the Fourier expansion (14). To do this we first express the wave equation as a set of linear algebraic equations in the C's. We substitute the Fourier expansion (14) into the wave equation (13). The kinetic energy term is

$$\frac{1}{2m} p^2 \psi(x) = \frac{1}{2m}\left(-i\hbar \frac{d}{dx}\right)^2 \psi(x) = -\frac{\hbar^2}{2m}\frac{d^2\psi}{dx^2} = \frac{\hbar^2}{2m}\sum_K K^2 C(K) e^{iKx} \; ;$$

(15a)

and the potential energy term is

$$\left(\sum_G U_G e^{iGx}\right)\psi(x) = \sum_G \sum_K U_G e^{iGx} C(K) e^{iKx} \; .$$

(15b)

The wave equation is obtained as the sum of (15a) and (15b):

$$\sum_K \frac{\hbar^2}{2m} K^2 C(K) e^{iKx} + \sum_G \sum_K U_G C(K) e^{i(K+G)x} = \epsilon \sum_K C(K) e^{iKx} \; .$$ (15c)

We clarify this equation by use of the orthogonality property of the different Fourier components:

$$\int_0^L dx\, e^{-iK'x} e^{iKx} = \begin{cases} \dfrac{1}{i(K - K')}\,[e^{i(K-K')L} - 1] = 0 \; , & \text{if } K' \neq K \; . \\[2ex] & \text{(15d)} \\[2ex] L & , & \text{if } K' = K \; . \end{cases}$$

We have used the forms $K = 2\pi n/L$ and $K' = 2\pi n'/L$, where n, n' are integers, whence $\exp[i(K - K')L] = \exp[2\pi i(n - n')] = 1$. We now multiply both sides of (15c) by $\exp(-iK'x)$ and integrate over dx to obtain the wave equation in the form

$$\boxed{\frac{\hbar^2}{2m} K'^2 C(K') + \sum_G U_G C(K' - G) = \epsilon C(K') \; ,}$$

(16)

because by (15d) the nonvanishing terms in the double sum in (15c) have K such that $K + G = K'$, or $K = K' - G$.

This is the most important equation of the band theory of solids. It may be written more compactly if we replace the arbitrary index symbol K' by K, as we are now free to do without confusion. We also use the notation

$$\lambda_K \equiv \frac{\hbar^2}{2m} K^2$$

(17)

for the kinetic energy of the Fourier component K. Thus (16) becomes

$$\boxed{(\lambda_K - \epsilon) C(K) + \sum_G U_G C(K - G) = 0 \; ,}$$

(18)

which is a very useful form of the wave equation (13). It appears unfamiliar[7] only because a set of algebraic equations has emerged in place of the usual differential equation. The differential equation itself is hard to handle in a physical or instructive way for the periodic potential energy functions of interest.

Once we determine the C's from (18), the wavefunction is given as

$$\psi_K(x) = \sum_G C(K - G)e^{i(K-G)x} . \tag{19}$$

Equation (18) connects the Fourier coefficient $C(K_0)$ of a plane wave component $\exp[iK_0 x]$ of any particular orbital with the set of Fourier coefficients $C(K_0 + G)$ of all other plane waves $\exp[i(K_0 + G)x]$ which are contained in the Fourier series for the orbital. There is no fixed rule to tell us how to label the orbital $\psi_K(x)$; one man's K is another man's $K + G$. A useful convention is to use as a label the wavevector K_0 of that component of ψ that lies inside the first Brillouin zone of the lattice. In one dimension this will be the wavevector that lies between $-\pi/a$ and $+\pi/a$. The zone construction rules then assure that every other component $K_0 + G$ will lie outside the first zone. (The subscript zero is put on the K_0 as a temporary crutch to mark one particular wavevector of the set under consideration.)

For any given K_0 included in the large set of values $2\pi n/L$ permitted by the periodic boundary conditions for the lattice of length L, we can always find an orbital of the Schrödinger equation by solving the set of equations (18). In fact, this is an infinite set of equations, and there will be an infinite number of orbitals for any K_0. The orbitals generally will have different energies, and those of lowest energy interest us most.

The entire argument will be clear if we write out the equations for an explicit problem. Let g denote the shortest G, and suppose that the potential energy contains only one Fourier component $U_g = U_{-g}$, denoted by U. Then (18) contains the equations (20), (21a), (21b), where the second and third of these equations are obtained from (18) by replacing K_0 with $K_0 + g$ and $K_0 - g$;

$$(\lambda_{K_0} - \epsilon)C(K_0) + U[C(K_0 + g) + C(K_0 - g)] = 0 ; \tag{20}$$

$$(\lambda_{K_0+g} - \epsilon)C(K_0 + g) + U[C(K_0 + 2g) + C(K_0)] = 0 ; \tag{21a}$$

$$(\lambda_{K_0-g} - \epsilon)C(K_0 - g) + U[C(K_0) + C(K_0 - 2g)] = 0 . \tag{21b}$$

We see that the coefficients $C(K_0 + 2g)$ and $C(K_0 - 2g)$ have become coupled into the problem, and two more equations of the set (18) are

$$(\lambda_{K_0+2g} - \epsilon)C(K_0 + 2g) + U[C(K_0 + 3g) + C(K_0 + g)] = 0 ; \tag{22}$$

$$(\lambda_{K_0-2g} - \epsilon)C(K_0 - 2g) + U[C(K_0 - g) + C(K_0 - 3g)] = 0 . \tag{23}$$

We can go on to write equations for $C(K_0 \pm 3g)$ and so on ad infinitum. In all

[7] This equation appears in the work of H. Bethe on the diffraction of electrons by crystals: Annalen der Physik **87**, 55 (1928). We call it the central equation.

these equations ϵ will be the same value of the energy (which is to be determined), and λ_{K_0+sg} denotes $\hbar^2(K_0 + sg)^2/2m$, where s is any integer. We see that an infinite set of equations is obtained even in a problem for which the lattice potential has only a single Fourier component g.

To summarize, Eq. (18) connects a given Fourier coefficient $C(K)$ with the infinite number of other Fourier coefficients for which the wavevector differs from K by a reciprocal lattice vector. We generate an infinite set of independent equations, all of the same form, by subtracting successive reciprocal lattice vectors from K whenever it appears in (18). If we subtract a particular G' from K, we have the equation:

$$(\lambda_{K-G'} - \epsilon)C(K - G') + \sum_G U_G C(K - G' - G) = 0 . \qquad (24)$$

In practice it is very fortunate that we can often approximate the infinite set successfully with one or two equations or, more important, with one or two potential energy coefficients U_G. (This is the secret of the practical usefulness of the method developed here.) With two equations, for example, we have only to solve the 2×2 determinant of the coefficients of the C's to find the two roots ϵ that are the energy eigenvalues. With these roots we can then solve for the ratio of the two C's. First let us give an argument to suggest that there are circumstances in which two components dominate in the wavefunction ψ.

It is helpful to write the central equation (18) in the form

$$C(K) = \frac{\displaystyle\sum_G U_G C(K - G)}{\epsilon - \dfrac{\hbar^2 K^2}{2m}} , \qquad (25)$$

with use of the definition $\lambda_K = \hbar^2 K^2/2m$. The form (25) emphasizes that a coefficient $C(K)$ may tend to be large if the kinetic energy $\hbar^2 K^2/2m$ of the plane wave component $\exp(iKx)$ is nearly equal to the correct energy ϵ of the orbital $\psi_K(x)$ under consideration. The situation is especially interesting if $C(K)$ is large and there exists another coefficient $C(K - G')$ whose plane wave component $\exp[i(K - G')x]$ has nearly the same kinetic energy:

$$\frac{\hbar^2(K - G')^2}{2m} \cong \frac{\hbar^2 K^2}{2m} \approx \epsilon . \qquad (26)$$

Then $C(K - G')$ may also be large because the new denominator is also small in the equation we obtain on rewriting (25) with $K - G'$ in place of K:

$$C(K - G') = \frac{\displaystyle\sum_G U_G C(K - G' - G)}{\epsilon - \dfrac{\hbar^2}{2m}(K - G')^2} . \qquad (27)$$

The condition for strong mixing of the two components K and $K - G'$ is therefore that

$$(K - G')^2 = K^2 \; ; \tag{28}$$

this is familiar from Chapter 2 as exactly the condition for Bragg reflection of x-rays or electrons or neutrons from a crystal lattice. By strong mixing we mean that both $\exp[iKx]$ and $\exp[i(K - G')x]$ are important components of the orbital $\psi_K(x)$.

Bloch Functions

We can derive starting from the central equation (18) an exceedingly important and useful exact result about the form of the orbitals of the periodic potential problem. First we need a method of labeling any particular orbital. Let us choose *any* wavevector that appears in the Fourier series (14) for ψ and call it k; this will be the label:

$$\psi_k(x) = \sum_K C(K)e^{iKx} \; . \tag{29}$$

It does not matter that the choice of label is not unique. Earlier we wrote K_0 as a label; now it is convenient to write k. By (19) we know that

$$\psi_k(x) = \sum_G C(k - G)e^{i(k-G)x} \; , \tag{30}$$

which may be rearranged as

$$\psi_k(x) = \left(\sum_G C(k - G)e^{-iGx} \right) e^{ikx} = e^{ikx}u_k(x) \; , \tag{31}$$

with the definition

$$u_k(x) \equiv \sum_G C(k - G)e^{-iGx} \; . \tag{32}$$

Because $u_k(x)$ is a Fourier series over the reciprocal lattice vectors, it is invariant[8] under a crystal lattice translation T, so that

$$u_k(x) = u_k(x + T) \; . \tag{33}$$

We verify this directly by evaluating $u_k(x + T)$:

$$u_k(x + T) = \Sigma C(k - G)e^{-iG(x+T)} = e^{-iGT}(\Sigma C(k - G)e^{-iGx}) \tag{33a}$$

Because $\exp(-iGT) = 1$ it follows that $u_k(x + T) = u_k(x)$, thereby establishing the periodicity of u_k.

The result (31) is a statement of the **Bloch theorem** which states that **the eigenfunctions of the wave equation for a periodic potential are of the form of**

---

[8] A similar argument was developed in Chapter 2.

the product of a plane wave $e^{i\mathbf{k}\cdot\mathbf{r}}$ times a function $u_\mathbf{k}(\mathbf{r})$ with the periodicity of the crystal lattice:

$$\psi_\mathbf{k}(\mathbf{r}) = e^{i\mathbf{k}\cdot\mathbf{r}}u_\mathbf{k}(\mathbf{r}) \ . \tag{34}$$

The subscript $\mathbf{k}$ indicates that the function $u_\mathbf{k}(\mathbf{r})$ depends on the wavevector $\mathbf{k}$. An orbital of the form (34) is known as a **Bloch function.** These solutions are composed of travelling waves, and they can be assembled into wave packets to represent electrons that propagate freely through the potential field of the ion cores.

Crystal Momentum of an Electron

What is the significance of the wavevector $\mathbf{k}$ used to label the Bloch function? It has several properties:

(a) Under a crystal lattice translation which carries $\mathbf{r}$ to $\mathbf{r} + \mathbf{T}$ we have

$$\psi_\mathbf{k}(\mathbf{r} + \mathbf{T}) = e^{i\mathbf{k}\cdot\mathbf{T}}e^{i\mathbf{k}\cdot\mathbf{r}}u_\mathbf{k}(\mathbf{r} + \mathbf{T}) = e^{i\mathbf{k}\cdot\mathbf{T}}\psi_\mathbf{k}(\mathbf{r}) \ , \tag{35}$$

because $u_\mathbf{k}(\mathbf{r} + \mathbf{T}) = u_\mathbf{k}(\mathbf{r})$ by (33). Thus $e^{i\mathbf{k}\cdot\mathbf{T}}$ is the phase factor[9] by which a Bloch function is multiplied when we make a crystal lattice translation $\mathbf{T}$.

(b) If the lattice potential vanishes, the central equation (18) reduces to $(\lambda_\mathbf{k} - \epsilon)C(\mathbf{k}) = 0$, so that all $C(\mathbf{k} - \mathbf{G})$ are zero except $C(\mathbf{k})$, and thus $u_\mathbf{k}(\mathbf{r})$ is constant. We have

$$\psi_\mathbf{k}(\mathbf{r}) = e^{i\mathbf{k}\cdot\mathbf{r}} \ , \tag{36}$$

just as for a free electron. (This assumes we have had the foresight to pick the "right" $\mathbf{k}$ as the label; for many purposes other choices of $\mathbf{k}$, as we shall see, are more convenient.)

(c) The value of $\mathbf{k}$ enters into the conservation laws for collision processes of electrons in crystals. For this reason $\hbar\mathbf{k}$ is called the **crystal momentum** of the electron. When an electron $\mathbf{k}$ collides with a phonon of wavevector $\mathbf{K}$, the selection rule is

$$\mathbf{k} + \mathbf{K} = \mathbf{k}' + \mathbf{G} \ , \tag{37}$$

if the phonon is absorbed in the collision. The electron has been scattered from a state $\mathbf{k}$ to a state $\mathbf{k}'$; and $\mathbf{G}$ is any reciprocal lattice vector.[10]

---

[9] We may also say that $\exp(i\mathbf{k}\cdot\mathbf{T})$ is the eigenvalue of the crystal translation operation $\mathfrak{I}$, and $\psi_\mathbf{k}$ is the eigenvector. That is, $\mathfrak{I}\,\psi_\mathbf{k}(x) = \psi_\mathbf{k}(x + \mathbf{T}) = \exp(i\mathbf{k}\cdot\mathbf{T})\psi_\mathbf{k}(x)$, so that $\mathbf{k}$ is a suitable label for the eigenvalue. Here we have used the Bloch theorem.

[10] The probability of the collision involves a matrix element roughly of the form

$$\int d^3x\,\psi_{\mathbf{k}'}^{\circ}(\mathbf{r})\exp(i\mathbf{K}\cdot\mathbf{r})\,\psi_\mathbf{k}(\mathbf{r}) = \int d^3x\,u_{\mathbf{k}'}^{\circ}(\mathbf{r})u_\mathbf{k}(\mathbf{r})\exp[i(\mathbf{k} + \mathbf{K} - \mathbf{k}')\cdot\mathbf{r}] \ .$$

The product $u_{\mathbf{k}'}u_\mathbf{k}$ is periodic in the crystal lattice and can be expressed as a Fourier series in the $\mathbf{G}$'s. The integral vanishes unless $\mathbf{k} + \mathbf{K} - \mathbf{k}' = $ a reciprocal lattice vector $\mathbf{G}$, because the integral over all space of the function $\exp(i\mathbf{Q}\cdot\mathbf{r})$ vanishes unless $\mathbf{Q} = 0$. The presence of $\mathbf{G}$ takes care of the arbitrariness of $\mathbf{k}$ and $\mathbf{k}'$.

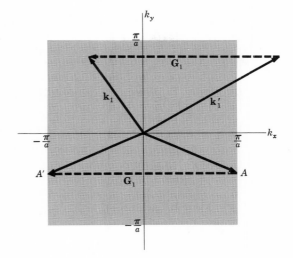

Figure 5   First Brillouin zone of a square lattice of side a. The wavevector $\mathbf{k}_1'$ can be carried into the first zone by forming $\mathbf{k}_1' + \mathbf{G}_1$. The wavevector at a point A on the zone boundary is carried by $\mathbf{G}_1$ to the point A' on the opposite boundary of the same zone. Shall we count both A and A' as lying in the first zone? We count them as *one* identical point in the zone.

Reduced Zone Scheme

It is always possible and often convenient to select the wavevector index $\mathbf{k}$ of the Bloch function so that it lies within the first Brillouin zone. This procedure is known as the **reduced zone scheme.** If we encounter a Bloch function written as

$$\psi_{\mathbf{k}'}(\mathbf{r}) = e^{i\mathbf{k}'\cdot\mathbf{r}}u_{\mathbf{k}'}(\mathbf{r}) \ , \tag{38}$$

with $\mathbf{k}'$ outside the first zone, as in Fig. 5, we may always find a suitable reciprocal lattice vector $\mathbf{G}'$ such that

$$\mathbf{k} = \mathbf{k}' - \mathbf{G}' \tag{39}$$

lies within the first Brillouin zone. Then (38) may be written as

$$\psi_{\mathbf{k}'}(\mathbf{r}) = e^{i\mathbf{k}'\cdot\mathbf{r}}u_{\mathbf{k}'}(\mathbf{r}) = e^{i\mathbf{k}\cdot\mathbf{r}}(e^{i\mathbf{G}'\cdot\mathbf{r}}u_{\mathbf{k}'}(\mathbf{r})) \equiv e^{i\mathbf{k}\cdot\mathbf{r}}u_{\mathbf{k}}(\mathbf{r}) = \psi_{\mathbf{k}}(\mathbf{r}) \ , \tag{40}$$

where we have defined

$$u_{\mathbf{k}}(\mathbf{r}) \equiv e^{i\mathbf{G}'\cdot\mathbf{r}}u_{\mathbf{k}'}(\mathbf{r}) \ . \tag{41}$$

308

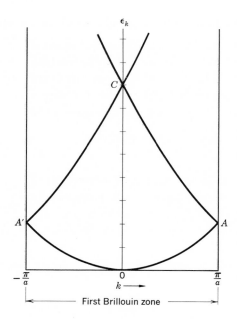

ϵ_k

C

A' A

$-\dfrac{\pi}{a}$ 0 $\dfrac{\pi}{a}$

$k \longrightarrow$

|←————— First Brillouin zone —————→|

Figure 6 Energy-wavevector relation $\epsilon_k = \hbar^2 k^2/2m$ for free electrons as drawn in the reduced zone scheme. This construction often gives a useful idea of the overall appearance of the band structure of a crystal. The branch AC if reflected in the vertical line at $k = \pi/a$ gives the usual free electron curve for positive k. The branch $A'C$ if reflected in $k = -\pi/a$ gives the usual curve for negative k. A crystal potential $U(x)$ will introduce band gaps at the edges of the zone (as at A and A') and at the center of the zone (as at C). The overall width and gross features of the band structure are often indicated properly by such free electron bands in the reduced zone scheme.

Both $e^{i\mathbf{G}'\cdot\mathbf{r}}$ and $u_{\mathbf{k}'}(\mathbf{r})$ are periodic in the crystal lattice, so $u_{\mathbf{k}}(\mathbf{r})$ is also, whence $\psi_{\mathbf{k}}(\mathbf{r})$ is of the Bloch form. Even with free electrons it may be useful to work in the reduced zone scheme, as in Fig. 6. It follows also that any energy $\epsilon_{\mathbf{k}'}$ for $\mathbf{k}'$ outside the first zone is equal to an $\epsilon_{\mathbf{k}}$ in the first zone, where $\mathbf{k}$ is related to $\mathbf{k}'$ by (39). Thus we need solve for the energy only in the first Brillouin zone, for each band. (An **energy band** is a single branch of the $\epsilon_{\mathbf{k}}$ versus $\mathbf{k}$ surface.)

In the reduced zone scheme we should not be surprised to find different energies at the same value of the wavevector. Each different energy characterizes a different band. Two wavefunctions at the same $\mathbf{k}$ but of different energies will be independent of each other: the wavefunctions will be made up of different combinations of the plane wave components $e^{i(\mathbf{k}-\mathbf{G})\cdot\mathbf{r}}$. Because the values of the coefficients $C(\mathbf{k} - \mathbf{G})$ are different for the two bands we really should add a symbol, say μ, to the C's to serve as a band index: $C_\mu(\mathbf{k} - \mathbf{G})$. Thus the Bloch function for a state of wavevector $\mathbf{k}$ in the band μ will be written as

$$\psi_{\mu,\mathbf{k}} = e^{i\mathbf{k}\cdot\mathbf{r}}u_{\mu,\mathbf{k}}(\mathbf{r}) = \sum_G C_\mu(\mathbf{k} - \mathbf{G})e^{i(\mathbf{k}-\mathbf{G})\cdot\mathbf{r}} \ .$$

Periodic Zone Scheme

There are problems for which it is helpful to imagine that we repeat a given Brillouin zone periodically through all of wavevector space. To repeat a zone, we translate the zone by a reciprocal lattice vector. If we can translate a band from other zones into the first zone, it follows that we can translate a band in the first zone into every other zone. In this scheme the energy ϵ_k of a band is a periodic function in the reciprocal lattice:

$$\epsilon_k = \epsilon_{k+G} \; . \tag{42}$$

Here ϵ_{k+G} is understood to refer to the same energy band as ϵ_k. The result of this construction is known as the **periodic zone scheme;** it will be especially helpful in Chapter 10 in exhibiting the connectivity of electron orbits in a magnetic field. The periodic property of the energy also can be seen easily from the central equation (18).

Consider as an example an energy band of a simple cubic lattice as calculated in the tight-binding approximation, Eq. (F.9):

$$\epsilon_k = E_0 - \alpha - 2\gamma \, (\cos k_x a + \cos k_y a + \cos k_z a) \; , \tag{43}$$

where E_0, α, and γ are constants. A reciprocal lattice vector of the sc lattice is $\mathbf{G} = (2\pi/a)\hat{x}$; if we add this vector to $\mathbf{k}$ the only change in (43) is in the term

$$\cos k_x a \rightarrow \cos \, (k_x + 2\pi/a)a = \cos \, (k_x a + 2\pi) \; , \tag{44}$$

but this is identically equal to $\cos k_x a$. We see that the energy is unchanged when the wavevector is increased by a reciprocal lattice vector, so that in the periodic zone scheme the energy is a periodic function of the wavevector.

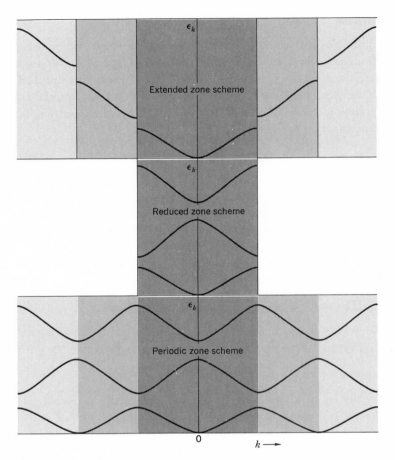

Figure 7 Three energy bands of a linear lattice plotted in the extended (Brillouin), reduced, and periodic zone schemes.

Three different zone schemes can be useful:

(a) the **extended zone scheme** in which different bands are drawn in different zones in wavevector space;

(b) the **reduced zone scheme** in which all bands are drawn in the first Brillouin zone;

(c) the **periodic zone scheme** in which every band is drawn in every zone.

These schemes are displayed in Fig. 7. The periodic zone scheme is also illustrated by Figs. 8 and 10.5.

APPROXIMATE SOLUTION NEAR A ZONE BOUNDARY

We are now equipped to understand the occurrence of band gaps, of light effective masses, and the role of holes as charge carriers. We suppose that the values of the Fourier components U_G of the potential energy are small in comparison with the kinetic energy $\hbar^2 k_F{}^2/2m$ of a free electron on the Fermi sphere. This assumption will allow us to restrict attention to simple wavefunctions represented approximately as the sum of two plane waves.

On Zone Boundary

We first consider the wavefunction with a wavevector exactly at the zone boundary at $\tfrac{1}{2}G_1$, that is, at π/a. Here

$$k^2 = (\tfrac{1}{2}G_1)^2 \; ; \qquad (k - G_1)^2 = (\tfrac{1}{2}G_1 - G_1)^2 = (\tfrac{1}{2}G_1)^2 \; , \quad (45)$$

so that at the zone boundary the kinetic energy of the two component waves $\exp[ikx]$ and $\exp[i(k - G_1)x]$ are equal:

$$\frac{\hbar^2}{2m} k^2 = \frac{\hbar^2}{2m} (k - G_1)^2 = \frac{\hbar^2}{2m} (\tfrac{1}{2}G_1)^2 \; . \quad (46)$$

If $C(\tfrac{1}{2}G_1)$ is an important coefficient in the orbital (19) at the zone boundary, then $C(-\tfrac{1}{2}G_1)$ is also an important coefficient in the orbital. This result follows from the discussion of (25).

We retain only those equations in the central equation that contain both coefficients $C(\tfrac{1}{2}G_1)$ and $C(-\tfrac{1}{2}G_1)$, and neglect all other coefficients. Then one equation of (18) becomes, with $K = \tfrac{1}{2}G_1$ and $\lambda_1 \equiv \hbar^2(\tfrac{1}{2}G_1)^2/2m$,

$$(\lambda_1 - \epsilon)C(\tfrac{1}{2}G_1) + U_1 C(-\tfrac{1}{2}G_1) = 0 \; . \quad (47)$$

We have written $U_1 = U_{G_1} = U_{-G_1}$. This result is of the form of (20). With $K = -\tfrac{1}{2}G_1$, another equation of (18) becomes

$$(\lambda_{-1} - \epsilon)C(-\tfrac{1}{2}G_1) + U_1 C(\tfrac{1}{2}G_1) = 0 \; , \quad (48)$$

of the form of (21b).

These two equations have nontrivial solutions for the two coefficients $C(\tfrac{1}{2}G_1)$, $C(-\tfrac{1}{2}G_1)$ if the energy ϵ satisfies

$$\begin{vmatrix} \lambda_1 - \epsilon & U_1 \\ U_1 & \lambda_{-1} - \epsilon \end{vmatrix} = 0 \; , \quad (49)$$

whence, with $\lambda_1 = \lambda_{-1}$,

$$(\lambda_1 - \epsilon)^2 = U_1{}^2 \; ; \qquad \epsilon = \lambda_1 \pm U_1 = \frac{\hbar^2}{2m} (\tfrac{1}{2}G_1)^2 \pm U_1 \; . \quad (50)$$

The energy has two roots, one lower than the free electron kinetic energy by U_1, and one higher by U_1. Thus the potential energy $2U_1 \cos G_1 x$ has created an energy gap of extent $2U_1$ at the zone boundary.

The ratio of the C's may be found from either (47) or (48):

$$\frac{C(-\tfrac{1}{2}G_1)}{C(\tfrac{1}{2}G_1)} = \frac{\epsilon - \lambda_1}{U_1} = \pm 1 \ , \tag{51}$$

where the last step uses (50). Thus the Fourier expansion of $\psi(x)$ has the two solutions

$$\psi(x) = e^{i\frac{1}{2}G_1 x} \pm e^{-i\frac{1}{2}G_1 x} \ , \tag{52}$$

apart from a normalization constant. These orbitals are identical to (5). One solution gives the wavefunction at the bottom of the energy gap; the other gives the wavefunction at the top of the gap. Which solution has the lower energy depends on the sign of U_1 in the potential energy.

Near Zone Boundary

We now solve for the orbitals with a wavevector k near the zone boundary $\tfrac{1}{2}G_1$. We use the same two-component approximation, now with a wavefunction of the form

$$\psi(x) = C(k)e^{ikx} + C(k - G_1)e^{i(k-G_1)x} \ . \tag{53}$$

As directed by the central equation (18), we solve the pair of equations

$$(\lambda_k - \epsilon)C(k) + U_1 C(k - G_1) = 0 \ ; \\
(\lambda_{k-G_1} - \epsilon)C(k - G_1) + U_1 C(k) = 0 \ , \tag{54}$$

with λ_k defined as $\hbar^2 k^2 / 2m$. These equations have a solution if the energy ϵ satisfies

$$\begin{vmatrix} \lambda_k - \epsilon & U_1 \\ U_1 & \lambda_{k-G_1} - \epsilon \end{vmatrix} = 0 \ , \tag{55}$$

whence

$$\epsilon^2 - \epsilon(\lambda_{k-G_1} + \lambda_k) + \lambda_{k-G_1}\lambda_k - U_1{}^2 = 0 \ . \tag{56}$$

The energy has two roots:

$$\epsilon = \tfrac{1}{2}(\lambda_{k-G_1} + \lambda_k) \pm [\tfrac{1}{4}(\lambda_{k-G_1} - \lambda_k)^2 + U_1{}^2]^{\frac{1}{2}} \ , \tag{57}$$

and each root describes an energy band. The two roots are plotted in Fig. 8a in the periodic zone scheme. In the extended zone scheme the first band would be plotted only in the first Brillouin zone, $-\tfrac{1}{2}G_1 \leq k \leq \tfrac{1}{2}G_1$, and the second band would be plotted only in the second Brillouin zone (which means in the two segments $-G_1 \leq k \leq -\tfrac{1}{2}G_1$ and $\tfrac{1}{2}G_1 \leq k \leq G_1$).

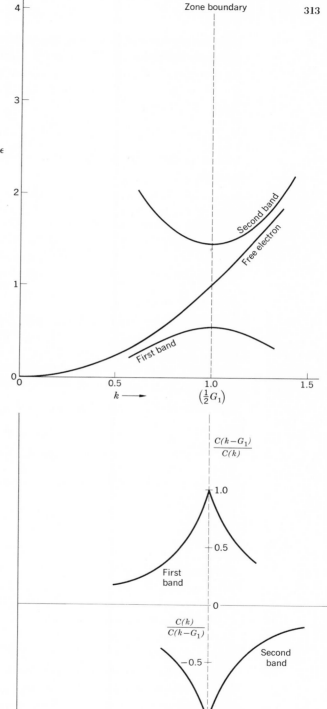

Figure 8a Solutions of (57) in the periodic zone scheme, in the region near a boundary of the first Brillouin zone. The units are such that $U_1 = -0.45$; $G_1 = 2$; and $\hbar^2/m = 1$. The free electron curve is drawn for comparison. The energy gap at the zone boundary is 0.90. The value of U_1 has deliberately been chosen large for this illustration; the value is too large for the two-term approximation to be accurate.

Figure 8b Ratio of the coefficients in $\psi(x) = C(k)e^{ikx} + C(k - G_1)e^{i(k-G_1)x}$ as calculated near the boundary of the first Brillouin zone.

It is convenient to expand the energy (57) in terms of a quantity δ which measures the difference in wavevector between the zone boundary and k:

$$\delta \equiv \tfrac{1}{2}G_1 - k \; ; \qquad k = \tfrac{1}{2}G_1 - \delta \; . \tag{58}$$

Then (57) becomes

$$
\begin{aligned}
\epsilon_k &= (\hbar^2/2m)(\tfrac{1}{4}G_1{}^2 + \delta^2) \pm [4\lambda_1(\hbar^2\delta^2/2m) + U_1{}^2]^{\frac{1}{2}} \\
&\cong (\hbar^2/2m)(\tfrac{1}{4}G_1{}^2 + \delta^2) \pm U_1[1 + 2(\lambda_1/U_1{}^2)(\hbar^2\delta^2/2m)] \; ,
\end{aligned}
\tag{59}
$$

in the region $\hbar^2 G_1 \delta / 2m \ll |U_1|$. This restricts δ to very small values, because our whole calculation is based on the assumption that $\hbar^2 G_1{}^2/2m \gg |U_1|$. Writing the two roots of (50) as $\epsilon_1(\pm)$, we may write (59) as

$$\epsilon_k(+) = \epsilon_1(+) + \frac{\hbar^2\delta^2}{2m}\left(1 + \frac{2\lambda_1}{U_1}\right) \; ; \tag{60}$$

$$\epsilon_k(-) = \epsilon_1(-) + \frac{\hbar^2\delta^2}{2m}\left(1 - \frac{2\lambda_1}{U_1}\right) \; . \tag{61}$$

These are the roots for the energy when the wavevector is very close to the zone boundary at $\tfrac{1}{2}G_1$. Note the quadratic dependence of the energy on the wavevector δ. For U_1 negative, the solution $\epsilon_k(-)$ corresponds to the upper of the two bands, and $\epsilon_k(+)$ to the lower of the two bands. Bear in mind our assumption $\lambda_1 \gg |U_1|$.

NUMBER OF ORBITALS IN A BAND

Consider a linear crystal constructed of primitive cells of lattice constant a. In order to count states we apply periodic boundary conditions to the wavefunctions over the length of the crystal. The allowed values of the electron wavevector k in the first Brillouin zone are given by (2):

$$k = 0 \; ; \quad \pm\frac{2\pi}{L} \; ; \quad \pm\frac{4\pi}{L} \; ; \quad \ldots \; ; \quad \frac{N\pi}{L} \; . \tag{62}$$

We have cut the series off at $k = N\pi/L \equiv \pi/a$, for this is the boundary of the zone. The point $-N\pi/L = -\pi/a$ is not to be counted as an independent point because it is connected by a reciprocal lattice vector with π/a. The total number of points given in (62) is exactly N, the number of primitive cells. We see that **each primitive cell contributes exactly one independent value of k to each energy band.** This result carries over into three dimensions. With account taken of the two independent orientations of the electron spin, **there are 2N independent orbitals in each energy band.**

If there is a single atom of valence one in each primitive cell, the band can be half-filled with electrons. If each atom contributes two valence electrons to the band, the band can be exactly filled. If there are two atoms of valence one in each primitive cell, the band can also be exactly filled.

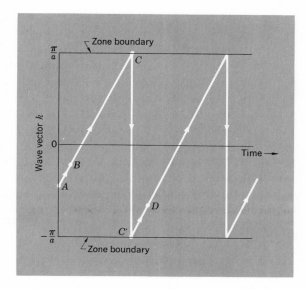

Figure 9 Motion in k space of the value of the wavevector of an electron in a linear crystal in an external electric field, with all collision processes neglected. An electron with wavevector initially at A will be accelerated by the field B and then to C at the zone boundary. But C is equivalent by a reciprocal lattice vector to C' at the opposite boundary of the zone. After a further interval the electron will reach D from C', and the process will repeat itself. There is some doubt about this theoretical possibility that an electron may oscillate in an energy band, because of the possibility of interband transitions induced by the electric field, according to estimates by A. Rabinovitch and J. Zak (to be published).

Metals and Insulators

If the valence electrons exactly fill one or more bands, leaving others empty, the crystal will be an insulator. An external electric field will not cause the flow of current.[11] Because a filled band is separated by an energy gap from the next higher band, there is no continuous way to change the total momentum of the electrons. Every accessible state is filled; nothing changes when the field is applied. (This is quite unlike the situation for free electrons pictured in Fig. 7.10).

Another way of seeing what happens is by use of the force equation derived in Chapter 10:

$$\hbar \, \frac{d\mathbf{k}}{dt} = \mathbf{F} \ . \tag{63}$$

Under the action of a constant force the wavevector of an electron will increase continuously with time. But when $\mathbf{k}(t)$ reaches the boundary of the zone the wavevector will be umklapped (Fig. 9) to the opposite boundary. The motion will then start all over again, but the net acceleration is zero when the motion

---

[11] We suppose that the electric field is not strong enough to disrupt the electronic structure, as in the Zener effect (see J. M. Ziman, *Principles of the theory of solids*, Cambridge University Press, 1964, sec. 6.8).

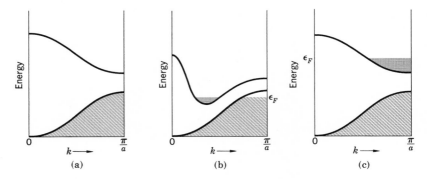

Figure 10 Occupied states and band structures giving (a) an insulator, (b) a metal or a semimetal because of band overlap, and (c) a metal because of electron concentration. In (b) the overlap need not occur along the same directions in the Brillouin zone. If the overlap is small, with relatively few states involved, we speak of a semimetal.

is averaged over all states of the band.[12]

A crystal can be an insulator only if the number of valence electrons in a primitive cell of the crystal is an even integer. (An exception must be made sometimes for electrons in tightly bound inner shells which cannot be treated by band theory.) If a crystal has an even number of valence electrons per primitive cell, it is necessary to consider whether or not the bands overlap in energy. If the bands overlap, then instead of one filled band giving an insulator, we can have two partly filled bands giving a metal (Fig. 10).

The alkali metals and the noble metals have one valence electron per primitive cell, so that they have to be metals. The alkaline earth metals have two valence electrons per primitive cell; they could be insulators, but the bands overlap in energy to give metals, but not very good metals.[13] Diamond, silicon, and germanium each have two atoms of valence four (or eight valence electrons) per primitive cell; the bands do not overlap, so that the pure crystals are insulators at absolute zero.

---

[12] According to this argument it would appear that there is no net current even if the band is not entirely filled. But the reasoning does not apply in the presence of the strong collision processes characteristic of real solids. Collision processes return electrons to thermal equilibrium before the wave-vector has increased in an important way. In equilibrium the high energy states in the conduction band of a metal are occupied less often than are the low energy states, and the electrical conductivity can be calculated as in Chapters 7 and 8. Thus the argument that led to zero net current is entirely irrelevant to the actual situation in metals.

[13] See the values of the electrical resistivity in Table 7.3.

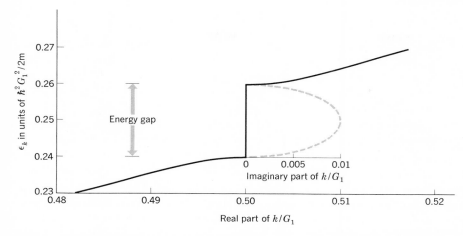

Figure 11 In the energy gap there exist solutions of the wave equation for complex values of the wavevector. At the boundary of the first zone the real part of the wavevector is $\frac{1}{2}G_1$. The imaginary part is plotted in the approximation of two plane waves, for $U_1 = 0.01\ \hbar^2 G_1^2/2m$. (Courtesy of R. Cahn.)

SUMMARY

1. The solutions of the wave equation in a periodic lattice are of the Bloch form $\psi_\mathbf{k}(\mathbf{r}) = e^{i\mathbf{k}\cdot\mathbf{r}}u_\mathbf{k}(\mathbf{r})$, where $u_\mathbf{k}(\mathbf{r})$ is invariant under a crystal lattice translation. The **K**'s which appear in a Fourier series expansion

$$\psi_\mathbf{k}(\mathbf{r}) = \sum_\mathbf{K} C(\mathbf{K})e^{i\mathbf{K}\cdot\mathbf{r}}$$

are all of the form $\mathbf{k} + \mathbf{G}$, where **G** runs over all reciprocal lattice vectors.

2. There are regions of energy for which no Bloch function solutions of the wave equation exist. These energies form forbidden regions in which the wavefunctions are damped in space and the values of the **K**'s are complex, as pictured in Fig. 11. The existence of insulators is due to the existence of forbidden regions of energy.

3. Energy bands may often be approximated by one or two plane waves: for example, $\psi_k(x) \cong C(k)e^{ikx} + C(k - G)e^{i(k-G)x}$ near the zone boundary at $\frac{1}{2}G$. The bands may usefully be pictured in three schemes: extended (Brillouin), reduced, and periodic.

4. The number of orbitals in a band is $2N$, where N is the number of primitive cells in the specimen.

Problems

1. *Square lattice, free electron energies.* (a) Show for a simple square lattice (two dimensions) that the kinetic energy of a free electron at a corner of the first zone is higher than that of an electron at midpoint of a side face of the zone by a factor of 2. (b) What is the corresponding factor for a simple cubic lattice (three dimensions)? (c) What bearing might the result of (b) have on the conductivity of divalent metals?

2. *Number of states.* A simple cubic crystal in the form of a cube has N^3 primitive cells. Counting carefully, show that the number of independent values of the wavevector **k** in the Brillouin zone is exactly N^3. *Hint:* If one point of the zone can be connected to p other points by reciprocal lattice vectors, all $p + 1$ points are to be counted as one point.

3. *Complex wavevectors in the energy gap.* Find an expression for the imaginary part of the wavevector in the energy gap at the boundary of the first Brillouin zone, in the approximation which led to Eq. (49). Give the result in the form of $\mathrm{Im}(k)$ versus σ, where σ will denote the energy as measured from the center of the energy gap. The result for small σ and small $\mathrm{Im}(k)$ is

$$\frac{\hbar^2}{2m}\,[\mathrm{Im}(k)]^2 \approx \frac{U_1{}^2 - \sigma^2}{\hbar^2 G^2/2m}\,,$$

as plotted in Fig. 11. This form is of importance in the theory of Zener tunneling from one band to another in the presence of a strong electric field; see Ziman, pp. 163–168. [Experimental confirmation of the predicted form of $\mathrm{Im}(k)$ is reported by G. H. Parker and C. A. Mead, "Energy-momentum relationship in InAs", Phys. Rev. Letters, **21**, 605 (1968); see also S. Kurtin, T. C. McGill, and C. A. Mead, Phys. Rev. **25**, 756 (1970).]

4. *Potential energy in the diamond structure.* (a) Show that for the diamond structure the Fourier component U_G of the crystal potential seen by an electron is equal to zero for **G** = 2**A**, where **A** is a basis vector in the reciprocal lattice referred to the conventional cubic cell. (b) Show that in the usual first-order approximation to the solutions of the wave equation in a periodic lattice the energy gap vanishes at the zone boundary plane normal to the end of the vector **A**, and show in the next order of approximation that this energy gap does not vanish.

References

J. M. Ziman, *Principles of the theory of solids,* Cambridge University Press, 1964.

N. F. Mott and H. Jones, *Theory of the properties of metals and alloys,* Oxford, 1936. (Dover paperback reprint.)

(Further references on band theory are given at the end of Chapter 10.)

10

Energy Bands II

Advanced Topics Relevant to This Chapter

 G. Particle motion in wavevector space and in real space
 H. Mott transition
 I. Vector potential, including field momentum, gauge transformation, and quantization of orbits

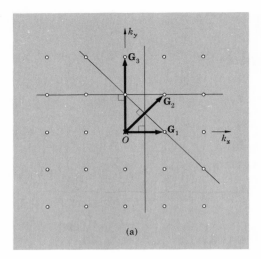

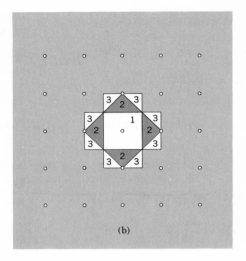

Figure 1 (a) Construction in **k** space of the first three Brillouin zones of a square lattice. The three shortest forms of the reciprocal lattice vectors are indicated as G_1, G_2, and G_3. The lines drawn are the perpendicular bisectors of these G's. (b) On constructing all lines equivalent by symmetry to the three lines in (a) we obtain the regions in **k** space which form the first three Brillouin zones. The numbers denote the zone to which the regions belong; the numbers here are ordered according to the length of the vector G involved in the construction of the outer boundary of the region.

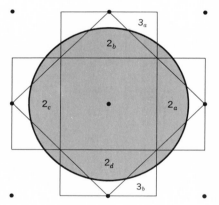

Figure 2 Brillouin zones of a square lattice in two dimensions. The circle shown is a surface of constant energy for free electrons; it will be the Fermi surface for some particular value of the electron concentration. The total area of the filled region in **k** space depends only on the electron concentration and is independent of the interaction of the electrons with the lattice. The shape of the Fermi surface depends on the lattice interaction, and the shape will not be an exact circle in an actual lattice. The labels within the sections of the second and third zones refer to Fig. 3.

1st zone

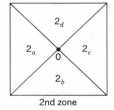

2nd zone

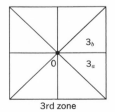

3rd zone

Figure 3 Mapping of the first, second, and third Brillouin zones in the reduced zone scheme. The sections of the second zone in Fig. 2 are put together into a square by translation through an appropriate reciprocal lattice vector. A different G is needed for each piece of a zone.

"It is interesting to establish which [electron] waves are especially influenced by these anomalies that arise from selective Bragg reflections. We construct the reciprocal lattice of the crystal . . . ."

(L. Brillouin, 1930)

CONSTRUCTION OF FERMI SURFACES

With metals we are particularly concerned with the shape of the **Fermi surface**, the surface of constant energy ϵ_F in wavevector or **k** space. The Fermi surface separates the unfilled orbitals from the filled orbitals, at absolute zero. The electrical properties of the metal are largely determined by the shape of the Fermi surface, because the current is due to changes in the occupancy of states near the Fermi surface. The shape may be very intricate and yet have a simple interpretation in terms of a spherical Fermi surface viewed in the reduced zone scheme.

Figure 9.6 shows the energy versus the wavevector for free electrons in one dimension in the reduced zone scheme. We extend the analysis to two dimensions in Fig. 1. The Bragg equation (2.40) for the zone boundaries is $2\mathbf{k} \cdot \mathbf{G} + G^2 = 0$, which is satisfied if **k** terminates on the plane normal to **G** at the midpoint of **G**. The first Brillouin zone of the square lattice is the area enclosed by the perpendicular bisectors of $\mathbf{G}_1$ and of the three reciprocal lattice vectors equivalent by symmetry to $\mathbf{G}_1$ in Fig. 1a. These four reciprocal lattice vectors are $\pm(2\pi/a)\hat{\mathbf{k}}_x$ and $\pm(2\pi/a)\hat{\mathbf{k}}_y$.

The second zone is constructed from $\mathbf{G}_2$ and the three vectors equivalent to it by symmetry, and similarly for the third zone. The pieces of the second and third zones are drawn in Fig. 1b. To determine the boundaries of some zones we have to consider sets of several nonequivalent reciprocal lattice vectors. Thus the boundaries of section 3_a of the third zone are formed from the perpendicular bisectors of three **G**'s, namely $(2\pi/a)\hat{\mathbf{k}}_x$; $(4\pi/a)\hat{\mathbf{k}}_y$; and $(2\pi/a)(\hat{\mathbf{k}}_x + \hat{\mathbf{k}}_y)$.

The free electron Fermi surface for an arbitrary electron concentration is shown in Fig. 2. It is inconvenient to have sections of the Fermi surface that belong to the same zone, such as the second zone, appear so detached from one another. The detachment can be repaired by a transformation to the reduced zone scheme as discussed in Eqs. (9.38) to (9.41). We take from Fig. 2 the triangle labeled 2_a and move it by a reciprocal lattice vector, in this instance by $\mathbf{G} = -(2\pi/a)\hat{\mathbf{k}}_x$, such that the triangle reappears in the area of the first Brillouin zone (Fig. 3). Other reciprocal lattice vectors will shift the triangles 2_b, 2_c, 2_d to other parts of the first zone, thereby completing the mapping of the second zone into the reduced zone scheme. The parts of the Fermi surface falling in the second zone are now connected, as shown in Fig. 4.

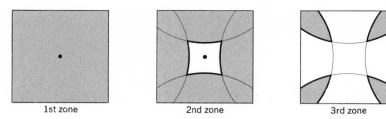

1st zone   2nd zone   3rd zone

Figure 4 The free electron Fermi surface of Fig. 2, as viewed in the reduced zone scheme. The shaded areas represent occupied electron states. Parts of the Fermi surface fall in the second, third, and fourth zones. The fourth zone is not shown. The first zone is shown entirely occupied.

A third zone has been assembled into a square in Fig. 4, but the parts of the Fermi surface falling in the third zone still appear disconnected. When we look at it in the periodic zone scheme (Fig. 5) the Fermi surface forms a rosette, or a lattice of rosettes.

How do we go from Fermi surfaces for free electrons to Fermi surfaces for nearly free electrons? We can make approximate constructions freehand by the use of four facts:

(a) The interaction of the electron with the periodic potential of the crystal causes energy gaps to appear at the zone boundaries.

(b) Almost always the Fermi surface will intersect zone boundaries perpendicularly.

(c) The crystal potential will round out sharp corners in the Fermi surfaces.

(d) The total volume enclosed by the Fermi surface depends only on the electron concentration and is independent of the details of the lattice interaction.

We cannot make quantitative statements without a detailed calculation, but qualitatively we expect the Fermi surfaces in the second and third zones of Fig. 4 to be changed in the direction of the surfaces shown in Fig. 6.

Freehand impressions of the Fermi surfaces derived from free electron surfaces are quite useful. Fermi surface constructions for free electrons are most easily carried out by a procedure due to W. A. Harrison, Fig. 7. The reciprocal lattice points are determined, and then a free-electron sphere of radius appropriate to the electron concentration is drawn around each reciprocal lattice point. Any point in **k** space which lies within at least one of the spheres corresponds to an occupied state in the first zone. Points lying within at least two spheres correspond to occupied states in the second zone, and similarly for points lying in three or more spheres.

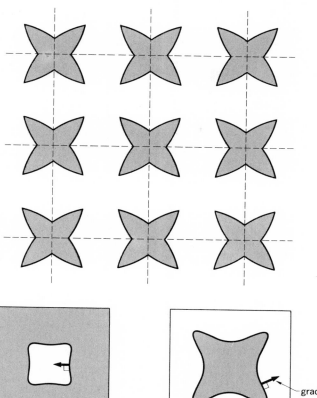

Figure 5 The Fermi surface in the third zone as drawn in the periodic zone scheme. The figure was constructed by repeating the third zone as shown in Fig. 4 in the reduced zone scheme.

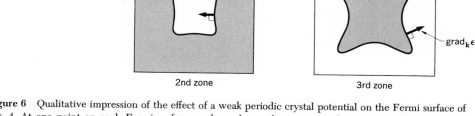

2nd zone

3rd zone

Figure 6 Qualitative impression of the effect of a weak periodic crystal potential on the Fermi surface of Fig. 4. At one point on each Fermi surface we have shown the vector $\text{grad}_{\mathbf{k}}\ \epsilon$. In the second zone the energy increases toward the interior of the figure, and in the third zone the energy increases toward the exterior. The shaded regions are filled with electrons and are lower in energy than the unshaded regions. We shall see that a Fermi surface like that of the third zone is electronlike, whereas one like that of the second zone is holelike.

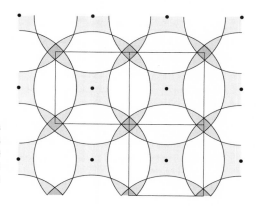

Figure 7 Harrison construction of free electron Fermi surfaces in the second, third, and fourth zones for a square lattice. The Fermi surface does not intercept the first zone, which therefore is filled with electrons. The darker the shading, the higher the zone number.

ELECTRONS, HOLES, AND OPEN ORBITS

We now derive the equation of motion of an electron in a crystal. We look first at the motion of a wave packet in a linear crystal in an applied electric field. Suppose that the wave packet is made up of wavefunctions in a single band with wavevectors near a particular wavevector $\mathbf{k}$. We assume from wave optics the general expression for the group velocity as $v_g = d\omega/dk$. The frequency associated with a wavefunction of energy ϵ is given by $\omega = \epsilon/\hbar$, and so

$$v_g = \hbar^{-1} \frac{d\epsilon}{dk} \;. \tag{1}$$

The effects of the crystal on the motion of the electron are all contained in the dispersion relation $\epsilon(\mathbf{k})$. The work $\delta\epsilon$ done on the electron by the electric field E in the time interval δt is

$$\delta\epsilon = -eEv_g \, \delta t \;. \tag{2}$$

We observe that
$$\delta\epsilon = \frac{d\epsilon}{dk} \delta k = \hbar v_g \, \delta k \;, \tag{3}$$

using (1). Thus on comparing (2) with (3) we have

$$\delta k = -(eE/\hbar)\delta t \;, \tag{4}$$

whence $\hbar \, dk/dt = -eE$.

In terms of the external force $\mathbf{F}$,

$$\boxed{\hbar \frac{d\mathbf{k}}{dt} = \mathbf{F}} \;. \tag{5}$$

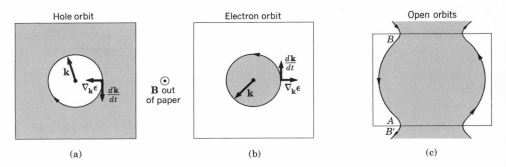

Figure 8 Motion in a magnetic field of the wavevector of an electron on the Fermi surface, in (a) and (b) for Fermi surfaces topologically equivalent to those of Fig. 6. In (a) the wavevector moves around the orbit in a clockwise direction; in (b) the wavevector moves around the orbit in a counterclockwise direction. The direction in (b) is what we expect for a free electron of charge $-e$: the smaller $\mathbf{k}$ values have the lower energy, so that the filled electron states lie inside the Fermi surface. We call the orbit in (b) **electronlike.** The sense of the motion in a magnetic field is opposite in (a) to that in (b), so that we refer to the orbit in (a) as **holelike.** A hole moves as a particle of positive charge e. In (c) for a rectangular zone we show the motion on an **open orbit** in the periodic zone scheme. This is topologically intermediate between a hole orbit and an electron orbit.

This is an important relation: In a crystal $\hbar \, d\mathbf{k}/dt$ is equal to the external force on the electron. In free space $d(m\mathbf{v})/dt$ is equal to the force. We have not overthrown Newton's second law of motion: the electron in the crystal is subject to forces from the crystal lattice as well as from external sources. If we choose to express the motion of the electron in terms of the external force alone, it is not surprising that the resulting equation of motion is not simply $\mathbf{F} = m\mathbf{a}$. Perhaps it is more surprising that any good at all comes out of an approach involving only external forces.

We shall assume the result of the argument (1) to (5) applies also to the Lorentz force on an electron in a magnetic field. Laborious calculations have justified this assumption under ordinary conditions where the magnetic field is not so strong that it breaks down the band structure. Thus the equation of motion of an electron of group velocity $\mathbf{v}$ in a constant magnetic field $\mathbf{B}$ is

(CGS) $$\hbar \frac{d\mathbf{k}}{dt} = -\frac{e}{c} \mathbf{v} \times \mathbf{B} \; ; \tag{6}$$

(SI) $$\hbar \frac{d\mathbf{k}}{dt} = -e\mathbf{v} \times \mathbf{B} \; ,$$

where the right-hand side is the Lorentz force on the electron. With the group velocity $\hbar\mathbf{v} = \mathrm{grad}_{\mathbf{k}}\epsilon$, the rate of change of the wavevector is

(CGS) $$\frac{d\mathbf{k}}{dt} = -\frac{e}{\hbar^2 c} \nabla_{\mathbf{k}}\epsilon \times \mathbf{B} \; ; \tag{7}$$

(SI) $$\frac{d\mathbf{k}}{dt} = -\frac{e}{\hbar^2} \nabla_{\mathbf{k}}\epsilon \times \mathbf{B} \; ,$$

where now both sides of the equation refer to the electron coordinates in $\mathbf{k}$ space.

We see from the vector cross-product in (7) that in a magnetic field an electron moves in $\mathbf{k}$ space in a direction normal to the direction of the gradient of the energy ϵ, so that the **electron moves on a surface of constant energy.** The value of the projection k_B of $\mathbf{k}$ on $\mathbf{B}$ is arbitrary, but constant during the motion. This component is the same as the initial crystal momentum component of the electron. The motion in $\mathbf{k}$ space is on a plane normal to the direction of $\mathbf{B}$, and the orbit is defined by the intersection of this plane with a surface of constant energy.

Three types of orbits[1] in a magnetic field are shown in Fig. 8. The closed orbits of (a) and (b) are traversed in opposite senses. Because particles of opposite charge circulate in a magnetic field in opposite senses it is natural to say that one orbit is **electronlike** and the other orbit is **holelike.** Electrons in holelike or-

---

[1] In Fig. 8 we discuss the motion of electrons at the Fermi surface, but we could as well discuss the motion on any surface of constant energy. Most experiments when analyzed in detail can be expressed in terms of the properties of orbits on the Fermi surface because experiments are concerned with changes in the occupancy of states in $\mathbf{k}$ space and these changes occur most easily at the Fermi surface.

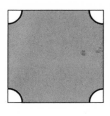

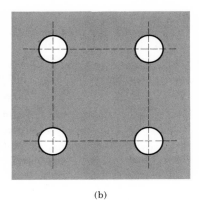

Figure 9 **Figure 9** (a) Vacant states at the corners of an almost-filled band, drawn in the reduced zone scheme. (b) In the periodic zone scheme the various parts of the Fermi surface are connected. Each circle forms a hole-like orbit. The different circles are entirely equivalent to each other, and the density of states is that of a single circle. (The orbits need not be true circles: for the lattice shown it is only required that the orbits have four-fold symmetry.)

(a)   (b)

Figure 10 Vacant states near the top of a filled band in a two-dimensional crystal. This figure is equivalent to Fig. 8a.

bits move in a magnetic field as if endowed with a positive charge. Thus holes give positive values of the Hall constant, thereby resolving the problem discussed in Chapter 8. In (c) the orbit is not closed: the particle on reaching the zone boundary at A is instantly umklapped back to B, where B is equivalent to B′ because they are connected by a reciprocal lattice vector. Such an orbit is called an **open orbit.** Open orbits have an important effect on the magnetoresistance (QTS, Chap. 12). Vacant orbitals near the top of an otherwise filled band give rise to holelike orbits, as in Figs. 9 and 10.

HOLES

The existence of holes is one of the most interesting features of the band theory of solids. It is also a feature of practical importance, because the operation of transistors depends directly on the coexistence of holes and electrons within semiconducting crystals.

Vacant orbitals in a band are commonly called hole orbitals or hole states. The clearest case is that of a single vacant orbital near the top of an otherwise filled energy band, a situation more likely to arise in a semiconductor than in a metal. The concept of a hole *orbit* is also well-defined even if there are more vacant orbitals than filled orbitals in the band, but the topological nature of the particular orbit may depend on the direction of the applied magnetic field. An intricate argument determines the properties of holes.

It has been established (see Chapter 11) by means of cyclotron resonance

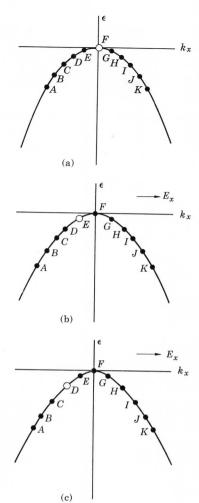

Figure 11 (a) At $t = 0$ all states are filled except F at the top of the band; the velocity v_x is zero at F because $d\epsilon/dk_x = 0$. (b) An electric field E_x is applied in the $+x$ direction. The force on the electrons is in the $-k_x$ direction and all electrons make transitions together in the $-k_x$ direction, moving the hole to the state E. (c) After a further interval the electrons move farther along in k space and the hole is now at D.

experiments with circularly polarized radiation on semiconductors that holes and electrons rotate in opposite senses in a magnetic field, as one would expect for charges of opposite sign. The radiation is absorbed by electrons for one sense of circular polarization and by holes for the opposite sense.

Consider the motion of a hole in an applied electric field, as in Fig. 11. Initially the band is filled except for the single vacant orbital F at the top of the band. An electric field E_x is now applied in the $+x$ direction. The motion of the electrons in the band is governed by the usual equation

$$\hbar \frac{dk_x}{dt} = -eE_x \; ; \tag{8}$$

each electron changes its k_x value at the same time. We see that Δk_x is negative in the figure. The vacancy initially at orbital F is displaced first to orbital E and subsequently to state D. That is, the hole moves along in **k** space together with the electrons in the direction of decreasing k_x.

The total wavevector of the electrons in a filled band is zero:

$$\Sigma \mathbf{k} = 0 \ . \tag{9}$$

This result follows from the geometrical symmetry of the Brillouin zone: every fundamental lattice type (Chapter 1) has symmetry under the inversion operation $\mathbf{r} \to -\mathbf{r}$ about any lattice point; and from the geometrical definition it follows that the Brillouin zone of the lattice also has inversion symmetry. Thus if the band is filled all pairs of state $\mathbf{k}$ and $-\mathbf{k}$ are necessarily filled and the total wavevector is zero.

Even in a filled band every electron changes its $\mathbf{k}$ value at a rate given by $\hbar \, d\mathbf{k}/dt = \mathbf{F}$, as derived above. The electrons are not blocked on reaching a zone boundary, but they are umklapped back to the opposite boundary to start life again, as in Fig. 9.9. (We can also picture the motion of the electrons in the periodic zone scheme, where the electrons advance continuously to higher k values in one band.)

If the band is filled except for an electron missing from the state E (Fig. 11b), we say that there is a hole in the state E. **The physical properties of the hole follow from those of the totality of electrons in the band.** This sentence is the key to the understanding of holes. The first application is to the wavevector of the hole: If the electron is missing from a state of wavevector $\mathbf{k}_e$, the total wavevector of the system is $-\mathbf{k}_e$; thus the wavevector to be attributed to the hole[2] is

$$\boxed{\mathbf{k}_h = -\mathbf{k}_e \ .} \tag{10}$$

This result is surprising: the electron is missing from $\mathbf{k}_e$ and the position of the hole is usually indicated graphically as situated at $\mathbf{k}_e$, as in the figure. But the true wavevector of the hole is $-\mathbf{k}_e$, which is the wavevector of the point G if the hole is at E. The wavevector $-\mathbf{k}_e$ enters into selection rules, as for photon absorption in Fig. 12.

The equation of motion of an electron in a crystal is $\hbar \, d\mathbf{k}_e/dt = \mathbf{F}_e$, where $\mathbf{F}_e$ is the force on an electron. Because $\mathbf{k}_h = -\mathbf{k}_e$, we have

$$\hbar \frac{d\mathbf{k}_h}{dt} = -\hbar \frac{d\mathbf{k}_e}{dt} = -\mathbf{F}_e \ , \tag{11}$$

whence

(CGS) $$\hbar \frac{d\mathbf{k}_h}{dt} = e\left(\mathbf{E} + \frac{1}{c}\mathbf{v}_e \times \mathbf{B}\right) \ ; \tag{12}$$

---

[2] Do not make the mistake of adding $-\mathbf{k}_e$ for the hole to $-\mathbf{k}_e$ for the band to obtain $-2\mathbf{k}_e$ for the wavevector of the band plus the hole. The hole is just an alternate and convenient description of a band with one missing electron, and we either say that the hole has wavevector $-\mathbf{k}_e$ or that the band (with one missing electron) has total wavevector $-\mathbf{k}_e$.

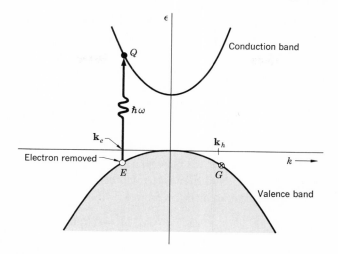

Figure 12   Absorption of a photon of energy $\hbar\omega$ and negligible wavevector takes an electron from E in the filled valence band to Q in the conduction band. If $\mathbf{k}_e$ was the wavevector of the electron at E, it becomes the wavevector of the electron at Q. The total wavevector of the valence band after the absorption is $-\mathbf{k}_e$, and this is the wavevector we must ascribe to the hole if we describe the valence band as occupied by one hole. Thus $\mathbf{k}_h = -\mathbf{k}_e$; the wavevector of the hole is the same as the wavevector of the electron which remains at G. For the entire system the total wavevector after the absorption of the photon is $\mathbf{k}_e + \mathbf{k}_h = 0$, so that the total wavevector is unchanged by the absorption of the photon and the creation of a free electron and free hole.

(SI)
$$\hbar \frac{d\mathbf{k}_h}{dt} = e(\mathbf{E} + \mathbf{v}_e \times \mathbf{B}) \ .$$

This is the equation of motion for a *positive* charge in an electric field; it also describes the motion of a positive charge in a magnetic field, provided that $\mathbf{v}_h = \mathbf{v}_e$. The proof of this relation for the velocity of the hole is given below. The reason the sign of $d\mathbf{k}_h/dt$ is opposite to $d\mathbf{k}_e/dt$ can be seen from Fig. 11: the vacant state moves along from $E \rightarrow D \rightarrow C \ldots$ just like the electrons in the adjacent states, but the corresponding unpaired electron moves backward from $G \rightarrow H \rightarrow I \ldots$, in $\mathbf{k}$ space. Here $\mathbf{v}_e$ is the velocity the missing electron would have.

The velocity (or group velocity) $\mathbf{v}_h$ of the hole is determined by the following argument. If an electron is missing from the state E of Fig. 11b, the net electric current carried by the band is that of the unpaired electron in the state G:

$$\mathbf{j} = -e\mathbf{v}(G) = e[-\mathbf{v}(G)] \ . \tag{13}$$

This current may be viewed as that of a positive charge e with the velocity $-\mathbf{v}(G)$. Now $-\mathbf{v}(G)$ is equal to velocity $\mathbf{v}(E)$ of the orbital at E, the orbital from which the electron was removed, so that the current may be written as

$$\mathbf{j} = e\mathbf{v}(E) \ . \tag{14}$$

We have $\mathbf{v}(E) = -\mathbf{v}(G)$ because $d\epsilon/d\mathbf{k}$ has opposite values at the points E and G. The current is consistent with a positive charge for the hole if the velocity of the hole is that of the missing electron.

The electron is missing from the state $\mathbf{k}_e$. If $\mathbf{v}_e$ is the velocity which an electron would have in the state $\mathbf{k}_e$, then

$$\boxed{\mathbf{v}_h = \mathbf{v}_e = \hbar^{-1}\nabla\epsilon(\mathbf{k}_e) \ ,} \tag{15}$$

where the symbol $\epsilon(\mathbf{k}_e)$ denotes the energy of an electron in the state $\mathbf{k}_e$.

For convenience we may take the zero of energy at the top of the filled or nearly filled band, and then $\epsilon(\mathbf{k}_e)$ will be negative in this band. The hole created by the removal of an electron from $\mathbf{k}_e$ will have the positive energy ϵ_h, where

$$\epsilon_h = -\epsilon(\mathbf{k}_e) \ . \tag{16}$$

If the band is symmetric[3] with $\epsilon(\mathbf{k}) = \epsilon(-\mathbf{k})$ we may interpret the hole energy ϵ_h as $\epsilon_h(\mathbf{k}_h)$ because $\mathbf{k}_h = -\mathbf{k}_e$, and now

$$\boxed{\epsilon_h(\mathbf{k}_h) = -\epsilon(\mathbf{k}_e) \ .} \tag{17}$$

The energy of the hole is opposite in sign to the energy of the missing electron—it takes more work to remove an electron from a low energy state than from a higher state in the band. For a symmetric band we have from (15) and (17) that

$$\mathbf{v}_h = \hbar^{-1}\nabla\epsilon_h(\mathbf{k}_h) \ , \tag{18}$$

where the gradient is taken with respect to $\mathbf{k}_h$. From (12) and (15) we have the equation of motion of a hole:

$$\text{(CGS)} \qquad \hbar\frac{d\mathbf{k}_h}{dt} = e\left(\mathbf{E} + \frac{1}{c}\mathbf{v}_h \times \mathbf{B}\right) = \mathbf{F}_h \ ; \tag{19}$$

$$\text{(SI)} \qquad \boxed{\hbar\frac{d\mathbf{k}_h}{dt} = e(\mathbf{E} + \mathbf{v}_h \times \mathbf{B}) = \mathbf{F}_h \ ,}$$

where $\mathbf{v}_h$ may be found from (15) or (18). The result (19) is the equation of motion of a positive charge, which immediately accounts for the difference in sign of the Hall effect of holes as compared with electrons or of a p type semiconductor as compared with an n type semiconductor.

The effective mass (see following section) of the hole is opposite in sign to that of the missing electron: The effective mass of an electron in the state $\mathbf{k}_e$ where it has velocity $\mathbf{v}_e$ is defined so that the equation of motion takes on the

[3] Bands are always symmetric under the inversion $\mathbf{k} \to -\mathbf{k}$ if the spin-orbit interaction is neglected. Even with spin-orbit interaction, bands are always symmetric if the crystal structure permits the inversion operation. Without a center of symmetry, but with spin-orbit interaction, the bands are symmetric if we compare subbands for which the spin direction is reversed: $\epsilon(\mathbf{k},\uparrow) = \epsilon(-\mathbf{k},\downarrow)$. See QTS, Chap. 9.

form of Newton's second law, $m_e \, dv_e/dt = -e\mathbf{E}$. The effective mass of a hole is defined from an equation of the same form, but for a positive charge: $m_h \, dv_h/dt = e\mathbf{E}$. We have seen from (15) that $v_e = v_h$, whence $dv_e/dt = dv_h/dt$ and from a comparison of the equations of motion

$$\boxed{m_h = -m_e \; .}$$
(20)

There is a big difference between the behavior of a *single* hole near the top of an otherwise filled band and a *single* electron near the top of an otherwise empty band. The single electron has negative charge and, by the argument of the following section, the effective mass of an electron near the top of a band is negative. The hole in the same position in the band acts as if it had positive charge and positive mass. Thus the ratio of charge to mass is the same for the single electron as for the corresponding single hole. It follows that the electron and hole will be accelerated in the same direction by an electric field, but the field will do work on the hole whereas the single electron does work on the field. The particles will circulate in the same sense in a static magnetic field. We can tell them apart by effects such as electric conductivity which depend on the ratio (charge)^2/mass. The conductivity of the hole will be positive, leading to power absorption by the specimen in an electric field. The conductivity of the isolated electron in a negative mass state will be negative,[4] leading to power emission (rather than absorption) by the specimen in an electric field. The electron is unstable in such a state. Notice that electrons at high ϵ_k tend to sink to lower energies in approaching thermal equilibrium, whereas holes at low ϵ_k tend to float to higher energies.

EFFECTIVE MASS OF ELECTRONS IN CRYSTALS

Going back to (9.61) and to Fig. 9.8 we see that an electron near the bottom of the second band has an energy that may be written in the form

$$\epsilon(\delta) = \epsilon_1(-) + \frac{\hbar^2}{2m^*}\delta^2 \; ,$$
(21)

where δ is the wavevector as measured from the zone boundary and

$$m^* \equiv \frac{m}{1 - (2\lambda_1/U_1)} \; ;$$
(22)

here $\lambda_1 = \hbar^2(\tfrac{1}{2}G_1)^2/2m$, and U_1 is taken such that $U(x) = 2U_1 \cos G_1 x$. We take U_1 as negative in order to give an attractive potential at $x = 0$. A small negative U_1 leads to a low value of the effective mass close to the energy gap.

The form of (21) suggests that an electron in a crystal may behave as if it had a mass different from the free electron mass. There are crystals in which the effective mass of the carriers is much larger or much smaller than m; the effective mass may be anisotropic, and it may even be negative. Effective masses of the order of $0.01 \, m$ have been observed (Chapter 11). The crystal does not weigh any less if m^* is smaller than m, nor is Newton's second law violated for

[4] See H. Kroemer, Phys. Rev. **109**, 1856 (1955); C. Kittel, Proc. Nat. Acad. Sci. **45**, 744 (1959).

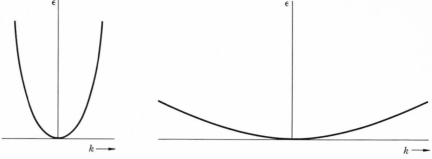

Figure 13   Energy band with light effective mass.

Figure 14   Energy band with heavy effective mass.

the crystal *taken as a whole*. The important point is that an electron in a periodic potential is accelerated relative to the lattice in an applied electric or magnetic field as if the mass of the electron were equal to an effective mass which we now define.

We differentiate the result (1) for the group velocity to obtain

$$\frac{dv_g}{dt} = \hbar^{-1}\frac{d^2\epsilon}{dk\,dt} = \hbar^{-1}\left(\frac{d^2\epsilon}{dk^2}\frac{dk}{dt}\right) . \tag{23}$$

Now we know from (5) that $dk/dt = F/\hbar$, whence

$$\frac{dv_g}{dt} = \left(\frac{1}{\hbar^2}\frac{d^2\epsilon}{dk^2}\right)F ; \qquad \text{or} \qquad F = \frac{\hbar^2}{d^2\epsilon/dk^2}\frac{dv_g}{dt} . \tag{24}$$

If we identify $\hbar^2/(d^2\epsilon/dk^2)$ as a mass, then (24) assumes the form of Newton's second law. We define the **effective mass** m^* as

$$\boxed{m^* = \frac{\hbar^2}{d^2\epsilon/dk^2} .} \tag{25}$$

If the energy is a quadratic function of k we may write $\epsilon = (\hbar^2/2m^*)k^2$. Portions of bands with light and heavy effective masses are sketched in Figs. 13 and 14.

It is easy to generalize (24) and (25) to take account of an anisotropic energy surface. We introduce the components of the reciprocal effective mass tensor

$$\left(\frac{1}{m^*}\right)_{\mu\nu} = \frac{1}{\hbar^2}\frac{d^2\epsilon_k}{dk_\mu\,dk_\nu} ; \qquad \frac{dv_\mu}{dt} = \left(\frac{1}{m^*}\right)_{\mu\nu}F_\nu , \tag{26}$$

where μ, ν are Cartesian coordinates.

Physical Basis of the Effective Mass

How can an electron of mass m when put into a crystal respond to applied fields as if the mass were m^*? It is helpful to think of the process of Bragg reflection of electron waves in a lattice.

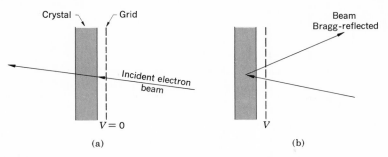

Figure 15 Explanation of light effective masses which occur near a Brillouin zone boundary. In (a) the energy of the electron beam incident on a thin crystal is slightly too low or too high to satisfy the condition for Bragg reflection and the beam is transmitted through the crystal. The application of a small voltage across the grid may, as in (b), cause the Bragg condition to be satisfied, and the electron beam will then be reflected from the appropriate set of crystal planes.

Consider the familiar weak interaction approximation for which the band structure was shown in Fig. 9.8b. Near the bottom of the lower band the state is represented quite adequately by a plane wave $\exp(ikx)$ with momentum $\hbar k$; the component $\exp[i(k - G_1)x]$ with momentum $\hbar(k - G_1)$ is small and increases only slowly as k is increased, and in this region $m^* \cong m$. An increase in the reflected component $\exp[i(k - G_1)x]$ as k is increased represents momentum transfer from the lattice to the electron. Near the boundary the reflected component is quite large; at the boundary it becomes equal in amplitude to $\exp(ikx)$, at which point the eigenfunctions are standing waves, rather than running waves. Here the component of momentum $\hbar(-\frac{1}{2}G_1)$ cancels the component of momentum $\hbar(\frac{1}{2}G_1)$.

It is not surprising to find negative values for m^* just below a zone boundary. A negative[5] effective mass means that on going from state k to state $k + \Delta k$ the momentum transfer from the lattice to the electron is opposite to and larger than the momentum transfer from the applied force to the electron. Although k is increased by Δk by the applied electric field, the consequent approach to Bragg reflection can result in an overall decrease in the forward momentum of the electron; if this happens the effective mass is described as negative.

As we proceed in the second band away from the boundary, the amplitude of $\exp[i(k - G_1)x]$ decreases rapidly and m^* assumes a small positive value. Here the increase in electron velocity resulting from a given impulse is larger than that which a free electron would experience. The lattice makes up the difference through the recoil it experiences when the amplitude of $\exp[i(k - G_1)x]$ is diminished. A pictorial discussion of this argument is given in Fig. 15. A small change in the energy of the electron beam has caused a very large change in the momentum of the beam; this situation corresponds to a light

[5] A *single electron* in an energy band may have positive or negative effective mass: the states of positive effective mass occur near the bottom of a band because positive effective mass means that the band has upward curvature ($\partial^2\epsilon/\partial k^2$ is positive). States of negative effective mass occur near the top of the band.

effective mass. The effective mass is positive or negative in this experiment, according to whether the initial energy is above or below the Bragg energy. The crystal itself experiences the usual classical recoil when an electron is reflected.

If the energy in a band depends only slightly on **k**, then the effective mass will be very large. That is, $m^*/m \gg 1$ because $d^2\epsilon/dk^2$ is very small. The tight-binding approximation discussed in Advanced Topic F gives quick insight into the formation of narrow bands. If the wavefunctions centered on neighboring atoms overlap very little, then the overlap integral γ defined by (F.6) will be small; the width of the band (F.9) will be narrow; and the effective mass (F.11) will be large. The overlap of wavefunctions centered on neighboring atoms is small for the inner or core electrons. The $4f$ electrons of the rare earth metals, for example, overlap very little. The overlap integral determines the rate of quantum tunneling of an electron from one ion to another. When the effective mass is heavy, the electron tunnels slowly from one ion to an adjacent ion in the lattice. The very narrow bands associated with the $1s$, $2s$, and $2p$ levels of sodium are shown on p. 240 of J. C. Slater, Rev. Mod. Phys. **6**, 209 (1934).

WAVEFUNCTIONS FOR ZERO WAVEVECTOR

There is no inconsistency between the complicated form of the electron wavefunctions in free atoms and the usefulness of a nearly free electron model of the overall band structure of a crystal. It is possible over most of a band for the energy to depend on the wavevector in approximately the same way as for a free electron, but at the same time the wavefunction may be quite unlike a plane wave and may pile up charge on the positive ion cores much as in the isolated atom.

The wave equation satisfied by a Bloch function is

$$\left(\frac{1}{2m}\mathbf{p}^2 + U(\mathbf{r})\right)e^{i\mathbf{k}\cdot\mathbf{r}}u_{\mathbf{k}}(\mathbf{r}) = \epsilon_{\mathbf{k}}e^{i\mathbf{k}\cdot\mathbf{r}}u_{\mathbf{k}}(\mathbf{r}) \ . \tag{27}$$

With $\mathbf{p} \equiv -i\hbar\,\mathrm{grad}$,

$$\mathbf{p}e^{i\mathbf{k}\cdot\mathbf{r}}u_{\mathbf{k}}(\mathbf{r}) = \hbar\mathbf{k}e^{i\mathbf{k}\cdot\mathbf{r}}u_{\mathbf{k}}(\mathbf{r}) + e^{i\mathbf{k}\cdot\mathbf{r}}\mathbf{p}u_{\mathbf{k}}(\mathbf{r}) \ ; \tag{28}$$

$$\mathbf{p}^2 e^{i\mathbf{k}\cdot\mathbf{r}}u_{\mathbf{k}}(\mathbf{r}) = (\hbar k)^2 e^{i\mathbf{k}\cdot\mathbf{r}}u_{\mathbf{k}}(\mathbf{r}) + e^{i\mathbf{k}\cdot\mathbf{r}}(2\hbar\mathbf{k}\cdot\mathbf{p})u_{\mathbf{k}}(\mathbf{r}) + e^{i\mathbf{k}\cdot\mathbf{r}}\mathbf{p}^2 u_{\mathbf{k}}(\mathbf{r}) \ ;$$

thus (27) may be written as

$$\left(\frac{1}{2m}(\mathbf{p} + \hbar\mathbf{k})^2 + U(\mathbf{r})\right)u_{\mathbf{k}}(\mathbf{r}) = \epsilon_{\mathbf{k}}u_{\mathbf{k}}(\mathbf{r}) \ . \tag{29}$$

At $\mathbf{k} = 0$ we have $\psi_0 = u_0(\mathbf{r})$, where $u_0(\mathbf{r})$ has the periodicity of the lattice and near the ion cores will look somewhat like the wavefunction of the free atom.

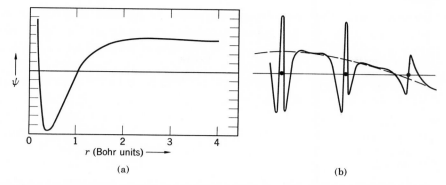

0   1   2   3   4
r (Bohr units) ———➤

(a)   (b)

Figure 16   (a) The lowest wavefunction $u_0(\mathbf{r})$ of metallic sodium, after Wigner and Seitz. The unit of length is 0.529×10^{-8} cm. (b) Schematic wavefunction $\psi_\mathbf{k}$ in sodium for a finite value of $\mathbf{k}$.

It is usually much easier to find a good solution at $\mathbf{k} = 0$ than for a general $\mathbf{k}$. We can use the solution $u_0(\mathbf{r})$ to construct the function

$$\psi = e^{i\mathbf{k}\cdot\mathbf{r}} u_0(r) \ . \tag{30}$$

This is of the Bloch form, but u_0 is not an exact solution of (29); it is a solution if we drop the term in $\mathbf{p}\cdot\mathbf{k}$. Because it takes some account of the ion cores the function is likely to be a much better approximation than a plane wave to the correct wavefunction. The energy of the approximate solution depends on $\mathbf{k}$ as $(\hbar k)^2/2m$, exactly as for the plane wave, even though the modulation represented by $u_0(\mathbf{r})$ may be very strong. Because u_0 is a solution of

$$\left(\frac{1}{2m}\mathbf{p}^2 + U(\mathbf{r})\right)u_0(\mathbf{r}) = \epsilon_0 u_0(\mathbf{r}) \ , \tag{31}$$

the function (30) has the energy $\epsilon_0 + (\hbar^2 k^2/2m)$.

There is considerable interest in reliable calculations of $u_0(\mathbf{r})$, as this function often will give us a good picture of the distribution of charge within a unit cell. Wigner and Seitz developed a simple and fairly accurate method of calculating $u_0(\mathbf{r})$. Figure 16a shows the Wigner-Seitz wavefunction for $\mathbf{k} = 0$ in the conduction (3*s*) band of metallic sodium. The function is practically constant over 90 percent of the atomic volume. To the extent that the solutions for higher $\mathbf{k}$ may be approximated by $e^{i\mathbf{k}\cdot\mathbf{r}} u_0(\mathbf{r})$ as in (30), it will be true that the wavefunctions in the conduction band are similar to plane waves over most of the atomic volume, but may oscillate and increase markedly in the region of the ion core.

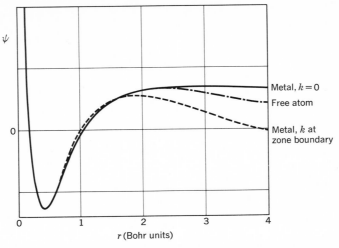

Figure 17 Radial wavefunctions for the 3s orbital of free sodium atom and for 3s conduction band in sodium metal. The wavefunctions are found by integrating the Schrödinger equation for an electron in the potential well of an Na⁺ ion core. For the free atom the wavefunction is integrated subject to the usual Schrödinger boundary condition $\psi(r) \to 0$ as $r \to \infty$; the energy eigenvalue is -5.15 eV. The wavefunction for wavevector $k = 0$ in the metal is subject to the Wigner-Seitz boundary condition that $d\psi/dr = 0$ when r is midway between neighboring atoms; the energy of this orbital is -8.2 eV, considerably lower than for the free atom. The orbitals at the zone boundary are not filled in sodium; their energy is $+2.7$ eV. [After E. Wigner and F. Seitz, Phys. Rev. 34, 804 (1933).]

Lattice Effects on Cohesive Energy of Metals

We now study the cohesive energy of simple metals. The stability of simple metals with respect to free atoms is due to the lowering of the energy of the Bloch orbital with $\mathbf{k} = 0$ in the metal as compared to the ground electronic orbital of the free atom. The effect is illustrated in Fig. 17 for sodium and in Fig. 18 for a linear periodic potential of attractive square wells. The ground orbital energy is much lower for atoms at the actual spacing in the metal than for isolated atoms.

A decrease in ground orbital energy corresponds to an increase in binding. The decrease in ground orbital energy when atoms are assembled in a periodic array is a consequence of the change in the boundary condition on the wavefunction: in the free atom the boundary condition on the wavefunction is $\psi(\mathbf{r}) \to 0$ as $r \to \infty$, with no discontinuities allowed in $d\psi/dr$. In the periodic crystal no discontinuities are allowed either: but the $\mathbf{k} = 0$ wavefunction $u_0(\mathbf{r})$ is periodic with the period of the lattice, and the only way it can do this without discontinuities is for the normal derivative of ψ to vanish across every plane midway between adjacent atoms. In a spherical approximation to the smallest Wigner-Seitz cell we require that

$$\left(\frac{d\psi}{dr}\right)_{r_0} = 0 \ , \tag{32}$$

where r_0 is the radius of a sphere equal in volume to a primitive cell of the lattice. In sodium $r_0 = 3.95$ Bohr units, or 2.08 Å; the half-distance to a nearest neighbor

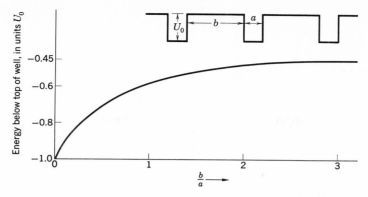

Figure 18 Ground orbital ($k = 0$) energy for an electron in a periodic square well potential of depth $|U_0| = 2\hbar^2/ma^2$. The energy is lowered as the wells come closer together. Here a is held constant and b is varied. Large b/a corresponds to separated atoms. (Courtesy of C. Y. Fong.)

Figure 19 Here a is the wavefunction of a particle in a parabolic potential well. Parts b, c, d are concerned with the wavefunction of a particle that moves along a linear array of parabolic potentials. The function b is obtained from direct juncture of functions of the form of a, but b is excluded because $d\psi/dx$ is discontinuous. In c we have rounded off the discontinuity, but the values of $d^2\psi/dx^2$ near the junctures will be rather high, and probably will give too high a kinetic energy for this function to be a solution of the wave equation with constant energy. The actual lowest energy solution could be more like d.

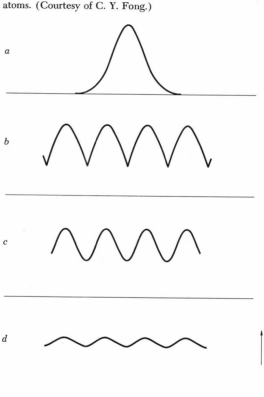

is 1.86 Å. The spherical approximation is not bad for fcc and bcc structures.

The boundary condition (32) on the ground state wavefunction is called the **Wigner-Seitz boundary condition.** The condition has a substantial effect on the wavefunction: it allows the ground orbital wavefunction to have much less curvature than the free atom boundary condition. Much less curvature means much less kinetic energy $-(\hbar^2/2m)\nabla^2\psi$. The new boundary conditions are explained further in Fig. 19. They apply strictly only to the orbital $\mathbf{k} = 0$.

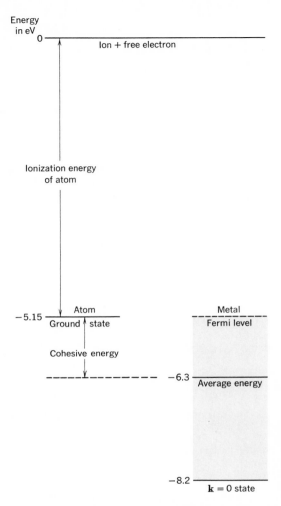

Figure 20 Cohesive energy of sodium metal is the difference between the average energy of an electron in the metal (-6.3 eV) and the ground state energy (-5.15 eV) of the valence $3s$ electron in the free atom.

In sodium the other filled orbitals in the conduction band can be represented in a rough approximation by wavefunctions of the form (30), with

$$\psi_{\mathbf{k}} = e^{i\mathbf{k}\cdot\mathbf{r}}u_0(\mathbf{r}) \; ; \qquad \epsilon_{\mathbf{k}} = \epsilon_0 + \frac{\hbar^2 k^2}{2m} \; . \tag{33}$$

The Fermi energy is 3.1 eV, from Table 7.1. The average kinetic energy per electron is 0.6 of the Fermi energy, or 1.9 eV. Because $\epsilon_0 = -8.2$ eV for the solution in Fig. 17, the average electron energy is

$$\langle \epsilon_{\mathbf{k}} \rangle = -8.2 + 1.9 = -6.3 \, \text{eV} \; , \tag{34}$$

compared with -5.15 eV for the valence electron of the free atom. We therefore estimate that sodium metal is stable by about 1.1 eV with respect to the free atom. This result agrees rather well with the experimental value of 1.13 eV given in Table 3.1. The relationship of the several energies is pictured in Fig. 20.

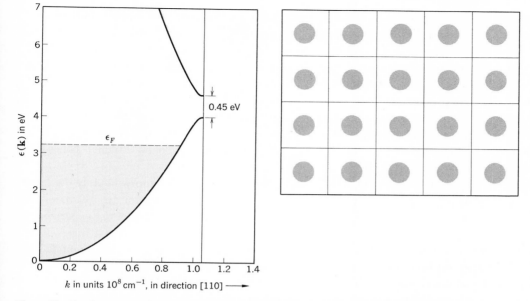

Figure 21   Band structure of sodium in direction [110], as calculated by A. W. Overhauser, Phys. Rev. **156**, 844 (1967).

Figure 22   In many metals the volume of the ion cores is a small part of the volume of the metal. For example, in sodium the volume of the ion core is approximately 4.0 Å³ and the atomic volume is 38 Å³. The potential energy is weaker and smoother outside the ion cores than within the cores, and the conduction electron wavefunctions are much smoother outside the cores.

Our discussion has neglected several corrections whose overall effect in sodium is small.

The structure of two bands of sodium is shown in Fig. 21, calculated in the nearly-free electron approximation. The value of the relevant Fourier component of the lattice potential was inferred from experiments on the de Haas-van Alphen effect as discussed below. The experiments are very sensitive to departures from sphericity of the Fermi surface, and they permit the accurate determination of the $U_{\mathbf{G}}$.

PSEUDOPOTENTIALS°

The orbitals of conduction electrons in metals are simple and smooth in the region between the ion cores, but the orbitals are complicated by a nodal structure in the region of the ion cores, as shown for the ground orbital of sodium in Fig. 17. Most of the volume of many metals lies between the ion cores (Fig. 22). In this outer region the potential energy of a conduction electron is relatively

---

° The appreciation of this section requires some grasp of quantum theory.

weak: it is the Coulomb potential of the positive charge of the ions, but reduced by the electrostatic screening as provided by the other conduction electrons. In the outer region the wavefunctions are somewhat like plane waves, neither complicated by the strong and rapidly varying potential found near the atomic nuclei nor by the requirement of orthogonality[6] to the wavefunctions of the electrons of the ion cores. The existence of nodes (zeros) in the wavefunction in the core region is caused by the orthogonality requirement: the $3s$ conduction band orbital of sodium has two nodes that enable it to be orthogonal to the $1s$ and $2s$ electrons of the ion core, and the $4s$ conduction band orbital of potassium has three nodes.

If the conduction orbitals in the outer region have approximately the form of plane waves, their energy must depend on the wavevector approximately as the relation $\epsilon_k = \hbar^2 k^2/2m$ for free electrons. The weak potential experienced by an electron in the outer region is treated as a perturbation that mixes plane wave components $\mathbf{k}$ and $\mathbf{k} + \mathbf{G}$ strongly only near Brillouin zone boundaries. But how do we handle the core region, where the orbitals are not like plane waves and the potential is strong? What goes on in the core is largely irrelevant to the dependence of ϵ on $\mathbf{k}$, for we can calculate the energy by applying the hamiltonian operator to an orbital at any point in space, and in the outer region this operation will give the free electron energy.

In fact we can replace the actual potential energy in the core region by an effective potential energy, the **pseudopotential**, that gives the same wavefunctions outside the core as does the actual potential.[7] The startling fact about pseudopotentials is that in the core region the pseudopotential is nearly zero. This conclusion is derived from a large amount of practical experience with pseudopotentials as well as from theoretical arguments sometimes referred to as the cancellation theorem.[8] If boldly we take the unscreened pseudopotential to be zero inside of some radius R_e, we have what is called the empty core model, which works surprisingly well:

$$U(r) = \begin{cases} 0 & , \quad \text{for } r < R_e \ ; \\ \\ -\dfrac{e^2}{r} & , \quad \text{for } r > R_e \ . \end{cases} \tag{35}$$

[6] An orbital $\psi(\mathbf{r})$ is orthogonal to another orbital $\psi_c(\mathbf{r})$ if $\int d^3x \psi^*(\mathbf{r})\psi_c(\mathbf{r}) = 0$.

[7] J. C. Phillips and L. Kleinman, Phys. Rev. **116**, 287 (1959); E. Antončik, J. Phys. Chem. Solids **10**, 314 (1959). The general theory of pseudopotentials is discussed by B. J. Austin, V. Heine, and L. J. Sham, Phys. Rev. **127**, 276 (1962).

[8] A full review of experience with different pseudopotential and model calculations is given by M. L. Cohen and V. Heine, *Solid state physics* **24**, 37 (1970). The utility of the empty core model has been known for many years; see H. Hellmann, Acta Physiochimica URSS **1**, 913 (1935) and H. Hellmann and W. Kassatotschkin, J. Chem. Phys. **4**, 324 (1936), who wrote "Since the field of the ion determined in this way runs a rather flat course, it is sufficient in the first approximation to set the valence electron in the lattice equal to a plane wave."

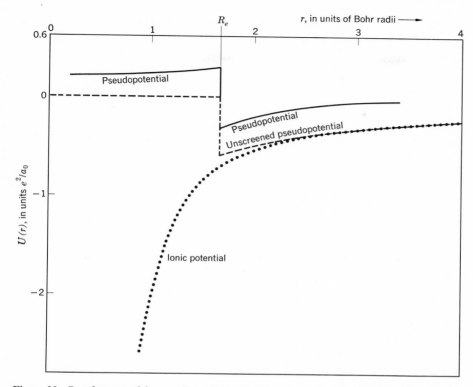

Figure 23 Pseudopotential for metallic sodium, based on the empty core model and screened by the Thomas-Fermi dielectric function (D. 13). The calculations were made for an empty core radius $R_e = 1.66a_0$, where a_0 is the Bohr radius, and for a screening parameter $\lambda a_0 = 0.79$, from Eq. (8.23). The dashed curve shows the assumed unscreened potential, as from (35). The dotted curve is the actual potential of the ion core; other values of $V(r)$ are -50.4, -11.6, and -4.6, for $r = 0.15$, 0.4, and 0.7, respectively. Thus the actual potential of the ion (chosen to fit the energy levels of the free atom) is very much larger than the pseudopotential, over 200 times larger at $r = 0.15$.

This potential is to be screened; that is, each Fourier component $U(\mathbf{K})$ of $U(r)$ must be divided by the dielectric function $\epsilon(\mathbf{K})$ of the electron gas. If we use (just as an example) the dielectric function (D. 13), we obtain the pseudopotential plotted in Fig. 23. The pseudopotential is much weaker than the true potential, but the wavefunctions in the outer regions are nearly identical for both potentials.

The calculation of the band structure of a crystal involves only the values $U(\mathbf{G})$ of the Fourier coefficients of the potential at the reciprocal lattice vectors. Often only a few coefficients are required for the determination of the band structure to good accuracy. These coefficients are sometimes calculated from model potentials, and sometimes they are taken from experiment. Values of $U(\mathbf{G})$ for the elements are tabulated and plotted in the review by Cohen and Heine.

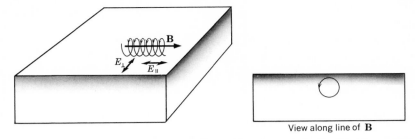

Figure 24 The Azbel-Kaner geometry is often employed for studies of cyclotron resonance in metals. The rf electric field **E** may be perpendicular or parallel to the static magnetic field **B**, but both **E** and **B** are parallel to the surface of the specimen. The penetration depth (skin depth) of the rf field is indicated by the shading. An electron orbit is shown. Near the top of each turn the electron enters the skin depth and experiences the rf electric field, gaining or losing energy from the field.

EXPERIMENTAL METHODS IN FERMI SURFACE STUDIES

Powerful experimental methods have been developed for studies of the Fermi surfaces of metals. These methods[9] include

a. Anomalous skin effect
b. Cyclotron resonance
c. Magnetoresistance
d. De Haas-van Alphen effect
e. Ultrasonic propagation in magnetic fields
f. Optical reflectivity

We shall discuss (b) and (d) for general Fermi surfaces. We have previously discussed (c), but only for free electrons.

Cyclotron Resonance in Metals

A geometry commonly employed in cyclotron resonance experiments in metals is sketched in Fig. 24. The radius of the orbit of an electron in a magnetic field of 10 kilogauss or 1 tesla is of the order of 10^{-3} cm, which is much larger than the skin depth at microwave frequencies in a pure metal at low temperatures. Electrons in the orbits drawn will see the rf electric field only for a small part of each cycle of their motion. Electrons obtain a net acceleration if on successive cycles they arrive in the skin depth in the same phase of the rf field. The resonance condition is that the period T of the electron in its orbit should equal an integral number n of periods $2\pi/\omega$ of the rf field:

$$T = \frac{2\pi n}{\omega} . \qquad (36)$$

Now $T = 2\pi/\omega_c = 2\pi m_c c/eB$, where m_c denotes the effective mass for cyclo-

[9] Extensive discussions are given in A. B. Pippard, *Dynamics of conduction electrons*, Gordon and Breach, 1965; in W. A. Harrison and M. B. Webb (eds.), *The Fermi surface*, Wiley, 1960; and in *QTS*.

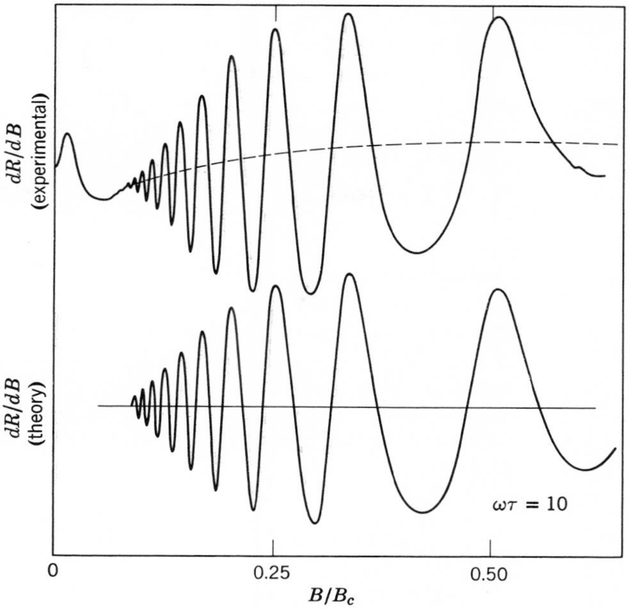

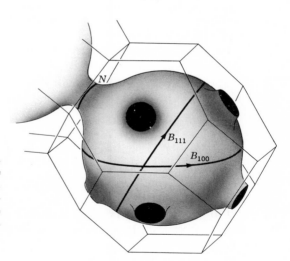

Figure 25 Cyclotron resonance in copper at 24 GHz. Comparison of calculations of the magnetic field dependence of the derivative of the surface resistivity with experimental results. Here B_c is the field (61) for $n = 1$. [After A. F. Kip, D. N. Langenberg, and T. W. Moore, Phys. Rev. **124**, 359 (1961). For results at 400 GHz, see P. Goy and G. Weisbuch, Phys. kondens. Materie **9**, 200 (1969).]

dR/dB (experimental)

dR/dB (theory)

$\omega\tau = 10$

0   0.25   0.50

B/B_c

Figure 26 Fermi surface of copper, after Pippard. The Brillouin zone of the fcc structure is the truncated octahedron, as derived in Chapter 2. The Fermi surface makes contact with the boundary at the center of the hexagonal faces of the zone, in the [111] directions in **k** space. Two "belly" extremal orbits are shown, denoted by B, the extremal "neck" orbit is denoted by N.

tron resonance, so that the electrons will be in resonance at values of the magnetic field given by $2\pi m_c c/eB = 2\pi n/\omega$, or

(CGS)
$$B = \frac{\omega m_c c}{en} \; ; \tag{37}$$

(SI)
$$B = \frac{\omega m_c}{en} \; .$$

Many subharmonics ($n > 1$) have been observed, as in Fig. 25 for pure copper at helium temperatures. The results for copper support the model of the Fermi surface (Fig. 26) originally deduced by Pippard in another way.

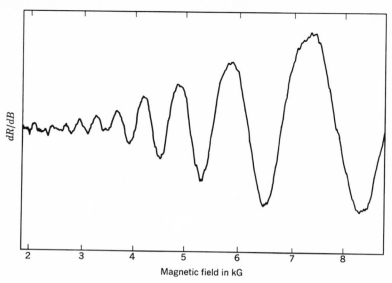

Figure 27 Experimental absorption derivative for cyclotron resonance in potassium at 68 GHz. The static magnetic field lies in a (110) plane; closely similar traces are obtained for all directions of **B** in this plane. [After C. C. Grimes and A. F. Kip, Phys. Rev. **132**, 1991 (1963).]

The Fermi surface of copper is distinctly nonspherical: eight necks make contact with the hexagonal faces of the first Brillouin zone of the fcc lattice. The electron concentration in a monovalent metal with an fcc structure is $n = 4/a^3$: there are 4 electrons in a cube of volume a^3. The radius of a free electron Fermi sphere is

$$k_F = (3\pi^2 n)^{\frac{1}{3}} = \left(\frac{12\pi^2}{a^3}\right)^{\frac{1}{3}} \cong \frac{4.90}{a} \; , \tag{38}$$

and the diameter is $9.8/a$. From Eq. (2.55) we see that the shortest distance across the Brillouin zone (the perpendicular distance between hexagonal faces) is $(2\pi/a)(3)^{\frac{1}{2}} = 10.90/a$, somewhat larger than $9.8/a$. The free electron sphere does not touch the zone boundary, but we know from Chapter 9 that the presence of a zone boundary tends to lower the band energy near the boundary. Thus it is plausible to find that the Fermi surface of copper necks out to meet the hexagonal faces of the zone. The square faces of the zone are more distant, with separation $12.57/a$.

Resonance peaks for potassium are shown in Fig. 27: the measurements indicate that the Fermi surface is very nearly spherical, with an anisotropy of less than 1 part in 100. The measured effective mass is $m_c = (1.24 \pm 0.02)m$.

How is the period T related to the Fermi surface? From (7)

(CGS) $$\frac{dk}{dt} = -\frac{eB}{\hbar^2 c}(\nabla \epsilon_{\mathbf{k}})_\perp \; , \tag{39}$$

where $(\nabla \epsilon_{\mathbf{k}})_\perp/\hbar$ is the velocity component in real space in a plane normal to **B**.

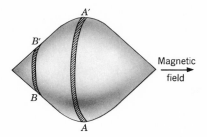

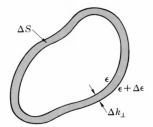

Figure 28 The orbits in the section AA' are extremal orbits: the cyclotron period is roughly constant over a reasonable section of the Fermi surface. Other orbits such as BB' vary in period over the section.

Figure 29 Orbits in **k** space at a constant value of k_B: one orbit is at energy ϵ, the other at energy $\epsilon + \Delta\epsilon$, where $\Delta\epsilon$ is constant. The spacing $\Delta k_\perp$ between the orbits may vary around an orbit. The shaded area between the orbits is ΔS.

The period is found by integration of (39) around one cycle of the motion:

$$\text{(CGS)} \qquad T = \oint dt = \frac{\hbar^2 c}{eB} \oint \frac{dk}{(\nabla_{\mathbf{k}}\epsilon)_\perp} \ . \tag{40}$$

The essential features of the Fermi surface can sometimes be deduced from (40). In SI, replace c by 1.

Extremal Orbits

One point in the interpretation of cyclotron resonance in metals is subtle and important. For a spherical or ellipsoidal Fermi surface it can easily be shown that orbits on the Fermi surface in any plane normal to **B** have the identical period, independent of the value of k_B, the projection of **k** on the direction of **B**. But for a Fermi surface of general shape the sections at different values of k_B will have different periods. The power absorption will be the sum of contributions from all sections or all orbits. *But the dominant response of the system comes from orbits whose periods are stationary with respect to small changes in k_B.* Such orbits are called **extremal orbits**. Thus in Fig. 28 the section AA' dominates the observed cyclotron period. The argument can be put in mathematical form, but we do not give the proof here. Essentially it is a question of phase cancellation: the contributions of different nonextremal orbits cancel, but near extrema the phase varies only slowly and there is a net signal from these orbits. Both theory and experiment agree that sharp resonances are obtained even from complicated Fermi surfaces because the extremal orbits are always accentuated.

The expression (40) for the cyclotron period can be cast in a form which establishes the connection with the area S (in **k** space) of the Fermi surface section enclosed by the orbit. Let ΔS in Fig. 29 be the area between orbits having the same k_B but separated in energy by $\Delta\epsilon$. Because $(\nabla_{\mathbf{k}}\epsilon)_\perp = (\Delta\epsilon)/(\Delta k)_\perp$, the integral in (40) may be written as

$$\oint \frac{dk}{(\nabla_{\mathbf{k}}\epsilon)_\perp} = \frac{1}{\Delta\epsilon} \oint (\Delta k)_\perp \ dk \ . \tag{41}$$

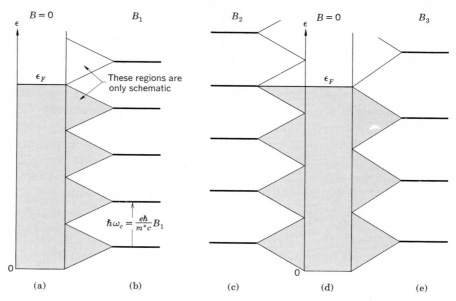

Figure 30 Explanation of the de Haas-van Alphen effect for a free electron gas in two dimensions in a magnetic field. The spin of the electron is neglected, in the interest of clarity. The filled orbitals of the Fermi sea in the absence of a magnetic field are shown shaded in a and d. The energy levels in a magnetic field are shown in b, c, and e. In b the field has a value B_1 such that the total energy of the electrons is the same as in the absence of a magnetic field: as many electrons have their energy raised as have their energy lowered by the orbital quantization in the magnetic field B_1. When we increase the field to B_2 the total electron energy is increased, because the uppermost electrons have their energy raised. In e for field B_3 the energy is again equal to that for the field $B = 0$. The total energy is a minimum at points such as B_1, B_3, B_5, . . . , and a maximum near points such as B_2, B_4, ...

Now

$$\oint (\Delta k)_\perp \, dk = \Delta S \tag{42}$$

is the area between the two orbits, so that (40) may be written as

$$\text{(CGS)} \quad T = \frac{\hbar^2 c}{eB} \frac{\partial S}{\partial \epsilon} \; ; \quad \omega_c = \frac{2\pi eB}{\hbar^2 c} \frac{\partial \epsilon}{\partial S} \equiv \frac{eB}{m_c c} \; ; \quad m_c \equiv \frac{\hbar^2}{2\pi} \frac{\partial S}{\partial \epsilon} \; . \tag{43}$$

This defines the **cyclotron effective mass** m_c in terms of the extremal sectional area of the Fermi surface.

De Haas-van Alphen Effect*

In a strong[10] magnetic field the states of a free electron gas are not plane waves, and the energy is no longer given simply by $\epsilon_{\mathbf{k}} = \hbar^2 k^2 / 2m$. Several physical properties of the metal are changed remarkably by the magnetic field. The **de Haas-van Alphen effect** is concerned with oscillations in the magnetic

---

* This section may be omitted on a first reading.

[10] A field is strong if an electron completes at least several cycles of a helical orbit before a collision, i.e. $\omega_c \tau \gg 1$, where ω_c is the cyclotron frequency. In practice we need low temperatures and pure specimens to satisfy this condition in the usual laboratory fields. The dHvA effect is not observable at room temperature. It is also necessary that $k_B T \ll \hbar \omega_c$, for otherwise the population oscillations are smudged out. Our analysis is for absolute zero.

moment as a function of magnetic field. The effect is much easier to discuss in two dimensions than in three dimensions; the two-dimensional model in most respects is not a bad approximation to reality because for real Fermi surfaces the extremal sections play the same dominant role as they played in cyclotron resonance. The de Haas-van Alphen effect arises from the periodic variation of the total electronic energy as a function of a static magnetic field. The energy variation is usually observed experimentally as a periodic variation in the magnetic moment of the metal. We neglect the spin of the electron in the present discussion. An elementary explanation of the effect is sketched in Fig. 30.

The energy levels of a free particle of mass m_c in two dimensions x, y in a magnetic field B normal to the xy plane are given by

$$\text{(CGS)} \qquad\qquad \epsilon_l = \frac{e\hbar B}{m_c c}\left(l + \frac{1}{2}\right) , \qquad\qquad (44)$$

where l is a quantum number that assumes integral values. This is the exact solution of a standard problem in elementary quantum mechanics. We do not give the derivation, but when (44) is combined with the selection rule $\Delta l = \pm 1$ for transitions induced by a uniform electric field we have the standard result for the cyclotron resonance frequency

$$\text{(CGS)} \qquad\qquad \hbar\omega_c = \epsilon_{l+1} - \epsilon_l = \frac{e\hbar B}{m_c c} . \qquad\qquad (45)$$

To obtain the above results in SI, replace c by 1.

We now assume that the energy of electrons on a *general* Fermi surface in two dimensions may be written as

$$\text{(CGS)} \qquad\qquad \epsilon_l = \frac{e\hbar B}{m_c c}\left(l + \frac{1}{2}\right) . \qquad\qquad (46)$$

In general the phase factor in (46) need not be $\frac{1}{2}$, but we write it as $\frac{1}{2}$ to be concrete and to agree with (44) for free electrons.

The relation (43) for the cyclotron effective mass,

$$m_c = \frac{\hbar^2}{2\pi}\frac{\partial S}{\partial \epsilon} , \qquad\qquad (47)$$

is satisfied if

$$\epsilon_l = \frac{\hbar^2}{2\pi m_c} S_l , \qquad\qquad (48)$$

where S_l is the area of the orbit l in $\mathbf{k}$ space. On comparison with (46),

$$\text{(CGS)} \qquad\qquad \boxed{S_l = \frac{2\pi e B}{\hbar c}\left(l + \frac{1}{2}\right) .} \qquad\qquad (49)$$

Thus in a magnetic field the area of the orbit in $\mathbf{k}$ space is quantized. The

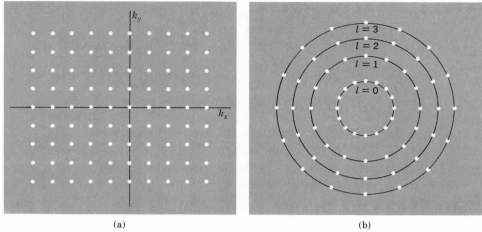

(a)   (b)

Figure 31 (a) Allowed electron orbitals in two dimensions in absence of a magnetic field. (b) In a sufficiently strong magnetic field the points which represent the orbitals of free electrons may be viewed as restricted to circles in the former $k_x k_y$ plane. The successive circles correspond to successive values of the quantum number l in the energy $(l + \frac{1}{2})\hbar\omega_c$. The area between successive circles is

(CGS) $$\pi\,\Delta(k^2) = 2\pi k(\Delta k) = (2\pi m/\hbar^2)\,\Delta\epsilon = 2\pi m\omega_c/\hbar = 2\pi eB/\hbar c.$$

The angular position of the points has no significance. The number of orbitals on a circle is constant and is equal to the area between successive circles times the number of orbitals per unit area in (a), or $(2\pi eB/\hbar c)(L/2\pi)^2 = L^2 eB/2\pi\hbar c$, neglecting electron spin. By a sufficiently strong magnetic field we mean one such that an electron may complete at least several cycles of motion before colliding with a lattice imperfection, a phonon, or with the boundary of the specimen.

expression for S is of general validity and is not restricted to free electrons or to spherical energy surfaces; the derivation is given in (I.34) to (I.48). The quantity $2\pi\hbar c/e$ is called the **flux quantum** and has the value 4.14×10^{-7} gauss-cm^2.

What happens in a magnetic field to the distribution of orbitals in **k** space? The orbitals are no longer described as in Fig. 31a by values of k_x and k_y, but instead are described by the quantum number l as in Fig. 31b. The number[11] of orbitals of given l may be quite high, for every allowed orbital in (a) in the absence of the field has to be accommodated somewhere in (b) when the field is turned on. It is not hard to show[12] that for a square specimen of side L there are $BL^2/(2\pi\hbar c/e)$ orbitals for each value of the quantum number l. This number of orbitals is what we expect if each ring in Fig. 31 sweeps up the states in **k** space lying between it and an adjacent ring. We write for the degeneracy of a level l:

(CGS) $$\boxed{\text{degeneracy} \equiv \xi B = (L/2\pi)^2(2\pi e/\hbar c)B\ ,}$$ (50)

[11] This number is called the degeneracy. For $B = 0$, the density of orbitals of a free electron gas in two dimensions is easily found by the arguments of Chap. 7 to be $\mathfrak{D}(\epsilon) = L^2 m/2\pi\hbar^2$, neglecting spin. The number of orbitals in an energy range of extent $\hbar\omega_c$ is $\mathfrak{D}(\epsilon)\hbar\omega_c = L^2 eB/2\pi\hbar c$, in agreement with (50). The application of a magnetic field collects the orbitals of the continuum throughout an energy range $\hbar\omega_c$.

[12] *QTS*, Chap. 11.

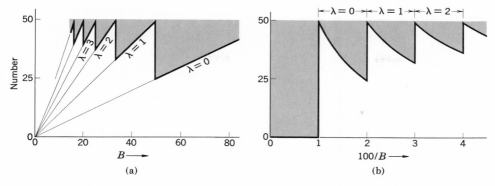

Figure 32   (a) The heavy line gives the number of particles in levels which are completely occupied in a magnetic field B, for a two-dimensional system with $N = 50$ and $\xi = 0.50$. The shaded area gives the number of particles in levels which are partially occupied. The value of λ denotes the quantum number of the highest level which is completely filled. Thus at $B = 40$ we have $\lambda = 1$: the levels $l = 0$ and $l = 1$ are filled and there are 10 particles in the level $l = 2$. At $B = 50$ the level $l = 2$ is empty. (b) The periodicity is $1/B$ is evident when the same points are plotted against $1/B$.

where the degeneracy parameter ξ is defined by this equation. The degeneracy is roughly the ratio of the flux BL^2 through the specimen to the flux quantum $2\pi\hbar c/e$. In SI units, delete the c.

Where does the Fermi level lie? For a system of N electrons at absolute zero the energy levels are filled up to some quantum number λ, and orbitals at the level $(\lambda + 1)$ will be partly filled to the extent needed to accommodate the electrons.[13] Because the degeneracy increases with B, the higher the field, the lower the value of λ. The distribution of electrons is shown in Fig. 32. As B is increased there occur certain values of B at which λ decreases abruptly by unity, as in Fig. 32. At these critical values B_λ no level is partly occupied, so that

$$\xi B_\lambda(\lambda + 1) = N \ . \tag{51}$$

This states that the degeneracy ξB_λ of a level times the number of filled levels $\lambda + 1$ is equal to the number of particles N. We recall that $l = 0$ is an allowed level, which accounts for the $+1$ in $\lambda + 1$. We rewrite (51) as

$$\frac{1}{B_\lambda} = \frac{\xi(\lambda + 1)}{N} \ . \tag{52}$$

It is now not surprising that certain properties of the system may be periodic in $1/B$, as in Fig. 32b, with the period ξ/N.

The energy of the electrons in *fully occupied* levels is

---

[13] That is, λ is defined as the maximum value of l in (46) for which all orbitals are filled with electrons.

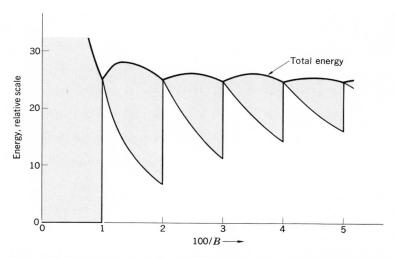

Figure 33 The upper curve is the total electronic energy versus $1/B$. The energy is an oscillatory function of $1/B$. The oscillations in the energy E may be detected by measurement of the magnetic moment, given by $-dE/dB$. The thermal and the transport properties of the metal also oscillate as successive orbital levels cut through the Fermi level when the field is increased. The shaded region in the figure gives the contribution to the energy from levels that are only partly filled. The parameters for the figure are the same as for Fig. 32, and we have taken the units of B such that $B = \hbar\omega_c$.

$$E_1 = \sum_{l=0}^{\lambda} (\xi B)(\hbar\omega_c)\left(l + \frac{1}{2}\right) = \frac{1}{2}(\xi B)(\hbar\omega_c)(\lambda + 1)^2 \; , \qquad (53)$$

where ξB is the number of electrons in each level, as in (50). The energy of the electrons in the *partly occupied* level $\lambda + 1$ is

$$E_2 = (\hbar\omega_c)\left(\lambda + \frac{3}{2}\right)[N - (\xi B)(\lambda + 1)] \; , \qquad (54)$$

where $(\xi B)(\lambda + 1)$ is the number of electrons in the lower filled levels. The total energy of the system of N electrons is the sum of E_1 and E_2: $E = E_1 + E_2$. The contributions to the energy are indicated in Fig. 33.

The magnetic moment μ of a system at absolute zero is given by $\mu = -\partial E/\partial B$. The moment for the system of Fig. 33 is an oscillatory function of $1/B$,

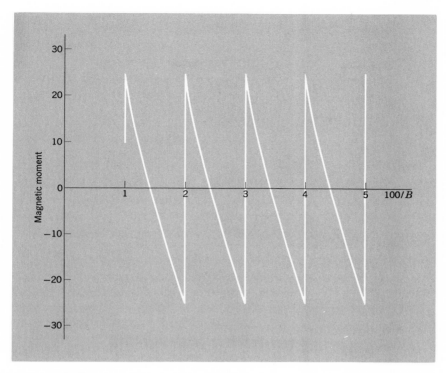

Figure 34 At absolute zero the magnetic moment is given by $-\partial E/\partial B$. The energy plotted in Fig. 33 leads to the magnetic moment shown here, which is an oscillatory function of $1/B$. The positive moments are paramagnetic (Chapter 15); the negative moments are diamagnetic. Effects of electron spin have not been included. In impure specimens or at finite temperatures the oscillations are reduced in amplitude or smudged out in part because the energy levels are no longer sharply defined.

as shown in Fig. 34. This oscillatory magnetic moment of the Fermi gas at low temperatures is the de Haas-van Alphen effect. From (49) we see that the oscillations occur at equal intervals of $1/B$ such that

$$\Delta\left(\frac{1}{B}\right) = \frac{2\pi e}{\hbar c S} \, , \tag{55}$$

where S is the extremal area of the Fermi surface normal to the direction of **B**. This period is identical with that of (52), for $\xi = L^2 e/2\pi\hbar c$ and $N = (L/2\pi)^2 S$. From measurements of $\Delta(1/B)$ for different directions of **B** we can deduce the corresponding extremal areas S; thereby much can be inferred about the shape and size of the Fermi surface.[14]

[14] For a discussion of experimental methods, see D. Shoenberg, Proc. Phys. Soc. (London) **79**, 1–9 (1962).

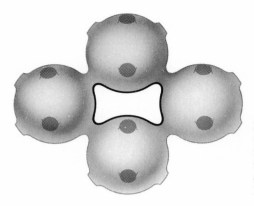

Figure 35 Dog's bone orbit of an electron on the Fermi surface of copper or gold in a magnetic field. This is a hole-like orbit because the energy increases toward the interior of the orbit. The Fermi surface is also shown in Fig. 26.

EXAMPLE: *Fermi Surface of Gold.* In gold for quite a wide range of field directions Shoenberg finds the magnetic moment has a period of 2×10^{-9} gauss^{-1}. From (55) we see that this period corresponds to an extremal orbit of area

$$S = \frac{2\pi e/\hbar c}{\Delta(1/B)} \cong \frac{9.55 \times 10^7}{2 \times 10^{-9}} \cong 4.8 \times 10^{16} \,\text{cm}^{-2} \ .$$

From Table 7.1 we have $k_F = 1.2 \times 10^8$ cm^{-1} for a free electron Fermi sphere for gold, which corresponds to an extremal area of 4.5×10^{16} cm^{-2}, in general agreement with the experimental value. The actual periods reported by Shoenberg are 2.05×10^{-9} gauss^{-1} for the orbit B_{111} of Fig. 35 and 1.95×10^{-9} gauss^{-1} for B_{100}. In the [111] direction in Au a large period of 6×10^{-8} gauss^{-1} is also found; the corresponding orbital area if 1.6×10^{15} cm^{-2}. This is the area of the "neck" orbit N of Fig. 26. Another extremal orbit, the "dog's bone," is shown in Fig. 35; its area in Au is about 0.4 of the belly area. Experimental results are shown in Figs. 36 and 37. To do the example in SI, drop c from the relation for S and use as the period 2×10^{-5} tesla^{-1}.

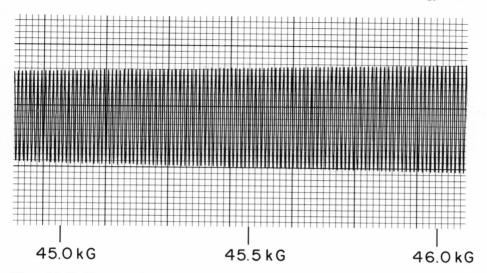

45.0 kG   45.5 kG   46.0 kG

Figure 36   De Haas-van Alphen effect in gold with **B** ∥ [110]. The oscillation is from the dog's bone orbit of Fig. 35. The signal is related to the second derivative of the magnetic moment with respect to field. The results were obtained by a field modulation technique in a high-homogeneity superconducting solenoid at about 1.2°K. (Courtesy of I. M. Templeton.)

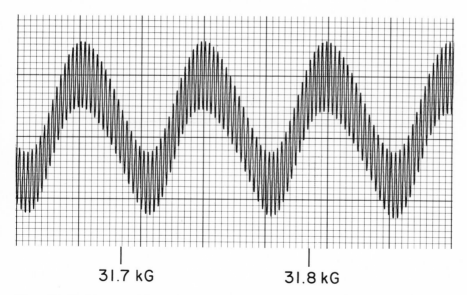

31.7 kG   31.8 kG

Figure 37   De Haas-van Alphen effect in gold with **B** ∥ [111], showing the belly (fine spacing) and neck (coarse spacing) oscillations. These orbits are shown in Fig. 26 as B_{111} and N, respectively. (Courtesy of I. M. Templeton.)

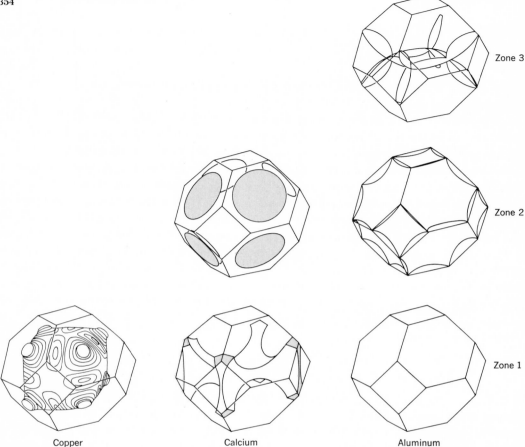

Zone 3

Zone 2

Zone 1

Copper   Calcium   Aluminum

Figure 38 Free electron Fermi surfaces for fcc metals with one (Cu), two (Ca), and three (Al) valence electrons per primitive cell. The Fermi surface shown for copper has been deformed from a sphere to agree with the experimental results. The second zone of calcium contains bubbles of electrons: the second zone of aluminum is nearly filled with electrons. The complex structure in the first zone of calcium and in the third zone of aluminum are shown only in part: for each structure shown, there are two similar structures in the same zone oriented so that the set of three structures has cubic symmetry. (Redrawn from A. R. Mackintosh, "Fermi surface of metals," Scientific American, July, 1963, pp. 110–120).

FERMI SURFACES OF Fcc METALS

We exhibit in Fig. 38 the free electron Fermi surfaces constructed for three metals that have the face-centered cubic crystal structure: copper, with one valence electron; calcium, with two valence electrons; and aluminum, with three. The drawings are of the Fermi surface in the reduced zone scheme. The free electron Fermi surfaces were developed from spheres of radius k_F

equal to

$$\text{Cu: } \frac{4.90}{a} \qquad \text{Ca: } \frac{7.8}{a} \qquad \text{Al: } \frac{10.2}{a} \;,$$

where a is the lattice parameter of the cubic cell. These values follow from (38) with appropriate changes for the valency.

The value of k_F for which a sphere just touches the boundary of the first Brillouin zone of the fcc lattice is $k_F = 5.45/a$. The normal from the origin to the hexagonal face of the zone is $(\pi/a)(\hat{x} + \hat{y} + \hat{z})$, of length $3^{\frac{1}{2}}\pi/a$. Thus the free electron Fermi spheres extend beyond the first zone for calcium and aluminum. We know from magnetoresistance measurements on calcium that there are electrons in the second zone, equal in number to the holes in the first zone.

The free electron Fermi sphere of aluminum fills the first zone entirely and has a large overlap into the second and third zones. The third zone Fermi surface is quite complicated, even though it is just made up of certain pieces of the surface of the free electron sphere. The free electron model also gives small pockets of holes in the fourth zone, but when the lattice potential is taken into account these empty out, the electrons being added to the third zone. The general features of the predicted Fermi surface for aluminum are quite well verified by experiment.[15]

SUMMARY
(In CGS units)

1. A Fermi surface is the surface in **k** space of constant energy equal to ϵ_F. The Fermi surface separates filled states from empty states at absolute zero. The form of the Fermi surface is usually exhibited best in the reduced zone scheme, but the connectivity of the surfaces is clearest in the periodic zone scheme.

2. The motion of a wave packet centered at wavevector **k** is described by $\mathbf{F} = \hbar \, d\mathbf{k}/dt$, where **F** is the applied force. The motion in real space is obtained from the group velocity $\mathbf{v}_g = \hbar^{-1}\nabla_\mathbf{k}\epsilon(\mathbf{k})$. The motion in **k** space in a magnetic field distinguishes orbits of three types: electron-like, hole-like, and open orbits. The effective mass m^* of an electron at **k** is given by

$$\left(\frac{1}{m^*}\right)_{\mu\nu} = \frac{1}{\hbar^2} \frac{\partial^2 \epsilon}{\partial k_\mu \partial k_\nu} \cdot$$

The smaller the energy gap, the smaller is $|m^*|$ near the gap.

3. A crystal with one hole has one empty electron state in an otherwise filled band. The properties of the hole are those of the $N - 1$ electrons:

(a) If the electron is missing from the state of wavevector $\mathbf{k}_e$, then the wavevector of the hole is

$$\mathbf{k}_h = -\mathbf{k}_e \;.$$

[15] W. A. Harrison, Phys. Rev. **118**, 1182 (1960).

(b) The rate of change of $\mathbf{k}_h$ in an applied field requires the assignment of a positive charge to the hole: $e_h = e = -e_e$, whence

$$\frac{d\mathbf{k}_h}{dt} = e\left(\mathbf{E} + \frac{1}{c}\mathbf{v}_h \times \mathbf{B}\right) .$$

(c) If $\mathbf{v}_e$ is the velocity an electron would have in the state $\mathbf{k}_e$, then the velocity to be ascribed to the hole of wavevector $\mathbf{k}_h = -\mathbf{k}_e$ is

$$\mathbf{v}_h = \mathbf{v}_e .$$

(d) The energy of the hole referred to zero for a filled band is positive and is

$$\epsilon_h(\mathbf{k}_h) = -\epsilon(\mathbf{k}_e) .$$

(e) The effective mass of a hole is opposite to the effective mass of an electron at the same point on an energy band:

$$m_h = -m_e .$$

4. The cohesion of simple metals is accounted for by the lowering of energy of the $\mathbf{k} = 0$ conduction band orbital when the boundary conditions on the wavefunction are changed from Schrödinger to Wigner-Seitz.

5. The periodicity in the de Haas-van Alphen effect measures the extremal cross-section area S in $\mathbf{k}$ space of the Fermi surface, the cross-section being taken perpendicular to $\mathbf{B}$:

$$\Delta\left(\frac{1}{B}\right) = \frac{2\pi e}{\hbar c S} .$$

Problems

1. *Brillouin zones of rectangular lattice.* Make a plot of the first two Brillouin zones of a primitive rectangular two-dimensional lattice with axes a, $b = 3a$.

2. *Brillouin zone, square lattice.* (a) Construct the fourth zone of the square lattice. (b) For the Fermi surface of Fig. 2, sketch in the reduced zone the form of that part of the Fermi surface which lies in the fourth zone. Indicate the region filled with electrons.

3. *Hexagonal close-packed structure.* Consider the first Brillouin zone of a crystal with a simple hexagonal lattice in three dimensions with lattice constants a and c, as treated in Problem 2.2. Let $\mathbf{G}_c$ denote the shortest reciprocal lattice vector parallel to the **c** axis of the crystal lattice. (a) Show that for a hexagonal close-packed crystal structure the Fourier component $U(\mathbf{G}_c)$ of the crystal potential $U(\mathbf{r})$ is zero. (b) Is $U(2\mathbf{G}_c)$ also zero? (c) Why is it possible in principle to obtain an insulator made up of divalent atoms at the lattice points of a simple hexagonal lattice? (d) Why is it not possible to obtain an insulator made up of monovalent atoms in a hexagonal-close-packed structure?

4. *Brillouin zones of two-dimensional divalent metal.* A two-dimensional metal in the

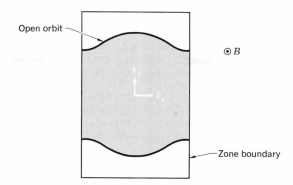

Figure 39 An open orbit in a tetragonal Brillouin zone.

form of a square lattice has two conduction electrons per atom. In the almost free electron approximation, sketch carefully the electron and hole energy surfaces. For the electrons choose a zone scheme such that the Fermi surface is shown as closed.

5. **Properties of holes.** The energy near a valence band edge is given by $\epsilon_{\mathbf{k}} = -1 \times 10^{-26} k^2$ ergs. An electron is removed from the orbital $\mathbf{k} = 1 \times 10^7 \, \hat{\mathbf{k}}_x$ cm^{-1}. The band is otherwise full. (a) Give the sign and magnitude of the effective mass of the hole. (b) What is the direction and magnitude of the wavevector of the hole? (c) The crystal momentum of the hole? (d) The velocity of the hole? (e) The energy of the hole, referred to the valence band edge? (f) The electric current carried by the hole?

6. **Open orbits.** An open orbit in a monovalent tetragonal metal connects opposite faces of the boundary of a Brillouin zone (Fig. 39). The faces are separated by $G = 2 \times 10^8$ cm^{-1}. A magnetic field $B = 10^3$ gauss $= 10^{-1}$ tesla is normal to the plane of the open orbit. (a) What is the order of magnitude of the period of the motion in $\mathbf{k}$ space? Take $v \approx 10^8$ cm/sec. (b) Describe in real space the motion of an electron on this orbit in the presence of the magnetic field.

7. **Cohesive energy for a square well potential.** (a) Find an expression for the binding energy of an electron in one dimension in a single square well of depth U_0 and width a. (This is the standard first problem in elementary quantum mechanics). Assume that the solution is symmetric about the midpoint of the well. (b) Find a numerical result for the binding energy in terms of U_0 for the special case $|U_0| = 2\hbar^2/ma^2$ and compare with the appropriate limit of Fig. 18.

8. **Cyclotron resonance on spherical Fermi surface.** Show from (40) that the period of cyclotron resonance for a particle on a spherical Fermi surface is independent of the value of k_B, the projection of $\mathbf{k}$ on the direction of $\mathbf{B}$.

9. **De Haas-van Alphen period of potassium.** (a) Calculate the period $\Delta(1/B)$ expected for potassium on the free electron model. (b) What is the area in real space of the extremal orbit, for $B = 10$ kG $= 1$ T?

References

Solid state physics **24** (1970) contains three fine articles on pseudopotentials: V. Heine, "The pseudo-potential concept," p. 1; M. L. Cohen and V. Heine "Fitting of pseudopotentials to experimental data and their subsequent application," p. 37; V. Heine and D. Weaire, "Pseudopotential theory of cohesion and structure," p. 249.

W. A. Harrison and M. B. Webb, ed., *The Fermi surface,* Wiley, 1960.

J. M. Ziman, *Principles of the theory of solids,* Cambridge University Press, 1964.

P. W. Anderson, *Concepts in solids,* Benjamin, 1963.

A. B. Pippard, *Dynamics of conduction electrons,* Gordon and Breach, 1965. (A geometrical discussion of Fermi surface phenomena.)

W. A. Harrison, *Solid state theory,* McGraw-Hill, 1970.

N. H. March, *Liquid metals,* Pergamon, 1968.

J. Ziman, ed., *Electrons,* Vol. 1 of *Physics of metals,* Cambridge, 1969.

C. B. Duke, *Tunneling in solids,* Academic, 1969.

J. F. Cochran and R. R. Haering, eds., *Electrons in metals,* Gordon and Breach, 1968.

CLASSIC TEXTS

N. F. Mott and H. Jones, *Theory of the properties of metals and alloys,* Oxford, 1936. (Dover paper-back reprint.)

R. Peierls, *Quantum theory of solids,* Oxford, 1955.

F. Seitz, *Modern theory of solids,* McGraw-Hill, 1940.

G. H. Wannier, *Elements of solid state theory,* Cambridge University Press, 1959.

A. H. Wilson, *Theory of metals,* 2nd ed., Cambridge University Press, 1959.

11

Semiconductor Crystals

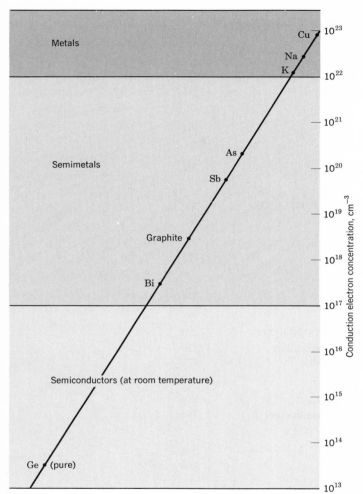

Figure 1 Carrier concentrations for metals, semimetals, and semiconductors. The semiconductor range may be extended upward by increasing the concentration of impurity atoms. (The horizontal axis of this plot is for display purposes and has no significance.)

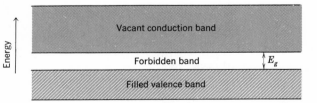

Figure 2 Band scheme for intrinsic conductivity in a semiconductor. At $0°K$ the conductivity is zero because all states in the valence band are filled and all states in the conduction band are vacant. As the temperature is increased, electrons are thermally excited from the valence band to the conduction band, where they become mobile.

There is an essential difference between a semiconductor, such as germanium, and a good conductor, such as silver .... The resistance of a good conductor falls rapidly with falling temperature, while that of a poor conductor rises and becomes very large as the temperature approaches the absolute zero.   (A. H. Wilson)

["The theory of electronic semiconductors," Proc. Roy. Soc. (London) **A133**, 458 (1933)].

At absolute zero a pure, perfect crystal of most semiconductors will be an insulator. The characteristic semiconducting properties are brought about by thermal excitation, impurities, lattice defects, or lack of stoichiometry (departure from nominal chemical composition). Semiconductors are electronic conductors with electrical resistivity values generally in the range of 10^{-2} to 10^9 ohm-cm at room temperature, intermediate between good conductors (10^{-6} ohm-cm) and insulators (10^{14} to 10^{22} ohm-cm). Carrier concentrations representative of metals, semimetals, and semiconductors are shown in Fig. 1.

The electrical resistivity of a semiconductor may be strongly dependent on temperature. Devices based on the properties of semiconductors include transistors, rectifiers, modulators, detectors, thermistors, and photocells. We discuss in this chapter the central physical features of semiconductor crystals, particularly silicon and germanium; other important crystals include cuprous oxide (Cu_2O), selenium, PbTe, PbS, SiC, InSb, GaAs, and graphite.

Some useful nomenclature: the semiconductor compounds of chemical formula AB, where A is a trivalent element and B is a pentavalent element, are called III-V (three-five) compounds. Examples are indium antimonide and gallium arsenide. Where A is divalent and B is hexavalent, the compound is called a II-VI compound; examples are zinc sulfide and cadmium sulfide. Silicon and germanium are sometimes called diamond-type semiconductors, because they have the crystal structure of diamond. Diamond itself is more an insulator rather than a semiconductor. Silicon carbide SiC is a IV-IV compound.

INTRINSIC CONDUCTIVITY

A highly purified semiconductor exhibits intrinsic conductivity, as distinguished from the impurity conductivity of less pure specimens. In the **intrinsic temperature range** the electrical properties of a semiconductor are not essentially modified by impurities in the crystal. An electronic band scheme leading to intrinsic conductivity is indicated in Fig. 2. The conduction band is vacant at absolute zero and is separated by an energy gap E_g from the filled valence band. The **band gap** is the difference in energy between the lowest point of

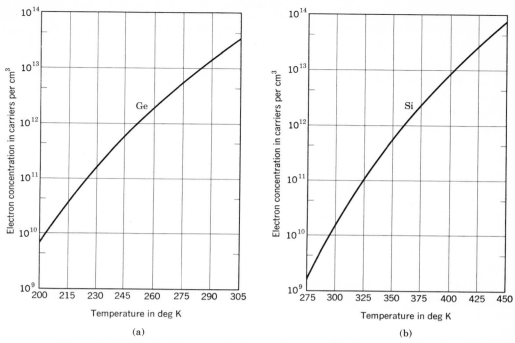

Figure 3 Intrinsic electron concentration as a function of temperature for (a) germanium and (b) silicon. Under intrinsic conditions the hole concentration is equal to the electron concentration. The intrinsic concentration at a given temperature is higher in Ge than in Si because the energy gap is narrower in Ge (0.67 eV) than in Si (1.14 eV). (After W. C. Dunlap.)

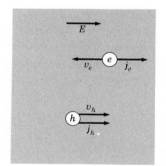

Figure 4 Motion of electrons and holes in the electric field E. The hole and electron velocities are in opposite directions, but their electric currents are in the same direction, the direction of the electric field.

the conduction band and the highest point of the valence band. The lowest point in the conduction band is called the **conduction band edge;** the highest point in the valence band is called the **valence band edge.**

As the temperature is increased, electrons are thermally excited from the valence band to the conduction band (Fig. 3). Both the electrons in the conduction band and the vacant orbitals or holes left behind in the valence band contribute to the electrical conductivity (Fig. 4). At temperatures below the intrinsic range the electrical properties are controlled by impurities, and here we speak of impurity conductivity or extrinsic conductivity. The intrinsic conductivity is dominant at the high temperatures at the left-hand half of Fig. 5.

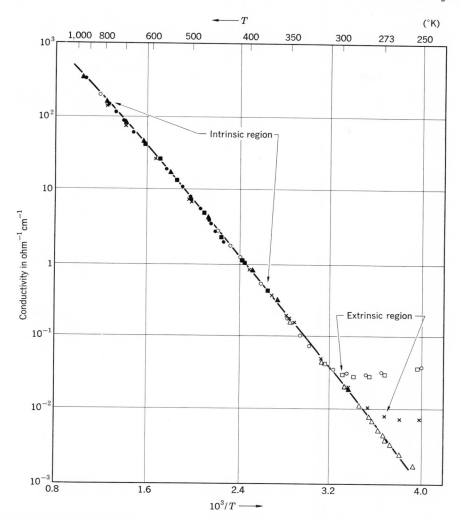

Figure 5 Logarithm of the conductivity versus the reciprocal temperature for crystals of germanium with different concentrations of trivalent and pentavalent impurities. The most temperature sensitive factor in the conductivity is the carrier concentration. The impurity content affects the concentration of carriers at low temperatures, but at higher temperatures the carrier concentration is determined by the intrinsic properties of the pure semiconductor crystal. In the intrinsic region the temperature dependence of the conductivity is dominated by a factor $\exp(-E_g/2k_BT)$ in the carrier concentration, where E_g is the energy gap, so that a plot of log (conductivity) versus $1/T$ gives a straight line in the intrinsic region. Use these data to estimate the energy gap of germanium. *Ans.* ∼0.7 eV. [After F. J. Morin and J. P. Maita, Phys. Rev. **96**, 28 (1954).]

Table 1 Values of the energy gap between the valence and conduction bands in semiconductors, at absolute zero and at room temperature

(i = indirect gap; d = direct gap)

Crystal	Gap	E_g, eV		Crystal	Gap	E_g, eV	
		0°K	300°K			0°K	300°K
Diamond	i	5.4		HgTe°	d	−0.30	
Si	i	1.17	1.14	PbS	d	0.29	0.34–0.37
Ge	i	0.74	0.67	PbSe	d	0.17	0.27
αSn	d	0.00	0.00	PbTe	d	0.19	0.30
InSb	d	0.23	0.18	CdS	d	2.58	2.42
InAs	d	0.36	0.35	CdSe	d	1.84	1.74
InP	d	1.29	1.35	CdTe	d	1.61	1.45
GaP	i	2.32	2.26	ZnO		3.44	3.2
GaAs	d	1.52	1.43	ZnS		3.91	3.6
GaSb	d	0.81	0.78	SnTe	d	0.3	0.18
AlSb	i	1.65	1.52	AgCl		—	3.2
SiC(hex)		3.0	—	AgI		—	2.8
Te	d	0.33	—	Cu$_2$O		2.17	—
ZnSb		0.56	0.56	TiO$_2$		3.03	—

° HgTe is a semimetal; the bands overlap.

General references: D. Long, *Energy bands in semiconductors*, Interscience, 1968; also the *AIP Handbook*.

BAND GAP

The intrinsic conductivity and intrinsic carrier concentrations are largely controlled by E_g/k_BT, the ratio of the band gap to the temperature. When this ratio is large, the concentration of intrinsic carriers will be low and the conductivity will be low. Band gaps of representative semiconductors are given in Table 1.

The best values of the band gap are obtained by optical absorption. The threshold of continuous optical absorption at frequency ω_g determines the band gap $E_g = \hbar\omega_g$ in Figs. 6a and 7a. In the **direct absorption process** a photon is absorbed by the crystal with the creation of an electron and a hole. In the **indirect absorption process** in Figs. 6b and 7b the minimum energy gap of the band structure involves electrons and holes separated by a substantial wavevector $\mathbf{k}_c$. Here a direct photon transition at the energy of the minimum gap cannot satisfy the requirement of conservation of wavevector, because photon wavevectors are negligible in comparison with $\mathbf{k}_c$ at the energy range (~ 1 eV) of interest. But if a phonon of wavevector $\mathbf{K}$ and frequency Ω is created in the process, then we can have

$$\mathbf{k}(\text{photon}) = \mathbf{k}_c + \mathbf{K} \cong 0 \; ; \qquad \hbar\omega = E_g + \hbar\Omega \; ,$$

as required by the conservation laws. The phonon energy $\hbar\Omega$ will generally

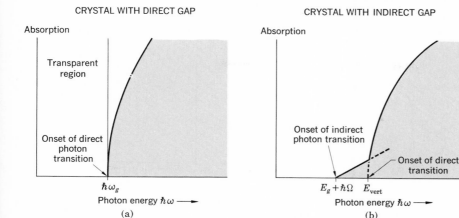

CRYSTAL WITH DIRECT GAP   CRYSTAL WITH INDIRECT GAP

(a)   (b)

Figure 6 Optical absorption as a function of photon energy in pure insulators at absolute zero. In (a) the absorption threshold has a simple variation with the photon energy; the threshold determines the energy gap as $E_g = \hbar\omega_g$. In (b) the optical absorption is weaker near the threshold: at the photon energy $\hbar\omega = E_g + \hbar\Omega$ a photon is absorbed with the creation of three particles: a free electron, a free hole, and a phonon of energy $\hbar\Omega$. In (b) the energy E_{vert} marks the threshold for the creation of a free electron and a free hole, with no phonon involved. Such a transition is called vertical; it is similar to the direct transition in (a). These plots do not show absorption lines, which sometimes are seen lying just to the low energy side of the threshold. These lines are due to the creation of a bound electron-hole pair, called an exciton. The absorption due to the production of excitons is shown in Chapter 18 for gallium arsenide, GaAs.

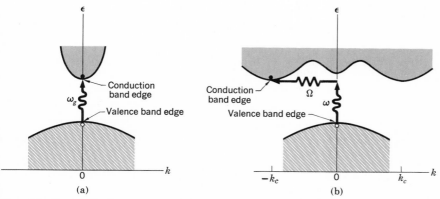

Figure 7 In (a) the lowest point of the conduction band occurs at the same value of **k** as the highest point of the valence band. A direct optical transition is drawn vertically with no significant change of **k**, because the absorbed photon has a very small wavevector. The threshold frequency ω_g for absorption by the direct transition determines the energy gap $E_g = \hbar\omega_g$. The indirect transition in (b) involves both a photon and a phonon because the band edges of the conduction and valence bands are widely separated in **k** space. The threshold energy for the indirect process in (b) is greater than the true band gap. The absorption threshold for the indirect transition between the band edges is at $\hbar\omega = E_g + \hbar\Omega$, where Ω is the frequency of an emitted *phonon* of wavevector $\mathbf{K} \cong -\mathbf{k}_c$. At higher temperatures phonons are already present; if a phonon is absorbed along with a photon, the threshold energy is $\hbar\omega = E_g - \hbar\Omega$. *Note:* the figure shows only the threshold transitions. Transitions occur generally between almost all points of the two bands for which the wavevectors and energy can be conserved.

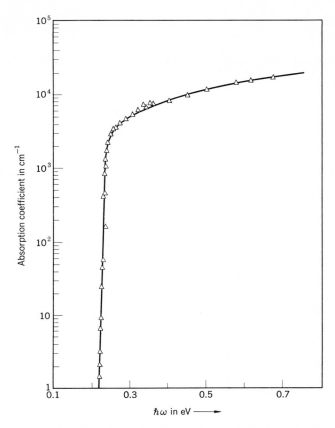

Figure 8 Optical absorption in pure indium antimonide, InSb. The transition is direct because both conduction and valence band edges are at the center of the Brillouin zone, **k** = 0. (After G. W. Gobeli and H. Y. Fan.)

be much less than E_g: a phonon even of high wavevector is an inexpensive source of crystal momentum because the phonon energies are characteristically small ($\sim$0.01 to 0.03 eV) in comparison with the energy gap. If the temperature is so high that the necessary phonon is already thermally excited in the crystal, it is possible also to have a photon absorption processs in which the phonon is absorbed.

The band gap may also be deduced from the temperature dependence of the conductivity or of the carrier concentration in the intrinsic range. The carrier concentration is obtained from measurements of the Hall voltage (Chapter 8), sometimes supplemented by conductivity measurements. Only optical measurements determine whether the gap is direct or indirect. The band edges in Ge and in Si are connected by indirect transitions; the band edges in InSb are connected by a direct transition (Fig. 8). The gap in αSn is direct and is exactly zero;[1] HgTe and HgSe are semimetals and have negative gaps—that is, the bands overlap.

[1] S. Groves and W. Paul, Phys. Rev. Letters **11**, 194 (1963).

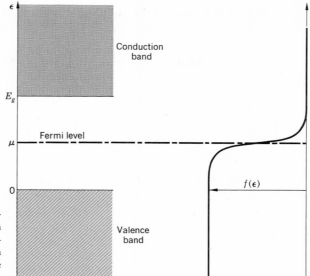

Figure 9 Energy scale for statistical calculations. The Fermi distribution function is shown on the same scale, for a temperature $k_B T \ll E_g$; the Fermi level μ is taken to lie in the band gap, as for an intrinsic semiconductor.

LAW OF MASS ACTION

We want to calculate the concentration of intrinsic carriers in terms of the band gap. We first calculate in terms of the chemical potential μ the number of electrons excited to the conduction band at temperature T. In semiconductor physics μ is called the **Fermi level**. We measure the energy ϵ from the top of the valence band, as in Fig. 9. At the temperatures of interest we may suppose for the conduction band that $\epsilon - \mu \gg k_B T$, and the Fermi-Dirac distribution function reduces to

$$f \cong e^{(\mu - \epsilon)/k_B T} \; . \tag{1}$$

This is the probability that a conduction electron orbital is occupied. We recall that μ is the energy at which $f = \frac{1}{2}$, but (1) is written in an approximation valid for $f \ll 1$. We suppose that the energy of an electron in the conduction band is

$$\epsilon_{\mathbf{k}} = E_g + \frac{\hbar^2 k^2}{2m_e} \; , \tag{2}$$

where m_e is the effective mass of an electron. Thus from (7.24) the number of orbitals with energy between ϵ and $\epsilon + d\epsilon$ is

$$\mathfrak{D}_e(\epsilon) \, d\epsilon = \frac{1}{2\pi^2} \left(\frac{2m_e}{\hbar^2} \right)^{\frac{3}{2}} (\epsilon - E_g)^{\frac{1}{2}} \, d\epsilon \; , \tag{3}$$

per unit volume. We combine (1) and (3) to obtain the concentration of electrons in the conduction band:

$$n = \int_{E_g}^{\infty} \mathfrak{D}_e(\epsilon) f_e(\epsilon) \, d\epsilon = \frac{1}{2\pi^2} \left(\frac{2m_e}{\hbar^2} \right)^{\frac{3}{2}} e^{\mu/k_B T} \int_{E_g}^{\infty} (\epsilon - E_g)^{\frac{1}{2}} e^{-\epsilon/k_B T} \, d\epsilon \ , \quad (4)$$

which integrates to give[2]

$$n = 2 \left(\frac{m_e k_B T}{2\pi \hbar^2} \right)^{\frac{3}{2}} e^{(\mu - E_g)/k_B T} \ . \tag{5}$$

The problem is not solved until μ is known. It is useful to calculate the equilibrium concentration of holes p. The distribution function f_h for holes is related to the electron distribution function f_e by $f_h = 1 - f_e$, because a hole is defined as the absence of an electron. We have

$$f_h = 1 - \frac{1}{e^{(\epsilon - \mu)/k_B T} + 1} = \frac{1}{e^{(\mu - \epsilon)/k_B T} + 1} \cong e^{(\epsilon - \mu)/k_B T} \ , \tag{6}$$

provided $(\mu - \epsilon) \gg k_B T$. If the holes near the top of the valence band behave as particles with effective mass m_h, the density of hole orbitals is given by

$$\mathfrak{D}_h(\epsilon) \, d\epsilon = \frac{1}{2\pi^2} \left(\frac{2m_h}{\hbar^2} \right)^{\frac{3}{2}} (-\epsilon)^{\frac{1}{2}} \, d\epsilon \ , \tag{7}$$

recalling that the energy is measured positive upwards from the top of the valence band. Proceeding as in (4) we obtain

$$p = \int_{-\infty}^{0} \mathfrak{D}_h(\epsilon) f_h(\epsilon) \, d\epsilon = 2 \left(\frac{m_h k_B T}{2\pi \hbar^2} \right)^{\frac{3}{2}} e^{-\mu/k_B T} \tag{8}$$

for the concentration p of holes in the valence band.

We multiply together the expressions for n and p to obtain the equilibrium relation:

$$\boxed{ np = 4 \left(\frac{k_B T}{2\pi \hbar^2} \right)^3 (m_e m_h)^{\frac{3}{2}} e^{-E_g/k_B T} \ . } \tag{9}$$

---

[2] In the notation of *TP*, Chapter 11, the result may be written as

$$n = \frac{2\lambda e^{-E_g/k_B T}}{V_Q(e)} \ , \tag{5a}$$

where

$$V_Q(e) \equiv \left(\frac{2\pi \hbar^2}{m_e k_B T} \right)^{\frac{3}{2}} \tag{5b}$$

is the quantum volume of a conduction electron and λ is the absolute activity $e^{\mu/k_B T}$. An alternative derivation of (9) is given by C. Kittel, American Journal of Physics **35**, 483 (1967).

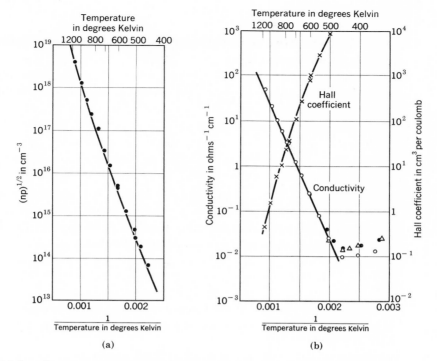

Figure 10 (a) Log of carrier concentration product $(np)^{1/2}$ in the intrinsic range of silicon as a function of reciprocal absolute temperature. The plotted points are derived from experiment. The solid curve is an empirical fit to the points below $700°$K: $np = 1.5 \times 10^{33}T^3 \exp(-1.21/k_BT)$, with the Boltzmann constant in eV/deg K. From this expression we find the energy gap $E_g(T = 0) = 1.21$ eV. (b) Conductivity and Hall coefficient in the intrinsic range of silicon. [After F. J. Morin and J. P. Maita, Phys. Rev. **96**, 28 (1954).]

This useful result does not involve the Fermi level μ. It is an expression of the law of mass action.[3] We have nowhere assumed in the derivation that the material is intrinsic: the result (9) holds in the presence of impurities as well. The only assumption made is that the distance of the Fermi level from the edge of both bands should be large in comparison with k_BT so that (1) and (6) are valid approximations. Experimental results[4] for silicon are shown in Fig. 10.

---

[3] A simple kinetic argument shows that the product np is constant at a given temperature. Suppose that the equilibrium population of electrons and holes is maintained by black-body photon radiation. The photons generate electron-hole pairs at a rate $A(T)$, while $B(T)np$ is the rate of the recombination reaction $e + h = $ photon. Then

$$\frac{dn}{dt} = A(T) - B(T)np = \frac{dp}{dt}.$$

In equilibrium $dn/dt = 0$; $dp/dt = 0$, whence $np = A(T)/B(T)$.

[4] The coefficient of $\exp(-E_g/k_BT)$ deduced from Fig. 10 is considerably larger than the value calculated from (9). If the gap depends on temperature as $E_g(T) = E_g(0)(1 - \alpha T)$, then a temperature-independent factor $\exp(\alpha E_g/k_B)$ is introduced in front of $\exp[-E_g(0)/k_BT]$, in a first approximation.

At $300°$K the value of np is 3.6×10^{27} cm^{-6} for Ge and 4.6×10^{19} cm^{-6} for Si, all calculated with $m_e = m_h = m$.

Because the product of the electron and hole concentrations is a constant independent of impurity concentration at a given temperature, the introduction of a small proportion of a suitable impurity to increase n, say, must decrease p. This result is important in practice—we can reduce the total carrier concentration $n + p$ in an impure crystal, sometimes enormously, by the controlled introduction of suitable impurities. Such a reduction is called **compensation** of one impurity type by adding another.

INTRINSIC CARRIER CONCENTRATION

In an intrinsic semiconductor the number of electrons is equal to the number of holes, because the thermal excitation of an electron leaves behind a hole in the valence band. Thus from (9) we have, letting the subscript i denote intrinsic,

$$n_i = p_i = 2\left(\frac{k_B T}{2\pi\hbar^2}\right)^{\frac{3}{2}} (m_e m_h)^{\frac{3}{4}} e^{-E_g/2k_B T} . \tag{10}$$

The intrinsic carrier excitation depends exponentially on $E_g/2k_B T$, where E_g is the energy gap. We set (4) equal to (8) to obtain

$$e^{2\mu/k_B T} = (m_h/m_e)^{\frac{3}{2}} e^{E_g/k_B T} \tag{11}$$

or, on solving for the chemical potential,

$$\mu = \tfrac{1}{2} E_g + \tfrac{3}{4} k_B T \log (m_h/m_e) . \tag{12}$$

If $m_h = m_e$, then $\mu = \tfrac{1}{2} E_g$ and the Fermi level is in the middle of the forbidden gap. This is not inconsistent with the general result that the average occupancy is one-half at the Fermi level: if there were an orbital at the Fermi level, its occupancy would indeed be one-half.

Mobility in the Intrinsic Region

The mobility[5] is the magnitude of the drift velocity per unit electric field:

$$\mu = \frac{|v|}{E} .$$

The mobility is defined to be positive for both electrons and holes, although their drift velocities are opposite.

In an ideal intrinsic semiconductor the mobility is determined by lattice scattering; that is, by collisions between electrons and phonons. In actual intrinsic specimens there are always some impurity atoms that may dominate the scattering of electrons at low temperatures when phonons are not present.

---

[5] By writing μ_e or μ_h for the electron or hole mobility we can avoid any confusion between μ as the chemical potential and μ as the mobility.

Table 2   Carrier mobilities at room temperature

(Most of the values are probably representative of lattice scattering. For further data see the solid state physics section by H. P. R. Frederikse in the A.I.P. Handbook, 3rd ed.).

Crystal	Mobility, cm²/volt-sec		Crystal	Mobility, cm²/volt-sec	
	Electrons	Holes		Electrons	Holes
Diamond	1800	1200	GaSb	4000	1400
Si	1300	500	PbS	550	600
Ge	4500	3500	PbSe	1020	930
InSb	77,000	750	PbTe	1620	750
InAs	33,000	460	AgCl	50	—
InP	4600	150	KBr(100°K)	100	—

The electrical conductivity in the presence of both electrons and holes is given by the sum of the separate contributions:

$$\sigma = (ne\mu_e + pe\mu_h) \; , \tag{13a}$$

where n and p are the concentrations of electrons and holes. On comparison with the expression $\sigma = ne^2\tau/m$ for the static conductivity we have

$$\mu_e = \frac{e\tau_e}{m_e} \; ; \qquad \mu_h = \frac{e\tau_h}{m_h} \; . \tag{13b}$$

The mobilities are likely to depend on temperature as a modest power law. The temperature dependence of the conductivity in the intrinsic region will be dominated by the exponential dependence $\exp\left(-E_g/2k_BT\right)$ of the carrier concentration. This is the basis for the use of conductivity data in the determination of the energy gap.

Experimental values of the mobility[6] at room temperature are given in Table 2. By comparison, the mobility in metallic copper is only 35 cm²/volt-sec at room temperature. In most substances the values quoted are probably representative of lattice scattering of carriers by phonons. There is a tendency for crystals with small energy gaps to have high values of the electron mobility. As discussed in Chapter 10, small gaps lead to small effective masses, which by (13b) favor high mobilities. The highest mobility observed in a semiconductor is 5×10^6 cm²/volt-sec for electrons in PbTe at 4°K.

The mobility in CGS is expressed in cm²/statvolt-sec and is 300 times higher than the mobility in practical units expressed in cm²/volt-sec.

---

[6] For electron mobilities in alkali halide crystals, see R. K. Ahrenkiel and F. C. Brown, Phys. Rev. **136**, A223 (1964).

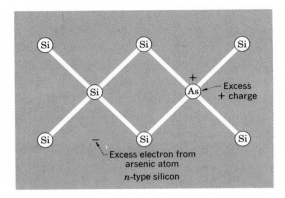 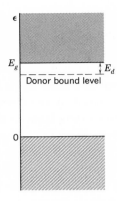

Figure 11 Charges associated with an arsenic impurity atom in silicon. Arsenic has five valence electrons, but silicon has only four valence electrons. Thus four electrons on the arsenic form tetrahedral covalent bonds similar to silicon, and the fifth electron is available for conduction. The arsenic atom is called a **donor** atom because when ionized it gives up an electron to the conduction band.

IMPURITY CONDUCTIVITY

Certain types of impurities and imperfections drastically affect the electrical properties of a semiconductor. The addition of boron to silicon in the proportion of 1 boron atom to 10^5 silicon atoms increases the conductivity of pure silicon by a factor of 10^3 at room temperature. In a compound semiconductor a stoichiometric deficiency of one constituent will act as an impurity; such semiconductors are known as **deficit semiconductors** (see Problem 6). The deliberate addition of impurities to a semiconductor is called **doping.**

We consider in particular the effect of impurities in silicon and germanium. These elements crystallize in the diamond structure, as shown in Fig. 1.29. Each atom forms four covalent bonds, one with each of its four nearest neighbors, corresponding to the chemical valence four. If an impurity atom of valence five, such as phosphorus, arsenic, or antimony, is substituted in the lattice in place of a normal atom, there will be one valence electron from the impurity atom left over after the four covalent bonds are established with the nearest neighbors, that is, after the impurity atom has been accommodated in the structure with as little disturbance as possible.

IMPURITY STATES

The structure in Fig. 11 has a positive charge on the impurity atom (which has lost one electron). Lattice constant studies have verified that the pentavalent impurities enter the lattice by substitution for normal atoms, and not in interstitial positions. Impurity atoms that can give up an electron are called

donors. The crystal as a whole remains neutral because the electron remains in the crystal.

The electron moves in the Coulomb potential $e/\epsilon r$ of the impurity ion, where ϵ in a covalent crystal is the static dielectric constant of the medium. The factor $1/\epsilon$ takes account of the reduction in the Coulomb force between charges caused by the electronic polarization of the medium. This treatment is valid for orbits large in comparison with the distance between atoms, and for slow motions of the electron such that the orbital frequency is low in comparison with the frequency ω_g corresponding to the energy gap. These conditions are satisfied quite well in Ge and Si by the donor electron of P, As, or Sb.

We now estimate the binding energy of the donor impurity. The Bohr theory of the hydrogen atom may readily be modified to take into account both the dielectric constant of the medium and the effective mass of an electron in the periodic potential of the crystal. The binding energy of atomic hydrogen is $-e^4m/2\hbar^2$ in CGS and $-e^4m/2(4\pi\epsilon_0\hbar)^2$ in SI. In the semiconductor we replace e^2 by e^2/ϵ and m by the effective mass m^* to obtain

$$\text{(CGS)} \quad E_d = \frac{e^4m^*}{2\epsilon^2\hbar^2} \; ; \qquad\qquad \text{(SI)} \quad E_d = \frac{e^4m^*}{2(4\pi\epsilon\epsilon_0\hbar)^2} \, , \qquad (14)$$

as the donor ionization energy of the semiconductor. The Bohr radius of the ground state of hydrogen is $\hbar^2/me^2$ in CGS or $4\pi\epsilon_0\hbar^2/me^2$ in SI. Thus the Bohr radius of the donor is

$$\text{(CGS)} \quad a_d = \frac{\epsilon\hbar^2}{m^*e^2} \; ; \qquad\qquad \text{(SI)} \quad a_d = \frac{4\pi\epsilon\epsilon_0\hbar^2}{m^*e^2} \, . \qquad (15)$$

The application of these results to germanium and silicon is complicated by the anisotropic effective mass of the conduction electrons, as discussed below. But the dielectric constant has the most important effect on the donor energy because it enters the energy (14) as the square, whereas the effective mass enters only as the first power. To obtain a general impression of the impurity levels we use $m^* \approx 0.1\,m$ for electrons in germanium and $m^* \approx 0.2\,m$ in silicon. The static dielectric constant has the value 5.5 in diamond, 11.7 in silicon, and 15.8 in germanium. In some semiconductor crystals the static dielectric constant is much larger: 205, 400, and 1770 in PbS, PbTe, and SnTe, respectively.

The ionization energy of the free hydrogen atom is 13.6 eV. For germanium the donor ionization energy E_d on our model is 0.006 eV, reduced with respect to hydrogen by the factor $m^*/m\epsilon^2 = 4 \times 10^{-4}$. The corresponding result for silicon is 0.02 eV. Calculations[7] using the correct anisotropic mass tensor predict 0.00905 eV for germanium and 0.0298 eV for silicon. Further corrections for

---

[7] J. M. Luttinger and W. Kohn, Phys. Rev. **98**, 915 (1955); M. Lampert, Phys. Rev. **97**, 352 (1955); C. Kittel and A. H. Mitchell, Phys. Rev. **96**, 1488 (1954); see also *QTS*, Chap. 14.

**Table 3   Donor ionization energies E_d
of pentavalent impurities in
germanium and silicon, in eV**

	P	As	Sb
Si	0.045	0.049	0.039
Ge	0.0120	0.0127	0.0096

silicon have been considered by Luttinger and Kohn. Observed values of donor ionization energies are given in Table 3.

The radius of the first Bohr orbit is increased by $\epsilon m/m^*$ over the value 0.53 Å for the free hydrogen atom. The corresponding radius is $(160)(0.53) \cong$ 80 Å in germanium and $(60)(0.53) \cong 30$ Å in silicon. These are large radii, so that impurity orbits overlap at relatively low impurity concentrations.

Just as an electron may be bound to a pentavalent impurity, a hole may be bound to a trivalent impurity in germanium or silicon (Fig. 12). Trivalent impurities such as B, Al, Ga, and In are called **acceptors** because they can accept electrons from the valence band, leaving mobile holes in the band. When we ionize an acceptor, we free a hole, which requires an input of energy. (The acceptor problem is similar in principle to that for donors, although the initial strain of visualization on the reader is greater. On the usual energy band diagram an electron rises when it gains energy, whereas a hole sinks in gaining energy.)

Experimental values of the ionization energies of acceptors in germanium and silicon are given in Table 4; they are not unlike those for donors. The modified Bohr model applies qualitatively for holes just as for electrons, but in germanium and silicon the degeneracy at the top of the valence band complicates the effective mass problem.[7a]

A glance at the tables shows that some donor and acceptor ionization energies are comparable with $k_B T$ at room temperature (0.026 eV), so that the thermal ionization of donors and acceptors is important in the electrical conductivity of germanium and silicon at room temperature. If, for example, donor atoms are present in considerably greater numbers than acceptors, the thermal ionization of donors will cause electrons to be freed in the conduction band. The conductivity of the specimen will be controlled by electrons (negative charges), and the material is said to be n type. If acceptors are dominant, holes will be freed in the valence band and the conductivity will be controlled by holes (positive charges): the material is p type. The sign of the Hall voltage (Chapter 8) is a rough test for n or p type. Another handy laboratory test is the sign of the thermoelectric potential: if the two ends of a specimen are at

[7a] See, however, N. O. Lipari and A. Baldereschi, Phys. Rev. Letters **25**, 1660 (1970).

Table 4   Acceptor ionization energies E_a of trivalent
impurities in germanium and silicon, in eV

	B	Al	Ga	In
Si	0.045	0.057	0.065	0.16
Ge	0.0104	0.0102	0.0108	0.0112

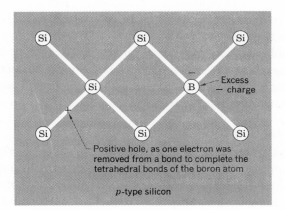
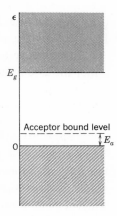

Figure 12   Boron has only three valence electrons; it can complete its tetrahedral bonds only by taking an electron from a Si-Si bond, leaving behind a hole in the silicon valence band. The positive hole is then available for conduction. The boron atom is called **acceptor** atom because when ionized it accepts an electron from the valence band. At 0°K the hole is bound.

different temperatures the carriers tend to concentrate at the colder end.[8] The extra concentration causes a voltage of sign determined by the sign of the charge carriers.

We recall that the numbers of holes and electrons are equal in the absence of impurities: such material is described as intrinsic. The intrinsic electron concentration n_i at 300°K is 6×10^{13} cm^{-3} in germanium and 7×10^9 cm^{-3} in silicon; the electrical resistivity of intrinsic material is 43 ohm-cm for germanium and 2.6×10^5 ohm-cm for silicon. The lowest impurity concentrations attained at present are of the order of 10^{12} impurity atoms per cm^3; we may obtain germanium intrinsic at room temperature, but not silicon. Impurities which do not ionize have no effect on the carrier concentration and may be present in higher proportions without being detected in electrical measurements.

[8] This conclusion follows from inspection of *TP*, Eq. (11.83).

THERMAL IONIZATION OF IMPURITIES

The calculation of the equilibrium concentration of conduction electrons from ionized donors is identical with the standard calculation in statistical mechanics of the thermal ionization of hydrogen atoms. If there are no acceptors present, the result in the low temperature limit $k_B T \ll E_d$ is

$$ n \cong (n_0 N_d)^{\frac{1}{2}} e^{-E_d/2k_B T} \ , \tag{16} $$

with $n_0 \equiv 2(m_e k_B T / 2\pi\hbar^2)^{\frac{3}{2}}$; here N_d is the concentration of donors. To obtain (16) we apply the laws of chemical equilibria to the concentrations $[e][N_d^+]/[N_d]$ = function of the temperature, and then set $[N_d^+] = [e] = n$.

Identical results hold for acceptors, appropriate changes being made, under the assumption that there are no donor atoms. If the donor and acceptor concentrations are comparable, then affairs are quite complicated and the equations are solved by numerical methods.

Mobility in the Presence of Impurities

Scattering by phonons determines the mobility of carriers when relatively few impurity atoms are present, or at high temperatures. Scattering by impurity atoms may be important at higher impurity concentrations. The scattering will depend on whether the impurity is neutral or ionized. The neutral atom problem is equivalent to the scattering of an electron by a hydrogen atom. We note that the area of the first Bohr orbit is increased by $(\epsilon m/m^*)^2$. An exact solution for the neutral scattering cross-section is quite difficult in the energy range of interest in semiconductors. The scattering of carriers by ionized donors or acceptors has been treated by Conwell and Weisskopf, who utilized the Rutherford scattering formula. The effect of impurity scattering in reducing the mobility is shown in Fig. 13 for electrons in AgCl.

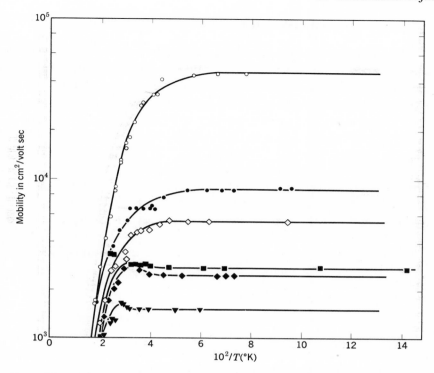

Figure 13   Mobility of electrons in crystals of AgCl of various purities. The highest mobility found is 45000 cm²/volt-sec. At low temperatures the mobility is limited by crystal purity and at high temperatures by scattering by optical phonons. (After T. Masumi, R. K. Ahrenkiel, and F. C. Brown.)

Analysis of Experimental Results

A good general picture of the physical behavior of a semiconductor can be obtained by measuring[9] the electrical conductivity and Hall coefficient as functions of temperature and impurity doping over wide ranges. In a simple metal the Hall coefficient is given by

$$R_H = -\frac{1}{nec} \, ,$$

where n is the electron concentration. In semiconductors the connection between R_H and n may vary slightly according to the form of the dependence of the mean free path on the electron velocity. Carrier concentration deduced from measurements of the Hall voltage versus temperature for a set of silicon crystals doped with arsenic donors are shown in Fig. 14a. Above room temperature the donors are essentially all ionized. Below 100°K the carrier concentration decreases as the donors become un-ionized.

If the current carriers are predominantly of one type only, we may obtain the mobility μ simply from the product of the conductivity and the Hall coefficient

(CGS) $$c|R_H|\sigma = c\left(\frac{1}{nec}\right)\left(\frac{ne^2\tau}{m^*}\right) = \frac{e\tau}{m^*} = \mu \; ; \qquad (17)$$

(SI) $$|R_H|\sigma = \left(\frac{1}{ne}\right)\left(\frac{ne^2\tau}{m^*}\right) = \frac{e\tau}{m^*} = \mu \, ,$$

apart from factors of the order of unity. The product $c|R_H|\sigma$ or $|R_H|\sigma$ is called the **Hall mobility.** The temperature dependence of the Hall mobility for the set of arsenic-doped silicon crystals is shown in Fig. 14b.

---

[9] For germanium see E. M. Conwell, Proc. I. R. E. **40**, 1327 (1952); P. P. Debye and E. M. Conwell, Phys. Rev. **93**, 693 (1954); for silicon see F. J. Morin and J. P. Maita, Phys. Rev. **96**, 28 (1954).

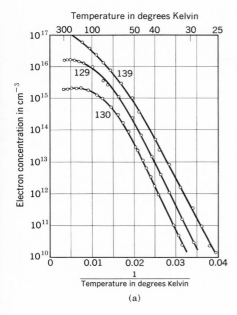

(a)

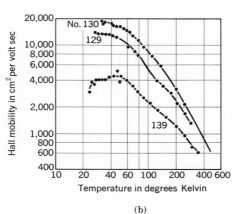

Figure 14 Silicon specimens containing arsenic. (a) Charge carrier concentration as a function of reciprocal absolute temperature; (b) Mobility as a function of temperature. (After Morin and Maita.)

(b)

ENERGY BANDS IN Si and Ge

The conduction and valence bands of germanium are shown in Fig. 15, as based on a combination of theoretical and experimental results. The valence band edge in both crystals is at $k = 0$ and is derived from $p_{\frac{3}{2}}$ and $p_{\frac{1}{2}}$ states of the free atoms. (This is clear from the tight-binding approximation to the wave-functions. The $p_{\frac{3}{2}}$ level is four-fold degenerate as in the atom; the four states correspond to m_J values $\pm\frac{3}{2}$ and $\pm\frac{1}{2}$. The $p_{\frac{1}{2}}$ level is doubly degenerate, with $m_J = \pm\frac{1}{2}$. The $p_{\frac{3}{2}}$ states are higher in energy than the $p_{\frac{1}{2}}$ states; the difference in energy is a measure of the spin-orbit interaction.) The conduction band edges are not at $k = 0$; this is known from cyclotron resonance and from the occurrence of indirect optical transitions between the band edges in Fig. 6b.

Cyclotron Resonance in Semiconductors

In a number of semiconductors it has been possible to determine by cyclotron resonance the form of the energy surfaces of the conduction and valence bands near the band edges.[10] The determination of the energy surface is equivalent to a determination of the effective mass tensor

$$\left(\frac{1}{m_{\mu\nu}^*}\right) = \hbar^{-2}\frac{\partial^2\epsilon}{\partial k_\mu \partial k_\nu}.$$

Cyclotron resonance in a semiconductor is somewhat different than in a metal because at low carrier concentration the rf field will penetrate the entire specimen. The entire orbit of a carrier in a semiconductor will therefore see a uniform rf field. The current carriers are accelerated in spiral orbits about the axis of a static magnetic field. The angular rotation frequency ω_c of the carriers is

$$(\text{CGS}) \quad \omega_c = \frac{eB}{m^*c}, \qquad\qquad (\text{SI}) \quad \omega_c = \frac{eB}{m^*},$$

where m^* is the appropriate effective mass. Resonant absorption of energy from an rf electric field perpendicular to the static magnetic field (Fig. 16) occurs when the frequency of the rf field is equal to the cyclotron frequency. Holes and electrons will rotate in opposite senses in a magnetic field.

We consider the order of magnitude of several physical quantities relevant to the experiment, for $m^*/m \cong 0.1$. For $f_c = 24$ GHz, or $\omega_c = 1.5 \times 10^{11}$ rad s^{-1}, we have $B \cong 860$ G at resonance. The line width is determined by the collision relaxation time τ, and to obtain a distinctive resonance it is necessary that $\omega_c\tau \geq 1$. In other words, the mean free path must be long enough to permit the average carrier to get one radian around a circle between

---

[10] Recent advances in the analysis of ultraviolet reflectance measurements on semiconductors have led to the approximate determination of the overall shape of the energy surfaces; the experimental developments involve the modulation of the reflectance by electric fields and by elastic stress. See M. Cardona, "Modulation spectroscopy," *Solid state physics* Suppl. 11, (1969).

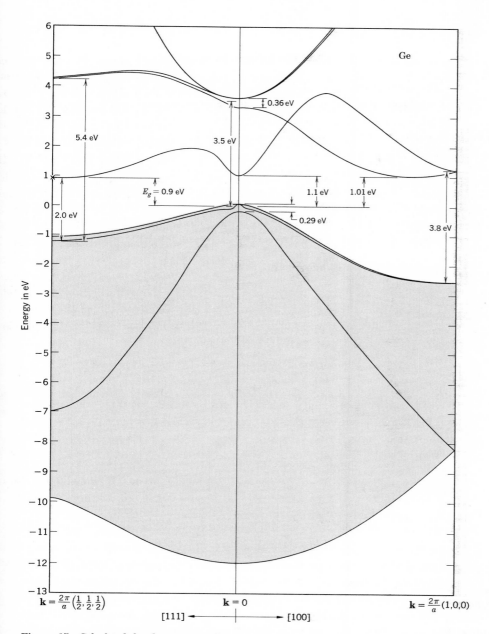

Figure 15 Calculated band structure of germanium, after C. Y. Fong. The general features are in good agreement with experiment. The four valence bands are shown in gray. The fine structure of the valence band edge is caused by spin-orbit splitting.

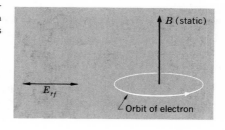

Figure 16 Arrangement of fields in a cyclotron resonance experiment in a semiconductor. The sense of the circulation is opposite for electrons and holes.

successive collisions. For $\omega_c = 1.5 \times 10^{11}$ rad s^{-1}, we require $\tau = 10^{-11}$ s or longer. At room temperature the relaxation times of carriers in crystals are too short, being commonly in the range 10^{-13} to 10^{-15} s. It is usually necessary to work with high-purity crystals in liquid helium to obtain long enough relaxation times. The requirements are relaxed with the use of higher frequency radiation, but higher magnetic fields are then needed. Pulsed fields up to 400 kG or more are available in several laboratories.

We derive the cyclotron resonance frequency for an electron on a spheroidal energy surface; the result is applicable to the conduction band edges in Si, Ge, and many other crystals. If we measure the wavevector from the band edge, then the energy will be written in the form

$$\epsilon(\mathbf{k}) = \hbar^2\left(\frac{k_x{}^2 + k_y{}^2}{2m_t} + \frac{k_z{}^2}{2m_l}\right) , \tag{18}$$

where m_t is the transverse mass parameter and m_l is the longitudinal mass parameter. A surface on which $\epsilon(\mathbf{k})$ is constant will be a spheroid. The velocity components in real space are related to $\mathbf{k}$ by $\mathbf{v} = \hbar^{-1}\nabla_{\mathbf{k}}\epsilon$:

$$v_x = \frac{\hbar k_x}{m_t} ; \qquad v_y = \frac{\hbar k_y}{m_t} ; \qquad v_z = \frac{\hbar k_z}{m_l} .$$

The equation of motion in $\mathbf{k}$ space is

(CGS) $$\hbar\frac{d\mathbf{k}}{dt} = -\frac{e}{c}\mathbf{v} \times \mathbf{B} . \tag{19}$$

The results (19) through (29) are in CGS, but can be converted to SI by replacing c by unity. If $\mathbf{B}$ lies in the equatorial plane of the spheroid and parallel to the k_x axis, then

$$\hbar\frac{dk_x}{dt} = 0 ;$$

$$\hbar\frac{dk_y}{dt} = -\frac{eB}{c}v_z = -\hbar\frac{eB}{m_l c}k_z , \tag{20}$$

or

$$\frac{dk_y}{dt} = -\omega_l k_z ; \qquad \omega_l \equiv \frac{eB}{m_l c} ; \tag{21}$$

further

$$\frac{dk_z}{dt} = \frac{eB}{c}v_y = \hbar\frac{eB}{m_t c}k_y , \tag{22}$$

or

$$\frac{dk_z}{dt} = \omega_t k_y ; \qquad \omega_t \equiv \frac{eB}{m_t c} . \tag{23}$$

We differentiate (21) with respect to time to obtain

$$\frac{d^2 k_y}{dt^2} = -\omega_l \frac{dk_z}{dt} \; ; \tag{24}$$

on substitution of (23) for dk_z/dt we have

$$\frac{d^2 k_y}{dt^2} + \omega_l \omega_t k_y = 0 \; , \tag{25}$$

the equation of motion of a simple harmonic oscillator of natural frequency

(CGS) $$\boxed{\omega_0 = (\omega_l \omega_t)^{\frac{1}{2}} = \frac{eB}{(m_l m_t)^{\frac{1}{2}} c}} \; . \tag{26}$$

This is the cyclotron frequency when **B** lies in the equatorial plane of the spheroidal energy surface.

If **B** lies parallel to the symmetry axis of the spheroid (the k_z axis), then the equations of motion are

$$\frac{dk_x}{dt} = -\omega_t k_y; \qquad \frac{dk_y}{dt} = \omega_t k_x \; , \tag{27}$$

whence

$$\frac{d^2 k_x}{dt^2} + \omega_t^2 k_x = 0 \; . \tag{28}$$

This describes an oscillator of natural frequency

(CGS) $$\boxed{\omega_0 = \omega_t = \frac{eB}{m_t c}} \; . \tag{29}$$

The effective mass determining the cyclotron frequency when the static magnetic field makes an angle θ with the longitudinal axis of the spheroidal energy surface (18) is

$$\left(\frac{1}{m_c}\right)^2 = \frac{\cos^2 \theta}{m_t^2} + \frac{\sin^2 \theta}{m_t m_l} \; , \tag{30}$$

which reduces to (26) and (29) for $\theta = \frac{1}{2}\pi$ and 0, respectively.

An early cyclotron resonance run in germanium is shown in Fig. 17. We see two hole masses and three electron masses; each electron mass originates from one or more spheroidal energy surfaces oriented differently with respect to the magnetic field. In Fig. 18 we plot results obtained for electrons in germanium as a function of the angle between the direction of the static magnetic field in a (110) plane and a [001] direction lying in the plane. Mass values derived from the theoretical fit to the experimental points[11] are $m_l = 1.59\,m$ and $m_t = 0.082\,m$. A set of crystallographically equivalent energy spheroids is oriented along all $\langle 111 \rangle$ directions in the Brillouin zone, with their center at the edge of the zone.

In silicon the conduction band edges are spheroids oriented along the equivalent $\langle 100 \rangle$ directions in the Brillouin zone, with mass parameters $m_l = 0.98\,m$ and $m_t = 0.19\,m$, as in Fig. 19.

The valence band edges in germanium and silicon are complicated, (Fig. 15). The holes observed in these crystals are characterized by two effective masses, and we speak of light and heavy holes. The two hole masses arise from the two bands formed from the $p_{\frac{3}{2}}$ level of the atom. The energy surfaces are of the form (QTS, p. 271):

$$\epsilon(\mathbf{k}) = Ak^2 \pm [B^2k^4 + C^2(k_x{}^2k_y{}^2 + k_y{}^2k_z{}^2 + k_z{}^2k_x{}^2)]^{\frac{1}{2}} , \qquad (31)$$

The choice of sign distinguishes the two masses. The experiments give, in units $\hbar^2/2m$,

Si: $A = -4.0$; $|B| = 1.1$; $|C| = 4.1$

Ge: $A = -13.3$; $|B| = 8.6$; $|C| = 12.5$.

Roughly, the holes in germanium have masses $0.04\,m$ and $0.3\,m$; in silicon $0.16\,m$ and $0.5\,m$. There is some anisotropy, easily apparent for the heavy hole (Fig. 20).

---

[11] B. W. Levinger and D. R. Frankel, J. Phys. Chem. Solids **20**, 281 (1961).

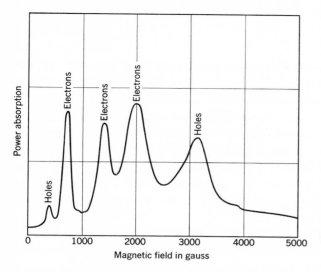

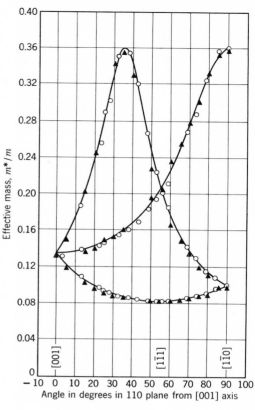

Figure 17 (*above*) Cyclotron resonance absorption in germanium near 24 GHg and 4°K. The static magnetic field is in a (110) plane at 60° from a [100] axis. The holes and electrons were produced by photons from a light bulb.

Figure 18 (*right*) Effective cyclotron mass of electrons in germanium at 4°K for magnetic field directions in a (110) plane. (After Dresselhaus, Kip, and Kittel.)

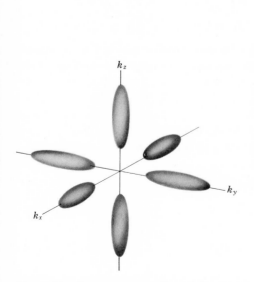

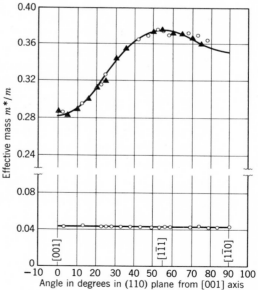

Figure 19 Constant energy ellipsoids for electrons in silicon, drawn for $m_l/m_t = 5$.

Figure 20 Effective cyclotron mass of holes in germanium at 4°K for magnetic field directions in a (110) plane.

385

Table 5   Effective masses of electrons and holes

(In cubic crystals with conduction and valence band edges at $\mathbf{k} = 0$.)

Crystal	Energy gap, in eV	Electron mass, m_e/m	Heavy hole mass, m_{hh}/m	Light hole mass, m_{lh}/m
InSb	0.23	0.0155	0.4	0.016
InAs	0.36	0.024	0.41	0.026
GaSb	0.81	0.042	—	0.052
GaAs	1.52	0.07	0.68	0.07

In Table 5 we present effective mass data on four III-V compounds with both conduction and valence band edges at the center of the Brillouin zone, $\mathbf{k} = 0$. The electron and light hole masses increase almost directly as the energy gap increases, in agreement with the theory (*QTS*, p. 186). The two hole surfaces are warped spheres, and the light and heavy hole masses m_{lh}, m_{hh} are defined only as averages.

p-n JUNCTIONS

What happens if we place in contact two pieces of a semiconductor, one p type and the other n type, as in Fig. 21? On the p side of the junction there are free holes and an equal concentration of ($-$) ionized acceptor impurity atoms that maintain charge neutrality. On the n side of the junction there are free electrons and an equal number of ($+$) ionized donor impurity atoms. Thus the majority carriers are holes on the p side and electrons on the n side, Fig. 22a. In thermal equilibrium with the majority carriers there are small concentrations of the minority carrier types, exaggerated somewhat in the figure.

The holes concentrated on the p side would like to diffuse to fill the crystal uniformly. Electrons would like to diffuse from the n side. But diffusion upsets electrical neutrality. As soon as a small charge transfer by diffusion has taken place, there is left behind on the p side an excess of ($-$) ionized acceptor atoms and on the n side an excess of ($+$) ionized donor atoms (Fig. 22b). This double

Figure 21   A *p-n* junction is made from a single crystal modified in two separate regions. Acceptor impurity atoms are incorporated into part of the crystal during growth to produce the *p* region in which the major part of the mobile charge is carried by holes. Donor impurity atoms are grown into the other part to produce the *n* region in which the majority carriers are electrons. The boundary region may be 10^{-4} cm thick.

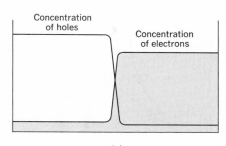

(a)

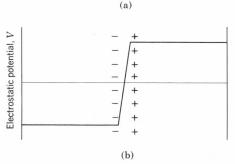

(b)

Figure 22 (a) Variation of the hole and electron concentrations across an unbiased (zero applied voltage) *p-n* junction. The carriers are in thermal equilibrium with the acceptor and donor impurity atoms. The product *pn* of the hole and electron concentrations will be constant throughout the crystal in conformity with the law of mass action. (b) Electrostatic potential from unbalanced acceptor (−) and donor (+) ions near the junction. The potential inhibits diffusion of holes from the *p* side to the *n* side, and it inhibits diffusion of electrons from the *n* side to the *p* side.

layer of charge creates an electric field directed from *n* to *p* that inhibits diffusion and maintains the separation of the two carrier types. Because of the double layer the electrostatic potential in the crystal takes a jump at the junction region.

The electrochemical potential[12] is constant across the crystal and the junction even though there is a jump in the electrostatic potential at the junction. The net current flow of holes or electrons is zero in thermal equilibrium, for the current is proportional to the gradient of the electrochemical potential and not to the gradient of the electrostatic potential alone. (The concentration gradient exactly cancels the electrostatic potential gradient.) A voltmeter measures the difference of electrochemical potential, so that a voltmeter connected across the crystal reads zero.

---

[12] The electrochemical potential of each carrier type is everywhere constant in thermal equilibrium; see *TP* Eq. (11.85). For holes

$$k_B T \log p(\mathbf{r}) + e\varphi(\mathbf{r}) = \text{constant},$$

where *p* is the hole concentration and φ is the electrostatic potential. We see *p* will be low where φ is high. Similarly, for electrons

$$k_B T \log n(\mathbf{r}) - e\varphi(\mathbf{r}) = \text{constant},$$

and *n* will be low where φ is low.

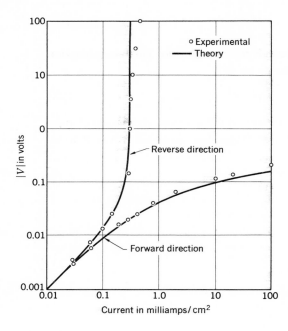

Figure 23 Rectification characteristic of a *p-n* junction in germanium, after Shockley. Note that voltage is plotted vertically and current horizontally.

If the ends are connected to a circuit, we get a current when we shine light on the junction. A photon when absorbed creates an electron and a hole. If the pair is produced at the junction, the electric field of the double layer drives the hole toward the *p* region and the electron toward the *n* region. Thus a current flows in the circuit from *n* to *p*. The energy of the photon is converted to electrical energy in the junction. **Solar cells** work on this principle, and they are used to convert sunlight to provide power for the operation of satellites.

Even in thermal equilibrium there will be a small flow of electrons J_{nr} from the *n* region into the *p* region where the electrons end their lives by recombination with holes. This recombination current is balanced by a current J_{ng} of electrons generated thermally in the *p* region and which diffuse to the *n* region:

$$J_{nr}(0) + J_{ng}(0) = 0 , \tag{32}$$

for otherwise electrons would pile up on one side of the barrier.

Rectification

A *p-n* junction can act as a rectifier. A large current will flow if we apply a voltage across the junction in one direction, but if the voltage is in the opposite direction only a very small current will flow. If an alternating voltage is applied across the junction the current will flow chiefly in one direction—the junction has rectified the current (Fig. 23).

For back voltage bias a negative voltage is applied to the *p* region and a positive voltage to the *n* region, thereby increasing the potential difference between the two regions. Now practically no electrons can climb the potential energy hill from the low side of the barrier to the high side. The recombination current is reduced by the Boltzmann factor:[13]

$$J_{nr}(V\text{ back}) = J_{nr}(0)\exp\left(-e|V|/k_B T\right) .\qquad(33)$$

The thermal generation current of electrons is not particularly affected by the back voltage because the generation electrons flow downhill (from p to n) anyway:

$$J_{ng}(V\text{ back}) = J_{ng}(0) .\qquad(34)$$

We saw in (32) that $J_{nr}(0) = -J_{ng}(0)$; thus by (33) the generation current dominates the recombination current for a back bias.

When a forward voltage is applied, the recombination current increases because the potential energy barrier is lowered, thereby enabling more electrons to flow from the n side to the p side:

$$J_{nr}(V\text{ forward}) = J_{nr}(0)\exp\left(e|V|/k_B T\right) .\qquad(35)$$

Again the generation current is unchanged:

$$J_{ng}(V\text{ forward}) = J_{ng}(0) .\qquad(36)$$

The hole current flowing across the junction behaves similarly to the electron current. The applied voltage which lowers the height of the barrier for electrons also lowers it for holes, so that large numbers of electrons flow from the n region to the p region under the same voltage conditions that produce large hole currents in the opposite direction.

The electrical currents due to holes and electrons add. The total forward electric current, including the effects of both holes and electrons, is given by

$$I = I_s(e^{eV/k_B T} - 1) ,\qquad(37)$$

where I_s is the sum of the two generation currents. As shown in Fig. 23, this equation is well satisfied for p-n junctions in germanium.

POLARONS

An electron in a crystal lattice interacts via its electrical charge with the ions or atoms of the lattice and creates a local deformation of the lattice. The deformation tends to follow the electron as it moves through the lattice. The combination of the electron and its strain field is known as a **polaron**.[14]

[13] The Boltzmann factor controls the number of electrons with enough energy to get over the barrier.

[14] See *QTS*, Chap. 7; also C. G. Kuper and G. D. Whitfield, ed., *Polarons and excitons*, Plenum Press, New York, 1963, particularly the article "Experiments on the polaron" by F. C. Brown; I. G. Austen and N. F. Mott, "Polarons in crystalline and non-crystalline materials," Advances in Physics **18**, 41 (1969).

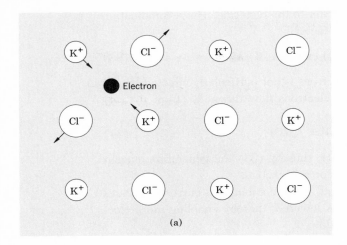

(a)

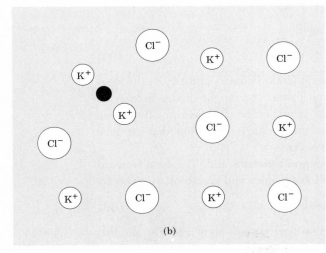

(b)

Figure 24 The formation of a polaron. (a) A conduction electron is shown in a rigid lattice of an ionic crystal, KCl. The forces on the ions adjacent to the electron are shown. (b) The electron is shown in an elastic or deformable lattice. The electron plus the associated strain field is called a polaron. The displacement of the ions increases the effective inertia and, hence, the effective mass of the electron; in KCl the mass is increased by a factor of 2.5 with respect to the band theory mass in a rigid lattice. In extreme situations, often with holes, the particle can become self-trapped (localized) in the lattice. In covalent crystals the forces on the atoms from the electron are weaker than in ionic crystals, so that polaron deformations are small in covalent crystals.

The most important effect of the lattice deformation is the increase in the effective mass of the electron because the heavy ion cores are set into motion when the electron moves. The electron drags the ions along with it (Fig. 24). The effect is large in ionic crystals because of the strong Coulomb interaction between ions and electrons. In covalent crystals the effect is weak because neutral atoms have only a weak interaction with electrons.

The strength of the electron-lattice interaction is measured by the dimensionless coupling constant α given by

$$\frac{1}{2}\alpha = \frac{\text{deformation energy}}{\hbar\omega_L} , \qquad (38)$$

Table 6 **Polaron coupling constants α, masses m_{pol}^*, and band masses m^***
for electrons in the conduction band

(In units of the mass of the free electron)

Crystal	KCl	KBr	AgCl	AgBr	ZnO	PbS	InSb	GaAs
α	3.97	3.52	2.00	1.69	0.85	0.16	0.014	0.06
m_{pol}^*	1.25	0.93	0.51	0.33	—	—	0.014	—
m^*	0.50	0.43	0.35	0.24	—	—	0.014	—
m_{pol}^*/m^*	2.5	2.2	1.5	1.4	—	—	1.0	—

where ω_L is the longitudinal optical phonon frequency near zero wavevector. We view $\frac{1}{2}\alpha$ as "the number of phonons which surround a slow-moving electron in a crystal." Values deduced from diverse experiments and theory are given in Table 6 after F. C. Brown. The values of α are high in ionic crystals and low in covalent crystals. The values of the effective mass m_{pol}^* of the polaron are from cyclotron resonance experiments.[15] The values given for the band effective mass m^* were calculated from m_{pol}^* by use of Eq. (39). The last row in the table gives the factor m_{pol}^*/m^* by which the band mass is increased by the deformation of the lattice.

Theory relates the effective mass of the polaron m_{pol}^* to the effective band mass m^* of the electron in the undeformed lattice by the relation[16]

$$m_{pol}^* \cong m^*\left(\frac{1 - 0.0008\alpha^2}{1 - \frac{1}{6}\alpha + 0.0034\alpha^2}\right) ; \tag{39}$$

for $\alpha \ll 1$ this is approximately $m^*(1 + \frac{1}{6}\alpha)$. Because the coupling constant α is always positive, the polaron mass is greater than the bare mass, as we expect from the inertia of the ions. For high α the mobility is low when the temperature is high enough to permit the excitation of optical phonons. The electron mobility is low in AgCl and KBr (Table 2).

It is common to speak of large and small polarons. The electron associated with a large polaron moves in a band, but the mass is slightly enhanced; these are the polarons we have discussed above. The electron associated with a small polaron[16a] spends most of its time trapped on a single ion. At high temperatures the electron moves from site to site by thermally activated hopping; at low temperatures the electron tunnels slowly through the crystal, as if in a band of large effective mass.

[15] J. W. Hodby, J. A. Borders, F. C. Brown, and S. Foner, Phys. Rev. Letters **19**, 952 (1967).
[16] D. C. Langreth, Phys. Rev. **159**, 717 (1967); see also *QTS*, p. 141.
[16a] The theory of small polarons is treated by T. Holstein, Annals of Physics **8**, 343 (1959); see also references given by J. Appel, *Solid state physics* **21**, 193 (1968).

**Table 7 Experimental values of electron and hole
concentrations in semimetals**
(Courtesy of L. Falicov)

Semimetal	n_e, in cm^{-3}	n_h, in cm^{-3}
Arsenic[a]	$(2.12 \pm 0.01) \times 10^{20}$	$(2.12 \pm 0.01) \times 10^{20}$
Antimony[b]	$(5.54 \pm 0.05) \times 10^{19}$	$(5.49 \pm 0.03) \times 10^{19}$
Bismuth[c]	2.88×10^{17}	3.00×10^{17}
Graphite[d]	2.72×10^{18}	2.04×10^{18}

[a] M. G. Priestley, L. R. Windmiller, J. B. Ketterson and Y. Eckstein, Phys. Rev. **154**, 671 (1967).

[b] L. R. Windmiller, Phys. Rev. **149**, 472 (1966).

[c] R. N. Bhargava, Phys. Rev. **156**, 785 (1967).

[d] J. W. McClure, Phys. Rev. **108**, 612 (1957).

SEMIMETALS

In semimetals the conduction band edge is very slightly lower in energy than the valence band edge. A small overlap in energy of the conduction and valence bands leads to small concentration of holes in the valence band and of electrons in the conduction band (Table 7). Three of the semimetals, arsenic, antimony, and bismuth, are in group V of the periodic table. Their atoms associate in pairs[17] in the crystal lattice, with two ions and ten valence electrons per primitive cell. Small overlaps of the fifth and sixth energy bands create equal numbers of holes and electrons in small nearly ellipsoidal pockets in the Brillouin zone. Like semiconductors, the semimetal elements may be doped with suitable impurities to vary the relative concentrations of holes and electrons. The absolute concentrations may also be varied by application of pressure, for the overlap of the band edges varies with pressure. The Fermi surfaces are described in detail in the references cited in the table.

MOBILITY OF PROTONS, PIONS, AND MUONS

We know from the observed mobilities that electrons propagate in crystals relatively freely as Bloch waves or wave packets in the most commonly studied metals and semiconductors. There is evidence that positrons and positronium atoms propagate similarly. What of heavier particles—protons, pions, or muons —will they also propagate as Bloch waves?

There exist meaningful experiments for protons. In some solids, protons are mobile, but not as wave packets in uniform motion. Instead, they travel by a random step-by-step or hopping process from one site in the structure to

[17] M. H. Cohen, L. M. Falicov, and S. Golin, IBM J. Res. Develop. **8**, 215 (1964), have shown that the qualitative features of the energy bands are determined by the crystal structures of these elements. This volume of the journal contains the proceedings of a conference on semimetals.

another. The motion may be by thermal activation of the proton over a potential energy barrier, or the proton may tunnel quantum-mechanically from one site to another, perhaps assisted by thermal fluctuations.[18] The motion of heavier ions may also require the existence of vacant lattice sites, Chapter 19.

The mobility of protons in ice[19] is 0.1 to 0.5 cm^2/volt-sec at $-10°$C, about 10^{-2} to 10^{-5} of the electron mobilities cited in Table 2. The mass of the proton is 1840 times the mass of a free electron; thus we infer from the mobilities that the relaxation times of electrons and protons may be comparable: a free proton in ice is quite lively.

We expect that a proton injected into a crystal will move by a random hopping process and not as a wave. The proton digs a localized trap for itself by polarizing or straining the lattice. The energy of the system is lower when the proton is localized in the trap than when it is spread out over the lattice as a wave. Self-trapping is more likely for protons than for electrons because the de Broglie wavelength of a proton is shorter than that of an electron of the same energy. The ratio of the wavelengths at equal energies is

$$\frac{\lambda_e}{\lambda_p} = \left(\frac{M_p}{m_e}\right)^{\frac{1}{2}} = (1840)^{\frac{1}{2}} \cong 43 \ . \tag{40}$$

The short wavelength suggests that the proton may easily be localized and bound to an individual ion, probably a negative ion, together with a local distortion of the lattice. Probably pions and muons are also easily bound.

There exist situations in which electrons or, more commonly, holes become self-trapped by inducing an asymmetric local deformation of the lattice. This is most likely to occur when the relevant band edge is degenerate and the crystal is polar (such as an alkali halide or silver halide), with strong coupling of the particle to the lattice. Here degenerate means that more than one band has the same energy at the same value of **k**. The valence band edge is more often degenerate than the conduction band edge. Holes appear to be self-trapped in all the alkali and silver halides, as discussed in Chapter 19 in connection with V_K centers.

Ionic solids at room temperature generally have very low conductivities for the motion of ions through the crystal, less than 10^{-6} (ohm-cm)$^{-1}$, but a family of compounds has been reported[20] with conductivities of 0.2 (ohm-cm)$^{-1}$ at 20°C. The compounds have the composition MAg_4I_5, where M denotes K, Rb, or NH_4. The Ag^+ ions occupy only a fraction of the lattice sites available for Ag^+, and the ionic conductivity proceeds by the hopping of a silver ion from one site to a nearby vacant site. The crystal structures have parallel open channels responsible for the high conductivity.

---

[18] J. A. Sussman, Phys. kondensierte Materie **2**, 146 (1964).

[19] An excellent survey is given by M. Eigen and L. De Maeyer, "Self-dissociation and protonic charge transport in water and ice," Proc. Roy. Soc. (London) **247A**, 565 (1958).

[20] See B. B. Owens and G. R. Argue, Science **157**, 308 (1967); S. Geller, Science **157**, 310 (1967).

AMORPHOUS SEMICONDUCTORS

Amorphous covalent semiconducting alloys are readily formed over broad ranges of composition, particularly from elements of groups IV, V, and VI of the periodic system. They behave as low-mobility intrinsic semiconductors with a temperature-dependent conductivity characterized by an activation energy, as for ionic conductivity, Eq. 19.12. It is a very interesting question to consider what happens to the band model in a disordered solid. The most striking results of band theory are consequences of the assumption of a regular crystalline order. But we know that Bragg reflection and the band gap do not disappear when the solid is disordered by thermal distortions, as we recall from the discussion of the Debye-Waller factor at the end of Chapter 2. In an amorphous alloy it is believed that the band gap is modified so that the tails of the valence and conduction bands overlap, which means that an electron in a valence band in some region of the specimen may have a higher energy than an extra electron in a non-bonding localized conduction state in another part of the specimen. But the mobility associated with states in the gap is very low, and may involve thermal activation between one localized state and another. This model of the amorphous solid is referred to as the Mott and Cohen-Fritzsche-Ovshinsky model.[21]

Problems

1. *Impurity orbits.* Indium antimonide has $E_g = 0.23$ eV; dielectric constant $\epsilon = 17$; electron effective mass $m_e = 0.015\ m$. Calculate (a) the donor ionization energy; (b) the radius of the ground state orbit. (c) At what minimum donor concentration will appreciable overlap effects between the orbits of adjacent impurity atoms occur? This overlap tends to produce an impurity band—a band of energy levels which permit conductivity presumably by a hopping mechanism in which electrons move from one impurity site to a neighboring ionized impurity site.

2. *Ionization of donors.* In a particular semiconductor there are 10^{13} donors/cm^3 with an ionization energy E_d of 1×10^{-3} eV and an effective mass of $1 \times 10^{-2}\ m$. (a) Estimate the concentration of conduction electrons at 4°K. (b) What is the value of the Hall constant? *Note:* Assume no acceptor atoms are present.

3. *Hall effect with two carrier types.* Assuming concentrations n, p; relaxation times τ_e, τ_h; and masses m_e, m_h; show that the Hall constant is

(CGS)
$$R_H = \frac{1}{ec} \cdot \frac{p - nb^2}{(p + nb)^2}\ ,$$

[21] N. F. Mott, Advan. Phys. **16**, 49 (1967). Phil. Mag. **17**, 1259 (1968); M. H. Cohen, H. Fritzsche, and S. R. Ovshinsky, Phys. Rev. Letters **22**, 1065 (1969).

where $b = \mu_e/\mu_h$ is the mobility ratio. In the derivation neglect terms of order $\mathbf{B}^2$. In SI we drop the c. *Hint:* In the presence of a longitudinal electric field, find the transverse electric field such that the transverse current vanishes. The algebra may seem tedious, but the result is worth the trouble. Use (8.39), but for two carrier types; neglect $(\omega_c\tau)^2$ in comparison with $\omega_c\tau$.

4. **Impurity compensation.** (a) Explain the phenomenon of impurity compensation, which is the reduction of carrier concentration and conductivity in a semiconductor initially of one conductivity type (n or p) by the addition of impurities of the other type; assume for simplicity that the mobilities of electrons and holes are equal. (b) Does a nearly intrinsic resistivity prove the specimen is pure?

5. **Cuprous oxide.** Cu_2O is usually a p-type semiconductor in the impurity range. This circumstance is attributed to a deficiency of one of the components. (a) Which component must be missing in order to account for the conductivity type? (b) Cuprous oxide in thin sections is red by transmitted light. Why?

*References**

R. L. Sproull, *Modern physics,* 2nd ed., Wiley, 1963. Contains a good elementary discussion of semiconductor devices.

J. L. Moll, *Physics of semiconductors,* McGraw-Hill, 1964. For the physics of semiconductor devices.

S. Wang, *Solid state electronics,* McGraw-Hill, 1966. Good introduction to devices.

R. A. Smith, *Semiconductors,* Cambridge University Press, 1959.

R. B. Adler, A. C. Smith, R. L. Longini, *Introduction to semiconductor physics,* Wiley, 1964.

R. A. Smith, ed., *Semiconductors* (Italian Physical Society Course **22**), Academic Press, 1963.

Selected constants related to semiconductors (Constantes sélectionées **12**), Pergamon, 1961.

F. Gutmann and L. E. Lyons, *Organic semiconductors,* Wiley, 1967.

D. Long, *Energy bands in semiconductors,* Interscience, 1968.

E. H. Putley, *Hall effect and related phenomena,* Butterworths, 1960.

A. E. Owen, "Semiconducting glasses," Contemp. Physics **11**, 227, 257 (1970).

A. F. Gibson, P. Aigrain, and R. E. Burgess, ed., *Progress in semiconductors,* Heywood, London; a continuing series of volumes starting in 1956.

R. K. Willardson and A. C. Beer, ed., *Semiconductors and semimetals,* Academic Press; a continuing series.

° For an excellent selective bibliography see P. Handler, "Resource letter Scr-1 on semiconductors," Am. J. Phys. **32**, 329 (1964).

12

Superconductivity

Advanced topics relevant to this chapter:

    I. Vector potential
    J. Flux quantization
    K. Josephson superconductor tunneling effects
    L. BCS theory of the superconducting energy gap

NOTATION: In this chapter B_a denotes the applied magnetic field. In the CGS system the critical value B_{ac} of the applied field will be denoted by the symbol H_c in accordance with the custom of workers in superconductivity. Values of B_{ac} are given in gauss in CGS units and in teslas in SI units. In SI we have $B_{ac} = \mu_0 H_c$.

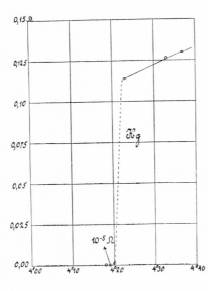

Figure 1 Resistance in ohms of a specimen of mercury versus absolute temperature. This plot by Kamerlingh Onnes marked the discovery of superconductivity.

The electrical resistivity of many metals and alloys drops suddenly to zero when the specimen is cooled to a sufficiently low temperature, often a temperature in the liquid helium range. This phenomenon was observed first by Kamerlingh Onnes[1] in Leiden in 1911, three years after he first liquified helium. His original measurements on mercury are shown in Fig. 1. Note the narrowness of the temperature interval in which the resistivity change occurs. We say that at a critical temperature T_c the specimen undergoes a phase transition from a state of normal electrical resistivity to a superconducting state.

EXPERIMENTAL SURVEY

In the superconducting state the dc electrical resistivity is exactly zero, or at least so close to zero that persistent electrical currents have been observed to flow without attenuation in superconducting rings for more than a year, until at last the experimentalist wearied of the experiment. The decay of supercurrents in a solenoid of $Nb_{0.75} Zr_{0.25}$ was studied by File and Mills[2] using precision nuclear magnetic resonance methods (Chapter 17) to measure the magnetic field associated with the supercurrent. They concluded that the decay time of the supercurrent is not less than 100,000 years. In some superconducting materials, particularly those used for superconducting magnets, finite decay times are observed because of an irreversible redistribution of magnetic flux in the material.

The magnetic properties exhibited by superconductors are as dramatic as their electrical properties. The magnetic properties cannot be accounted for by the assumption that the superconducting state is characterized properly by zero electrical resistivity. It is an experimental fact that a bulk superconductor in a weak magnetic field will act as a perfect diamagnet, with zero magnetic induction in the interior of a bulk superconductor. When a specimen is placed in a magnetic field and is then cooled through the transition temperature for superconductivity, the magnetic flux originally present is ejected from the specimen.

---

[1] H. Kamerlingh Onnes, Akad. van Wetenschappen (Amsterdam) **14**, 113, 818 (1911): "The value of the mercury resistance used was 172.7 ohms in the liquid condition at 0°C; extrapolation from the melting point to 0°C by means of the temperature coefficient of solid mercury gives a resistance corresponding to this of 39.7 ohms in the solid state. At 4.3°K this had sunk to 0.084 ohms, that is, to 0.0021 times the resistance which the solid mercury would have at 0°C. At 3°K the resistance was found to have fallen below 3×10^{-6} ohms, that is to one ten-millionth of the value which it would have at 0°C. As the temperature sank further to 1.5°K this value remained the upper limit of the resistance." Historical references are given by C. J. Gorter, Rev. Mod. Phys. **36**, 1 (1964).

[2] J. File and R. G. Mills, Phys. Rev. Letters **10**, 93 (1963).

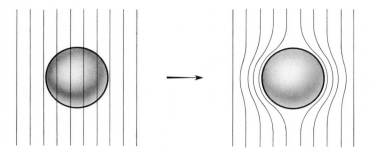

Figure 2 Meissner effect in a superconducting sphere cooled in a constant applied magnetic field; on passing below the transition temperature the lines of induction **B** are ejected from the sphere.

This is called the **Meissner effect.** The sequence of events is shown in Fig. 2. The unique magnetic properties of superconductors are of central importance to the characterization of the superconducting state.

The superconducting state is known to be an ordered state of the conduction electrons of the metal. The order is in the formation of loosely associated pairs of electrons. The electrons are ordered at temperatures below the transition temperature, and they are disordered above the transition temperature. The nature and origin of the ordering was explained first in 1957 by Bardeen, Cooper, and Schrieffer.[3] In the present chapter we develop as far as we can in an elementary way the physics of the superconducting state. We shall also discuss the basic physics of the materials used for superconducting magnets, but not their technology.

Occurrence of Superconductivity

Superconductivity occurs[4] in many metallic elements of the periodic system and also in alloys, intermetallic compounds, and semiconductors.[5] The range of transition temperatures at present extends from about $21°K$ for the alloy $Nb_3(Al_{0.8}Ge_{0.2})$ to $0.01°K$ for some semiconductors. In many metals superconductivity has not been found down to the lowest temperatures at which the metal was examined, usually well below $1°K$. Thus Li, Na, and K have been investigated for superconductivity down to $0.08°K$, $0.09°K$, and $0.08°K$, respectively, where they were still normal conductors. Similarly, Cu,

---

[3] J. Bardeen, L. N. Cooper, and J. R. Schrieffer, Phys. Rev. **106**, 162 (1957); **108**, 1175 (1957).

[4] A review of data on the occurrence of superconductivity is given by B. T. Matthias, T. H. Geballe, and V. B. Compton, Rev. Mod. Phys. **35**, 1–22 (1963).

[5] Superconductivity in certain semiconductors was predicted theoretically by M. L. Cohen, Phys. Rev. **134**, 511 (1964); Rev. Mod. Phys. **36**, 240 (1964). For experiments on oxygen-deficient $SrTiO_3$, see J. F. Schooley et al., Phys. Rev. Letters **14**, 305 (1965); the lowest carrier concentration studied was 2×10^{18} cm^{-3}, and for this specimen $T_c \approx 0.01°K$.

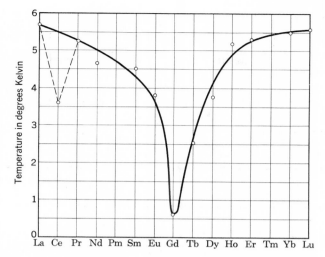

Figure 3 Superconducting transition temperatures of 1 at. percent solid solutions of rare earth elements in lanthanum, after Matthias, Suhl, and Corenzwit. Most of the rare earth elements in solid solution in lanthanum have electronic magnetic moments.

Ag, and Au have been investigated down to 0.05°K, 0.35°K, and 0.05°K, where they are still normal conductors. It has been predicted[6] by theoretical calculation that if sodium and potassium superconduct at all, the transition temperature will be much less than 10^{-5}°K. These statements refer to atmospheric pressure. Cesium becomes a superconductor ($T_c = 1.5$°K) at a pressure of 110 kbar, after several phase transformations.

Will every nonmagnetic metallic element become a superconductor at sufficiently low temperatures? We do not know. In experimental searches for superconductors with ultralow transition temperatures it is important to eliminate from the specimen even trace quantities of foreign paramagnetic elements, because they can lower the transition temperature severely. A few parts per million of Fe will destroy the superconductivity of Mo, which when pure has $T_c = 0.92$°K; and 1 at. percent of gadolinium lowers the transition temperature (Fig. 3) of lanthanum[7] from 5.6°K to 0.6°K. Nonmagnetic impurities have no very marked effect on the transition temperature, although they may affect the behavior of the superconductor in strong magnetic fields.

[6] J. P. Carbotte and R. C. Dynes, Phys. Rev. **172**, 476 (1968); for the experiments on cesium, see J. Wittig, Phys. Rev. Letters **24**, 812 (1970).

[7] H. Suhl and B. T. Matthias, Phys. Rev. Letters **2**, 5 (1959).

Table 1 Superconductivity parameters of the elements.

(An asterisk denotes an element superconducting only in thin films or under high pressure in a crystal modification not normally stable. Related values of the Debye θ and the electronic γ are given in Tables 6.1 and 7.2, respectively.)

Transition temperature in deg K.
Critical magnetic field at absolute zero in gauss (10^{-4} tesla).

Li	Be												B	C	N	O	F	Ne
Na	Mg												Al 1.180 / 105	Si*	P	S	Cl	Ar
K	Ca	Sc	Ti 0.39 / 100	V 5.38 / 1420	Cr	Mn	Fe	Co	Ni	Cu	Zn 0.875 / 53	Ga 1.091 / 51	Ge*	As	Se*	Br	Kr
Rb	Sr	Y*	Zr 0.546 / 47	Nb 9.20 / 1980	Mo 0.92 / 95	Tc 7.77 / 1410	Ru 0.51 / 70	Rh	Pd	Ag	Cd 0.56 / 30	In 3.4035 / 293	Sn (w) 3.722 / 309	Sb*	Te*	I	Xe
Cs*	Ba*	La fcc 6.00 / 1100	Hf	Ta 4.483 / 830	W 0.012 / 1.07	Re 1.698 / 198	Os 0.655 / 65	Ir 0.14 / 19	Pt	Au	Hg (α) 4.153 / 412	Tl 2.39 / 171	Pb 7.193 / 803	Bi*	Po	At	Rn
Fr	Ra	Ac															

Ce*	Pr	Nd	Pm	Sm	Eu	Gd	Tb	Dy	Ho	Er	Tm	Yb	Lu
Th 1.368 / 1.62	Pa 1.4	U (α) 0.68	Np	Pu	Am	Cm	Bk	Cf	Es	Fm	Md	No	Lw

Table 2   Superconductivity of selected compounds

Compound	T_c, in °K	Compound	T_c, in °K
Nb_3Sn	18.05	V_3Ga	16.5
$Nb_3(Al_{0.8}Ge_{0.2})$	20.9	V_3Si	17.1
Nb_3Al	17.5	UCo	1.70
Nb_3Au	11.5	Ti_2Co	3.44
NbN	16.0	La_3In	10.4
MoN	12.0	InSb°	1.9

° Metallic phase under pressure.

The elements known to be superconducting are displayed in Table 1, together with their transition temperatures. None of the monovalent metals (except Cs under pressure) is known to be a superconductor; none of the ferromagnetic metals; and none of the rare earth elements except lanthanum (which has an entirely empty $4f$ electronic shell). The transition temperatures of a number of interesting superconducting compounds are listed in Table 2.

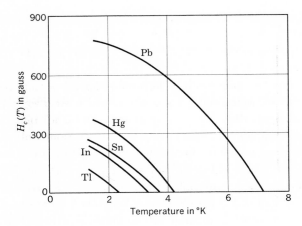

Figure 4 Threshold curves of the critical field $H_c(T)$ versus temperature for several superconductors. A specimen is superconducting below the curve and normal above the curve.

Destruction of Superconductivity by Magnetic Fields

A sufficiently strong magnetic field will destroy superconductivity. The threshold or critical value of the applied magnetic field for the destruction of superconductivity is denoted by $H_c(T)$ and is a function of the temperature. At the critical temperature the critical field is zero: $H_c(T_c) = 0$. The variation of the critical field with temperature for several superconducting elements is shown in Fig. 4. The threshold curves separate the superconducting state in the lower left of the figure from the normal state in the upper right. The dependence of $H_c(0)$ on T_c is shown in Fig. 5 for a number of superconductors. *Note:* We should denote the critical value of the applied magnetic field as B_{ac}, but this is not common practice among workers in superconductivity. In the CGS system we shall always understand that $H_c \equiv B_{ac}$, and in the SI we have $H_c \equiv B_{ac}/\mu_0$. The symbol B_a denotes the applied magnetic field.

Meissner Effect

Meissner and Ochsenfeld[8] found that if a superconductor is cooled in a magnetic field to below the transition temperature, then at the transition the lines of induction B are pushed out (Fig. 2). This phenomenon is called the Meissner effect. It shows that a bulk superconductor behaves in an applied external field B_a as if inside the specimen $B = 0$. We obtain a particularly useful form of this result if we limit ourselves to long thin specimens with long axes

[8] W. Meissner and R. Ochsenfeld, Naturwiss. **21**, 787 (1933).

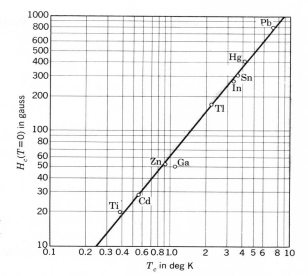

Figure 5 Log-log plot of critical field H_c at $T = 0$ versus transition temperature T_c for superconductors in bulk form.

parallel to B_a; now the demagnetizing field contribution [see Eqs. (13.15) and (17.43)] to B will be negligible, whence:[9]

(CGS) $B = B_a + 4\pi M = 0$;   or $\dfrac{M}{B_a} = -\dfrac{1}{4\pi}$;   (1)

(SI) $B = B_a + \mu_0 M = 0$;   or $\dfrac{M}{B_a} = -\dfrac{1}{\mu_0}$.

This important result cannot be derived merely from the characterization of a superconductor as a medium of zero resistivity. From Ohm's law $\mathbf{E} = \rho\mathbf{j}$ we see that if the resistivity ρ goes to zero while $\mathbf{j}$ is held finite, then $\mathbf{E}$ must be zero. By a Maxwell equation $d\mathbf{B}/dt$ is proportional to curl $\mathbf{E}$, so that zero resistivity implies $d\mathbf{B}/dt = 0$. This argument is not entirely transparent, but the result predicts that the flux through the metal cannot change on cooling through the transition. The Meissner effect contradicts this result and suggests that perfect diamagnetism is an essential property of the superconducting state.[10]

---

[9] Diamagnetism, the magnetization M, and the magnetic susceptibility are defined in Chapter 15. The magnitude of the apparent diamagnetic susceptibility of bulk superconductors is very much larger than in typical diamagnetic substances. In (1), M is the magnetization equivalent to the superconducting currents in the specimen.

[10] We expect another difference between a superconductor and a perfect conductor, defined as a conductor in which there is nothing to scatter the electrons. When the field penetration problem in a perfect conductor is solved in detail, it turns out that a perfect conductor when placed in a magnetic field cannot produce a permanent eddy current screen: the field will penetrate about 1 cm in an hour. See A. B. Pippard, *Dynamics of conduction electrons*, Gordon and Breach, 1965.

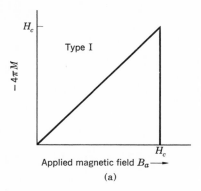

H_c

$-4\pi M$

Type I

H_c

Applied magnetic field B_a ⟶

(a)

Figure 6a Magnetization versus applied magnetic field for a bulk superconductor exhibiting a complete Meissner effect (perfect diamagnetism). A superconductor with this behavior is called a type I superconductor. Above the critical field H_c the specimen is a normal conductor and the magnetization is too small to be seen on this scale. Note that *minus* $4\pi M$ is plotted on the vertical scale: the negative value of M corresponds to diamagnetism. The quantity $4\pi M$ is simply the magnetic field produced by the superconducting currents induced when an external field is applied. (CGS units in all parts of this figure.)

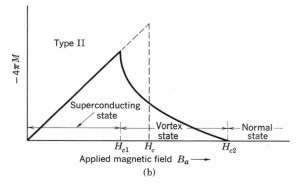

$-4\pi M$

Type II

Superconducting state

Vortex state

Normal state

H_{c1} H_c H_{c2}

Applied magnetic field B_a ⟶

(b)

Figure 6b Superconducting magnetization curve of a type II superconductor. The flux starts to penetrate the specimen at a field H_{c1} lower than the thermodynamic critical field H_c. The specimen is in a **vortex state** between H_{c1} and H_{c2}, and it has superconducting electrical properties up to H_{c2}. Above H_{c2} the specimen is a normal conductor in every respect, except for possible surface effects. For given H_c the area under the magnetization curve is the same for a type II superconductor as for a type I.

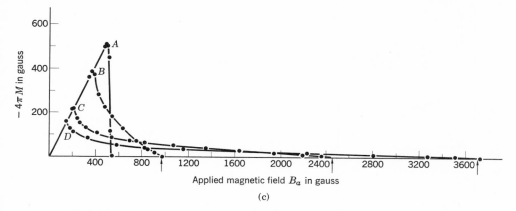

Figure 6c   Superconducting magnetization curves of annealed polycrystalline lead and lead-indium alloys at 4.2°K. (*A*) lead; (*B*) lead–2.08 wt. percent indium; (*C*) lead–8.23 wt. percent indium; (*D*) lead–20.4 wt. percent indium. (After Livingston.)

The magnetization curve expected for a superconductor under the conditions of the Meissner-Ochsenfeld experiment is sketched in Fig. 6a. This applies quantitatively to a specimen in the form of a long solid cylinder[11] placed in a longitudinal magnetic field. Pure specimens of many materials exhibit this behavior; they are called **type I superconductors,** or, formerly, soft superconductors. The values of H_c are always too low for type I superconductors to have any useful technical application in coils for superconducting magnets.

Other materials exhibit a magnetization curve of the form of Fig. 6b and are known as **type II superconductors.** They tend to be alloys (as in Fig. 6c) or transition metals with high values of the electrical resistivity in the normal state: that is, the electronic mean free path in the normal state is short. We shall see later why the mean free path is involved in the "magnetization" of superconductors. Type II superconductors have superconducting electrical properties up to a field denoted by H_{c2}. Between the **lower critical field** H_{c1} and the **upper critical field** H_{c2} the flux density $B \neq 0$ and the Meissner effect is said to be incomplete. The value of H_{c2} may be 100 times or more higher than the value of the critical field H_c calculated from the thermodynamics of the transition. In the region between H_{c1} and H_{c2} the superconductor is threaded by flux

[11] In other geometries the field may not be homogeneous around the specimen and the field may penetrate below H_c; for a sphere the field penetrates at $\frac{2}{3}H_c$ as a consequence of the nonzero demagnetization factor of a sphere.

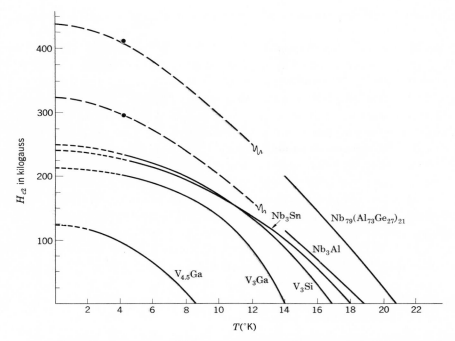

Figure 7 Upper critical field versus temperature, for various high field superconductors. (After S. Foner, E. J. McNiff, Jr., B. T. Matthias, T. H. Geballe, R. H. Willens, and E. Corenzwit.)

lines and is said to be in the **vortex state,** as in Fig. 36 below. A field H_{c2} of 410 kG (41 teslas) has been attained[12] in an alloy of Nb, Al, and Ge at the boiling point of helium (Fig. 7). Commercial solenoids (Fig. 8) wound with a hard superconductor produce high steady fields, some over 100 kG. A hard superconductor is a type II superconductor with a large amount of magnetic hysteresis or flux pinning induced by mechanical treatment.

[12] For detailed data on high H_{c2} materials, see T. G. Berlincourt and R. R. Hake, Phys. Rev. **131**, 140 (1963); Y. Shapira and L. J. Neuringer, Phys. Rev. **140**, A1638 (1965); S. Foner, E. J. McNiff, Jr., B. T. Matthias, T. H. Geballe, R. H. Willens and E. Corenzwit, Physics Letters **31A**, 349 (1970).

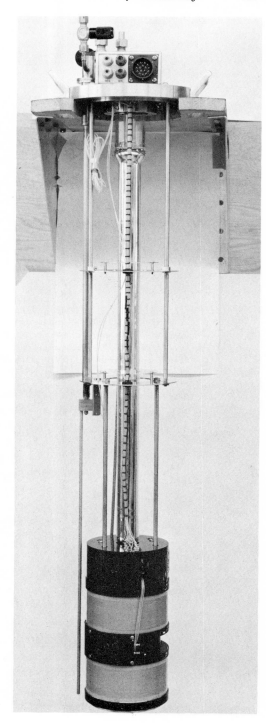

Figure 8 A superconducting magnet assembly for immersion in a liquid He cryostat. (Courtesy of Varian Associates.)

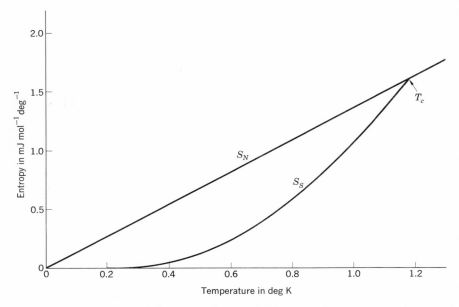

Figure 9 Entropy S of aluminum in the normal and superconducting states as a function of the temperature. The entropy is lower in the superconducting state because the electrons are more ordered here than in the normal state. At any temperature below the critical temperature T_c the specimen can be put in the normal state by application of a magnetic field stronger than the critical field. (Courtesy of N. E. Phillips)

Heat Capacity

In all superconductors the entropy decreases markedly on cooling below the critical temperature T_c. Measurements for aluminum are plotted in Fig. 9. The decrease in entropy between the normal state and the superconducting state tells us that the superconducting state is more ordered than the normal state, for the entropy is a measure of the disorder of a system. Some or all of the electrons thermally excited in the normal state are ordered in the superconducting state. The change in entropy is small, in aluminum of the order of 10^{-4} k_B per atom.

The heat capacity of gallium is plotted in Fig. 10: (a) compares the normal and superconducting states; (b) shows that the electronic contribution to the heat capacity in the superconducting state is an exponential form with an argument proportional to $-1/T$, suggestive of excitation of electrons across an energy gap. An energy gap (Fig. 11) is a characteristic, but not universal, feature of the superconducting state. The gap is accounted for by the Bardeen-Cooper-Schrieffer (BCS) theory of superconductivity.

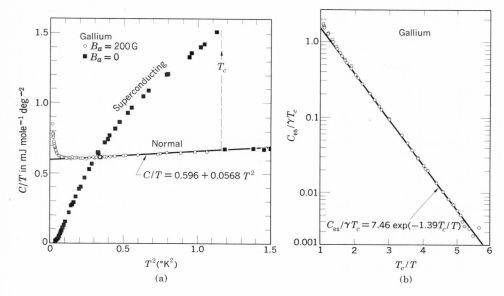

Figure 10 (a) The heat capacity of gallium in the normal and superconducting states. The normal state (which is restored by a 200 G field) has electronic, lattice, and (at low temperatures) nuclear quadrupole contributions. In (b) the electronic part C_{es} of the heat capacity in the superconducting state is plotted on a log scale versus T_c/T: the exponential dependence on $1/T$ is evident. Here $\gamma = 0.60$ mJ mole^{-1} deg^{-2}. [After N. E. Phillips, Phys. Rev. **134**, 385 (1964).]

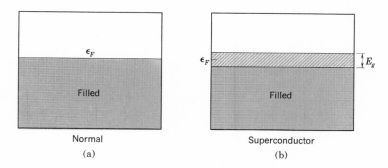

Figure 11 (a) Conduction band in the normal state; (b) energy gap at the Fermi level in the superconducting state. Electrons in excited states above the gap behave as normal electrons in rf fields: they cause resistance; at dc they are shorted out by the superconducting electrons. At absolute zero there are no electrons above the gap. The gap E_g is exaggerated in the figure: typically $E_g \sim 10^{-4} \, \epsilon_F$.

Table 3   Energy gaps in superconductors, at $T = 0$

											Al	Si
$E_g(0)$ in 10^{-4}eV.											3.4	
$E_g(0)/k_B T_c$.											3,3	
Sc	**Ti**	**V**	**Cr**	**Mn**	**Fe**	**Co**	**Ni**	**Cu**	**Zn**	**Ga**	**Ge**	
		16.							2.4	3.3		
		3.4							3.2	3.5		
Y	**Zr**	**Nb**	**Mo**	**Tc**	**Ru**	**Rh**	**Pd**	**Ag**	**Cd**	**In**	**Sn** (w)	
		30.5	2.7						1.5	10.5	11.5	
		3.80	3.4						3.2	3.6	3.5	
La fcc	**Hf**	**Ta**	**W**	**Re**	**Os**	**Ir**	**Pt**	**Au**	**Hg** (α)	**Tl**	**Pb**	
19.		14.							16.5	7.35	27.3	
3.7		3.60							4.6	3.57	4.38	

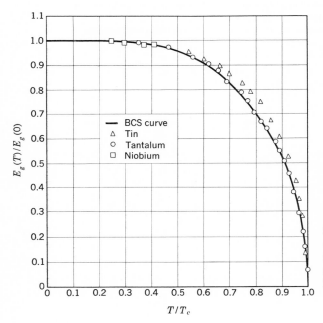

Figure 12   Reduced values of the observed energy gap $E_g(T)/E_g(0)$ as a function of the reduced temperature T/T_c, after Townsend and Sutton. The solid curve is drawn for the BCS theory.

Energy Gap

The energy gap in superconductors is of an entirely different nature than the energy gap in insulators.[13] The argument of the exponential in the heat capacity is found[14] to be one-half of the gap. In Fig. 10b we see that the heat capacity of gallium varies as $e^{-\Delta/k_B T}$, with $\Delta \cong 1.4 k_B T_c$. Thus the gap is $E_g \equiv 2\Delta = 2(1.4 k_B T_c)$, or $E_g/k_B T_c = 2.8$. This value of the ratio $E_g/k_B T_c$ is representative. Values of the energy gaps[15] in several superconductors are given in Table 3; the values were obtained by the electron tunneling method to be described. The quantity Δ is often called the **energy gap parameter.**

The transition in zero magnetic field from the superconducting state to the normal state is observed to be a second-order phase transition. At a second-order transition there is no latent heat, but there is a discontinuity in the heat capacity, evident in Fig. 10a. Further, the energy gap decreases continuously to zero as the temperature is increased to the transition temperature T_c, as in Fig. 12. A first-order transition would be characterized by a latent heat and by a discontinuity in the energy gap.

---

[13] In an insulator the gap is tied to the lattice; in a superconductor the gap is tied to the Fermi gas.

[14] From theory and from comparison with optical and other determinations of the gap.

[15] For a review, see D. H. Douglass, Jr., and L. M. Falicov, *Progress in low temperature physics* **4**, 97–193 (1964).

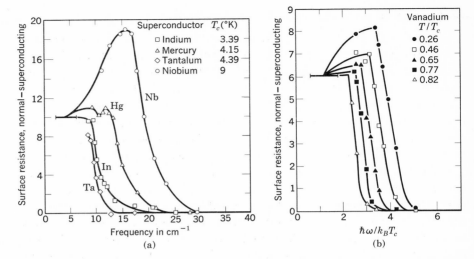

Figure 13 At high frequencies the surface resistivity of a superconductor approaches that of a normal metal. (a) Low temperature results on relative scale of (normal state surface resistance) minus (superconducting state surface resistance), as a function of frequency. The curves have been normalized so that the value at the lowest frequency is the same for each metal. (b) The same quantity for vanadium, as a function of temperature. (After Richards and Tinkham.)

Microwave and Infrared Properties

The existence of an energy gap in superconductors means that photons of energy less than the gap energy are not absorbed. This has been established experimentally by Glover and Tinkham, and by several other workers. Nearly all the photons incident are reflected as for any metal because of the impedance mismatch at the boundary between vacuum and metal, but for a very thin ($\sim$20 Å) film more photons are transmitted in the superconducting state than in the normal state.

For photon energies less than the energy gap, the resistivity of a super-conductor vanishes at absolute zero. Experimental results in the far infrared[16] are shown in Fig. 13. Results in the microwave region[17] at different temperatures are shown in Fig. 14. At $T \ll T_c$ the resistance in the superconducting state has a sharp threshold at the gap energy. Photons of lower energy see a resistanceless surface. Photons of higher energy see a resistance which approaches that of the normal state because such photons cause transitions to unoccupied "normal" energy levels above the gap. As the temperature is

[16] P. L. Richards and M. Tinkham, Phys. Rev. **119**, 575 (1960).
[17] M. A. Biondi and M. P. Garfunkel, Phys. Rev. Letters **2**, 143 (1959).

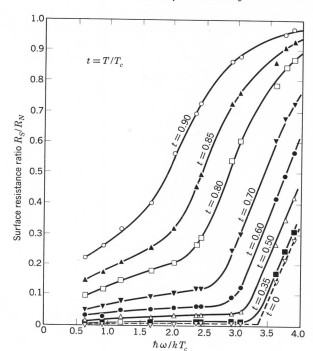

Figure 14 Isotherms of surface resistance ratio of aluminum versus photon frequency. (After Biondi and Garfunkel.)

increased not only does the gap decrease in energy (as in Fig. 12) but the resistivity for photons below the gap no longer vanishes, except at zero frequency. At zero frequency the superconducting electrons short-circuit any normal electrons that have been thermally excited above the gap. At finite frequencies the inertia of the superconducting electrons prevents them from completely screening the electric field, so that thermally excited normal electrons now can absorb energy.

Isotope Effect

It has been observed that the critical temperature of superconductors varies with isotopic mass. The first observations were made by Maxwell[18] and by Reynolds and co-workers.[19] In mercury T_c varies from 4.185°K to 4.146°K as the average atomic mass M varies from 199.5 to 203.4 atomic mass units. The transition temperature changes smoothly when we mix different isotopes of the same element.

The experimental results within each series of isotopes may be fitted by a relation of the form

$$M^\alpha T_c = \text{constant}, \tag{2}$$

[18] E. Maxwell, Phys. Rev. **78**, 477 (1950).
[19] Reynolds, Serin, Wright, and Nesbitt, Phys. Rev. **78**, 487 (1950).

Table 4   Isotope effect in superconductors

Experimental values of α in $M^{\alpha}T_c$ = constant, where M is the isotopic mass. (After a tabulation by J. W. Garland, Jr., Phys. Rev. Letters **11**, 114 (1963), with revisions suggested by Dr. V. Compton.)

Substance	α	Substance	α
Zn	0.45 ± 0.05	Ru	0.00 ± 0.05
Cd	0.32 ± 0.07	Os	0.15 ± 0.05
Sn	0.47 ± 0.02	Mo	0.33
Hg	0.50 ± 0.03	Nb_3Sn	0.08 ± 0.02
Pb	0.49 ± 0.02	Mo_3Ir	0.33 ± 0.03
Tl	0.61 ± 0.10	Zr	0.00 ± 0.05

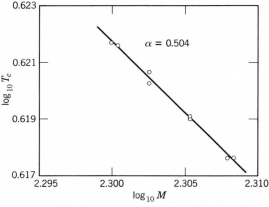

Figure 15 Log-log plot of transition temperature versus average mass number for separated isotopes of mercury. [After Reynolds, Serin, and Nesbitt, Phys. Rev. **84**, 691 (1951).]

as shown for mercury in Fig. 15. Observed values of α are given in Table 4. From the dependence of T_c on the isotopic mass we learn that lattice vibrations and hence electron-lattice interactions are deeply involved in superconductivity.[20] There is no other reason for the superconducting transition temperature to depend on the number of neutrons in the nucleus. The original simple BCS model gave the result $T_c \propto \theta_{\text{Debye}} \propto M^{-\frac{1}{2}}$, so that $\alpha = \frac{1}{2}$ in (2), but the inclusion of Coulomb interactions between the electrons changes the relation. There is nothing sacred about $\alpha = \frac{1}{2}$.

THEORETICAL SURVEY

A theoretical understanding of the phenomena associated with superconductivity has been reached in several ways. Certain results follow directly from thermodynamics. Many important results can be described by phenomeno-

---

[20] The absence of an isotope effect in Ru and Os has been accounted for in terms of the band structure of these metals. J. W. Garland, Jr., Phys. Rev. Letters **11**, 114 (1963); see also W. I. McMillan, Phys. Rev. **167**, 331 (1968).

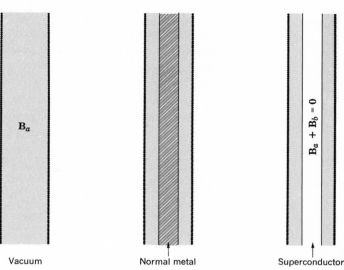

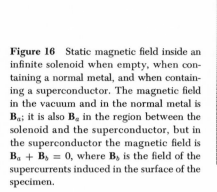

Figure 16 Static magnetic field inside an infinite solenoid when empty, when containing a normal metal, and when containing a superconductor. The magnetic field in the vacuum and in the normal metal is $\mathbf{B}_a$; it is also $\mathbf{B}_a$ in the region between the solenoid and the superconductor, but in the superconductor the magnetic field is $\mathbf{B}_a + \mathbf{B}_b = 0$, where $\mathbf{B}_b$ is the field of the supercurrents induced in the surface of the specimen.

logical equations: the London equations and the Landau-Ginzburg equations. A successful quantum theory of superconductivity was given by Bardeen, Cooper and Schrieffer, and has provided the basis for subsequent work. Our survey will have to be sketchy because of the advanced level of the theory.

Thermodynamics of the Superconducting Transition

It was demonstrated experimentally by van Laer and Keesom[21] that the transition between the normal and superconducting states is thermodynamically reversible, just as the transition between liquid and vapor phases of a substance is reversible under conditions of slow evaporation. The Meissner effect also suggests that the transition is reversible. Thus we may apply thermodynamics[22] to the transition, and we thereby obtain an expression for the entropy difference between normal and superconducting states in terms of the critical field curve H_c versus T. We treat only a type I superconductor with a complete Meissner effect, so that $\mathbf{B} = 0$ inside the superconductor (Fig. 16). We shall see that the critical field[23] H_c is a valuable quantitative measure of the energy difference between the superconducting and normal states at absolute zero.

The stabilization energy of the superconducting state with respect to the normal state can be determined calorimetrically or magnetically. Direct measurements of the heat capacity are made for the normal metal in a magnetic field

[21] P. H. van Laer and W. H. Keesom, Physica **5**, 993 (1938).

[22] C. Gorter and H. B. G. Casimir, Physica **1**, 306 (1934). The result is analogous to the vapor pressure equation for dp/dT.

[23] The symbol H_c will always refer to a bulk specimen, never to a thin film. For type II superconductors H_c is understood to be the **thermodynamic critical field** which can be determined from the stabilization energy.

and for the superconductor in zero magnetic field. From their difference we compute the energy difference at absolute zero, the **stabilization energy** of the superconducting state.[24] It is also possible to obtain the stabilization energy and the free energy simply from the value of the applied magnetic field that will destroy the superconducting state and will carry the specimen into the normal state. The argument depends on the important property of nearly perfect diamagnetism exhibited by type I superconductors. According to the Meissner effect the magnetic induction $\mathbf{B}$ in a bulk superconductor is zero, so that a superconductor simulates a perfect diamagnetic material. In (1) we obtained results in a limit that applies to a thin film or a long needle oriented parallel to the applied field:

$$(\text{CGS}) \qquad \mathbf{B} \equiv \mathbf{B}_a + 4\pi\mathbf{M} = 0 \; ; \qquad \mathbf{M} = -\frac{\mathbf{B}_a}{4\pi} \; ; \tag{3}$$

$$(\text{SI}) \qquad \mathbf{B} \equiv \mathbf{B}_a + \mu_0\mathbf{M} = 0 \; ; \qquad \mathbf{M} = -\frac{\mathbf{B}_a}{\mu_0} \, .$$

The simplest way to understand the effect of an applied magnetic field $\mathbf{B}_a$ on the transition from superconductor to normal is to consider the work done on a superconductor when it is brought from a position at infinity (where the applied field is zero) to a position $\mathbf{r}$ in the field of a permanent magnet. The work done in the process (Fig. 17) is

$$W = -\int_0^{\mathbf{B}_a} \mathbf{M} \cdot d\mathbf{B}_a \; , \tag{4}$$

per unit volume of specimen.[25] This work appears in the energy of the magnetic field. The thermodynamic identity for the process is

$$dU = T\,dS - \mathbf{M} \cdot d\mathbf{B}_a \; , \tag{5}$$

or, for a superconductor with $\mathbf{M}$ related to $\mathbf{B}_a$ by (3) we have

$$(\text{CGS}) \qquad dU_S = T\,dS + \frac{1}{4\pi}B_a\,dB_a \; ; \tag{6}$$

$$(\text{SI}) \qquad dU_S = T\,dS + \frac{1}{\mu_0}B_a\,dB_a \, .$$

---

[24] We assume in using the heat capacity measurements that the thermodynamic properties of the normal state are approximately independent of the field.

[25] This work is discussed carefully in Chapters 22 and 23 of C. Kittel, *Thermal physics*, Wiley, 1969.

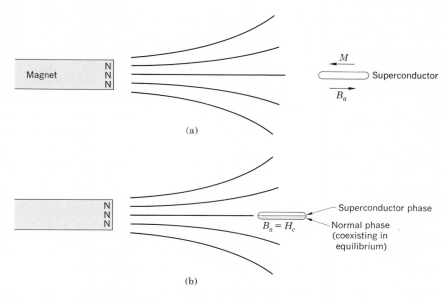

(a)

(b)

Figure 17   (a) A superconductor in which the Meissner effect is complete has $B = 0$, as if the magnetization were $M = -B_a/4\pi$, in CGS units. The work done on the superconductor in a displacement from infinity to the position where the field of the permanent magnet is B_a is

(CGS) $$W = -\int \mathbf{M} \cdot d\mathbf{B}_a = \frac{1}{8\pi} B_a^2 \; ,$$

per unit volume. (b) When the applied field reaches the value B_{ac}, the normal state can coexist in equilibrium with the superconducting state. In coexistence the free energy densities are equal: $F_N(T, B_{ac}) = F_S(T, B_{ac})$.

Thus at absolute zero, where $T\,dS = 0$, the increase in the energy density of the superconductor is

(CGS) $$U_S(B_a) - U_S(0) = \frac{1}{8\pi} B_a^2 \; ; \tag{7}$$

(SI) $$U_S(B_a) - U_S(0) = \frac{1}{2\mu_0} B_a^2 \; ,$$

on being brought from a position where the applied field is zero to a position where the applied field is B_a.

Now consider a normal nonmagnetic metal. If we neglect the small susceptibility[26] of a metal in the normal state, then $M = 0$ and the energy of the normal metal is independent of field. In particular at the critical field we

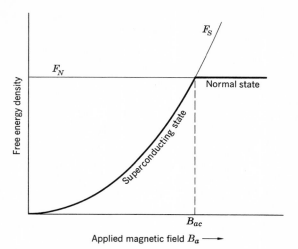

Applied magnetic field B_a —→

Figure 18 The free energy density F_N of a nonmagnetic normal metal is approximately independent of the intensity of the applied magnetic field B_a. At a temperature $T < T_c$ the metal is a superconductor in zero magnetic field, so that $F_S(T,0)$ is lower than $F_N(T,0)$. An applied magnetic field increases F_S by $B_a{}^2/8\pi$, in CGS units, so that

(CGS) $$F_S(T, B_a) = F_S(T, 0) + \frac{1}{8\pi} B_a{}^2 \ .$$

If B_a is larger than the critical field B_{ac} the free energy density is lower in the normal state than in the superconducting state, and now the normal state is the stable state. The origin of the vertical scale in the drawing is at $F_S(T, 0)$. The figure applies equally to U_S and U_N at $T = 0$.

have

$$U_N(B_{ac}) = U_N(0) \ . \tag{8}$$

The results (7) and (8) are all we need to determine the stabilization energy of the superconducting state at absolute zero. At the critical value B_{ac} of the applied magnetic field the energies are equal in the normal and superconducting states:

(CGS) $$U_N(B_{ac}) = U_S(B_{ac}) = U_S(0) + \frac{1}{8\pi} B_{ac}{}^2 \ ; \tag{9}$$

(SI) $$U_N(B_{ac}) = U_S(B_{ac}) = U_S(0) + \frac{1}{2\mu_0} B_{ac}{}^2 \ .$$

In SI units $H_c \equiv B_{ac}/\mu_0$, whereas in CGS units $H_c \equiv B_{ac}$. The specimen is stable in either state when the applied field is equal to the critical field. Now by (8)

[26] This is an adequate assumption for type I superconductors. In type II superconductors $H_c \ll H_{c2}$ and in high fields the change in spin paramagnetism of the conduction electrons (Chap. 15) lowers the energy of the normal phase significantly with respect to the superconducting phase. In some type II superconductors the upper critical field is limited by this effect. Clogston has suggested that $H_{c2}(\text{max}) = 18{,}400\, T_c$, where H_{c2} is in gauss and T_c in deg K. See A. M. Clogston, Phys. Rev. Letters **9**, 266 (1962); B. Chandrasekhar, Appl. Phys. Letters **1**, 7 (1962).

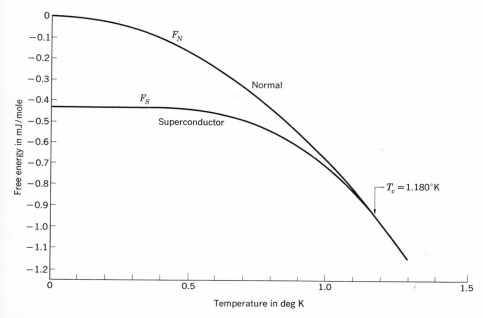

Figure 19 Experimental values of the free energy as a function of temperature for aluminum in the superconducting state and in the normal state. Below the transition temperature $T_c = 1.180°K$ the free energy is lower in the superconducting state. The two curves merge at the transition temperature, so that the phase transition is second order (there is no latent heat of transition at T_c). The curve F_S is measured in zero magnetic field, and F_N is measured in a magnetic field sufficient to put the specimen in the normal state; however, the value of F_N is essentially independent of the magnetic field intensity. (Courtesy of N. E. Phillips)

it follows that

$$\text{(CGS)} \quad U_N(0) = U_S(0) + \frac{1}{8\pi} B_{ac}^2 \; ;$$

$$\Delta U \equiv U_N(0) - U_S(0) = \frac{1}{8\pi} B_{ac}^2 \; , \quad (10)$$

where ΔU is the stabilization energy of the superconducting state at absolute zero, per unit volume of specimen. As an example, we see from Table 1 that B_{ac} for aluminum at absolute zero is 105 gauss, so that

$$\text{(CGS)} \quad \Delta U = \frac{(105)^2}{8\pi} \cong 440 \text{ ergs/cm}^3 \; , \quad (11)$$

in excellent agreement with the result of thermal measurements, 430 ergs/cm³.

At a finite temperature the normal and superconducting phases are in equilibrium when their free energies $F = U - TS$ are equal. The free energies of the two phases are sketched in Fig. 18 as a function of the magnetic field. Experimental curves of the free energies of the two phases in aluminum as a function of the temperature are given in Fig. 19. The slopes dF/dT are equal at the transition so that there is no latent heat.

London Equation

We saw that the Meissner effect implies a magnetic susceptibility $\chi = -1/4\pi$ in CGS in the superconducting state or, in SI, $\chi = -1$. This sweeping assumption tends to cut off further discussion, and it does not account for the flux penetration observed in thin films. Can we modify a constitutive equation of electrodynamics (such as Ohm's law) in some way to obtain the Meissner effect? We do not want to modify the Maxwell equations themselves.

Electrical conduction in the normal state of a metal is described by Ohm's law:

$$\mathbf{j} = \sigma \mathbf{E} \ . \tag{12}$$

We need to modify this drastically to describe conduction and the Meissner effect in the superconducting state. Let us make a *postulate* and see what happens. We postulate that in the superconducting state the current density is directly proportional to the vector potential $\mathbf{A}$ of the local magnetic field. In CGS units we write the constant of proportionality as $-c/4\pi\lambda_L^2$ for reasons that will become clear. Here c is the speed of light and λ_L is a constant with the dimensions of length. In SI units we write $-1/\mu_0\lambda_L^2$. Thus

(CGS)
$$\mathbf{j} = -\frac{c}{4\pi\lambda_L^2} \mathbf{A} \ ; \tag{13}$$

(SI)
$$\mathbf{j} = -\frac{1}{\mu_0\lambda_L^2} \mathbf{A} \ .$$

This is the London equation.[27] The properties of the vector potential are derived in Advanced Topic I. We express (13) in another way by taking the curl of both sides to obtain

(CGS)
$$\operatorname{curl} \mathbf{j} = -\frac{c}{4\pi\lambda_L^2} \mathbf{B} \ ; \tag{14}$$

(SI)
$$\operatorname{curl} \mathbf{j} = -\frac{1}{\mu_0\lambda_L^2} \mathbf{B} \ .$$

First we show that the London equation leads to the Meissner effect. By a Maxwell equation we know that

(CGS) $\quad \operatorname{curl} \mathbf{B} = \dfrac{4\pi}{c} \mathbf{j} \ ; \qquad$ (SI) $\quad \operatorname{curl} \mathbf{B} = \mu_0 \mathbf{j}_0 \ ; \tag{15}$

[27] F. London and H. London, Proc. Roy. Soc. (London) **A149**, 72 (1935); Physica **2**, 341 (1935). The London equation (13) is understood to be written with the vector potential in the London gauge in which div $\mathbf{A} = 0$, and $\mathbf{A}_n = 0$ on any external surface through which no external current is fed. The subscript n denotes the component normal to the surface. Thus div $\mathbf{j} = 0$ and $\mathbf{j}_n = 0$, the actual physical boundary conditions. The form (13) applies to a simply-connected superconductor; additional terms may be present in a ring or cylinder, but (14) holds true independent of geometry.

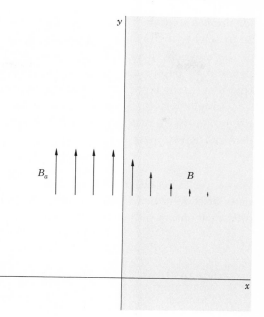

Figure 20 Penetration of an applied magnetic field into a semi-infinite super-conductor. The penetration depth λ is defined as the distance in which the field decreases by the factor e^{-1}. Typically, $\lambda \approx 500 \text{ Å}$ in a pure super-conductor.

under static conditions. We take the curl of both sides to obtain

(CGS) $$\text{curl curl } \mathbf{B} = -\nabla^2 \mathbf{B} = \frac{4\pi}{c} \text{ curl } \mathbf{j} \; ; \tag{16}$$

(SI) $$\text{curl curl } \mathbf{B} = -\nabla^2 \mathbf{B} = \mu_0 \text{ curl } \mathbf{j} \; ,$$

which may be combined with (14) to give for a superconductor

$$\nabla^2 \mathbf{B} = \frac{1}{\lambda_L{}^2} \mathbf{B} \; . \tag{17}$$

This equation accounts for the Meissner effect because it does not allow a solution uniform in space, so that **a uniform magnetic field cannot exist in a superconductor.** That is, $\mathbf{B}(\mathbf{r}) = \mathbf{B}_0 = $ constant is not a solution of (17) unless the constant field $\mathbf{B}_0$ is identically zero. The result follows because $\nabla^2 \mathbf{B}_0$ is always zero, but $\mathbf{B}_0/\lambda_L{}^2$ is not zero unless $\mathbf{B}_0$ is zero.

In the pure superconducting state the only field allowed is exponentially damped as we go in from an external surface. Let a semi-infinite superconductor occupy the space on the positive side of the x axis, as in Fig. 20. If $B(0)$ is the field at the plane boundary, then the field inside the superconductor is

$$B(x) = B(0) \exp(-x/\lambda_L) \; , \tag{18}$$

for this is a solution[28] of (17). In this example the magnetic field is assumed to be parallel to the boundary. Thus we see λ_L measures the depth of penetration of the magnetic field; it is known as the **London penetration depth.** Actual penetration depths are not described precisely by λ_L alone, for the London equation is now known to be somewhat oversimplified.

Insight into the physical basis of the London equation and the order of magnitude of λ_L is obtained from a simple observation. The electric current density is given quite generally by

$$\mathbf{j} = nq\mathbf{v} \; , \tag{19}$$

where n is the concentration of carriers of charge q. In the presence of a magnetic field described by the vector potential $\mathbf{A}$, the velocity $\mathbf{v}$ is related to the total momentum $\mathbf{p}$ by

$$(\text{CGS}) \qquad \mathbf{p} = m\mathbf{v} + \frac{q}{c}\mathbf{A} \; ; \qquad \mathbf{v} = \frac{1}{m}\left(\mathbf{p} - \frac{q}{c}\mathbf{A}\right) \, , \tag{20}$$

according to (I.16). Thus (19) may be written as

$$(\text{CGS}) \qquad \mathbf{j} = \frac{nq}{m}\mathbf{p} - \frac{nq^2}{mc}\mathbf{A} \; . \tag{21}$$

In SI the quantity c is replaced by unity.

The London equation is a consequence of (21) if $\mathbf{p} = 0$ in the superconducting state, in the London gauge for $\mathbf{A}$. The accomplishment of the quantum theory of superconductivity is to explain why the total momentum $\mathbf{p}$ is zero in the superconducting state, although it is not equal to zero in the normal state. With $\mathbf{p} = 0$, Eq. (21) reduces to

$$(\text{CGS}) \quad \mathbf{j} = -\frac{nq^2}{mc}\mathbf{A} \; ; \qquad\qquad (\text{SI}) \quad \mathbf{j} = -\frac{nq^2}{m}\mathbf{A} \, , \tag{22}$$

which is the London equation (13) with

$$(\text{CGS}) \quad \lambda_L{}^2 = \frac{mc^2}{4\pi nq^2} \; ; \qquad\qquad (\text{SI}) \quad \lambda_L{}^2 = \frac{m}{\mu_0 nq^2} \; . \tag{23}$$

If the effective carriers are pairs of electrons with charge $q = -2e$, the concentration n in (23) is one-half of the conduction electron concentration. The mass m is then twice the electron mass. A typical experimental value[29] of the penetration depth in a metal is 500Å, of the expected order of magnitude.

---

[28] This expression and the London equation cannot be entirely correct because they neglect the mean free path of the electrons, and they neglect restrictions on localization imposed by the uncertainty principle. A coherence length is introduced in the theory for these reasons. The London equation relation between $\mathbf{j}(\mathbf{r})$ and $\mathbf{A}(\mathbf{r})$ is not strictly valid for fields that vary rapidly in space; instead, some form of local average of $\mathbf{A}$ over a coherence length should determine $\mathbf{j}(\mathbf{r})$.

[29] J. M. Lock, Proc. Roy. Soc. (London) **A208**, 391 (1951).

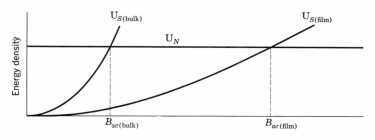

Figure 21 Increase of the critical field in a thin film as compared with a bulk specimen of superconductor. In the film the applied magnetic field is incompletely screened: the reduced screening in turn reduces the magnetic field dependence of the free energy as compared with the bulk value. At absolute zero the state of lowest energy density is the stable state at the applied magnetic field considered.

An applied magnetic field B_a will penetrate a thin film fairly uniformly if the thickness is much less than λ_L; thus in a thin film the Meissner effect is not complete. In this situation the induced field is much less than B_a, and there is little effect of B_a on the energy density of the superconducting state, so that (7) does not apply. From our thermodynamic argument (Fig. 21) it follows that the critical field H_c of thin films in parallel magnetic fields should be very high, as in Fig. 33 below.

Coherence Length

The London penetration depth λ_L is a fundamental length that characterizes a superconductor. Another independent length of equal importance is the **coherence length** ξ. The coherence length is a measure of the distance within which the gap parameter cannot change drastically in a spatially-varying magnetic field. The London equation is a *local* equation: it relates the current density at a point $\mathbf{r}$ to the vector potential at the same point. So long as $\mathbf{j}(\mathbf{r})$ is given as a constant times $\mathbf{A}(\mathbf{r})$, the current is required to follow exactly any variation in the vector potential. But the coherence length ξ is a measure of the range over which we should average $\mathbf{A}$ to obtain $\mathbf{j}$. Actually, two different coherence lengths are involved in the theory, but we shall not go into this.

Any spatial variation in the state of an electronic system requires extra kinetic energy.[30] It is reasonable to restrict the spatial variation of $\mathbf{j}(\mathbf{r})$ in such a way that the extra energy is less than the stabilization energy of the superconducting state. A suggestive argument (based on the uncertainty principle) for the coherence length at absolute zero follows: We compare the plane wave $\psi(x) = e^{ikx}$ with the strongly modulated wavefunction:

$$\varphi(x) = 2^{-\frac{1}{2}}(e^{i(k+q)x} + e^{ikx}) \ . \tag{24}$$

[30] A modulation of an eigenfunction increases the kinetic energy because the modulation will increase the integral of $d^2\varphi/dx^2$.

The probability density associated with the plane wave is uniform in space:

$$\psi^*\psi = e^{-ikx}e^{ikx} = 1 \ , \tag{25}$$

whereas $\varphi^*\varphi$ is modulated with the wavevector q:

$$\varphi^*\varphi = \tfrac{1}{2}(e^{-i(k+q)x} + e^{-ikx})(e^{i(k+q)x} + e^{ikx})$$
$$= \tfrac{1}{2}(2 + e^{iqx} + e^{-iqx}) = 1 + \cos qx \ . \tag{26}$$

The kinetic energy of the wave $\psi(x)$ is

$$\epsilon = \frac{\hbar^2}{2m}k^2 \ ; \tag{27}$$

the kinetic energy of the modulated density distribution is higher, for

$$\int dx \, \varphi^* \left(-\frac{\hbar^2}{2m}\frac{d^2}{dx^2}\right)\varphi = \frac{1}{2}\left(\frac{\hbar^2}{2m}\right)[(k+q)^2 + k^2] \cong \frac{\hbar^2}{2m}k^2 + \frac{\hbar^2}{2m}kq \ , \tag{28}$$

where we neglect q^2 on the assumption that $q \ll k$.

The increase of energy required to modulate is $\hbar^2 kq/2m$. If this increase exceeds the energy gap E_g, superconductivity will be destroyed. The critical value q_0 of the modulation wavevector is defined by

$$\frac{\hbar^2}{2m}k_F q_0 = E_g \ . \tag{29}$$

We define an **intrinsic coherence length** ξ_0 related to the ciritical modulation by $\xi_0 = 2\pi/q_0$. From (29) we have

$$\xi_0 = \frac{2\pi\hbar^2 k_F}{2mE_g} = \frac{\pi\hbar v_F}{E_g} \ , \tag{30}$$

where v_F is the electron velocity at the Fermi surface. On the BCS theory[31] a similar result is found:

$$\boxed{\xi_0 = \frac{2\hbar v_F}{\pi E_g} \ .} \tag{31}$$

Calculated values of ξ_0 from (31) are given in Table 5.

The intrinsic coherence length ξ_0 is characteristic of a pure superconductor. In impure materials and in alloys the coherence length ξ is shorter than ξ_0. This may be understood qualitatively: in impure material the electron eigenfunctions already have wiggles in them. We can construct a given localized variation of current density with less energy from wavefunctions with wiggles than from smooth wavefunctions.

The coherence length ξ first appeared in the solutions of a pair of phenomenological equations known as the Landau-Ginzburg equations; these equations

---

[31] See *QTS*, pp. 173–174.

Table 5   Calculated intrinsic coherence length and London penetration depth, at absolute zero

(After R. Meservey and B. B. Schwartz)

Metal	Intrinsic Pippard coherence length ξ_0, in 10^{-6} cm	London penetration depth λ_L, in 10^{-6} cm	$\dfrac{\xi_0}{\lambda_L}$
Sn	23.	3.4	6.2
Al	160.	1.6	100.
Pb	8.3	3.7	2.2
Cd	76.	11.0	6.9
Nb	3.8	3.9	0.98

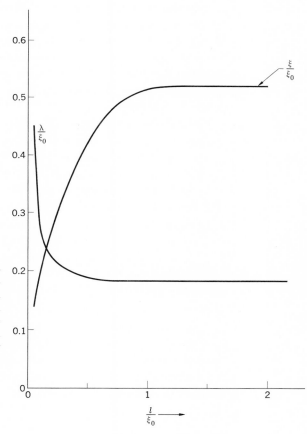

Figure 22   A schematic variation of the penetration depth λ and the coherence length ξ as functions of the mean free path l of the conduction electrons in the normal state. All lengths are in units of ξ_0, the intrinsic coherence length. The curves are sketched for $\xi_0 = 10\lambda_L$. For short mean free paths the coherence length becomes shorter and the penetration depth becomes longer. The increase in the ratio λ/ξ favors type II superconductivity.

also follow from the BCS theory. They describe the structure of the transition layer between normal and superconducting phases in contact. The coherence length and the actual penetration depth λ are found theoretically to depend on the mean free path of the electrons measured in the normal state; the relationships are indicated in Fig. 22.

BCS Theory of Superconductivity

We have given above a simple unified discussion of the rich experimental and phenomenological knowledge of superconductors. There seems at present to be little need for fundamentally different theories of superconductivity for the different rows and columns of the periodic table, for pure metals and alloys, or for different crystal structures. These considerations affect the quantitative details of the superconducting properties, but apparently only within the framework of a general theory which we now discuss. The basis of a general quantum theory of superconductivity was laid by the classic 1957 papers of Bardeen, Cooper, and Schrieffer.[32]

The accomplishments of the BCS theory include:

1. An attractive interaction[33] between electrons can lead to a ground state of the entire electronic system which is separated from excited states by an energy gap. The critical field, the thermal properties,[34] and most of the electromagnetic properties are consequences of the energy gap. The mathematics that leads to the energy gap and to the special BCS ground state is discussed in Advanced Topic L. In special circumstances, superconductivity may occur without an actual energy gap.

2. The electron-lattice-electron interaction is attractive and leads to an energy gap of the observed magnitude. The indirect interaction proceeds roughly as follows: one electron interacts with the lattice and deforms it; a second electron sees the deformed lattice and adjusts itself to take advantage of the deformation to lower its energy. Thus the second electron interacts with the first electron via the lattice deformation field or phonon field. The interactions are dynamic, so that the atomic mass enters the theory of the interaction in a natural way and causes an isotope effect.

3. The penetration depth and the coherence length emerge as natural consequences of the BCS theory. The London equation (13) is obtained for magnetic fields that vary slowly in space. Thus the central phenomenon in superconductivity, the Meissner effect, is obtained in a natural way.[35]

4. The criterion for the occurrence of superconductivity and for the magnitude of the transition temperature in an element or alloy involves the electron density of orbitals $\mathfrak{D}(\epsilon_F)$ at the Fermi level and the electron-lattice interaction U, which can be estimated from the electrical resistivity. For $U\mathfrak{D}(\epsilon_F) \ll 1$ the BCS theory predicts

$$T_c = 1.14\,\theta \exp\left[-1/U\mathfrak{D}(\epsilon_F)\right]\ ,\tag{32}$$

[32] J. Bardeen, L. N. Cooper, and J. R. Schrieffer, Phys. Rev. **106**, 162 (1957); **108**, 1175 (1957).

[33] Strictly speaking, the net interaction does not have to be attractive, but it has to be less repulsive for the superconducting state than for the normal state.

[34] The observed ratios of $E_g(0)/k_BT_c$ listed in Table 3 are close to the prediction of the BCS theory. The character of the normal/superconductor phase transition is described correctly.

[35] The value of the penetration depth is no longer given simply by Eq. (23), but the qualitative results are the same.

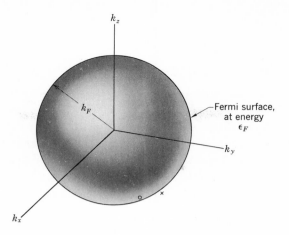

Figure 23 Ground state of a noninteracting Fermi gas; all one-particle states with $k \leq k_F$ are occupied; for $k > k_F$ all states are vacant. An excited state of arbitrarily low excitation energy can be formed by removing an electron from the point o just inside the Fermi surface and placing it at x just outside the Fermi surface.

where θ is the Debye temperature. This relation assumes that U is an attractive interaction (positive as used here); otherwise the ground state is not superconducting. The result for T_c is satisfied at least qualitatively by the experimental data. There is an interesting apparent paradox: the higher the resistivity at room temperature,[36] the more likely it is that a metal will be a superconductor when cooled. This holds only if we compare metals with comparable densities of conduction electrons. Another simple conclusion is that elements with an even number of valence electrons per atom are less likely to be superconductors than elements with an odd number of valence electrons: this is just another way of saying that an even number of valence electrons may favor filling a Brillouin zone, so that $\mathfrak{D}(\epsilon_F)$ will be small.

5. Magnetic flux through a superconducting ring is quantized and the effective unit of charge is $2e$ rather than e. The BCS ground state involves pairs of electrons; thus flux quantization[37] in terms of the pair charge $2e$ is a consequence of the theory.

BCS Ground State

We saw in Chapter 7 that the ground state of a Fermi gas of noninteracting electrons is just the filled Fermi sea (Fig. 23). This state, which we call the Fermi state, allows arbitrarily small excitations—we can form an excited state by taking an electron from the Fermi surface and raising it just above the Fermi surface. The BCS theory shows that with an appropriate attractive interaction between electrons the ground state differs from the Fermi state and is separated by a finite energy E_g from its lowest excited state.

[36] Because the resistivity at room temperature is a measure of the electron-phonon interaction. In the BCS relation quoted for T_c we should take $\mathfrak{D}(\epsilon_F)$ as the density of orbitals of one spin.

[37] Early experiments on flux quantization are reported by H. S. Deaver and W. M. Fairbank, Phys. Rev. Letters **7**, 43 (1961); R. Doll and M. Näbauer, Phys. Rev. Letters **7**, 51 (1961). An elementary account of the theory of flux quantization is given in Advanced Topic J; see also K.

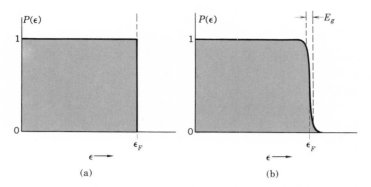

Figure 24 (a) Probability P that a one-electron state of energy ϵ is occupied in the ground state of the noninteracting Fermi gas; (b) the BCS ground state differs from the Fermi state in a region of width of the order of the energy gap E_g. (Both curves are for absolute zero.)

The formation of the BCS ground state is indicated from one point of view in Fig. 24. The BCS state in (b) contains admixtures of one-electron orbitals from above as well as below the Fermi energy ϵ_F. At first sight the BCS state appears therefore to have a higher energy than the Fermi state. The comparison of (b) and (a) shows that the kinetic energy of the BCS state is higher than that of the Fermi state. But the attractive potential energy of the BCS state, although not represented in the figure, acts to lower the total energy of the BCS state with respect to the Fermi state. One-particle orbitals or pairs of orbitals somewhat above the Fermi energy are brought into the BCS state because the value of the energy gap tends to be proportional to the number of orbitals that participate in forming the BCS state. This point is explained in Advanced Topic L.

If the BCS ground state of a many-electron system is described in terms of the occupancy of one-particle orbitals, then those near ϵ_F are filled somewhat like a Fermi-Dirac distribution for some finite temperature. A central feature of the BCS state is that the one-particle orbitals are occupied in pairs: if an orbital with wavevector $\mathbf{k}$ and spin up is occupied, then the orbital with wavevector $-\mathbf{k}$ and spin down is also occupied. If $\mathbf{k}_{1\uparrow}$ is vacant, then $-\mathbf{k}_{1\downarrow}$ is also vacant.

Persistent Currents

Several arguments[38] can be given for the stability of persistent currents in a superconductor. The easiest, due to Landau, concerns the spectrum of elementary excitations in Figs. 25 and 26; this particular argument does not apply to gapless superconductors.[39] Consider a crystal of mass M that contains an imperfection, such as a phonon or an impurity. When current flows in the superconducting state the electron gas moves collectively with respect to the lattice. Let the lattice flow with velocity $\mathbf{v}$ relative to the electron gas. Friction will decrease the velocity only if the relative motion can generate excitations in the electron gas. To create an excitation of energy $E_\mathbf{k}$ and momentum $\hbar\mathbf{k}$ in a collision event we must have

$$\tfrac{1}{2}Mv^2 = \tfrac{1}{2}Mv'^2 + E_\mathbf{k} \; ; \qquad M\mathbf{v} = M\mathbf{v}' + \hbar\mathbf{k} \; , \tag{33}$$

from energy and momentum conservation.

If we combine these two equations we have

$$0 = \hbar\mathbf{k} \cdot \mathbf{v} + \frac{\hbar^2 k^2}{2M} + E_\mathbf{k} \; . \tag{34}$$

For $M \to \infty$ we can neglect the term in $1/M$. The lowest value of $\mathbf{v}$ for which the equation $E_\mathbf{k} = -\hbar\mathbf{k} \cdot \mathbf{v}$ can be satisfied is the **critical velocity**

$$v_c = \text{minimum value of } \frac{E_\mathbf{k}}{\hbar k} \; . \tag{35}$$

[38] The position has been very well stated by G. Rickayzen, *Theory of superconductivity,* Interscience, 1965: "The most difficult property of superconductors to understand is their infinite conductivity. How is it that the scattering mechanisms, impurities, phonons, etc., which are so effective in reducing currents in a normal metal, are powerless when the metal becomes superconducting? We can be sure that we shall never be able to show theoretically that superconductors do possess an infinite conductivity. We can never show that no scattering mechanism can reduce the current because there will always be the possibility that we have neglected some weak scattering mechanism. It is probably true that we will not be able to show that the resistance is less than the experimental maximum of 10^{-20} ohm-cm because this means listing and studying all mechanisms whose strength in normal metals is 10^{-11} times that of phonon scattering. The best we can hope for is to show that some of the more obvious scatterers, which certainly limit the conductivity in the normal state, do not (at least in some approximation) limit the conductivity in the superconductive state . . . .

"We shall also never be able to show experimentally that the conduction is infinite. In fact, the maximum we have quoted was obtained from the observation of the current in a ring. The upper limit for the conductivity of a simply connected superconductor is very much less . . . .

"Theoretically, the conductivity of rings and wires also differs. In wires we have only the ineffectiveness of the scatterers to explain the absence of resistance. In rings we have also the quantization of flux."

[39] Gapless superconductors may arise from the presence of magnetic impurities; see E. B. Hansen, "Infinite conductivity of ordinary and gapless superconductors," Physica **39**, 271 (1968); K. Maki, "Gapless superconductivity," in R. D. Parks, ed., *Superconductivity,* Dekker, 1969, p. 1035.

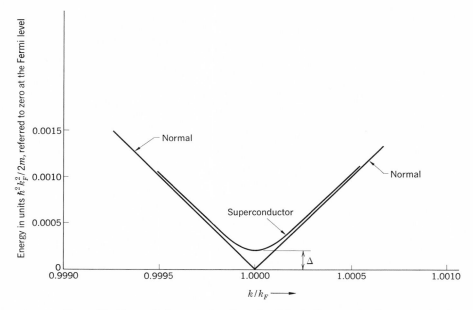

Figure 25 The excitation energies of quasiparticles in the normal and superconducting states, as a function of the wavevector. The zero of energy is taken for the ground state of a Fermi gas. An electron added to the system in the normal state gives an excitation with $k > k_F$ for which the energy is

$$\epsilon_k = \frac{\hbar^2}{2m}(k^2 - k_F{}^2) \cong \frac{\hbar^2}{m}k_F(k - k_F)$$

for $k - k_F \ll k_F$. An electron removed from the system in the normal state gives a hole-like excitation with $k < k_F$; the excitation energy is

$$\epsilon_k = \left(\frac{\hbar^2}{2m}\right)(k_F{}^2 - k^2) \cong \frac{\hbar^2}{m}k_F(k_F - k).$$

The quasiparticle excitation energy in the superconducting state is $E_k = (\epsilon_k{}^2 + \Delta^2)^{\frac{1}{2}}$, where Δ is the energy gap parameter. The graph is drawn for $\Delta = 0.0002\epsilon_F$.

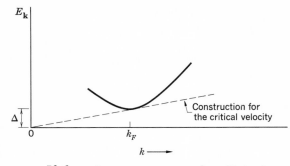

Figure 26 Spectrum of elementary excitations in a superconductor. The vertical axis is the energy above the ground state of one member of a pair of two excited particles. The horizontal axis is the magnitude of the wavevector. The dashed line has a slope equal to $\hbar v_c$, where v_c is the critical velocity.

If there is an energy gap then $E_\mathbf{k} > 0$, so that $v_c > 0$. Thus superconducting currents can flow with velocities less than v_c without risk of dissipation of energy by excitation of electrons from the superconducting state to the normal state. The values of the critical current density are quite high (Problem 2). Essentially the same argument applies to the excitation of a pair of electrons.

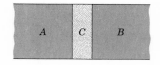

Figure 27   Two metals, *A* and *B*, separated by a thin layer of an insulator *C*.

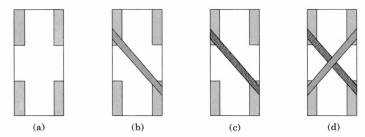

(a)   (b)   (c)   (d)

Figure 28   Preparation of an Al/Al₂O₃/Sn sandwich. (a) Glass slide with indium contacts. (b) An aluminum strip 1 mm wide and 1000 to 3000 Å thick has been deposited across the contacts. (c) The aluminum strip has been oxidized to form an Al₂O₃ layer 10 to 20 Å in thickness. (d) A tin film has been deposited across the aluminum film, forming an Al/Al₂O₃/Sn sandwich. The external leads are connected to the indium contacts; two contacts are used for the current measurement and two for the voltage measurement. The critical temperatures of Sn and Al are 3.7 and 1.2°K, respectively; between these two temperatures the Sn strip is superconducting and the Al strip is normal. The Al₂O₃ layer is an insulator. (After Giaever and Megerle.)

Single Particle Tunneling

Consider two metals separated by an insulator, as in Fig. 27. The insulator normally acts as a barrier to the flow of conduction electrons from one metal to the other. If the barrier is sufficiently thin (less than 10 or 20 Å) there is a significant probability that an electron which impinges on the barrier will pass from one metal to the other: this is called **tunneling.** The concept that particles can tunnel through potential barriers is as old as quantum mechanics. In many experiments the insulating layer is simply a thin oxide layer formed on one of two evaporated metal films, as in Fig. 28.

434

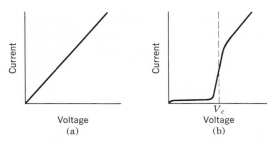

Figure 29 (a) Linear current-voltage relation for junction of normal metals separated by oxide layer; (b) current-voltage relation with one metal normal and the other metal superconducting.

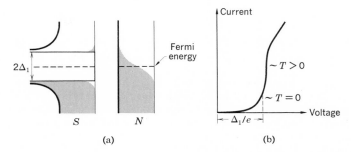

Figure 30 The density of orbitals and the current-voltage characteristic for a tunneling junction. In (a) the energy is plotted on the vertical scale and the density of orbitals on the horizontal scale. One metal is in the normal state and one in the superconducting state. (b) I versus V; the dashes indicate the expected break at $T = 0$. (After Giaever and Megerle.)

When both metals are normal conductors, the current-voltage relation of the sandwich or tunneling junction is ohmic at low voltages,[40] with the current directly proportional to the applied voltage (Fig. 29a).

Giaever[41] discovered that if one of the metals becomes superconducting the current-voltage characteristic changes from the straight line of Fig. 29a to the curve shown in Fig. 29b. Figure 30a contrasts the electron density of orbitals in the superconductor with that in the normal metal. In the superconductor there is an energy gap centered at the Fermi level. At absolute zero no current can flow until the applied voltage is $V = E_g/2e = \Delta/e$. The gap E_g corresponds to the break-up of a pair of electrons in the superconducting state, with the formation of two electrons, or an electron and a hole, in the normal state. The current starts when $eV = \Delta$. At temperatures different from zero there is a small current flow even at low voltages, because of electrons in the superconductor that are thermally excited across the energy gap.

[40] See, for example, I. Giaever and K. Megerle, Phys. Rev. **122**, 1101 (1961); J. Bardeen, Phys. Rev. Letters **6**, 57 (1961).

[41] I. Giaever, Phys. Rev. Letters **5**, 147, 464 (1960).

Another representation of the threshold voltage for tunneling is given in Fig. 25. One curve gives the excitation energy to add an electron or to add a hole to the Fermi sea of the normal metal, referred to the Fermi level as the zero of energy. The other curve gives the excitation energy of a quasiparticle in a superconductor referred to the same Fermi level. The threshold voltage to take an electron from the normal metal to the superconductor is given by $eV = \Delta$. The study of superconductors by single-electron tunneling has been highly successful. The results are in good agreement with the BCS theory. We discuss in Advanced Topic K the remarkable effects that arise in the tunneling of pairs of superconducting electrons, known as Josephson tunneling.

Type II Superconductors

There is no difference in the fundamental mechanism of superconductivity in type I and type II superconductors. In both types the mechanism is the electron-phonon-electron interaction. Both types have similar thermal properties at the superconductor-normal transition in zero magnetic field. But the Meissner effect is entirely different in the two types (Fig. 6). A good type I superconductor excludes a magnetic field until superconductivity is destroyed suddenly and completely, and then the field penetrates completely. A good type II superconductor excludes the field completely only for relatively weak fields, up to a field H_{c1}. Above H_{c1} the field is partially excluded (Fig. 6c), but the specimen remains electrically superconducting. At a much higher field, sometimes 100 kG or more, the flux penetrates completely and all superconductivity vanishes; this field is called H_{c2}. (An outer surface layer of the specimen may remain superconducting up to a still higher field H_{c3}.)

An important difference in the physical constitution of a type I and a type II superconductor is in the value of the mean free path of the conduction electrons in the normal state at low temperatures. If the coherence length is longer than the penetration depth, the superconductor will be type I. Most pure metals are type I. But if the mean free path is short, the coherence length is short and the penetration depth is great (Fig. 22); then the superconductor will be type II. We can sometimes change a metal from type I to type II by a modest addition of an alloying element. In Figure 6c the addition of 2 wt. percent of indium changes lead from type I to type II, although the transition temperature is scarcely changed at all. Neither would we expect the energy gap to be changed. Nothing fundamental has been done to the intrinsic electronic structure or to the superconducting nature of lead by this amount of alloying, but the behavior in a magnetic field has been changed drastically. The theory of type II superconductors was developed by Ginzburg, Landau, Abrikosov, and Gorkov. Later Kunzler and co-workers observed that Nb_3Sn wires can carry large supercurrents in fields approaching 100 kG.

Consider the interface between a region in the superconducting state and a region in the normal state in a metal specimen. The interface has an extra energy associated with it. This **surface energy** may be positive or negative; examples of both are plentiful. The surface energy decreases as the applied magnetic field is increased. A superconductor is type I if the surface energy is always positive, and type II if the surface energy becomes negative as the magnetic field is increased. To refer to Fig. 6b, the surface energy is negative for fields above H_{c1}.

The consequences of the sign of the surface energy are spectacular with respect to critical fields, although of no importance for the transition temperature. The sign of the surface energy is shown below to depend on the ratio of the penetration depth λ to the coherence length ξ, as in Fig. 31. The results are usually expressed in terms of the Ginzburg-Landau parameter

$$\kappa = \frac{\lambda}{\xi\sqrt{2}} \; .$$

In a type I superconductor $\kappa < 1/\sqrt{2}$, and in a type II superconductor $\kappa > 1/\sqrt{2}$. That is, type II behavior requires that the magnetic field penetrate farther than a coherence length into the superconductor. The factor $\sqrt{2}$ is historical.

The consequence of a negative interface energy took a long time to be understood. The total energy is lower the more interfaces are packed into the specimen, until the interfaces start pressing against each other. Many interfaces actually do form, and the normal regions are circular cores surrounded by superconducting phase. The induced superconducting currents flow in vortices around cores arranged in a fairly regular pattern. This condition of the superconductor is called the **vortex state.** It is quite different from the intermediate state,[42] which only can exist in a finite geometry.

---

[42] There is no special problem in imagining the nature of the structure of a type I superconductor when both superconducting and normal states coexist. Slabs in one state are intercalated between slabs of the other state. Because the interface energy is positive, the electronic energy would be lowered if the interfaces were removed. But in a finite geometry, for example a sphere, the magnetic energy of the system will increase if the interfaces are swept out and the whole specimen becomes superconducting. There is a range of field strengths in which the total energy of the system is lower with interfaces than without them. It would lead us astray to analyze the problem here, but the experimental and theoretical result is that for a sphere of type I material the pure superconducting state is stable when the applied field is less than $\frac{2}{3}H_c$; the normal state is stable when the applied field is greater than H_c; and slabs of the two states coexist in the **intermediate state** when the applied field is between $\frac{2}{3}H_c$ and H_c. The phrase intermediate state is applied only to superconductors with positive surface energy; it was discussed on pages 470 to 473 of the second edition of this text. The problem is like that of the origin of ferromagnetic domains, discussed in Chapter 16.

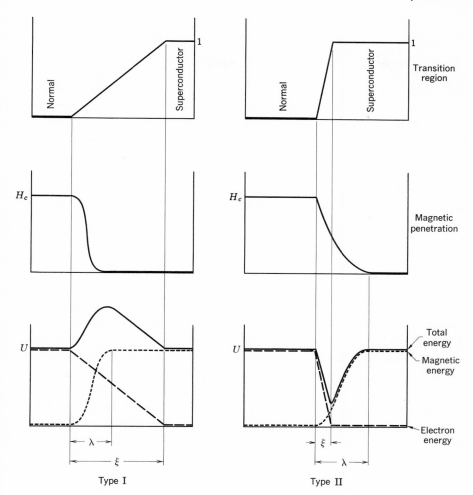

Figure 31 Energy relationships at a normal-superconductor interface in a type I superconductor and in a type II superconductor. Notice the increase in total energy for the interface in type I and the decrease in total energy for the interface in type II; thus the surface energy is positive for type I and negative for type II. (After J. L. Olsen and E. Fischer.)

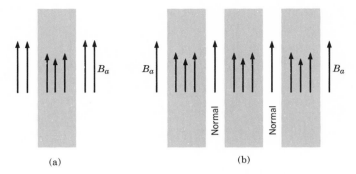

Figure 32 (a) Magnetic field penetration into a thin film of thickness equal to the penetration depth λ. The arrows indicate the intensity of the magnetic field. (b) Magnetic field penetration in a homogeneous bulk structure in the mixed or vortex state, with alternate layers in normal and superconducting states. The superconducting layers are thin in comparison with λ. The laminar structure is shown for convenience; the actual structure consists of rods of the normal state surrounded by the superconducting state. The N regions in the vortex state are not exactly normal, but are described by low values of Δ as in Fig. 38.

The energy of a bulk superconductor is increased in a magnetic field, provided that the field does not penetrate the specimen. The field penetration in films is the subject of Problems 1 and 4. A parallel field can penetrate a very thin film nearly uniformly (Fig. 32a), and then the energy of the superconducting film will increase only slowly as the external magnetic field is increased, causing a large increase in the field intensity required for the destruction of superconductivity (Figs. 21 and 33). In a thin film in the superconducting state the magnitude of the apparent magnetic susceptibility may be much smaller than $1/4\pi$ (or 1 in SI) because only part of the flux is expelled, but the film has the usual energy gap and will be resistanceless. The persistence of superconductivity in films has been observed in magnetic fields 100 times or more higher than the critical field H_c of a bulk superconductor of the same material. A thin film is not classed as a type II superconductor, but the behavior of a film shows that superconductivity can exist in high magnetic fields under suitable conditions.

Vortex State. The results for thin films suggest an important question: Do there exist stable configurations of a homogeneous bulk superconductor in a magnetic field in which regions in the form of thin rods (or plates) are in the normal state, with each normal region surrounded by a superconducting region? In such a mixed state the external magnetic field will penetrate the thin normal regions uniformly, and the field will also penetrate somewhat into the surrounding superconducting material, as in Fig. 31b. The term **vortex state** is descriptive of the distribution of superconducting currents in vortices throughout the bulk specimen, as in Fig. 36 below. There is no chemical or crystallographic dif-

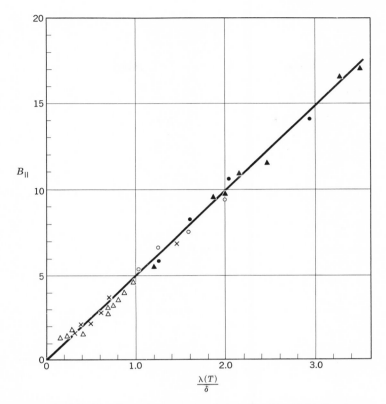

Figure 33   Parallel critical field for thin films of tin, in units of the bulk critical field. Results are given for five different films from 850 Å to 4500 Å in thickness. For each film the penetration depth is varied by varying the temperature. The horizontal scale is the ratio of the penetration depth λ to the film thickness δ. [After B. K. Sevastyonov, Soviet Phys. JETP **13**, 35 (1961).]

ference between the normal and the superconducting regions in the vortex state. The vortex state is stable because the penetration of the applied field into the superconducting material causes the surface energy to become negative. **A type II superconductor is characterized by a vortex state stable over a certain range of magnetic field strength;** namely, between H_{c1} and H_{c2}.

Estimation of H_{c1}. The vortex state in a type II superconductor first forms in a field H_{c1}. The value of H_{c1} will be smaller than the thermodynamic critical field H_c defined so that $H_c^2/8\pi$ equals the difference of free energy density between the normal and superconducting states in zero magnetic field. The difference is determined by calorimetric measurements, because now there is no transition in magnetic properties at H_c (recall Fig. 6c).

To estimate H_{c1} we consider the stability of the vortex state at absolute zero in the impure limit $\xi < \lambda$; here the coherence length is short in comparison with the penetration depth. We estimate in the vortex state the energy of a fluxoid

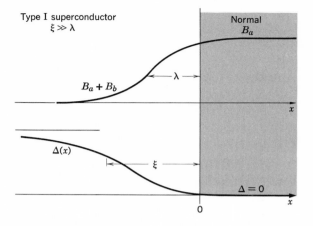

Type I superconductor
$\xi \gg \lambda$

Normal
B_a

$B_a + B_b$

λ

x

$\Delta(x)$

ξ

$\Delta = 0$

x

0

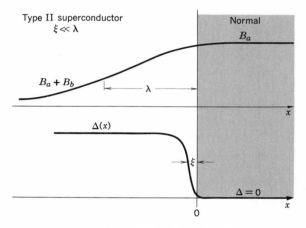

Type II superconductor
$\xi \ll \lambda$

Normal
B_a

$B_a + B_b$

λ

x

$\Delta(x)$

ξ

$\Delta = 0$

x

0

Figure 34 Variation of the magnetic field and energy gap parameter $\Delta(x)$ at the interface of superconducting and normal regions, for type I and type II superconductors.

or cylinder of normal metal which carries an average magnetic field B_a over its area. The core of the fluxoid is essentially in the normal phase; the radius will be of the order of the coherence length, which is the thickness of the boundary between N and S phases. The energy of the normal core referred to the energy of a pure superconductor is given by the product of the stabilization energy times the area of the core:

(CGS)
$$f_{\text{core}} \approx \frac{1}{8\pi} H_c^2 \times \pi \xi^2 \ , \tag{36}$$

per unit length. But there is also a decrease in energy because of the penetration of the applied field B_a into the superconducting material around the core (Fig. 34):

(CGS)
$$f_{\text{mag}} \approx -\frac{1}{8\pi} B_a^2 \times \pi \lambda^2 \ . \tag{37}$$

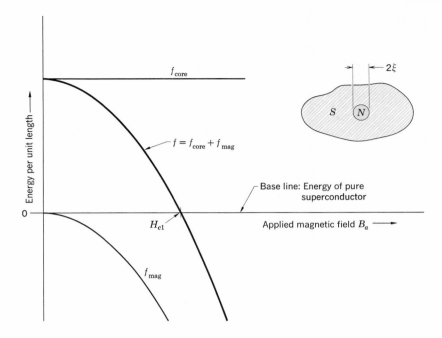

Figure 35   The determination of the lower critical field H_{c1} for the formation of the vortex state. Here f is the energy per unit length of a rod of radius ξ in the normal state N and surrounded by the superconducting state S, plotted as a function of the applied magnetic field B_a. The contributions to f of the N and S regions are shown as f_{core} and f_{mag}. The curves are drawn for $\lambda = 3\xi$. The base line is the energy of the specimen when in the pure superconducting state. The vortex state is stable when $B_a > H_{c1}$, for then f is negative.

For a single fluxoid we add these two contributions to obtain

(CGS) $$f = f_{\text{core}} + f_{\text{mag}} \approx \tfrac{1}{8}(H_c{}^2\xi^2 - B_a{}^2\lambda^2) \; , \tag{38}$$

and the core is stable if $f < 0$; a statement about the sign of f is equivalent to a statement about the sign of the surface energy. The threshold value H_{c1} of the applied field for a fluxoid to be stable is given when $f = 0$, or, with H_{c1} written for B_a,

$$\boxed{\; \frac{H_{c1}}{H_c} \approx \frac{\xi}{\lambda} \; \cdot \;} \tag{39}$$

The threshold field divides the region of positive surface energy $(B_a < H_{c1})$ from the region of negative surface energy $(B_a > H_{c1})$. The energetic relationships are indicated in Fig. 35. Abrikosov[43] has carried out a more careful calculation of H_{c1}.

[43] A. A. Abrikosov, Soviet Phys.-JETP **5**, 1174 (1957).

Figure 36 Contour diagram of the local spatial variation of the superconducting energy gap in a type II superconductor just below the upper critical field H_{c2}, after W. H. Kleiner, L. M. Roth, and S. H. Autler, Phys. Rev. **133**, A1226 (1964). At the center of each fluxoid the energy gap goes to zero. Such a triangular lattice has been found experimentally.

Near the upper critical field the fluxoids are packed close together and the external field infiltrates the specimen almost uniformly, with small ripples on the scale of the fluxoid lattice. The upper critical field marks the destruction of bulk superconductivity. The calculation for the upper critical field H_{c2} shows that

$$\boxed{(H_{c1}H_{c2})^{\frac{1}{2}} \approx H_c \ .}$$
(40)

The geometric mean of the upper and lower critical fields is equal to the thermodynamic critical field.

A theoretical plot of the spatial variation of the energy gap parameter Δ in a type II superconductor is given in Fig. 36. The lattice parameter of the fluxoid array is determined by the coherence length ξ and may be of the order of 10^{-5} cm. The fluxoid array has been observed by neutron diffraction[44] and by a magnetic powder pattern technique,[45] as in Fig. 37. The structure of a fluxoid is sketched in Fig. 38.

[44] D. Cribier et al., Proceedings International Conference on Magnetism, Institute of Physics, London, 1965, pp. 285–7.

[45] H. Träuble and U. Essmann, J. Appl. Phys. **39**, 4052 (1968); for a description of the method see Jahrbuch Akad. Wiss. Göttingen (1967) 18.

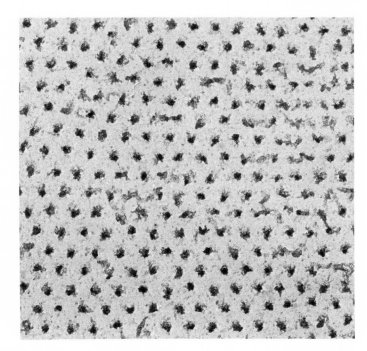

Figure 37 Triangular lattice of flux lines through top surface of a superconducting cylinder. The points of exit of the flux lines are decorated with fine ferromagnetic particles. The electron microscope image is at a magnification of 8300. (Courtesy of U. Essmann and H. Träuble.)

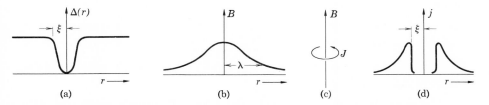

Figure 38 Structure of fluxoid in the mixed state is supported by a circulating supercurrent. The core of the flux line of size $\approx \xi$ is in the normal state, and contains most of the flux, but the field extends into the superconducting region over a distance $\approx \lambda$. The current density is j; the energy gap parameter is $\Delta(r)$. Each flux line contains one flux quantum $hc/2e = 2 \times 10^{-7}$ gauss-cm^2. (After Y. B. Kim, Physics Today, September 1964, pp. 21–30.)

SUMMARY

(In CGS units)

1. A superconductor exhibits infinite conductivity.

2. A bulk specimen of metal in the superconducting state exhibits perfect diamagnetism, with the magnetic induction $\mathbf{B} = 0$. This is the Meissner effect. The external magnetic field will penetrate the surface of the specimen over a distance determined by the penetration depth λ.

3. There are two types of superconductors, I and II. In a bulk specimen of type I superconductor the superconducting state is destroyed and the normal state is restored by application of an external magnetic field in excess of a critical value H_c. A type II superconductor has two critical fields, $H_{c1} < H_c < H_{c2}$; a vortex state exists in the range between H_{c1} and H_{c2}. The stabilization energy density of the pure superconducting state is $H_c^2/8\pi$ in both type I and II superconductors.

4. In the superconducting state an energy gap $E_g \approx 4k_B T_c$ separates superconducting electrons below from normal electrons above the gap. The gap is detected in experiments on heat capacity, infrared absorption, and tunneling.

5. The London equation

$$\mathbf{j} = -\frac{c}{4\pi\lambda_L^2}\mathbf{A} \qquad \text{or} \qquad \text{curl } \mathbf{j} = -\frac{c}{4\pi\lambda_L^2}\mathbf{B}$$

leads to the Meissner effect through the penetration equation $\nabla^2 B = B/\lambda_L^2$, where $\lambda_L \approx (mc^2/4\pi ne^2)^{\frac{1}{2}}$ is the London penetration depth.

6. In the London equation $\mathbf{A}$ or $\mathbf{B}$ should be a weighted average over the coherence length ξ. The intrinsic coherence length $\xi_0 = 2\hbar v_F/\pi E_g$.

7. The BCS theory accounts for a superconducting state formed from pairs of electrons $\mathbf{k}\uparrow$ and $-\mathbf{k}\downarrow$.

8. Three important lengths enter the theory of superconductivity: the London penetration depth λ_L; the intrinsic coherence length ξ_0; and the normal electron mean free path ℓ.

9. Type II superconductors have $\xi < \lambda$. The critical fields are related by $H_{c1} \approx (\xi/\lambda)H_c$ and $H_{c2} \approx (\lambda/\xi)H_c$. Values of H_{c2} are as high as 400 kG.

Problems

1. *Magnetic field penetration in a plate.* The penetration equation may be written as $\lambda^2\nabla^2 B = B$, where λ is the penetration depth. (a) Show that $B(x)$ inside a superconducting plate perpendicular to the x axis and of thickness δ is given by

$$B(x) = B_a \frac{\cosh (x/\lambda)}{\cosh (\delta/2\lambda)} ,$$

where B_a is the field outside the plate and parallel to it; here $x = 0$ is at the center of the plate. (b) The effective magnetization $M(x)$ in the plate is defined by $B(x) - B_a = 4\pi M(x)$. Show that

(CGS) $$4\pi M(x) = -B_a(1/8\lambda^2)(\delta^2 - 4x^2) ,$$

for $\delta \ll \lambda$. In SI we replace the 4π by μ_0.

2. *Critical velocity.* (a) Find the critical velocity v_c of Eq. (35) if the elementary excitation spectrum is given by

$$E_{\mathbf{k}} = [\Delta^2 + \epsilon_{\mathbf{k}}^2]^{\frac{1}{2}} ,$$

for $\Delta = 1 \times 10^{-16}$ erg and m equal to the free electron mass. Here $\epsilon_{\mathbf{k}}$ is the free electron energy referred to the Fermi level. Take $k_F = 0.66 \times 10^8$ cm^{-1}. (b) Estimate the critical current density, using the value of v_c found in part (a) and with $n = 1 \times 10^{22}$ electrons per cm^3. Express the result in amperes per cm^2.

3. *Superconductor parameters.* Consider a volume of 1 cm^3 of a metal with a conduction electron concentration $n = 1 \times 10^{23}$ electrons per cm^3; Debye temperature $\theta = 300°$K; and superconducting transition temperature $0.3°$K. (a) From the BCS relation (32), find the electron-electron interaction U in the specimen. (b) The BCS theory predicts $E_g \cong 3.5 k_B T_c$ for the energy gap and $\mathfrak{D}(\epsilon_F)E_g{}^2$ for the stabilization energy density of the superconducting state. Find the value of the critical field H_c at $T = 0°$K. (c) Find the intrinsic coherence length ξ_0.

4. *Critical field of thin films.* (a) Using the result of Problem 1b, show that the energy density at $T = 0°$K within a superconducting film of thickness δ in an external magnetic field B_a is given by, for $\delta \ll \lambda$,

(CGS) $$U_S(x, B_a) = U_S(0) + \frac{1}{64\pi\lambda^2}(\delta^2 - 4x^2)B_a{}^2 .$$

In SI the factor π is replaced by $\frac{1}{4}\mu_0$. We neglect a kinetic energy contribution to the problem.

(b) Show that the magnetic contribution to U_S when averaged over the thickness of the film is

$$\frac{1}{96\pi}B_a{}^2\left(\frac{\delta}{\lambda}\right)^2 .$$

(c) Show that the critical field of the thin film is proportional to $(\lambda/\delta)H_c$, where H_c is the bulk critical field, if we consider only the magnetic contribution to U_S. Experimental results are shown in Fig. 33.

5. *Two-fluid model of a superconductor.* On the two-fluid model of a superconductor we assume that at temperatures $0 < T < T_c$ the current density may be written as the sum of the contributions of normal and superconducting electrons: $\mathbf{j} = \mathbf{j}_N + \mathbf{j}_S$, where $\mathbf{j}_N = \sigma_0\mathbf{E}$ and $\mathbf{j}_S$ is given by the London equation. Here σ_0 is an ordinary normal conductivity, decreased by the reduction in the number of normal electrons at temperature T as compared to the normal state. Neglect inertial effects on both j_N and j_S.

(a) Show from the Maxwell equations that the dispersion relation connecting wave-vector $\mathbf{k}$ and frequency ω for electromagnetic waves in the superconductor is

(CGS) $$k^2c^2 = 4\pi\sigma_0\omega i - c^2\lambda_L^{-2} + \omega^2 ; \qquad \text{or}$$

(SI)
$$k^2 = \mu_0 \sigma_0 \omega i - \lambda_L^{-2} + \omega^2 \mu_0 \epsilon_0 \ ,$$

where $\lambda_L{}^2$ is given by (23) with n replaced by n_S. Recall that curl curl $\mathbf{B} = -\nabla^2 \mathbf{B}$. (b) If τ is the relaxation time of the normal electrons and n_N is their concentration, show by use of the expression $\sigma_0 = n_N e^2 \tau / m$ that at frequencies ω below $1/\tau$ the dispersion relation does not involve the normal electrons in an important way, so that the motion of the electrons is described by the London equation alone. The supercurrent short-circuits the normal electrons. The London equation itself only holds true if $\hbar\omega$ is small in comparison with the energy gap. *Note:* The frequencies of interest are such that $\omega \ll \omega_p$, where ω_p is the plasma frequency, Chapter 8.

References

E. A. Lynton, *Superconductivity*, Methuen, 1969, 3rd ed. An excellent short introduction.

R. D. Parks, ed., *Superconductivity*, Dekker, 1969. Thorough, useful collection of review articles; 1412 pp.

G. Rickayzen, *Theory of superconductivity*, Interscience, 1965. Excellent introduction to the theory.

M. Tinkham, "Superconductivity," in de Witt et al., ed., *Low temperature physics*, Gordon and Breach, 1962.

J. R. Schrieffer, *Theory of superconductivity*, Benjamin, 1964.

D. Shoenberg, *Superconductivity*, Cambridge University Press, 1960, 2nd ed.

B. W. Roberts, "Superconducting materials," *Progress in cryogenics* 4, 161 (1964); Nat. Bur. Standards Tech. Note 408 (1966). Tables of properties.

J. M. Blatt, *Theory of superconductivity*, Academic Press, 1964.

P. Nozières, "Superfluidity in Bose and Fermi fluids," D. F. Brewer, ed., *Quantum fluids*, North-Holland, 1966. Especially good discussion of the Meissner effect on pp. 1–9.

P. R. Wallace, ed., *Superconductivity*, Gordon and Breach, 1969, 2 vols.

C. G. Kuper, *Introduction to the theory of superconductivity*, Oxford, 1968.

TYPE II SUPERCONDUCTIVITY

D. Saint-James, G. Sarma, E. J. Thomas, *Type II superconductivity*, Pergamon, 1969.

P. G. de Gennes, *Superconductivity of metals and alloys*, Benjamin, 1966.

DEVICES

V. L. Newhouse, *Applied superconductivity*, Wiley, 1964.

P. F. Chester, "Superconducting magnets," Repts. Prog. Physics **30**, pt. II, 561–614 (1967).

HISTORICAL

K. Mendelssohn, *Quest for absolute zero*, McGraw-Hill, 1966, paperback.

F. London, *Superfluids*, Vol. I., Wiley, 1950.

13

Dielectric Properties

NOTATION: $\epsilon_0 = 10^7/4\pi c^2$;

(CGS) $D = E + 4\pi P = \epsilon E = (1 + 4\pi\chi)E$; $\alpha = p/E_{\text{local}}$;

(SI) $D = \epsilon_0 E + P = \epsilon\epsilon_0 E = (1 + \chi)\epsilon_0 E)$; $\alpha = p/E_{\text{local}}$;

$\epsilon_{\text{CGS}} = \epsilon_{\text{SI}}$; $4\pi\chi_{\text{CGS}} = \chi_{\text{SI}}$

(Our definition of the polarizability in SI units is the most agreeable definition; the strict rules, seldom applied, would require $4\pi\epsilon_0\alpha = p/E_{\text{local}}$.)

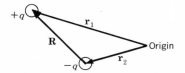

Figure 1 The dipole moment of the pair of charges $\pm q$ is $\mathbf{p} = q\mathbf{r}_1 - q\mathbf{r}_2 = q\mathbf{R}$ and is directed from the negative charge toward the positive charge.

$p = 1.9 \times 10^{-18}$ esu·cm

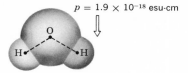

Figure 2 The permanent dipole moment of a molecule of water has the magnitude 1.9×10^{-18} esu-cm and is directed from O^{--} ion toward the midpoint of the line connecting the H^+ ions. (To convert to SI units, multiply by $\frac{1}{3} \times 10^{11}$.)

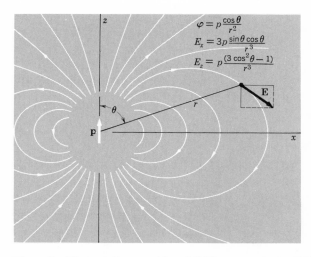

$$\varphi = p\frac{\cos\theta}{r^2}$$
$$E_x = 3p\frac{\sin\theta\cos\theta}{r^3}$$
$$E_z = p\frac{(3\cos^2\theta - 1)}{r^3}$$

Figure 3 Electrostatic potential and field components in CGS at position r, θ for a dipole $\mathbf{p}$ directed along the z axis. For $\theta = 0$, we have $E_x = E_y = 0$ and $E_z = 2p/r^3$; for $\theta = \pi/2$ we have $E_x = E_y = 0$ and $E_z = -p/r^3$. To convert to SI, replace p by $p/4\pi\epsilon_0$. (After E. M. Purcell.)

We first derive the relationship between the applied electric field and the internal electric field in a dielectric crystal. We then discuss the dielectric polarizability of atoms, molecules and crystals, in static fields and at high frequencies. The study of the electric field within dielectric matter arises when we look for the answers to two questions:

(a) What is the relation in the material between the dielectric polarization **P** and the macroscopic electric field **E** which occurs in the Maxwell equation? The equations in the usual form are:

(CGS)	(SI)	
$\text{curl } \mathbf{H} = \dfrac{4\pi}{c}\mathbf{j} + \dfrac{1}{c}\dfrac{\partial}{\partial t}(\mathbf{E} + 4\pi\mathbf{P})$;	$\text{curl } \mathbf{H} = \mathbf{j} + \dfrac{\partial}{\partial t}(\epsilon_0\mathbf{E} + \mathbf{P})$;	(1)
$\text{curl } \mathbf{E} = -\dfrac{1}{c}\dfrac{\partial \mathbf{B}}{\partial t}$;	$\text{curl } \mathbf{E} = -\dfrac{\partial \mathbf{B}}{\partial t}$;	(2)
$\text{div}(\mathbf{E} + 4\pi\mathbf{P}) = 4\pi\rho$;	$\text{div}(\epsilon_0\mathbf{E} + \mathbf{P}) = \rho$;	(3)
$\text{div } \mathbf{B} = 0$;	$\text{div } \mathbf{B} = 0$.	(4)

(b) What is the relation between the dielectric polarization and the local electric field which acts at the site of an atom in the lattice? It is this local field which determines the dipole moment of the atom.

Polarization

The **polarization P** is defined as the **dipole moment per unit volume,** averaged over the volume of a crystal cell. The total dipole moment is defined (Fig. 1) as

$$\mathbf{p} = \Sigma q_n \mathbf{r}_n , \tag{5a}$$

where $\mathbf{r}_n$ is the position vector of the charge q_n. The value of the sum will be independent of the origin chosen for the position vectors, provided that the system is neutral. The dipole moment of a water molecule is shown in Fig. 2.

The electric field at a point **r** from a point of moment **p** is given by a standard result of elementary electrostatics:

$$(\text{CGS}) \quad \mathbf{E}(\mathbf{r}) = \frac{3(\mathbf{p}\cdot\mathbf{r})\mathbf{r} - r^2\mathbf{p}}{r^5} ; \quad (\text{SI}) \quad \mathbf{E}(\mathbf{r}) = \frac{3(\mathbf{p}\cdot\mathbf{r})\mathbf{r} - r^2\mathbf{p}}{4\pi\epsilon_0 r^5} . \tag{5b}$$

The lines of force of a dipole pointing along the z axis are shown in Fig. 3.

MACROSCOPIC ELECTRIC FIELD

One contribution to the electric field inside a body is that of the applied electric field, defined as

$$\mathbf{E}_0 \equiv \text{field produced by fixed charges external to the body.} \qquad (6)$$

The other contribution to the electric field is the sum of the fields of all charges that constitute the body. If the body is neutral, the contribution to the average field may be expressed in terms of the sum of the fields of atomic dipoles of the form of (5).

We define the average electric field $\mathbf{E}(\mathbf{r}_0)$ as the **average field over the volume of the crystal cell** that contains the lattice point $\mathbf{r}_0$:

$$\mathbf{E}(\mathbf{r}_0) = \frac{1}{V_c} \int dV \, \mathbf{e}(\mathbf{r}) \ , \qquad (7)$$

where $\mathbf{e}(\mathbf{r})$ is the microscopic electric field at the point $\mathbf{r}$. The field $\mathbf{E}$ is a much smoother quantity[1] than the microscopic field $\mathbf{e}$. We could well have written the dipole field (5b) as $\mathbf{e}(\mathbf{r})$ because it is the microscopic unsmoothed field.

We call $\mathbf{E}$ the **macroscopic electric field**; it is adequate for all problems in the electrodynamics of crystals provided that we know the connection between $\mathbf{E}$, the polarization $\mathbf{P}$, and the current density $\mathbf{j}$ in Eq. (1), and provided that the wavelengths of interest are long in comparison with the lattice spacing.[2]

To find the contribution of the polarization to the macroscopic field, we can simplify the sum over all the dipoles in the specimen. By a famous theorem of electrostatics[3] the macroscopic electric field caused by a uniform polariza-

---

[1] If the specimen is not crystalline we must average over a volume large enough to be representative of the structure of the material.

[2] A detailed derivation of the Maxwell equations for the macroscopic fields $\mathbf{E}$ and $\mathbf{B}$, starting from the Maxwell equations in terms of the microscopic fields $\mathbf{e}$ and $\mathbf{h}$, is given in several textbooks; see J. H. Van Vleck, *Electric and magnetic susceptibilities*, Oxford, 1932, pp. 1–13. The clearest elementary exposition is by E. M. Purcell, *Electricity and magnetism*, McGraw-Hill, 1965, Chaps. 9 and 10.

[3] The electrostatic potential in CGS units of a dipole $\mathbf{p}$ is

$$\varphi(\mathbf{r}) = \mathbf{p} \cdot \operatorname{grad} \frac{1}{r} \ . \qquad (8a)$$

For a volume distribution of polarization $\mathbf{P}$ we have

$$\varphi(\mathbf{r}) = \int dV \left(\mathbf{P} \cdot \operatorname{grad} \frac{1}{r} \right) , \qquad (8b)$$

which by a vector identity becomes

$$\varphi(\mathbf{r}) = \int dV \left(-\frac{1}{r} \operatorname{div} \mathbf{P} + \operatorname{div} \frac{\mathbf{P}}{r} \right) . \qquad (8c)$$

If $\mathbf{P}$ is constant, then $\operatorname{div} \mathbf{P} = 0$ and by the Gauss theorem we have

$$\varphi(\mathbf{r}) = \int dS \, \frac{P_n}{r} = \int dS \, \frac{\sigma}{r} \ , \qquad (8d)$$

where dS is an element of area on the surface of the body. This completes the proof.

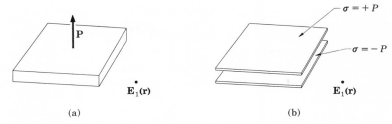

Figure 4   (a) A uniformly polarized dielectric slab, with the polarization vector **P** normal to the plane of the slab. (b) A pair of uniformly charged parallel plates which give rise to the identical electric field $\mathbf{E}_1$ as in (a). The upper plate has the surface charge density $\sigma = +P$, and the lower plate has $\sigma = -P$.

tion is equal to the electric field in vacuum of a fictitious surface charge density

$$\sigma = \hat{\mathbf{n}} \cdot \mathbf{P} \tag{8}$$

on the surface of the body. Here $\hat{\mathbf{n}}$ is the unit normal to the surface, drawn outward from the polarized matter.

We apply the result (8) to a thin dielectric slab (Fig. 4a) with a uniform volume polarization **P**. The electric field $\mathbf{E}_1(\mathbf{r})$ produced by the polarization is equal to the field produced by the fictitious surface charge density $\sigma = \hat{\mathbf{n}} \cdot \mathbf{P}$ on the surface of the slab. On the upper boundary the unit vector $\hat{\mathbf{n}}$ is directed upward and on the lower boundary $\hat{\mathbf{n}}$ is directed downward. The upper boundary bears the fictitious charge $\sigma = \hat{\mathbf{n}} \cdot \mathbf{P} = P$ per unit area, and the lower boundary bears $-P$ per unit area.

The electric field $\mathbf{E}_1$ due to these charges has a simple form at any point between the plates, but comfortably removed from their edges. By Gauss's law

(CGS) $$E_1 = -4\pi|\sigma| = -4\pi P \; ; \tag{9}$$

(SI) $$E_1 = -\frac{|\sigma|}{\epsilon_0} = -\frac{P}{\epsilon_0} \; .$$

We add $\mathbf{E}_1$ to the applied field $\mathbf{E}_0$ to obtain the total macroscopic field inside the slab:

(CGS) $$\mathbf{E} = \mathbf{E}_0 + \mathbf{E}_1 = \mathbf{E}_0 - 4\pi P \hat{\mathbf{z}} \; ; \tag{10}$$

(SI) $$\mathbf{E} = \mathbf{E}_0 + \mathbf{E}_1 = \mathbf{E}_0 - \frac{P}{\epsilon_0} \hat{\mathbf{z}} \; ,$$

where $\hat{\mathbf{z}}$ is the unit vector normal to the plane of the slab.

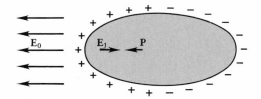

Figure 5 The depolarization field E_1 is opposite to $\mathbf{P}$. The fictitious surface charges are indicated: the field of these charges is E_1 within the ellipsoid.

We define

$$
\boxed{E_1 \equiv \text{field of the surface charge density } \hat{n} \cdot \mathbf{P} \text{ on the boundary of a simply-connected body.}} \tag{11}
$$

This field is smoothly-varying in space inside and outside the body and satisfies the Maxwell equations (1) to (3) as written for the macroscopic field $\mathbf{E}$. The reason E_1 is a smooth function when viewed on an atomic scale is that we have replaced the discrete lattice of dipoles $\mathbf{p}_j$ with the smoothed polarization $\mathbf{P}$.

Depolarization field, E_1

The geometry in many of our problems is such that the polarization is uniform within the body, and then the only contributions to the macroscopic field are from $\mathbf{E}_0$ and $\mathbf{E}_1$:

$$
\boxed{\mathbf{E} = \mathbf{E}_0 + \mathbf{E}_1 \,.} \tag{12}
$$

Here $\mathbf{E}_0$ is the applied field and $\mathbf{E}_1$ is the field due to the uniform depolarization.

The field $\mathbf{E}_1$ is called the **depolarization field,** for within the body it tends to oppose the applied field $\mathbf{E}_0$ as in (Fig. 5). Specimens in the shape of ellipsoids, a class that includes spheres, cylinders, and discs as limiting forms, have an advantageous property: a uniform polarization produces a uniform depolarization field. This is a famous mathematical result demonstrated in classic texts on electricity and magnetism.[4] If P_x, P_y, P_z are the components of the polarization $\mathbf{P}$ referred to the principal axes of an ellipsoid, then the components of the depolarization field are written

$$
\text{(CGS)} \quad E_{1x} = -N_x P_x \;; \qquad E_{1y} = -N_y P_y \;; \qquad E_{1z} = -N_z P_z \;; \tag{13}
$$

$$
\text{(SI)} \quad E_{1x} = -\frac{N_x P_x}{\epsilon_0} \;; \qquad E_{1y} = -\frac{N_y P_y}{\epsilon_0} \;; \qquad E_{1z} = -\frac{N_z P_z}{\epsilon_0} \,.
$$

Here N_x, N_y, N_z are the **depolarization factors;** their values depend on the ratios of the principal axes of the ellipsoid. The N's are positive and satisfy the sum rule $N_x + N_y + N_z = 4\pi$ in CGS, and $N_x + N_y + N_z = 1$ in SI.

[4] R. Becker, *Electromagnetic fields and interactions*, Blaisdell, 1964, pp. 102–107.

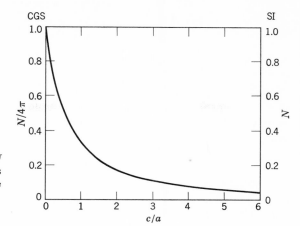

Figure 6 Depolarization factor N parallel to the figure axis of ellipsoids of revolution, as a function of the axial ratio c/a.

Values of N parallel to the figure axis of ellipsoids of revolution are plotted in Fig. 6; additional cases have been calculated by Osborn[5] and by Stoner. In limiting cases N has the values:

Shape	Axis	N (CGS)	N (SI)
Sphere	any	$\dfrac{4\pi}{3}$	$\dfrac{1}{3}$
Thin slab	normal	4π	1
Thin slab	in plane	0	0
Long circular cylinder	longitudinal	0	0
Long circular cylinder	transverse	2π	$\dfrac{1}{2}$

It is possible to reduce the depolarization field to zero in two ways, either by working with a long fine specimen or by making an electrical connection between electrodes deposited on the opposite surfaces of a thin slab.

A uniform applied field $\mathbf{E}_0$ will induce uniform polarization in an ellipsoid. We introduce the **dielectric susceptibility** χ such that

$$(CGS) \qquad\qquad \mathbf{P} = \chi\mathbf{E} \; ; \qquad\qquad (14)$$

$$(SI) \qquad\qquad \mathbf{P} = \epsilon_0\chi\mathbf{E} \; ,$$

connect the macroscopic field $\mathbf{E}$ inside the ellipsoid with the polarization $\mathbf{P}$. If $\mathbf{E}_0$ is uniform and parallel to a principal axis of the ellipsoid, then

$$(CGS) \qquad\qquad E = E_0 + E_1 = E_0 - NP \; ; \qquad\qquad (15)$$

$$(SI) \qquad\qquad E = E_0 - \frac{NP}{\epsilon_0} \; ,$$

[5] J. A. Osborn, Physical Review **67**, 351 (1945); E. C. Stoner, Philosophical Magazine **36**, 803 (1945).

by (13), whence

(CGS) $$P = \chi(E_0 - NP) \; ; \qquad P = \frac{\chi}{1 + N\chi} E_0 \; ; \qquad (16)$$

(SI) $$P = \chi(\epsilon_0 E_0 - NP) \; ; \qquad P = \frac{\chi \epsilon_0}{1 + N\chi} E_0 \; .$$

The value of the polarization depends on the depolarization factor N. If the susceptibility χ is very large in comparison with N, then

(CGS) $\quad P \approx \dfrac{E_0}{N} \; ;$ \qquad\qquad (SI) $\quad P \approx \dfrac{\epsilon_0 E_0}{N} \; .$ \qquad (17)

In this limit the polarization is essentially determined by the shape of the specimen. Such a situation must be avoided if we are interested in the determination of the dielectric susceptibility χ of the material.

LOCAL ELECTRIC FIELD AT AN ATOM

The value of the local electric field at the site of an atom is significantly different from the value of the macroscopic electric field. We can convince ourselves of this by consideration of the local field at a site with a cubic arrangement of neighbors[6] in a crystal of spherical shape. The macroscopic electric field in a sphere is

(CGS) $$\mathbf{E} = \mathbf{E}_0 + \mathbf{E}_1 = \mathbf{E}_0 - \frac{4\pi}{3} \mathbf{P} \; ; \qquad (18)$$

(SI) $$\mathbf{E} = \mathbf{E}_0 + \mathbf{E}_1 = \mathbf{E}_0 - \frac{1}{3\epsilon_0} \mathbf{P} \; ,$$

by (15). But consider the field that acts on the atom at the center of the sphere (this atom is not unrepresentative, as we shall see below). If all dipoles are parallel to the z axis and have magnitude p, the z component of the field at the center due to all other dipoles is, from (5b),

(CGS) $$E_{\text{dipole}} = p \sum_i \frac{3z_i{}^2 - r_i{}^2}{r_i{}^5} = p \sum_i \frac{2z_i{}^2 - x_i{}^2 - y_i{}^2}{r_i{}^5} \; . \qquad (19)$$

In SI we replace p by $p/4\pi\epsilon_0$. The x, y, z directions are equivalent because of

---

[6] Atom sites in a cubic crystal do not necessarily have cubic symmetry: thus the O^{--} sites in the barium titanate structure of Fig. 14.2 do not have a cubic environment. However, the Na^+ and Cl^- sites in the NaCl structure and the Cs^+ and Cl^- sites in the CsCl structure have cubic symmetry. By the field at an atom site, we mean the field that acts on an atom at that site.

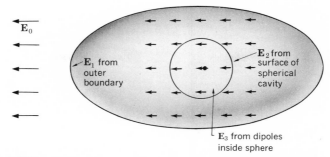

Figure 7 The internal electric field on an atom in a crystal is the sum of the external applied field $\mathbf{E}_0$ and of the field due to the other atoms in the crystal. The standard method of summing the dipole fields of the other atoms is first to sum individually over a moderate number of neighboring atoms inside an imaginary sphere concentric with the reference atom: this defines the field $\mathbf{E}_3$, which vanishes at a reference site with cubic symmetry. The atoms outside the sphere can be treated as a uniformly polarized dielectric. Their contribution to the field at the reference point is $\mathbf{E}_1 + \mathbf{E}_2$, where $\mathbf{E}_1$ is the depolarization field associated with the other boundary and $\mathbf{E}_2$ is the field associated with the surface of the spherical cavity.

the symmetry of the lattice and of the sphere; thus

$$\sum_i \frac{z_i{}^2}{r_i{}^5} = \sum_i \frac{x_i{}^2}{r_i{}^5} = \sum_i \frac{y_i{}^2}{r_i{}^5} \; , \tag{20}$$

whence

$$E_{\text{dipole}} = 0 \; . \tag{21}$$

The correct local field is just equal to the applied field,

$$\mathbf{E}_{\text{local}} = \mathbf{E}_0 \; , \tag{22}$$

for an atom site with a cubic environment in a spherical specimen. Thus the local field is not the same as the macroscopic average field.

We now develop an expression for the local field at a general lattice site, not necessarily of cubic symmetry. The local field at an atom is the sum of the electric field $\mathbf{E}_0$ from external sources and of the field of the dipoles within the specimen. It is convenient to decompose the dipole field so that part of the summation over dipoles may be replaced by integration. We write

$$\mathbf{E}_{\text{local}} = \mathbf{E}_0 + \mathbf{E}_1 + \mathbf{E}_2 + \mathbf{E}_3 \; . \tag{23}$$

Here

$\mathbf{E}_0$ = field produced by fixed charges external to the body;

$\mathbf{E}_1$ = depolarization field, from a surface charge density $\hat{n} \cdot \mathbf{P}$ on the outer surface of the specimen;

$\mathbf{E}_2$ = Lorentz cavity field: field from polarization charges on inside of a spherical cavity cut (as a mathematical fiction) out of the specimen with the reference atom as center, as in Fig. 7;

$\mathbf{E}_3$ = field of atoms inside cavity.

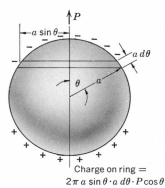

Charge on ring =
$2\pi a \sin\theta \cdot a \, d\theta \cdot P\cos\theta$

Figure 8 Calculation of the field in a spherical cavity in a uniformly polarized medium.

The contribution $\mathbf{E}_1 + \mathbf{E}_2 + \mathbf{E}_3$ to the local field is the total field at one atom caused by the dipole moments of all the other atoms in the specimen:

$$\text{(CGS)} \qquad \mathbf{E}_1 + \mathbf{E}_2 + \mathbf{E}_3 = \sum_i \frac{3(\mathbf{p}_i \cdot \mathbf{r}_i)\mathbf{r}_i - r_i^2 \mathbf{p}_i}{r_i^5} \, , \qquad (24)$$

and in SI we replace $\mathbf{p}_i$ by $\mathbf{p}_i/4\pi\epsilon_0$. Dipoles at distances greater than perhaps ten lattice constants from the reference site make a smoothly varying contribution to this sum, a contribution which may be replaced by two surface integrals (footnote 3). One surface integral is taken over the outer surface of the ellipsoidal specimen and defines $\mathbf{E}_1$, as in Eq. (11). The second surface integral defines $\mathbf{E}_2$ and may be taken over any interior surface that is a suitable distance (say 50 Å) from the reference site, for we shall count in $\mathbf{E}_3$ any dipoles not included in the volume bounded by the inner and outer surfaces. It is most convenient to let the interior surface be spherical.

Lorentz Field, $\mathbf{E}_2$

The field $\mathbf{E}_2$ due to the polarization charges on the surface of the fictitious cavity was calculated by Lorentz in 1878. If θ is the polar angle (Fig. 8) referred to the polarization direction as axis, the surface charge density on the surface of the cavity is $-P\cos\theta$. The electric field at the center of the spherical cavity of radius a is

$$\text{(CGS)} \quad \mathbf{E}_2 = \int_0^\pi (a^{-2})(2\pi a \sin\theta)(a \, d\theta)(\mathbf{P}\cos\theta)(\cos\theta) = \frac{4\pi}{3}\mathbf{P} \; ; \qquad (25)$$

$$\text{(SI)} \qquad \qquad \mathbf{E}_2 = \frac{1}{3\epsilon_0}\mathbf{P} \; .$$

Field of Dipoles Inside Cavity, $\mathbf{E}_3$

The field $\mathbf{E}_3$ due to the dipoles within the spherical cavity is the only term that depends on the crystal structure. We showed in Eqs. (19) to (21) for a

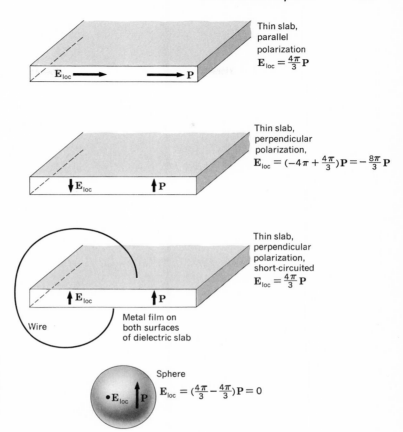

[7] H. Mueller, Phys. Rev. **47**, 947 (1935); **50**, 547 (1936); see also L. W. McKeehan, Phys. Rev. **43**, 1022, 1025 (1933).

Figure 9   (a) Local electric fields due to polarization alone, as seen at atom sites with cubic local surroundings, for four important geometries. (To obtain the equations in SI, multiply **P** by $1/4\pi\epsilon_0$.)

reference site with cubic surroundings in a sphere that

$$\mathbf{E}_3 = 0 \qquad (26)$$

if all the atoms may be replaced by point dipoles *parallel* to each other. This confirms the result (23) for the local field. Values of $\mathbf{E}_3$ for tetragonal and simple hexagonal lattices have been given by Mueller.[7]

The total local field at a cubic site is

(CGS) $$\mathbf{E}_{local} = \mathbf{E}_0 + \mathbf{E}_1 + \frac{4\pi}{3}\mathbf{P} = \mathbf{E} + \frac{4\pi}{3}\mathbf{P} \; ; \qquad (27)$$

(SI) $$\mathbf{E}_{local} = \mathbf{E} + \frac{1}{3\epsilon_0}\mathbf{P} \; ,$$

from (23) and (26). Local fields for several geometries are shown in Fig. 9. Equation (27) is the **Lorentz relation:** the field acting at an atom in a cubic

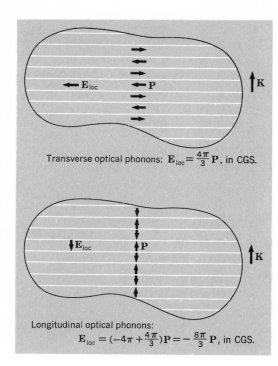

Transverse optical phonons: $\mathbf{E}_{loc} = \frac{4\pi}{3}\mathbf{P}$, in CGS.

Longitudinal optical phonons:
$\mathbf{E}_{loc} = (-4\pi + \frac{4\pi}{3})\mathbf{P} = -\frac{8\pi}{3}\mathbf{P}$, in CGS.

Figure 10 The local electric field (the long range part of the interaction) tends to assist the lattice distortion associated with a transverse optical phonon (a), but tends to resist the deformation for a longitudinal optical phonon (b). Thus $\omega_L > \omega_T$. The values of the local fields shown are for ions in cubic surroundings. The rulings indicate the lines (or planes) of nodes.

site is the macroscopic field $\mathbf{E}$ of Eq. (18) plus the contribution $4\pi\mathbf{P}/3$ or $\mathbf{P}/3\epsilon_0$ from the polarization of the other atoms in the specimen. Experimental data for cubic ionic crystals[8] support the Lorentz relation. The local fields for optical phonon modes in ionic crystals depend on the polarization type, as in Fig. 10.

DIELECTRIC CONSTANT AND POLARIZABILITY

The **dielectric constant** ϵ of an isotropic or cubic medium relative to vacuum is defined as

(CGS)
$$\epsilon \equiv \frac{E + 4\pi P}{E} = 1 + 4\pi\chi \; ; \qquad (28)$$

(SI)
$$\epsilon = \frac{\epsilon_0 E + P}{\epsilon_0 E} = 1 + \chi \;.$$

The susceptibility is related to the dielectric constant by

(CGS)
$$\chi = \frac{P}{E} = \frac{\epsilon - 1}{4\pi} \; ; \qquad (29)$$

(SI)
$$\chi = \frac{P}{\epsilon_0 E} = \epsilon - 1 \;.$$

[8] J. Tessman, A. Kahn, and W. Shockley, Phys. Rev. **92**, 890 (1953); E. Kartheuser and J. Deltour, Physics Letters **19**, 548 (1965).

Here E is the macroscopic electric field. In a noncubic crystal the dielectric response is described by the components of the susceptibility tensor or of the dielectric constant tensor:

(CGS) $\qquad P_\mu = \chi_{\mu\nu} E_\nu \ ; \qquad \epsilon_{\mu\nu} = 1 + 4\pi\chi_{\mu\nu} \ .$ (30)

(SI) $\qquad P_\mu = \chi_{\mu\nu}\epsilon_0 E_\nu \ ; \qquad \epsilon_{\mu\nu} = 1 + \chi_{\mu\nu} \ .$

The **polarizability** α of an atom is defined in terms of the local electric field at the atom:

$$p = \alpha E_{\text{local}} \ ,$$ (31)

where p is the dipole moment. This definition applies in CGS and in SI, although some authors may define α in SI as $p = \alpha\epsilon_0 E_{\text{local}}$. The polarizability is an atomic property, but the dielectric constant will depend on the manner in which the atoms are assembled to form a crystal. The polarizability in CGS units has the dimensions of $[\text{length}]^3$, for the dipole moment is $[\text{charge}] \times [\text{length}]$ and the field is $[\text{charge}] \times [\text{length}]^{-2}$.

The polarization in a crystal may be expressed approximately as the product of the polarizabilities of the atoms times the local electric field:

$$P = \sum_j N_j p_j = \sum_j N_j \alpha_j E_{\text{loc}}(j) \ ,$$ (32)

where N_j is the concentration and α_j the polarizability of atoms j, and $E_{\text{loc}}(j)$ is the local field at atom sites j. We next want to relate the dielectric constant to the polarizabilities; the result will depend on the relation that holds between the macroscopic electric field and the local electric field. We give the derivation in CGS units and state the result in both systems of units.

If the local field is given by the Lorentz relation (27), then

(CGS) $\qquad P = (\Sigma N_j \alpha_j)\left(E + \dfrac{4\pi}{3}P\right) \ ;$ (33)

and we solve for P to find the susceptibility

(CGS) $\qquad \chi = \dfrac{P}{E} = \dfrac{\Sigma N_j \alpha_j}{1 - \dfrac{4\pi}{3}\Sigma N_j \alpha_j} \ .$ (34)

By definition $\epsilon = 1 + 4\pi\chi$ in CGS; we may rearrange (34) to obtain

(CGS) $\quad \dfrac{\epsilon - 1}{\epsilon + 1} = \dfrac{4\pi}{3}\Sigma N_j \alpha_j \ ;$ $\qquad$ (SI) $\quad \dfrac{\epsilon - 1}{\epsilon + 2} = \dfrac{1}{3\epsilon_0}\Sigma N_j \alpha_j \ ,$ (35)

the **Clausius-Mossotti relation.** This relates the dielectric constant to the electronic polarizability, but only for crystal structures for which the Lorentz local field obtains.

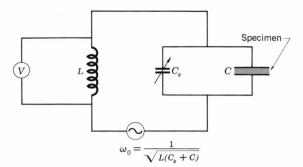

Figure 11 Schematic diagram of apparatus for the measurement of dielectric constants.

Measurement of Dielectric Constant

The usual methods of measuring the dielectric constant are based on a comparison of the capacity C'' of a capacitor filled with the substance and the capacity C' of the empty capacitor. The ratio $C''/C' = \epsilon$, the dielectric constant. The determination of the value of the capacitance may in principle be accomplished by an LC resonant circuit as shown in Fig. 11, where C_s is a calibrated variable capacitor and C is the capacitor in which the specimen may be placed. By varying the calibrated capacitor so as to keep the resonance frequency $\omega_0 = [L(C_s + C)]^{-\frac{1}{2}}$ constant when C is inserted and then filled, we may determine C' and C'', and thus ϵ.

Descriptions of the actual circuits employed are abundant in the literature. At microwave frequencies the technique of measurement is altered somewhat, and here one often measures essentially the wavelength λ of the microwave radiation in the specimen, obtaining the dielectric constant from the relation λ (vacuum)$/\lambda$ (specimen) $= (\epsilon\mu)^{\frac{1}{2}}$, where μ is the permeability relative to vacuum.

Electronic Polarizability

The total polarizability may usually be separated into three parts:[9] electronic, ionic, and dipolar, as in Fig. 12. The electronic contribution arises from the displacement of the electron shell relative to a nucleus. The ionic contribution comes from the displacement of a charged ion with respect to other ions. The dipolar polarizability arises from molecules with a permanent electric dipole moment that can change orientation in an applied electric field. The separate contributions are indicated in Fig. 13.

---

[9] In heterogeneous materials there is usually also an *interfacial polarization* arising from the accumulation of charge at structural interfaces. This is of little fundamental interest, but it is of considerable practical interest because commercial insulating materials are usually heterogeneous.

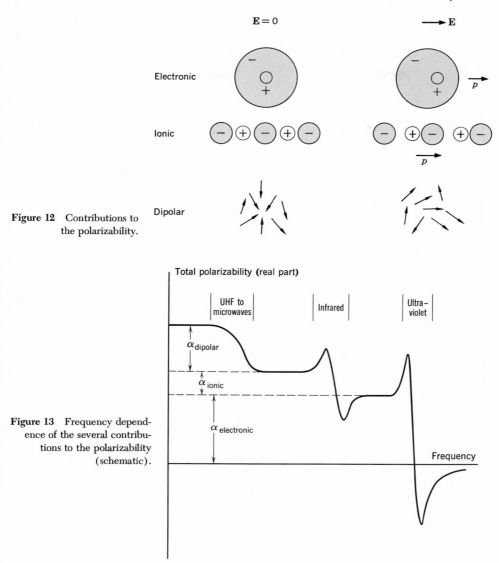

Figure 12 Contributions to the polarizability.

Figure 13 Frequency dependence of the several contributions to the polarizability (schematic).

The dielectric constant at optical frequencies arises almost entirely from the electronic polarizability. The dipolar and ionic contributions are small at high frequencies because of the inertia of the molecules and ions. In the optical range (35) reduces to

(CGS) $$\frac{n^2 - 1}{n^2 + 2} = \frac{4\pi}{3} \Sigma N_j \alpha_j (\text{electronic});$$ (36)

here we have used the relation $n^2 = \epsilon$, where n is the refractive index. By

Table 1   Electronic polarizabilities of ions in 10^{-24} cm^3

Values from L. Pauling, Proc. Roy. Soc. (London) **A114**, 181 (1927); J. Pirenne and E. Kartheuser, Physica **20**, 2005 (1964); and J. Tessman, A. Kahn, and W. Shockley, Phys. Rev. **92**, 890 (1953). The PK and TKS polarizabilities are at the frequency of the D lines of sodium. The values are in CGS; to convert to SI, multiply by $\frac{1}{9} \times 10^{-15}$.

			He	Li$^+$	Be^{2+}	B^{3+}	C^{4+}
Pauling			0.201	0.029	0.008	0.003	0.0013
PK				0.029			
	O^{2-}	F$^-$	Ne	Na$^+$	Mg^{2+}	Al^{3+}	Si^{4+}
Pauling	3.88	1.04	0.390	0.179	0.094	0.052	0.0165
PK-(TKS)	(2.4)	0.87		0.312			
	S^{2-}	Cl$^-$	Ar	K$^+$	Ca^{2+}	Sc^{3+}	Ti^{4+}
Pauling	10.2	3.66	1.62	0.83	0.47	0.286	0.185
PK-(TKS)	(5.5)	3.06		1.14	(1.1)		(0.19)
	Se^{2-}	Br$^-$	Kr	Rb$^+$	Sr^{2+}	Y^{3+}	Zr^{4+}
Pauling	10.5	4.77	2.46	1.40	0.86	0.55	0.37
PK-(TKS)	(7.)	4.28		1.76	(1.6)		
	Te^{2-}	I$^-$	Xe	Cs$^+$	Ba^{2+}	La^{3+}	Ce^{4+}
Pauling	14.0	7.10	3.99	2.42	1.55	1.04	0.73
PK-(TKS)	(9.)	6.52		3.02	(2.5)		

applying (36) to large numbers of crystals we determine in Table 1 empirical values of the electronic polarizabilities that are reasonably consistent with the observed values of the refractive index. The scheme is not entirely self-consistent, because the electronic polarizability of an ion depends somewhat on the environment in which it is placed. The negative ions are highly polarizable because they are large in size (Table 3.7).

Classical Theory of Electronic Polarizability. An electron bound harmonically to an atom will show resonance absorption at a frequency $\omega_0 = (\beta/m)^{\frac{1}{2}}$, where β is the force constant. The displacement x of the electron occasioned by the application of a field E_{loc} is given by

$$-eE_{\mathrm{loc}} = \beta x = m\omega_0^2 x ,\qquad (37)$$

so that the static electronic polarizability is

$$\alpha \text{ (electronic)} = \frac{p}{E_{\mathrm{loc}}} = -\frac{ex}{E_{\mathrm{loc}}} = \frac{e^2}{m\omega_0^2} .\qquad (38)$$

The electronic polarizability will depend on frequency, and it is shown in the following example that for frequency ω

$$\alpha \text{ (electronic)} = \frac{e^2/m}{\omega_0^2 - \omega^2} ,\qquad (39)$$

but in the visible region the frequency dependence (dispersion) is not usually very important in most transparent materials.

EXAMPLE: *Frequency dependence.* Find the frequency dependence of the electronic polarizability of an electron having the resonance frequency ω_0, treating the system as a simple harmonic oscillator.

The equation of motion in the local electric field $E_{\text{loc}} \sin \omega t$ is

$$m \frac{d^2x}{dt^2} + m\omega_0^2 x = -eE_{\text{loc}} \sin \omega t \ , \tag{40}$$

so that, for $x = x_0 \sin \omega t$,

$$m(-\omega^2 + \omega_0^2)x_0 = -eE_{\text{loc}} \ .$$

The dipole moment has the amplitude

$$p_0 = -ex_0 = \frac{e^2 E_{\text{loc}}}{m(\omega_0^2 - \omega^2)} \ , \tag{41}$$

from which (39) follows.

In quantum theory the expression corresponding to (39) is

$$\alpha(\text{electronic}) = \frac{e^2}{m} \sum_j \frac{f_{ij}}{\omega_{ij}^2 - \omega^2} \ , \tag{42}$$

where f_{ij} is called the **oscillator strength** of the electric dipole transition between the atomic states i and j. Equation (42) is written for atoms and should be modified slightly for solids.

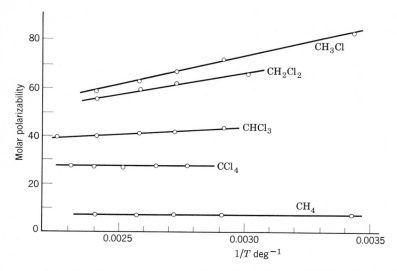

Figure 14 Plot of $[(\epsilon - 1)/(\epsilon + 1)]$ times the molar volume, which is known as the molar polarizability, for polar and nonpolar substituted methane compounds in gaseous form. [After R. Sanger, *Physik Z.* **27**, 556 (1926).]

Orientational Polarizability

The static dielectric constant of water is 81 at room temperature, but the dielectric constant at optical frequencies is only 1.77. The difference is caused chiefly by the orientational polarization which is effective at low frequencies, but is damped out for frequencies above 10^{10} Hz. The characteristic temperature dependence of the orientational polarizability of polar molecules is shown in Fig. 14; CH_3Cl has a permanent electric dipole moment; CH_4 and CCl_4 are symmetrical and do not.

The orienting tendency of the electric field on a permanent dipole is opposed by thermal agitation. The potential energy U of a molecule of permanent moment $\mathbf{p}$ in a field $\mathbf{E}$ is

$$U = -\mathbf{p} \cdot \mathbf{E} = -pE \cos \theta \; , \tag{45}$$

where θ is the angle between the moment and the field direction. Then

$$P = Np \langle \cos \theta \rangle \; , \tag{46}$$

where N is the concentration of molecules and $\langle \cos \theta \rangle$ is the thermal average.

According to the Boltzmann distribution law the relative probability of finding a molecule in an element of solid angle $d\Omega$ is proportional to $\exp(-U/k_BT)$, and

$$\langle \cos \theta \rangle = \frac{\int e^{-\beta U} \cos \theta \, d\Omega}{\int e^{-\beta U} \, d\Omega} \; , \tag{47}$$

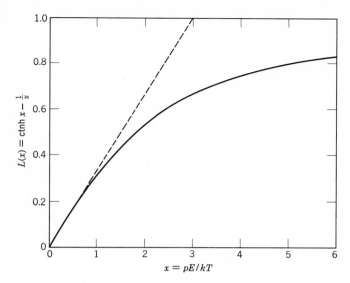

Figure 15 Plot of Langevin function $L(x)$ versus $x = pE/k_BT$. The initial slope is shown by the dashed line. The polarization is 80 percent of saturation when $pE/k_BT = 5$.

where $\beta \equiv 1/k_BT$. The integration is to be carried out over all solid angles, so that

$$\langle \cos \theta \rangle = \frac{\int_0^\pi 2\pi \sin \theta \cos \theta \, e^{\beta pE \cos \theta} \, d\theta}{\int_0^\pi 2\pi \sin \theta \, e^{\beta pE \cos \theta} \, d\theta} \, . \tag{48}$$

We let $s \equiv \cos \theta$ and $x \equiv pE/k_BT$, so that

$$\langle \cos \theta \rangle = \int_{-1}^1 e^{sx} s \, ds \Big/ \int_{-1}^1 e^{sx} \, ds = \frac{d}{dx} \log \int_{-1}^1 e^{sx} \, ds$$

$$= \frac{d}{dx} \log (e^x - e^{-x}) - \frac{d}{dx} \log x = \operatorname{ctnh} x - \frac{1}{x} \equiv L(x) \, . \tag{49}$$

This defines the **Langevin function** $L(x)$. The function[10] is plotted in Fig. 15, where the saturation property for $pE \gg k_BT$ is clearly seen.

The most important situation experimentally is when $pE \ll k_BT$. In CGS units, molecular dipole moments are of the order of 10^{-18} esu-cm, so in an electric field $E = 3000$ V/cm $= 10$ statV/cm we have $pE \approx 10^{-17}$ erg. In SI units, dipole moments are of the order 10^{-29} C m. At room temperature $k_BT \approx 4 \times 10^{-14}$ erg, so that $x \equiv pE/k_BT \approx 1/4000$. In the limit of $x \ll 1$, we have

$$\operatorname{ctnh} x = \frac{1}{x} + \frac{x}{3} - \frac{x^3}{45} + \cdots \, ; \qquad L(x) \cong \frac{x}{3} = \frac{pE}{3k_BT} \, , \tag{50}$$

[10] The Langevin function is tabulated in Jahnke-Emde-Lösch, *Tables of higher functions*, McGraw-Hill, 6th ed., 1960, table 51.

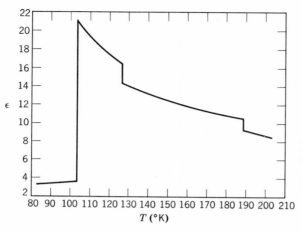

Figure 16 Dielectric constant temperature curve of hydrogen sulfide at 5 kHz. The solid-liquid transition occurs near 190 K. [After Smyth and Hitchcock, J. Am. Chem. Soc. **56**, 1084 (1934).]

and the polarization is

$$P = Np\langle \cos\theta \rangle = \frac{Np^2E}{3k_BT} \ . \tag{51}$$

The orientational polarizability per molecule is

$$\alpha(\text{dipolar}) = \frac{p^2}{3k_BT} \ . \tag{52}$$

At room temperature this is of the same order of magnitude as the electronic polarizability.

The dipole moment p is determined in practice by plotting either α or $(\epsilon - 1)/(\epsilon + 2)$ from (35) as a function of $1/T$ as in Fig. 14; the slope is simply related to p. In this way one obtains, for example, $p(H_2O) = 1.87 \times 10^{-18}$ esu-cm. The moments are often expressed in **Debye units** equal to 10^{-18} esu, which is of the order of the electronic charge times an interatomic distance.

Dipole Orientation in Solids

The ability of a molecule to reorient in a solid depends very much on its shape and on the strength of its interactions with the environment. The nearer to sphericity and the lower the dipole moment, the more easily and faster the molecule will reorient in a changing electric field. Solid methane (CH_4), a symmetrical nonpolar molecule, rotates fairly freely in the solid state; the molecules in solid hydrogen also rotate quite freely. Less symmetrical molecules such as HCl and H_2O have several stable orientations and change direction relatively slowly from one stable orientation to another. The average time between changes is called the **relaxation time.**

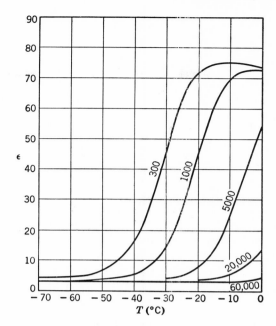

Figure 17 Variation of the dielectric constant of ice with temperature and frequency (in Hz). [After Smyth and Hitchcock, J. Am. Chem. Soc. **54**, 4631 (1932).]

The dielectric constant of solid H_2S as a function of temperature is shown in Fig. 16. Notice the sharp increase in the dielectric constant as the temperature is lowered; the sudden drop below 105°K is thought to mark the transition to an ordered state in which the directions of the H_2S molecules form a regular array. The variation of the dielectric constant above the transition temperature is suggestive of free rotation, but a similar variation with temperature arises when there are only a discrete number of allowed orientations for each dipole, so that the temperature variation is not evidence of free rotation.

DIELECTRIC RELAXATION AND LOSS

In solids and liquids composed of polar molecules the principal part of the difference between the low frequency dielectric constant and the optical frequency dielectric constant (the square of the optical refractive index) is due to the orientational contribution to the dielectric constant. We have introduced the relaxation time as the time interval characterizing the restoration of a disturbed system to its equilibrium configuration; the **relaxation frequency** is the reciprocal of the relaxation time. When the frequency of an applied field is higher than the relaxation frequency, the system will not follow or respond to the field. The orientation relaxation frequencies vary over a wide range and may be strongly dependent on the temperature. In water at room temperature, relaxation occurs at about 3×10^{10} Hz. In ice at −20°C the relaxation frequency is of the order of 1kHz, as we see from Fig. 17.

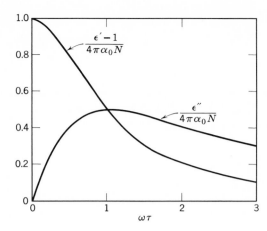

Figure 18 Frequency dependence of real and imaginary parts of the dielectric constant $\epsilon = \epsilon' + i\epsilon''$, for orientational relaxation. (CGS units.)

Debye Relaxation Time

Debye[11] has given an elegant discussion of dielectric relaxation of polar molecules in liquids. His central result is that the orientational part of the polarizability depends on the applied frequency ω as, for $E \propto \exp(-i\omega t)$,

$$\alpha(\omega) = \frac{\alpha_0}{1 - i\omega\tau} \; , \tag{53}$$

where τ is the relaxation time and α_0 is the static orientational polarizability. In liquids the relaxation time is related to the viscosity η by the approximate relation

$$\tau = \frac{4\pi\eta a^3}{k_B T} \; , \tag{54}$$

where a is the radius of the molecule, supposed to be spherical. For water at room temperature we obtain $\tau \approx 10^{-11}$ sec, using $a \approx 10^{-8}$ cm and $\eta = 0.01$ poise; this relaxation frequency is in approximate agreement with experiment.

Relaxation times in solids are usually much longer than in liquids, because a solid presents a more rigid barrier to internal motion. Breckenridge[12] has related the observed dielectric losses in alkali halide crystals to the motion of lattice defects in the crystals. Dielectric losses due to electrons hopping from one lattice site to another in transition metal oxides are discussed by S. van Houten and A. J. Bosman, Proc. 1963 Buhl Int. Conf. Transition Metal Compounds, pp. 123–136.

Shephard and Feher[13] have reported that OH^- ions substituted for Cl^- ions in a KCl crystal have a relaxation time less than 10^{-10} sec near $1°K$. The OH^- ion has six equally likely orientations in the solid, with the dipole moment along the direction of each cube edge.

[11] P. Debye, *Polar molecules,* Chemical Catalog Co., New York, 1929, Chap. V. For a discussion of the transition from resonance to relaxation-type behavior, see J. H. Van Vleck and V. F. Weisskopf, Revs. Modern Phys. **17**, 227 (1945).

[12] R. G. Breckenridge, in *Imperfections in nearly perfect crystals,* eds., Shockley, Hollomon, Maurer, and Seitz, Wiley, 1952; J. Volger, *Prog. semiconductors* **4**, 207 (1959).

[13] I. Shepherd and G. Feher, Phys. Rev. Let. **15**, 194 (1965).

Complex Dielectric Constant

In the presence of relaxation effects the dielectric constant may conveniently be written as complex. For the polarizability (53) the dielectric constant is, with the local field taken as equal to the applied field,

$$(\text{CGS}) \quad \epsilon = \epsilon' + i\epsilon'' = 1 + \frac{4\pi\alpha_0 N}{1 - i\omega\tau} = 1 + \frac{4\pi\alpha_0 N}{1 + \omega^2\tau^2} + i\frac{4\pi\alpha_0\omega\tau N}{1 + \omega^2\tau^2}, \quad (55)$$

as plotted in Fig. 18. The decrease of ϵ' occurs near the peak of ϵ''; this is an example of the general result that if $\epsilon'(\omega)$ changes with frequency, then $\epsilon''(\omega)$ also changes[14] with frequency. In SI replace 4π by $1/\epsilon_0$.

SUMMARY
(In CGS units)

1. The electric field averaged over the volume of the specimen defines the electric field $\mathbf{E}$ of the Maxwell equations. It is known as the macroscopic electric field.

2. The electric field at the site $\mathbf{r}_j$ of an atom j is the local electric field, $\mathbf{E}_{\text{loc}}$. It is calculated as a sum over all charges, grouped in terms as $\mathbf{E}_{\text{loc}}(\mathbf{r}_j) = \mathbf{E}_0 + \mathbf{E}_1 + \mathbf{E}_2 + \mathbf{E}_3(\mathbf{r}_j)$, where only $\mathbf{E}_3$ varies rapidly within a cell of the crystal. Here:

$\mathbf{E}_0 = $ external electric field;
$\mathbf{E}_1 = $ depolarization field associated with the boundary of the specimen;
$\mathbf{E}_2 = $ field from polarization outside a sphere centered about $\mathbf{r}_j$;
$\mathbf{E}_3(\mathbf{r}_j) = $ field at $\mathbf{r}_j$ due to all atoms inside the sphere.

3. The macroscopic field $\mathbf{E}$ of the Maxwell equations is equal to $\mathbf{E}_0 + \mathbf{E}_1$, which, in general, is not equal to $\mathbf{E}_{\text{loc}}(\mathbf{r}_j)$.

4. The depolarization field in an ellipsoid is $E_{1\mu} = -N_{\mu\nu}P_\nu$, where $N_{\mu\nu}$ is the depolarization tensor; the polarization $\mathbf{P}$ is the dipole moment per unit volume. In a sphere

$$N = \frac{4\pi}{3}.$$

5. The Lorentz field is

$$\mathbf{E}_2 = \frac{4\pi}{3}\mathbf{P}.$$

6. The polarizability α of an atom is defined in terms of the local electric field as $\mathbf{p} = \alpha\mathbf{E}_{\text{loc}}$.

---

[14] The connection between $\epsilon'(\omega)$ and $\epsilon''(\omega)$ is known as the Kramers-Kronig relation; see, for example, C. Kittel, *Elementary statistical physics*, Wiley, 1958, pp. 206–210.

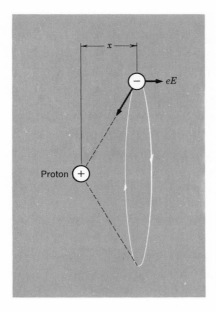

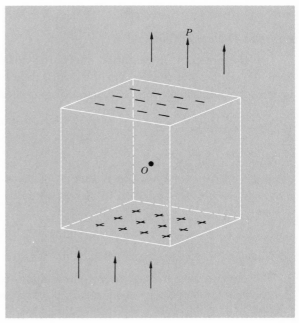

Figure 19 An electron in a circular orbit of radius a_H is displaced a distance x on application of an electric field E in the $-x$ direction. The force on the electron due to the nucleus is e^2/a_H^2 in CGS or $e^2/4\pi\epsilon_0 a_H^2$ in SI.

Figure 20 In CGS the field is $4\pi P/3$ at the center O of a cubical cavity bearing charges of uniform surface density $\pm P$ on the upper and lower faces. Solve this problem by finding the vertical component of the field due to a rectangular strip of width dx and integrating over one face. (In SI the field is $P/3\epsilon_0$ at the center O.)

7. The dielectric susceptibility χ and dielectric constant ϵ are defined in terms of the macroscopic electric field E as $\mathbf{D} = \mathbf{E} + 4\pi\mathbf{P} = \epsilon\mathbf{E} = (1 + 4\pi\chi)\mathbf{E}$, or $\chi = P/E$. In SI, we have $\chi = P/\epsilon_0 E$.

8. An atom at a site with cubic symmetry has $\mathbf{E}_{\text{loc}} = \mathbf{E} + (4\pi/3)\mathbf{P}$ and satisfies the Clausius-Mossotti relation (35).

9. The Langevin-Debye equation $\alpha = \alpha_0 + p^2/3k_B T$ gives the contribution of a permanent electric dipole moment p to the polarizability, in the limit $pE \ll k_B T$.

Problems

1. *Polarizability of atomic hydrogen.* Consider a semiclassical model of the ground state of the hydrogen atom in an electric field normal to the plane of the orbit (Fig. 19), and show that for this model $\alpha = a_H^3$, where a_H is the radius of the unperturbed orbit. *Note:* If the applied field is in the x direction, then the x component of the field of the nucleus at the displaced position of the electron orbit must be equal to the applied field. The correct quantum-mechanical result is larger than this by the factor $\frac{9}{2}$. (We are speaking of α_0 in the expansion $\alpha = \alpha_0 + \alpha_1 E + \cdots .$)

Figure 22 An air gap of thickness qd is in series in a capacitor with a dielectric slab of thickness d.

Figure 21 The total field inside a conducting sphere is zero. If a field $\mathbf{E}_0$ is applied externally, then the field $\mathbf{E}_1$ due to surface charges on the sphere must just cancel $\mathbf{E}_0$, so that $\mathbf{E}_0 + \mathbf{E}_1 = 0$ within the sphere. But $\mathbf{E}_1$ can be simulated by the depolarization field $-4\pi\mathbf{P}/3$ of a uniformly polarized sphere of polarization $\mathbf{P}$. Relate $\mathbf{P}$ to $\mathbf{E}_0$ and calculate the dipole moment $\mathbf{p}$ of the sphere. In SI the depolarization field is $-P/3\epsilon_0$.

2. **Field of cubical cavity.** In the local field problem the cavity need not be chosen as spherical, but may be a cube (Fig. 20) with a face normal to the polarization. In this case the polarization charge density on the upper and lower faces of the cube is uniform and equal to $\mp P$, whereas the other faces do not carry any charge. Show that at the center of this cavity $E_2 = 4\pi P/3$, just as for the spherical cavity. (In SI the result is $E_2 = P/3\epsilon_0$.) Contrast this result with (9) for an infinite slab.

3. **Polarizability of conducting sphere.** Show that the polarizability of a conducting metallic sphere of radius a is $\alpha = a^3$. This result is most easily obtained by noting that $E = 0$ inside the sphere and then using the depolarization factor $4\pi/3$ for a sphere (Fig. 21). The result gives values of α of the order of magnitude of the observed polarizabilities of atoms. A lattice of N conducting spheres per unit volume has dielectric constant $\epsilon = 1 + 4\pi Na^3$, for $Na^3 \ll 1$. The suggested proportionality of α to the cube of the ionic radius is satisfied quite well for alkali and halogen ions. To do the problem in SI, use $\frac{1}{3}$ as the depolarization factor; the result is the same.

4. **Effect of air gap.** Discuss the effect of an air gap (Fig. 22) between condenser plates and dielectric on the measurement of high dielectric constants. What is the highest apparent dielectric constant possible if the air gap thickness is 10^{-3} of the total thickness?

5. **Interfacial polarization.** Show that a parallel-plate capacitor made up of two parallel layers of material—one layer with dielectric constant ϵ, zero conductivity, and thickness d, and the other layer with $\epsilon = 0$ for convenience, finite conductivity σ, and thickness qd—behaves as if the space between the condenser plates were filled with a homogeneous dielectric with dielectric constant

(CGS)
$$\epsilon_{\text{eff}} = \frac{\epsilon(1 + q)}{1 - (i\epsilon\omega q/4\pi\sigma)} \; ,$$

where ω is the angular frequency [K. W. Wagner, Arch. Elektrotech. **2**, 371 (1941)]. Values of ϵ_{eff} as high as 10^4 or 10^5 caused largely by this Maxwell-Wagner mechanism are sometimes found, but the high values are always accompanied by large losses.

6. **Refractive indexes.** Estimate using Table 1 the refractive index of solid xenon.

7. **Dielectric constant of damped oscillator.** Suppose that a damped oscillator of mass m and charge q has the equation of motion

$$m\left(\frac{d^2x}{dt^2} + \frac{1}{\tau}\frac{dx}{dt} + \omega_0^2 x\right) = qE_0 e^{-i\omega t} \; ,$$

where ω_0 is the resonant frequency and τ, the relaxation frequency, represents the damping effects. (a) Neglecting interactions among the oscillators, show that the contribution of the N such oscillators per unit volume to the dielectric constant is

(CGS)
$$\epsilon = 1 + \frac{4\pi Nq^2/m}{\omega_0^2 - \omega^2 - i\omega/\tau} \; .$$

(b) Assuming $\omega\tau \gg 1$, sketch roughly the dependence of ϵ' and ϵ'' on ω/ω_0, where ϵ' is the real part and ϵ'' is the imaginary part of ϵ.

8. **Dielectric constant of PbTe, using LST relation.** Experiments by R. S. Allgaier and W. W. Scanlon, Phys. Rev. **111**, 1029 (1958), on the mobility at low temperatures of electrons in the semiconductor lead telluride suggest that the static dielectric constant may be very large. The electrical conductivity interferes with low frequency measurements of the dielectric constant. It is found by neutron diffraction [W. Cochran, Phys. Letters **13**, 193 (1964)] that the low wavevector limit of the frequencies of the longitudinal and transverse optical phonons are $\nu_L = 3.2 \times 10^{12}$ Hz and $\nu_T = 1.0 \times 10^{12}$ Hz. With $\epsilon(\infty) = 28$, use the Lyddane-Sachs-Teller relation to show that $\epsilon(0) \approx 300$.

9. **Polarization of sphere.** A sphere of dielectric constant ϵ is placed in a uniform external electric field E_0. (a) What is the volume average electric field E in the sphere? (b) Show that the polarization in the sphere is $P = \chi E_0/[1 + (4\pi\chi/3)]$, where $\chi = (\epsilon - 1)/4\pi$. *Hint:* You do not need to calculate E_{loc} in this problem; in fact it is confusing to do so, because ϵ and χ are defined so that $P = \chi E$. We require E_0 to be unchanged by insertion of the sphere. We can produce a fixed E_0 by placing positive charges on one thin plate of an insulator and negative charges on an opposite plate. If the plates are always far from the sphere, the field of the plates will remain unchanged when the sphere is inserted between them. The results above are in CGS.

References

H. Fröhlich, *Theory of dielectrics: dielectric constant and dielectric loss,* Oxford, 1958, 2nd ed.

C. P. Smyth, *Dielectric behavior and structure,* McGraw-Hill, 1955.

J. H. Van Vleck, *Theory of electric and magnetic susceptibilities,* Oxford, 1932.

A. R. von Hippel, *Dielectrics and waves,* Wiley, 1954.

A. R. von Hippel, ed., *Dielectric materials and applications,* Wiley, 1954.

W. F. Brown, Jr., "Dielectrics," *Encyclo. of physics* **17,** 1–263 (1956).

V. Daniel, *Dielectric relaxation,* Academic Press, 1967.

J. B. Birks, ed., *Progress in dielectrics,* Chemical Rubber. A series.

14

Ferroelectric Crystals

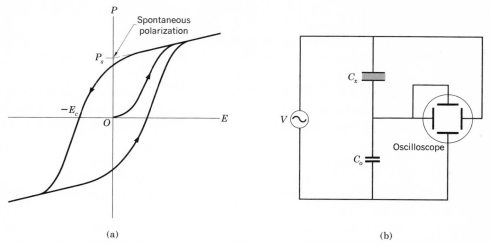

(a)   (b)

Figure 1 (a) Ferroelectric hysteresis loop; the coercive force is E_c. (b) Circuit for display of loop. The voltage across the ferroelectric crystal C_x is applied to the horizontal plates of the oscilloscope. The linear capacitor C_0 is in series with C_x. The voltage across C_0 is proportional to the polarization of C_x and is applied to the vertical plates. [After C. B. Sawyer and C. H. Tower, Phys. Rev. **35**, 269 (1930).]

Table 1   Ferroelectric crystals°

To obtain P_s in the SI unit of C m^{-2}, divide the value given in CGS by 3×10^5. To obtain P_s in μcoul/cm^2, divide the value in CGS by 3×10^3.

		T_c, °K	P_s in esu cm^{-2}, at T °K	
KDP type	KH$_2$PO$_4$	123	16,000	[96]
	KD$_2$PO$_4$	213	13,500	—
	RbH$_2$PO$_4$	147	16,800	[90]
	RbH$_2$AsO$_4$	111	—	—
	KH$_2$AsO$_4$	96	15,000	[80]
	KD$_2$AsO$_4$	162	—	—
	CsH$_2$AsO$_4$	143	—	—
	CsD$_2$AsO$_4$	212	—	—
TGS type	Tri-glycine sulfate	322	8,400	[293]
	Tri-glycine selenate	295	9,600	[273]
Perovskites	BaTiO$_3$	393	78,000	[296]
	SrTiO$_3$†	32	(9,000)	[4]
	WO$_3$	223	—	—
	KNbO$_3$	712	90,000	[523]
	PbTiO$_3$	763	>150,000	[300]
	LiTaO$_3$	—	70,000	[720]
	LiNbO$_3$	1470	900,000	

° A table of 76 ferroelectric crystals (excluding solid solutions) is given in App. A of F. Jona and G. Shirane, *Ferroelectric crystals*, Pergamon, 1962. This reference is an excellent source of data on the crystal structure of ferroelectrics. For further data, see E. Nakumura, T. Mitsui, and J. Furuichi, J. Phys. Soc. Japan **18**, 1477 (1963).

† It is not at all certain that SrTiO$_3$ has a ferroelectric phase; it may be paraelectric down to at least 1 °K. Possibly a ferroelectric phase can be induced by an electric field at low temperatures.

A ferroelectric crystal exhibits an electric dipole moment even in the absence of an external electric field. In the **ferroelectric state** the center of positive charge of the crystal does not coincide with the center of negative charge.

A typical plot of polarization versus electric field[1] for the ferroelectric state is shown in Fig. 1. The loop is called a **hysteresis loop;** it is a sign of a ferroelectric state. A crystal in a normal dielectric state usually does not show perceptible hysteresis when the electric field is increased and reversed slowly.

Ferroelectricity usually disappears above a certain temperature called the **transition temperature;** above the transition the crystal is said to be in a **paraelectric state.** The term paraelectric suggests an analogy with paramagnetism (Chapter 15) and implies a rapid decrease in the dielectric constant as the temperature increases.

CLASSIFICATION OF FERROELECTRIC CRYSTALS

We list in Table 1 some of the crystals commonly considered to be ferroelectric, along with the transition temperature or Curie point T_c at which the crystal changes from the low temperature polarized state to the high temperature unpolarized state. Thermal motion tends to destroy the ferroelectric order. Some ferroelectric crystals have no Curie point because they melt before leaving the ferroelectric phase. The table also includes values of the spontaneous polarization P_s.

Ferroelectric crystals may be classified into two main groups, order-disorder or displacive, according to whether the transition is associated with the ordering of ions or is associated with the displacement of a whole sublattice of ions of one type relative to another sublattice.

The order-disorder class of ferroelectrics includes crystals with hydrogen bonds in which the motion of the protons is related to the ferroelectric properties, as in potassium dihydrogen phosphate[2] (KH_2PO_4) and isomorphous salts.

---

[1] In some crystals the ferroelectric dipole moment may not be changed by an electric field of the maximum intensity which it is possible to apply without causing electrical breakdown of the crystal. In these crystals we are often able to observe a change in the spontaneous moment when they are heated: changing the temperature changes the value of the dipole moment. Such crystals are called **pyroelectric,** whereas crystals such that the direction of the spontaneous moment can be altered by an electric field are called ferroelectric. Lithium niobate, $LiNbO_3$, is pyroelectric at room temperature; it has a high transition temperature ($T_c = 1470°K$) and a very high saturation polarization. It can only be given a remanent polarization by an electric field applied over 1000°C.

[2] G. Busch and P. Scherrer, Naturwiss. **23,** 737 (1935); R. J. Mayer and J. L. Bjorkstam, J. Phys. Chem. Solids **23,** 619 (1962).

The behavior of crystals in which the hydrogen has been replaced by deuterium is interesting:

	KH_2PO_4	KD_2PO_4	KH_2AsO_4	KD_2AsO_4
Curie temperature	123°K	213°K	96°K	162°K

The substitution of deuterons for protons nearly doubles T_c, although the fractional change in the molecular weight of the compound is less than 2 percent. This extraordinarily large isotope shift is believed to be a quantum effect[3] involving the mass-dependence of the de Broglie wavelength.

Neutron diffraction data[4] show that above the Curie temperature the proton distribution along the hydrogen bond is symmetrically elongated. Below the Curie temperature the distribution is more concentrated and asymmetric with respect to neighboring ions, so that one end of the hydrogen bond is preferred by the proton over the other end.

The displacive class of ferroelectrics includes ionic crystal structures closely related to the perovskite and ilmenite structures. The simplest ferroelectric crystal[5] is GeTe with the sodium chloride structure. We shall devote ourselves primarily to crystals with the perovskite structure (Figs. 2 and 3).

[3] J. Pirenne, Physica **15**, 1019 (1949). A theory of the transition is given by J. C. Slater, J. Chem. Phys. **9**, 16 (1941); see also T. Nagamiya, Prog. Theor. Phys. **7**, 275 (1952); H. B. Silsbee, E. A. Uehling, and V. H. Schmidt, Phys. Rev. **133**, A165 (1964).

[4] B. C. Frazer and R. Pepinsky, Acta Cryst. **6**, 273 (1953); R. S. Pease and G. E. Bacon, Proc. Roy. Soc. (London) **A220**, 397 (1953); **A236**, 359 (1955).

[5] G. S. Pawley, W. Cochran, R. A. Cowley, and G. Dolling, Phys. Rev. Letters **17**, 753 (1966).

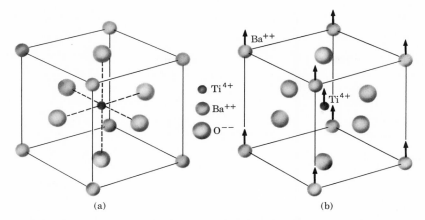

Figure 2   (a) The perovskite crystal structure of barium titanate. The prototype crystal is calcium titanate (perovskite). The structure is cubic, with Ba^{2+} ions at the cube corners, O^{2-} ions at the face centers, and a Ti^{4+} ion at the body center. (b) Below the Curie temperature the structure is slightly deformed, with Ba^{++} and Ti^{4+} ions displaced relative to the O^{2-} ions, thereby developing a dipole moment. It is possible that the upper and lower oxygen ions move downward slightly.

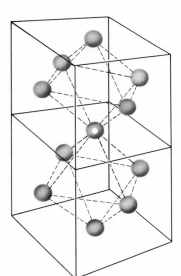

Figure 3   Two cubic cells of barium titanate, with only O^{2-} ions shown. These ions form interconnected octahedra with Ti^{4+} ions at the centers. The local symmetry about each oxygen ion is not cubic: thus the O^{2-}, Ti^{4+} or Ba^{2+} ions are not cubic with respect to the ion marked with a white spot.

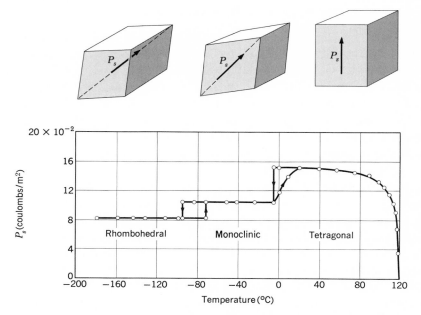

Figure 4 Spontaneous polarization projected on cube edge of barium titanate, as a function of temperature. The discontinuities near $0°C$ and $-80°C$ are caused by small changes in the crystal structure. The direction of spontaneous polarization is parallel to a cube edge above $0°C$, becomes parallel to a face diagonal below $0°C$ and parallel to a body diagonal below $-80°C$. [After W. J. Merz, Phys. Rev. **76**, 1221 (1949).]

Ferroelectricity in Ionic Crystals

Consider first the order of magnitude of the ferroelectric effects in barium titanate: The observed saturation polarization P_s at room temperature (Fig. 4) is 8×10^4 esu cm^{-2}. The volume of a unit cell is $(4 \times 10^{-8})^3 = 64 \times 10^{-24}$ cm^3, so that the dipole moment of a unit cell is

(CGS) $\qquad p \cong (8 \times 10^4 \text{ esu cm}^{-2})(64 \times 10^{-24} \text{ cm}^3) \cong 5 \times 10^{-18}$ esu cm ; $\qquad$ (1)

(SI) $\qquad p \cong (3 \times 10^{-1} \text{ C m}^{-2})(64 \times 10^{-30} \text{ m}^3) \cong 2 \times 10^{-29}$ C m .

If in the ideal perovskite structure the positive ions (Ba^{2+} and Ti^{4+}) were moved[6] by $\delta = 0.1$ Å with respect to the negative O^{2-} ions, the dipole moment of a unit cell would be $6e\delta \cong 3 \times 10^{-18}$ esu cm.

---

[6] The actual shifts are not accurately known. G. Shirane, H. Danner, and R. Pepinsky [Phys. Rev. **105**, 856 (1957)] report the following displacements in the tetragonal phase referred to the central four O^{2-} ions: δ(Ba) = 0.09 Å; δ(Ti) = 0.15 Å; δ(O$_I$) = -0.03 Å. In LiNbO$_3$ the displacements are considerably larger, being 0.9 Å and 0.5 Å for the lithium and niobium ions respectively.

THE POLARIZATION CATASTROPHE

Two related viewpoints contribute to an understanding of the occurrence of ferroelectricity. We may speak in terms of a polarization catastrophe in which for some critical condition the polarization becomes very large; or we may speak in terms of a transverse optical phonon of very low frequency.[7] The two viewpoints are related by the Lyddane-Sachs-Teller relation derived in Chapter 5.

In a polarization catastrophe the local electric field caused by the polarization increases faster than the elastic restoring force on an ion in the crystal, thereby leading to an asymmetrical shift in ionic positions. The shift is ultimately limited to a finite displacement by higher order restoring forces. The occurrence of ferroelectricity in an appreciable number of crystals with the perovskite structure suggests that this structure is favorably disposed to a polarizability catastrophe. Calculations of local fields by Slater[8] and others have made clear the physical reason for the favored position of the perovskite structure.

We give first the simple form of the catastrophe theory, supposing that the local field at all atoms is equal to $\mathbf{E} + 4\pi\mathbf{P}/3$ in CGS or $\mathbf{E} + \mathbf{P}/3\epsilon_0$ in SI, as derived in Chapter 13. The theory as given now leads to a second-order transition,[9] but the physical ideas can be carried over to a first-order transition.

We may rewrite (13.35) for the dielectric constant in the form

(CGS)
$$\epsilon = \frac{1 + \dfrac{8\pi}{3} \Sigma N_i\alpha_i}{1 - \dfrac{4\pi}{3} \Sigma N_i\alpha_i} , \tag{2}$$

where α_i is the electronic plus ionic polarizability of an ion of type i and N_i is the number of ions i per unit volume. The numerical factors multiplying $\Sigma N_i\alpha_i$ are the consequence of the Lorentz local field $E + 4\pi P/3$, in CGS.

The dielectric constant becomes infinite, corresponding to a finite polarization for zero applied field, when

(CGS)
$$\Sigma N_i\alpha_i = \frac{3}{4\pi} . \tag{3}$$

This is the condition for a polarization catastrophe.

The value of ϵ in (2) is sensitive to small departures of $\Sigma N_i\alpha_i$ from the critical value $3/4\pi$. If we write

(CGS)
$$\frac{4\pi}{3} \Sigma N_i\alpha_i = 1 - 3s , \tag{4}$$

---

[7] P. W. Anderson, Moscow dielectric conference, 1960; W. Cochran, Adv. in Physics **9**, 387 (1960); V. L. Ginzburg, Sov. Physics (Solid State) **2**, 1824 (1960).

[8] J. C. Slater, Phys. Rev. **78**, 748 (1950). We note that the O^{2-} ions in the perovskite structure do not have cubic surroundings. The local field factors at these ions turn out to be large.

[9] In a second-order transition there is no latent heat; the order parameter (in this instance, the polarization) is not discontinuous at the transition temperaure. In a first-order transition there is a latent heat; the order parameter changes discontinuously at the transition temperature.

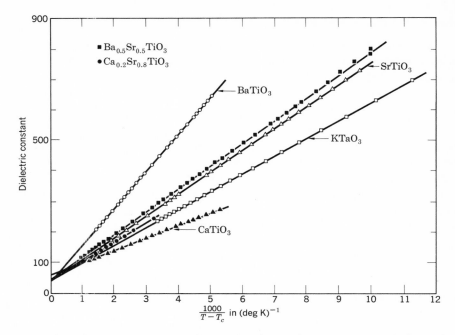

Figure 5 Dielectric constant versus $1/(T - T_c)$ in the paraelectric state $(T > T_c)$ of perovskites, after G. Rupprecht and R. O. Bell, Phys. Rev. **135**, A748 (1964). A plot versus $1/(T - T_0)$ would also be of interest.

where $s \ll 1$, we have for the dielectric constant in (2)

$$\epsilon \cong \frac{1}{s} \ . \tag{5}$$

Let us suppose that near the critical temperature the value of s in (4) varies with temperature in a linear fashion:

$$s \cong \frac{T - T_c}{\xi} \ , \tag{6}$$

where ξ is a constant. Such a variation of $\Sigma N_i \alpha_i$ might come about from the normal thermal expansion of the lattice. The dielectric constant has the form

$$\epsilon \cong \frac{\xi}{T - T_c} \ . \tag{7}$$

This is close to the observed temperature variation in the paraelectric state, Fig. 5.

Nature of the Phase Transition[10]

We can obtain a consistent formal thermodynamic theory of the behavior of a ferroelectric crystal by considering the form of the expansion of the energy as a function of the polarization P. We assume that the Landau[11] free energy density $\hat{F}$ in one dimension may be expanded formally as

$$\hat{F}(P;T,E) = -EP + g_0 + \tfrac{1}{2}g_2 P^2 + \tfrac{1}{4}g_4 P^4 + \tfrac{1}{6}g_6 P^6 + \cdots , \qquad (8a)$$

where the coefficients g_n depend on the temperature. The series does not contain terms in odd powers of P if the unpolarized crystal has a center of inversion symmetry. The value of P in thermal equilibrium is given by the minimum of $\hat{F}$ as a function of P; the value of $\hat{F}$ at this minimum defines the Helmholtz free energy $F(T,E)$. The equilibrium polarization in an electric field E satisfies

$$\frac{\partial \hat{F}}{\partial P} = 0 = -E + g_2 P + g_4 P^3 + g_6 P^5 + \cdots . \qquad (8b)$$

To obtain a ferroelectric state we must suppose that the coefficient of the term in P^2 in (8a) passes through zero at some temperature T_0:

$$g_2 = \gamma(T - T_0) , \qquad (9)$$

where γ is taken as a positive constant and T_0 may be equal to or lower than the transition temperature. A small positive value of g_2 means that the lattice is "soft" and is close to instability. A negative value of g_2 means that the unpolarized lattice is unstable. The variation of g_2 with temperature can be accounted for by thermal expansion and other effects of anharmonic lattice interactions.

---

[10] An example of a first-order phase transition is given by the liquid-vapor transition at constant pressure (the boiling of water). A ferroelectric with a first-order phase transition between the ferroelectric and the paraelectric state is distinguished by a discontinuous change (as in Fig. 10a) of the saturation polarization at the transition temperature. An example of a second-order phase transition is given by the transition between the ferromagnetic and paramagnetic states (Chapter 16). A good discussion of second-order phase transitions is given in Chap. 18 of J. C. Slater, *Introduction to Chemical physics*, McGraw-Hill, 1939; and in Chap. 14 of L. Landau and E. Lifshitz, *Statistical physics*, Pergamon, 1958. A power series expansion such as (8a) cannot always be made. For example, the transition in KH_2PO_4 appears to have a logarithmic singularity in the heat capacity at the transition. Such a singularity is not classifiable as either first- or second-order.

[11] In *TP*, see pp. 300 to 304 for a discussion of the Landau function and p. 361 for the term $-EP$. The Helmholtz free energy $F(T, E)$ is in the A scheme of *TP* Chapter 22. The possibility of crystals with reorientable polarization at angles different from 90° and the related occurrence of odd powers of P_s in the Landau function have been discussed in particular by L. A. Shuvalov and K. Aizu, and H. Schmid has reported a realization of this situation in the boracites; for references see the proceedings of the Kyoto conference cited at the end of the chapter.

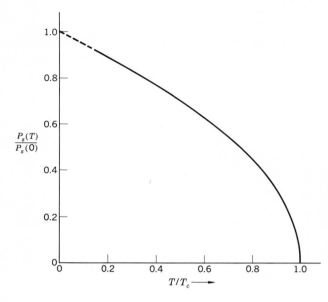

Figure 6 Spontaneous polarization versus temperature, for a second-order phase transition. (The curve is not realistic at low temperatures: the third law of thermodynamics requires that $dP_s/dT \to 0$ as $T \to 0$. Thus Eq. (9) cannot be a valid assumption near $T = 0$.)

Second-order Transition

If g_4 is positive, nothing new is added by the term in g_6, and this may then be neglected. The polarization for zero applied electric field is found from (8b):

$$\gamma(T - T_0)P_s + g_4 P_s{}^3 = 0 , \tag{10}$$

so that either $P_s = 0$ or

$$P_s{}^2 = (\gamma/g_4)(T_0 - T) . \tag{11}$$

For $T \geq T_0$ the only real root of (10) is at $P_s = 0$, because γ and g_4 are positive. Thus T_0 is the Curie temperature.

For $T < T_0$ the minimum of the Landau free energy in zero applied field is at

$$|P_s| = (\gamma/g_4)^{\frac{1}{2}}(T_0 - T)^{\frac{1}{2}} , \tag{12}$$

as plotted in Fig. 6. The phase transition is a second-order phase transition because the polarization (12) goes continuously to zero at the transition temperature. Plots of the Landau free energy versus P^2 are given in Fig. 7 for three representative temperatures. The transition in triglycine sulfate is an example (Fig. 8) of a second-order transition; the observed exponent[12] of $(T_0 - T)$ in the saturation polarization P_s is 0.51 $\pm$0.05.

[12] J. A. Gonzalo, Phys. Rev. **144**, 662 (1966); P. P. Craig, Physics Letters **20**, 140 (1966).

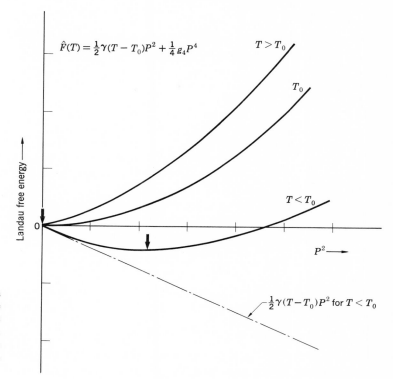

$$\hat{F}(T) = \tfrac{1}{2}\gamma(T - T_0)P^2 + \tfrac{1}{4}g_4P^4$$

$T > T_0$

T_0

$T < T_0$

$P^2 \longrightarrow$

$-\tfrac{1}{2}\gamma(T - T_0)P^2$ for $T < T_0$

Landau free energy

0

Figure 7 Landau free energy function versus (polarization)^2, at representative temperatures. As the temperature drops below T_0 the equilibrium polarization gradually increases, as defined by the position of the minimum of the free energy.

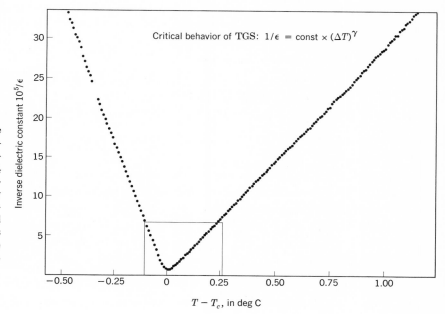

Critical behavior of TGS: $1/\epsilon = \text{const} \times (\Delta T)^\gamma$

Inverse dielectric constant $10^5/\epsilon$

$T - T_c$, in deg C

Figure 8 Plot of inverse dielectric constant of triglycine sulfate versus temperature difference $\Delta T = T - T_c$ from the transition temperature, 49.92°C. The linearity is as predicted for a second-order phase transition, and the difference in slopes is also accounted for by the theory. (After J. A. Gonzalo.)

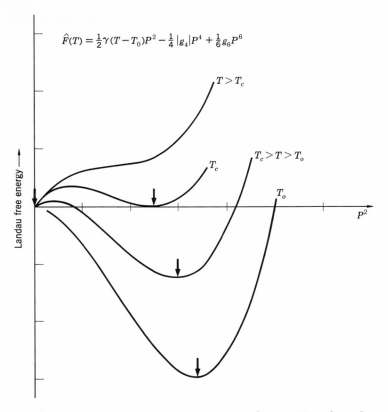

Figure 9 Landau free energy function versus (polarization)^2 in a first-order transition, at representative temperatures. At T_c the Landau function has equal minima at $P = 0$ and at a finite P as shown. For T below T_c the absolute minimum is at larger values of P; as T passes through T_c there is a discontinuous change in the position of the absolute minimum. The arrows mark the minima.

First-order Transition

The transition is first-order if g_4 is negative. We must now retain g_6 and take it positive in order to restrain $\hat{F}$ from going to minus infinity (Fig. 9). The equilibrium condition for $E = 0$ is given by (8b):

$$\gamma(T - T_0)P_s - |g_4|P_s{}^3 + g_6 P_s{}^5 = 0 \ , \tag{13}$$

so that either $P_s = 0$ or

$$\gamma(T - T_0) - |g_4|P_s{}^2 + g_6 P_s{}^4 = 0 \ . \tag{14}$$

At the transition temperature T_c the free energies of the paraelectric and ferroelectric phases will be equal. That is, the value of F for $P_s = 0$ will be equal to the value of F at the minimum given by (14). In Fig. 10 we show the characteristic variation with temperature of P_s for a first-order phase transition; contrast this with the variation shown in Fig. 6a for a second-order phase transition. The transition in $BaTiO_3$ is first order.

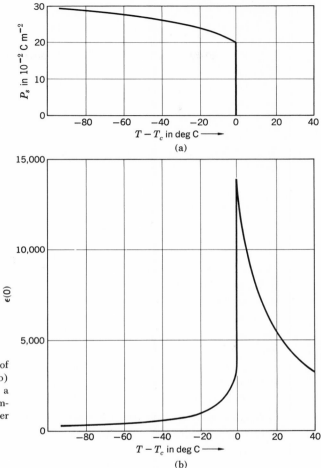

Figure 10 Calculated values (a) of the spontaneous polarization and (b) of the static dielectric constant as a function of temperature, with parameters as for barium titanate. (After W. Cochran.)

The dielectric constant is calculated from the equilibrium polarization in an applied electric field E, and is given by (8b). In equilibrium at temperatures over the Curie temperature the terms in P^4 and P^6 may usually be neglected; thus

$$E = \gamma(T - T_0)P \; , \qquad (15)$$

or

(CGS) $\qquad \epsilon(T > T_c) = 1 + 4\pi \dfrac{P}{E} = 1 + \dfrac{4\pi/\gamma}{T - T_0} \; . \qquad (16)$

This has the form of (7). The result (16) applies whether the transition is of the first or second order, but if second order we have $T_0 = T_c$; if first order, then $T_0 < T_c$. We recall that T_0 is defined by (9), while T_c is the temperature at which the transition actually occurs.

Table 2   Ferroelectric constants

Crystal	C, in 10^4 °K	T_c, in °K	T_0, in °K
$BaTiO_3$	17	381	370
$KNbO_3$	27	683	623
$PbTiO_3$	11	763	693
KH_2PO_4	0.3	123	123

If we write the temperature-dependent part of the dielectric constant as $C/(T - T_0)$, then selected experimental data may be represented as in Table 2.

LOW FREQUENCY OPTICAL PHONONS

The Lyddane-Sachs-Teller relation (Chapter 5) says that

$$\frac{\omega_T{}^2}{\omega_L{}^2} = \frac{\epsilon(\infty)}{\epsilon(0)} \ . \tag{17}$$

Here ω_T is the frequency of a transverse optical phonon near zero wavevector. It is important for ferroelectricity that (17) requires the static dielectric constant to become infinite when the transverse optical phonon frequency approaches zero. Whenever the static dielectric constant $\epsilon(0)$ is observed to have a very high value, such as 100 to 10,000, we expect that ω_T will have a very low value. When $\omega_T = 0$ the crystal is unstable because there is no effective restoring force. The ferroelectric $BaTiO_3$ at 24°C has a low frequency optical mode[13] centered at 12 cm^{-1}: this is quite a low frequency for an optical mode.

If the transition to a ferroelectric state is first order, we do not find $\omega_T = 0$ or $\epsilon(0) = \infty$ at the transition. The LST relation suggests only that $\epsilon(0)$ extrapolates to a singularity at a temperature T_0. In the usual first-order transition the transition occurs at a temperature T_c higher than T_0; then ω_T never becomes zero and $\epsilon(0)$ never becomes infinite.

Experiments with Strontium Titanate

The association of a high static dielectric constant with a low frequency optical mode is supported very well by experiments on strontium titanate, $SrTiO_3$. The dielectric constant (Fig. 11) takes on very large values near 30°K and below.[14] Neutron diffraction experiments (Fig. 12) show a large decrease in the frequency of the lowest optical phonon between room temperature and 90°K.[15] The extrapolation in Fig. 13 of the phonon frequency goes through zero near 32°K, in approximate agreement with the extrapolation of the reciprocal of the dielectric constant. According to the LST relation, if the recipro-

[13] J. M. Ballantyne, Phys, Rev. **136**, A429 (1964).

[14] At present definite proof of the state of the crystal below 30°K is lacking.

[15] For a discussion of the lattice dynamics of $SrTiO_3$, see R. A. Cowley, Phys. Rev. **134**, A981 (1964). It is possible to obtain large changes in the frequency of the lowest transverse optical phonon by only slight changes in the parameters which describe the interatomic forces.

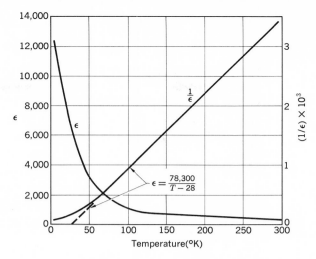

Figure 11 Dielectric constant of a SrTiO$_3$ crystal versus temperature, after T. Mitsui and W. B. Westphal, Phys. Rev. **124**, 1354 (1961).

$$\epsilon = \frac{78{,}300}{T - 28}$$

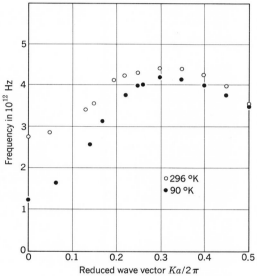

Figure 12 The dispersion curve of the lowest frequency transverse optic branch in strontium titanate along [100] at 90°K and 296°K, after R. A. Cowley, Phys. Rev. Letters 9, 159 (1962).

○ 296 °K
● 90 °K

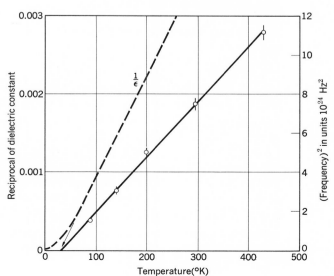

Figure 13 Plot of the square of the frequency of the zero wavevector transverse optic mode against temperature, for SrTiO$_3$, as observed in neutron diffraction experiments by Cowley. The solid line gives a Curie temperature of 32 ± 5°K. The broken line represents the reciprocal of the dielectric constant from the measurements of Mitsui and Westphal.

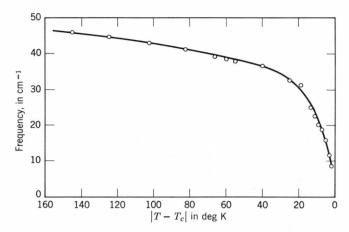

Figure 14 Decrease of a transverse phonon frequency as the Curie temperature is approached from below, in the ferroelectric crystal antimony sulpho-iodide (SbSI). [After Raman scattering experiments by C. H. Perry and D. K. Agrawal, Solid State Comm. **8**, 225 (1970).]

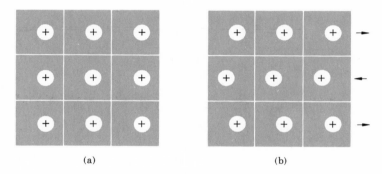

Figure 15 Comparison of (a) ferroelectric and (b) antiferroelectric deformations. Only the displacement of positive ions is shown.

cal of the static dielectric constant has a temperature dependence $1/\epsilon(0) \propto (T - T_0)$ over some region of temperature, then the square of the optical mode frequency will have a similar temperature dependence: $\omega_T{}^2 \propto (T - T_0)$, if ω_L is independent of temperature. The result for $\omega_T{}^2$ is very well confirmed by Fig. 13. Measurements of ω_T versus T for another ferroelectric crystal, SbSI, are shown in Fig. 14.

Antiferroelectricity

A ferroelectric displacement is not the only type of instability which may develop in a dielectric crystal. In the perovskite structure other deformations occur. These deformations, even if they do not give a spontaneous polarization, may be accompanied by changes in the dielectric constant. One type of deformation is called **antiferroelectric** and has neighboring lines of ions displaced in opposite senses, as in Fig. 15.

Table 3   Antiferroelectric crystals

(From a compilation by Walter J. Merz)

Crystal	Transition temperature to antiferroelectric state, in °K.
WO_3	1010
$NaNbO_3$	793, 911
$PbZrO_3$	506
$PbHfO_3$	488
$NH_4H_2PO_4$	148
$ND_4D_2PO_4$	242
$NH_4H_2AsO_4$	216
$ND_4D_2AsO_4$	304
$(NH_4)_2H_3IO_6$	254

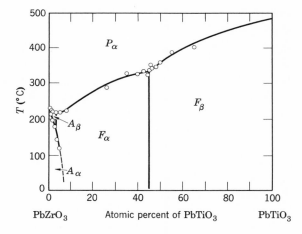

Figure 16   Phase diagram of $PbZrO_3$–$PbTiO_3$ system; *A*, *F*, *P* denote antiferroelectric, ferroelectric, and paraelectric, respectively. [After F. Sawaguchi. J. Phys. Soc. Japan **8**, 615 (1953).]

The perovskite structure appears to be susceptible to many types of deformation, often with little difference in energy between them. The phase diagrams of mixed perovskite systems, such as the $PbZrO_3$–$PbTiO_3$ system in Fig. 16, show transitions between para-, ferro-, and antiferroelectric states.

Ordered antiferroelectric arrangements of permanent electric dipole moments occur at low temperatures in ammonium salts and in hydrogen halides. Several crystals believed to have an ordered nonpolar state are listed in Table 3.

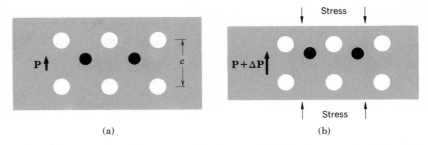

Figure 17 (a) Unstressed ferroelectric crystal and (b) stressed ferroelectric crystal. The stress changes the polarization by Δ**P**, the induced piezoelectric polarization.

Piezoelectricity

All crystals in a ferroelectric state are also piezoelectric: a stress Z applied to the crystal will change the electric polarization (Fig. 17). Similarly, an electric field E applied to the crystal will cause the crystal to become strained. In schematic one-dimensional notation, the piezoelectric equations are

$$(\text{CGS}) \qquad P = Zd + E\chi \; ; \qquad e = Zs + Ed \; , \qquad (18)$$

where P is the polarization, Z the stress, d the **piezoelectric strain constant**, E the electric field, χ the dielectric susceptibility, e the elastic strain, and s the elastic compliance constant. To obtain (18) in SI, replace χ by $\epsilon_0\chi$.

These relations exhibit the development of polarization by an applied stress and the development of elastic strain by an applied electric field. The first aspect is exploited in strain gauges and in the detection of ultrasonic waves; the second aspect is exploited in ultrasonic generators.

A crystal may be piezoelectric without being ferroelectric: a schematic example of such a structure is given[16] in Fig. 18. Quartz is piezoelectric, but not ferroelectric; barium titanate is both. For order of magnitude, in quartz $d \approx 10^{-7}$ cm/statvolt and in barium titanate $d \approx 10^{-5}$ cm/statvolt.

The general definition of the piezoelectric strain constants is

$$d_{ik} = (\partial e_k/\partial E_i)_Z \; , \qquad (19)$$

where $i \equiv x, y, z$ and $k \equiv xx, yy, zz, yz, zx, xy$. To convert to cm/statvolt from values of d_{ik} given in meters/volt, multiply by 3×10^4.

Ferroelectric Domains

Consider a ferroelectric crystal (such as barium titanate in the tetragonal phase) in which the spontaneous polarization may be either up or down the c axis of the crystal. A ferroelectric crystal generally consists of regions called **domains** within each of which the polarization is in the same direction,

---

[16] See also Chap. 11 of A. Holden and P. Singer, *Crystals and crystal growing*, Doubleday Anchor, 1960.

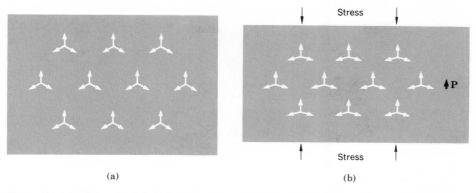

Figure 18 (a) The unstressed crystal has a three-fold symmetry axis. The arrows represent dipole moments, each set of three arrows represents a planar group of ions denoted by $A_3^+B^{3-}$, with a B^{3-} ion at each vertex. The sum of the three dipole moments at each vertex is zero. (b) The crystal when stressed develops a polarization in the direction indicated. The sum of the dipole moments about each vertex is no longer zero.

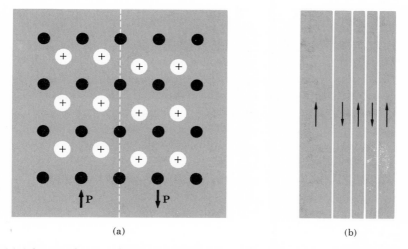

Figure 19 (a) Schematic drawing of atomic displacements on either side of a boundary between domains polarized in opposite directions in a ferroelectric crystal; (b) view of a domain structure, showing 180° boundaries between domains polarized in opposite directions.

but in adjacent domains the polarization is in different directions. In our example in Fig. 19 the polarization is in opposite directions. The net polarization of the crystal will depend on the difference in the volumes of the upward- and downward-directed domains. The crystal as a whole will appear to be unpolarized, as measured by the charge on electrodes covering the ends, when the volumes of domains in opposite senses are equal. The total dipole moment of the crystal may be changed by the movement of the walls between domains or by the nucleation of new domains.

494

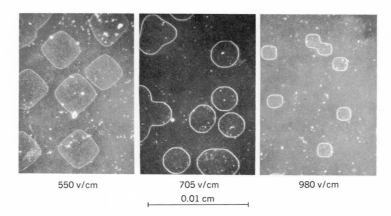

550 v/cm   705 v/cm   980 v/cm

0.01 cm

Figure 20 Ferroelectric domains on the face of a single crystal of barium titanate. The face is normal to the tetragonal or c axis. The net polarization of the crystal as judged by domain volumes is increased markedly as the electric field intensity parallel to the axis is increased from 550 volts/cm to 980 volts/cm. The domain boundaries are made visible by etching the crystal in a weak acid solution. (Courtesy of R. C. Miller.)

We exhibit in Fig. 20 a series of photomicrographs of a single crystal of barium titanate in an electric field normal to the plane of the photographs and parallel to the tetragonal axis. The closed curves are boundaries between domains polarized into and out of the plane of the photographs. The domain boundaries move sidewise and change size and shape when the intensity of the electric field is altered. In $BaTiO_3$ the domain widths are characteristically 10^{-4} to 10^{-2} cm; the thickness of the boundary between domains may be only one lattice constant.

The motion of domain walls in ferroelectrics is not simple: it is known[17] that in an electric field a 180° wall[18] in $BaTiO_3$ does not move as a whole perpendicular to itself, but the wall motion appears to result from the repeated nucleation of steps by thermal fluctuations along the parent wall. The nucleation rate[19] is the controlling factor in the propagation of the wall. This is not unlike the usual situation with ferromagnetic domain walls (Chapter 16).

---

[17] W. J. Merz, Phys. Rev. **95**, 690 (1954); R. Landauer, J. Appl. Phys. **28**, 227 (1957); R. C. Miller and A. Savage, Phys. Rev. **115**, 1176 (1959).

[18] A "180° wall" is the boundary between regions having opposite polarization directions.

[19] R. C. Miller and G. Weinreich, Phys. Rev. **117**, 1460 (1960).

Problems

1. **Ferroelectric criterion for atoms.** Consider a system of two neutral atoms separated by a fixed distance a, each atom having a polarizability α. Find the relation between a and α for such a system to be ferroelectric. *Hint:* The dipolar field is strongest along the axis of the dipole.

2. **Saturation polarization at Curie point.** In a first-order transition the equilibrium condition (14) with T set equal T_c gives one equation for the polarization $P_s(T_c)$. A further condition at the Curie point is that $\hat{F}(P_s, T_c) = \hat{F}(0, T_c)$. (a) Combining these two conditions, show that $P_s^2(T_c) = 3|g_4|/4g_6$. (b) Using this result, show that $T_c = T_0 + 3g_4^2/16\gamma g_6$.

3. **Piezoelectricity.** (a) Consider a flat plate of a piezoelectric crystal with the plane of the plate normal to the direction of E and Z considered in (18). Find expressions for the effective elastic compliance constant when the surfaces of the plate are short-circuited and when they are electrically open. (b) Consider a longitudinal sound wave of strain $e(x, t) = A \cos(kx - \omega t)$. Find an expression for the polarization $P(x, t)$. Using the Maxwell equation div $(\mathbf{E} + 4\pi\mathbf{P}) = 0$, find $E(x, t)$. This electric field is important in determining the interaction of electrons with elastic waves in piezoelectric semiconductors. In SI the relevant Maxwell equation is div $(\epsilon_0\mathbf{E} + \mathbf{P}) = 0$.

References

J. C. Burfoot, *Ferroelectrics, an introduction to the physical principles,* Van Nostrand, 1967.

E. Fatuzzo and W. J. Merz, *Ferroelectricity,* North-Holland (Interscience), 1967.

E. F. Weller, ed., *Ferroelectricity,* Elsevier, 1967. Valuable symposium survey.

"Ferroelectrics," IEEE Transactions on Electron Devices, vol. ED-16, No. 6, June 1969. Survey of ferroelectric technology.

A. F. Devonshire, "Theory of ferroelectrics," Advances in Physics **3**, 85–130 (1954).

W. P. Mason, *Piezoelectric crystals,* Van Nostrand, 1950.

F. Jona and G. Shirane, *Ferroelectric crystals,* Pergamon, 1962.

W. Cochran, "Crystal stability and the theory of ferroelectricity," Advances in Physics **9**, 387 (1960); **10**, 401–420 (1961).

W. Känzig, "Ferroelectrics and antiferroelectrics," *Solid state physics* **4**, 1–197 (1957).

P. W. Forsbergh, "Piezoelectricity, electrostriction, and ferroelectricity," *Encyclo. of physics* **17**, 264–392 (1956).

Proceedings of the Second International Conference on Ferroelectricity, Kyoto, 1969. J. Phys. Soc. Japan **28**, Suppl. (1970).

15

Diamagnetism and Paramagnetism

NOTATION: In the problems treated in this chapter the magnetic field B is always closely equal to the applied field B_a, so that we write B for B_a in most instances.

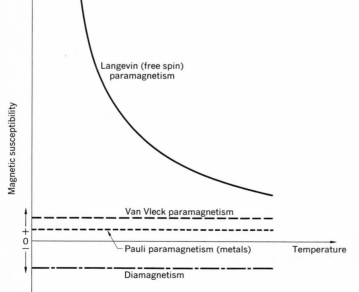

Figure 1 Characteristic magnetic susceptibilities of diamagnetic and paramagnetic substances.

This is the first of three chapters on magnetism. Chapter 16 treats ferromagnetism and antiferromagnetism, and Chapter 17 treats magnetic resonance.

Magnetism is inseparable from quantum mechanics, for a strictly classical system in thermal equilibrium can display no magnetic moment, even in a magnetic field. This theorem, although not obvious, is true[1]; and were the value of $\hbar$ to go to zero, the loss of the science of magnetism is one of the catastrophes that would overwhelm the universe.

The magnetic moment of a free atom has three principal sources: the spin with which electrons are endowed; their orbital angular momentum about the nucleus; and the change in the orbital moment induced by an applied magnetic field. We shall see that the first two effects give paramagnetic contributions to the magnetization, and the third effect gives a diamagnetic contribution. In the ground $1s$ state of the hydrogen atom the orbital moment is zero, and the magnetic moment is chiefly that of the electron spin along with a small induced diamagnetic moment. In the $1s^2$ state of helium the spin and orbital moments are both zero, and there is only an induced moment. Atoms with filled electron shells have zero spin and orbital moments: these moments are associated with unfilled shells.

The magnetization M is defined as the magnetic moment per unit volume. The **magnetic susceptibility** per unit volume is defined as

$$(\text{CGS}) \quad \chi = \frac{M}{B} ; \qquad\qquad (\text{SI}) \quad \chi = \frac{\mu_0 M}{B} , \qquad (1)$$

where B is the macroscopic magnetic field intensity. In both systems of units χ is dimensionless. We shall sometimes for convenience refer to M/B as the susceptibility without specifying the system of units. Quite frequently a susceptibility is defined referred to unit mass or to a mole of the substance. The molar susceptibility is written as χ_M; the magnetic moment per gram is sometimes written as σ.

Substances with a negative magnetic susceptibility are called **diamagnetic.** Substances with a positive susceptibility are called **paramagnetic,** as in Fig. 1. Ordered arrays of magnetic moments are discussed in Chapter 16; the arrays may be ferromagnetic, ferrimagnetic, antiferromagnetic, helical, or more complex in form. Nuclear magnetic moments give rise to **nuclear paramagnetism.** Magnetic moments of nuclei are of the order of 10^{-3} times smaller than the magnetic moment of the electron.

---

[1] The theorem, due to J. H. van Leeuwen, is treated by J. H. Van Vleck, *Electric and magnetic susceptibilities,* Oxford, 1932, pp. 94–104.

LANGEVIN DIAMAGNETISM EQUATION

Diamagnetism is associated with the tendency of electrical charges partially to shield the interior of a body from an applied magnetic field. In electromagnetism we are familiar with Lenz's law: when the flux through an electrical circuit is changed, an induced current is set up in such a direction as to oppose the flux change. In a resistanceless circuit, in a superconductor, or in an electron orbit within an atom, the induced current persists as long as the field is present. The magnetic field produced by the induced current is opposite to the applied field, and the magnetic moment associated with the current is a diamagnetic moment. Even in a normal metal there is a diamagnetic contribution from the conduction electrons, and this diamagnetism is not destroyed by collisions of the electrons.

The usual treatment of the diamagnetism of atoms and ions employs the Larmor theorem[2]: in a magnetic field the motion of the electrons around a central nucleus is, to the first order in B, the same as a possible motion in the absence of B except for the superposition of a precession of the electrons with angular frequency

$$\text{(CGS)} \quad \omega = \frac{eB}{2mc} \; ; \qquad\qquad \text{(SI)} \quad \omega = \frac{eB}{2m} \; . \qquad (2)$$

If the field is applied slowly, the motion in the rotating reference system will be the same as the original motion in the rest system before the application of the field. If the average electron current around the nucleus is zero initially, the application of the magnetic field will cause a finite average current around the nucleus. The current thus established is equivalent to a magnetic moment. The direction of the moment is opposite to the direction of the applied field.

The Larmor precession of Z electrons is equivalent to an electric current

$$\text{(SI)} \quad I = (\text{charge})\,(\text{revolutions per unit time}) = (-Ze)\left(\frac{1}{2\pi} \cdot \frac{eB}{2m}\right) . \qquad (3)$$

The magnetic moment μ of a current loop is given by the product (current) $\times$ (area of the loop). The area of a loop of radius ρ is $\pi\rho^2$. We have

$$\text{(SI)} \quad \mu = -\frac{Ze^2 B}{4m} \langle \rho^2 \rangle \; ; \qquad\qquad \text{(CGS)} \quad \mu = -\frac{Ze^2 B}{4mc^2} \langle \rho^2 \rangle \; . \qquad (4)$$

Here

$$\langle \rho^2 \rangle = \langle x^2 \rangle + \langle y^2 \rangle \qquad (5)$$

is the mean square of the perpendicular distance of the electron from the field

[2] H. Goldstein, *Classical mechanics*, Addison-Wesley, 1953, pp. 176–178. The Larmor frequency is one-half the cyclotron frequency of a free electron in the magnetic field.

axis through the nucleus. The mean square distance of the electrons from the nucleus is

$$\langle r^2 \rangle = \langle x^2 \rangle + \langle y^2 \rangle + \langle z^2 \rangle \ . \tag{6}$$

For a spherically symmetrical distribution of charge we have $\langle x^2 \rangle = \langle y^2 \rangle = \langle z^2 \rangle$, so that

$$\langle r^2 \rangle = \tfrac{3}{2}\langle \rho^2 \rangle \ . \tag{7}$$

From (4) and (7) the diamagnetic susceptibility per unit volume is, if N is the number of atoms per unit volume,

(CGS) $$\chi = \frac{N\mu}{B} = -\frac{NZe^2}{6mc^2}\langle r^2 \rangle \ ; \tag{8}$$

(SI) $$\chi = \frac{\mu_0 N\mu}{B} = -\frac{\mu_0 NZe^2}{6m}\langle r^2 \rangle \ .$$

This is the classical Langevin result. A quantum-theoretical derivation is given in Advanced Topic M.

The problem of calculating the diamagnetic susceptibility of an isolated atom is reduced to the calculation of $\langle r^2 \rangle$ for the electron distribution within the atom. The distribution can be calculated by quantum mechanics. Experimental values for neutral atoms are most easily obtained for the inert gases. Typical experimental values of the molar susceptibilities are the following:

	He	Ne	Ar	Kr	Xe
χ_M in CGS in 10^{-6} cm^3/mole:	-1.9	-7.2	-19.4	-28.0	-43.0

In dielectric solids the diamagnetic contribution of the ion cores is described roughly by the Langevin result. The contribution of conduction electrons is more complicated, as is evident from the de Haas–van Alphen effect discussed in Chapter 10.

Diamagnetism of Molecules

The derivation of the Larmor equation assumes implicitly that the field direction is an axis of symmetry of the system. In most molecular systems this condition is not satisfied, and the general theory of Van Vleck must be applied. For a polyatomic molecule with spin quantum number zero we have, according to Advanced Topic M, the total molar susceptibility

(CGS) $$\chi_M = -\frac{N_0 e^2}{6mc^2}\sum \langle r^2 \rangle + 2N_0 \sum_s \frac{|\langle s|\mu_z|0\rangle|^2}{E_s - E_0} \ , \tag{9}$$

where N_0 is the Avogadro number, $\langle s|\mu_z|0\rangle$ is the matrix element of the z component of the orbital magnetic moment connecting the ground state 0 with the excited state s, and $E_s - E_0$ is the energy separation of the two states. The material is diamagnetic or paramagnetic according to whether the first or second term of (9) is greater. The second term is known as **Van Vleck paramagnetism.**

For the normal state of the H_2 molecule Van Vleck and Frank[3] calculate

$$\chi_M = -4.71 \times 10^{-6} + 0.51 \times 10^{-6} = -4.20 \times 10^{-6} \text{ cm}^3/\text{mole.}$$

The experimental values are between -3.9 and -4.0×10^{-6}.

PARAMAGNETISM

Electronic paramagnetism (positive contribution to χ) is found in:

a. Atoms, molecules, and lattice defects possessing an odd number of electrons, as here the total spin of the system cannot be zero. Examples: free sodium atoms; gaseous nitric oxide (NO); organic free radicals such as triphenylmethyl, $C(C_6H_5)_3$; F centers in alkali halides.

b. Free atoms and ions with a partly filled inner shell: transition elements; ions isoelectronic with transition elements; rare earth and actinide elements. Examples: Mn^{2+}, Gd^{3+}, U^{4+}. Paramagnetism is exhibited by many of these ions when incorporated into solids, but not invariably.

c. A few compounds with an even number of electrons, including molecular oxygen and organic biradicals.

d. Metals.

We shall consider classes (b) and (d).

LANGEVIN PARAMAGNETISM EQUATION AND THE CURIE LAW

We treat a medium containing N atoms per unit volume, each bearing a magnetic moment $\boldsymbol{\mu}$. Magnetization results from the orientation of the magnetic moments in an applied magnetic field, but thermal disorder resists the tendency of the field to orient the moments. The energy of interaction of the moment $\boldsymbol{\mu}$ with an applied magnetic field $\mathbf{B}$ is

$$U = -\boldsymbol{\mu} \cdot \mathbf{B} . \tag{10}$$

The magnetization in thermal equilibrium is calculated by following exactly the steps (13.46) to (13.49) in the derivation of the Debye orientation polarizability, with $\boldsymbol{\mu}$ written for $\mathbf{p}$ and $\mathbf{B}$ for $\mathbf{E}$. The magnetization is then given by the Langevin equation

$$M = N\mu L(x) , \tag{11}$$

[3] J. H. Van Vleck and A. Frank, Proc. Natl. Acad. Sci. U.S. **15**, 539 (1929).

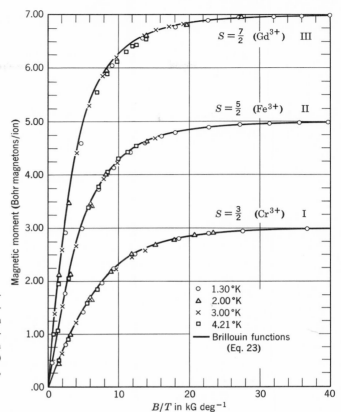

Figure 2 Plot of magnetic moment versus B/T for spherical samples of (I) potassium chromium alum. (II) ferric ammonium alum, and (III) gadolinium sulfate octahydrate. Over 99.5% magnetic saturation is achieved at 1.3°K and about 50,000 gauss. [After W. E. Henry, Phys. Rev. 88, 559 (1952).]

where $x \equiv \mu B/k_B T$, and the Langevin function $L(x)$ is

$$L(x) \equiv \operatorname{ctnh} x - \frac{1}{x} . \qquad (12)$$

For $x \ll 1$, we have $L(x) \cong x/3$ from (13.50), so that the magnetization is

$$M \cong \frac{N\mu^2 B}{3k_B T} = \frac{C}{T} B , \qquad (13)$$

where the **Curie constant**

$$C \equiv \frac{N\mu^2}{3k_B} . \qquad (14)$$

The result (13) is known as the **Curie law** and applies in the limit $\mu B \ll k_B T$. For an electron $\mu = 0.927 \times 10^{-20}$ erg/gauss or 0.927×10^{-23} joule/tesla. At room temperature in a field of 10^4 gauss we have $\mu B/k_B T \approx 2 \times 10^{-3}$, so that under these conditions we may safely approximate the Langevin function by $\mu B/3k_B T$. At low temperatures saturation effects are observed, as shown in Fig. 2.

QUANTUM THEORY OF PARAMAGNETISM

The magnetic moment of an atom or ion in free space is given by

$$\boxed{\mu = \gamma\hbar\mathbf{J} = -g\mu_B\mathbf{J} \ ,}$$ (15)

where the total angular momentum $\hbar\mathbf{J}$ is the sum of the orbital $\hbar\mathbf{L}$ and spin $\hbar\mathbf{S}$ angular momenta. The constant γ is the ratio of the magnetic moment to the angular momentum; γ is called the **gyromagnetic ratio** or **magnetogyric ratio.**

For electronic systems a quantity g is defined as in (15) by

$$g\mu_B \equiv -\gamma\hbar \ .$$ (15a)

The quantity g is called the **g factor** or the **spectroscopic splitting factor.** It represents the ratio of the number of Bohr magnetons to the units of $\hbar$ of angular momentum. For an electron spin $g = 2.0023$, usually taken as 2.00. For a free atom with an orbital angular momentum the g factor is given by the Landé equation[4]

$$g = 1 + \frac{J(J+1) + S(S+1) - L(L+1)}{2J(J+1)} \ .$$ (16)

The **Bohr magneton** μ_B is defined as $e\hbar/2mc$ in CGS and $e\hbar/2m$ in SI. It is closely equal to the spin magnetic moment of a free electron.

The energy levels of the system in a magnetic field are

$$E = m_J g\mu_B B \ ,$$ (17)

where m_J is the azimuthal quantum number and has the values $J, J - 1, \cdots,$ $-J$. For a single spin with no orbital moment we have $m_J = \pm\frac{1}{2}$ and $g = 2$, whence

$$E = \pm\mu_B B \ .$$ (18)

---

[4] For a derivation see Born, pp. 164 and 370.

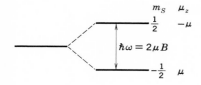

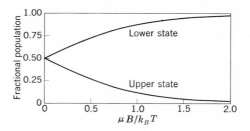

Figure 3 Energy levels splitting scheme for one electron, with only spin angular momentum, in a magnetic field B directed along the positive z axis. For an electron the magnetic moment μ is opposite in sign to the spin S, so that $\mu = -g\mu_B S$. In the low energy state the magnetic moment is parallel to the magnetic field.

Figure 4 Fractional populations of a two-level spin system in thermal equilibrium at temperature T in a magnetic field B. The magnetic moment is proportional to the difference between the two curves.

This splitting is shown in Fig. 3.

If a system has only two levels the equilibrium populations are, with $\tau \equiv k_B T$,

$$\frac{N_1}{N} = \frac{\exp{(\mu B/\tau)}}{\exp{(\mu B/\tau)} + \exp{(-\mu B/\tau)}} \; ;$$

$$\frac{N_2}{N} = \frac{\exp{(-\mu B/\tau)}}{\exp{(\mu B/\tau)} + \exp{(-\mu B/\tau)}} \; ; \tag{19}$$

here N_1, N_2 are the populations of the lower and upper levels, and $N = N_1 + N_2$ is the total number of atoms. The fractional populations are plotted in Fig. 4.

The projection of the magnetic moment of the upper state along the field direction is $-\mu$ and of the lower state is μ. The resultant magnetization for N atoms per unit volume is

$$M = (N_1 - N_2)\mu = N\mu \cdot \frac{e^x - e^{-x}}{e^x + e^{-x}} = N\mu \tanh x \; , \tag{20}$$

where $x \equiv \mu B/k_B T$. Note that the function L in (11) and tanh in (20) are different because of the difference between continuous orientation and quantized orientation. The low field expansions of the functions are also different.

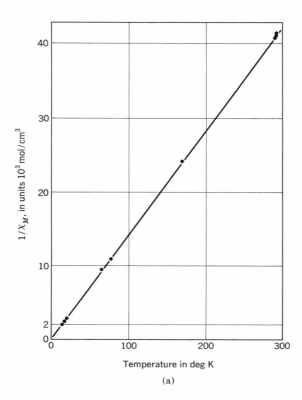

Figure 5a Plot of $1/\chi$ vs T for a gadolinium salt, $Gd(C_2H_5SO_4)_3 \cdot 9H_2O$. The straight line is the Curie law. (After L. C. Jackson and H. Kamerlingh Onnes, Leiden Communications 168a.)

For $x \ll 1$, $\tanh x \cong x$, and we have

$$M \cong N\mu(\mu B/k_B T) \ , \tag{21}$$

of the Curie form. Results for the paramagnetic ions in a gadolinium salt are shown in Fig. 5a, and results for the He³ nuclei in solid He³ are shown in Fig. 5b.

In a magnetic field an atom with angular momentum quantum number J has $2J + 1$ equally spaced energy levels. The magnetization is given by

$$M = NgJ\mu_B B_J(x) \ , \qquad (x \equiv gJ\mu_B B/k_B T) \tag{22}$$

where the **Brillouin function** B_J is defined by

$$B_J(x) = \frac{2J + 1}{2J} \operatorname{ctnh}\left(\frac{(2J + 1)x}{2J}\right) - \frac{1}{2J} \operatorname{ctnh}\left(\frac{x}{2J}\right) . \tag{23}$$

Equation (20) is a special case of (22) for $J = \frac{1}{2}$. For $x \ll 1$, we have

$$\operatorname{ctnh} x = \frac{1}{x} + \frac{x}{3} - \frac{x^3}{45} + \cdots \ , \tag{24}$$

and the susceptibility is given from (23) and (24) by

$$\frac{M}{B} \cong \frac{NJ(J + 1)g^2\mu_B{}^2}{3k_B T} = \frac{Np^2\mu_B{}^2}{3k_B T} = \frac{C}{T} \ . \tag{25}$$

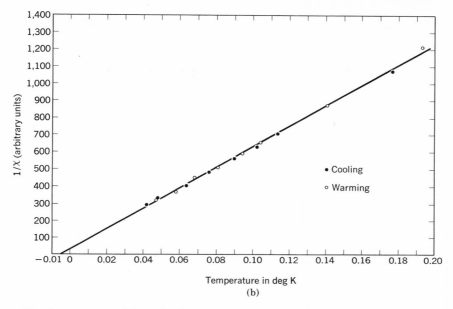

Figure 5b Inverse susceptibility of solid He^3 versus temperature, at a molar volume of 23.6 cm^3/mole^{-1}. The magnetic susceptibility is due to the He^3 nuclei. [After P. B. Pipes and W. M. Fairbanks, Phys. Rev. Letters **23**, 520 (1969).]

Here p is the **effective number of Bohr magnetons,** defined as

$$p \equiv g[J(J + 1)]^{\frac{1}{2}} . \tag{26}$$

Rare Earth Ions

The ions of the rare-earth elements have closely similar chemical properties, and their chemical separation in tolerably pure form was accomplished only long after their discovery. Their magnetic properties are fascinating: the ions exhibit a systematic variety and intelligible complexity. The chemical properties of the trivalent ions are similar because the outermost electron shells are identically in the $5s^2$ $5p^6$ configuration, like neutral xenon. In lanthanum, just before the rare earth group begins, the $4f$ shell is empty; at cerium there is one $4f$ electron, and the number of $4f$ electrons increases steadily through the group until we have $4f^{13}$ at ytterbium and the filled shell $4f^{14}$ at lutecium. (See the front endpapers of this volume.) The radii of the trivalent ions contract fairly smoothly as we go through the group from 1.11 Å at cerium to 0.94 Å at ytterbium. This is the famous "lanthanide contraction." What distinguishes the magnetic behavior of one ion species from another is the number of $4f$ electrons secreted in the inner shell with a radius of perhaps 0.3 Å. Even in the metals the $4f$ core retains its integrity and its atomic properties: no other group of elements in the periodic table is as interesting.

Table 1   Effective magneton numbers p for trivalent lanthanide group ions
(Near room temperature)

Ion	Configuration	Basic level	$p(\text{calc}) =$ $g[J(J+1)]^{\frac{1}{2}}$	$p(\text{exp})$, approximate
Ce^{3+}	$4f^{1}5s^{2}p^{6}$	$^{2}F_{5/2}$	2.54	2.4
Pr^{3+}	$4f^{2}5s^{2}p^{6}$	$^{3}H_{4}$	3.58	3.5
Nd^{3+}	$4f^{3}5s^{2}p^{6}$	$^{4}I_{9/2}$	3.62	3.5
Pm^{3+}	$4f^{4}5s^{2}p^{6}$	$^{5}I_{4}$	2.68	—
Sm^{3+}	$4f^{5}5s^{2}p^{6}$	$^{6}H_{5/2}$	0.84	1.5
Eu^{3+}	$4f^{6}5s^{2}p^{6}$	$^{7}F_{0}$	0	3.4
Gd^{3+}	$4f^{7}5s^{2}p^{6}$	$^{8}S_{7/2}$	7.94	8.0
Tb^{3+}	$4f^{8}5s^{2}p^{6}$	$^{7}F_{6}$	9.72	9.5
Dy^{3+}	$4f^{9}5s^{2}p^{6}$	$^{6}H_{15/2}$	10.63	10.6
Ho^{3+}	$4f^{10}5s^{2}p^{6}$	$^{5}I_{8}$	10.60	10.4
Er^{3+}	$4f^{11}5s^{2}p^{6}$	$^{4}I_{15/2}$	9.59	9.5
Tm^{3+}	$4f^{12}5s^{2}p^{6}$	$^{3}H_{6}$	7.57	7.3
Yb^{3+}	$4f^{13}5s^{2}p^{6}$	$^{2}F_{7/2}$	4.54	4.5

The preceding discussion of paramagnetism applies to atoms which have a $(2J+1)$-fold degenerate ground state, the degeneracy being lifted by a magnetic field. The influence of all higher energy states of the system is neglected. These assumptions appear to be satisfied by a number of rare-earth ions, Table 1. The calculated magneton numbers are obtained with g values from the Landé result (16) and the ground-state level assignment predicted by the Hund theory of spectral terms.

The discrepancy between the experimental magneton numbers and those calculated on these assumptions is quite marked for Eu^{3+} and Sm^{3+} ions. For these ions it is necessary to consider the influence of the higher states of the $L - S$ multiplet,[5] as the intervals between successive states of the multiplet are not large compared to $k_B T$ at room temperature.

The full theoretical expression for the susceptibility may be quite complicated if higher states are considered. In Advanced Topic M we consider two limiting cases, when the level splitting is larger or smaller than $k_B T$. Levels higher than $k_B T$ above the ground state may contribute a Van Vleck term (9) to the susceptibility, independent of temperature over the appropriate range.

---

[5] A multiplet is the set of levels of different J values arising out of a given L and S. The levels of a multiplet are split by the spin-orbit interaction.

Hund Rules

The Hund rules as applied to electrons in a given shell of an atom affirm that electrons will occupy orbitals in such a way that the ground state is characterized by the following:

1. The maximum value of the total spin S allowed by the exclusion principle;
2. The maximum value of the orbital angular momentum L consistent with this value of S;
3. The value of the total angular momentum J is equal to $|L - S|$ when the shell is less than half full and to $L + S$ when the shell is more than half full. (When the shell is just half full, the application of the first rule gives $L = 0$, so that $J = S$.)

The first Hund rule has its origin in the exclusion principle and the Coulomb repulsion between electrons. The exclusion principle prevents two electrons of the same spin from being at the same place at the same time. Thus electrons of the same spin are kept apart, further apart than electrons of opposite spin. Because of the Coulomb interaction the energy of electrons of the same spin is lower—the average potential energy is less positive for parallel spin than for antiparallel spin.

A good example is the ion Mn^{2+}. This ion has five electrons in the $3d$ shell, which is therefore half-filled. The spins can all be parallel if each electron enters a different orbital, and there are exactly five different orbitals available, characterized by the orbital quantum numbers $m_L = 2, 1, 0, -1, -2$. These will be occupied by one electron each. We expect $S = \frac{5}{2}$, and because $\Sigma m_L = 0$ the only possible value of L is 0, as observed.

The second Hund rule is best approached by model calculations. Pauling and Wilson,[6] for example, give a calculation of the spectral terms that arise from the configuration p^2. The third Hund rule is a consequence of the sign of the spin-orbit interaction: For a single electron the energy is lowest when the spin is antiparallel to the orbital angular momentum. But the low energy pairs m_L, m_S are progressively used up as we add electrons to the shell; by the exclusion principle when the shell is more than half full the state of lowest energy necessarily has the spin parallel to the orbit.

Consider two examples of the Hund rules: The ion Ce^{3+} has a single f electron; an f electron has $l = 3$ and $s = \frac{1}{2}$. Because the f shell is less than half full, the J value by the preceding rule is $|L - S| = L - \frac{1}{2} = \frac{5}{2}$. The ion Pr^{3+} has two f electrons: one of the rules tells us that the spins add to give $S = 1$. Both f electrons cannot have $m_l = 3$ without violating the Pauli exclusion principle, so that the maximum L consistent with the Pauli principle is not 6, but 5. The J value is $|L - S| = 5 - 1 = 4$.

[6] L. Pauling and E. B. Wilson, *Introduction to quantum mechanics*, McGraw-Hill, 1935, pp. 239–246.

Table 2   Effective magneton numbers for iron group ions

Ion	Config-uration	Basic level	$p(\text{calc}) = g[J(J+1)]^{\frac{1}{2}}$	$p(\text{calc}) = 2[S(S+1)]^{\frac{1}{2}}$	$p(\text{exp})°$
Ti^{3+}, V^{4+}	$3d^1$	$^2D_{3/2}$	1.55	1.73	1.8
V^{3+}	$3d^2$	$^3F_2$	1.63	2.83	2.8
Cr^{3+}, V^{2+}	$3d^3$	$^4F_{3/2}$	0.77	3.87	3.8
Mn^{3+}, Cr^{2+}	$3d^4$	$^5D_0$	0	4.90	4.9
Fe^{3+}, Mn^{2+}	$3d^5$	$^6S_{5/2}$	5.92	5.92	5.9
Fe^{2+}	$3d^6$	$^5D_4$	6.70	4.90	5.4
Co^{2+}	$3d^7$	$^4F_{9/2}$	6.63	3.87	4.8
Ni^{2+}	$3d^8$	$^3F_4$	5.59	2.83	3.2
Cu^{2+}	$3d^9$	$^2D_{5/2}$	3.55	1.73	1.9

° Representative values.

Iron Group Ions

Table 2 shows that the experimental magneton numbers for salts of the iron transition group of the periodic table are in poor agreement with (26). The values often agree quite well with magneton numbers $p = 2[S(S+1)]^{\frac{1}{2}}$ calculated as if the orbital moment were not there at all. One expresses this situation by saying that the orbital moments are "quenched."

Crystal Field Splitting

The basic reason for the difference in behavior of the rare earth and the iron group salts is that the $4f$ shell responsible for paramagnetism in the rare earth ions lies deep inside the ions, within the $5s$ and $5p$ shells, whereas in the iron group ions the $3d$ shell responsible for paramagnetism is the outermost shell. The $3d$ shell experiences the intense inhomogeneous electric field produced by neighboring ions. This inhomogeneous electric field is called the **crystal field.** The interaction of the paramagnetic ions with the crystal field has two major effects: the coupling of **L** and **S** vectors is largely broken up, so that the states are no longer specified by their J values; further, the $2L+1$ sublevels belonging to a given L which are degenerate in the free ion may now be split up by the crystal field, as in Fig. 6. This splitting diminishes the contribution of the orbital motion to the magnetic moment.

Quenching of the Orbital Angular Momentum

In an electric field directed toward a fixed nucleus, the plane of a classical orbit is fixed in space, so that all the orbital angular momentum components L_x, L_y, L_z are constant. In quantum theory one angular momentum component, usually taken as L_z, and the square of the total orbital angular momentum L^2 are constant in a central field. In a noncentral field the plane of the orbit will move about; the angular momentum components are no longer

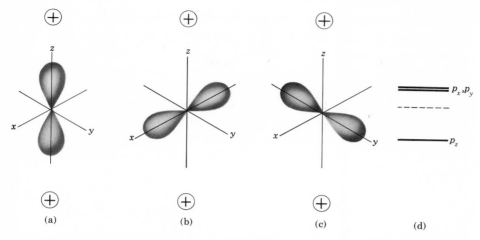

Figure 6 Consider an atom with orbital angular momentum $L = 1$ placed in the uniaxial crystalline electric field of the two positive ions along the z axis. In the free atom the states $m_L = \pm 1, 0$ have identical energies—they are degenerate. In the crystal the atom has a lower energy when the electron cloud is close to positive ions as in (a) than when it is oriented midway between them, as in (b) and (c). The wavefunctions that give rise to these charge densities are of the form $zf(r)$, $xf(r)$ and $yf(r)$ and are called the p_z, p_x, p_y orbitals, respectively. In an axially symmetric field as shown the p_x and p_y orbitals are degenerate. The energy levels referred to the free atom (dotted line) are shown in (d). If the electric field does not have axial symmetry, all three states will have different energies.

constant and may average to zero. In a crystal, as shown in detail in Advanced Topic M, L_z will no longer be a constant of the motion, although to a good approximation L^2 may continue to be constant. When L_z averages to zero the orbital angular momentum is said to be **quenched.**

The magnetic moment of a state is given by the average value of the magnetic moment operator $\mu_B(\mathbf{L} + 2\mathbf{S})$. In a magnetic field along the z direction the orbital contribution to the magnetic moment is proportional to the quantum expectation value of L_z; the orbital magnetic moment is quenched if the mechanical moment L_z is quenched.

When the spin-orbit interaction energy is introduced as an additional perturbation on the system, the spin may drag some orbital moment along with it. If the sign of the interaction favors parallel orientation of the spin and orbital magnetic moments, the total magnetic moment will be larger than for the spin alone, and the g value will be larger than 2. The experimental results are in agreement with the known variation of sign of the spin-orbit interaction: $g > 2$ when the $3d$ shell is more than half full, $g = 2$ when the shell is half full, and $g < 2$ when the shell is less than half full.

Nuclear Paramagnetism

Magnetic moments of nuclei are smaller than the magnetic moment of the electron by a factor $\sim m/M_p \sim 10^{-3}$, where M_p is the proton mass. According to (14) the susceptibility of a nuclear paramagnetic system for the same number of particles will be smaller by a factor $\sim 10^{-6}$ than that of an electronic paramagnetic system. The susceptibility of solid hydrogen, which is diamagnetic with respect to electrons but paramagnetic with respect to protons, was measured at very low temperatures by Lasarew and Schubnikow.[7] Results for solid He³ were shown in Fig. 5b. Nuclear magnetism is discussed in Chapter 17.

COOLING BY ADIABATIC DEMAGNETIZATION
OF A PARAMAGNETIC SALT

The first method for attaining temperatures much below 1°K was that of adiabatic demagnetization.[8] By its use temperatures of 10^{-3}°K and lower have been reached. The method rests on the fact that at a fixed temperature the entropy of a system of magnetic moments is lowered by the application of a magnetic field. (The entropy is a measure of the disorder of a system: the greater the disorder, the higher is the entropy.) In the magnetic field the moments will be partly lined up (partly ordered), so that the entropy is lowered by the field. The entropy is also lowered if the temperature is lowered, as more of the moments line up.

If the magnetic field can then be removed without changing the entropy of the spin system, the order of the spin system will look like a lower temperature than the same degree of order in the presence of the field. When the specimen is demagnetized adiabatically, entropy can flow into the spin system only from the system of lattice vibrations, as in Fig. 7. At the temperatures of interest the entropy of the lattice vibrations is usually negligible; thus the entropy of the spin system will be essentially constant during adiabatic demagnetization of the specimen. Magnetic cooling is a one-shot operation, not cyclic.

We first find an expression for the spin entropy of a system of N ions, each of spin S, at a temperature sufficiently high that the spin system is entirely disordered. That is, T is supposed to be much higher than some temperature Δ which characterizes the energy of the interactions ($E_{\text{int}} \equiv k_B \Delta$) tending to orient the spins preferentially. Some of these interactions are discussed in

[7] B. Lasarew and L. Schubnikow, Physik. Z. Sowjetunion 11, 445 (1937); see also D. F. Evans, Phil. Mag. 1, 370 (1956).

[8] The method was suggested independently by P. Debye, Ann. Physik 81, 1154 (1926); and W. F. Giauque, J. Am. Chem. Soc. 49, 1864 (1927). For many purposes the method has been supplanted by the He³-He⁴ dilution refrigerator which operates in a continuous cycle. The He³ atoms in solution in liquid He⁴ play the role of atoms in a gas, and cooling is effected by "vaporization" of He³. References are given by J. C. Wheatley, Amer. J. Phys. 36, 181 (1968); see especially pp. 198–210.

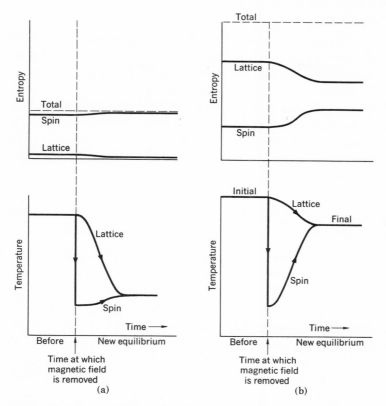

Figure 7 During adiabatic demagnetization the total entropy of the specimen is constant, but situation (a) is favorable for cooling, whereas (b) is not. In (a) the initial entropy of the lattice is small in comparison with the entropy of the spin system, whereas in (b) the initial lattice entropy is large, so that the low spin temperature reached just after the magnetic field is switched off is chiefly wasted in a slight cooling off of the lattice.

Chapter 17. The definition of the entropy σ of a system of G accessible states is

$$\sigma = k_B \log G \ . \tag{27}$$

At a temperature so high that all of the $2S + 1$ states of each ion are nearly equally populated, G is the number of ways of arranging N spins in $2S + 1$ states. Thus

$$G = (2S + 1)^N \ , \tag{28}$$

whence the spin entropy[9] is

$$\sigma_S = k_B \log (2S + 1)^N = N k_B \log (2S + 1) \ . \tag{29}$$

It is this spin entropy which is reduced by a magnetic field. The field separates the $2S + 1$ states in energy, and the lower levels gain in population.

[9] Here σ_S denotes the entropy of the spin system.

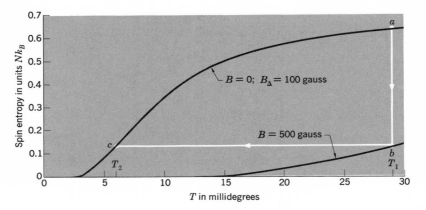

Figure 8 Entropy for a spin $\frac{1}{2}$ system as a function of temperature, assuming an internal random magnetic field B_Δ of 100 gauss. The specimen is magnetized isothermally along ab, and is then insulated thermally. The external magnetic field is turned off along bc. In order to keep the figure on a reasonable scale the initial temperature T_1 is lower than would be used in practice, as is also the external magnetic field.

The steps carried out in the cooling process are shown in Figs. 8 and 9. The field is applied at temperature T_1 with the specimen in good thermal contact with the surroundings, giving the isothermal path ab. The specimen is then insulated ($\Delta\sigma = 0$) and the field removed; the specimen follows the constant entropy path bc, ending up at temperature T_2. The thermal contact at T_1 is provided by helium gas, and the thermal contact is broken by removing the gas with a pump.

EXAMPLE. Consider a unit volume of a crystal containing N unpaired electrons of spin $\frac{1}{2}$ and magnetic moment μ. Let the heat capacity of the lattice be $C_{\text{lat}} = 3AT^3$, where A is a constant. The lattice entropy is

$$\sigma_{\text{lat}} = \int_0^T (C_{\text{lat}}/T)\, dT = AT^3 \ . \tag{30}$$

The entropy of the spin system in a magnetic field B is found readily. The partition function per spin is

$$Z = e^{-\beta\delta} + e^{\beta\delta} \equiv 2\cosh\beta\delta \ , \tag{31}$$

where $\beta \equiv 1/k_B T$ and $\delta \equiv \mu B$. The free energy of N spins of $S = \frac{1}{2}$ is

$$F_S = -Nk_B T \log Z = -Nk_B T \log(2\cosh\beta\delta) \ , \tag{32}$$

and the spin entropy σ_S is

$$\sigma_S = -(\partial F_S/\partial T)_B = Nk_B[\log(2\cosh\beta\delta) - \beta\delta\tanh\beta\delta] \ . \tag{33}$$

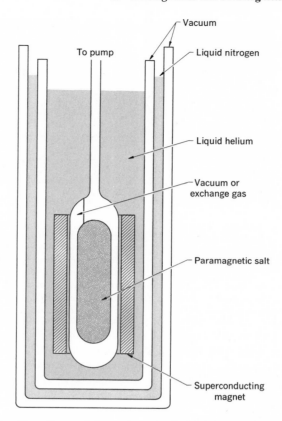

Figure 9 Apparatus for magnetic cooling. From *Heat and thermodynamics,* by M. W. Zemansky, 5th ed. (Copyright © 1968 by McGraw-Hill, Inc. Used with permission of McGraw-Hill Book Company.)

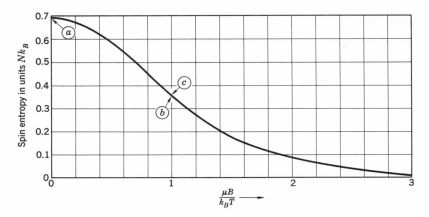

Figure 10 Entropy of a spin $\frac{1}{2}$ system as a function of $\mu B/k_B T$. In a realistic experiment the specimen is initially at $1°K$ and the internal random magnetic field B_Δ is of the order of 100 gauss. Then $\mu_B B_\Delta / k_B T_1 \approx 10^{-18}/10^{-16} \approx 10^{-2}$, and the specimen is at the point a in the figure. If we magnetize the specimen isothermally in a 10,000 gauss field, then $\mu_B B/k_B T_1 \approx 1$ and the specimen is brought to point b. The total lattice entropy is assumed to be much smaller than the amount by which the entropy of the spin system has been lowered. If now we remove the external magnetic field the entropy remains constant, so that $\mu_B B_\Delta / k_B T$ must be ≈ 1, whence $T_2 \approx 0.01°K$ at point c.

This is plotted in Fig. 10 as a function of $\mu B/k_B T$.

The constant A in the expression (30) for the lattice entropy may be of the order of $10^{-6} N_a k_B$ erg deg^{-4} for a representative solid, as we see from Chapter 6. Here N_a is the number of atoms, which may be 10 to 100 times larger than the number of spins in the paramagnet, according to the chemical composition and dilution. When we compare σ_{lat} at $1°K$ with σ_S, we see that at and below this temperature the lattice entropy is negligible in comparison with the spin entropy. This is usually the situation in practice, provided the initial temperature is not too high. If initially the lattice entropy is higher than the spin entropy, the cooling effect following demagnetization will be negligible.

In Fig. 10, $T_1 = 1°K$ and $B = 10$ kG; the specimen will be cooled to $0.01°K$. The ultimate temperature reached by adiabatic demagnetization is limited by the intrinsic or **zero-field splitting** of the spin energy levels—the splitting which occurs in the absence of external magnetic fields. The zero-field splitting may be caused by electrostatic interaction with the other ions in the crystal, by the interaction of the magnetic moments with each other, or by interactions of nuclear moments. In Fig. 10 the zero-field splitting of the spin levels is equivalent to an internal magnetic field B_Δ of 100 gauss. The zero-field splitting makes the entropy at a and at c in Fig. 8 less than for a smaller zero-field splitting; thus the final temperature is not as low as for a smaller splitting.

Nuclear Demagnetization

The result of the preceding discussion of cooling by adiabatic demagnetization of a paramagnetic salt is that T_2, the final temperature reached, is given as in Fig. 10 by $B/T_1 = B_\Delta/T_2$, or

$$T_2 = T_1(B_\Delta/B) \ , \qquad (34)$$

where B_Δ corresponds to the zero-field splitting and T_1 is the starting temperature. Because nuclear magnetic moments are weak, nuclear magnetic interactions are much weaker than similar electronic interactions. We expect to reach a temperature 100 times lower with a nuclear paramagnet than with an electron paramagnet.

The initial temperature T_1 of the nuclear stage in a nuclear spin-cooling experiment must be lower than in an electron spin-cooling experiment. If we start at $B = 50$ kG and $T_1 = 0.01°$K, then $\mu B/k_B T_1 \approx 0.5$, and the entropy decrease on magnetization is over 10 percent of the maximum spin entropy. This is sufficient to overwhelm the lattice[10] and from (34) we estimate a final temperature $T_2 \approx 10^{-7}°$K.

The first[11] nuclear cooling experiment was carried out on Cu nuclei in the metal, starting from a first stage at about 0.02°K as attained by electronic

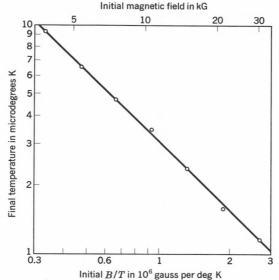

Figure 11 Nuclear demagnetizations of copper nuclei in the metal, starting from 0.012°K and various fields. [After M. V. Hobden and N. Kurti, Phil. Mag. **4**, 1092 (1959).]

Initial magnetic field in kG

Final temperature in microdegrees K

Initial B/T in 10^6 gauss per deg K

---

[10] If the nuclei are in a metal, we must consider also the entropy of the electrons.

[11] N. Kurti, F. N. H. Robinson, F. E. Simon, and D. A. Spohr, Nature **178**, 450 (1956); R. E. Walstedt, E. L. Hahn, C. Froidevaux, and E. Geissler, Proc. Roy. Soc. (London) **A284**, 499 (1965); for reviews see N. Kurti, Cryogenics **1**, 2 (1960); Adv. in Cryogenic Engineering **8**, 1 (1963).

cooling. The lowest temperature reached was $1.2 \times 10^{-6}°K$. The results in Fig. 11 fit a line of the form of (34): $T_2 = T_1(3.1/B)$ with B in gauss, so that $B_\Delta = 3.1$ gauss. This is the effective interaction field of the magnetic moments of the Cu nuclei. The motivation for using nuclei in a metal is that conduction electrons help ensure rapid thermal contact of lattice and nuclei at the temperature of the first stage.

PARAMAGNETIC SUSCEPTIBILITY OF CONDUCTION ELECTRONS

"We are going to try to show how on the basis of these statistics the fact that many metals are diamagnetic or only weakly paramagnetic, can be brought into agreement with the existence of a magnetic moment of the electrons." (W. Pauli, 1927)

Classical free electron theory gives an unsatisfactory account of the paramagnetic susceptibility of the conduction electrons. An electron has associated with it a magnetic moment of one Bohr magneton, μ_B. One might expect that the conduction electrons would make a Curie-type paramagnetic contribution (25) to the magnetization of the metal:

$$M = \frac{N\mu_B^2}{k_B T} B \ .$$
(35)

Instead it is observed that the magnetization of most normal nonferromagnetic metals is *independent of temperature;* the magnitude is perhaps only 0.01 of that expected from (35) at room temperature.

Pauli[12] showed that the application of the Fermi-Dirac distribution (Chapter 7) would correct the theory as required. We first give a qualitative explanation of the situation. The result (21) tells us that the probability an atom will be lined up parallel to the field B exceeds the probability of the antiparallel orientation by roughly $\mu B/k_B T$. For N atoms per unit volume, this gives a net magnetization $\sim N\mu^2 B/k_B T$, the standard result. Most conduction electrons in a metal, however, have zero probability of turning over when a field is applied, because most orbitals in the Fermi sea with parallel spin are already occupied. Only the electrons within a range $k_B T$ of the top of the Fermi distribution have a chance to turn over in the field; thus only the fraction T/T_F of the total number of electrons contribute to the susceptibility. Hence

$$M \approx \frac{N\mu^2 B}{k_B T} \cdot \frac{T}{T_F} = \frac{N\mu^2}{kT_F} B \ ,$$

which is independent of temperature and of the observed order of magnitude.

We now calculate the expression for the paramagnetic susceptibility of a free electron gas at $T \ll T_F$. We follow the method of calculation suggested by Fig. 12a. An alternate derivation is the subject of Problem 8. The concentra-

[12] W. Pauli, Z. Physik **41**, 81 (1927).

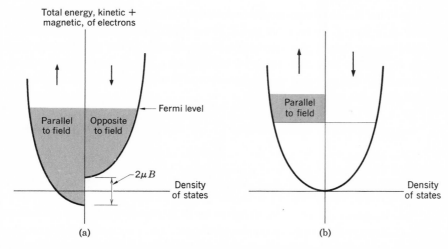

Figure 12   Pauli paramagnetism at absolute zero; the orbitals in the shaded regions in (a) are occupied. The numbers of electrons in the "up" and "down" band will adjust to make the energies equal at the Fermi level. The chemical potential (Fermi level) of the spin up electrons is equal to that of the spin down electrons. In (b) we show the excess of spin up electrons in the magnetic field.

tion of electrons with magnetic moments parallel to the magnetic field is

$$N_+ = \frac{1}{2} \int_{\mu B}^{\epsilon_F} d\epsilon \, f(\epsilon) \, \mathfrak{D}(\epsilon + \mu B) \;\cong\; \frac{1}{2} \int_0^{\epsilon_F} d\epsilon \, f(\epsilon) \, \mathfrak{D}(\epsilon) + \frac{1}{2} \mu B \, \mathfrak{D}(\epsilon_F) \; ,$$

$$(36)$$

where $f(\epsilon)$ is the Fermi-Dirac distribution function and $\frac{1}{2}\mathfrak{D}(\epsilon + \mu B)$ is the density of orbitals of one spin orientation, with allowance for the downward shift of energy by $-\mu B$. (Make a drawing for yourself of the integrands in the expressions for N_+ and N_-.) The approximation (36) is written for $k_B T \ll \epsilon_F$. The concentration of electrons with magnetic moments antiparallel to the magnetic field is

$$N_- = \frac{1}{2} \int_{\mu B}^{\epsilon_F} d\epsilon \, f(\epsilon) \, \mathfrak{D}(\epsilon - \mu B) \;=\; \frac{1}{2} \int_0^{\epsilon_F} d\epsilon \, f(\epsilon) \, \mathfrak{D}(\epsilon) - \frac{1}{2} \mu B \, \mathfrak{D}(\epsilon_F) \; .$$

$$(37)$$

The magnetization is given by

$$M = \mu(N_+ - N_-) \; , \qquad (38)$$

so that

$$M \cong \mu^2 \, \mathfrak{D}(\epsilon_F) B = \frac{3N\mu^2}{2k_B T_F} B \; , \qquad (39)$$

with $\mathfrak{D}(\epsilon_F) = 3N/2\epsilon_F = 3N/2k_B T_F$ from Chapter 7. The result (39) gives the **Pauli spin magnetization** of the conduction electrons.

In deriving the paramagnetic susceptibility, we have supposed that the spatial motion of the electrons is not affected by the magnetic field. We saw in Chapter 10 that the wavefunctions are modified by the magnetic field; Landau[13] has shown that for free electrons this causes a diamagnetic moment equal to $-\frac{1}{3}$ of the paramagnetic moment. Thus the total magnetization of a free electron gas is

$$M = \frac{N\mu_B{}^2}{k_B T_F} B \ . \tag{40}$$

Before comparing (40) with the experiment we must take account of the diamagnetism of the ionic cores, of band effects, and of electron-electron interactions. In sodium the interaction effects increase the spin susceptibility by perhaps 75 percent. Detailed calculations and comparison with experiment for the alkali metals are reported by Silverstein.[14]

The magnetic susceptibility is considerably higher for most transition metals (with unfilled inner electron shells) than for the alkali metals (Fig. 13). The high values suggest that the density of electron orbitals in (39) is unusually high for transition metals, in agreement with measurements of the electronic heat capacity. We saw in Chapter 10 how this arises from band theory.

[13] L. Landau, Z. Physik **64**, 629 (1930).

[14] S. D. Silverstein, Phys. Rev. **130**, 1703 (1963). For measurements of the spin susceptibility of sodium see R. T. Schumacher and W. E. Vehso, J. Phys. Chem. Solids **24**, 297 (1963).

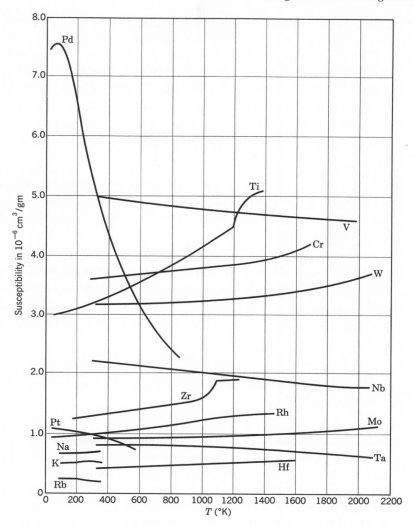

Figure 13 Temperature dependence of the magnetic susceptibility of metals. (Courtesy of C. J. Kriessman.)

SUMMARY
(CGS units)

1. The diamagnetic susceptibility of N atoms of atomic number Z is $\chi = -Ze^2N\langle r^2\rangle/6mc^2$, where $\langle r^2\rangle$ is the mean square atomic radius. (Langevin)

2. Atoms with a permanent magnetic moment μ have a paramagnetic susceptibility $\chi = N\mu^2/3k_BT$, for $\mu B \ll k_BT$. (Curie-Langevin)

3. For a system of spins $S=\frac{1}{2}$, the exact magnetization is $M=N\mu\tanh(\mu B/k_BT)$, where $\mu = \frac{1}{2}g\mu_B$. (Brillouin)

4. The ground state of electrons in the same shell have the maximum value of S allowed by the Pauli principle and the maximum L consistent with this S. The J value is $L + S$ if the shell is more than half full and $|L - S|$ if the shell is less than half full.

5. A cooling process operates by demagnetization of a paramagnetic salt at constant entropy. The final temperature reached is of the order of $(B_\Delta/B)\,T_{\text{initial}}$, where B_Δ is the effective local field and B is the initial applied magnetic field.

6. The paramagnetic susceptibility of a Fermi gas of conduction electrons is $\chi = 3N\mu^2/2\epsilon_F$, independent of temperature for $k_BT \ll \epsilon_F$. (Pauli)

Problems

1. *Diamagnetic susceptibility of atomic hydrogen.* The wavefunction of the hydrogen atom in its ground state ($1s$) is

$$\psi = (\pi a_0{}^3)^{-\frac{1}{2}}e^{-r/a_0} \; ,$$

where $a_0 = \hbar^2/me^2 = 0.529 \times 10^{-8}$ cm. The charge density is $\rho(x, y, z) = -e|\psi|^2$, according to the statistical interpretation of the wave function. Show that for this state $\langle r^2\rangle = 3a_0{}^2$, and calculate the molar diamagnetic susceptibility of atomic hydrogen $(-2.36 \times 10^{-6}$ cm^3/mole$)$.

2. *Langevin paramagnetism.* Show that on the Langevin theory the first two terms in a series expansion of the differential susceptibility are

(CGS) $$\chi = \frac{dM}{dB} = \left(\frac{N\mu^2}{3k_BT}\right)\left[1 - \frac{1}{5}\left(\frac{\mu B}{k_BT}\right)^2 + \cdots\right] .$$

3. *Hund rules.* Apply the Hund rules to find the ground state (the basic level in the notation of Table 1) of (a) Eu^{++}, in the configuration $4f^7\,5s^2p^6$; (b) Yb^{3+}; (c) Tb^{3+}. The results for (b) and (c) are in Table 1, but you should give the separate steps in applying the rules.

4. *Triplet excited states.* Some organic molecules have a triplet ($S = 1$) excited state at an energy $k_B\Delta$ above a singlet ($S = 0$) ground state. (a) Find an expression for the magnetic moment $\langle\mu\rangle$ in a field B. (b) Show that the susceptibility for $T \gg \Delta$ is approximately independent of Δ. (c) With the help of a diagram of energy levels versus field and a rough sketch of entropy versus field, explain how this system might be cooled by adiabatic magnetization (not demagnetization).

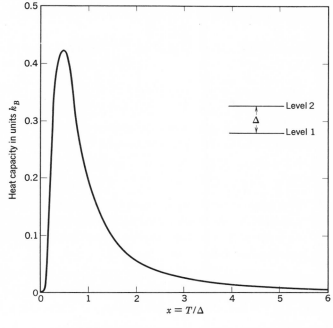

Figure 14 Heat capacity of a two-level system as a function of T/Δ, where Δ is the level splitting. The Schottky anomaly is a very useful tool for determining energy-level splittings of ions in rare-earth and transition-group metals, compounds, and alloys.

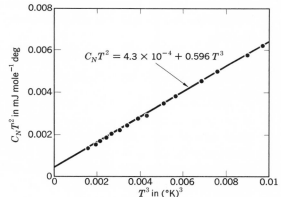

Figure 15 The normal-state heat capacity of gallium at $T < 0.21°K$. The nuclear quadrupole ($C \propto T^{-2}$) and conduction electron ($C \propto T$) contributions dominate the heat capacity at very low temperatures. [After N. E. Phillips, Phys. Rev. 134, 385 (1964).]

5. **Heat capacity from internal degrees of freedom.** (a) Consider a two-level system with an energy splitting $k_B\Delta$ between upper and lower states; the splitting may arise from a magnetic field or in other ways. Show that the heat capacity per system is

$$C = \left(\frac{\partial E}{\partial T}\right)_\Delta = k_B \frac{(\Delta/T)^2 \, e^{\Delta/T}}{(1 + e^{\Delta/T})^2} \; .$$

The function is plotted in Fig. 14. Peaks of this type in the heat capacity are often known as Schottky anomalies. The maximum heat capacity is quite high, but for $T \ll \Delta$ and for $T \gg \Delta$ the heat capacity is low. (b) Show that for $T \gg \Delta$ we have $C \cong k_B(\Delta/2T)^2 + \cdots$. The hyperfine interaction between nuclear and electronic magnetic moments in paramagnetic salts (and in systems having electron spin order) causes splittings with $\Delta \approx 0.001$ to $0.1°K$. These splittings are often detected experimentally by the presence of a term in $1/T^2$ in the heat capacity in the region $T \gg \Delta$. Nuclear electric quadrupole interactions (see Chapter 17) with crystal fields also cause splittings, as in Fig. 15.

6. **Spin entropy and lattice entropy.** Making *rough* calculations, compare the entropy (for $B = 0$) at $2°K$ of 1 cm^3 of the salt iron ammonium alum, FeNH$_4$(SO$_4$)$_2 \cdot$ 12H$_2$O, with that of 1 cm^3 of sodium at the same temperature. (The result shows that one may use the salt to cool other substances). At $2°K$, $T/\Delta > 10$ for iron ammonium alum, where $k_B\Delta$ is the zero field splitting. Neglect nuclear spin effects.

7. **Paraelectric cooling.** It is known that OH$^-$ ions substitute for Cl$^-$ ions in KCl crystals. Suppose that the electric dipole moment of the OH$^-$ ion may be oriented freely. The electric dipole moment is $\sim 4 \times 10^{-18}$ esu-cm. (a) What is the overall splitting in ergs for $E = 60$ kV/cm parallel to [100]? (b) For what temperature is $k_B T$ equal to this overall splitting? *Note:* Paraelectric cooling was first observed for this system by W. Känzig, H. R. Hart, Jr., and S. Roberts, Phys. Rev. Letters **13**, 543 (1964). I. Sheperd and G. Feher, Phys. Rev. Letters **15**, 194 (1965) cooled a crystal with 2.9×10^{18} OH$^-$ ions/cm^3 from $1.27°K$ to about $0.4°K$ by adiabatic depolarization with an initial electric field of 75 kV/cm.

8. **Pauli spin susceptibility.** The spin susceptibility of a conduction electron gas at absolute zero may be discussed by another method. Let

$$N^+ = \tfrac{1}{2}N(1 + \zeta) \; ; \qquad N^- = \tfrac{1}{2}N(1 - \zeta)$$

be the concentrations of spin-up and spin-down electrons.
(a) Show that in a magnetic field H the total energy of the spin-up band in a free electron gas is

$$E^+ = E_0(1 + \zeta)^{\frac{5}{3}} - \tfrac{1}{2}N\mu B(1 + \zeta) \; ,$$

where $E_0 = \tfrac{3}{10}N\epsilon_F$, in terms of the Fermi energy ϵ_F in zero magnetic field. Find a similar expression for E^-.
(b) Minimize $E_{\text{total}} = E^+ + E^-$ with respect to ζ and solve for the equilibrium value of ζ in the approximation $\zeta \ll 1$. Go on to show that the magnetization is $M = 3N\mu^2B/2\epsilon_F$, in agreement with Eq. (39).

9. **Conduction electron ferromagnetism.** We approximate the effect of exchange interactions among the conduction electrons if we assume that electrons with parallel spins interact with each other with energy $-V$, where V is positive, while electrons with antiparallel spins do not interact with each other.
(a) Show with the help of Problem 8 that the total energy of the spin-up band is

$$E^+ = E_0(1 + \zeta)^{\frac{5}{3}} - \tfrac{1}{8}VN^2(1 + \zeta)^2 - \tfrac{1}{2}N\mu B(1 + \zeta) \; ;$$

find a similar expression for E^-.
(b) Minimize the total energy and solve for ζ in the limit $\zeta \ll 1$. Show that the magnetization is

$$M = \frac{3N\mu^2}{2\epsilon_F - \tfrac{3}{2}VN}B \; ,$$

so that the exchange interaction enhances the susceptibility.
(c) Show that with $B = 0$ the total energy is unstable at $\zeta = 0$ when $V > 4\epsilon_F/3N$. If this is satisfied a ferromagnetic state ($\zeta \neq 0$) will have a lower energy than the paramagnetic state. Because of the assumption $\zeta \ll 1$, this is a sufficient condition for ferromagnetism, but it may not be a necessary condition.

References

L. F. Bates, *Modern magnetism*, Cambridge University Press, 4th ed., 1961.

P. W. Selwood, *Magnetochemistry*, Interscience, 2nd ed., 1956.

J. H. Van Vleck, *The theory of electric and magnetic susceptibilities*, Oxford, 1932. Superb derivations of basic theorems.

L. Orgel, *Introduction to transition metal chemistry*, Wiley, 2nd ed., 1966.

H. B. G. Casimir, *Magnetism and very low temperatures*, Cambridge University Press, 1940. A classic work.

N. Kurti, Nuovo Cimento (Supplement) **6**, 1101–1139 (1957).

D. de Klerk, "Adiabatic demagnetization," *Encyclo. of physics* **15**, 38–209 (1956). This reference contains detailed entropy data for many paramagnetic salts.

W. A. Little, "Magnetic cooling," *Prog. in cryogenics* **4**, 101 (1964).

H. Weinstock, "Thermodynamic and statistical aspects of magnetic cooling," Amer. J. Physics **36**, 36 (1968). Elementary survey.

G. K. White, *Experimental techniques in low temperature physics*, Oxford, 1968, 2nd ed.

R. L. Sanford and I. L. Cooter, "Basic magnetic quantities and the measurement of the magnetic properties of materials," NBS Monograph **47** (1962).

R. M. White, *Quantum theory of magnetism*, McGraw-Hill, 1970.

H. Knoepfel, *Pulsed high magnetic fields*, North Holland, 1970.

A. Abragam and B. Bleaney, *Electron paramagnetic resonance of transition ions*, Clarendon, 1970.

16

Ferromagnetism and Antiferromagnetism

NOTATION: (CGS) $B = H + 4\pi M$;   (SI) $B = \mu_0(H + M)$.

We call B_a the applied magnetic field in both systems of units: in CGS we have $B_a = H_a$ and in SI we have $B_a = \mu_0 H_a$. The susceptibility is $\chi = M/B_a$ in CGS and $\chi = M/H_a = \mu_0 M/B_a$ in SI.

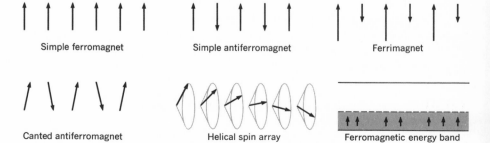

Figure 1 Possible ordered arrangements of electron spins. Note that the particular helix shown has a magnetic moment along the axis.

FERROMAGNETIC ORDER

A ferromagnet has a spontaneous magnetic moment—a magnetic moment even in zero applied magnetic field. A spontaneous moment suggests that electron spins and magnetic moments are arranged in a regular manner. The order need not be simple: all of the spin arrangements sketched in Fig. 1 except the simple antiferromagnet (and except the helix if the spin directions lie in a plane) have a spontaneous magnetic moment, usually called the saturation moment.

Curie Point and the Exchange Integral

Consider a paramagnet with a concentration of N ions of spin S. Given an internal interaction tending to line up the magnetic moments parallel to each other, we shall have a ferromagnet. Let us postulate such an interaction and call it the **exchange field.**[1] The orienting effect of the exchange field is opposed by thermal agitation, and at elevated temperatures the spin order is destroyed.

We treat the exchange field as equivalent to a magnetic field $\mathbf{B}_E$. The magnitude of the exchange field may be as high as 10^7 gauss. We assume that $\mathbf{B}_E$ is proportional to the magnetization $\mathbf{M}$. The magnetization is defined as the magnetic moment per unit volume; unless otherwise specified it is understood to be the value in thermal equilibrium in the field $\mathbf{B}_E$ at the temperature T. If domains (regions magnetized in different directions) are present, the magnetization refers to the value within a domain. In the **mean field approximation** we assume each magnetic atom experiences a field proportional to the magnetization:

$$\mathbf{B}_E = \lambda \mathbf{M} , \qquad (1)$$

where λ is a constant, independent of temperature. According to (1) each spin sees the average magnetization of all the other spins. In truth, it may see only near neighbors, but our oversimplification of the situation is good for a first look at the problem.

The **Curie temperature** T_c is the temperature above which the spontaneous magnetization vanishes; it separates the disordered paramagnetic phase at $T > T_c$ from the ordered ferromagnetic phase at $T < T_c$. We can find T_c in terms of λ. Consider the paramagnetic phase: an applied field B_a will cause a

---

[1] Also called the molecular field or the Weiss field, after Pierre Weiss who was the first to imagine such a field. The exchange field B_E simulates a real magnetic field in the expressions for the energy $-\boldsymbol{\mu} \cdot \mathbf{B}_E$ and the torque $\boldsymbol{\mu} \times \mathbf{B}_E$ on a magnetic moment $\boldsymbol{\mu}$. But B_E is not really a magnetic field and therefore does not enter into the Maxwell equations: for example, there is no current density $\mathbf{j}$ related to $\mathbf{B}_E$ by curl $\mathbf{H} = 4\pi\mathbf{j}/c$. The magnitude of B_E is typically 10^4 larger than the average magnetic field of the magnetic dipoles of the ferromagnet.

finite magnetization and this in turn will cause a finite exchange field B_E. If χ is the susceptibility,

(CGS) $$M = \chi(B_a + B_E) \ ;$$ (2)

(SI) $$\mu_0 M = \chi(B_a + B_E) \ .$$

We saw in Chapter 15 that we can write the magnetization as a constant susceptibility times a field only if the fractional alignment is small: this is where the assumption enters that the specimen is in the paramagnetic phase. The susceptibility χ of a paramagnet is given by the Curie law $\chi = C/T$, where C is the Curie constant. Using (1) and (2), $MT = C(B_a + \lambda M)$ and

(CGS) $$\chi = \frac{M}{B_a} = \frac{C}{(T - C\lambda)} \ .$$ (3)

The susceptibility has a singularity at $T = C\lambda$. At this temperature (and below) there exists a spontaneous magnetization, because if χ is infinite we can have a finite M for zero B_a. From (3) we have the **Curie-Weiss law**

(CGS) $$\chi = \frac{C}{T - T_c} \ ; \qquad T_c = C\lambda \ .$$ (4)

This expression describes fairly well the observed susceptibility variation in the paramagnetic region above the Curie point. Detailed calculations[2] predict

$$\chi \propto \frac{1}{(T - T_c)^{1.33}}$$

at temperatures close to T_c, in general agreement with the experimental data summarized in Table 1. The reciprocal susceptibility of nickel is plotted in Fig. 2.

From (4) and the definition (15.25) of the Curie constant C we may determine the value of the mean field constant λ:

(CGS) $$\lambda = \frac{T_c}{C} = \frac{3k_B T_C}{Ng^2 S(S + 1)\mu_B^2} \ .$$ (5)

For iron $T_c \approx 1000°K$, $g \approx 2$, and[3] $S \approx 1$; from (5) we have $\lambda \approx 5000$. With $M_s \approx 1700$ we have $B_E \approx \lambda M \approx (5000)(1700) \approx 10^7$ G. The exchange field in iron is very much stronger than the real magnetic field due to the other magnetic ions in the crystal: a magnetic ion produces a field $\approx \mu_B/a^3$ or about 10^3 G at a neighboring lattice point.

---

[2] Experimentally the susceptibility for $T \gg T_c$ is given quite accurately by $C/(T - \theta)$, where θ is appreciably greater than the actual transition temperature T_c. See the review by C. Domb in *Magnetism*, Vol. 2A, G. T. Rado and H. Suhl, eds., Academic Press, 1965. For experimental references see R. W. Kedzie and D. H. Lyons, Phys. Rev. Letters **15**, 632 (1965).

[3] We assume $S = 1$ for iron; Table 2 shows that the moment per atom in the metal is close to $2\mu_B$.

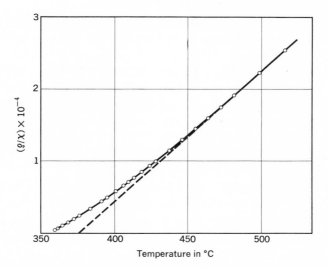

Figure 2 Reciprocal of the susceptibility per gram of nickel in the neighborhood of the Curie temperature (358°C). The density is ρ. The dashed line is a linear extrapolation from high temperatures. After P. Weiss and R. Forrer. An analysis of the curve near T_c has been given by J. S. Kouvel and M. E. Fisher, Phys. Rev. **136**, A1626 (1964).

Table 1   Critical point exponents for ferromagnets

As $T \to T_c$ from above, the susceptibility χ becomes proportional to $(T - T_c)^{-\gamma}$; as $T \to T_c$ from below, the magnetization M_s becomes proportional to $(T_c - T)^{\beta}$. In the mean field approximation $\gamma = 1$ and $\beta = \frac{1}{2}$. The experimental data in this table were collected by H. E. Stanley.

	γ	β	T_c in deg K
Fe	1.33 ± 0.015	0.34 ± 0.04	1043
Co	1.21 ± 0.04	—	1388
Ni	1.35 ± 0.02	0.42 ± 0.07	627.2
Gd	1.3 ± 0.1	—	292.5
CrO_2	1.63 ± 0.02	—	386.5
$CrBr_3$	1.215 ± 0.02	0.368 ± 0.005	32.56
EuS	—	0.33 ± 0.015	16.50

The exchange field gives an approximate representation of the quantum-mechanical exchange interaction. On certain assumptions it can be shown[4] that the energy of interaction of atoms i, j bearing spins $\mathbf{S}_i$, $\mathbf{S}_j$ contains a term

$$U = -2J\mathbf{S}_i \cdot \mathbf{S}_j \ , \tag{6}$$

[4] See most texts on quantum theory; also J. H. Van Vleck, Revs. Modern Phys. **17**, 27 (1945). The origin of exchange in insulators is reviewed by P. W. Anderson in Rado and Suhl, Vol. I, 25 (1963); in metals by C. Herring in Vol. IV (1966).

where J is the exchange integral and is related to the overlap of the charge distributions of the atoms i, j. Equation (6) is called the **Heisenberg model.**

The charge distribution of a system of two spins depends on whether the spins are parallel or antiparallel[5], for the Pauli principle excludes two electrons of the same spin from being at the same place at the same time. It does not exclude two electrons of opposite spin. Thus the electrostatic energy of a system will depend on the relative orientation of the spins: the difference in energy defines the **exchange energy.** The exchange energy of two electrons may be written in the form $-2J s_1 \cdot s_2$ as in (6), just as if there were a direct coupling between the directions of the two spins.[6]

We establish an approximate connection between the exchange integral J in (6) and the exchange constant λ in (1). We suppose that the atom under consideration has z nearest neighbors, each connected with the central atom by the interaction J. For more distant neighbors we take J as zero. The energy required to reverse the spin under consideration in the presence of all other spins may be written, neglecting components of $\mathbf{S}$ perpendicular to the average magnetization, as

$$U = 4J z S^2 = 2\mu B_E = 2\mu(\lambda M_s) = 2\mu(\lambda\mu/\Omega) \ , \tag{7}$$

where S is the average value of $\mathbf{S}$ in the direction of the magnetization and Ω is the volume per atom. The average magnetic moment of a spin is $\mu = gS\mu_B$, and the saturation magnetization is $M_s = \mu/\Omega$. Thus from (7)

$$\lambda = \frac{2Jz\Omega}{g^2{\mu_B}^2} \ . \tag{8}$$

Using (5) and $\Omega = 1/N$, we have the mean field theory result:

$$J = \frac{3k_B T_c}{2zS(S+1)} \ . \tag{9}$$

Better approximations to the quantum-statistical problem give somewhat different results. For the sc, bcc, and fcc structures with $S = \frac{1}{2}$, Rushbrooke and Wood[7] give $k_B T_c/zJ = 0.28$; 0.325; and 0.346, respectively, as compared with 0.500 from (9) for all three structures. If iron is represented by the Heisenberg model (6) with $S = 1$, then the observed Curie temperature corresponds to $J = 1.19 \times 10^{-2}$ eV.

[5] If two spins are antiparallel, the wavefunctions of the two electrons must be symmetric, as in the combination $u(\mathbf{r}_1)v(\mathbf{r}_2) + u(\mathbf{r}_2)v(\mathbf{r}_1)$. If the two spins are parallel, the Pauli principle requires that the orbital part of the wavefunction be antisymmetric, as in $u(\mathbf{r}_1)v(\mathbf{r}_2) - u(\mathbf{r}_2)v(\mathbf{r}_1)$, for here if we interchange the coordinates $\mathbf{r}_1$, $\mathbf{r}_2$ the wavefunction changes sign. If we set the positions equal so that $\mathbf{r}_1 = \mathbf{r}_2$, then the antisymmetric function vanishes: for parallel spins there is zero probability of finding the two electrons at the same position. See also Fig. 3.6.

[6] Equation (6) is an equation in the spin operators $\mathbf{S}_i$, $\mathbf{S}_j$; for many purposes in ferromagnetism it is a good approximation to treat the spins as classical angular momentum vectors.

[7] G. S. Rushbrooke and P. J. Wood, Molecular Physics **1**, 257 (1958).

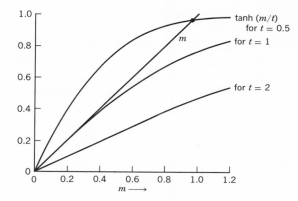

Figure 3 Graphical solution of Eq. (11) for the reduced magnetization m as a function of temperature. The reduced magnetization is defined as $m = M/N\mu$. The left-hand side of Eq. (11) is plotted as a straight line m with unit slope. The right-hand side is tanh (m/t) and is plotted vs. m for three different values of the reduced temperature $t = k_BT/N\mu^2\lambda = T/T_c$. The three curves correspond to the temperatures $2T_c$, T_c, and $0.5T_c$. The curve for $t = 2$ intersects the straight line m only at $m = 0$, as appropriate for the paramagnetic region (there is no external applied magnetic field). The curve for $t = 1$ (or $T = T_c$) is tangent to the straight line m at the origin; this temperature marks the onset of ferromagnetism. The curve for $t = 0.5$ is in the ferromagnetic region and intersects the straight line m at about $m = 0.94N\mu$. As $t \to 0$ the intercept moves up to $m = 1$, so that all magnetic moments are lined up at absolute zero.

Temperature Dependence of the Saturation Magnetization

We can also use the mean field approximation below the Curie temperature to find the magnetization as a function of temperature. We proceed as before, but instead of the Curie law we use the complete Brillouin expression (15.23) for the magnetization. For spin $\frac{1}{2}$ this is[8] $M = N\mu$ tanh $(\mu B/k_BT)$, according to (15.20). If we omit the applied magnetic field and replace B by the molecular field $B_E = \lambda M$, then

$$M = N\mu \tanh (\mu\lambda M/k_BT) \ . \tag{10}$$

We shall see that solutions of this equation with nonzero M exist in the temperature range between 0 and T_c.

To solve (10) we write it in terms of the reduced magnetization $m \equiv M/N\mu$ and the reduced temperature $t \equiv k_BT/N\mu^2\lambda$, whence

$$m = \tanh (m/t) \ . \tag{11}$$

We then plot the right and left sides of this equation separately as functions of m, as in Fig. 3. The intercept of the two curves gives the value of m at the temperature of interest. The critical temperature is $t = 1$, or $T_c = N\mu^2\lambda/k_B$, in agreement with (5) for $S = \frac{1}{2}$.

[8] We often write M_s for the spontaneous or saturation magnetization, but where no ambiguity is possible we shall use M.

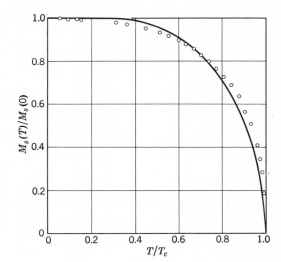

Figure 4 Saturation magnetization of nickel as a function of temperature, together with the theoretical curve for $S = \frac{1}{2}$ on the mean field theory. Experimental values by P. Weiss and R. Forrer, Ann. phys. **5**, 153 (1926).

The curves of M versus T obtained in this way reproduce roughly the features of the experimental results, as shown in Fig. 4 for nickel. As T increases the magnetization decreases smoothly to zero at $T = T_c$. This behavior classifies the usual ferromagnetic/paramagnetic transition as second-order.

The mean field theory does not give a good description of the variation of M at low temperatures. For $T \ll T_c$ the argument of tanh in (11) is large, and

$$\tanh \xi \cong 1 - 2e^{-2\xi} \ldots \tag{12}$$

To lowest order the magnetization deviation $\Delta M \equiv M(0) - M(T)$ is

$$\Delta M \cong 2N\mu \exp\left(-2\lambda N\mu^2/k_B T\right) . \tag{13}$$

The argument of the exponential is equal to $-2T_c/T$. For $T = 0.1T_c$ we have $\Delta M/N\mu \cong 2 \exp(-20) \cong 4 \times 10^{-9}$.

The experimental results show a much more rapid dependence of ΔM on

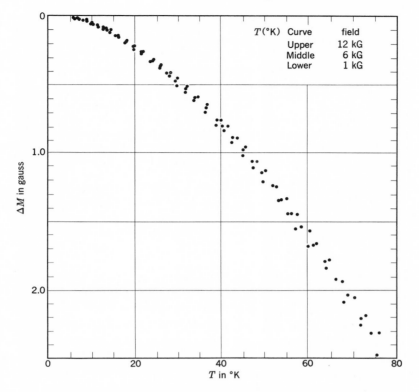

Figure 5 Decrease in magnetization of nickel with temperature above 4.2°K, after Argyle, Charap, and Pugh. In the plot $\Delta M \equiv 0$ at 4.2°K.

temperature at low temperatures than predicted by (13). At $T = 0.1T_c$ we have $\Delta M/M \cong 2 \times 10^{-3}$ from the data of Fig. 5. The leading term in ΔM is observed from experiment to have the form

$$\frac{\Delta M}{M(0)} = C_{\frac{3}{2}} T^{\frac{3}{2}} , \tag{14}$$

where[9] the constant $C_{\frac{3}{2}}$ has the experimental value $(7.5 \pm 0.2) \times 10^{-6}$ deg$^{-\frac{3}{2}}$ for Ni and $(3.4 \pm 0.2) \times 10^{-6}$ deg$^{-\frac{3}{2}}$ for Fe. The result (14) finds a natural explanation in terms of spin wave theory, as discussed below.

[9] B. E. Argyle, S. Charap, and E. W. Pugh, Phys. Rev. **132**, 2051 (1963).

Table 2 Ferromagnetic crystals

(Data selected with the assistance of R. M. Bozorth. General references:
A.I.P. Handbook, 1963, Sec. 5g; Landolt-Bornstein **2**, pt. 9, 6th ed., 1962)

Substance	Saturation magnetization M_s, in gauss		$n_B(0°K)$, per formula unit	Ferromagnetic Curie temperature, in °K
	Room temperature	0°K		
Fe	1707	1740	2.22	1043
Co	1400	1446	1.72	1388
Ni	485	510	0.606	627
Gd	—	2010	7.10	292
Dy	—	2920	10.0	85
Cu_2MnAl	500	(550)	(4.0)	710
MnAs	670	870	3.4	318
MnBi	620	680	3.52	630
Mn_4N	183	—	1.0	743
MnSb	710	—	3.5	587
MnB	152	163	1.92	578
CrTe	247	—	2.5	339
$CrBr_3$	—	—	—	33
CrO_2	515	—	2.03	386
$MnOFe_2O_3$	410	—	5.0	573
$FeOFe_2O_3$	480	—	4.1	858
$CoOFe_2O_3$	400	—	3.7	793
$NiOFe_2O_3$	270	—	2.4	858
$CuOFe_2O_3$	135	—	1.3	728
$MgOFe_2O_3$	110	—	1.1	713
UH_3	—	230	0.90	180
EuO	—	1920	6.8	69
$GdMn_2$	—	215	2.8	303
$Gd_3Fe_5O_{12}$	0	605	16.0	564
$Y_3Fe_5O_{12}$ (YIG)	130	200	5.0	560

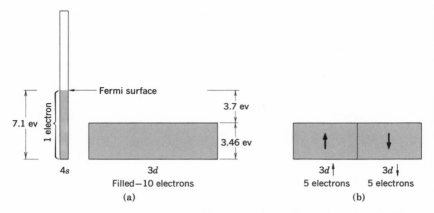

Figure 6a   Schematic relationship of 4s and 3d bands in metallic copper. The 3d band holds ten electrons per atom and is filled in copper. The 4s band can hold two electrons per atom; it is shown half-filled, as copper has one valence electron outside the filled 3d shell. The energies shown are from calculations by Howarth; it is coincidental that the bottoms of both bands fall nearly at the same energy.

Figure 6b   The filled 3d band of copper shown as two separate sub-bands of opposite electron spin orientation, each band holding five electrons. With both sub-bands filled as shown, the net spin (and hence the net magnetization) of the d band is zero.

Saturation Magnetization at Absolute Zero

Table 2 gives representative values of the saturation magnetization M_s, the effective magneton number n_B, and the ferromagnetic Curie temperature. The effective magneton number of a ferromagnet is defined by $M_s(0) = n_B N \mu_B$, where N is the number of formula units of the element or compound per unit volume. Do not confuse n_B with the paramagnetic effective magneton number p defined by (15.26).

Observed values of n_B are often nonintegral. There are many possible causes. One is the spin-orbit interaction which adds or subtracts some orbital magnetic moment. Another cause in ferromagnetic metals is the conduction electron magnetization induced locally about a paramagnetic ion core. A third cause is suggested by the drawing in Fig. 1 of the spin arrangement in a ferrimagnet: if there is one atom of spin projection $-S$ for every two atoms $+S$, the average spin is $\frac{1}{3}S$.

A band model[10] is the most likely model for the ferromagnetism of the transition metals Fe, Co, Ni. The approach is indicated in Figs. 6 and 7. The relationship of 4s and 3d bands is shown in Fig. 6 for copper, which is not ferromagnetic. If we remove one electron from copper, we obtain nickel which has

[10] E. C. Stoner, Repts. Prog. Phys. **11**, 43 (1948); Proc. Roy. Soc. (London) **A165**, 372 (1938); C. Herring and C. Kittel, Phys. Rev. **81**, 869 (1951); E. C. Mattis, *Theory of magnetism,* Harper and Row, 1965, Chap. 7; C. Herring in Rado and Suhl, Vol. IV (1966).

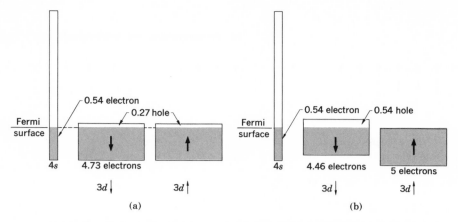

Figure 7a Band relationships in nickel above the Curie temperature. The net magnetic moment is zero, as there are equal numbers of holes in both $3d\downarrow$ and $3d\uparrow$ bands.

Figure 7b Schematic relationship of bands in nickel at absolute zero. The energies of the $3d\uparrow$ and $3d\downarrow$ sub-bands are separated by an exchange interaction. The $3d\uparrow$ band is filled; the $3d\downarrow$ band contains 4.46 electrons and 0.54 hole. The $4s$ band is usually thought to contain approximately equal numbers of electrons in both spin directions, and so we have not troubled to divide it into sub-bands. The net magnetic moment of 0.54 μ_B per atom arises from the excess population of the $3d\uparrow$ band over the $3d\downarrow$ band. It is often convenient to speak of the magnetization as arising from the 0.54 hole in the $3d\downarrow$ band.

the possibility of a vacant state in the $3d$ band. In the band structure of nickel shown in Fig. 7a for $T > T_c$ we have taken $2 \times 0.27 = 0.54$ of an electron away from the $3d$ band and 0.46 away from the $4s$ band, as compared with copper. The band structure of nickel at absolute zero is shown in Fig. 7b. Nickel is ferromagnetic and at absolute zero $n_B = 0.60$ Bohr magnetons per atom. After allowance[11] for the magnetic moment contribution of orbital electronic motion, nickel has an excess of 0.54 electron per atom having spin preferentially oriented in one direction. The exchange enhancement of the susceptibility of metals was the subject of Problem 15.9.

Are there in fact any simple ferromagnetic insulators, with all ionic spins parallel in the ground state? The few simple ferromagnets known at present include [12,13] $CrBr_3$, EuO, and EuS.

Spin Waves

The ground state of a simple ferromagnet has all spins parallel, as in Fig. 8a. Consider N spins each of magnitude S on a line or a ring, with nearest neighbor spins coupled by the Heisenberg interaction (6):

$$U = -2J \sum_{p=1}^{N} \mathbf{S}_p \cdot \mathbf{S}_{p+1} \ . \tag{15}$$

[11] P. Argyres and C. Kittel, Acta Met. **1**, 241 (1953). The number of effective ferromagnetic electrons n_e is just n_B corrected for the orbital contribution. We have $n_e = 2n_B/g$, where g is the spectroscopic splitting factor defined in Chapters 15 and 17. In metallic Ni, $g = 2.20$.

[12] A review of the properties of ferromagnetic europium compounds is given by T. R. McGuire and M. W. Shafer, J. Appl. Phys. **35**, 984 (1964).

[13] Numerous references to work on $CrBr_3$ are given by H. L. Davis and A. Narath, Phys. Rev. **134**, A433 (1964).

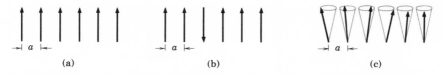

(a)   (b)   (c)

Figure 8 (a) Classical picture of the ground state of a simple ferromagnet; all spins are parallel. (b) A possible excitation; one spin is reversed. (c) The low-lying elementary excitations are spin waves. The ends of the spin vectors precess on the surfaces of cones, with successive spins advanced in phase by a constant angle.

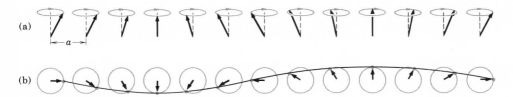

Figure 9 A spin wave on a line of spins. (a) The spins viewed in perspective. (b) Spins viewed from above, showing one wavelength. The wave is drawn through the ends of the spin vectors.

Here J is the exchange integral and $\hbar \mathbf{S}_p$ is the angular momentum of the spin at site p. If we treat the spins $\mathbf{S}_p$ as classical vectors, then in the ground state $\mathbf{S}_p \cdot \mathbf{S}_{p+1} = S^2$ and the exchange energy of the system is $U_0 = -2NJS^2$. What is the energy of the first excited state? Consider an excited state with one particular spin reversed, as in Fig. 8b. We see from (15) that this increases the energy by $8JS^2$, so that $U_1 = U_0 + 8JS^2$.

We can form an excitation of much lower energy if we let all the spins share the reversal, as in Fig. 8c. The elementary excitations of a spin system have a wavelike form and are called **spin waves** or, when quantized, **magnons** (Fig. 9). These are analogous to lattice vibrations or phonons. Spin waves are oscillations in the relative orientations of spins on a lattice; lattice vibrations are oscillations in the relative positions of atoms on a lattice.

We now give a classical derivation of the magnon dispersion relation for the problem described by the interaction (15). The terms in (15) which involve the pth spin are

$$-2J\mathbf{S}_p \cdot (\mathbf{S}_{p-1} + \mathbf{S}_{p+1}) \ . \tag{16a}$$

We write the magnetic moment at site p as

$$\boldsymbol{\mu}_p = -g\mu_B\mathbf{S}_p \ , \tag{16b}$$

as in (15.15). Then (16a) becomes

$$-\boldsymbol{\mu}_p \cdot [(-2J/g\mu_B)(\mathbf{S}_{p-1} + \mathbf{S}_{p+1})] \ , \tag{16c}$$

which is of the form

$$-\boldsymbol{\mu}_p \cdot \mathbf{B}_p \ . \tag{16d}$$

The effective magnetic field or exchange field that acts on the pth spin is

$$\mathbf{B}_p = (-2J/g\mu_B)(\mathbf{S}_{p-1} + \mathbf{S}_{p+1}) \ , \tag{17}$$

by comparison of (16d) with (16c).

From elementary mechanics the rate of change of the angular momentum $\hbar\mathbf{S}_p$ is equal to the torque $\boldsymbol{\mu}_p \times \mathbf{B}_p$ which acts on the spin:

$$\hbar\frac{d\mathbf{S}_p}{dt} = \boldsymbol{\mu}_p \times \mathbf{B}_p \ , \tag{18}$$

or

$$\frac{d\mathbf{S}_p}{dt} = \left(-\frac{g\mu_B}{\hbar}\right)\mathbf{S}_p \times \mathbf{B}_p = \left(\frac{2J}{\hbar}\right)(\mathbf{S}_p \times \mathbf{S}_{p-1} + \mathbf{S}_p \times \mathbf{S}_{p+1}) \ . \tag{19}$$

In Cartesian components

$$dS_p{}^x/dt = (2J/\hbar)\,[S_p{}^y(S_{p-1}^z + S_{p+1}^z) - S_p{}^z(S_{p-1}^y + S_{p+1}^y)] \ , \tag{20}$$

and similarly for $dS_p{}^y/dt$ and $dS_p{}^z/dt$. These equations involve products of spin components and are nonlinear.

If the amplitude of the excitation is small (if $S_p{}^x$, $S_p{}^y \ll S$), we may obtain an approximate set of linear equations by taking all $S_p{}^z = S$, and by neglecting terms in the product of S^x and S^y which appear in the equation for dS^z/dt. The linearized equations are

$$\frac{dS_p{}^x}{dt} = (2JS/\hbar)(2S_p{}^y - S_{p-1}^y - S_{p+1}^y) \ ; \tag{21a}$$

$$\frac{dS_p{}^y}{dt} = -(2JS/\hbar)(2S_p{}^x - S_{p-1}^x - S_{p+1}^x) \ ; \tag{21b}$$

$$\frac{dS_p{}^z}{dt} = 0 \ . \tag{22}$$

By analogy with the phonon problems in Chapter 5 we look for traveling wave solutions of (21) of the form

$$S_p{}^x = ue^{i(pka-\omega t)} \ ; \qquad S_p{}^y = ve^{i(pka-\omega t)} \ , \tag{23}$$

where u, v are constants, p is an integer, and a is the lattice constant. On substitution into (21) we have

$$-i\omega u = \left(\frac{2JS}{\hbar}\right)(2 - e^{-ika} - e^{ika})v = \left(\frac{4JS}{\hbar}\right)(1 - \cos ka)v \ ;$$

$$-i\omega v = -\left(\frac{2JS}{\hbar}\right)(2 - e^{-ika} - e^{ika})u = -\left(\frac{4JS}{\hbar}\right)(1 - \cos ka)u \ .$$

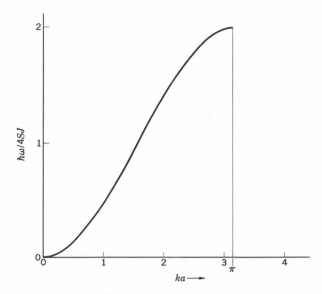

Figure 10 Dispersion relation for spin waves in a ferromagnet in one dimension with nearest-neighbor interactions.

These equations have a solution for u and v if the determinant of the coefficients is equal to zero:

$$\begin{vmatrix} i\omega & \left(\dfrac{4JS}{\hbar}\right)(1-\cos ka) \\ -\left(\dfrac{4JS}{\hbar}\right)(1-\cos ka) & i\omega \end{vmatrix} = 0 \; , \qquad (24)$$

whence

$$\hbar\omega = 4JS(1-\cos ka) \; . \qquad (25)$$

This result is plotted in Fig. 10. With this solution we find that $v = -iu$, corresponding to circular precession[14] of each spin about the z axis.

Equation (25) is the dispersion relation $\omega(k)$ for spin waves in one dimension with nearest-neighbor interactions.[15] At long wavelengths $ka \ll 1$, so that $(1-\cos ka) \cong \frac{1}{2}(ka)^2$. In this limit (25) becomes

$$\hbar\omega \cong (2JSa^2)k^2 \; . \qquad (26)$$

Notice that the frequency is proportional to k^2, whereas the frequency of a phonon in the same limit is proportional to k.

[14] We see this on taking real parts of (23), with v set equal to $-iu$. Then

$$S_p{}^x = u \cos(pka - \omega t); \; S_p{}^y = u \sin(pka - \omega t) \; .$$

[15] Precisely the same result is obtained from the quantum-mechanical solution; see *QTS*, Chap. 4.

The dispersion relation for a ferromagnetic cubic lattice (sc, bcc, or fcc) with nearest-neighbor interactions may be shown to be (see Problem 1)

$$\hbar\omega = 2JS\left[z - \sum_{\delta} \cos (\mathbf{k} \cdot \boldsymbol{\delta})\right] , \qquad (27)$$

where the summation is over the z vectors denoted by $\boldsymbol{\delta}$ which join the central atom to its nearest neighbors. The leading terms in the expansion of (27) for $ka \ll 1$ is

$$\hbar\omega = (2JSa^2)k^2 \qquad (28)$$

for all three cubic lattices, where a is the lattice constant. The coefficient of k^2 often may be determined accurately by spin wave resonance[16] in thin films.

Quantization of Spin Waves

The values of the total spin quantum number of a system of N spins S are NS, $NS - 1$, $NS - 2, \ldots$, according to the quantum mechanics of angular momenta. In the ferromagnetic ground state the total spin quantum number has the value NS: all spins are parallel in the ground state. The excitation of a spin wave lowers the total spin because the spins are no longer parallel. We look for the relation between the amplitude of the spin wave and the reduction in the z component of the total spin quantum number.

Consider the spin wave (23) with $v = -iu$, as we found earlier:

$$S_p^x = ue^{i(pka-\omega t)} \; ; \qquad S_p^y = -iue^{i(pka-\omega t)} . \qquad (29)$$

The spin component perpendicular to the z direction is u, independent of the site position p and of the time. The z component of a spin is

$$S_z = (S^2 - u^2)^{\frac{1}{2}} \cong S - \frac{u^2}{2S} , \qquad (30)$$

for small amplitudes $u/S \ll 1$.

Quantum theory allows only integral values for $S - S_z$. If N is the total number of spins and $NS - n_k$ is the z component of the total spin when a spin wave k is excited, then by (30) we have the quantization condition for the spin wave amplitude u_k:

$$n_k \cong \frac{Nu_k^2}{2S} \; ; \qquad \text{or} \qquad u_k^2 \cong \frac{2Sn_k}{N} . \qquad (31)$$

Here n_k is an integer equal to the number of **magnons** of wavevector $\mathbf{k}$ that are excited. Each magnon lowers the z component of the total spin by one.

---

[16] See Chapter 17.

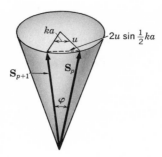

Figure 11   Construction relating the angle φ between two successive spin vectors to the spin wave amplitude u and the phase angle ka. The length of the dashed line is $2u \sin \frac{1}{2}ka$; if the length of a spin is S, then $S \sin \frac{1}{2}\varphi = u \sin \frac{1}{2}ka$.

Does the energy ϵ_k satisfy the same quantum condition

$$\epsilon_k = n_k \hbar \omega_k \ , \tag{32}$$

as for photons and phonons? The exchange energy (15) depends on the cosine of the angle between the spins at sites p and $p + 1$. The *phase difference* at the same time t between successive spins is ka radians, according to (29). The tips of the two spin vectors in Fig. 11 are separated by a distance $2u \sin \frac{1}{2}ka$, so that the angle φ between the spin vectors is given by

$$\sin \tfrac{1}{2}\varphi = (u/S) \sin \tfrac{1}{2}ka \ . \tag{33}$$

For $u/S \ll 1$ the cosine of φ is

$$\cos \varphi = 1 - 2(u/S)^2 \sin^2 \tfrac{1}{2}ka \ . \tag{34}$$

Thus the energy (15) is

$$U \cong -2JNS^2 + 4JNu^2 \sin^2 \tfrac{1}{2}ka = -2JNS^2 + 2JNu^2(1 - \cos ka) \ , \tag{35}$$

and the excitation energy of a spin wave of amplitude u_k and wavevector k is $\epsilon_k = 2JNu_k{}^2(1 - \cos ka)$. With the quantization condition (31) we have a result of the form (32):

$$\epsilon_k = 4JS(1 - \cos ka)n_k = n_k \hbar \omega_k \ , \tag{36}$$

by (25).

Thermal Excitation of Magnons

In thermal equilibrium the average value of $n_\mathbf{k}$ is given by the Planck distribution:[17]

$$\langle n_\mathbf{k} \rangle = \frac{1}{\exp\,(\hbar\omega_\mathbf{k}/k_B T)\,-\,1}\,. \tag{37}$$

The total number of magnons excited at a temperature T is

$$\sum_\mathbf{k} n_\mathbf{k} = \int\,d\omega\,\mathfrak{D}(\omega)\langle n(\omega)\rangle\,, \tag{38}$$

where $\mathfrak{D}(\omega)$ is the number of magnon modes per unit frequency range, as in Chapter 6 for phonons. The integral is taken over the allowed range of $\mathbf{k}$, which is the first Brillouin zone. At sufficiently low temperatures we may carry the integral between 0 and ∞ because $\langle n(\omega)\rangle \to 0$ exponentially as $\omega \to \infty$.

Magnons have only a single polarization for each value of $\mathbf{k}$. In three dimensions the number of modes of wavevector less than k is $(1/2\pi)^3(4\pi k^3/3)$ per unit volume, whence the number of magnons $\mathfrak{D}(\omega)d\omega$ with frequency in $d\omega$ at ω is $(1/2\pi)^3(4\pi k^2)\,(dk/d\omega)\,d\omega$. In the approximation (28)

$$\frac{d\omega}{dk} = \frac{4JSa^2 k}{\hbar} = 2\Big(\frac{2JSa^2}{\hbar}\Big)^{\frac{1}{2}}\omega^{\frac{1}{2}}\,.$$

Thus the density of modes for magnons is

$$\mathfrak{D}(\omega) = \frac{1}{4\pi^2}\cdot\Big(\frac{\hbar}{2JSa^2}\Big)^{\frac{3}{2}}\omega^{\frac{1}{2}}\,, \tag{39}$$

so that (38) becomes

$$\sum_\mathbf{k} n_\mathbf{k} = \frac{1}{4\pi^2}\Big(\frac{\hbar}{2JSa^2}\Big)^{\frac{3}{2}}\int_0^\infty d\omega\,\frac{\omega^{\frac{1}{2}}}{e^{\beta\hbar\omega}-1} = \frac{1}{4\pi^2}\Big(\frac{k_B T}{2JSa^2}\Big)^{\frac{3}{2}}\int_0^\infty dx\,\frac{x^{\frac{1}{2}}}{e^x-1}\,. \tag{40}$$

The definite integral is found in tables and has the value $(0.0587)\,(4\pi^2)$.

The number N of atoms per unit volume is Q/a^3, where $Q = 1,\,2,\,4$ for sc, bcc, fcc lattices respectively. Now $(\Sigma n_\mathbf{k})/NS$ is equal to the fractional change of magnetization $\Delta M/M(0)$, whence

$$\boxed{\frac{\Delta M}{M(0)} = \frac{0.0587}{SQ}\cdot\Big(\frac{k_B T}{2JS}\Big)^{\frac{3}{2}}\,.} \tag{41}$$

This result[18] due to Felix Bloch[19] is known as the **Bloch $T^{\frac{3}{2}}$ law**; it has the form found experimentally. The physical basis of the $T^{\frac{3}{2}}$ law is the subject of Problem 8.

---

[17] The argument is exactly as for phonons or photons. The Planck distribution follows for any problem where the energy levels are identical with those of a harmonic oscillator or collection of harmonic oscillators.

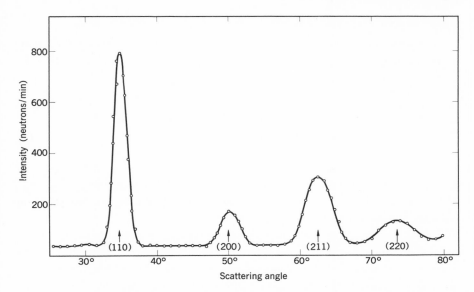

Figure 12   Neutron diffraction pattern for iron. The observed reflections satisfy the index rule for a body-centered cubic structure: the sum of the reflection indices is an even integer. [After C. G. Shull, E. O. Wollan, and W. C. Koehler, Phys. Rev. **84**, 912 (1951).]

Neutron Scattering: Elastic and Inelastic

In Chapter 5 we discussed the determination of phonon spectra by inelastic x-ray or neutron scattering. An x-ray photon sees the spatial distribution of electronic charge, whether or not the charge density is magnetized or unmagnetized. But a neutron sees two aspects of a crystal: the distribution of nuclei and the distribution of electronic magnetization. The neutron diffraction pattern for iron is shown in Fig. 12.

The magnetic moment of the neutron interacts with the magnetic moment of the electron. The cross-section for the neutron-electron interaction is of the same order of magnitude as for the neutron-nuclear interaction. Diffraction of neutrons by a magnetic crystal allows the determination of the distribution, direction, and order of the magnetic moments. Further, a neutron can be inelastically scattered by the magnetic structure, with the creation or annihilation of a magnon; such events make possible the experimental determination of magnon spectra.

[18] The result (41) is the correct leading term on the Heisenberg model. To go further we need to use the full dispersion relation; to carry out the integration over the first Brillouin zone; to take account of magnon-magnon interactions; and to take account of external magnetic fields, anisotropy fields, and dipolar interactions among the spins. An elementary method for taking account of magnon-magnon interactions is given by M. Bloch, Phys. Rev. Letters **9**, 286 (1962); J. Appl. Phys. **34**, 1151 (1963).

[19] F. Bloch, Z. Physik **61**, 206 (1931).

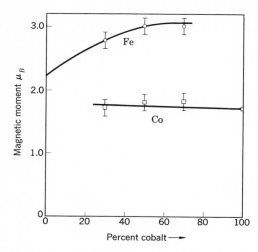

Figure 13 Moments attributable to $3d$ electrons in Fe-Co alloys as a function of composition, after M. F. Collins and J. B. Forsyth, Phil. Mag. **8**, 401 (1963).

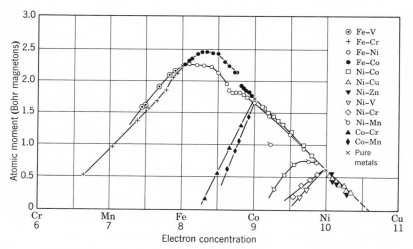

Figure 14 Average atomic moments of binary alloys of the elements in the iron group, after Bozorth.

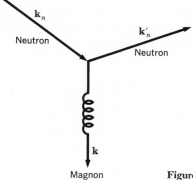

Figure 15 Scattering of a neutron with creation of a magnon.

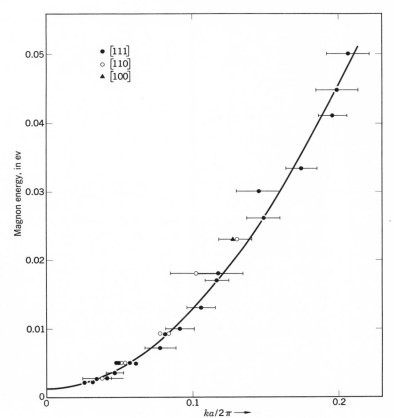

Figure 16 Magnon spectrum of a fcc cobalt alloy (92 Co 8 Fe) at room temperature, after R. N. Sinclair and B. N. Brockhouse, Phys. Rev. **120**, 1638 (1960). The solid line is the best-fit theoretical curve. The results shown do not extend very far out into the Brillouin zone.

The magnetic moments associated with particular components of alloys may be investigated by neutron diffraction. Results for the Fe-Co binary alloy system (which is ferromagnetic) are shown in Fig. 13. Notice that the magnetic moment on the cobalt atom does not appear to be affected by alloying, but that on the iron atom increases to about $3\mu_B$ as the cobalt concentration increases. The magnetizations are shown in Fig. 14, a famous plot.

In an inelastic scattering event a neutron may create or destroy a magnon (Fig. 15). If the incident neutron has wavevector $\mathbf{k}_n$ and is scattered to $\mathbf{k}_n'$ with the creation of a magnon of wavevector $\mathbf{k}$, then by conservation of crystal momentum

$$\mathbf{k}_n = \mathbf{k}_n' + \mathbf{k} + \mathbf{G} \,, \tag{42}$$

where $\mathbf{G}$ is a reciprocal lattice vector. By conservation of energy

$$\frac{\hbar^2 k_n{}^2}{2M_n} = \frac{\hbar^2 k_n'{}^2}{2M_n} + \hbar\omega_{\mathbf{k}} \,, \tag{43}$$

where $\hbar\omega_{\mathbf{k}}$ is the energy of the magnon created in the process. The observed magnon spectrum for a cobalt-rich alloy is shown in Fig. 16.

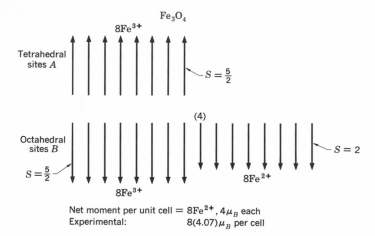

Figure 17 Schematic spin arrangements in magnetite, FeO · Fe₂O₃, showing that the moments of the Fe^{3+} ions cancel out, leaving only the moments of the Fe^{2+} ions. The types of interstices are defined in Fig. 19.

The magnon dispersion curves can be distinguished experimentally from the phonon dispersion curves in the same crystal by two features: (1) The magnons vanish (or at least their mean free paths become very short) at temperatures somewhat above the Curie temperature. (2) The intensity of neutron scattering by a magnon is proportional to the square of that component of the spin of the specimen which is normal to the neutron scattering vector $\mathbf{k}_n' - \mathbf{k}_n$, as proved in *QTS*, pp. 380–382. Thus the relative intensities of phonon and magnon inelastic scattering may be varied by a suitable rotation of the specimen, thereby permitting an unambiguous identification of the magnons.

FERRIMAGNETIC ORDER

In many ferromagnetic crystals the saturation magnetization at $T = 0°\text{K}$ does not correspond to parallel alignment of the magnetic moments of the constituent paramagnetic ions, even in crystals for which there is strong evidence that the individual paramagnetic ions have their normal magnetic moments. The most familiar example is magnetite, Fe_3O_4 or $FeO · Fe_2O_3$. From Table 15.2 we see that ferric (Fe^{3+}) ions are in a state with spin $S = \frac{5}{2}$ and zero orbital moment. Thus each ion should contribute $5\mu_B$ to the saturation moment. The ferrous (Fe^{2+}) ions have a spin of 2 and should contribute $4\mu_B$, apart from any residual orbital moment contribution. Thus the effective number of Bohr magnetons per Fe_3O_4 formula unit should be about $2 \times 5 + 4 = 14$ if all spins were parallel. The observed value (Table 2) is 4.1. The discrepancy[20] is accounted for if the moments of the Fe^{3+} ions are antiparallel to each other: then the observed moment arises only from the Fe^{2+} ion, as in Fig. 17. The neutron diffraction results in Fig. 18 agree with this model.

[20] L. Néel, Ann. phys. **3**, 137 (1948).

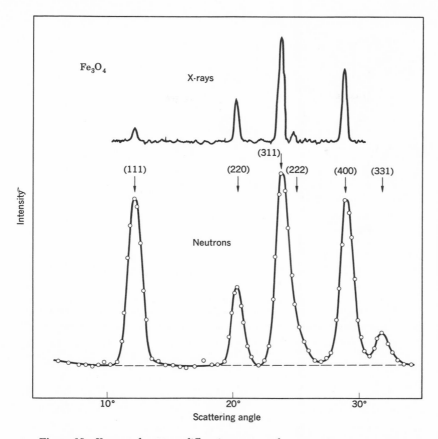

Figure 18 X-ray and neutron diffraction patterns for magnetite at room temperature. A pronounced magnetic scattering contribution is found in the neutron pattern: note the strength of the (111) neutron line. The relative intensities of the neutron diffraction lines are sensitive to the order of the Fe^{2+} and Fe^{3+} electronic magnetic moments. The intensities calculated for the Néel ferrimagnetic structure agree well with the observed intensities:

Reflection:	(111)	(220)	(311) + (222)	(400)	(331)
Calculated intensity	934	343	1060	765	110
Observed intensity	860	360	1070	780	135

(After Shull, Wollan, and Koehler.)

A systematic discussion of the consequences of this type of spin order was given by L. Néel with reference to an important class of magnetic oxides known as ferrites.[21] The term **ferrimagnetic** was coined originally to describe the ferrite-type ferromagnetic spin order such as Fig. 17, and by extension the term covers almost any compound in which some ions have a moment antiparallel to other ions. Many ferrimagnets are poor conductors of electricity, a quality which is exploited in device applications.

[21] The usual chemical formula of a **ferrite** is $MO \cdot Fe_2O_3$, where M is a divalent cation, often Zn, Cd, Fe, Ni, Cu, Co, or Mg.

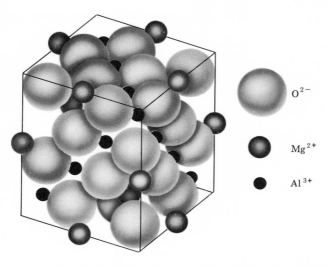

Figure 19 Crystal structure of the mineral spinel $MgAl_2O_4$; the Mg^{2+} ions occupy tetrahedral sites, each surrounded by four oxygen ions; the Al^{3+} occupy octahedral sites, each surrounded by six oxygen ions. This is a **normal spinel** arrangement: the divalent metal ions occupy the tetrahedral sites. In the **inverse spinel** arrangement the tetrahedral sites are occupied by trivalent metal ions, while the octahedral sites are occupied half by divalent and half by trivalent metal ions.

The cubic ferrites have the **spinel** crystal structure shown in Fig. 19. There are eight occupied tetrahedral (or A) sites and sixteen occupied octahedral (or B) sites in a unit cube. The lattice constant is about 8 Å. A remarkable feature of the spinels is that all exchange interactions (AA, AB, and BB) are believed to favor *antiparallel* alignment of the spins connected by the interaction. But the AB interaction is the strongest, so that the A spins are parallel to each other and the B spins are parallel to each other, just in order that the A spins may be antiparallel to the B spins. We believe that all exchange integrals J_{AA}, J_{AB}, and J_{BB} are negative.[22]

We now prove that three antiferromagnetic interactions can result in ferrimagnetism. The mean exchange fields acting on the A and B spin lattices may be written

$$\mathbf{B}_A = -\lambda\mathbf{M}_A - \mu\mathbf{M}_B \; ; \qquad \mathbf{B}_B = -\mu\mathbf{M}_A - \nu\mathbf{M}_B \; ; \qquad (44)$$

taking all constants λ, μ, ν to be positive. The minus sign then corresponds to an antiparallel interaction. The interaction energy density is

$$U = -\tfrac{1}{2}(\mathbf{B}_A \cdot \mathbf{M}_A + \mathbf{B}_B \cdot \mathbf{M}_B) = \tfrac{1}{2}\lambda M_A^2 + \mu\mathbf{M}_A \cdot \mathbf{M}_B + \tfrac{1}{2}\nu M_B^2 \; ; \quad (45)$$

this is lower when $\mathbf{M}_A$ is antiparallel to $\mathbf{M}_B$ than when $\mathbf{M}_A$ is parallel to $\mathbf{M}_B$. The energy when antiparallel should be compared with zero, because a possible solution is $M_A = M_B = 0$. Thus when

$$\mu M_A M_B > \tfrac{1}{2}(\lambda M_A^2 + \nu M_B^2) \; , \qquad (46)$$

[22] If J in $U = -2J\mathbf{S}_i \cdot \mathbf{S}_j$ is positive, we say that the exchange integral is ferromagnetic; if J is negative, the exchange integral is antiferromagnetic.

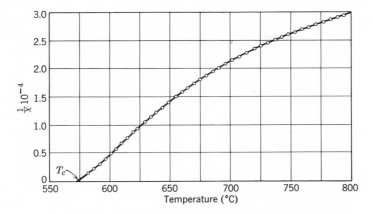

Figure 20   Reciprocal susceptibility of magnetite, $FeO \cdot Fe_2O_3$, above the Curie temperature.

the ground state will have M_A directed oppositely to M_B. Under certain conditions there may be noncollinear spin arrays of still lower energy.

Curie Temperature and Susceptibility of Ferrimagnets

We define separate Curie constants C_A and C_B for the A and B sites. We need separate C's for the two lattices because the numbers and types of paramagnetic ions will usually be different on the two lattices. Let all interactions be zero except for an antiparallel interaction between the A and B sites: $\mathbf{B}_A = -\mu\mathbf{M}_B$; $\mathbf{B}_B = -\mu\mathbf{M}_A$, where μ is positive. In analogy to the argument of (1) to (4) we have in the mean field approximation

$$\text{(CGS)} \quad M_A T = C_A(B_a - \mu M_B) \; ; \qquad M_B T = C_B(B_a - \mu M_A) \; , \qquad (47)$$

where B_a is the applied field. These equations have a nonzero solution for M_A and M_B in zero applied field if

$$\begin{vmatrix} T & \mu C_A \\ \mu C_B & T \end{vmatrix} = 0 \; , \qquad (48)$$

so that the ferrimagnetic Curie temperature is given by $T_c = \mu(C_A C_B)^{\frac{1}{2}}$.

We solve (47) for M_A and M_B to obtain the susceptibility at $T > T_c$:

$$\text{(CGS)} \qquad \chi = \frac{M_A + M_B}{B_a} = \frac{(C_A + C_B)T - 2\mu C_A C_B}{T^2 - T_c{}^2} \; , \qquad (49)$$

a result more complicated than (4). Experimental values for Fe_3O_4 are plotted in Fig. 20. The curvature of the plot of $1/\chi$ versus T is a characteristic feature of a ferrimagnet. In (51) below we consider the case antiferromagnetic $C_A = C_B$.

Iron Garnets. The iron garnets are cubic ferrimagnetic insulators with the general formula $M_3Fe_5O_{12}$, where M is a trivalent metal ion and the Fe is the trivalent ferric ion ($S = \frac{5}{2}, L = 0$). An example is yttrium iron garnet $Y_3Fe_5O_{12}$, known as YIG. Here Y^{3+} is diamagnetic.

The net magnetization of YIG is due to the resultant of two oppositely magnetized lattices of Fe^{3+} ions. At absolute zero each ferric ion contributes $\pm 5\mu_B$ to the magnetization, but in each formula unit the three Fe^{3+} ions on sites denoted as d sites are magnetized in one sense and the two Fe^{3+} ions on a sites are magnetized in the opposite sense, giving a resultant of $5\mu_B$ per formula unit in good agreement with the measurements of Geller et al.[23] The mean field at an a site due to the ions on the d sites is $B_a = -(1.5 \times 10^4)M_d$. The observed Curie temperature 559°K of YIG is due to the a-d interaction.

The only magnetic ions in YIG are the ferric ions. Because these are in an $L = 0$ state with a spherical change distribution their interaction with lattice deformations and phonons is weak. As a result YIG is characterized by very narrow linewidths in ferromagnetic resonance experiments (Chapter 17).

In the rare-earth iron garnets[24] the ions M^{3+} are paramagnetic trivalent rare-earth ions. Magnetization curves are given in Fig. 21. The rare-earth ions occupy sites labeled c; the magnetization M_c of the ions on the c lattice is opposite to the net magnetization of the ferric ions on the $a + d$ sites. At low temperatures (Fig. 22) the combined moments of the three rare-earth ions in a formula unit may dominate the net moment of the Fe^{3+} ions, but because of the weak c-a and c-d coupling the rare-earth lattice loses its magnetization rapidly with increasing temperature. The total moment can pass through zero and then increase again as the Fe^{3+} moment starts to be dominant. In GdIG the mean field at a c site can be represented by $\mathbf{B}_c \cong -(2 \times 10^3)(\mathbf{M}_a + \mathbf{M}_d)$, much weaker than the field at an a site as given above.

[23] S. Geller, H. J. Williams, R. C. Sherwood, J. P. Remeika, and G. P. Espinosa, Phys. Rev. **131**, 1080 (1963).

[24] A review of their properties is given by L. Néel, R. Pauthenet, and B. Dreyfus, *Prog. in low temperature physics* **4**, 344–383 (1964).

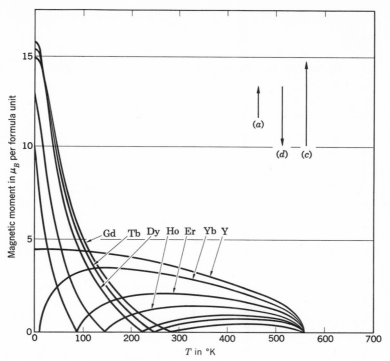

Figure 21 Experimental values of the saturation magnetization versus temperature of various iron garnets, after R. Pauthenet. The formula unit is $M_3Fe_5O_{12}$, where M is a trivalent metal ion. The temperature at which the magnetization crosses zero is called the compensation temperature; here the magnetization of the M sublattice is equal and opposite to the net magnetization of the ferric ion sublattices. Per formula unit there are 3 Fe^{3+} ions on tetrahedral sides d; 2 Fe^{3+} ions on octahedral sites a; and 3 M^{3+} ions on sites denoted by c. The ferric ions contribute $(3 - 2)5\mu_B = 5\mu_B$ per formula unit. The ferric ion coupling is strong and determines the Curie temperature. If the M^{3+} ions are rare earth ions they are magnetized opposite to the resultant of the Fe^{3+} ions. The M^{3+} contribution drops rapidly with increasing temperature because the M-Fe coupling is weak. Measurements on single crystal specimens are reported by Geller et al., Phys. Rev. **137**, 1034 (1965).

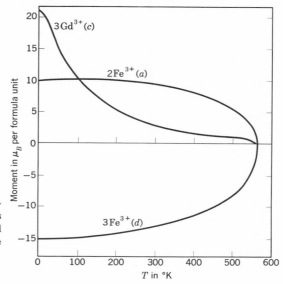

Figure 22 Magnetization of the sublattices in gadolinium iron garnet, as calculated by R. Pauthenet. The total moment is zero near 280°K; the Curie temperature is near 560°K.

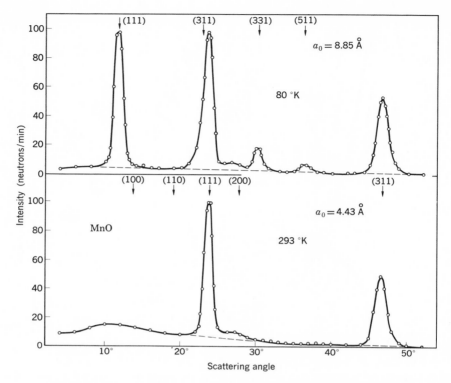

Figure 23 Neutron diffraction patterns for MnO below and above the spin-ordering temperature of 120°K, after C. G. Shull, W. A. Strauser, and E. O. Wollan. Phys. Rev. **83**, 333 (1951). The reflection indices are based on an 8.85 Å cell at 80°K and on a 4.43 Å cell at 293°K. At the higher temperature the Mn^{2+} ions are still magnetic, but they are no longer ordered.

ANTIFERROMAGNETIC ORDER

A classical example of magnetic structure determination by neutrons is shown in Fig. 23 for MnO, which has the NaCl structure. At 80°K there are extra neutron reflections not present at 293°K. The reflections at 80°K may be classified in terms of a cubic unit cell of lattice constant 8.85 Å. At 293°K the reflections correspond to an fcc unit cell of lattice constant 4.43 Å. But the lattice constant determined by x-ray reflection is 4.43 Å at *both* temperatures, 80°K and 293°K. We conclude that the chemical unit cell has the 4.43 Å lattice parameter, but that at 80°K the electronic magnetic moments of the Mn^{2+} ions are ordered in some nonferromagnetic arrangement. If the ordering were ferromagnetic, the chemical and magnetic cells would give the same reflections. The spin arrangement shown in Fig. 24 is consistent with the neutron diffraction results and with magnetic measurements. The spins in a single [111] plane are parallel, but adjacent [111] planes are antiparallel. Thus MnO is an antiferromagnet, as in Fig. 25.

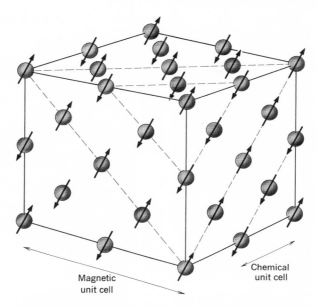

Figure 24   Ordered arrangements of spins of the Mn^{2+} ions in manganese oxide. MnO, as determined by neutron diffraction. The O^{2-} ions are not shown.

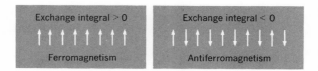

Figure 25   Spin ordering in ferromagnets $(J > 0)$ and antiferromagnets $(J < 0)$.

556

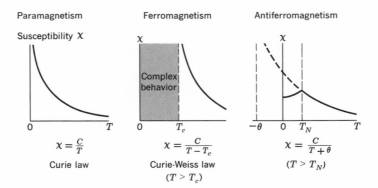

Figure 26 Temperature dependence of the magnetic susceptibility in paramagnets, ferromagnets, and antiferromagnets. Below the Néel temperature of an antiferromagnet the spins have antiparallel orientations; the susceptibility attains its maximum value at T_N where there is a well-defined kink in the curve of χ versus T. The transition is also marked by peaks in the heat capacity and the thermal expansion coefficient.

In an **antiferromagnet** the spins are ordered in an antiparallel arrangement with zero net moment at temperatures below the ordering or **Néel temperature.** The susceptibility of an antiferromagnet is not infinite at $T = T_N$, but has a weak cusp, as in Fig. 26.

An antiferromagnet is a special case of a ferrimagnet for which both sublattices A and B have equal saturation magnetizations. Thus $C_A = C_B$ in (47), and the Néel temperature in the mean field approximation is given by

$$T_N = \mu C , \tag{50}$$

where C refers to a single sublattice. The susceptibility in the paramagnetic region $T > T_N$ is obtained from (49):

$$\chi = \frac{2CT - 2\mu C^2}{T^2 - (\mu C)^2} = \frac{2C}{T + \mu C} = \frac{2C}{T + T_N} . \tag{51}$$

The experimental results at $T > T_N$ are of the form

(CGS) $$\chi = \frac{2C}{T + \theta} . \tag{52}$$

Experimental values of θ/T_N listed in Table 3 often differ substantially from the value unity expected from (51). Values of θ/T_N of the observed magnitude may be obtained when next-nearest-neighbor interactions are provided for, and when sublattice arrangements[25] are considered. It is shown in Problem 3 that if a mean field constant $-\epsilon$ is introduced to describe interactions within a sublattice, then $\theta/T_N = (\mu + \epsilon)/(\mu - \epsilon)$.

[25] P. W. Anderson, Phys. Rev. **79**, 350, 705 (1950); J. M. Luttinger, Phys. Rev. **81**, 1015 (1951).

Table 3   Antiferromagnetic crystals

The Néel temperatures T_N often vary considerably between samples, and in some cases there is large thermal hysteresis. For a bibliography relating to experimental data on antiferromagnetic substances, see T. Nagamiya, K. Yosida, and R. Kubo, Advances in Physics **4**, 1–112 (1955); and the A.I.P. Handbook. The value of θ is obtained by fitting an expression of the form $\chi = C/(T + \theta)$ to the susceptibility above the actual transition temperature T_N.

Substance	Paramagnetic ion lattice	Transition temperature, T_N in °K	Curie-Weiss θ in °K	$\dfrac{\theta}{T_N}$	$\dfrac{\chi(0)}{\chi(T_N)}$
MnO	fcc	116	610	5.3	$\frac{2}{3}$
MnS	fcc	160	528	3.3	0.82
MnTe	hex. layer	307	690	2.25	
MnF$_2$	bc tetr	67	82	1.24	0.76
FeF$_2$	bc tetr	79	117	1.48	0.72
FeCl$_2$	hex. layer	24	48	2.0	<0.2
FeO	fcc	198	570	2.9	0.8
CoCl$_2$	hex. layer	25	38.1	1.53	
CoO	fcc	291	330	1.14	
NiCl$_2$	hex. layer	50	68.2	1.37	
NiO	fcc	525	~2000	~4	
α-Mn	complex	~100			
Cr	bcc	308			
CrSb	hex. layer	723	550	0.76	$\sim\frac{1}{4}$
Cr$_2$O$_3$	complex	307	485	1.58	
FeCO$_3$	complex	35	14	0.4	$\sim\frac{1}{4}$

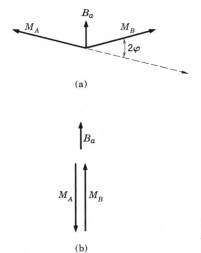

Figure 27 Calculation of (a) perpendicular and (b) parallel susceptibilities at $0°K$, in the mean field approximation.

Susceptibility below the Néel Temperature

There are two situations: with the applied magnetic field perpendicular to the axis of the spins; and with the field parallel to the axis of the spins. At the Néel temperature the susceptibility is nearly independent of the direction of the field relative to the spin axis.

For $\mathbf{B}_a$ perpendicular to the axis of the spins we can calculate the susceptibility by elementary considerations. The energy density in the presence of the field is, with $M = |M_A| = |M_B|$,

$$U = \mu\mathbf{M}_A \cdot \mathbf{M}_B - \mathbf{B}_a \cdot (\mathbf{M}_A + \mathbf{M}_B) \cong -\mu M^2[1 - \tfrac{1}{2}(2\varphi)^2] - 2B_a M\varphi \ , \tag{53a}$$

where 2φ is the angle the spins make with each other (Fig. 27a). The energy is a minimum when

$$\frac{dU}{d\varphi} = 0 = 4\mu M^2\varphi - 2B_a M \ ; \qquad \varphi = \frac{B_a}{2\mu M} \ , \tag{53b}$$

so that

(CGS) $$\qquad\qquad\qquad \chi_\perp = \frac{2M\varphi}{B_a} = \frac{1}{\mu} \ . \tag{53c}$$

In the parallel orientation (Fig. 27b) the magnetic energy is not changed if the spin systems A and B make equal angles with the field. Thus the susceptibility at $T = 0°K$ is zero:

$$\chi_\parallel(0) = 0 \ . \tag{54}$$

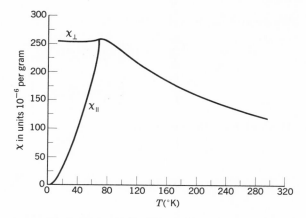

Figure 28 Magnetic susceptibility of manganese fluoride, MnF_2, parallel and perpendicular to the tetragonal axis. (After S. Foner.)

The parallel susceptibility increases smoothly with temperature up to T_N. Measurements on MnF_2 are shown in Fig. 28. In very strong fields the spin systems will flop from the parallel orientation to the perpendicular orientation, because the energy is lower here.

Antiferromagnetic Magnons

We obtain simply an expression for the dispersion relation of magnons in a one-dimensional antiferromagnet by making the appropriate substitutions in the treatment of the ferromagnetic line. Let spins with even indices $2p$ compose sublattice A, that with spins up ($S_z = S$); and let spins with odd indices $2p + 1$ compose sublattice B, that with spins down ($S_z = -S$). We consider only nearest-neighbor interactions, with J negative. Then (21) written for A becomes, with a careful look at (20),

$$\frac{dS_{2p}^x}{dt} = \left(\frac{2JS}{\hbar}\right)(-2S_{2p}^y - S_{2p-1}^y - S_{2p+1}^y) \; ; \tag{55a}$$

$$\frac{dS_{2p}^y}{dt} = -\left(\frac{2JS}{\hbar}\right)(-2S_{2p}^x - S_{2p-1}^x - S_{2p+1}^x) \; . \tag{55b}$$

The corresponding equations for a spin on B are

$$\frac{dS_{2p+1}^x}{dt} = \left(\frac{2JS}{\hbar}\right)(2S_{2p+1}^y + S_{2p}^y + S_{2p+2}^y) \; ; \tag{56a}$$

$$\frac{dS_{2p+1}^y}{dt} = -\left(\frac{2JS}{\hbar}\right)(2S_{2p+1}^x + S_{2p}^x + S_{2p+2}^x) \; . \tag{56b}$$

We form $S^+ = S_x + iS_y$; then

$$\frac{dS_{2p}^+}{dt} = \left(\frac{2iJS}{\hbar}\right)(2S_{2p}^+ + S_{2p-1}^+ + S_{2p+1}^+) \; ; \tag{57}$$

$$\frac{dS_{2p+1}^+}{dt} = -\left(\frac{2iJS}{\hbar}\right)(2S_{2p+1}^+ + S_{2p}^+ + S_{2p+2}^+) \; . \tag{58}$$

We look for solutions of the form

$$S_{2p}^+ = ue^{i[2pka-\omega t]} \; ; \qquad S_{2p+1}^+ = ve^{i[(2p+1)ka-\omega t]} \; , \tag{59}$$

so that (57) and (58) become, with $\omega_{ex} \equiv -\dfrac{4JS}{\hbar} = \dfrac{4|J|S}{\hbar}$,

$$\omega u = \tfrac{1}{2}\omega_{ex}(2u + ve^{-ika} + ve^{ika}) \; ; \tag{60a}$$

$$-\omega v = \tfrac{1}{2}\omega_{ex}(2v + ue^{-ika} + ue^{ika}) \; . \tag{60b}$$

Equations (60ab) have a solution if

$$\begin{vmatrix} \omega_{ex} - \omega & \omega_{ex}\cos ka \\ \omega_{ex}\cos ka & \omega_{ex} + \omega \end{vmatrix} = 0 \; . \tag{61}$$

Thus $\qquad \omega^2 = \omega_{ex}^2(1 - \cos^2 ka) \; ; \qquad \omega = \omega_{ex}|\sin ka| \; . \tag{62}$

The dispersion relation (62) for magnons in an antiferromagnet is quite different from (25) for magnons in a ferromagnet. For $ka \ll 1$ we see that (62) is linear[26] in k:

$$\omega \cong \omega_{ex}|ka| \; . \tag{63}$$

The magnon spectrum of $RbMnF_3$ is shown in Fig. 29, as determined by inelastic neutron scattering experiments. There is a large region in which the magnon frequency is directly proportional to the wavevector. Well-resolved magnons have been observed in MnF_2 at specimen temperatures up to 0.93 of the Néel temperature. Thus even at high temperatures the magnon approximation is useful. Further details concerning antiferromagnetic magnons are given in QTS, Chapter 4.

[26] A physical discussion of the difference between the dispersion relations for ferromagnetic and antiferromagnetic magnons is given by F. Keffer, H. Kaplan, and Y. Yafet, Am. J. Phys. **21**, 250 (1953).

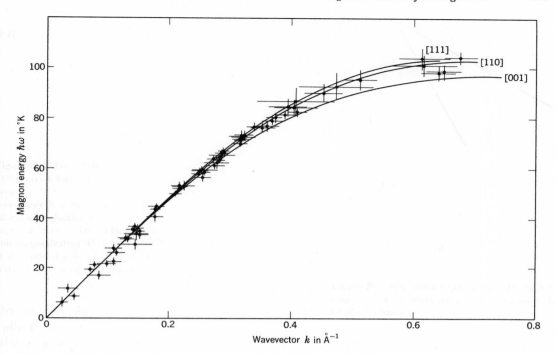

Figure 29 Magnon dispersion relation in the simple cubic antiferromagnet $RbMnF_3$ as determined at 4.2°K by inelastic neutron scattering. The observed points are for wavevectors anywhere in a (110) plane; the curves are calculated in three indicated directions for a nearest neighbor exchange interaction $J/k_B = 3.4°$K. [After C. G. Windsor and R. W. H. Stevenson, Proc. Phys. Soc. (London) **87**, 501 (1966).]

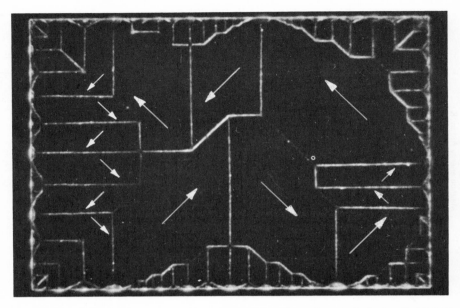

Figure 30 Ferromagnetic domain pattern on the surface of a single crystal platelet of nickel. The domain boundaries are made visible by the Bitter technique. The direction of magnetization within a domain is determined by observing growth or contraction of the domain in an applied magnetic field, as in Fig. 31a. (Courtesy of R. W. De Blois.)

FERROMAGNETIC DOMAINS

At temperatures well below the Curie point the electronic magnetic moments of a ferromagnet are essentially all lined up when regarded on a microscopic scale. Yet, looking at a specimen as a whole, the magnetic moment may be very much less than the saturation moment, and the application of an external magnetic field may be required to saturate the specimen. The behavior observed in polycrystalline specimens is similar to that in single crystals.

Weiss explained this behavior by assuming that actual specimens are composed of a number of small regions called domains, within each of which the local magnetization is saturated. The directions of magnetization of different domains, however, need not necessarily be parallel. An arrangement of domains with approximately zero resultant magnetic moment is shown in Fig. 30. Domains form also in antiferromagnetics, ferroelectrics, antiferroelectrics, ferroelastics, superconductors, and sometimes in metals under conditions of a strong de Haas-van Alphen effect.

The increase in the magnetic moment of the specimen under the action of an applied magnetic field takes place by two independent processes: (1) in weak applied fields the volume of domains (Fig. 31) which are favorably oriented with respect to the field increases at the expense of unfavorably oriented domains; (2) in strong applied fields the magnetization rotates toward the direction of the field.

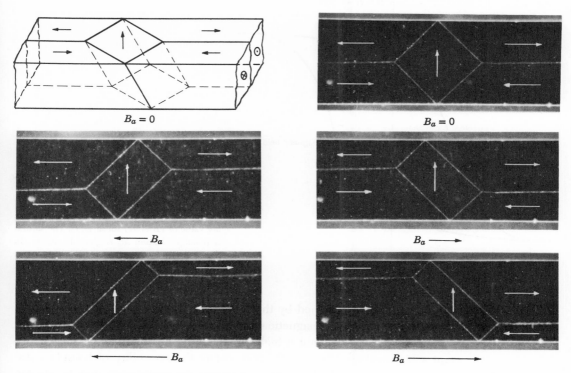

$B_a = 0$

$\longleftarrow B_a$

$\longleftarrow B_a$

$B_a = 0$

$B_a \longrightarrow$

$B_a \longrightarrow$

Figure 31a Smooth reversible domain wall motion in an iron crystal. The domains oriented in the direction of the applied field grow at the expense of the other domains. The field is in a [001] direction; the surface is a (100) plane; the maximum applied field is about 10 G. The crystal is a whisker (Chapter 20) about 10^{-2} cm on a side. (Courtesy of R. W. De Blois and C. D. Graham.)

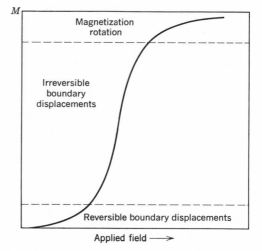

Figure 31b Representative magnetization curve, showing the dominant magnetization processes in the different regions of the curve.

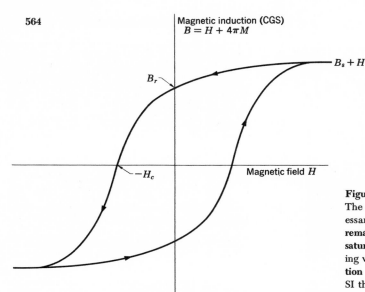

Figure 32 The technical magnetization curve. The **coercive force** H_c is the reverse field necessary to bring the induction B to zero; the **remanence** B_r is the value of B at $H = 0$; the **saturation induction** B_s is defined as the limiting value of $(B - H)$ for large H. The **saturation magnetization** M_s is given by $B_s/4\pi$. In SI the vertical axis is $B = \mu_0(H + M)$. (The vertical scale has been greatly compressed.)

Technical terms defined by the hysteresis loop are shown in Fig. 32. The domain structure of ferromagnetic materials affects their practical properties. In a transformer core we want a high permeability; in a permanent magnet we want a high coercive force.[27] By suppressing the possibility of boundary displacement we may achieve a high coercivity; the suppression may be accomplished by using very fine particles or by precipitating a second metallurgical phase so that the specimen is heterogeneous on a very fine scale. By making the material pure, homogeneous, and well oriented we facilitate boundary displacement and thereby attain high permeability; values of the relative permeability up to 3.8×10^6 have been reported.

Anisotropy Energy

There is an energy in a ferromagnetic crystal which directs the magnetization along certain definite crystallographic axes called directions of easy magnetization. This energy is called the **magnetocrystalline** or **anisotropy energy.** It does not come about from the pure isotropic exchange interaction considered thus far. Cobalt is a hexagonal crystal. The hexagonal axis is the direction of easy magnetization at room temperature, as shown in Fig. 33.

One origin of the anisotropy energy[28] is illustrated by Fig. 34. The magnetization of the crystal sees the crystal lattice through orbital overlap of the electrons: the spin interacts with the orbital motion by means of the spin-orbit

---

[27] The **coercive force** is defined as the reverse field needed to reduce the induction B or the magnetization M to zero, starting in a saturated condition. Usually the definition is understood to refer to B, except in theoretical work. When referred to M, one writes $_IH_c$ or $_MH_c$.

[28] The theory of anisotropy is reviewed by J. Kanamori in Rado and Suhl, Vol. I, 127 (1963).

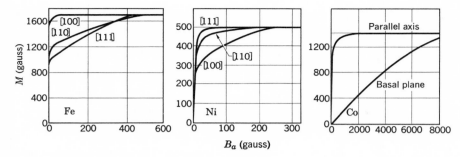

Figure 33 Magnetization curves for single crystals of iron, nickel, and cobalt. From the curves for iron we see that the [100] directions are easy directions of magnetization and the [111] directions are hard directions. The applied field is B_a. (After Honda and Kaya.)

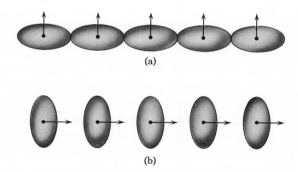

Figure 34 Asymmetry of the overlap of electron distributions on neighboring ions provides one mechanism of magnetocrystalline anisotropy. Because of spin-orbit interaction the charge distribution is spheroidal and not spherical. The asymmetry is tied to the direction of the spin, so that a rotation of the spin directions relative to the crystal axes changes the exchange energy and also changes the electrostatic interaction energy of the charge distributions on pairs of atoms. Both effects give rise to an anisotropy energy. The energy of (a) is not the same as the energy of (b).

coupling, and the orbital motion in turn interacts with the crystal structure by means of the electrostatic fields and overlapping wavefunctions associated with neighboring atoms in the lattice.

In cobalt the anisotropy energy density is given by

$$U_K = K_1' \sin^2 \theta + K_2' \sin^4 \theta , \tag{64}$$

where θ is the angle the magnetization makes with the hexagonal axis. At room temperature $K_1' = 4.1 \times 10^6$ ergs/cm^3; $K_2' = 1.0 \times 10^6$ ergs/cm^3.

Iron is a cubic crystal, and the cube edges are the directions of easy magnetization. To represent the anisotropy energy of iron magnetized in an arbitrary direction with direction cosines α_1, α_2, α_3 referred to the cube edges, we are guided by cubic symmetry. The expression for the anisotropy energy must be an even power[29] of each α_i, and it must be invariant under interchanges of the α_i among themselves. The lowest order combination satisfying the symmetry requirements is $\alpha_1{}^2 + \alpha_2{}^2 + \alpha_3{}^2$, but this is identically equal to unity and does not describe anisotropy effects. The next combination is of the fourth degree: $\alpha_1{}^2\alpha_2{}^2 + \alpha_1{}^2\alpha_3{}^2 + \alpha_3{}^2\alpha_2{}^2$, and then of the sixth degree: $\alpha_1{}^2\alpha_2{}^2\alpha_3{}^2$. Thus

$$U_K = K_1(\alpha_1{}^2\alpha_2{}^2 + \alpha_2{}^2\alpha_3{}^2 + \alpha_3{}^2\alpha_1{}^2) + K_2\alpha_1{}^2\alpha_2{}^2\alpha_3{}^2 \; ; \tag{65}$$

at room temperature $K_1 = 4.2 \times 10^5$ ergs/cm^3 and $K_2 = 1.5 \times 10^5$ ergs/cm^3. Results for iron at other temperatures are shown in Fig. 35; note that $K \to 0$ as $T \to T_c$. For nickel at room temperature $K_1 = -5 \times 10^4$ ergs/cm^3.

Transition Region between Domains

A **Bloch wall** in a crystal is the transition layer which separates adjacent regions (domains) magnetized in different directions. The entire change in spin direction between domains does not occur in one discontinuous jump across a single atomic plane, but takes place in a gradual way over many atomic planes (Fig. 36). The exchange energy is lower when the change is distributed over many spins.

This behavior may be understood by interpreting the Heisenberg equation (6) classically; we also replace cos φ by $1 - \frac{1}{2}\varphi^2$. Then $w_{ex} = JS^2\varphi^2$ is the exchange energy between two spins making a small angle φ with each other. Here J is the exchange integral and S is the spin quantum number; w_{ex} is referred to the energy for parallel spins. If a total change of π occurs in N equal steps, the angle between neighboring spins is π/N, and the exchange energy per pair of neighboring atoms is $w_{ex} = JS^2(\pi/N)^2$. The total exchange energy of a line of $N + 1$ atoms is

$$Nw_{ex} = JS^2\pi^2/N \; . \tag{66}$$

The wall would thicken without limit were it not for the anisotropy energy, which acts to limit the width of the transition layer. The spins contained within the wall are largely directed away from the axes of easy magnetization, so there is an anisotropy energy associated with the wall, roughly proportional to the thickness.

Let us consider a wall parallel to the cube face of a simple cubic lattice and separating domains magnetized in opposite directions. We wish to determine the number N of atomic planes contained within the wall.

The energy per unit area of wall is the sum of contributions from exchange and anisotropy energies: $\sigma_w = \sigma_{ex} + \sigma_{\text{anis}}$. The exchange energy is given approximately by (66) for each line of atoms normal to the plane of the wall. There are $1/a^2$ such lines per unit area, where a is the lattice constant. Thus

---

[29] Because opposite ends of a crystal axis are equivalent magnetically.

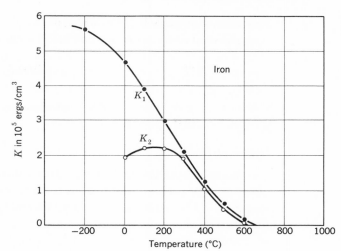

Figure 35 Temperature dependence of anisotropy constants of iron.

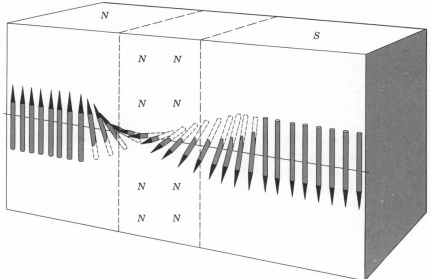

Figure 36 The structure of the Bloch wall separating domains. In iron the thickness of the transition region is about 300 lattice constants.

$\sigma_{ex} = \pi^2 J S^2 / N a^2$. The anisotropy energy is of the order of the anisotropy constant times the thickness Na, or $\sigma_{anis} \approx KNa$; therefore

$$\sigma_w \approx (\pi^2 J S^2 / N a^2) + KNa \ . \tag{67}$$

This is a minimum with respect to N when

$$\partial \sigma_w / \partial N = 0 = -(\pi^2 J S^2 / N^2 a^2) + Ka \ ; \tag{68}$$

or

$$N = \left(\frac{\pi^2 J S^2}{K a^3} \right)^{\frac{1}{2}} \ . \tag{69}$$

For order of magnitude, $N \approx 300$ in iron.

The total wall energy per unit area on our model is

$$\sigma_w = 2\pi \left(\frac{KJS^2}{a} \right)^{\frac{1}{2}} ; \tag{70}$$

in iron $\sigma_w \approx 1$ erg/cm². Accurate calculation for a 180° wall in a (100) plane gives $\sigma_w = 2(2K_1JS^2/a)^{\frac{1}{2}}$.

Origin of Domains

Landau and Lifshitz[30] showed that domain structure is a natural consequence of the various contributions to the energy—exchange, anisotropy, and magnetic—of a ferromagnetic body. Direct evidence of domain structure is furnished by photomicrographs of domain boundaries obtained by the technique of magnetic powder patterns and by optical studies using Faraday rotation. The powder pattern method developed by F. Bitter consists in placing a drop of a colloidal suspension of finely divided ferromagnetic material, such as magnetite, on the carefully prepared surface of the ferromagnetic crystal under study. The colloid particles in the suspension concentrate strongly about the boundaries between domains where strong local magnetic fields exist which attract the magnetic particles. The discovery of transparent ferromagnetic compounds has encouraged the use of optical rotation for domain studies.

We may understand the origin of domains by considering the structures shown in Fig. 37, each representing a cross section through a ferromagnetic single crystal. In (a) we have a single domain; as a consequence of the magnetic "poles" formed on the surfaces of the crystal this configuration will have a high value of the magnetic energy $(1/8\pi) \int B^2 \, dV$. The magnetic energy density for the configuration shown will be of the order of $M_s{}^2 \approx 10^6$ ergs/cm³; here M_s denotes the saturation magnetization, and the units are CGS.

In (b) the magnetic energy is reduced by roughly one-half by dividing the crystal into two domains magnetized in opposite directions. In (c) with N domains the magnetic energy is reduced to approximately $1/N$ of the magnetic energy of (a), because of the reduced spatial extension of the field.

In domain arrangements such as (d) and (e) the magnetic energy is zero. Here the boundaries of the triangular prism domains near the end faces of the crystal make equal angles—45°—with the magnetization in the rectangular domains and with the magnetization in the domains of closure. The component of magnetization normal to the boundary is continuous across the boundary and there is no magnetic field associated with the magnetization. The flux circuit is completed within the crystal—thus giving rise to the phrase **domains of closure** for surface domains that complete the flux circuit, as in Fig. 38.

[30] See the review by Kittel and Galt.

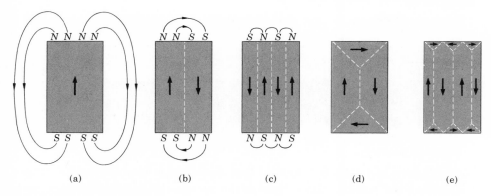

Figure 37   The origin of domains.

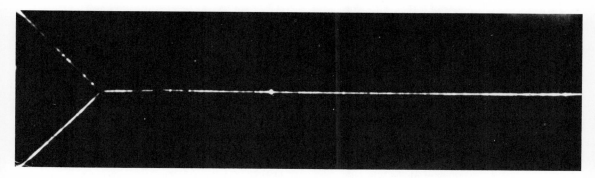

Figure 38   Domain of closure at the end of a single crystal iron whisker. The face is a (100) plane; the whisker axis is [001]. (Courtesy of R. V. Coleman, C. G. Scott, and A. Isin.)

Domain structures are often more complicated than our simple examples, but *domain structure always has its origin in the possibility of lowering the energy of a system by going from a saturated configuration with high magnetic energy to a domain configuration with a lower energy.*

Coercive Force and Hysteresis

The coercive force is the field $-H_c$ required to reduce the induction B to zero (Fig. 32). It is the most sensitive property of ferromagnetic materials which is subject to control. The coercive force may range from the value of 600 G in a loudspeaker permanent magnet (Alnico V) and 20,000 G in a special high stability magnet (Fe-Pt) to 0.5 G in a commercial power transformer (Si-Fe) or 0.004 G in a pulse transformer (Supermalloy). In a transformer low hysteresis is desired; this means a low coercive force.

Figure 39 Microstructure of Alnico V in its optimum state as a permanent magnet. The composition of Alnico V is, by weight, 8 Al, 14 Ni, 24 Co, 3 Cu, 51 Fe. As a permanent magnet it is a two-phase system, with fine particles of one phase embedded in the other phase. The precipitation is carried out in a magnetic field, and the particles are oriented with their long axis parallel to the field direction. (Courtesy of F. E. Luborsky.)

The coercive force decreases as the impurity content decreases and also as internal strains are removed by annealing (slow cooling). Alloys which contain a precipitated phase may have a high coercivity, as in Fig. 39.

The high coercivity of materials composed of very small grains or fine powders is well understood. A sufficiently small particle, with diameter less than 10^{-5} or 10^{-6} cm, is always magnetized to saturation as a single domain because the formation of a flux-closure configuration is energetically unfavorable[31] (Problem 6).

---

[31] L. Néel, Compt. rend. **224**, 1488 (1947), C. Kittel, Phys. Rev. **70**, 965 (1946). A review of permanent magnet technology is given by F. E. Luborsky, Electro-Technology **70** (1962)—see No. 1, p. 100; No. 2, p. 94; No. 3, p. 107. Careful calculations of critical radii for single domain behavior are given by W. F. Brown, Jr., Annals N.Y. Acad. Sci. **147**, 461 (1969); he finds that spheres of iron and nickel will be stable as single domains below radii of 167 Å and 382 Å, respectively.

In a sufficiently small single domain particle it will not be possible for magnetization reversal to take place by means of the process of boundary displacement, which usually requires relatively weak fields. Instead the magnetization of the particle must rotate as a whole, a process which may require large fields depending on the anisotropy energy of the material and the shape of the particle.

The coercive force of fine iron particles is expected theoretically to be about 500 gauss on the basis of rotation opposed by the crystalline anisotropy energy, and this is of the order of the value reported by several observers. Higher coercivities have been reported for elongated iron particles, the rotation here being opposed by the shape anisotropy of the demagnetization energy.

SUMMARY

(In CGS units)

1. The susceptibility of a ferromagnet above the Curie temperature has the form $\chi = C/(T - T_c)$ in the mean field approximation.

2. In the mean field approximation the effective magnetic field seen by a magnetic moment in a ferromagnet is $\mathbf{B}_a + \lambda\mathbf{M}$, where $\lambda = T_c/C$ and $\mathbf{B}_a$ is the applied magnetic field.

3. The elementary excitations in a ferromagnet are magnons. Their dispersion relation for $ka \ll 1$ has the form $\hbar\omega \approx Jk^2a^2$ in zero external magnetic field. The thermal excitation of magnons leads at low temperatures to a heat capacity and to a fractional magnetization change both proportional to $T^{\frac{3}{2}}$.

4. In an antiferromagnet two spin lattices are equal, but antiparallel. In a ferrimagnet two lattices are antiparallel, but the magnetic moment of one is larger than the magnetic moment of the other.

5. In an antiferromagnet the susceptibility above the Néel temperature has the form $\chi = 2C/(T + \theta)$.

6. The magnon dispersion relation in an antiferromagnet has the form $\hbar\omega \approx Jka$. The thermal excitation of magnons leads at low temperatures to a term in T^3 in the heat capacity, in addition to the phonon term in T^3.

7. A Bloch wall separates domains magnetized in different directions. The thickness of a wall is $\approx (J/Ka^3)^{\frac{1}{2}}$ lattice constants, and the energy per unit area is $\approx (KJ/a)^{\frac{1}{2}}$, where K is the anisotropy energy density.

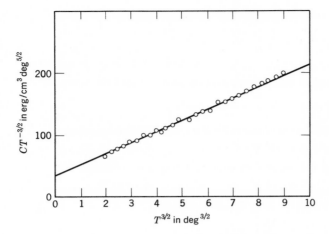

Figure 40 Heat capacity of yttrium iron garnet, $Y_3Fe_5O_{12}$, after S. Shinozaki, Phys. Rev. **122**, 388 (1961). A plot of $C/T^{3/2}$ versus $T^{3/2}$ will be a straight line if the heat capacity is of the form $C = aT^{3/2} + bT^3$, where $aT^{3/2}$ is the magnon contribution and bT^3 is the phonon contribution.

Problems

1. *Magnon dispersion relation.* Derive the magnon dispersion relation (27) for a spin S on a simple cubic lattice, $z = 6$. Hint: Show first that (21a) is replaced by

$$\frac{dS_\rho{}^x}{dt} = \left(\frac{2JS}{\hbar}\right)\left(6S_\rho{}^y - \sum_\delta S_{\rho+\delta}{}^y\right) \; ,$$

where the central atom is at ρ and the six nearest neighbors are connected to it by six vectors $\boldsymbol{\delta}$. Look for solutions of the equations for $dS_\rho{}^x/dt$ and $dS_\rho{}^y/dt$ of the form $\exp(i\mathbf{k} \cdot \boldsymbol{\rho} - i\omega t)$.

2. *Heat capacity of magnons.* Use the approximate magnon dispersion relation $\omega = Ak^2$ to find the leading term in the heat capacity of a three-dimensional ferromagnet at low temperatures $k_BT \ll J$. The result is $0.113\, k_B(k_BT/\hbar A)^{\frac{3}{2}}$, per unit volume. An experimental curve is given in Fig. 40; the heat capacity here is made up of a magnon contribution proportional to $T^{\frac{3}{2}}$ and a phonon contribution proportional to T^3. The heat capacity measurements on YIG are reviewed by A. J. Henderson, Jr., D. G. Onn, H. Meyer, and J. P. Remeika, Phys. Rev. **185**, 1218 (1969).

3. *Néel temperature.* Taking the effective fields on the two-sublattice model of an antiferromagnetic as

$$B_A = B_a - \mu M_B - \epsilon M_A \; ; \qquad B_B = B_a - \mu M_A - \epsilon M_B \; ,$$

show that

$$\frac{\theta}{T_N} = \frac{\mu + \epsilon}{\mu - \epsilon} \; .$$

4. **Magnetoelastic coupling.** In a cubic crystal the elastic energy density is

$$U_{el} = \tfrac{1}{2}C_{11}(e_{xx}^2 + e_{yy}^2 + e_{zz}^2) + \tfrac{1}{2}C_{44}(e_{xy}^2 + e_{yz}^2 + e_{zx}^2)$$
$$+ C_{12}(e_{yy}e_{zz} + e_{xx}e_{zz} + e_{xx}e_{yy}) \; ,$$

and the leading term in the magnetic anisotropy energy density is, from (65),

$$U_K \cong K_1(\alpha_1^2\alpha_2^2 + \alpha_2^2\alpha_3^2 + \alpha_3^2\alpha_1^2) \; .$$

Coupling between elastic strain and magnetization direction may be taken formally into account by including in the total energy density a term

$$U_c \cong B_1(\alpha_1^2 e_{xx} + \alpha_2^2 e_{yy} + \alpha_3^2 e_{zz}) + B_2(\alpha_1\alpha_2 e_{xy} + \alpha_2\alpha_3 e_{yz} + \alpha_3\alpha_1 e_{zx})$$

arising from the strain dependence of U_K; here B_1 and B_2 are called **magnetoelastic coupling constants.** Show that the total energy is a minimum when

$$e_{ii} = \frac{B_1[C_{12} - \alpha_i^2(C_{11} + 2C_{12})]}{[(C_{11} - C_{12})(C_{11} + 2C_{12})]} \; ;$$

$$e_{ij} = -\frac{B_2\alpha_i\alpha_j}{C_{44}} \qquad (i \neq j) \; .$$

This explains the origin of magnetostriction, the change of length on magnetization. For higher order magnetoelastic effects, see D. E. Eastman, Phys. Rev. **148**, 530 (1966) and references cited there.

5. **Coercive force of a small particle.** Consider a small spherical single-domain particle of a uniaxial ferromagnet. Show that the reverse field along the axis required to reverse the magnetization is

(CGS) $$B_a = \frac{2K}{M_s} \; .$$

The coercive force of single-domain particles is observed to be of this magnitude. Take $U_K = K \sin^2 \theta$ as the anisotropy energy density and $U_M = -B_a M \cos \theta$ as the interaction energy density with the external field H; here θ is the angle between $\mathbf{B}_a$ and $\mathbf{M}$. *Hint:* Expand the energies for small angles about $\theta = \pi$, and find the value of B_a for which $U_K + U_M$ does not have a minimum near $\theta = \pi$.

6. **Criterion for single domain particles.** Show that the magnetic energy of a saturated sphere of diameter d is $\approx M_s^2 d^3$. An arrangement with appreciably less magnetic energy has a single wall in an equatorial plane. The domain wall energy will be $\pi\sigma_w d^2/4$, where σ_w is the wall energy per unit area. *Estimate* for cobalt the critical radius below which the particles are stable as single domains, taking the value of JS^2/a as for iron.

7. **Saturation magnetization near T_c.** Show that in the mean field approximation the saturation magnetization just below the Curie temperature has the dominant temperature dependence $(T_c - T)^{\frac{1}{2}}$. Assume the spin is $\frac{1}{2}$. The result is the same as that for a second-order transition in a ferroelectric crystal, as discussed in Chapter 14. The experimental data for ferromagnets (Table 1) suggest that the exponent is closer to 0.33.

8. **Bloch $T^{\frac{3}{2}}$ law.** In Fig. 6.15 we constructed a qualitative argument for the Debye T^3 law for the heat capacity of phonons. Construct a similar argument for the Bloch $T^{\frac{3}{2}}$ law for the magnetization. Use the approximate magnon dispersion relation $\omega = Ak^2$.

References

G. T. Rado and H. Suhl, eds., *Magnetism*, Academic Press, 1963, several volumes. An encyclopedic work.

S. Chikazumi, *Physics of magnetism*, Wiley, 1964. Elementary account of ferromagnetism.

A. H. Morrish, *Physical principles of magnetism*, Wiley, 1965.

R. M. Bozorth, *Ferromagnetism*, Van Nostrand, 1951.

C. Kittel and J. K. Galt, "Ferromagnetic domains," *Solid state physics* **3**, 437 (1956).

D. J. Craik and R. S. Tebble, *Ferromagnetism and ferromagnetic domains*, North Holland, 1966.

J. Smit and H. P. J. Wijn, *Ferrites*, Wiley, 1959.

F. Keffer, "Spin waves," *Encyclo. of physics* **18/2** (1966).

T. Nagamiya, K. Yosida, and R. Kubo, "Antiferromagnetism," Adv. in Physics **4**, 1 (1955).

J. B. Goodenough, *Magnetism and the chemical bond*, Interscience, 1963.

D. C. Mattis, *Theory of magnetism*, Harper & Row, 1965.

A. Herpin, *Théorie du magnétisme*, Presses universitaires, 1968.

B. Lax and K. J. Button, *Microwave ferrites and ferrimagnetics*, McGraw-Hill, 1962.

17

Magnetic Resonance

NOTATION: In this chapter the symbols B_a and B_0 refer to the applied field, and B_i is the applied field plus the demagnetizing field. In particular we write $\mathbf{B}_a = B_0\hat{\mathbf{z}}$. For CGS readers it may be simpler to read H for B wherever it occurs in this chapter.

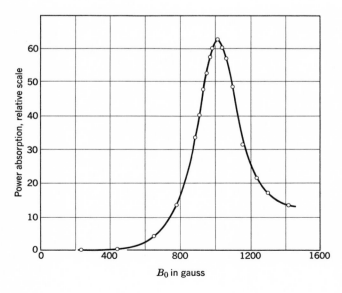

Figure 1 Electron spin resonance absorption in $MnSO_4$ at $298°K$ at 2.75 GHz, after Zavoisky.

In this chapter we discuss dynamical magnetic effects associated with the spin angular momentum of nuclei and of electrons. Applications of magnetic spin resonance in solid state physics are of great importance. The principal phenomena are often identified in the literature by their initial letters, such as

> NMR: nuclear magnetic resonance
> NQR: nuclear quadrupole resonance
> EPR: electron paramagnetic resonance (electron spin resonance)
> FMR: ferromagnetic resonance
> SWR: spin wave resonance (ferromagnetic films)
> AFMR: antiferromagnetic resonance

Zavoisky[1] performed the earliest magnetic resonance experiments in a solid. He observed strong electron spin resonance absorption (Fig. 1) in several paramagnetic salts. Spin resonance experiments on nuclei in liquids and solids were carried out first by Purcell, Torrey, and Pound[2] and by Bloch, Hansen, and Packard.[3]

The kinds of information that can be obtained about solids by resonance studies may in part be categorized as follows:

1. Electronic structure of single defects, as revealed by the fine structure of the absorption.

2. Motion of the spin or of the surroundings, as revealed by changes in the line width.

3. Internal magnetic fields sampled by the spin, as revealed by the position of the resonance line.

4. Collective spin excitations.

It is best to discuss NMR as a basis for a brief account of the other resonance experiments. However, the greatest impact of NMR has not been in solids, but in organic chemistry, where NMR provides a powerful tool for the identification and the structure determination of complex molecules. This success is due to the extremely high resolution attainable in diamagnetic liquids. The references listed at the end of the chapter are exceptionally useful sources of further information.

[1] E. Zavoisky, J. Phys. USSR **9**, 211, 245, 447 (1945).
[2] E. M. Purcell, H. C. Torrey, and R. V. Pound, Phys. Rev. **69**, 37 (1946).
[3] F. Bloch, W. W. Hansen, and M. Packard, Phys. Rev. **69**, 127 (1946).

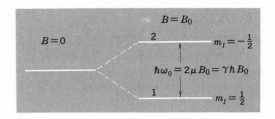

Figure 2 Energy level splitting of a nucleus of spin $I = \frac{1}{2}$ in a static magnetic field B_0.

NUCLEAR MAGNETIC RESONANCE

We consider a nucleus that possesses a magnetic moment μ and an angular momentum $\hbar\mathbf{I}$. The two quantities are parallel, and we may write

$$\boldsymbol{\mu} = \gamma\hbar\mathbf{I} \; ; \tag{1}$$

the magnetogyric ratio γ is constant. By convention $\mathbf{I}$ denotes the nuclear angular momentum measured in units of $\hbar$. The energy of interaction with the applied magnetic field is

$$U = -\boldsymbol{\mu} \cdot \mathbf{B}_a \; ; \tag{2}$$

if $\mathbf{B}_a = B_0\hat{z}$, then

$$U = -\mu_z B_0 = -\gamma\hbar B_0 I_z \; . \tag{3}$$

The allowed values of I_z are $m_I = I, I - 1, \ldots, -I$, and $U = -m_I\gamma\hbar B_0$.

In a magnetic field a nucleus with $I = \frac{1}{2}$ has two energy levels corresponding to $m_I = \pm\frac{1}{2}$, as in Fig. 2. If $\hbar\omega_0$ denotes the energy difference between the two levels, then $\hbar\omega_0 = \gamma\hbar B_0$ or

$$\omega_0 = \gamma B_0 \; . \tag{4}$$

This is the fundamental condition for magnetic resonance absorption. For the proton[4] $\gamma = 2.675 \times 10^4$ rad sec^{-1} gauss^{-1} = 2.675×10^8 rad sec^{-1} tesla^{-1}, so that

$$\boxed{\nu(\text{MHz}) = 4.258 \, B_0(\text{kilogauss}) = 42.58 \, B_0(\text{tesla})} \tag{4a}$$

where ν is the frequency. One tesla is precisely 10^4 gauss. For the electron spin

$$\boxed{\nu(\text{GHz}) = 2.80 \, B_0(\text{kilogauss}) = 28.0 \, B_0(\text{tesla}) \; .} \tag{4b}$$

Magnetic data for selected nuclei are given in Table 1.

[4] The magnetic moment μ_p of the proton is 1.4106×10^{-23} erg G^{-1} or 1.4106×10^{-26} JT^{-1}, and $\gamma \equiv 2\mu_p/\hbar$. The **nuclear magneton** μ_n is defined as $e\hbar/2M_p c$ and is equal to 5.0509×10^{-24} erg G^{-1} or 5.0509×10^{-27} JT^{-1}; thus $\mu_p = 2.793$ nuclear magnetons.

Table 1 Nuclear magnetic resonance data

For every element the most abundant magnetic isotope is shown.
After Varian Associates NMR Table, 4th ed., 1964.

Legend for each cell:
- Most abundant isotope with nonzero nuclear spin.
- Nuclear spin; in units of $\hbar$.
- Natural abundance of isotope, in percent.
- Nuclear magnetic moment, in units of $e\hbar/2M_p c$.

Each cell: Element^{A} / spin I / abundance (%) / magnetic moment μ

1	2	3	4	5	6	7	8	9	10	11	12	13	14	15	16	17	18
H^1 1/2; 99.98; 2.792																	He^3 1/2; 10^{-6}; -2.127
Li 3/2; 92.57; 3.256	Be 3/2; 100.; -1.177											B^{11} 3/2; 81.17; 2.688	C^{13} 1/2; 1.108; 0.702	N^{14} 1; 99.64; 0.404	O^{17} 5/2; 0.04; -1.893	F^{19} 1/2; 100.; 2.627	Ne^{21} 3/2; 0.257; -0.662
Na^{23} 3/2; 100.; 2.216	Mg^{25} 5/2; 10.05; 0.855											Al^{27} 5/2; 100.; 3.639	Si^{29} 1/2; 4.70; 0.555	P^{31} 1/2; 100.; 1.131	S^{33} 3/2; 0.74; 0.643	Cl^{35} 3/2; 75.4; 0.821	Ar
K^{39} 3/2; 93.08; 0.391	Ca^{43} 7/2; 0.13; -1.315	Sc^{45} 7/2; 100.; 4.749	Ti^{47} 5/2; 7.75; 0.787	V^{51} 7/2; ~100.; 5.139	Cr^{53} 3/2; 9.54; 0.474	Mn^{55} 5/2; 100.; 3.461	Fe^{57} 1/2; 2.245; 0.090	Co^{59} 7/2; 100.; 4.639	Ni^{61} 3/2; 1.25; 0.746	Cu^{63} 3/2; 69.09; 2.221	Zn^{67} 5/2; 4.12; 0.874	Ga^{69} 3/2; 60.2; 2.011	Ge^{73} 9/2; 7.61; 0.877	As^{75} 3/2; 100.; 1.435	Se^{77} 1/2; 7.50; 0.533	Br^{79} 3/2; 50.57; 2.099	Kr^{83} 9/2; 11.55; -0.967
Rb^{85} 5/2; 72.8; 1.348	Sr^{87} 9/2; 7.02; 1.089	Y^{89} 1/2; 100.; 0.137	Zr^{91} 5/2; 11.23; 1.298	Nb^{93} 9/2; 100.; 6.144	Mo^{95} 5/2; 15.78; 0.910	Tc	Ru^{101} 5/2; 16.98; -0.69	Rh^{103} 1/2; 100.; 0.088	Pd^{105} 5/2; 22.23; -0.57	Ag^{107} 1/2; 51.35; -0.113	Cd^{111} 1/2; 12.86; -0.592	In^{115} 9/2; 95.84; 5.507	Sn^{119} 1/2; 8.68; -1.041	Sb^{121} 5/2; 57.25; 3.342	Te^{125} 1/2; 7.03; -0.882	I^{127} 5/2; 100.; 2.794	Xe^{129} 1/2; 26.24; -0.773
Cs^{133} 7/2; 100.; 2.564	Ba^{137} 3/2; 11.32; 0.931	La^{139} 7/2; 99.9; 2.761	Hf^{177} 7/2; 18.39; 0.61	Ta^{181} 7/2; 100.; 2.340	W^{183} 1/2; 14.28; 0.115	Re^{187} 5/2; 62.93; 3.176	Os^{189} 3/2; 16.1; 0.651	Ir^{193} 3/2; 61.5; 0.17	Pt^{195} 1/2; 33.7; 0.600	Au^{197} 3/2; 100.; 0.144	Hg^{199} 1/2; 16.86; 0.498	Tl^{205} 1/2; 70.48; 1.612	Pb^{207} 1/2; 21.11; 0.584	Bi^{209} 9/2; 100.; 4.039	Po	At	Rn
Fr	Ra	Ac															

Lanthanides:

Ce^{141}*	Pr^{141}	Nd^{143}	Pm	Sm^{147}	Eu^{153}	Gd^{157}	Tb^{159}	Dy^{163}	Ho^{165}	Er^{167}	Tm^{169}	Yb^{173}	Lu^{175}
7/2; —; 0.16	5/2; 100.; 3.92	7/2; 12.20; -1.25		7/2; 15.07; -0.68	5/2; 52.23; 1.521	3/2; 15.64; -0.34	3/2; 100.; 1.52	5/2; 24.97; -0.53	7/2; 100.; 3.31	7/2; 22.82; 0.48	1/2; 100.; -0.20	5/2; 16.08; -0.677	7/2; 97.40; 2.9

Actinides:

Th	Pa	U	Np	Pu	Am	Cm	Bk	Cf	Es	Fm	Md	No	Lw

Equations of Motion

The rate of change of angular momentum of a system is equal to the torque which acts on the system. The torque on a magnetic moment μ in a magnetic field $\mathbf{B}$ is $\mu \times \mathbf{B}$ so that we have the gyroscopic equation

$$\hbar \frac{d\mathbf{I}}{dt} = \mu \times \mathbf{B}_a \; ; \qquad (5)$$

or

$$\frac{d\mu}{dt} = \gamma \mu \times \mathbf{B}_a \; . \qquad (6)$$

The nuclear magnetization $\mathbf{M}$ is the sum $\Sigma \mu_i$ over all the nuclei in a unit volume. If only a single isotope is important, we consider only a single value of γ, so that

$$\boxed{\frac{d\mathbf{M}}{dt} = \gamma \mathbf{M} \times \mathbf{B}_a \; .} \qquad (7)$$

We place the nuclei in a static field $\mathbf{B}_a = B_0 \hat{\mathbf{z}}$. In thermal equilibrium at temperature T the magnetization will be along $\hat{\mathbf{z}}$:

$$M_x = 0 \; ; \qquad M_y = 0 \; ; \qquad M_z = M_0 = \chi_0 B_0 = CB_0/T \; , \qquad (8)$$

where the Curie constant $C = N\mu^2/3k_B$ as defined by (15.14). The magnetization of a system of spins with $I = \frac{1}{2}$ is related to the population difference $N_1 - N_2$ of the lower and upper levels in Fig. 2: $M_z = (N_1 - N_2)\mu$, where the N's refer to a unit volume. The population ratio in thermal equilibrium is just given by the Boltzmann factor for the energy difference $2\mu B_0$:

$$\left(\frac{N_2}{N_1}\right)_0 = e^{-2\mu B_0/k_B T} \; . \qquad (9)$$

The equilibrium magnetization is given by $M = N\mu \tanh(\mu B/k_B T)$, according to (15.20).

When the magnetization component M_z is not in thermal equilibrium, we suppose that it approaches equilibrium at a rate proportional to the departure from the equilibrium value M_0:

$$\frac{dM_z}{dt} = \frac{M_0 - M_z}{T_1} \; . \qquad (10)$$

In the standard notation T_1 is called the **longitudinal relaxation time** or the **spin-lattice relaxation time**.

If at $t = 0$ an unmagnetized specimen is placed in a magnetic field $B_0 \hat{\mathbf{z}}$, the magnetization will increase from the initial value[5] $M_z = 0$ to a final value

[5] Before and just after the specimen is placed in the field the population N_1 will be equal to N_2, as appropriate to thermal equilibrium in zero magnetic field. It is necessary to reverse some spins to establish the new equilibrium distribution in the field B_0.

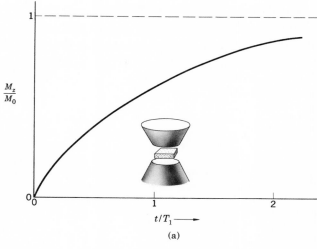

Figure 3a   At time $t = 0$ an unmagnetized specimen $M_z(0) = 0$ is placed in a static magnetic field B_0. The magnetization increases with time and approaches the new equilibrium value $M_0 = \chi_0 B_0$. This experiment defines the longitudinal relaxation time T_1.

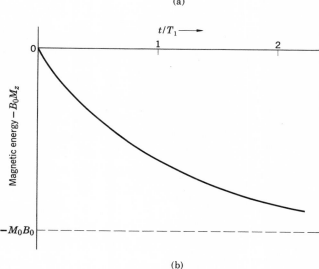

Figure 3b   In the experiment of Fig. 3a the magnetic energy density $-\mathbf{M} \cdot \mathbf{B}$ decreases as part of the spin population moves into the lower level. The asymptotic value at $t \gg T_1$ is $-M_0 B_0$. The energy flows from the spin system to the system of lattice vibrations; thus T_1 is also called the spin-lattice relaxation time.

$M_z = M_0$. On integrating (10):

$$\int_0^{M_z} \frac{dM_z}{M_0 - M_z} = \frac{1}{T_1} \int_0^T dt \ , \tag{11}$$

or

$$\log \frac{M_0}{M_0 - M_z} = \frac{t}{T_1} \ ; \qquad M_z(t) = M_0(1 - e^{-t/T_1}) \ , \tag{12}$$

as in Fig. 3a. The magnetic energy $-\mathbf{M} \cdot \mathbf{B}_a$ decreases as M_z approaches its new equilibrium value (Fig. 3b). Typical processes whereby the magnetization

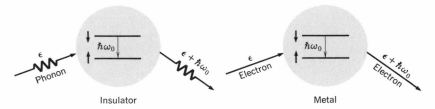

Figure 4a Some important processes that contribute to longitudinal magnetization relaxation in an insulator and in a metal. For the insulator we show a phonon scattered inelastically by the spin system. The spin system moves to a lower energy state, and the emitted phonon has higher energy by $\hbar\omega_0$ than the absorbed phonon. For the metal we show a similar inelastic scattering process in which a conduction electron is scattered.

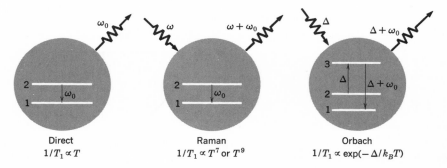

Figure 4b Spin relaxation from $2 \rightarrow 1$ by phonon emission, phonon scattering, and a two-stage phonon process. The temperature dependence of the longitudinal relaxation time T_1 is shown for the several processes.

approaches equilibrium are indicated in Fig. 4.

The dominant spin-lattice interaction of paramagnetic ions in crystals is by the phonon modulation of the crystalline electric field. Relaxation proceeds by three principal processes (Fig. 4b): direct (emission or absorption of a phonon); Raman (scattering of a phonon); and Orbach[6] (intervention of a third state). A thorough experimental analysis of spin-lattice relaxation in several rare-earth salts at helium temperatures has been given by Scott and Jeffries;[7] they discuss evidence for the three processes and give useful references to the literature of paramagnetic relaxation. An example of the results is given in Fig. 5.

[6] C. B. P. Finn, R. Orbach, and W. P. Wolf, Proc. Phys. (London) **77**, 261 (1961).

[7] P. L. Scott and C. D. Jeffries, Phys. Rev. **127**, 32 (1962).

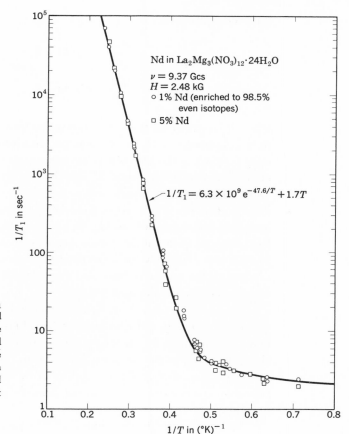

Figure 5 Log of relaxation rate $1/T_1$ versus reciprocal temperature for 1 and 5 percent Nd in lanthanum double nitrate, after observations by Scott and Jeffries between 1.4 and 4.3°K. The results are clear evidence for the Orbach process at the high temperatures, and for the direct one-phonon process at low temperatures.

Taking account of (10), the z component of the equation of motion (7) becomes

$$\frac{dM_z}{dt} = \gamma(\mathbf{M} \times \mathbf{B}_a)_z + \frac{M_0 - M_z}{T_1} , \qquad (13a)$$

where $(M_0 - M_z)/T_1$ is an extra term in the equation of motion, arising from interactions not included in the magnetic field $\mathbf{B}_a$. That is, besides precessing about the magnetic field, $\mathbf{M}$ will relax to the equilibrium value $\mathbf{M}_0$.

If in a static field $B_0\hat{\mathbf{z}}$ the transverse magnetization component M_x is not zero, then M_x will decay to zero, and similarly for M_y. The decay occurs because in thermal equilibrium the transverse components are zero. We can provide for transverse relaxation:

$$\frac{dM_x}{dt} = \gamma(\mathbf{M} \times \mathbf{B}_a)_x - \frac{M_x}{T_2} \; ; \qquad (13b)$$

$$\frac{dM_y}{dt} = \gamma(\mathbf{M} \times \mathbf{B}_a)_y - \frac{M_y}{T_2} \; , \qquad (13c)$$

where T_2 is called the **transverse relaxation time**. The magnetic energy $-\mathbf{M} \cdot \mathbf{B}_a$ does not change as M_x or M_y changes, provided that $\mathbf{B}_a$ is along $\hat{\mathbf{z}}$. No energy need flow out of the spin system for relaxation of M_x or M_y to occur, so that the conditions which determine T_2 may be less strict than for T_1. Sometimes the two times are nearly equal, and sometimes $T_1 \gg T_2$, depending on local conditions. The time T_2 is a measure of the time during which the individual moments that contribute to M_x, M_y remain in phase with each other. Different local magnetic fields at the different spins will cause them to precess at different frequencies. If initially the spins have a common phase, the phases will become random in the course of time and the values of M_x, M_y will become zero. We can think of T_2 as a dephasing time.

The set of equations (13) are called the **Bloch equations.** They are not symmetrical in x, y, and z because we have biased the system with a static magnetic field along $\hat{\mathbf{z}}$. In the experiments an rf magnetic field is usually applied along the $\hat{\mathbf{x}}$ or $\hat{\mathbf{y}}$ axes. Our main interest is in the behavior of the magnetization in the combined rf and static fields, as in Fig. 6. The Bloch equations are plausible, but not exact; they do not describe all spin phenomena, particularly in solids.

We determine the frequency of free precession of the spin system in a static field $\mathbf{B}_a = B_0\hat{\mathbf{z}}$ and with $M_z = M_0$. The Bloch equations reduce to

$$\frac{dM_x}{dt} = \gamma B_0 M_y - \frac{M_x}{T_2} \; ; \qquad \frac{dM_y}{dt} = -\gamma B_0 M_x - \frac{M_y}{T_2} \; ; \qquad \frac{dM_z}{dt} = 0 \; . \qquad (14)$$

We look for damped oscillatory solutions of the form

$$M_x = m \cos \omega t \, e^{-t/T'} \; ; \qquad M_y = -m \sin \omega t \, e^{-t/T'} \; ; \qquad (15)$$

on substitution in (14) we have for the left-hand equation

$$-\omega \sin \omega t - \frac{1}{T'} \cos \omega t = -\gamma B_0 \sin \omega t - \frac{1}{T_2} \cos \omega t \; , \qquad (16)$$

so that the free precession is characterized by

$$\omega_0 = \gamma B_0 \; ; \qquad T' = T_2 \; . \qquad (17)$$

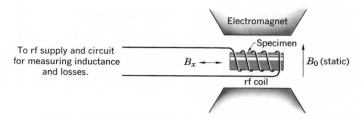

Figure 6   Schematic arrangement for magnetic resonance experiments.

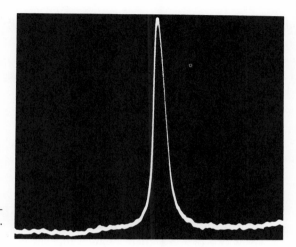

Figure 7   Proton resonance absorption in water. (Courtesy of E. L. Hahn.)

The motion is similar to that of a damped harmonic oscillator in two dimensions. The analogy suggests correctly that the spin system will show resonance absorption of energy from a driving field near the frequency $\omega_0 = \gamma B_0$, and the frequency width of the response of the system to the driving field will be $\Delta \omega \approx 1/T_2$. Figure 7 shows the resonance of protons in water.

The Bloch equations may be solved to give the power absorption from a rotating magnetic field of amplitude B_1:

$$B_x = B_1 \cos \omega t \; ; \qquad B_y = -B_1 \sin \omega t \; . \tag{18}$$

After a routine calculation one finds that the power absorption is

$$\text{(CGS)} \qquad \mathcal{P}(\omega) = \frac{\omega \gamma M_z T_2}{1 + (\omega_0 - \omega)^2 T_2{}^2} B_1{}^2 \; . \tag{19}$$

Notice that the half-width of the resonance at half-maximum power is

$$(\Delta \omega)_{\frac{1}{2}} = \frac{1}{T_2} \; . \tag{20}$$

In the steady state the energy absorbed from the rf field usually ends up as lattice vibrations.

LINE WIDTH

The magnetic dipolar interaction is usually the most important cause of line broadening in a rigid lattice of magnetic dipoles. The magnetic field $\Delta \mathbf{B}$ seen by a magnetic dipole $\boldsymbol{\mu}_1$ due to a magnetic dipole $\boldsymbol{\mu}_2$ at a point $\mathbf{r}_{12}$ from the first dipole is

$$\text{(CGS)} \qquad \Delta \mathbf{B} = \frac{3(\boldsymbol{\mu}_2 \cdot \mathbf{r}_{12})\mathbf{r}_{12} - \boldsymbol{\mu}_2 \mathbf{r}_{12}^2}{r_{12}^5} \ , \qquad (21)$$

by a fundamental result of magnetostatics. The order of magnitude of the interaction is, with B_i written for ΔB,

$$\text{(CGS)} \qquad B_i \approx \frac{\mu}{r^3} \ . \qquad (22)$$

The strong dependence on r suggests that close neighbor interactions will be dominant, so that

$$\text{(CGS)} \qquad B_i \approx \frac{\mu}{a^3} \ , \qquad (23)$$

where a is the separation of nearest neighbors. This result gives us a measure of the width of the spin resonance line, assuming random orientation of the neighbors. For protons at 2Å separation,

$$B_i \approx \frac{1.4 \times 10^{-23} \text{ G cm}^3}{8 \times 10^{-24} \text{ cm}^3} \approx 2 \text{ gauss} = 2 \times 10^{-4} \text{ tesla} \ . \qquad (24)$$

To express (21), (22), and (23) in SI, multiply the right-hand sides by $\mu_0/4\pi$.

Motional Narrowing

Experimentally it is found that the line width decreases for nuclei in rapid motion relative to one another. The effect in solids is illustrated by Fig. 8: diffusion[8] resembles a random walk as atoms jump from one crystal site to another. An atom remains in one site for an average time τ that decreases markedly as the temperature increases. The motional effects on the line width in liquids are usually even more spectacular than in solids, because atoms are much more mobile in a liquid. The width of the proton resonance line in water is only 10^{-5} of the width expected for water molecules frozen in position.

The effects of nuclear motion on T_2 and on the line width can be understood by an elementary argument. We know from the Bloch equations that T_2

---

[8] A detailed study of T_1 and T_2 in alkali metals has been made by D. F. Holcomb and R. E. Norberg, Phys. Rev. 98, 1074 (1955). Diffusion is discussed in Chapter 19.

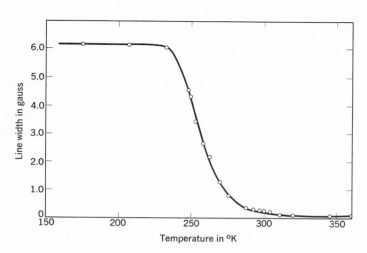

Figure 8 Effect of diffusion of nuclei on the Li^7 NMR line width in metallic lithium. At low temperatures the width agrees with the theoretical value for a rigid lattice. As the temperature increases the diffusion rate increases and the line width decreases. The abrupt decrease in line width above $T = 230°K$ is not the result of a phase transition in the metal; the decrease occurs when the diffusion hopping time τ becomes shorter than $1/\gamma B_i$. Thus the experiment gives a direct measure of the hopping time for an atom to change lattice sites. [After H. S. Gutowsky and B. R. McGarvey, J. Chem. Phys. **20**, 1472 (1952).]

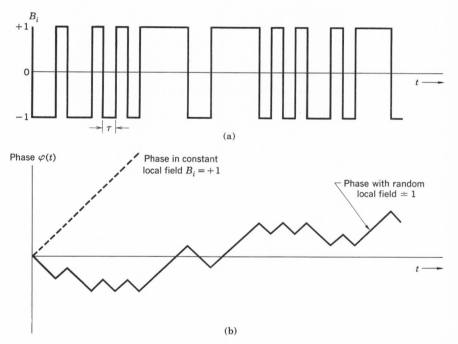

Figure 9 Phase of a spin in a constant local field ------, as compared with dephasing of a spin which after fixed time intervals τ hops at random among sites having local fields ±1. The dephasing is measured relative to the phase of a spin in the applied field B_0.

is a measure of the time in which an individual spin becomes dephased by one radian because of a local perturbation in the magnetic field intensity. Let $(\Delta\omega)_0 \approx \gamma B_i$ denote the local frequency deviation due to a perturbation B_i in a rigid lattice. The local field may be caused by dipolar interactions with other spins. If the atoms are in rapid relative motion, the local field B_i seen by a given spin will fluctuate rapidly in time. We suppose that the local field has a value $+B_i$ for an average time τ and then changes to $-B_i$, as in Fig. 9b. Such a random change could be caused by the relative motion of other atoms, as by a change of the angle between μ and $\mathbf{r}$ in (21). In the time τ the spin will precess by an extra phase angle $\delta\varphi = \pm\gamma B_i\tau$ relative to the phase angle of the steady precession in the applied field B_0. The motional narrowing effect arises for short τ such that $\delta\varphi \ll 1$. After n intervals of duration τ the mean square dephasing angle in the field B_0 will be

$$\langle\varphi^2\rangle = n(\delta\varphi)^2 = n\gamma^2 B_i{}^2\tau^2 \ , \tag{25}$$

by analogy with a random walk process: the mean square displacement from the initial position after n steps of length ℓ in random directions is $\langle r^2\rangle = n\ell^2$.

The average number of steps necessary to dephase a spin by one radian is $n = 1/\gamma^2 B_i{}^2\tau^2$. Spins dephased by much more than one radian do not contribute to the absorption signal. This number of steps takes place in a time

$$T_2 = n\tau = \frac{1}{\gamma^2 B_i{}^2\tau} \ , \tag{26}$$

quite different from the rigid lattice result $T_2 \cong 1/\gamma B_i$. From (26) we obtain as the line width for rapid motion with a characteristic time τ:

$$\Delta\omega = \frac{1}{T_2} = (\gamma B_i)^2\tau \ , \tag{27}$$

or

$$\Delta\omega = \frac{1}{T_2} = (\Delta\omega)_0{}^2\tau \ , \tag{28}$$

where $(\Delta\omega)_0$ is the line width in the rigid lattice.

The argument assumes that $(\Delta\omega)_0\tau \ll 1$, as otherwise $\delta\varphi$ will not be $\ll 1$. Thus $\Delta\omega \ll (\Delta\omega)_0$. The shorter is τ, the narrower is the resonance line! This remarkable effect is known as **motional narrowing**.[9] The rotational relaxation time of water molecules at room temperature is known from dielectric constant measurements to be of the order of 10^{-10} sec; if $(\Delta\omega)_0 \approx 10^5$ sec^{-1}, then $(\Delta\omega)_0\tau \approx 10^{-5}$ and $\Delta\omega \approx (\Delta\omega)_0{}^2\tau \approx 1$ sec^{-1}. Thus the motion narrows the proton resonance line to about 10^{-5} of the static width.

HYPERFINE SPLITTING

The hyperfine interaction is the magnetic interaction between the magnetic moment of a nucleus and the magnetic moment of an electron. To an observer stationed on the nucleus, the interaction is caused by the magnetic

---

[9] The physical ideas are due to N. Bloembergen, E. M. Purcell, and R. V. Pound, Phys. Rev. **73**, 679 (1948). The result differs from the theory of optical line width caused by *strong* collisions between atoms (as in a gas discharge), where a short τ gives a broad line. In the nuclear spin problem the collisions are weak. In most optical problems the collisions of atoms are strong enough to interrupt the phase of the oscillation. In nuclear resonance the phase may vary smoothly in a collision, although the frequency may vary suddenly from one value to another nearby value.

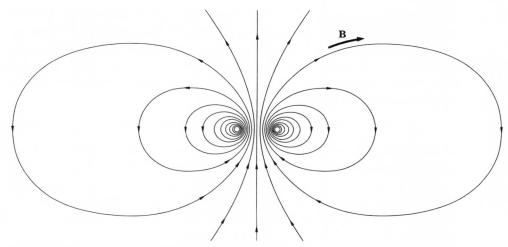

Figure 10 Magnetic field **B** produced by a charge moving in a circular loop. The contact part of the hyperfine interaction with a nuclear magnetic moment arises from the region within or near to the current loop. The field in the region outside the loop if averaged over a spherical shell gives zero. Thus for an *s* electron ($L = 0$) only the contact part contributes to the interaction.

field produced by the motion of the electron about the nucleus. There is an electron current about the nucleus if the electron is in a state with orbital angular momentum about the nucleus. But even if the electron is in a state of zero orbital angular momentum, there is an electron spin current about the nucleus, and this current gives rise to the **contact hyperfine interaction,** of particular importance in solids. We can understand the origin of the contact by a qualitative physical argument, given in CGS.

The results of the Dirac theory of the electron suggest that the magnetic moment of $\mu_B = e\hbar/2mc$ of the electron arises from the circulation of an electron with velocity c in a current loop of radius approximately the electron Compton wavelength, $\lambda_e = \hbar/mc \sim 10^{-11}$ cm. The electric current associated with the circulation is

$$I \sim e \times \text{(turns per unit time)} \sim \frac{ec}{\lambda_e} \; , \tag{29}$$

and the magnetic field (Fig. 10) produced by the current is

$$\text{(CGS)} \qquad\qquad B \sim \frac{I}{\lambda_e c} \sim \frac{e}{\lambda_e{}^2} \; . \tag{30}$$

The observer on the nucleus has the probability

$$P \approx |\psi(0)|^2 \lambda_e{}^3 \; . \tag{31}$$

of finding himself inside the electron, that is, within a sphere of volume $\lambda_e{}^3$

about the electron. Here $\psi(0)$ is the value of the electron wavefunction at the nucleus. Thus the average value of the magnetic field seen by the nucleus is

$$\overline{B} \approx e|\psi(0)|^2 \lambda_e \approx \mu_B|\psi(0)|^2 \ , \tag{32}$$

where $\mu_B = e\hbar/2mc = \frac{1}{2}e\lambda_e$ is the Bohr magneton. The contact part of the hyperfine interaction energy is

$$U = -\boldsymbol{\mu}_I \cdot \overline{\mathbf{B}} \approx -\boldsymbol{\mu}_I \cdot \mu_B|\psi(0)|^2 \approx \gamma\hbar\mu_B|\psi(0)|^2 \mathbf{I} \cdot \mathbf{S} \ , \tag{33}$$

where I is the nuclear spin in units of $\hbar$.

The contact interaction in an atom has the form

$$U = a\mathbf{I} \cdot \mathbf{S} \ . \tag{34}$$

Values of the hyperfine constant a for the ground states of several free atoms are:

nucleus	H¹	Li⁷	Na²³	K³⁹	K⁴¹
I	$\frac{1}{2}$	$\frac{3}{2}$	$\frac{3}{2}$	$\frac{3}{2}$	$\frac{3}{2}$
a in gauss	507	144	310	83	85
a in MHz	1420	402	886	231	127

The value of a in gauss as seen by an electron spin is defined as $a/2\mu_B$.

In a strong magnetic field the energy level scheme of a free atom or ion is dominated by the Zeeman energy splitting of the electron levels; the hyperfine interaction gives an additional splitting that in strong fields is $U' \cong am_S m_I$, where m_S, m_I are the magnetic quantum numbers. For the energy level diagram of Fig. 11 the two electronic transitions have $\Delta m_S = \pm 1$, $\Delta m_I = 0$; the frequencies are $\omega = \gamma H_0 \pm a/2\hbar$. The nuclear transitions are not marked; they have $\Delta m_S = 0$, so that $\omega_{\text{nuc}} = a/2\hbar$. The frequency of the nuclear transition $1 \rightarrow 2$ is equal to that of $3 \rightarrow 4$.

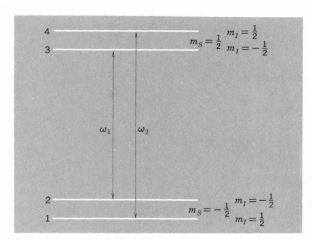

Figure 11 Energy levels in a magnetic field of a system with $S = \frac{1}{2}$, $I = \frac{1}{2}$. The diagram is drawn for the strong field approximation $\mu_B B \gg a$, where a is the hyperfine coupling constant, taken to be positive. The four levels are labeled by the magnetic quantum numbers m_S, m_I. The strong electronic transitions have $\Delta m_I = 0$, $\Delta m_S = \pm 1$.

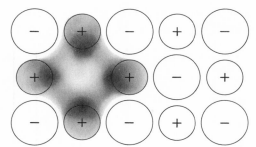

Figure 12 An *F* center is a negative ion vacancy with one excess electron bound at the vacancy. The distribution of the excess electron is largely on the positive metal ions adjacent to the vacant lattice site.

The hyperfine interaction in an atom splits the ground energy level. The splitting in hydrogen is 1420 MHz; and this is the radio frequency line of interstellar atomic hydrogen.

Examples: Paramagnetic Point Defects

The hyperfine splitting of the electron spin resonance furnishes valuable structural information about paramagnetic point defects, such as the *F* centers in alkali halide crystals and the donor atoms in semiconductor crystals.

F Centers in Alkali Halides. An *F* center is a negative ion vacancy with one excess electron bound at the vacancy (Fig. 12), according to the evidence discussed here and in Chapter 19. The wavefunction of the trapped electron is shared chiefly among the six alkali ions adjacent to the vacant lattice site, with smaller amplitudes on the twelve halide ions that form the shell of second nearest neighbors. The counting applies to crystals with the NaCl structure. If $\varphi(\mathbf{r})$ is the wavefunction of the valence electron on a single alkali ion, then in the first approximation[10]

$$\psi(\mathbf{r}) = C \sum_p \varphi(\mathbf{r} - \mathbf{r}_p) \; , \tag{36}$$

where in the NaCl structure the six values of $\mathbf{r}_p$ mark the alkali ion sites that bound the lattice vacancy. Here C is a normalization constant. Because the wavefunction also overlaps somewhat the nearby halogen ions, (36) is not a complete description.

The width of the electron spin resonance line of an *F* center is determined essentially by the hyperfine interaction of the trapped electron with the nuclear magnetic moments of the alkali ions adjacent to the vacant lattice site. The observed line width is evidence for the simple picture of the wavefunction of

[10] Kip, Kittel, Levy and Portis, Phys. Rev. **91**, 1066 (1953); G. Feher, Phys. Rev. **105**, 1122 (1957).

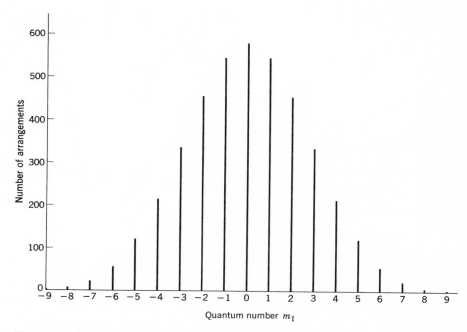

Figure 13 The 4096 arrangements of the six nuclear spins of K^{39} as distributed into 19 hyperfine components. Each component will be split further into a very large number of components by virtue of the residual hyperfine interaction with the 12 neighbor Cl nuclei, which may be Cl^{35} (75 percent) or Cl^{37} (25 percent). The envelope of the pattern is approximately gaussian in form.

the electron. By line width we mean the width of the envelope of the possible hyperfine structure components. As an example, consider an F center in KCl. Natural potassium is 93 percent K^{39} with nuclear spin $I = \frac{3}{2}$. The total spin of the six potassium nuclei at the F center is $I_{max} = 6 \times \frac{3}{2} = 9$, so that the number of hyperfine components is $2I_{max} + 1 = 19$: this is the number of possible values of the quantum number m_I. There are $(2I + 1)^6 = 4^6 = 4096$ independent arrangements of the six spins distributed into the 19 components, as in Fig. 13. Most often we observe only the envelope of the absorption line of an F center, as in Fig. 14 for RbBr, but occasionally the splitting due to the metal ions is so dominant that the separate components of Fig. 13 emerge, as in LiF, NaF, and RbCl. A review of the experimental results is included in the survey by Markham.[11] The most spectacular application to color centers of electron spin resonance has been to the structure of the V_K center, Chapter 19.

Donor Atoms in Silicon. Phosphorus, arsenic, and antimony act as donors when present in silicon. Each donor atom has five electrons, of which four enter diamagnetically into the covalent bond network of the crystal, and the fifth bound electron acts as a paramagnetic center of spin $S = \frac{1}{2}$. The experimental

[11] J. J. Markham, *F centers in alkali halides*, Academic, 1966.

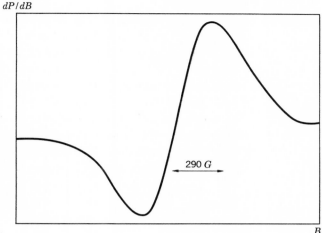

Figure 14 Electron spin resonance of *F* centers in RbBr. The derivative dP/dB of the power absorption with respect to magnetic field is plotted as a function of the static magnetic field. [After H. C. Wolf and K. H. Hausser, Naturwissenschaften **46**, 646 (1959).]

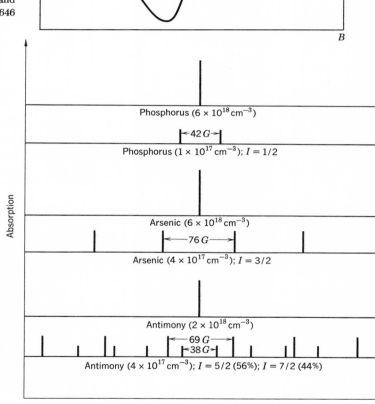

Figure 15 Electron spin resonance lines of P, As, and Sb donor atoms in silicon. The hyperfine splitting components are equal in number to $2I + 1$, where I is the spin of the nucleus. At the higher donor concentrations (that is, 6×10^{18} atoms cm^{-3} for phosphorous) a donor electron can hop from site to site so rapidly that the hyperfine structure is suppressed. [After R. C. Fletcher, W. A. Yager, G. L. Pearson, and F. R. Merritt, Phys. Rev. **95**, 844 (1954).]

hyperfine splittings that result from the interaction in the strong field limit,

$$U' = am_S m_I ; \qquad (\Delta m_S = \pm 1, \Delta m_I = 0) , \qquad (37)$$

are shown in Fig. 15. The donors P and As each have only one natural isotope, and the number of hyperfine components is $2I + 1 = 2$ for P and 4 for As. But Sb has two significant natural isotopes, Sb^{121} (56 percent) with $I = \frac{5}{2}$ and Sb^{123}

(44 percent) with $I = \frac{7}{2}$. The total number of lines observed is $6 + 8 = 14$, as expected. With all three donors it is found that when the concentration exceeds about 1×10^{18} donors cm^{-3}, the multiplicity of lines is replaced by a single narrow line. It is believed this is a motional narrowing effect [Eq. 28] caused by the rapid hopping of the donor electrons among many donor atoms. The rapid hopping acts to average out the hyperfine splitting. The rate of hopping increases at the higher concentrations as the overlap of the donor electron wavefunctions is increased, a view supported by conductivity measurements.

The donor electron wavefunction (Chapter 11) extends not only over the central donor atom, but significantly over some hundreds of silicon atoms. Most (92 percent) of these are Si^{28} with zero nuclear spin and give no hyperfine splitting, but 5 percent are Si^{29} with $I = \frac{1}{2}$. The Si^{29} nuclear spins give additional hyperfine splittings first studied by Feher[12] with a powerful electron nuclear double resonance technique; the studies give information on the conduction electron wavefunctions and the location in the Brillouin zone of the conduction band edge in silicon.

Knight Shift

At a fixed frequency the resonance of a nuclear spin is observed at a slightly different magnetic field in a metal than in a diamagnetic solid. The effect is known as the **Knight shift** and is valuable as a tool for the study of conduction electrons. The interaction energy of a nucleus of spin $\mathbf{I}$ and magnetogyric ratio γ_I is

$$U = (-\gamma_I \hbar B_0 + a\langle S_z \rangle)I_z \ , \tag{38}$$

where the first term is the interaction with the applied magnetic field B_0 and the second is the average hyperfine interaction of the nucleus with the conduction electrons. The average value of the conduction electron spin $\langle S_z \rangle$ is related to the Pauli spin susceptibility χ_s of the conduction electrons:

$$M_z = gN\mu_B\langle S_z \rangle = \chi_s B_0 \ , \tag{39}$$

whence the interaction may be written as

$$U = \left(-\gamma_I \hbar + \frac{a\chi_s}{gN\mu_B}\right)B_0 I_z = -\gamma_I \hbar B_0 \left(1 + \frac{\Delta B}{B_0}\right)I_z \ . \tag{40}$$

The Knight shift is defined as

$$K = -\frac{\Delta B}{B_0} = \frac{a\chi_s}{gN\mu_B\gamma_I\hbar} \ . \tag{41}$$

[12] G. Feher, Phys. Rev. **114**, 1219 (1959). The technique is known as ENDOR. The results are reviewed by W. Kohn, *Solid state physics* **5**, 257 (1957).

Table 2   Knight shifts in NMR in metallic elements°

(At room temperature)

Nucleus	Knight shift in percent	Nucleus	Knight shift in percent
Li^7	0.0261	Cu^{63}	0.237
Na^{23}	0.112	Rb^{87}	0.653
Al^{27}	0.162	Pd^{105}	−3.0
K^{39}	0.265	Pt^{195}	−3.533
V^{51}	0.580	Au^{197}	1.4
Cr^{53}	0.69	Pb^{207}	1.47

° After the compilation by L. E. Drain[13].

Experimental values are given in Table 2. As an example, the shift is 2.6×10^{-4} in metallic Li^7. The value of the hyperfine coupling constant a is somewhat different in the metal than in the free atom because the wave functions at the nucleus are different. From the Knight shift of metallic Li it is deduced that the value of $|\psi(0)|^2$ in the metal is 0.44 of the value in the free atom; a calculated value of the ratio is 0.49. It is only in rare instances that the absolute value of the spin contribution χ_s to the magnetic susceptibility can be determined, and then usually by very careful conduction electron spin resonance experiments. Most commonly we extract physical information from Knight shift experiments by making a reasonable estimate of the coupling constant in the metal and by deducing a value of χ_s. The Knight shift has been of value in the study of metals, alloys, soft and intermetallic superconductors, and unusual electronic systems such as Na_xWO_3. The field is reviewed by Drain.[13]

NUCLEAR QUADRUPOLE RESONANCE

Nuclei of spin $I \geq 1$ have an electric quadrupole moment. The quadrupole moment Q is a measure of the ellipticity of the distribution of charge in the nucleus. The quantity of interest is defined classically by[14]

$$eQ = \tfrac{1}{2} \int (3z^2 - r^2)\rho(\mathbf{r})\, d^3x \ , \tag{42}$$

where $\rho(\mathbf{r})$ is the charge density. An egg-shaped nucleus has Q positive; a saucer-shaped nucleus has Q negative.

---

[13] L. E. Drain, "Nuclear magnetic resonance in metals," Metallurgical Reviews, Review 119 (1967).

[14] See Slichter, Chap. 6.

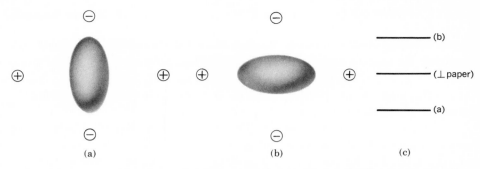

Figure 16 (a) Lowest energy orientation of a nuclear electric quadrupole moment ($Q > 0$) in the local electric field of the four ions shown. The electrons of the ion itself are not shown. (b) Highest energy orientation. (c) The energy level splitting for $I = 1$.

The nucleus when placed in a crystal will see the electrostatic field of its environment, as in Fig. 16. If the symmetry of this field is lower than cubic, then the nuclear quadrupole moment will lead to a set of energy levels split by the interaction of the quadrupole moment with the local electric field. Precisely this effect is discussed in Appendix M for an electronic quadrupole moment. There we do not use the term quadrupole moment; however, an electron in a p state ($L = 1$) has a quadrupole moment that is responsible for the crystal field splitting in the example treated.

The states that are split are the $2I + 1$ states of a spin I. The quadrupole splittings can often be observed directly because an rf magnetic field of the appropriate frequency can cause transitions between the levels. The term **nuclear quadrupole resonance** refers to observations of nuclear quadrupole splittings in the absence of a static magnetic field. The quadrupole splittings are particularly large in covalently bonded molecules such as Cl_2, Br_2, and I_2; the splittings are of the order 10^7 or 10^8 Hz.

FERROMAGNETIC RESONANCE

Spin resonance at microwave frequencies in ferromagnets[15] is similar in principle to nuclear spin resonance. The total magnetic moment of the specimen precesses about the direction of the static magnetic field, and energy is absorbed strongly from the rf transverse field when its frequency is equal to the precessional frequency. We may equally well think of the macroscopic vector representing the total spin of the ferromagnet as quantized in the static magnetic field, with energy levels separated by the usual Zeeman frequencies; the selection rule $\Delta m_S = \pm 1$ allows transitions only between adjacent levels.

The unusual features of ferromagnetic resonance include the following:

a. The transverse susceptibility components χ' and χ'' are very large

---

[15] Observed first by J. H. E. Griffiths, Nature **158**, 670 (1946).

because the magnetization of a ferromagnet in a given static field is very much larger than the magnetization of electronic or nuclear paramagnets in the same field.

b. The shape of the specimen plays an important role. Because the magnetization is large, the demagnetization field[16] is large.

c. The strong exchange coupling between the ferromagnetic electrons tends to suppress the dipolar contribution to the line width, so that the ferromagnetic resonance lines can be quite sharp (<1 G) under favorable conditions. The exchange-narrowing effect in the paramagnetic region is treated below under EPR.

d. Saturation effects are found to occur at relatively low rf power levels. It is not possible, as it is with nuclear spin systems, to drive a ferromagnetic spin system so hard that the magnetization M_z is reduced to zero or reversed. The ferromagnetic resonance excitation breaks down into spin wave modes before the magnetization vector can be rotated appreciably from its initial direction.

Shape Effects in FMR

We treat the effects of specimen shape on the resonance frequency.[17] Consider a specimen of a cubic ferromagnetic insulator in the form of an ellipsoid with principal axes parallel to x, y, z axes of a Cartesian coordinate system. The **demagnetization factors** N_x, N_y, N_z are identical with the depolarization factors defined in Chapter 13. The components of the internal magnetic field $\mathbf{B}_i$ in the ellipsoid are related to the applied field by

$$B_x{}^i = B_x{}^0 - N_x M_x \; ; \qquad B_y{}^i = B_y{}^0 - N_y M_y \; ; \qquad B_z{}^i = B_z{}^0 - N_z M_z \; . \tag{43}$$

The Lorentz field $(4\pi/3)\mathbf{M}$ and the exchange field $\lambda\mathbf{M}$ do not contribute to the torque because their vector product with $\mathbf{M}$ vanishes identically. In SI we replace the components of $\mathbf{M}$ in (43) by $\mu_0\mathbf{M}$, with the appropriate redefinition of the N's.

With (43) the components of the spin equation of motion $\dot{\mathbf{M}} = \gamma(\mathbf{M} \times \mathbf{B}^i)$ become, for an applied static field $B_0\hat{\mathbf{z}}$,

$$\frac{dM_x}{dt} = \gamma(M_y B_z{}^i - M_z B_y{}^i) = \gamma[M_y(B_0 - N_z M) - M(-N_y M_y)]$$

$$= \gamma[B_0 + (N_y - N_z)M]M_y \; ; \tag{44}$$

$$\frac{dM_y}{dt} = \gamma[M(-N_x M_x) - M_x(B_0 - N_z M)] = -\gamma[B_0 + (N_x - N_z)M]M_x \; .$$

[16] See depolarization field, Chapter 13.

[17] C. Kittel, Phys. Rev. **71**, 270 (1947); **73**, 155 (1948).

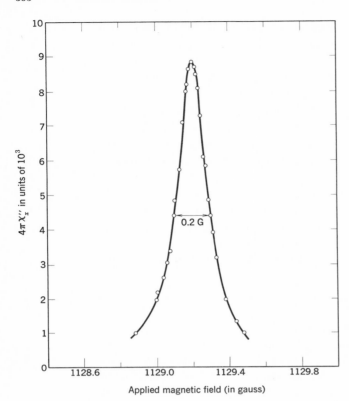

Figure 17 FMR in a polished sphere of yttrium iron garnet at 3.33 GHz and 300°K. for $B_0 \parallel [111]$. The total line width at half-power is only 0.2 G. (After R. C. LeCraw and E. Spencer, unpublished.)

To first order we may set $dM_z/dt = 0$ and $M_z = M$. Solutions of (44) with time dependence $e^{-i\omega t}$ will exist if

$$\begin{vmatrix} i\omega & \gamma[B_0 + (N_y - N_z)M] \\ - \gamma[B_0 + (N_x - N_z)M] & i\omega \end{vmatrix} = 0 , \quad (45)$$

so that the ferromagnetic resonance frequency in the applied field B_0 is

(CGS) $\qquad \omega_0^2 = \gamma^2[B_0 + (N_y - N_z)M][B_0 + (N_x - N_z)M] ; \qquad (46)$

(SI) $\qquad \omega_0^2 = \gamma^2[B_0 + (N_y - N_z)\mu_0 M][B_0 + (N_x - N_z)\mu_0 M] .$

The frequency ω_0 is called the frequency of the **uniform mode,** in distinction to the frequencies of magnon and other nonuniform modes. In the uniform mode all the moments precess together in phase with the same amplitude.

For a sphere $N_x = N_y = N_z$, so that

$$\omega_0 = \gamma B_0 . \qquad (47)$$

A very sharp resonance line in this geometry is shown in Fig. 17.

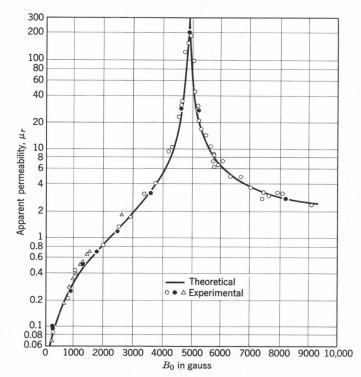

Figure 18 Resonance curve for Supermalloy, according to Yager and Bozorth; the vertical scale is that combination of the real and imaginary parts of the permeability which determines the eddy current losses.

For a flat plate with B_0 perpendicular to the plate $N_x = N_y = 0$; $N_z = 4\pi$, whence

(CGS) $\omega_0 = \gamma(B_0 - 4\pi M)$;   (SI) $\omega_0 = \gamma(B_0 - \mu_0 M)$.   (48)

If B_0 is parallel to the plane of the plate and the plate lies in the xz plane, then $N_x = N_z = 0$; $N_y = 4\pi$, and

(CGS) $\omega_0 = \gamma[B_0(B_0 + 4\pi M)]^{\frac{1}{2}}$;   (SI) $\omega_0 = \gamma[B_0(B_0 + \mu_0 M)]^{\frac{1}{2}}$.   (49)

A resonance line for this geometry is shown in Fig. 18.

The experiments determine γ which is related to the spectroscopic splitting factor g by $-\gamma \equiv g\mu_B/\hbar$, as in Chapter 15. Values of g for metallic Fe, Co, Ni at room temperature are 2.10, 2.18, and 2.21, respectively.

Spin Wave Resonance

Uniform rf magnetic fields can excite long wavelength spin waves in thin ferromagnetic films[18] if the electron spins on the surfaces of the film see dif-

---

[18] C. Kittel, Phys. Rev. **110**, 1295 (1958); M. H. Seavey, Jr., and P. E. Tannenwald, Phys. Rev. Letters **1**, 108 (1958); P. Pincus, Phys. Rev. **118**, 658 (1960); for a study of the angular dependence see M. Okochi, J. Phys. Soc. Japan **28**, 897 (1970).

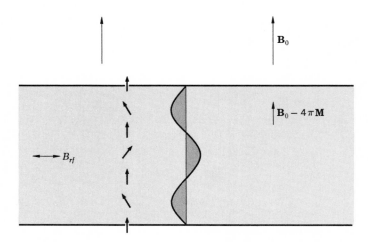

Figure 19 Spin wave resonance in a thin film. The plane of the film is normal to the applied magnetic field B_0. A cross section of the film is shown here. The internal magnetic field is $B_0 - 4\pi M$. The spins on the surfaces of the film are assumed to be held fixed in direction by surface anisotropy forces. A uniform rf field will excite spin wave modes having an odd number of half-wavelengths. The wave shown is for $n = 3$ half-wavelengths.

ferent anisotropy fields than the spins within the films. In effect, the surface spins may be pinned, as shown in Fig. 19. If the rf field is uniform, it can excite waves with an odd number of half-wavelengths within the thickness of the film. Waves with an even number of half-wavelengths have no net interaction energy with the field.

The condition for **spin wave resonance** (SWR) for the applied magnetic field normal to the film is obtained from (48) by adding to the right-hand side the exchange contribution to the frequency. The exchange contribution may be written as Dk^2, where D is a constant that in the model of Eq. (16.26) is equal to $2JSa^2$. The assumption $ka \ll 1$ is valid for the SWR experiments. Thus in an applied field B_0:

$$(\text{CGS}) \quad \omega_0 = \gamma(B_0 - 4\pi M) + Dk^2 = \gamma(B_0 - 4\pi M) + D\left(\frac{n}{\pi L}\right)^2 . \qquad (50)$$

where the wavevector for a mode of n half-wavelengths in a film of thickness L is $k = n\pi/L$.

Experimental results for two films of Permalloy (82Ni 18Fe) are plotted in Fig. 20. The experiments were done at a constant frequency, so that as the mode number n in (50) is increased the field at resonance is correspondingly decreased. The graph of B_0 versus n^2 is expected to be a straight line if the surface pinning condition is satisfied. We have selected a set of experiments for which the linearity is excellent; this appears to be largely a matter of the uniformity of the film. An experimental absorption curve is shown in Fig. 21.

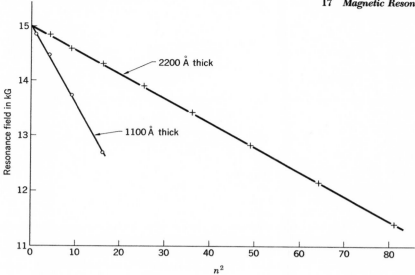

Figure 20 Spin wave resonances in Permalloy films at 12.33 GHz. The magnetic field at resonance is plotted versus the square of the mode number n. [After G. I. Lykken, Phys. Rev. Letters **19**, 1431 (1967).]

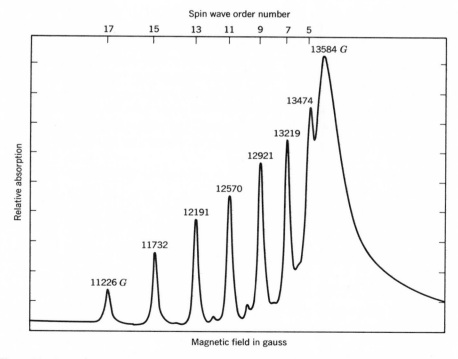

Figure 21 Spin wave resonance spectrum in a Permalloy (80Ni20Fe) film at 9 GHz. The order number is the number of half-wavelengths in the thickness of the film. (After R. Weber, IEEE Transactions on Magnetics, MAG-4:1, March 1968.)

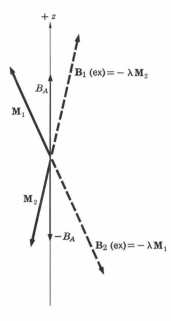

+z

$\mathbf{B}_1$ (ex) $= -\lambda \mathbf{M}_2$

B_A

$\mathbf{M}_1$

$\mathbf{M}_2$

$-B_A$

$\mathbf{B}_2$ (ex) $= -\lambda \mathbf{M}_1$

Figure 22 Effective fields in antiferromagnetic resonance. The magnetization $\mathbf{M}_1$ of sublattice 1 sees a field $-\lambda \mathbf{M}_2 + B_A \hat{\mathbf{z}}$; the magnetization $\mathbf{M}_2$ sees $-\lambda \mathbf{M}_1 - B_A \hat{\mathbf{z}}$. (Both ends of the crystal axis are "easy axes" of magnetization.)

ANTIFERROMAGNETIC RESONANCE

We consider a uniaxial antiferromagnetic with spins on two sublattices, 1 and 2. We suppose that the magnetization $\mathbf{M}_1$ on sublattice 1 is directed along the $+z$ direction by an anisotropy field $B_A \hat{\mathbf{z}}$; the anisotropy field (Chapter 16) results from an anisotropy energy density $U_K(\theta_1) = K \sin^2 \theta_1$, where θ_1 is the angle between $\mathbf{M}_1$ and the z axis. As in Problem 4, $B_A = 2K/M$, where $M = |\mathbf{M}_1| = |\mathbf{M}_2|$. The magnetization $\mathbf{M}_2$ is directed along the $-z$ direction by an anisotropy field[19] $- B_A \hat{\mathbf{z}}$.

The exchange interaction between $\mathbf{M}_1$ and $\mathbf{M}_2$ is treated in the mean field approximation. The exchange fields are

$$\mathbf{B}_1(\text{ex}) = -\lambda \mathbf{M}_2 \; ; \qquad \mathbf{B}_2(\text{ex}) = -\lambda \mathbf{M}_1 \; , \qquad (51)$$

where λ is positive. Here $\mathbf{B}_1$ is the field that acts on the spins of sublattice 1, and $\mathbf{B}_2$ acts on sublattice 2. In the absence of an external magnetic field the total field acting on $\mathbf{M}_1$ is $\mathbf{B}_1 = -\lambda \mathbf{M}_2 + B_A \hat{\mathbf{z}}$; the total field on $\mathbf{M}_2$ is $\mathbf{B}_2 = -\lambda \mathbf{M}_1 - B_A \hat{\mathbf{z}}$, as in Fig. 22. In what follows we set $M_1{}^z = M$; $M_2{}^z = -M$.

The linearized equations of motion are

$$\frac{dM_1{}^x}{dt} = \gamma[M_1{}^y(\lambda M + B_A) - M(-\lambda M_2{}^y)] \; ;$$

$$\frac{dM_1{}^y}{dt} = \gamma[M(-\lambda M_2{}^x) - M_1{}^x(\lambda M + B_A)] \; ; \qquad (52)$$

---

[19] If $+z$ is an easy direction of magnetization, so is $-z$. If one sublattice is directed along $+z$, the other will be directed along $-z$.

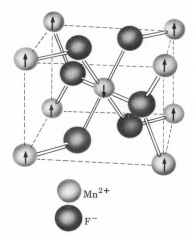

Figure 23 Chemical and magnetic structure of MnF_2. The arrows indicate the direction and arrangement of the magnetic moments assigned to the manganese atoms.

Mn^{2+}

F^-

$$\frac{dM_2^x}{dt} = \gamma[M_2^y(-\lambda M - B_A) - (-M)(-\lambda M_1^y)] \; ;$$

$$\frac{dM_2^y}{dt} = \gamma[(-M)(-\lambda M_1^x) - M_2^x(-\lambda M - B_A)] \; .$$

$$(53)$$

We define $M_1^+ = M_1^x + iM_1^y$; $M_2^+ = M_2^x + iM_2^y$. Then (52) and (53) become, for time dependence exp $(-i\omega t)$,

$$-i\omega M_1^+ = -i\gamma[M_1^+(B_A + \lambda M) + M_2^+(\lambda M)] \; ; \qquad (54a)$$

$$-i\omega M_2^+ = i\gamma[M_2^+(B_A + \lambda M) + M_1^+(\lambda M)] \; . \qquad (54b)$$

These equations have a solution if, with $B_E \equiv \lambda M$,

$$\begin{vmatrix} \gamma(B_A + B_E) - \omega & \gamma B_E \\ \gamma B_E & \gamma(B_A + B_E) + \omega \end{vmatrix} = 0 \; . \qquad (55)$$

Thus the AFMR frequency[20] is given by

$$\omega_0^2 = \gamma^2 B_A(B_A + 2B_E) \; . \qquad (56)$$

Commonly the exchange field is much larger than the anisotropy field, so that

$$\omega_0 \cong \gamma(2B_A B_E)^{\frac{1}{2}} \; . \qquad (57)$$

MnF_2 is an extensively studied antiferromagnet. The structure is shown in Fig. 23. The observed variation of ω_0 with temperature is shown[21] in Fig. 24.

[20] T. Nagamiya, Prog. Theor. Phys. **6**, 342 (1951); C. Kittel, Phys. Rev. **82**, 565 (1951); F. Keffer and C. Kittel, Phys. Rev. **85**, 329 (1952).

[21] F. M. Johnson and A. H. Nethercot, Jr., Phys. Rev. **104**, 847 (1956); **114**, 705 (1959); S. Foner, Phys. Rev. **107**, 683 (1957).

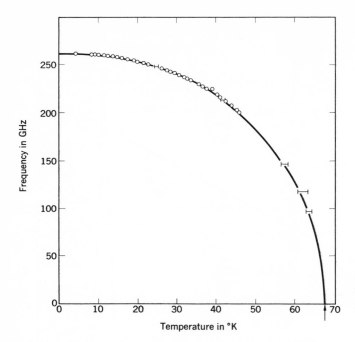

Figure 24 Antiferromagnetic resonance frequency for MnF_2 versus temperature, after Johnson and Nethercot.

Long before the first observations, careful estimates from empirical data were made by Keffer of B_A and B_E for MnF_2. He estimated $B_E = 540$ kG and $B_A = 8.8$ kG at $0°K$, whence $(2B_A B_E)^{\frac{1}{2}} = 100$ kG. The observed value is 93 kG.

Richards[22] has made a compilation of AFMR frequencies as extrapolated to $0°K$:

Crystal	CoF_2	NiF_2	MnF_2	FeF_2	MnO	NiO
Frequency in 10^{10} Hz	85.5	93.3	26.0	158.	82.8	109

ELECTRON PARAMAGNETIC RESONANCE

Electron spin resonance is now a vast area of research. We have space only to mention several isolated topics of considerable interest.

Exchange Narrowing[23]

We consider a paramagnet with an exchange interaction J among nearest-neighbor electron spins. The temperature of the specimen is assumed to be well above any spin-ordering temperature T_c. It is observed under these conditions that the width of the spin resonance line is usually much narrower than expected for the dipole-dipole interaction. The effect is called **exchange narrowing**; there is a close analogy with motional narrowing. We interpret the exchange frequency

$$\omega_{ex} \approx \frac{J}{\hbar} \tag{58}$$

---

[22] P. L. Richards, J. Appl. Phys. **34**, 1237 (1963).

[23] J. H. Van Vleck, Phys. Rev. **74**, 1168 (1948); F. Keffer, Phys. Rev. **88**, 686 (1952).

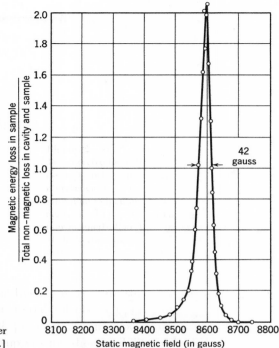

Figure 25   EPR in $NiSiF_6 \cdot 6H_2O$ at 24.45 GHz. [After Holden, Kittel, and Yager, Phys. Rev. **75**, 1443 (1949).]

as a hopping frequency $1/\tau$. Then by generalization of the motional-narrowing result (28) we have for the width of the exchange-narrowed line:

$$\Delta\omega \approx \frac{(\Delta\omega)_0{}^2}{\omega_{ex}} , \qquad (59)$$

where $(\Delta\omega)_0{}^2 = \gamma^2 \langle B_i{}^2 \rangle$ is the square of the static dipolar width in the absence of exchange.

A useful and striking example of exchange narrowing is the paramagnetic organic crystal known as the g marker or DPPH.[24] This free radical has a 1.35 G half-width of the resonance line at half-power, only a few percent of the pure dipole width.

Zero-field Splitting

A number of paramagnetic ions have crystal field splittings of their magnetic ground state energy levels in the range of $10^{10} - 10^{11}$ Hz which is conveniently accessible by microwave techniques.[25] An example is given in Fig. 25: the observed spectrum of the Ni^{2+} ion can be interpreted to give a

---

[24] Diphenyl picryl hydrazyl. The g value is 2.0036 ± 0.0002. Because of the sharp line width it is often used for calibration of the magnetic field; see A. N. Holden et al., Phys. Rev. **77**, 147 (1950).

[25] Much of the early work is due to B. Bleaney and co-workers at Oxford.

zero-field splitting $\Delta = 1.5 \times 10^{10}$ Hz at room temperature.

The Mn^{2+} ion is popular and has been studied in many crystals as an additive impurity. A ground state splitting in the range $10^7 - 10^9$ Hz is observed, according to the environment.

SUMMARY
(In CGS Units)

1. The resonance frequency of a free spin is $\omega_0 = \gamma B_0$, where $\gamma = \mu/\hbar I$ is the magnetogyric ratio.

2. The Bloch equations are

$$\frac{dM_x}{dt} = \gamma(\mathbf{M} \times \mathbf{B})_x - M_x/T_2 \; ;$$

$$\frac{dM_y}{dt} = \gamma(\mathbf{M} \times \mathbf{B})_y - M_y/T_2 \; ;$$

$$\frac{dM_z}{dt} = \gamma(\mathbf{M} \times \mathbf{B})_z + (M_0 - M_z)/T_1 \; .$$

3. The half-width of the resonance at half-power is $(\Delta\omega)_{\frac{1}{2}} = 1/T_2$.

4. Saturation effects at high rf power enter when $\gamma^2 B_1{}^2 T_1 T_2$ exceeds unity.

5. The dipolar line width in a rigid lattice is $(\Delta B)_0 \sim \mu/a^3$.

6. If the magnetic moments are ambulatory, with a characteristic time $\tau \ll 1/(\Delta\omega)_0$, the line width is reduced by the factor $(\Delta\omega)_0\tau$. In this limit $1/T_1 \approx 1/T_2 \approx (\Delta\omega)_0{}^2\tau$. With exchange coupling in a paramagnet the line width becomes $\approx (\Delta\omega)_0{}^2/\omega_{\text{ex}}$.

7. The ferromagnetic resonance frequency in an ellipsoid of demagnetization factors N_x, N_y, N_z is

$$\omega_0{}^2 = \gamma^2[B_0 + (N_y - N_z)M][B_0 + (N_x - N_z)M] \; .$$

8. The antiferromagnetic resonance frequency is $\omega_0{}^2 = \gamma^2 B_A(B_A + 2B_E)$, in a spherical specimen with zero applied field. Here B_A is the anisotropy field and B_E is the exchange field.

Problems

1. *Equivalent electrical circuit.* Consider an empty coil of inductance L_0 in a series with a resistance R_0; show if the coil is completely filled with a spin system characterized by the susceptibility components $\chi'(\omega)$ and $\chi''(\omega)$ that the inductance at frequency ω becomes

$$L = [1 + 4\pi\chi'(\omega)]L_0 ,$$

in a series with an effective resistance

$$R = 4\pi\omega\chi''(\omega)L_0 + R_0 .$$

In this problem χ', χ'' are defined for a linearly polarized rf field. *Hint:* Consider the impedance of the circuit. (CGS units.)

2. *Rotating coordinate system.* We define the vector $\mathbf{F}(t) = F_x(t)\hat{\mathbf{x}} + F_y(t)\hat{\mathbf{y}} + F_z(t)\hat{\mathbf{z}}$. Let the coordinate system of the unit vectors $\hat{\mathbf{x}}$, $\hat{\mathbf{y}}$, $\hat{\mathbf{z}}$ rotate with an instantaneous angular velocity Ω, so that $d\hat{\mathbf{x}}/dt = \Omega_y\hat{\mathbf{z}} - \Omega_z\hat{\mathbf{y}}$, etc. (a) Show that $d\mathbf{F}/dt = (d\mathbf{F}/dt)_R + \Omega \times \mathbf{F}$, where $(d\mathbf{F}/dt)_R$ is the time derivative of $\mathbf{F}$ as viewed in the rotating frame R. (b) Show that (7) may be written $(d\mathbf{M}/dt)_R = \gamma\mathbf{M} \times (\mathbf{B}_a + \Omega/\gamma)$. This is the equation of motion of $\mathbf{M}$ in a rotating coordinate system. The transformation to a rotating system is extraordinarily useful; it is exploited widely in the literature. (c) Let $\Omega = -\gamma B_0\hat{\mathbf{z}}$; thus in the rotating frame there is no static magnetic field. Still in the rotating frame, we now apply a dc pulse $B_1\hat{\mathbf{x}}$ for a time t. If the magnetization is initially along $\hat{\mathbf{z}}$, find an expression for the pulse length t such that the magnetization will be directed along $-\hat{\mathbf{z}}$ at the end of the pulse. (Neglect relaxation effects.) (d) Describe this pulse as viewed from the laboratory frame of reference.

3. *Hyperfine effects on ESR in metals.* We suppose that the electron spin of a conduction electron in a metal sees an effective magnetic field from the hyperfine interaction of the electron spin with the nuclear spin. Let the z component of the field seen by the conduction electron be written

$$B_i = \left(\frac{a}{N}\right)\sum_{j=1}^{N} I_j^z ,$$

where I_j^z is equally likely to be $\pm\frac{1}{2}$. (a) Show that $\langle B_i{}^2\rangle = (a/2N)^2 N$. (b) Show that $\langle B_i{}^4\rangle = 3(a/2N)^4 N^2$, for $N \gg 1$.

4. *FMR in the anisotropy field.* Consider a spherical specimen of a uniaxial ferromagnetic crystal with an anisotropy energy density of the form $U_K = K \sin^2 \theta$, where θ is the angle between the magnetization and the z axis. We assume that K is positive. Show that the ferromagnetic resonance frequency in an external magnetic field $B_0\hat{\mathbf{z}}$ is $\omega_0 = \gamma(B_0 + B_A)$, where $B_A \equiv 2K/M_s$.

5. *Exchange frequency resonance.* Consider a ferrimagnet with two sublattices A and B of magnetizations $\mathbf{M}_A$ and $\mathbf{M}_B$, where $\mathbf{M}_B$ is opposite to $\mathbf{M}_A$ when the spin system is at rest. The gyromagnetic ratios are γ_A, γ_B and the molecular fields are $\mathbf{B}_A = -\lambda\mathbf{M}_B$; $\mathbf{B}_B = -\lambda\mathbf{M}_A$. Show that there is a resonance at

$$\omega_0{}^2 = \lambda^2(\gamma_A|M_B| - \gamma_B|M_A|)^2 .$$

This is called the exchange frequency resonance.[26]

[26] See the review by S. Foner listed in the references.

References

INTRODUCTION

C. P. Slichter, *Principles of magnetic resonance, with examples from solid state physics*, Harper and Row, 1963. (An excellent introduction.)

G. E. Pake, "Nuclear magnetic resonance," *Solid state physics* **2**, 1–91 (1956).

G. E. Pake, *Paramagnetic resonance*, Benjamin, 1962.

A. A. Manenkov and R. Orbach, ed., *Spin-lattice relaxation in ionic solids*, Harper & Row, 1966. (Collected classical papers.)

N. Bloembergen, E. M. Purcell, and R. V. Pound, "Relaxation effects in nuclear magnetic resonance absorption," Phys. Rev. **73**, 679–712 (1948).

NUCLEAR MAGNETIC RESONANCE

A. Abragam, *Nuclear magnetism*, Oxford, 1961. (Definitive and comprehensive.)

T. P. Das and E. L. Hahn, "Nuclear quadrupole resonance spectroscopy," *Solid state physics*, Supplement **1**, (1958).

V. Jaccarino, "Nuclear resonance in antiferromagnets," *Magnetism*[27] IIA.

A. M. Portis and R. H. Lindquist, "Nuclear resonance in ferromagnetic materials," *Magnetism* IIA, 357.

E. L. Hahn, "Spin echoes," Physics Today **6**, 4 (Nov. 1953).

ELECTRON RESONANCE

B. Bleaney and K. W. H. Stevens, "Paramagnetic resonance," Repts. Prog. Phys. **16**, 108 (1953).

K. D. Bowers and J. Owen, "Paramagnetic resonance II," Repts. Prog. Phys. **18**, 304 (1955).

J. W. Orton, "Paramagnetic resonance data," Repts. Prog. Phys, **22**, 204 (1959).

W. Low, "Paramagnetic resonance in solids," *Solid state physics*, Supplement **2**, 1960.

S. A. Al'tshuler and B. M. Kozyrev, *Electron paramagnetic resonance*, Academic Press, 1964.

K. J. Standley and R. A. Vaughan, *Electron spin relaxation in solids*, Hilger, London, 1969.

FERRO- AND ANTIFERROMAGNETIC RESONANCE

M. Sparks, *Ferromagnetic relaxation theory*, McGraw-Hill, 1964.

S. Foner, "Antiferromagnetic and ferrimagnetic resonance," *Magnetism* I, 384.

C. W. Haas and H. B. Callen, "Ferromagnetic relaxation and resonance line widths," *Magnetism* I, 450.

R. W. Damon, "Ferromagnetic resonance at high power," *Magnetism* I, 552.

[27] Edited by G. T. Rado and H. Suhl, Academic Press.

18

Optical Phenomena in Insulators

COLOR OF CRYSTALS

Crystals that are electrical insulators at room temperature usually are transparent. A section 1 cm thick of such a single crystal appears clear to the eye, although rarely as clear as plate glass. To be clear a crystal can have no strong electronic or vibronic transitions in the visible spectral range from 7400 to 3600 Å, or 1.7 to 3.5 eV. We consider briefly the origin of the color exhibited by common solids. If the absorption is not strong, the color presented by a powder of small crystals is usually the color of light transmitted through the crystal.

(1) A pure perfect diamond is clear. The energy gap of diamond is 5.4 eV, so that electronic transitions from the valence band to the conduction band do not occur in the visible range. Diamonds can be colored by irradiation which produces lattice defects (Chapter 19).

(2) Cadmium sulfide is yellow-orange. The energy gap is 2.42 eV, so that the blue region of the spectrum is absorbed by the crystal.

(3) Silicon has a metallic luster because the band gap of 1.14 eV is below the visible region in energy: all wavelengths in the visible region excite electronic transitions[1] from the valence band to the conduction band. But a thin (<0.01 cm) section of silicon transmits weakly in the red, because the absorption process in silicon for frequencies near the gap involves a phonon as well as a photon and is not very intense. The threshold energy for direct absorption lies at 2.5 eV in the middle of the visible spectral region. Tin oxide is a semiconductor transparent in thin layers; it is often used as an electrode when transparency is required.

(4) Ruby is a dark red gem; sapphire is a blue gem. Both are impure crystals of Al_2O_3, which is colorless when pure. The color of ruby is caused by perhaps 0.5 percent of Cr^{3+}, which enters lattice sites normally occupied by an Al^{3+} ion. The blue color of sapphire is due to Ti^{3+} ions present as impurities in Al_2O_3.

(5) Many compounds containing transition elements are colored even though the crystals do not have energy gaps in the visible region. It is characteristic of many transition element ions to have electronic excited states at energies in the visible region. The excited state may be localized near or on the transition ion.[2]

---

[1] That is, the metallic appearance of silicon persists to 0°K and is not caused by any free charge carriers which may be present.

[2] The sense in which such an excitation is localized is discussed later in the section on tightly bound excitons.

(6) Some crystals can be colored by radiation damage, that is, by bombardment with energetic particles, gamma rays, or ultraviolet light. Electrons or holes trapped at lattice defects often have absorption lines in the visible region, as discussed in Chapter 19.

(7) The precipitation of metallic impurities as fine colloidal particles throughout a crystal will cause coloration by virtue of the wavelength dependence of the scattering cross-section of the particles. The production of ruby-colored glass by the controlled precipitation of gold is the classic example.

EXCITONS

We saw in Chapter 11 that an electron-hole pair is produced when a photon of energy greater than the energy gap E_g is absorbed in a crystal. The electron and hole produced in this way (Fig. 1) are free and may move independently through the crystal. But because an electron and hole have an attractive Coulomb interaction for each other it is possible for stable bound states of the two particles to be formed. The photon energy required to create a bound pair starting from a filled valence band will be less than the gap energy E_g.

The bound electron-hole pair (Fig. 2) is known as an **exciton;** it may move through the crystal transporting excitation energy but not charge. Thus an exciton[3] is a neutral excited mobile state of a crystal: an exciton can travel through a crystal, giving up its energy of formation on recombination.[4] Be-

---

[3] Important original papers on excitons include J. Frenkel, Phys. Rev. **37**, 17, 1276 (1931); Physik. Z. Sowjetunion **9**, 158 (1936); R. Peierls, Ann. Physik **13**, 905 (1932); J. C. Slater and W. Shockley, Phys. Rev. **50**, 705 (1936); G. H. Wannier, Phys. Rev. **52**, 191 (1937); W. R. Heller and A. Marcus, Phys. Rev. **84**, 809 (1951); N. F. Mott, Trans. Faraday Soc. **34**, 500 (1938); D. L. Dexter and W. R. Heller, Phys. Rev. **84**, 377 (1951); A. S. Davydov, J.E.T.P. **18**, 210 (1948).

[4] An exciton is said to recombine when the electron drops into the hole state from which it came.

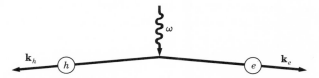

Figure 1   Absorption of a photon of frequency above the energy gap. The photon excites an electron from the valence band to the conduction band, leaving behind a hole in the valence band. The photon wavevectors are of negligible magnitude in the energy range of interest; thus $\mathbf{k}_h \cong -\mathbf{k}_e$.

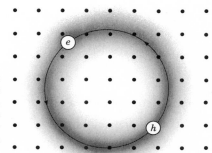

Figure 2   An exciton is a bound electron-hole pair, usually free to move together through the crystal. In some respects it is similar to an atom of positronium, which is formed from a positron and an electron. The exciton shown here is a Mott exciton: it is weakly bound, with an average electron-hole distance large in comparison with a lattice constant.

cause of its charge neutrality it does not contribute directly to the electrical conductivity.

We consider excitons in two different limiting approximations, one due to Frenkel in which the exciton is considered as tightly bound, and the other due to Mott and Wannier in which the exciton is weakly bound, with an electron-hole interparticle distance large in comparison with a lattice constant.

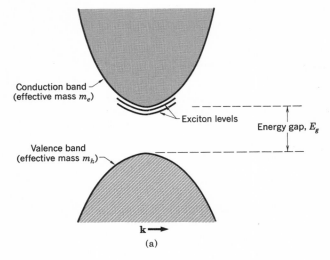

Conduction band
(effective mass m_e)

Exciton levels

Energy gap, E_g

Valence band
(effective mass m_h)

$\mathbf{k} \longrightarrow$

(a)

Figure 3a Exciton levels in relation to the conduction band edge, for a simple band structure with both conduction and valence band edges at $\mathbf{k} = 0$. An exciton can have translational kinetic energy, but if the translational energy is greater than the binding energy of the exciton, then the exciton is metastable with respect to decay into a free hole and free electron. All excitons are potentially unstable with respect to radiative recombination in which the electron drops down into the hole state in the valence band, accompanied by the emission of a photon or phonons.

Weakly Bound Excitons

We consider an electron in the conduction band of a crystal and a hole in the valence band. The electron and hole are attracted to each other by the attractive Coulomb potential

(CGS) $$U(r) = -\frac{e^2}{\epsilon r} , \tag{1}$$

where r is the distance between the particles and ϵ is the appropriate dielectric constant.[5] There will be bound states (Figs. 3 and 4) of the exciton system having total energies lower than the bottom of the conduction band. The problem is much like the hydrogen atom problem if the energy surfaces for the electron and hole are spherical and nondegenerate. The energy levels referred to the top of the valence band are given by a modified Rydberg equation

(CGS) $$E_n = E_g - \frac{\mu e^4}{2\hbar^2 \epsilon^2 n^2} . \tag{2}$$

Here n is the principal quantum number and μ is the reduced mass

$$\frac{1}{\mu} = \frac{1}{m_e} + \frac{1}{m_h} \tag{3}$$

formed from the effective masses m_e, m_h of the electron and hole. For general

---

[5] The lattice polarization contribution to the dielectric constant should not be included if the frequency of motion of the exciton is higher than the optical phonon frequencies; this is the most common situation. Examples are known, however, for which the exciton frequency is lower than the optical phonon frequencies: see R. Z. Bachrach and F. C. Brown, Phys. Rev. Letters **21**, 685 (1968); S. D. Mahanti and C. M. Varma, Phys. Rev. Letters **25**, 1115 (1970).

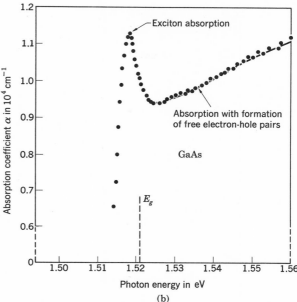

Figure 3b The observed effect of an exciton level on the optical absorption of a semiconductor for photons of energy near the band gap E_g. The optical absorption edge and the exciton absorption peak in gallium arsenide at 21°K, after M. D. Sturge, Phys. Rev. **127**, 768 (1962). The vertical scale is the intensity absorption coefficient, as in $I(x) = I_0 \exp(-\alpha x)$. The energy gap and exciton binding energy are deduced from the shape of the absorption curve: the gap E_g is 1.521 ev and the exciton binding energy is 0.0034 ev.

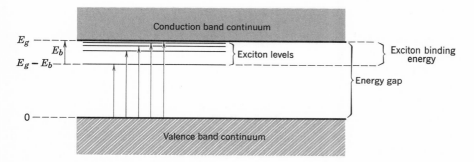

Figure 4 Energy levels of an exciton whose center of mass is at rest. Optical transitions from the top of the valence band are shown by the arrows; the longest arrow corresponds to the ionization of the exciton and therefore to the energy gap between the edges of the conduction and valence bands. The binding energy of the exciton is E_b, referred to a free electron and free hole. The lowest frequency absorption line of the crystal at absolute zero is not E_b, but is $E_g - E_b$.

energy surfaces the exciton problem may be quite complicated, but there is little doubt that bound exciton states[6] almost always exist in insulators.

The exciton ground state energy is obtained on setting $n = 1$ in (2); this energy corresponds to the ionization energy required to break up the exciton from its lowest state. It is difficult to produce excitons in sufficient concentration to observe directly transitions among the exciton levels, but it is possible to

---

[6] It may not always be possible to separate internal and center-of-mass coordinates, although this can be done for ellipsoidal energy surfaces.

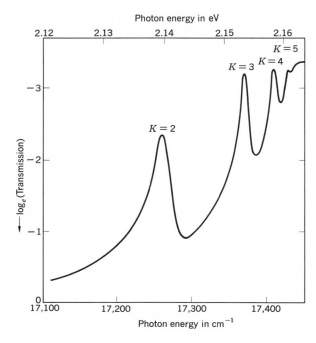

Figure 5 Logarithm of the optical transmission versus photon energy in cuprous oxide at 77°K, showing a series of exciton lines. [After P. W. Baumeister, Phys. Rev. **121**, 359 (1961).] Note that on the vertical axis the log is plotted decreasing upward; thus a peak corresponds to absorption. The band gap E_g is about 2.17 eV.

observe in optical absorption the transitions at $\hbar\omega_n = E_n$ between the valence band edge and an exciton level.

One crystal is known whose exciton spectrum satisfies (2) fairly accurately. Gross[7] and others have studied optical absorption lines in cuprous oxide Cu$_2$O at low temperatures, with results for the spacing of the exciton levels in surprisingly good agreement with the Rydberg equation (2), particularly for levels with $n > 2$. The results and experimental arrangement are shown in Figs. 5 and 6. An empirical fit to the lines as observed by Apfel and Hadley is obtained with the relation $\nu(\text{cm}^{-1}) = 17{,}508 - (800/n^2)$. Taking $\epsilon = 10$, we find $\mu \cong 0.7\ m$ from the coefficient of $1/n^2$. Unfortunately no adequate independent values of m_e and m_h are available[8] for Cu$_2$O. The constant term 17,508 cm^{-1} corresponds to an energy gap $E_g = 2.17$ eV.

[7] E. F. Gross, B. P. Zakharchenya, and N. M. Reinov, Doklady Akad. Nauk S.S.S.R. **92**, 265 (1953); **97**, 57, 221 (1954); **99**, 231, 527 (1954); S. Nikitine, G. Perny, and M. Sieskind, Compt. rend. (Paris) **238**, 67 (1954); J. H. Apfel and L. N. Hadley, Phys. Rev. **100**, 1689 (1955); A. G. Samoilovich and L. L. Kornblit, Doklady Akad. Nauk S.S.S.R. **100**, 43 (1955); S. Nikitine, Helv. Phys. Acta **28**, 307 (1955).

[8] For further details of the interpretation of the exciton series in Cu$_2$O, see the articles by R. J. Elliott and by M. Grosmann in *Polarons and excitons*, Plenum, New York, 1963. The article by T. P. McLean in the same volume gives an excellent account of excitons in germanium. An exhaustive analysis of excitations in CdS has been given by D. G. Thomas and J. Hopfield, Phys. Rev. **124**, 657 (1961); their work includes a remarkable demonstration of the motion of excitons through the lattice.

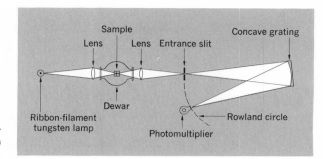

Figure 6   Apparatus used by Baumeister to measure the absorption coefficients of exciton lines in Cu_2O.

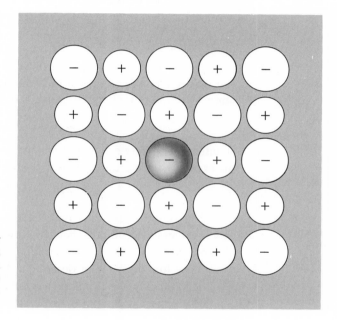

Figure 7   Schematic representation of a tightly bound or Frenkel exciton localized on one atom in the (100) plane of an alkali halide crystal. An ideal Frenkel exciton will travel as a wave throughout the crystal, but the electron is always close to the hole. Contrast with Fig. 2.

Tightly Bound Excitons

In a **Frenkel exciton** the excitation is localized on or near a single atom in the sense that the hole is usually on the same atom as the electron, although the pair may be anywhere in the crystal. A Frenkel exciton is essentially an excited state of a single atom, but the excitation can hop from one atom to another by virtue of the coupling between neighbors. The excitation wave travels through the crystal much as the reversed spin of a magnon travels through the crystal.

In alkali halide crystals (Fig. 7) the excitons of lowest energy are localized on the negative halogen ions: the negative ions have lower electronic excitation levels than the positive ions. Alkali halide crystals when pure are transparent in the visible region of the spectrum, but in the vacuum ultraviolet region they show a considerable structure[9] in spectral absorption (Fig. 8). It is known (particularly from the work of L. Apker and E. Taft) that free electrons and holes are not produced when light is absorbed in the region of the lowest-energy absorption peak. It is very likely that this absorption results in the creation of excitons.

A doublet structure is particularly evident in the bromides in Fig. 8. This structure, as noted by Mott, is similar to the doublet structure of the lowest excited states of the kryton atom, which has the same number of electrons as the Br^- ion. The resemblance of the doublet structures supports our identification of the lowest energy states in the alkali bromides as arising from an exciton centered on the Br^- ions.[10] The splitting is caused by the spin-orbit interaction.

Exciton waves *

We can easily show that the states of an exciton have the form of propagating waves. Consider a crystal of N atoms on a line or ring. If u_j is the ground state wavefunction of the jth atom, the ground state of the crystal is

$$\psi_g = u_1 u_2 \cdots u_{N-1} u_N ,\tag{4}$$

if interactions between the atoms are neglected. If now a single atom j is in an excited state v_j, the system is described by the function

$$\varphi_j = u_1 u_2 \cdots u_{j-1} v_j u_{j+1} \cdots u_N .\tag{5}$$

This function has the same energy as the corresponding function φ_l with any other atom l excited. If there is an interaction between an excited atom and a nearby unexcited atom, the excitation energy can be transferred from atom to atom. The functions φ that describe a definite single excited atom (and $N - 1$ unexcited atoms) are not stationary states of the problem.

The eigenfunctions are easily found. We can describe the effect of the interaction by the following statement: When the hamiltonian of the system

[9] R. Hilsch and R. W. Pohl, Z. Physik **57**, 145 (1929); **59**, 812 (1930); N. F. Mott and R. W. Gurney, *Electronic processes in ionic crystals,* Oxford, 1948.

[10] For identification of the other peaks in the absorption spectra of alkali halides see J. C. Phillips, Phys. Rev. Letters **12**, 142 (1964); also R. S. Knox and N. Inchauspé, Phys. Rev. **116**, 1093 (1959); A. W. Overhauser, Phys. Rev. **101**, 1702 (1956).

* This section requires a knowledge of quantum mechanics.

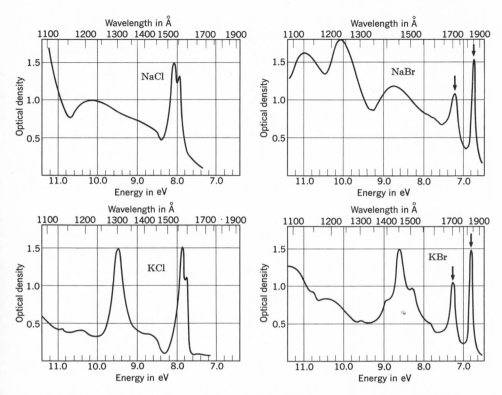

Figure 8 Optical absorption spectra of thin films of alkali halides at 80°K, after J. E. Eby, K. J. Teegarden, and D. B. Dutton, Phys. Rev. **116**, 1099 (1959). The vertical axes give the absorption on a relative scale. A doublet structure in the bromides is marked by arrows. The lowest excitations of the halogen ions are doublets: in Br$^-$ the ground state is a $^1S_0$ state arising from the electron configuration $4p^6$; the lowest excited states arise from the configuration $4p^55s$. The $4p^5$ core may have total angular momentum $J = \frac{3}{2}$ or $\frac{1}{2}$, split by about 0.5 eV; the observed doublet structure results from this splitting.

operates on the function φ_j with the jth atom excited, we obtain

$$\mathcal{H}\varphi_j = \epsilon\varphi_j + T(\varphi_{j-1} + \varphi_{j+1}) \; , \tag{6}$$

where ϵ is the free atom excitation energy, and the interaction T measures the rate of transfer of the excitation from j to its nearest neighbors, $j-1$ and $j+1$. The solutions of (6) are waves of the Bloch form:

$$\psi_k = \sum_j e^{ikj}\varphi_j \; . \tag{7}$$

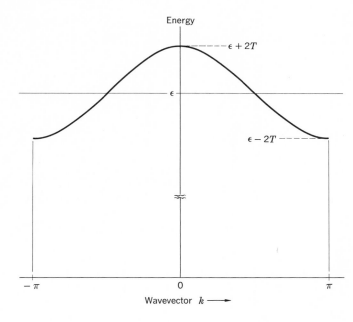

Figure 9 Energy versus wavevector for a Frenkel exciton, calculated with positive nearest neighbor interaction T.

To see this we form

$$\mathcal{H}\psi_k = \sum_j e^{ikj}\,\mathcal{H}\varphi_j = \sum_j e^{ikj}[\epsilon\varphi_j + T(\varphi_{j-1} + \varphi_{j+1})] \; , \qquad (8)$$

from (6). We can rearrange the terms on the right-hand side to obtain

$$\mathcal{H}\psi_k = \sum_j e^{ikj}[\epsilon + T(e^{ik} + e^{-ik})]\varphi_j = (\epsilon + 2T\cos k)\psi_k \; , \qquad (9)$$

so that the energy eigenvalues of the problem are

$$E_k = \epsilon + 2T\cos k \; ; \qquad (10)$$

this is sketched in Fig. 9. The application of periodic boundary conditions determines the allowed values of the wavevector k:

$$k = \frac{2\pi}{N}s \; ; \qquad s = -\tfrac{1}{2}N, \; -\tfrac{1}{2}N + 1, \ldots, \tfrac{1}{2}N - 1 \; . \qquad (11)$$

Excitons in Molecular Crystals

Molecular crystals also furnish examples of the Frenkel or tight-binding model of excitons. In molecular crystals the covalent binding within a molecule is strong in comparison with the van der Waals binding between molecules. Electronic excitation lines of an individual molecule will appear in the crystalline solid as an exciton, often with little shift in frequency. At low temperatures the lines in the solid are quite sharp, although there may be more structure

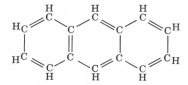

Figure 10 Chemical structure of anthracene.

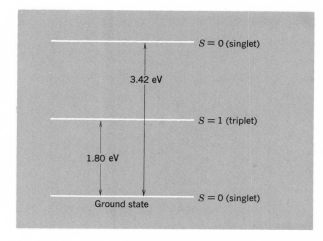

Figure 11 Low-lying exciton states of anthracene. The electron spin is denoted by S.

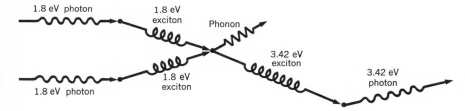

Figure 12 Detection of excitons in anthracene, after Avakian, Kepler, et al.

to the lines in the solid than in the molecule.[11] In such solids the exciton energies are much more closely related to the spectroscopic properties of the isolated molecule than to the Mott-Wannier model discussed earlier.

Considerable interesting work has been done on excitons in crystals of anthracene (Fig. 10). The low-lying exciton levels of anthracene are indicated in Fig. 11. The optical transition between the ground state (singlet) and the lowest electronic excited state (triplet) is not an allowed electric dipole transition, but the transition probability is nonzero and a significant concentration of triplet excitons can be produced by irradiation of the crystal with an intense laser beam of photons of energy 1.79 eV. Two triplet excitons can combine[12] to form one singlet exciton of energy 3.14 eV, the excess energy being carried off by phonons. The 3.14 eV exciton is connected with the ground state by an allowed transition, and the emitted photon at 3.14 eV is detected in the experiments (Fig. 12).

---

[11] See, for example, the discussion of solid benzene by D. Fox and O. Schnepp, J. Chem. Phys. **23,** 767 (1955); the general theory is discussed by A. S. Davydov, J. Exptl. Theoret. Phys. (U.S.S.R.) **18,** 210 (1948); H. Winston, J. Chem. Phys. **19,** 156 (1951). With two nonequivalent molecules in a cell there will be two branches to the dispersion relation E versus k. The separation of the two branches at $k = 0$ is called the **Davydov splitting.**

[12] R. G. Kepler, J. C. Caris, P. Avakian, and E. Abramson, Phys. Rev. Letters **10,** 400 (1963); P. Avakian and R. E. Merrifield, Phys. Rev. Letters **13,** 541 (1964).

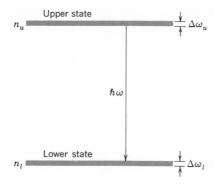

Figure 13 A two-level system, to explain maser operation. The populations of the upper and lower states are n_u and n_l, respectively. The frequency of the emitted radiation is ω; the combined width of the states is $\Delta\omega = \Delta\omega_u + \Delta\omega_l$.

Several systems, particularly certain ion radical salts of tetracyanoquino-methane, exist for which the molecular solid possesses triplet electronic excited states close enough in energy to a singlet ground state for the triplet exciton states to be densely populated at room temperature.[13] Electron spin resonance studies have been carried out on the triplet excitons. The absence of hyperfine broadening of the exciton resonance line (Problem 17.3) suggests that the excitons move fairly freely throughout the crystals.

SOLID STATE QUANTUM ELECTRONICS

Principle of Maser Action

Crystals can be used as microwave and light amplifiers and as sources of coherent radiation. A **maser** amplifies microwaves by the stimulated emission of radiation; a **laser** amplifies light by the same method. The principle, due to Townes, may be understood from the two-level magnetic system of Fig. 13. There are n_u atoms in the upper state and n_l atoms in the lower state.

We immerse the system in radiation at frequency ω; the amplitude of the magnetic component of the radiation field is B_{rf}. The probability per atom per unit time of a transition between the upper and lower states is

$$P = \left(\frac{\mu B_{\mathrm{rf}}}{\hbar}\right)^2 \frac{1}{\Delta\omega} \; ; \tag{12}$$

[13] D. B. Chesnut and W. D. Phillips, J. Chem. Phys. **35**, 1002 (1961); D. B. Chesnut, J. Chem. Phys. **40**, 405 (1964).

here μ is the magnetic moment, and $\Delta\omega$ is the combined width of the two levels. The result (12) is the standard result of quantum mechanics, called Fermi's golden rule. The net energy emitted from atoms in both upper and lower states is

$$\mathcal{P} = \left(\frac{\mu B_{\mathrm{rf}}}{\hbar}\right)^2 \frac{1}{\Delta\omega} \cdot \hbar\omega \cdot (n_u - n_l) \ , \tag{13}$$

per unit time. Here $\mathcal{P}$ denotes the power out; $\hbar\omega$ is the energy per photon; and $n_u - n_l$ is the excess of the number of atoms n_u initially able to emit a photon to the number of atoms n_l able to absorb a photon. In thermal equilibrium $n_u < n_l$, so we cannot get net emission of radiation, but in a nonequilibrium condition with $n_u > n_l$ there will be emission. In fact, if we start with $n_u > n_l$ and reflect the emitted radiation back onto the system, we increase B_{rf} and thereby stimulate a higher rate of emission. The enhanced stimulation continues until the population in the upper state decreases and becomes equal to the population in the lower state.

We can build up the intensity of the radiation field by placing the crystal in an electromagnetic cavity. There will be some power loss in the walls of the cavity: the rate of power loss is

$$(\text{CGS}) \quad \mathcal{P}_L = \frac{B_{\mathrm{rf}}^2 V}{8\pi} \cdot \frac{\omega}{Q} \ ; \qquad\qquad (\text{SI}) \quad \mathcal{P}_L = \frac{B_{\mathrm{rf}}^2 V}{2\mu_0} \cdot \frac{\omega}{Q} \ , \tag{14}$$

where V is the volume and Q is the Q factor of the cavity. We understand B_{rf}^2 to be a volume average. The condition for maser action is that the emitted power (13) exceed the power loss (14); that is, $\mathcal{P} > \mathcal{P}_L$. Both quantities involve B_{rf}^2. The maser condition can be expressed in terms of the population excess in the upper state:

$$(\text{CGS}) \quad n_u - n_l > \frac{V \Delta B}{8\pi\mu Q} \ , \qquad\qquad (\text{SI}) \quad n_u - n_l > \frac{V \Delta B}{2\mu_0\mu Q} \ , \tag{15}$$

where μ is the magnetic moment and the line width ΔB is defined in terms of the combined line width $\Delta\omega$ of the upper and lower states as

$$\mu \, \Delta B = \hbar \, \Delta\omega \ . \tag{16}$$

The central problem of the maser or laser is to obtain a suitable excess population in the upper state. This is accomplished in various ways in various devices.

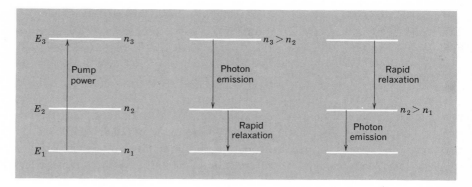

Figure 14 Three-level maser system. Two possible modes of operation are shown, starting from rf saturation of the states 3 and 1 to obtain $n_3 = n_1$.

Three-Level Maser

The three-level maser system (Fig. 14) is a clever solution to the excess population problem. Such a system may derive its energy levels from paramagnetic ions in a crystal.[14] Rf power is applied at the frequency $\hbar\omega_p = E_3 - E_1$ in sufficient intensity to maintain the population of level 3 substantially equal to the population of level 1. Now consider the rate of change of the population n_2 of level 2 due to normal thermal relaxation processes. In terms of the indicated transition rates P,

$$\frac{dn_2}{dt} = -n_2 P(2 \rightarrow 1) - n_2 P(2 \rightarrow 3) + n_3 P(3 \rightarrow 2) + n_1 P(1 \rightarrow 2) \ . \quad (17)$$

In the steady state $dn_2/dt = 0$, and by virtue of the saturation rf power we have $n_3 = n_1$, whence

$$\frac{n_2}{n_1} = \frac{P(3 \rightarrow 2) + P(1 \rightarrow 2)}{P(2 \rightarrow 1) + P(2 \rightarrow 3)} \ . \quad (18)$$

The transition rates are affected by many details of the paramagnetic ion and its environment, but one can hardly lose with this system, for either $n_2 > n_1$ and we get maser action between levels 2 and 1, or else $n_2 < n_1 = n_3$, and we get maser action between levels 3 and 2.

Ruby is an excellent crystal for a three-level maser. Ruby is Al_2O_3 with Cr^{3+} impurity. The Cr^{3+} ions have spin $S = \frac{3}{2}$; the ground level splits into four states in a magnetic field. Three of the four levels are utilized in the maser. Good low noise amplifiers at microwave frequencies have been made from ruby and these find application in radio astronomy and space communication.

[14] N. Bloembergen, Phys. Rev. **104**, 329 (1956).

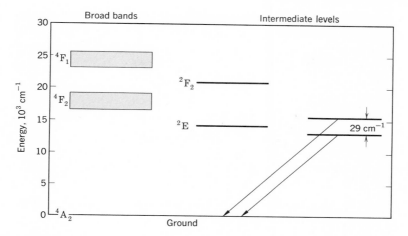

Figure 15 Energy level diagram of Cr^{3+} in ruby, as used in laser operation. The initial excitation takes place to the broad bands; they decay to the intermediate levels by the emission of phonons, and the intermediate levels radiate photons as the ion makes the transition to the ground level.

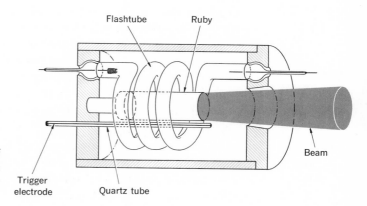

Figure 16 Apparatus for pulsed excitation of ruby, as used in the original work of Maiman. The actual length is 5 cm and the outer diameter is 2.5 cm.

Ruby Laser

The same crystal, ruby, used in the microwave maser was also the first crystal to exhibit optical maser action,[15] but a different set of energy levels of Cr^{3+} are involved (Fig. 15). About 15000 cm^{-1} above the ground state there lie a pair of states labeled ^2E, spaced 29 cm^{-1} apart. Above ^2E lie two broad bands of states, labeled $^4F_1$ and $^4F_2$. Because the bands are broad they can be populated efficiently by optical absorption from broadband light sources (Fig. 16) such as xenon flash lamps.

[15] T. H. Maiman, Phys. Rev. Letters **4**, 564 (1960); Phys. Rev. **123**, 1145 (1961); for the principles of infrared and optical lasers, see A. L. Schawlow and C. H. Townes, Phys. Rev. **112**, 1940 (1958).

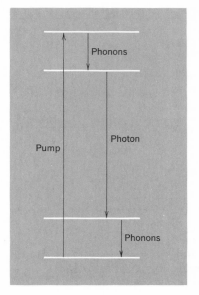

Figure 17 Four-level laser system, as in the neodymium glass laser.

In operation of a ruby laser both of the broad ^4F bands are populated by broadband light. Atoms thus excited will decay in 10^{-7} sec by radiationless processes with the emission of phonons to the states ^2E. Photon emission from the lower of the states ^2E to the ground state occurs slowly, in about 5×10^{-3} sec, so that a large excited population can pile up in ^2E. For laser action this population must exceed that in the ground state. The stored energy in ruby is 10^8 erg cm^{-3} if 10^{20} Cr^{3+} ions per cm^3 are in an excited state. The ruby laser can emit at a very high power level if all this stored energy comes out in a short burst. The overall efficiency of conversion of a ruby laser from input electrical energy to output laser light is about one percent.

Another popular solid state laser is the neodymium glass laser, made of calcium tungstate glass doped with Nd^{3+} ions. This operates as a four level system (Fig. 17). Here it is not necessary to empty out the ground state before laser action can occur.

Semiconductor Junction Lasers

Stimulated emission of radiation can occur at a p-n semiconductor junction from the radiation emitted when electrons recombine with holes (Fig. 18). The pumping action is provided by the dc voltage. The diode wafer provides its own electromagnetic cavity, for the reflectivity at the crystal-air interface is high. Crystals are usually polished or cleaned to provide two flat parallel surfaces; the radiation is emitted in the plane of the junction.

Crystals with direct band gaps (both band edges at the same point in k space) have high recombination probabilities and are required normally for junction lasers. Indirect gaps involve phonons as well as photons; carriers re-

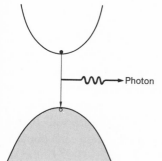

Figure 18a Direct photon emission from the recombination of an electron with a hole.

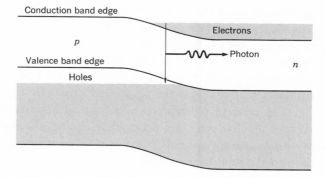

Figure 18b Biased semiconductor junction between a degenerate *p*-type region and a degenerate *n*-type region. Recombination radiation is obtained by biasing the *p-n* junction diode in the forward direction. This injection of minority carriers across the junction permits recombination, with the consequent emission of photons.

combine less efficiently because of competing processes. Gallium arsenide[16] was the first to work and has been widely studied; it emits in the near infrared at about 8383 Å (1.48 eV), with the exact wavelength tunable by variation of temperature or pressure. The gap is direct, with both conduction and valence band edges at $k = 0$. The system is very efficient; the ratio of light energy output to dc electrical energy input is near 50 percent. Other semiconductor junction lasers include GaP ($\lambda = 0.65\mu$) and InSb ($\lambda = 5.3\mu$).

---

[16] For early experiments see R. N. Hall, G. E. Fenner, J. D. Kingsley, T. J. Soltys, and R. O. Carlson, Phys. Rev. Letters **9**, 366 (1962); M. I. Nathan, W. P. Dumke, G. Burns, F. H. Dill, Jr., and G. Lasher, Appl. Phys. Letters **1**, 62 (1962); T. M. Quist, R. H. Rediker, R. J. Keyes, W. E. Krag, B. Lax, A. L. McWhorter, and H. J. Zeigler, Appl. Phys. Letters **1**, 91 (1962).

PHOTOCONDUCTIVITY

Photoconductivity is the increase in electrical conductivity of an insulating crystal caused by radiation incident on the crystal. Much of the pioneering work in the field was done by Gudden, Pohl, and Rose. The photoconductive effect finds practical application in television cameras, infrared detectors, light meters, and indirectly in the photographic process. The direct effect of illumination is to increase the number of mobile charge carriers in the crystal. If the energy of the incident photon is higher than the energy gap E_g, then each photon absorbed in the crystal will produce a free electron-hole pair. The photon is absorbed by raising to the conduction band an electron originally in the valence band. Both the hole in the valence band and the electron in the conduction band may contribute to the conductivity.

The hole and electron will eventually recombine with each other, but they may have quite different histories before recombination. Each may spend various amounts of time trapped locally, perhaps on impurities and imperfections, in the crystal.[17] It is not usual to find that holes and electrons make comparable contributions to the photoconductivity in a given specimen.

The concept of trapping is of central importance in understanding the photoconductive response of a crystal. The mechanism of the atomic processes occurring in traps are not always well understood, but we cannot understand the experimental facts of photoconductivity without invoking the presence of traps. Their role is treated in the following section.

If the energy of the incident photon is below the threshold for the production of pairs of holes and electrons, the photon may be able to cause ionization of donor and acceptor impurity atoms and in this way produce mobile electrons or holes, according to the nature of the impurity.

We discuss first the simplest possible model of a photoconductor. There exist few, if any, realizations of this model in actual crystals, but from the failure of the predictions to apply to real crystals we shall learn how to improve the model. The model (Fig. 19) supposes that electron-hole pairs are produced uniformly throughout the volume of the crystal by irradiation with an external light source. We suppose that recombination occurs by direct annihilation of electrons with holes. We suppose that electrons leaving the crystal at one elec-

---

[17] Crystals (such as AgBr and AgCl) are known in which the hole is trapped immediately after an electron-hole pair is produced. It is believed that the hole may be trapped by any halide ion to form a stable V_K center (Chapter 19).

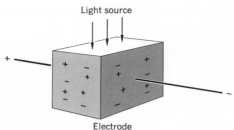

Light source

Electrode

Figure 19 Model of an ideal photoconductor: electron-hole pairs are produced throughout the volume of the crystal by an external light source. Recombination occurs by direct annihilation of electrons with holes. Electrons leaving at one electrode are replaced by others entering from the opposite electrode.

trode are replaced by electrons flowing in from the opposite electrode. It is convenient to neglect the mobility of the holes in comparison with the mobility of the electrons. In many photoconducting substances the mobility of the holes may often be neglected.

On this model the rate of change of the electron concentration n is given by

$$\frac{dn}{dt} = L - Anp = L - An^2 \ , \tag{19}$$

using $n = p$. Here L is the number of photons absorbed per unit volume of specimen per unit time. The term Anp is a bimolecular recombination rate, proportional to the product of hole and electron concentrations.

In the steady state $dn/dt = 0$, so that the steady-state electron concentration is

$$n_0 = (L/A)^{\frac{1}{2}} \ ; \tag{20}$$

the associated conductivity is

$$\sigma = n_0 e\mu = (L/A)^{\frac{1}{2}} e\mu \ , \tag{21}$$

where μ is the electron mobility. This relation predicts that at a given voltage the photocurrent will vary with light level L as $L^{0.5}$; the actual exponents observed run between 0.5 and 1.0, or higher.

The decay of carriers if the light is switched off suddenly is described by

$$\frac{dn}{dt} = -An^2 \ , \tag{22}$$

which has the solution

$$n = \frac{n_0}{1 + Atn_0} \ , \tag{23}$$

where n_0 is the concentration at $t = 0$ when the light was turned off. The carrier concentration drops to $\frac{1}{2}n_0$ in the time

$$t_0 = \frac{1}{An_0} = (LA)^{-\frac{1}{2}} = \frac{n_0}{L} \ . \tag{24}$$

Thus the elementary theory predicts that the **response time** t_0 should be directly proportional to the photoconductivity at a given illumination level: sensitive photoconductors should have long response times. Photographers know that the response time of a very sensitive cadmium sulfide light meter is several seconds. The precise details of the predicted association of these properties is rarely observed in practice.

We define the **sensitivity** or gain factor G as the ratio of the number of carriers crossing the specimen to the number of photons absorbed in the specimen. If the thickness of the specimen is d and the cross-section area is unity, then a potential V produces the particle flux

$$J_N = \frac{n_0 \mu V}{d} = \frac{V\mu}{d^2 (AL)^{\frac{1}{2}}} (Ld) \; , \tag{25}$$

using (20). The sensitivity $G \equiv J_N / Ld$, or

$$G = \frac{V\mu}{d^2 (AL)^{\frac{1}{2}}} \; . \tag{26}$$

The transit time T_d of a carrier between the electrodes is given by

$$T_d = \frac{d}{V\mu/d} = \frac{d^2}{V\mu} \; . \tag{27}$$

The lifetime T_e of an electron before recombination is given by t_0 in (24):

$$T_e = (LA)^{-\frac{1}{2}} \; . \tag{28}$$

The gain (26) may be expressed as

$$G = \frac{T_e}{T_d} \; ; \tag{29}$$

that is, the gain is equal to the ratio of the carrier lifetime to the transit time of a carrier between electrodes. This expression for the gain is quite general and is not limited to the specific model just discussed. But if the lifetime T_e is taken as equal to the observed response time, the gains calculated from (29) are very much larger than observed experimentally, in some instances by a factor of 10^8. A new mechanism must be added to our picture of the photoconductive process. Traps supply the missing mechanism.

Traps

A trap is an atom or imperfection in the crystal capable of capturing an electron or hole. The captured carrier may be re-emitted at a subsequent time. It is convenient to discuss models in which at any instant the holes are entirely trapped and a fraction of the electrons nominally in the conduction band are trapped. We shall discuss only very simple models.

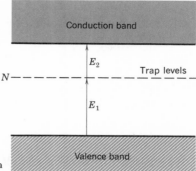

Figure 20   Model for photoconductivity with electron traps in concentration N.

There is an operational distinction between two types of traps. One type acts principally as a recombination center, helping electrons and holes to recombine and thereby assisting in the restoration of thermal equilibrium. Another type of trap affects principally the freedom of motion of charge carriers of one sign. It is this latter variety of trap which we treat.

We consider first a crystal with N electron trap levels per unit volume (Fig. 20). We suppose that the temperature is sufficiently low in relation to the relevant ionization energies so that the concentration of thermal carriers may be neglected. For simplicity we assume that the recombination coefficient A is the same for electron-hole recombination as for electron-trap capture. Then

$$\frac{dn}{dt} = L - An(n + N) \ , \tag{30}$$

where n is the electron concentration in the conduction band. In (30) we have omitted the effect of thermal ionization of carriers from traps back into the conduction band. In the steady state

$$n_0(n_0 + N) = \frac{L}{A} \ . \tag{31}$$

There are two limiting cases to discuss. It is difficult to grow crystals with trap concentrations N much less than 10^{14} cm^{-3}. At low current levels the carrier concentration n_0 may be very much less than this, perhaps only 10^8 or 10^{10} cm^{-3}. In the limit $n_0 \ll N$ we have the result

$$n_0 = \frac{L}{AN} \ , \tag{32}$$

in place of (20). The photocurrent is now directly proportional to the illumination L. At high levels of illumination with $n_0 \gg N$ the response is given by $n_0 = (L/A)^{\frac{1}{2}}$, just as found earlier in the absence of traps. The experimental results indicate just such a change in the response with illumination level.

The decay of the system on switching off the light is given by the solution of the rate solution (30) with $L = 0$:

$$\log \frac{n + N}{n} - \log \frac{n_0 + N}{n_0} = NAt \; . \tag{33}$$

If $N \gg n_0$, the solution reduces to $n = n_0 \exp(-NAt)$, and so the time for the signal to fall to e^{-1} of its initial value is

$$t_0 = \frac{1}{NA} \; , \tag{34}$$

in contrast with the earlier result (9) in the absence of traps. The presence of traps reduces the conductivity and the response time.

The model can be improved by taking account of the ionization of trapped carriers, and one obtains as observed a response time much longer than the carrier lifetime. For details consult the book by Rose (see references).

Space Charge Effects

When the illumination is not uniform througout the crystal or when the electrodes cannot supply or drain off charge carriers freely in the crystal, space charges build up which may reduce the photocurrents severely. Suppose that 300 volts is applied across a crystal slab 1 cm thick by electrodes not in contact with the crystal. This electric field is equivalent to that produced by about 2×10^8 carriers/cm^2 drawn to opposite faces of the slab.

After this charge has collected on the crystal surfaces, the current will stop flowing because the electric field of the surface charges cancels the field applied by the electrodes. The currents and times involved in the production of the surface charge are not large. Such polarization effects are a major obstacle to measurements of photoconductivity; thus pulse methods are often used. The crystal counter using a pulse of carriers is a useful tool in the investigation of mobility and trapping in crystals.

Crystal Counters

The crystal counter detects single ionizing particles by means of the pulse of charge carriers produced on passage of the particle through a crystal slab. The first practical crystal counter detected beta rays passing through a silver chloride crystal. The basic circuit is illustrated in Fig. 21.

The mechanism of counting is simple: the charge carriers produced by the ionizing particle drift under the influence of an applied electric field until they reach the electrodes or are trapped. The net displacement of charge induces a proportional charge on the electrodes. The electrode signal is then amplified.

We now analyze the voltage pulse induced by an alpha particle. The range of natural alpha particles in crystals is usually very small, of the order of 10^{-3} cm. Suppose that n free electrons are produced when one alpha particle pene-

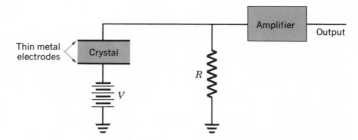

Figure 21   Schematic diagram of a crystal counter.

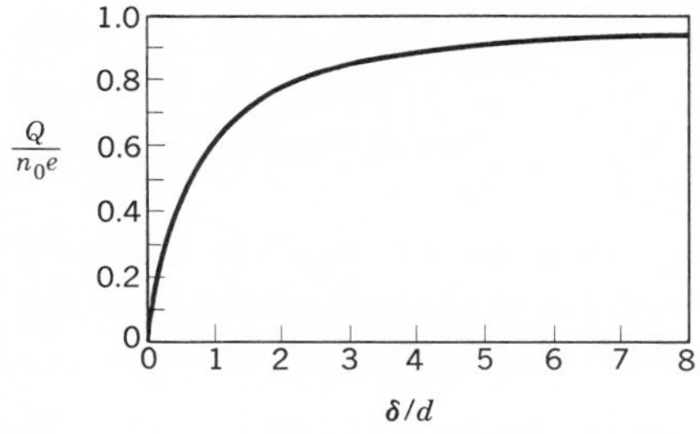

Figure 22   Induced charge on electrodes for an alpha particle incident on a crystal counter as a function of the ratio of electron range δ to crystal thickness d.

trates the negative electrode and stops shortly thereafter. The contribution of the holes to the signal will be neglected, as they are swept up by the cathode near where they are produced; thus they are displaced less than the electrons. An electron moving a distance x across the crystal induces a charge $Q = ex/d$ on the electrodes, where d is the thickness of the crystal. Electrons drifting toward the anode will be trapped along their path with a trapping time T:

$$n = n_0 \exp(-t/T) = n_0 \exp(-x/\mu ET) \;, \tag{35}$$

because $x = \mu Et$, where μ is the mobility and E the electric field intensity. The number trapped in dx at x is $dn = dx\,(n_0/\mu ET) \exp(-x/\mu ET)$.

The total charge appearing on the capacitance is

$$Q = \frac{e}{d} \int x\,dn = \frac{n_0 e}{d\mu\,ET} \int_0^d xe^{-x/\mu ET}\,dx + n_0 e \exp(-d/\mu ET) \;. \tag{36}$$

The second term on the right arises from electrons which get all the way across the crystal to the anode. On integrating we find

$$Q = (n_0 e\mu ET/d)(1 - e^{-d/\mu ET}) \;. \tag{37}$$

The quantity $\delta = \mu ET$ is called the range of the carriers. A plot of Q versus δ/d is given in Fig. 22.

Measurements of the rise time give us the trapping time T, and measurements of the pulse height give us $\delta = \mu ET$; on combining the results we can determine the mobility.

LUMINESCENCE

Luminescence denotes the absorption of energy in matter and its re-emission as visible or near visible radiation. The initial excitation may be by light, particle bombardment, mechanical strain, chemical reaction, or heat. If the emission occurs during excitation, or within 10^{-8} sec of excitation, the process is commonly called **fluorescence.** The interval 10^{-8} sec is chosen as of the order of the lifetime of an atomic state for an allowed electric dipole transition in the visible region of the spectrum. If the emission occurs after excitation has ceased, the process may be called **phosphorescence** or **afterglow.** The delay period may be of the order of microseconds to hours. Crystalline luminescent solids are known as **phosphors.**

Many solids are luminescent with low efficiency for the conversion of other forms of energy into radiation. The ability of a given material to luminesce with high efficiency is frequently related to **activators,** which are special impurity atoms present in small proportions. Luminescent crystals may be divided into two classes: photoconductors (of which Cu-activated ZnS is the prototype), and crystals in which photoconductivity is incidental to the luminescent process.

Thallium-activated Potassium Chloride

Thallium-activated alkali halide phosphors have been studied extensively[18] and provide good examples of phosphors which are not photoconducting. The phosphor KCl:Tl consists of an ionic lattice containing 0.01 percent or less of Tl^+ ions substituted for K^+ ions. Optical absorption of the pure KCl crystal (Fig. 8) begins at 1650 Å and extends to shorter wavelengths. The thallium introduces two bell-shaped absorption bands centered about 1960 Å and 2490 Å, and a broad emission band centered about 3050 Å. The absorption and emission bands are associated with excited states of the Tl^+ ion.

The electronic configuration of the ground state of Tl^+ is $6s^2$, and the state is $^1S_0$. (The spins of the two s electrons are antiparallel.) The lowest excited states arise from the configuration $6s6p$ and consist (Fig. 23) of $^3P_0$, $^3P_1$, $^3P_2$, and $^1P_1$, each separated by the order of 1 volt. The spectroscopic selection rule against transitions from $J = 0$ to $J' = 0$ suggests that the transition $^1S_0 \leftrightarrow {}^3P_0$ does not occur, and the transition $^1S_0 \leftrightarrow {}^3P_2$ is excluded by the general selection rule $\Delta J = 0, \pm 1$. The selection rule $\Delta S = 0$ is not very effective and the transitions $^1S_0 \rightarrow {}^3P_1$ and $^1S_0 \rightarrow {}^1P_1$ have comparable intensities, the former leading to the 2490 Å absorption band and the latter to the 1960 Å absorption band. The emission at 3050 Å is associated with the downward transition $^3P_1 \rightarrow {}^1S_0$, as in Fig. 24.

---

[18] The work is reviewed by F. E. Williams, *Advances in Electronics* **5**, 137 (1953); an early discussion was given by F. Seitz, J. Chem. Phys. **6**, 150 (1938).

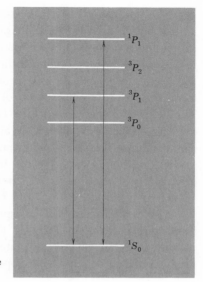

Figure 23 Ground and lowest excited states of the free Tl⁺ ion, with transitions indicated.

Figure 24 Two energy levels of Tl⁺ in KCl plotted as a function of the coordinate representing the symmetric displacement of the six neighboring Cl⁻ ions. The thallium ion in the ground state is close to point A, with some spread about this point caused by the thermal motion of the lattice. When the crystal is irradiated with light near 2490 Å, a transition to the upper state at B may take place. According to the Franck-Condon principle the absorption occurs with the lattice configuration characteristic of the ground state: thus the absorption occurs from A to B, rather than from A to C. After the transition a rearrangement of the neighboring ions takes place and the system assumes the equilibrium position C. The energy difference $B - C$ is dissipated by generation of lattice phonons. From C the ion emits light in a band around 3050 Å, passing to D, and, after giving energy to the lattice, returns to the equilibrium position A. (After F. E. Williams.)

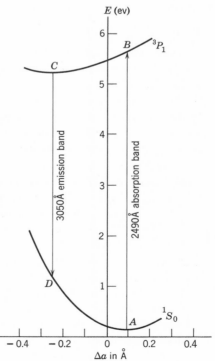

The absorption from the ground state occurs with the neighboring ions in the lattice roughly fixed in position, according to the Franck-Condon[19] principle. In different absorption acts, different positions of the neighbors are involved. The width of the absorption band is accounted for quite well by zero-point and thermal motions of the lattice; these motions cause transitions to occur over a certain spread of values of the configurational coordinate.[20] The lifetime of the excited state (say $^3P_1$) is 10^5 times longer than the period of a lattice vibration; thus after absorption and before re-emission the system can come to thermal equilibrium in the $^3P_1$ state. Luminescent emission occurs from the $^3P_1$ state to the $^1S_0$ state at the value of the configuration coordinate near the minimum energy position of the excited state. The re-arrangement energy $B \to C$ and $D \to A$ is emitted as phonons.

The luminescence of phosphors activated with divalent manganese is somewhat parallel to that of the thallium-activated phosphors. Divalent manganese is an efficient activator in many crystals and finds application in fluorescent lamps and oscilloscope screens.

Problems

1. *Orbital g factor of exciton.* What is the orbital spectroscopic splitting factor of a Mott exciton composed of an electron and hole of effective masses m_e and m_h? (Find the center of mass; calculate the orbital angular momentum about the c.m. for a given angular velocity; calculate the total orbital magnetic moment, bearing in mind that the hole and electron act as if oppositely charged; finally, find g. Do the whole calculation classically.)

2. *Stark effect of Mott exciton.* (a) Taking $m_e = m_h = m$ and $\epsilon = 10$, estimate the electric field intensity required for an exciton to have a first-order Stark splitting of 1 cm^{-1} in the level $n = 2$. (b) Compare this field with that required for atomic hydrogen. (Make use of the *results* of the theory of the Stark effect as given in most elementary texts on quantum theory.)

3. *Sign of photoconducting carriers.* You are given a crystal slab in which the photoconductivity is known to be associated with one carrier type, holes or electrons, but just which is not known. Describe an experiment (using light of a wavelength strongly absorbed in a short distance in the crystal) which will identify the carrier type, without utilizing the Hall effect.

4. *Activator concentration.* Thallous ions Tl^+ are activators for luminescence in KCl. The thallous ions are believed to occupy at random alkali metal sites in the lattice. Let the ratio of concentrations $[Tl^+]/[K^+] \equiv c$. Only those Tl ions which do not have other Tl ions among their nearest cation neighbors are effective activators. Derive and plot an expression for the variation in the concentration c^* of effective activators as a function of c.

---

[19] The Franck-Condon principle states that atoms in molecules do not change their internuclear distances during an electronic transition.

[20] A configurational coordinate is any convenient linear combination of the position vectors of the nuclei of ions in the neighborhood of interest. In our problem Δa is defined as the symmetric displacement from the perfect KCl lattice positions of the six Cl^- ions bounding the Tl^+ ion.

References

R. S. Knox, "Theory of excitons," *Solid state physics,* Supplement 5, 1963.

C. G. Kuper and G. D. Whitfield, ed., *Polarons and excitons,* Plenum, New York, 1963.

D. S. McClure, "Electronic spectra of molecules and ions in crystals," *Solid state physics* 8, 1 (1959).

H. C. Wolf, "The electronic spectra of aromatic molecular crystals," *Solid state physics* 9, 1 (1959).

A. S. Davydov, *Theory of molecular excitons,* McGraw-Hill, 1962.

A. Rose, *Concepts in photoconductivity,* Interscience, 1963.

C. C. Klick and J. S. Schulman, "Luminescence in solids," *Solid state physics* 5, 97 (1957).

D. Fox, M. M. Davis, and A. Weissburger, *Physics and chemistry of the organic solid state,* Interscience, Vol. I (1963), Vol. II (1965), Vol. III (1967).

R. F. Wallis, ed., *Localized excitations in solids,* Plenum Press, 1968.

J. Tauc, ed., *Optical properties of solids,* Academic, 1966.

A. Yariv, *Quantum electronics,* Wiley, 1967.

D. Ross, *Lasers, light amplifiers, and oscillators,* Academic, 1969.

W. V. Smith and P. P. Sorokin, *The laser,* McGraw-Hill, 1966.

C. H. Gooch, ed., *Gallium arsenide lasers,* Wiley-Interscience, 1969.

J. W. Orton, D. H. Paxman, and J. C. Walling, *Solid state maser,* Pergamon, 1970.

19
Point Defects and Alloys

From *Go fly a kite, Charlie Brown,* by kind permission of C. M. Schulz, © 1960 by United Features Syndicate, Inc.

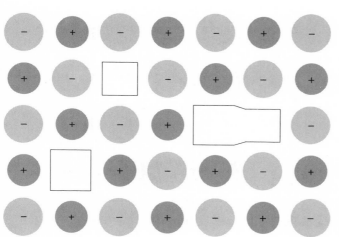

Figure 1 A plane of a pure alkali halide crystal, showing a vacant positive ion site, a vacant negative ion site, and a coupled pair of vacant sites of opposite sign.

Any deviation in a crystal from a perfect periodic lattice or structure is an imperfection. The common point imperfections are chemical impurities, vacant lattice sites, and interstitial atoms (extra atoms not in regular lattice positions). A point imperfection is localized near a point or atom in the structure, in contrast with a line or plane of imperfections. Line imperfections are considered in Chapter 20. Planar imperfections may occur in the initial stages of the formation of a new crystal structure within an existing crystal.

Real crystals are always imperfect in some respect. The nature of the imperfections are fairly well understood for some solids. Much work is concerned with the alkali and silver halides, germanium, silicon, copper, and alloys in general. An alloy represents a high concentration of point imperfections.

Many important properties of solids are controlled as much by imperfections as by the nature of the host crystal, which may act only as a vehicle or solvent or matrix for the imperfections. The conductivity of some semiconductors is due entirely to trace amounts of chemical impurities. The color of many crystals arises from imperfections. The luminescence of crystals is nearly always connected with the presence of impurities. Diffusion of atoms through solids may be accelerated enormously by impurities or imperfections. The mechanical and plastic properties of solids are usually controlled by imperfections.

LATTICE VACANCIES

The simplest imperfection is a **lattice vacancy,** which is a missing atom or ion, known as a **Schottky defect.** A lattice vacancy is often indicated in illustrations and in chemical equations by a square (Fig. 1). We create a Schottky defect in a perfect crystal by transferring an atom from a lattice site in the interior to a lattice site on the surface of the crystal.

In thermal equilibrium in an otherwise perfect crystal a certain number of lattice vacancies are always present, because the entropy is increased by the presence of disorder in the structure. At a finite temperature the equilibrium condition of a crystal is the state of minimum free energy[1] $F = E - TS$. In metals with close-packed structures the proportion of lattice sites vacant at temperatures just below the melting point is of the order of 10^{-3} to 10^{-4}. But in some alloys, in particular the very hard transition metal carbides such as TiC, the proportion of vacant sites of one component can be as high as 50 percent.

---

[1] For simplicity we assume the system is held at constant volume; it is not difficult to treat the system at constant pressure, for which $G = E - TS + pV$ is the appropriate thermodynamic potential.

The probability that a given lattice site is vacant is given simply by the Boltzmann factor for thermal equilibrium at temperature T:

$$P = e^{-E_V/k_BT} , \tag{1}$$

where E_V is the energy required to take an atom from a lattice site inside the crystal to a lattice site on the surface. The energy of a nearest-neighbor bond in a solid is typically of the order of 1 eV. If there are N atoms, the equilibrium number n of vacancies is given by the ratio of vacant to filled sites:

$$\frac{n}{N-n} = e^{-E_V/k_BT} . \tag{2}$$

If $n \ll N$, then

$$\frac{n}{N} \cong e^{-E_V/k_BT} . \tag{3}$$

If $E_V \sim 1$ eV and $T \sim 1000°$K, then $n/N \sim e^{-12} \sim 10^{-5}$. The equilibrium concentration of vacancies decreases as the temperature decreases. The actual concentration of vacancies will be higher than the equilibrium value if the crystal is grown at an elevated temperature and then cooled suddenly, thereby freezing in the vacancies (see the discussion of diffusion below). The thermal generation of lattice vacancies in aluminum is shown in Fig. 2.

In ionic crystals it is usually favorable energetically to form roughly equal numbers of positive and negative ion vacancies. The formation of such pairs of vacancies keeps the crystal electrostatically neutral on a local scale. On carrying out a statistical calculation we obtain

$$n \cong N \exp\left(-E_p/2k_BT\right) , \tag{4}$$

for the number of pairs, where E_p is the energy of formation of a pair.

Another type of vacancy defect is the **Frenkel defect** (Fig. 3) in which an atom is transferred from a lattice site to an **interstitial position,** a position not normally occupied by an atom. The calculation of the equilibrium number of Frenkel defects proceeds along the lines[2] of Problem 1. If the number n of Frenkel defects is much smaller than the number of lattice sites N and the number of interstitial sites N', the result is

$$n \cong (NN')^{\frac{1}{2}} \exp\left(-E_I/2k_BT\right) , \tag{5}$$

where E_I is the energy necessary to remove an atom from a lattice site to an interstitial position.

[2] Another method is given by C. Kittel, Am. J. Physics **35**, 483 (1967).

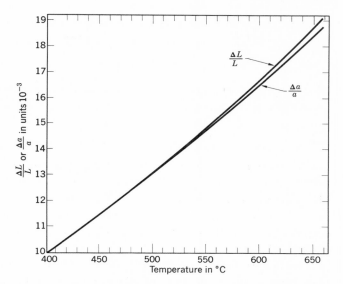

Figure 2 The *difference* between the linear dilation $\Delta L/L$ of the specimen on heating and the fractional lattice parameter change $\Delta a/a$ determined by x-ray diffraction is a measure of the concentration of vacant lattice sites. (The diffraction experiment is not significantly affected by a vacant site, but the length of the specimen is increased when atoms are taken out of lattice sites in the interior of the specimen and placed on the surface.) The graph gives results for aluminum, after R. O. Simmons and R. W. Balluffi, Phys. Rev. **117**, 52 (1960). The vertical scale is normalized to zero at 20°C.

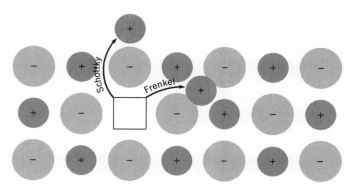

Figure 3 Schottky and Frenkel defects in an ionic crystal. The arrows indicate the displacement of the ions. In a Schottky defect the ion ends up on the surface of the crystal; in a Frenkel defect it is removed to an interstitial position.

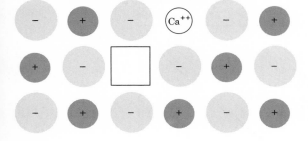

Figure 4 Production of lattice vacancy by the solution of $CaCl_2$ in KCl: to ensure electrical neutrality a positive ion vacancy is introduced into the lattice with each divalent cation Ca^{++}. The two Cl^- ions of $CaCl_2$ enter normal negative ion sites.

It is believed from ionic conductivity studies and density measurements that in pure alkali halides the most common lattice vacancies are Schottky defects; in pure silver halides the most common vacancies are Frenkel defects. The production of Schottky defects lowers the density of the crystal because the volume is increased with no increase in mass. The production of Frenkel defects does not change the volume[3] of crystal; thus the density remains unchanged.

Lattice vacancies in controlled concentrations are present in alkali halides containing additions of divalent elements. If a crystal of KCl is grown with controlled amounts of $CaCl_2$, the density varies as if a K^+ lattice vacancy were formed for each Ca^{2+} ion in the crystal. The Ca^{2+} enters the lattice in a normal K^+ site and the two Cl^- ions enter two Cl^- sites in the KCl crystal (Fig. 4). Demands of charge neutrality results in a vacant metal ion site. The experimental results (Fig. 5) show that the addition of $CaCl_2$ to KCl lowers the density of the crystal. The density would increase if no vacancies were produced, because Ca^{2+} is a heavier and smaller ion than K^+.

The mechanism of electrical conductivity in alkali and silver halide crystals is usually the motion of ions and not the motion of electrons. This has been established by comparing the transport of charge with the transport of mass as measured by the material plated out on electrodes in contact with the crystal. Conductivity by the motion of ions is called **ionic conductivity**.

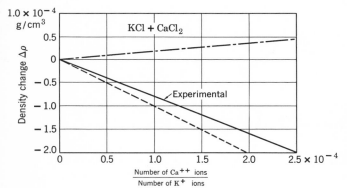

Figure 5 The change in density of KCl with controlled amounts of $CaCl_2$. The full line is experimental; the lower dashed line is the expected shift if the calcium ion plus vacancy occupies the same volume as two potassium ions. The upper hyphenated curve is the expected density if the densities of the two salts are additive.

[3] An interstitial atom takes up space, but the rest of the lattice is compressed. Under certain conditions elasticity theory predicts that the net volume change is zero.

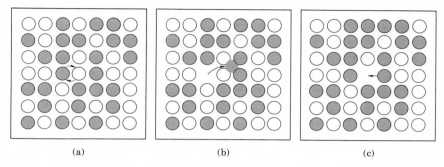

Figure 6 Three basic mechanisms of diffusion: (a) Interchange by rotation about a midpoint. More than two atoms may rotate together. (b) Migration through interstitial sites. (c) Atoms exchange position with vacant lattice sites. (From Seitz.)

The study of ionic conductivity is an important tool in the investigation of lattice defects. Work on alkali and silver halides containing known additions of divalent metal ions such as Cd, Ca, Sr, Ba, Mg shows that at not too high temperatures the ionic conductivity is directly proportional to the amount of divalent addition. This is not because the divalent ions are intrinsically highly mobile; it is predominantly the monovalent metal ion which deposits at the cathode, rather than the divalent addition. Thus **the lattice vacancies introduced with the divalent ions are responsible for the enhanced diffusion** (Fig. 6c). The diffusion of a vacancy in one direction is equivalent to the diffusion of an atom in the opposite direction.

When lattice defects are generated thermally their energy of formation gives an extra contribution to the heat capacity of the crystal, as shown in Fig. 7.

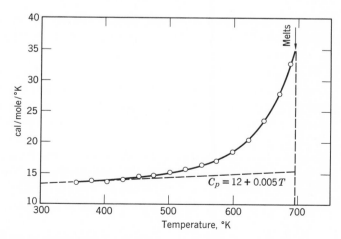

$$C_p = 12 + 0.005\,T$$

Figure 7 Heat capacity of silver bromide at constant pressure, exhibiting the excess heat capacity from the formation of lattice defects, after R. W. Christy and A. W. Lawson, J. Chem. Phys. **19**, 517 (1951).

Breckenridge[4] has noted that an associated pair of vacancies of opposite sign exhibits an electric dipole moment. He has observed contributions to the dielectric constant and dielectric loss at various frequencies in alkali halides which he attributes to the motion of pairs of vacancies; the dielectric relaxation time (Chapter 13) is a measure of the time required for one of the vacant sites to jump by one atomic position with respect to the other. (The dipole moment can change at low frequencies, but not at high.) In sodium chloride the relaxation frequency is 1000 Hz at 85°C.

DIFFUSION

If there is a concentration gradient of impurity atoms or vacancies in a solid, there will be a flux of these moving through the solid. In equilibrium the impurities or vacancies will be distributed uniformly. The net flux J_N of atoms of one species in a solid is related to the gradient of the concentration N of this species by a phenomenological relation called **Fick's law:**

$$\mathbf{J}_N = -D \operatorname{grad} N \ . \tag{6}$$

Here J_N is the number of atoms crossing unit area in unit time; the constant D is the **diffusion constant** or **diffusivity** and has the units cm^2/sec. The minus sign means that diffusion occurs away from regions of high concentration.

The form (6) of the law of diffusion is often adequate, but rigorously the gradient of the chemical potential is the driving force for diffusion and not the concentration gradient alone.[5]

The diffusion constant is often found to vary with temperature as

$$D = D_0 \exp \left(-E/k_B T \right) \ ; \tag{7}$$

here E is the **activation energy** for the process. Experimental results on the diffusion of carbon in alpha iron are shown in Fig. 8. The data are represented by $E = 0.87$ eV, $D_0 = 0.020$ cm^2/sec.

In order to diffuse, an atom must surmount the potential energy barrier presented by its neighbors. We consider the diffusion of impurity atoms in interstitial sites; identical results apply to the diffusion of vacancies. If the barrier is of height E, the atom will have sufficient thermal energy to pass over the barrier only a fraction $\exp \left(-E/k_B T \right)$ of the time. Quantum tunneling through a barrier is usually important only for the lightest nuclei, because for a given energy the de Broglie wavelength increases as the mass of the particle decreases. If ν is a characteristic atomic vibrational frequency, then the prob-

[4] R. G. Breckenridge, J. Chem. Phys. **16,** 959 (1948).

[5] See J. Bardeen and C. Herring, "Diffusion in alloys and the Kirkendall effect," *Imperfections in nearly perfect crystals,* Wiley, 1952.

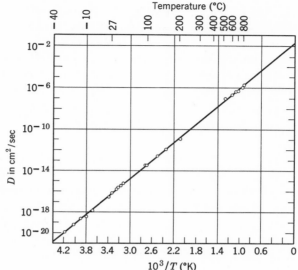

Figure 8 Diffusion coefficient of carbon in iron, after Wert. The logarithm of D is directly proportional to $1/T$. For the diffusion of nitrogen in alpha iron, see A. E. Lord, Jr., and D. N. Beshers, Acta Metal. **14**, 1659 (1966).

ability p that sometime during one second the atom will have enough thermal energy to pass over the barrier is

$$p \approx \nu \exp\left(-E/k_BT\right) \ . \tag{8}$$

In one second the atom makes ν passes at the barrier, with a probability $\exp\left(-E/k_BT\right)$ of surmounting the barrier on each try. The quantity p is also called the *jump frequency*. Values of ν are of the order of 10^{14} Hz.

We consider two parallel planes of impurity atoms in interstitial sites. The planes are separated by lattice constant a. There are S impurity atoms on one plane and $(S + a\, dS/dx)$ on the other. The net number of atoms crossing between the planes in one second is $\approx -pa\, dS/dx$. If N is the total concentration of impurity atoms, then $S = aN$ per cm^2 of a plane. The diffusion flux may now be written as

$$J_N \approx -pa^2 \frac{dN}{dx} \ . \tag{9}$$

On comparison with (6) we have the result

$$\boxed{D = \nu a^2 \exp\left(-E/k_BT\right) \ ,} \tag{10}$$

of the form of (7) with $D_0 = \nu a^2$.

Table 1   Activation energy E_+ for motion of a positive ion vacancy

Values of the energy of formation of a vacancy pair, E_f, are also given. The numbers given in parentheses for the silver salts refer to interstitial silver ions.

Crystal	E_+(eV)	E_f(eV)	Workers
NaCl	0.86	2.02	Etzel and Maurer
LiF	0.65	2.68	Haven
LiCl	0.41	2.12	Haven
LiBr	0.31	1.80	Haven
LiI	0.38	1.34	Haven
KCl	0.89	2.1–2.4	Wagner; Kelting and Witt
AgCl	0.39(0.10)	1.4°	Teltow
AgBr	0.25(0.11)	1.1°	Compton

° For Frenkel defect.

If the impurities are charged, we may find the ionic mobility μ and the conductivity σ from the diffusivity by using the Einstein relation $k_B T\mu = qD$:

$$\mu = (qva^2/k_BT) \exp{(-E/k_BT)} \; ; \tag{11}$$

$$\sigma = Nq\mu = (Nq^2\,va^2/k_BT) \exp{(-E/k_BT)} \; , \tag{12}$$

where now N is the concentration of impurity ions of charge q.

The proportion of vacancies is independent of temperature in the temperature range in which the number of vacancies is determined by the number of divalent metal ions. In this range the slope of a plot of log σ versus $1/k_BT$ gives E_+, the barrier activation energy for the jumping of positive ion vacancies (Table 1). At room temperature the jump frequency is of the order of 1 sec^{-1}, and at 100°K it is of the order of 10^{-25} sec^{-1}. We see that diffusion is very slow at low temperatures.

The proportion of vacancies in the temperature range in which the concentration of defects is determined by thermal generation is given by

$$f \cong \exp{(-E_f/2k_BT)} \; , \tag{13}$$

where E_f is the energy of formation of a vacancy pair, according to the theory of Schottky or Frenkel defects. Here the slope of a plot of log σ versus $1/k_BT$ will be $E_+ + \frac{1}{2}E_f$, according to (11) and (13). From measurements in different temperature ranges we determine the energy of formation of a vacancy pair E_f and the jump activation energy E_+.

The diffusion constant can be measured directly by radioactive tracer techniques. The diffusion of a known initial distribution of radioactive ions is followed as a function of time or distance. Values of the diffusion constant thus determined may be compared with values from ionic conductivities. The two sets of values do not usually agree within the experimental accuracy, so that there may also be present a diffusion mechanism which does not involve the transport of charge. For example, the diffusion of pairs of positive and negative ion vacancies or the diffusion of an associated complex of a divalent ion and a vacancy do not involve the transport of charge.

For the vacancy diffusion mechanism there should be a small difference between the diffusion constant as measured by radioactive atoms and as deduced from the ionic conductivity. Such differences were observed by Johnson.[6]

Metals

Self-diffusion in monatomic metals most commonly proceeds by lattice vacancies. **Self-diffusion** means the diffusion of atoms of the metal itself, and not of impurities. According to calculations by Huntington[7] the activation energy for self-diffusion in copper is expected to be in the range 2.4 to 2.7 eV for diffusion through vacancies and 5.1 to 6.4 eV for diffusion through interstitial sites. Observed values of the activation energy are 1.7 to 2.1 eV.

Activation energies for diffusion in Li and Na can be determined from measurements of the temperature dependence of the nuclear resonance line width. As discussed in Chapter 17, the resonance line width narrows when the jump frequency of an atom between sites becomes rapid in comparison with the frequency corresponding to the static line width. The values 0.57 eV and 0.45 eV were determined by NMR for Li and Na by Holcomb and Norberg.[8] Self-diffusion measurements for sodium also give 0.45 eV.

COLOR CENTERS

Pure alkali halide crystals are transparent throughout the visible region of the spectrum. The crystals may be colored in a number of ways: (a) by the introduction of chemical impurities; (b) by introducing an excess of the metal ion (we may heat the crystal in the vapor of the alkali metal and then cool it quickly—an NaCl crystal heated in the presence of sodium vapor becomes yellow; a KCl crystal heated in potassium vapor becomes magenta); (c) by x-ray, γ-ray, neutron, and electron bombardment; and (d) by electrolysis.

A **color center** is a lattice defect which absorbs visible light. An ordinary lattice vacancy does not color alkali halide crystals, although it affects the absorption in the ultraviolet.

---

[6] W. A. Johnson, Trans. AIME **147**, 331 (1943). The diffusion constant is lower for diffusion of a tracer atom than for a diffusion of a vacancy. Suppose that a tracer atom has jumped forward by trading places with a vacancy. The tracer atom now is not in a random position with respect to the vacancy, for the vacancy is directly behind the tracer. There is consequently a probability that the vacancy will be filled by the tracer atom jumping backwards.

[7] Vacancy formation in Cu has been reviewed by W. M. Lomer, *Prog. metal physics* 8 (1959).

[8] D. F. Holcomb and R. E. Norberg, Phys. Rev. **93**, 919 (1954).

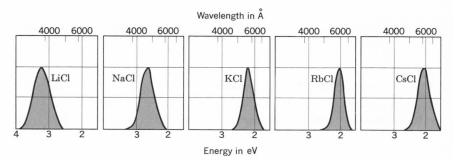

Figure 9 The *F* bands for several alkali halides: optical absorption versus wavelength for crystals which contain *F* centers.

F Centers

The simplest color center is an *F* center. The name comes from the German word for color, *Farbe*. We usually produce *F* centers by heating the crystal in excess alkali vapor or by x-irradiation. The central absorption band (*F* band) associated with *F* centers in several alkali halides are shown in Fig. 9, and the quantum energies are listed in Table 2. Experimental properties of *F* centers have been investigated in detail, originally by Pohl.

Table 2   Experimental *F* center
absorption energies in eV

LiCl	3.1	NaBr	2.3
NaCl	2.7	KBr	2.0
KCl	2.2	RbBr	1.8
RbCl	2.0	LiF	5.0
CsCl	2.0	NaF	3.6
LiBr	2.7	KF	2.7

As discussed in Chapter 17, the *F* center has been identified by electron spin resonance as an electron bound at a negative ion vacancy (Fig. 10) in agreement with a model suggested by de Boer. When excess alkali atoms are added to an alkali halide crystal, a corresponding number of negative ion vacancies are created. The valence electron of the alkali atom is not bound to the atom; the electron migrates in the crystal and becomes bound to a vacant

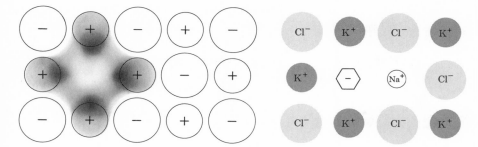

Figure 10   An *F* center is a negative ion vacancy with one excess electron bound at the vacancy. The distribution of the excess electron is largely on the positive metal ions adjacent to the vacant lattice site.

Figure 11   An *F_A* center in KCl: one of the six K⁺ ions which bound an *F* center is replaced by another alkali ion, here Na⁺. The hexagon is used here to denote a vacant negative ion site with a trapped electron.

negative ion site. (A negative ion vacancy in a perfect periodic lattice has the effect of an isolated positive charge:[9] it attracts and binds an electron.)

The model is consistent with the experimental facts:

a. The *F* band optical absorption is characteristic of the crystal and not of the alkali metal used in the vapor; that is, the band in potassium chloride is almost exactly the same whether the crystal is heated in potassium or sodium vapor. (The role of the alkali vapor is to produce an *F* center in the host crystal.)

b. Crystals colored by heating in alkali vapor are found by chemical analysis to contain an excess of alkali metal atoms, typically 10^{16} to 10^{19} per cm³. The integrated spectral absorption in the *F* band corresponds quantitatively to that expected from the known amount of excess alkali metal.

c. Colored crystals are less dense than uncolored crystals. This agrees with the elementary picture that the introduction of vacancies should lower the density.

Other Centers in Alkali Halides

The *F* center is the simplest trapped-electron center in alkali halide crystals. The optical absorption of an *F* center arises from an electric dipole transition to a bound excited state[10] of the *F* center.

Two absorption bands identify the *F_A* center, in contrast to the single absorption band of the *F* center. In the *F_A* center one of the six nearest neighbors of the *F* center has been replaced by a different[11] alkali ion, as in Fig. 11.

---

[9] We can simulate the electrostatic effect of a negative ion vacancy by adding a positive charge q to the normal charge $-q$ of an occupied negative ion site.

[10] R. K. Swank and F. C. Brown, Phys. Rev. **130**, 34 (1963).

[11] F. Lüty and H. Pick, J. Phys. Soc. Japan **18**, Suppl. II 240 (1963).

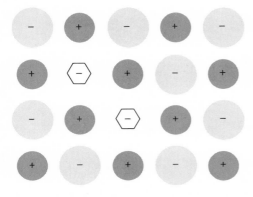

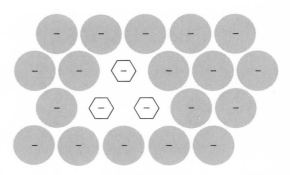

Figure 12   An M center consists of two adjacent F centers.

Figure 13   An R center consists of three adjacent F centers; that is, a group of three negative ion vacancies in a [111] plane of the NaCl structure, with three electrons associated with the group. (The [111] planes above and below this are made up of positive ions.)

Complex trapped-electron centers are formed by groups of F centers, as in Figs. 12 and 13. Two adjacent F centers form an M center; three F centers form an R center. These and other centers are usually identified by their optical properties.

Holes may be trapped also to form color centers. Hole centers differ somewhat from electron centers: a hole in the filled p^6 shell of a halogen ion leaves the halogen in a p^5 electron configuration, whereas an electron added to the filled p^6 shell of an alkali ion leaves the ion in a p^6s configuration. The chemistry of the two configurations is different. The antimorph to the F center is a hole trapped at a positive ion vacancy, but no such center has been identified in experiments.

The best-known trapped-hole center is the V_K center (Fig. 14). It is believed that a hole may be trapped by any halogen ion in a perfect alkali halide crystal to form a V_K center. The structure of the V_K center is known from electron spin resonance experiments[12] to resemble a negative halogen molecule ion: in KCl the V_K center is like a Cl_2^- ion. Results are shown in Fig. 15.

ALLOYS

The band structure of solids developed in Chapters 9 and 10 is based on the assumption that the crystal is periodic under the translations of the primitive lattice. What happens to the band structure if the crystal is not periodic, but contains impurities which occupy lattice sites at random, or if the crystal is an alloy of two elements? The translational symmetry of the lattice is no longer

---

[12] W. Känzig, Phys. Rev. **99**, 1890 (1955); T. G. Castner and W. Känzig, J. Phys. Chem. Solids **3**, 178 (1957). M. H. Cohen, Phys. Rev. **101**, 1432 (1958). The mechanism responsible for the trapping is the Jahn-Teller effect (Advanced Topic M). Calculations on the nature of the V_K center are given by T. P. Das, A. N. Jette, and R. S. Knox, Phys. Rev. **134**, A1079 (1964), and by D. F. Daly and R. L. Mieher, Phys. Rev. Letters **19**, 637 (1967).

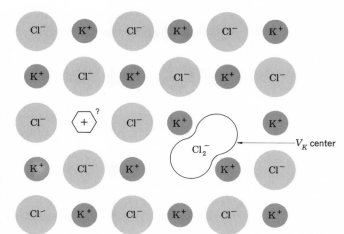

Figure 14   A V_K center is formed when a hole is trapped by a pair of negative ions. The stable condition resembles a negative halogen molecule-ion, which is Cl_2^- in KCl. No lattice vacancies or extra atoms are involved in a V_K center. The center at the left of the figure probably is not stable: the hexagon represents a hole trapped near a positive ion vacancy; such a center would be the antimorph to an F center. Holes have a lower energy trapped in a V_K center than in an anti-F center.

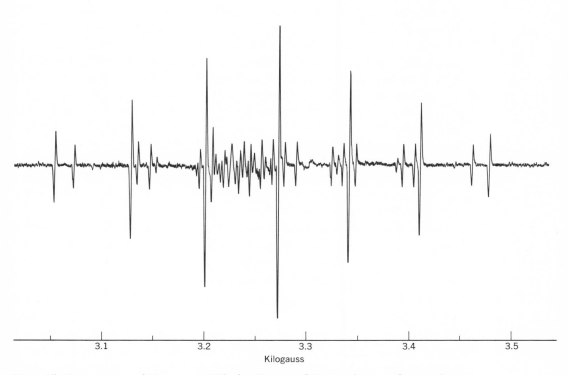

Kilogauss

Figure 15   Spin resonance of V_K centers in KCl, after Castner and Känzig. The seven dominant lines are the hyperfine components of an electron interacting with the nuclear magnetic moments of two Cl^{35} nuclei in a Cl_2^- molecule. Each nucleus has a spin of $\frac{3}{2}$; there are seven possible arrangements of the spins of the two nuclei. The spectrum is analyzed in detail in Sec. 7.4 of C. P. Slichter, *Principles of magnetic resonance*, Harper and Row, 1963.

perfect. Can we expect that all consequences of band theory, such as the Fermi surface and energy gaps, will no longer exist? Will insulators become conductors?

Experiment and theory[13] agree that the consequences of the destruction of the translational symmetry are much weaker than one expects at first sight. If the impurity belongs to the same column of the periodic table as the host element it replaces, then the effects are particularly small, in part because the average number of valence electrons remains constant.

One measure of the effect of alloying is given by the residual resistivity. One atomic percent of copper dissolved in silver increases the residual electrical resistivity by 0.07 μohm-cm, which corresponds to a scattering cross-section of perhaps 0.03 of the area of the impurity atom. Similarly, the mobility of electrons in Si–Ge alloys is much higher than one would expect from the simple geometrical argument that a germanium atom (32 electrons) is very unlike a silicon atom (14 electrons) and therefore a Si atom in Ge or a Ge atom in Si should act as an efficient scattering center for charge carriers. The theory shows us that the effective scattering potential of an impurity may be very small.

There is no experimental evidence for an intrinsic reduction in band gap due to the random aspects of alloying. For example, silicon and germanium form solid solutions over the entire composition range; the energies of the band edges in the alloys vary continuously with composition (Fig. 16). It is believed that the density of orbitals near the band edges are somewhat smudged out by alloying.

We now discuss substitutional solid solutions of one metal A in another B of different valence. We suppose that atoms A and B occupy equivalent lattice positions at random. The distinct effects which occur when the occupancies are regular and not random are considered under the heading of the order-disorder transformation.

Hume-Rothery has discussed empirical requirements for solid solutions to occur. One requirement is size. It is difficult to form solid solutions if the atomic diameters[14] of A and B differ by more than 15 percent. The sizes are favorable in the Cu (2.55 Å)–Zn (2.66 Å) system: zinc dissolves in copper as a fcc solid solution up to 38 atomic percent zinc. The sizes are somewhat unfavorable in the Cu (2.55 Å)–Cd (2.97 Å) system: only 1.7 atomic percent

---

[13] These questions can be discussed using the method of orthogonalized plane waves and effective potentials described in Chapter 10. A low concentration of impurity atoms cannot have much effect on the Fourier components U_G of the potential $U(\mathbf{r})$ responsible for the band gaps and for the behavior of the energy surfaces near the band gap: An impurity will introduce Fourier components of $U(\mathbf{r})$ at wavevectors which are not reciprocal lattice vectors, but such components are never large if the impurity atoms are located at random. Thus one obtains sharp x-ray diffraction lines from random alloys. See also the references toward the end of Chapter 11.

The atomic diameter is taken as the closest distance of approach in the crystal structure of the element (Table 1.5).

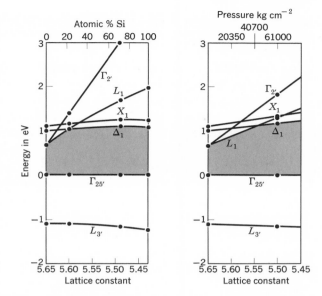

Figure 16 (a) Calculated variation of the major band edges in Ge-Si alloys as a function of the silicon concentration. The point $\Gamma_{25'}$ in **k** space is the valence band edge in both Ge and Si; L_1 is the conduction band edge in Ge. (Compare with Fig. 11.15.) The calculations are by F. Bassani and D. Brust, Phys. Rev. **131**, 1524 (1963), who give extensive references to the experimental data. (b) The calculated variation as a function of pressure for pure Ge. Qualitatively the variation of the bands with pressure is similar to the variation with alloying, as we see by comparing energy values at equal values of the lattice constant.

cadmium is soluble in copper. The atomic diameters referred to copper are 1.04 for zinc and 1.165 for cadmium.

Although the sizes may be favorable, solid solutions will not form if there is a strong tendency for A and B to form stable compounds of definite chemical proportions. If A is strongly electronegative and B strongly electropositive, it is likely that compounds such as AB and A_2B will precipitate out of solution. Although the atomic diameter ratio is favorable (1.02) for As in Cu, only 6 percent As is soluble. The ratio is also favorable (1.09) for Sb in Mg, yet the solubility of Sb in Mg is very small.

Several aspects of the electronic structure of alloys can be discussed in terms of the average number[15] of conduction electrons per atom, denoted by n. The value of n in the alloy 50 percent Cu–50 percent Zn is 1.50; in 50 Cu–50 Al, $n = 2.00$. Some of the principal effects of alloying elements of different valence come from the change in electron concentration.[16] Hume-Rothery first drew attention to the importance of the average electron concentration as determining structural changes in certain alloy systems.

[15] This number is often called the **electron concentration.**
[16] J. Friedel, Advances in Physics **3**, 446 (1954).

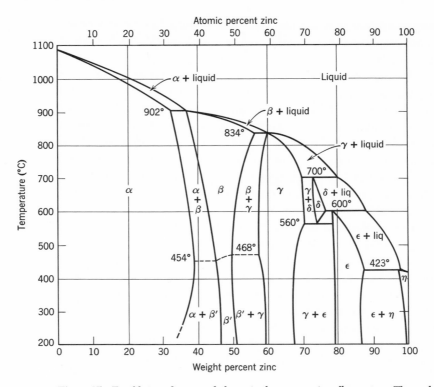

Figure 17 Equilibrium diagram of phases in the copper-zinc alloy system. The α phase is fcc; β and β' are bcc; γ is a complex structure; ϵ and η are both hcp, but ϵ has a c/a ratio near 1.56 and η (for pure Zn) has $c/a = 1.86$. The β' phase is ordered bcc, by which we mean that most of the Cu atoms occupy sites on one sc sublattice and most of the Zn atoms occupy sites on a second sc sublattice which interpenetrates the first sublattice. The β phase is disordered bcc: any site is equally likely to be occupied by a Cu or a Zn atom, almost irrespective of what atoms are in the neighboring sites.

The phase diagram of the copper-zinc system[17] is shown in Fig. 17. The fcc structure of pure copper ($n = 1$) persists on the addition of zinc ($n = 2$) until the electron concentration reaches 1.38. A bcc structure occurs at a minimum electron concentration of about 1.48. The γ phase exists for the approximate range of n between 1.58 and 1.66, and the hcp phase ϵ occurs near 1.75.

The term **electron compound** denotes an intermediate phase (such as the β phase of CuZn) whose crystal structure is determined by a fairly well-defined electron to atom ratio. The values of the ratio are called the **Hume-Rothery rules;** they are 1.50 for the β phase, 1.62 for the γ phase, and 1.75 for

---

[17] The phases of interest are usually denoted by metallurgists by Greek characters: in the Cu-Zn system we have α (fcc), β (bcc), γ (complex cubic cell of 52 atoms), ϵ (hcp) and η (hcp); ϵ and η differ considerably in c/a ratio. The meaning of the characters depends on the alloy system.

the ϵ phase. Representative experimental values are collected in Table 3, based on the usual chemical valence of 1 for Cu and Ag; 2 for Zn and Cd; 3 for Al and Ga; 4 for Si, Ge, and Sn.

The Hume-Rothery rules find a simple expression in terms of the band theory of nearly free electrons. The observed limit of the fcc phase occurs close to the electron concentration of 1.36 at which an inscribed Fermi sphere makes contact with the Brillouin zone boundary for the fcc lattice. The observed electron concentration of the bcc phase is close to the concentration 1.48 at which an inscribed Fermi sphere makes contact with the zone boundary for the bcc lattice. Contact of the Fermi sphere with the zone boundary for the γ phase is at the concentration 1.54. Contact for the hcp phase is at the concentration 1.69 for the ideal c/a ratio.

Why is there a connection between the electron concentration at which a new phase appears and the electron concentration at which the Fermi surface makes contact with the Brillouin zone boundary? It is costly in energy to add further electrons to an alloy once the filled states reach the zone boundary. Additional electrons can be accommodated only in states above the energy gap which occurs at the boundary or in the states of high energy near the corners of the lower zone. It may therefore be energetically favorable for the crystal structure to change to one which can contain a larger Fermi surface before contact. In this way the sequence fcc, bcc, γ, hcp was made plausible by H. Jones.

Table 3 Electron/atom ratios of electron compounds

Alloy	fcc phase boundary	Minimum bcc phase boundary	γ-phase boundaries	hcp phase boundaries
Cu–Zn	1.38	1.48	1.58–1.66	1.78–1.87
Cu–Al	1.41	1.48	1.63–1.77	
Cu–Ga	1.41			
Cu–Si	1.42	1.49		
Cu–Ge	1.36			
Cu–Sn	1.27	1.49	1.60–1.63	1.73–1.75
Ag–Zn	1.38		1.58–1.63	1.67–1.90
Ag–Cd	1.42	1.50	1.59–1.63	1.65–1.82
Ag–Al	1.41			1.55–1.80

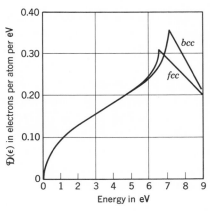

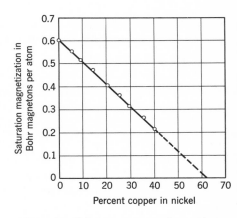

Figure 18 Number of free electron orbitals per unit energy range for the first Brillouin zone of the fcc and bcc lattices, as a function of energy.

Figure 19 Bohr magneton numbers of ferromagnetic nickel-copper alloys.

The transformation from fcc to bcc is illustrated by Fig. 18; this shows the number of orbitals per unit energy range as a function of energy, for the fcc and bcc structures. As the number of electrons is increased, a point is reached where it is easier to accommodate additional electrons in the Brillouin zone of the bcc lattice rather than in the Brillouin zone of the fcc lattice. The figure is drawn for free electrons, and it may perhaps be objected that the actual density of orbitals in pure fcc metals is somewhat different—see the Fermi surface of Cu in Figs. 10.26, 10.35, and 10.38.

We consider the relationship of the s and d bands in pure nickel at $0°K$ as suggested by Fig. 16.7b. There is a certain arbitrariness in the proposed distribution: we can transfer electrons from the d sub-bands to the s band provided we take 0.54 more from one d sub-band than from the other d sub-band. Evidence that our particular choice may correspond to reality is provided by Fig. 19 which shows the effect on the magneton number of adding copper to nickel. We add one extra electron with each copper atom because the atomic number of copper is larger by one than the atomic number of nickel. The density of electron orbitals in the d band is over ten times greater than in the s band, so that the extra electron goes at least 90 percent into the d band and less than 10 percent into the s band. The ferromagnetic magneton number is observed to go to zero at about 60 atomic percent copper.

At 60 atomic percent copper we have added about 0.54 electron per atom to the d band and about 0.06 electron to the s band. But 0.54 electron added to the d band of Fig. 16.7b will just fill both d sub-bands and will bring the magnetization to zero, in excellent agreement with observation. The distribution of electrons for 60Cu40Ni is shown in Fig. 20. At low Cu concentrations the decrease in magneton number from that of pure Ni should be a linear function of the concentration of copper, in agreement with Fig. 19.

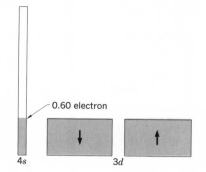

Figure 20 Distribution of electrons in the alloy 60Cu40Ni. The extra 0.6 valence electron provided by the copper has filled the d band entirely and increased slightly the number of electrons in the s band with respect to Fig. 16.7b.

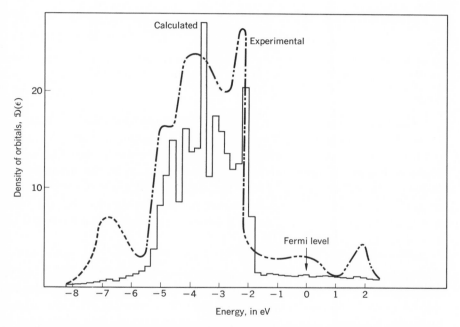

Figure 21 Density of orbitals in copper $3d$ and $4s$ bands. Calculations after C. Y. Fong and M. L. Cohen (unpublished); experimental, after W. E. Spicer, in *International colloquium on optical properties and electronic structure of metals and alloys*, F. Abeles, ed., North-Holland, 1966.

For simplicity the block drawings show the density of orbitals as uniform in energy. The actual density may be quite far from uniform: Fig. 21 gives the results of a calculation for copper. The d band is characterized by a high density of orbitals. The density of orbitals at the Fermi surface determines the electronic heat capacity and the Pauli paramagnetic susceptibility; in the transition metals they have higher values than in monovalent metals.

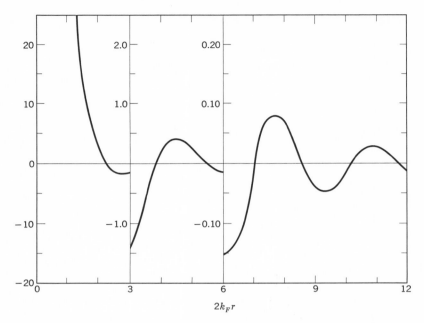

Figure 22 Magnetization of a free electron gas in neighborhood of a point magnetic moment at the origin $r = 0$, according to the RKKY theory. The horizontal axis is $2k_F r$, where k_F is the wavevector on the Fermi sphere. (After de Gennes.)

Magnetic Alloys and the Kondo Effect

In dilute solid solutions of a magnetic ion in a nonmagnetic metal crystal (such as Mn in Cu) the exchange coupling between the ion and the conduction electrons has important consequences. The conduction electron gas is magnetized in the vicinity of the magnetic ion, with the spatial dependence shown in Fig. 22. This magnetization causes an indirect exchange interaction[18] between two magnetic ions, because a second ion perceives the magnetization induced by the first ion. The interaction, known as the **RKKY interaction,** also plays a role in the magnetic spin order of the rare earth metals, where the spins of the $4f$ ion cores are coupled together by the magnetization induced in the conduction electron gas.

A dramatic consequence of the magnetic ion-conduction electron interaction is the **Kondo effect.** A minimum in the resistivity-temperature curve of dilute magnetic alloys at low temperatures has been observed in alloys of

---

[18] A review of indirect exchange interactions in metals is given by C. Kittel, *Solid state physics* **22**, 1 (1968); a review of the Kondo effect is given by J. Kondo, "Theory of dilute magnetic alloys," *Solid state physics* **23**, 184 (1969) and A. J. Heeger, "Localized moments and nonmoments in metals: the Kondo effect," *Solid state physics* **23**, 248 (1969).

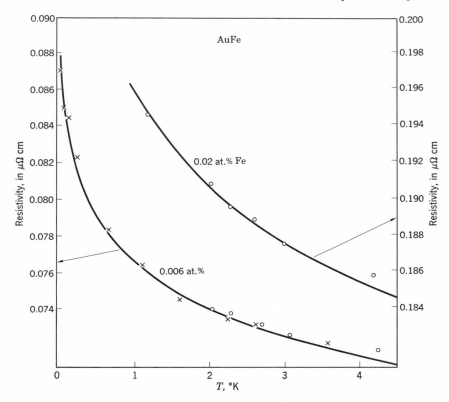

Figure 23   A comparison of experimental and theoretical results for the increase of electrical resistivity at low temperatures in dilute alloys of iron in gold. The resistance minimum lies to the right of the figure, for the resistivity increases at high temperatures because of scattering of electrons by thermal phonons. The experiments are due to D. K. C. MacDonald, W. B. Pearson, and I. M. Templeton, Proc. Roy. Soc. (London) **A266**, 161 (1962); the theory is by J. Kondo, Prog. Theo. Physics **32**, 37 (1964).

Cu, Ag, Au, Mg, Zn with Cr, Mn, Fe, Mo, Re, and Os as impurities, among others. The occurrence of a resistance minimum is connected with the existence of localized magnetic moments on the impurity atoms. Where a resistance minimum is found, there is inevitably a local moment. Kondo showed that the anomalously high scattering probability of magnetic ions at low temperatures is an esoteric consequence of the dynamic nature of the scattering and of the sharpness of the Fermi surface at low temperatures. The temperature region in which the Kondo effect is important is shown in Fig. 23. No simple physical explanation of the effect has appeared, but the original paper is quite accessible.

The central result is that the spin-dependent contribution to the resistivity is

$$\rho_{\text{spin}} = c\rho_M\left[1 + \frac{3zJ}{\epsilon_F}\log T\right] , \qquad (14)$$

where J is the exchange energy; z the number of nearest neighbors; c the concentration; and ρ_M is a measure of the strength of the exchange scattering. We see that the spin resistivity increases toward low temperatures if J is negative. If the phonon resistivity goes as T^5 in the region of interest, then the total resistivity has the form

$$\rho = aT^5 + c\rho_0 - c\rho_1 \log T , \qquad (15)$$

with a minimum at

$$\frac{d\rho}{dT} = 5aT^4 - \frac{c\rho_1}{T} = 0 , \qquad (16)$$

whence

$$T_{\min} = \left(\frac{c\rho_1}{5a}\right)^{\frac{1}{5}} . \qquad (17)$$

The temperature at which the resistivity is a minimum varies as the one-fifth power of the concentration of the magnetic impurity, in agreement with experiment.

ORDER-DISORDER TRANSFORMATION

The dashed horizontal line in the β phase region of the phase diagram of the Cu–Zn (Fig. 17) represents the transition temperature between ordered and disordered states of the alloy. Consider an alloy composed of equal numbers of two types of metal atoms, A and B. The alloy is said to be **ordered** if the A and B atoms have a regular periodic arrangement with respect to one another, as in Fig. 24a. The alloy is **disordered** if the A and B are randomly arranged, as in Fig. 24b. Many properties of an alloy are sensitive to the degree of order.

In the common ordered arrangement of an AB alloy with a bcc structure all the nearest-neighbor atoms of a B atom are A atoms, and vice versa. This arrangement results when the dominant interaction among the atoms is an attraction between A and B atoms. (If the dominant interaction is a repulsion between A and B atoms, a two-phase system is formed in which some crystallites are largely A and other crystallites are largely B.)

The alloy is completely ordered at absolute zero. It becomes less ordered as the temperature is increased, until a transition temperature is reached above which the structure is disordered. The transition temperature marks the dis-

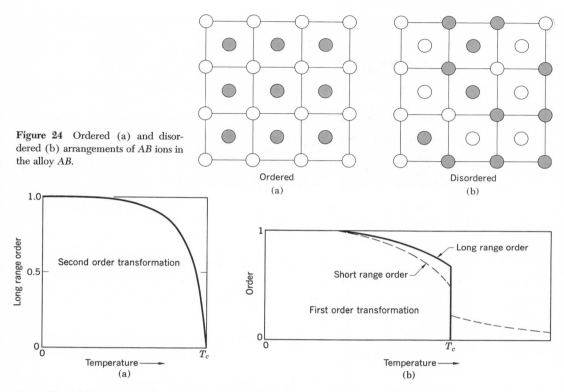

Figure 24 Ordered (a) and disordered (b) arrangements of *AB* ions in the alloy *AB*.

Ordered
(a)

Disordered
(b)

Figure 25 (a) Long-range order versus temperature for an *AB* alloy. The transformation is second order in the sense of Chapter 14. (b) Long-range and short-range order for an AB_3 alloy, after Nix and Shockley. The transformation for this composition is first order.

appearance of **long-range order,** which is order over many interatomic distances, but some **short-range order** or correlation among near neighbors may persist above the transition. The long-range order in an *AB* alloy is shown in Fig. 25a. Long- and short-range order for an alloy of composition AB_3 is given in Fig. 25b. The degree of order is defined below.

If an alloy is cooled rapidly from high temperatures to a temperature below the transition, a metastable condition may be produced in which a non-equilibrium disorder is frozen in the structure. The reverse effect occurs when an ordered specimen is disordered at constant temperature by heavy irradiation with nuclear particles.

The degree of order may be investigated experimentally by x-ray diffraction. The disordered structure in Fig. 24b has diffraction lines at the same positions as if the lattice points were all occupied by only one type of atom, because the effective scattering power of each plane is equal to the average of the *A* and *B* scattering powers. The ordered structure in Fig. 24a has extra diffraction lines not possessed by the disordered structure. The extra lines are called **superstructure lines.**

Figure 26 X-ray powder photographs in AuCu₃ alloy. (a) Disordered by quenching from $T > T_c$; (b) ordered by annealing at $T < T_c$. (Courtesy of G. M. Gordon.)

The structure of the ordered CuZn alloy is the cesium chloride structure (Fig. 1.26). The space lattice is simple cubic, and the basis has one Cu atom at 000 and one Zn atom at $\frac{1}{2}\frac{1}{2}\frac{1}{2}$. The diffraction structure factor (2.61) becomes

$$\mathcal{S}(hkl) = f_{Cu} + f_{Zn}e^{-i\pi(h+k+l)} . \tag{18}$$

This cannot vanish because $f_{Cu} \neq f_{Zn}$; therefore all reflections of the simple cubic space lattice will occur.

In the disordered structure the situation is different: the basis is equally likely to have either Zn or Cu at 0 0 0 and either Zn or Cu at $\frac{1}{2}\frac{1}{2}\frac{1}{2}$. Then the average structure factor is

$$\langle \mathcal{S}(hkl) \rangle = \langle f \rangle + \langle f \rangle e^{-i\pi(h+k+l)} , \tag{19}$$

where $\langle f \rangle = \frac{1}{2}(f_{Cu} + f_{Zn})$. Equation (19) is exactly the form of the result for the bcc lattice; the reflections vanish when $h + k + l$ is odd. We see that the ordered lattice has reflections (the superstructure lines) not present in the disordered lattice (Fig. 26).

Elementary Theory of Order

We give a simple statistical treatment of the dependence of order on temperature for an AB alloy with a bcc structure. The case A_3B differs[19] from AB, the former having a first-order transition marked by a latent heat and the latter having a second-order transition marked by a discontinuity in the heat capacity (Fig. 27).

We introduce a measure of the long-range order. We call one simple cubic lattice a and the other b: the bcc structure is composed of the two interpenetrating sc lattices, and the nearest neighbors of an atom on one lattice lie on

---

[19] The connection between composition and the order of the phase transition is discussed by S. Strässler and C. Kittel, Phys. Rev. **139**, A758 (1965).

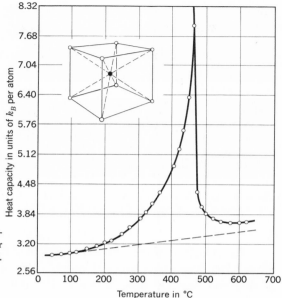

Figure 27 Heat capacity versus temperature of CuZn (β-brass) alloy, after F. C. Nix and W. Shockley, Revs. Mod. Physics **10**, 1 (1938).

the other lattice. If there are N atoms A and N atoms B in the alloy, the **long-range order parameter** P is defined so that the number of A's on the lattice a is equal to $\frac{1}{2}(1 + P)N$. The number of A's on lattice b is equal to $\frac{1}{2}(1 - P)N$. When $P = \pm 1$, the order is perfect and each lattice contains only one type of atom. When $P = 0$, each lattice contains equal numbers of A and B atoms and there is no long-range order.

We consider that part of the internal energy associated with the bond energies of AA, AB, and BB nearest-neighbor pairs, although this concept is an oversimplification. The total bond energy is

$$E = N_{AA}U_{AA} + N_{BB}U_{BB} + N_{AB}U_{AB} \; , \tag{20}$$

where N_{ij} is the number of nearest-neighbor ij bonds, and U_{ij} is the energy of an ij bond.

The probability that an atom A on lattice a will have an AA bond is equal to the probability that an A occupies a particular nearest-neighbor site on b, times the number of nearest-neighbor sites, which is 8 for the bcc structure. We assume that the probabilities are independent. Thus, by the preceding expressions for the number of A's on a and b,

$$N_{AA} = 8[\tfrac{1}{2}(1 + P)N][\tfrac{1}{2}(1 - P)] = 2(1 - P^2)N \; ;$$
$$N_{BB} = 8[\tfrac{1}{2}(1 + P)N][\tfrac{1}{2}(1 - P)] = 2(1 - P^2)N \; ; \tag{21}$$
$$N_{AB} = 8N[\tfrac{1}{2}(1 + P)]^2 + 8N[\tfrac{1}{2}(1 - P)]^2 = 4(1 + P^2)N \; .$$

The energy (20) becomes

$$E = E_0 + 2NP^2U \; , \tag{22}$$

where

$$E_0 = 2N(U_{AA} + U_{BB} + 2U_{AB}) \ ; \qquad U = 2U_{AB} - U_{AA} - U_{BB} \ . \quad (23)$$

We now calculate the entropy of this distribution of atoms. There are $\frac{1}{2}(1 + P)N$ atoms A and $\frac{1}{2}(1 - P)N$ atoms B on lattice a; there are $\frac{1}{2}(1 - P)N$ atoms A and $\frac{1}{2}(1 + P)N$ atoms B on lattice b. The number of arrangements σ of these atoms is

$$\sigma = \left[\frac{N!}{[\frac{1}{2}(1 + P)N]! \, [\frac{1}{2}(1 - P)N]!} \right]^2 \ . \quad (24)$$

From the Boltzmann definition of the entropy as $S = k_B \log \sigma$, we have, using Stirling's approximation,

$$S = 2Nk_B \log 2 - Nk_B[(1 + P) \log (1 + P) + (1 - P) \log (1 - P)].$$

For $P = \pm 1$, $S = 0$; for $P = 0$, $S = 2Nk_B \log 2$.

The equilibrium order is determined by the requirement that the free energy $F = E - TS$ be a minimum with respect to the order parameter P. On differentiating F with respect to P, we have as the condition for the minimum

$$4NPU + Nk_BT \log \frac{1 + P}{1 - P} = 0 \ . \quad (25)$$

The transcendental equation for P may be solved graphically; we find the smoothly decreasing curve shown in Fig. 25a. Near the transition we may expand (25), finding $4NPU + 2Nk_BTP = 0$. At the transition temperature $P = 0$, so that

$$T_c = -\frac{2U}{k_B} \ . \quad (26)$$

For a transition to occur, the effective interaction U must be negative (attractive).

The **short-range order parameter** r is a measure of the fraction of the average number q of nearest-neighbor bonds which are AB bonds. When completely disordered an AB alloy has an average of four AB bonds about each atom A. The total possible is eight. We may define[20]

$$r = \frac{(q - 4)}{4} \ , \quad (27)$$

so that $r = 1$ in complete order and $r = 0$ in complete disorder. Observe that r is a measure only of the local order about an atom, whereas the long-range order parameter P refers to the purity of the entire population on a given sublattice. Above the transition temperature T_c the long-range order is rigorously zero, but the short-range order is not zero, as in Fig. 25b.

[20] A careful discussion of long- and short-range order is given by H. A. Bethe, Proc. Roy. Soc. (London) **A150**, 552 (1935).

Problems

1. **Frenkel defects.** Show that the number n of interstitial atoms in equilibrium with n lattice vacancies in a crystal having N lattice points and N' possible interstitial positions is given by the equation

$$E_I = k_B T \log \left[(N - n)(N' - n)/n^2 \right] \ ,$$

whence, for $n \ll N, N'$,

$$n \cong (NN')^{1/2} \exp \left(-E_I/2k_B T \right) \ .$$

Here E_I is the energy necessary to remove an atom from a lattice site to an interstitial position.

2. **Schottky vacancies.** Suppose that the energy required to remove a sodium atom from the inside of a sodium crystal to the boundary is 1 eV. (a) Calculate the concentration of Schottky vacancies at 300°K. (b) If a sodium atom next to a vacancy has to move over a potential hill of 0.5 eV, and the atomic vibration frequency is 10^{12} Hz, estimate the diffusion coefficient at room temperature for radioactive sodium in normal sodium.

3. **F Center.** (a) Treat an F center as a free electron of mass m moving in the field of a point charge e in a medium of dielectric constant $\epsilon = n^2$; what is the $1s$-$2p$ energy difference of F centers in NaCl? (b) Compare from Table 2 the F center excitation energy in NaCl with the $3s$-$3p$ energy difference of the free sodium atom.

4. **Superlattice lines in Cu_3Au.** Cu_3Au alloy (75% Cu, 25% Au) has an ordered state below 400°C, in which the gold atoms occupy the 0 0 0 positions and the copper atoms the $\frac{1}{2} \frac{1}{2} 0$, $\frac{1}{2} 0 \frac{1}{2}$, and $0 \frac{1}{2} \frac{1}{2}$ positions in a face-centered cubic lattice. Give the indices of the new x-ray reflections which appear when the alloy goes from the disordered to the ordered state. List all new reflections with indices $\leqq 2$.

5. **Configurational heat capacity.** Derive an expression in terms of $P(T)$ for the heat capacity associated with order/disorder effects in an AB alloy. (The entropy associated with (24) is called the configurational entropy).

References

J. H. Schulman and W. D. Compton, *Color centers in solids,* Pergamon, 1962.

L. A. Girifalco, *Atomic migration in crystals,* Blaisdell, New York, 1964. (Elementary treatment.)

P. G. Shewmon, *Diffusion in solids,* McGraw-Hill, 1963.

D. Lazarus, "Diffusion in metals," *Solid state physics* **10**, 71 (1960).

C. C. Klick, "Point defects in insulators," Science **150**, 451 (1965).

J. Friedel, "Electronic structure of primary solid solutions in metal," Advances in Physics 3, 446 (1954).

W. Hume-Rothery, *Electrons, atoms, metals, and alloys,* 3rd ed., Dover, 1963.

L. Guttman, "Order-disorder phenomena in metals," *Solid state physics* 3, 145 (1956).

J. J. Markham, *F centers in alkali halides,* Academic, 1966.

N. B. Hannay, *Solid state chemistry,* Prentice-Hall, 1967.

Calculations of the properties of vacancies and interstitials, National Bureau of Standards Misc. Publ. 287.

Y. Quéré, *Défauts ponctuels dans les métaux,* Masson, 1967.

R. R. Hasiguti, ed., *Lattice defects in semiconductors,* Pennsylvania State University Press, 1967.

20
Dislocations

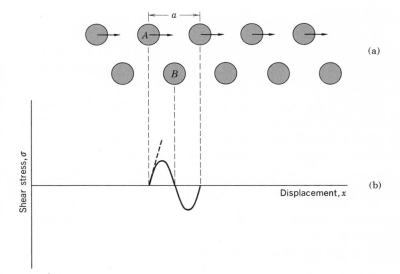

Figure 1 (a) Relative shear of two planes of atoms (shown in cross-section) in a uniformly strained crystal; (b) shear stress as a function of the relative displacement of the planes from their equilibrium position. The heavy broken line is drawn at the initial slope corresponding to the shear modulus G.

This chapter is concerned with the interpretation of the plastic mechanical properties of crystalline solids in terms of the theory of dislocations. Plastic properties are irreversible deformations, such as yield and slip; elastic properties are reversible. We shall also see that dislocations are involved in crystal growth.

The ease with which pure single crystals of many solids deform plastically is striking. This intrinsic weakness of crystals is exhibited in various ways. Pure silver chloride melts at 455°C, yet at room temperature it has a cheeselike consistency and can be rolled into sheets. Pure aluminum crystals are elastic (follow Hooke's law) only to a strain of about 10^{-5}, after which they deform plastically. Theoretical estimates of the elastic limit of perfect crystals give values 10^3 or 10^4 higher than the lowest observed values, although a factor 10^2 is more usual. There are few exceptions[1] to the rule that pure crystals are plastic and are not strong.

SHEAR STRENGTH OF SINGLE CRYSTALS

Frenkel[2] gave a simple method of estimating the theoretical shear strength of a perfect crystal. We consider in Fig. 1 the force needed to make a shear displacement of two planes of atoms past each other. For small elastic strains the stress σ is related to the displacement x by

$$\sigma = \frac{Gx}{d} , \tag{1}$$

as in Chapter 4. Here d is the interplanar spacing, and G denotes the appropriate shear modulus; for example, $G = C_{44}$ for shear in a $\langle 100 \rangle$ direction on a $\{100\}$ plane in a cubic crystal.

When the displacement is large and has proceeded to the point that atom A is directly over atom B in the figure, the two planes of atoms are in a configuration of unstable equilibrium and the stress is zero. As a first approximation we might represent the stress-displacement relation by a sine function:

$$\sigma = \left(\frac{Ga}{2\pi d}\right) \sin \left(\frac{2\pi x}{a}\right) , \tag{2}$$

where a is the interatomic spacing in the direction of shear. This relation reduces to (1) for small values of x/a.

The critical shear stress σ_c at which the lattice becomes unstable is given

[1] It seems that exceptions really exist; for example, crystals of high purity germanium and silicon are not plastic at room temperature and fail or yield only by fracture. Glass fails only by fracture at room temperature, but it is not crystalline. The fracture of glass is caused by stress concentration at minute cracks, as proposed by A. A. Griffith, Phil. Trans. Roy. Soc. (London) **A221**, 163 (1921).

[2] J. Frenkel, Z. Physik **37**, 572 (1926).

Table 1 Comparison of shear modulus and elastic limit

(After Mott)

	Shear modulus G, in dynes/cm^2	Observed elastic limit σ_c, in dynes/cm^2	G/σ_c
Sn, single crystal	1.9×10^{11}	1.3×10^7	15,000
Ag, single crystal	2.8×10^{11}	6×10^6	45,000
Al, single crystal	2.5×10^{11}	4×10^6	60,000
Al, pure, polycrystal	2.5×10^{11}	2.6×10^8	900
Al, commercial drawn	$\sim2.5 \times 10^{11}$	9.9×10^8	250
Duralumin	$\sim2.5 \times 10^{11}$	3.6×10^9	70
Fe, soft, polycrystal	7.7×10^{11}	1.5×10^9	500
Heat-treated carbon steel	$\sim8 \times 10^{11}$	6.5×10^9	120
Nickel-chrome steel	$\sim8 \times 10^{11}$	1.2×10^{10}	65

by the maximum value of σ, or

$$\sigma_c = \frac{Ga}{2\pi d} \ . \tag{3}$$

If $a \approx d$, then $\sigma_c \approx G/2\pi$ or $G/\sigma_c \approx 2\pi$. This calculation leads to the result that the ideal critical shear stress is of the order of $\frac{1}{6}$ of the shear modulus.

The observations in Table 1 show the experimental values of the elastic limit are much smaller than (3) would suggest. The theoretical estimate (3) may be improved by consideration of the actual form of the intermolecular forces and by consideration of other configurations of mechanical stability available to the lattice as it is sheared. Mackenzie has shown that these two effects may reduce the theoretical ideal shear strength to about $G/30$, corresponding to a critical shear strain angle of about 2 degrees.[3]

The observed low values of the shear strength cannot be explained without the presence of imperfections that can act as sources of mechanical weakness in real crystals. It is now known that special crystal imperfections called dislocations exist in almost all crystals, and their movement is responsible for slip at very low applied stresses.

Slip

Plastic deformation in many crystals occurs by slip, an example of which is shown in Fig. 2. In slip one part of the crystal slides as a unit across an adjacent part. The surface on which slip takes place is often a plane, and this is known as the slip plane. The direction of motion in slip is known as the slip

[3] J. K. Mackenzie, thesis, Bristol, 1949. W. L. Bragg and W. M. Lomer, Proc. Roy. Soc. (London) **196**, 171 (1949) have shown that the observed shear strength of a perfect two-dimensional raft of bubbles is of this order.

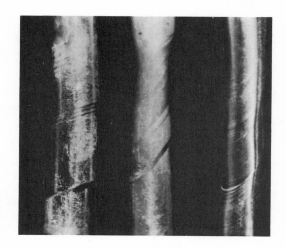

Figure 2 Translational slip in zinc single crystals. (Courtesy of E. R. Parker.)

direction. The great importance of lattice properties for plastic strain is indicated by the highly anisotropic nature of slip. Displacement takes place along crystallographic planes with a set of small Miller indices, such as the {111} planes in fcc metals and the {110}, {112}, and {123} planes in bcc metals. Under most conditions the slip direction is observed to lie in the line of closest atomic packing, ⟨110⟩ in fcc metals and ⟨111⟩ in bcc metals.

To maintain the crystal structure after slip, the displacement or slip vector must equal a lattice translation vector. The shortest lattice translation vector expressed in terms of the lattice constant a in a fcc structure is of the form $(a/2)(\hat{x} + \hat{y})$; in a bcc structure it is $(a/2)(\hat{x} + \hat{y} + \hat{z})$. But in fcc crystals one also observes partial displacements which upset the regular sequence $ABCABC\ldots$ of closest-packed planes (Chapter 1), to produce a **stacking fault** such as $ABCABABC.\ldots$ The result is then a "mixture" of fcc and hcp stacking.

Deformation of a crystal by slip is observed to be inhomogeneous. Large shear displacements occur on a few widely separated slip planes, while parts of the crystal lying between slip planes remain essentially undeformed. An incidental property of slip is the Schmid law of the critical shear stress: slip takes place along a given slip plane and direction when the corresponding *component* of shear stress reaches the critical value.

Slip is one mode of plastic deformation. A second mode of plastic deformation, called **twinning,** is observed in some crystals, particularly in hcp and bcc structures. During the slip process a considerable amount of displacement occurs on a few widely separated slip planes. During twinning a partial displacement occurs successively on each of many neighboring crystallographic planes. After deformation by twinning, the deformed part of the crystal is a mirror image of the undeformed part. Although both slip and twinning are caused by the motion of dislocations, we shall be concerned primarily with slip.

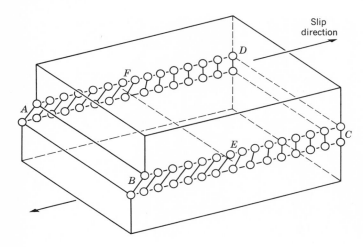

Slip
direction

Figure 3 An edge dislocation *EF* in the glide plane *ABCD*. The figure shows the slipped region *ABEF* in which the atoms have been displaced by more than half a lattice constant and the unslipped region *FECD* with displacement less than half a lattice constant.

DISLOCATIONS

The low observed values of the critical shear stress can be explained in terms of the motion through the lattice of a particular type of line imperfection known as a dislocation. The idea that slip propagates by the motion of dislocations was published in 1934 independently by Taylor, Orowan, and Polanyi; the concept of dislocations was introduced into physics somewhat earlier by Prandtl and Dehlinger.

There are several basic types of dislocations. We first describe an **edge dislocation**. Figure 3 shows a simple cubic crystal in which slip of one atom distance has occurred over the left half of the slip plane but not over the right half. The boundary between the slipped and unslipped regions is called the dislocation. Its position is marked by the termination of an extra vertical half-plane of atoms crowded into the upper half of the crystal as shown in Fig. 4. Near the dislocation the crystal is highly strained. The simple edge dislocation extends indefinitely in the slip plane in a direction normal to the slip direction. In Fig. 5 we show a photograph of a dislocation in a two-dimensional soap bubble raft obtained by the method of Bragg and Nye.[4]

The mechanism responsible for the mobility of a dislocation is shown in Fig. 6. The motion of an edge dislocation through a crystal is analogous to the passage of a ruck or wrinkle across a rug: the ruck moves more easily than the whole rug. If atoms on one side of the slip plane are moved with respect to those on the other side, atoms at the slip plane will experience repulsive forces from some neighbors and attractive forces from others across the slip plane. These forces cancel to a first approximation.

[4] W. L. Bragg and J. F. Nye, Proc. Roy. Soc. (London) **A190**, 474 (1947); W. L. Bragg and W. M. Lomer, Proc. Roy. Soc. (London) **A196**, 171 (1949).

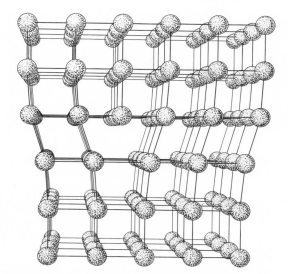

Figure 4 Structure of an edge dislocation. The deformation may be thought of as caused by inserting an extra plane of atoms on the upper half of the *y* axis. Atoms in the upper half-crystal are compressed by the insertion; those in the lower half are extended.

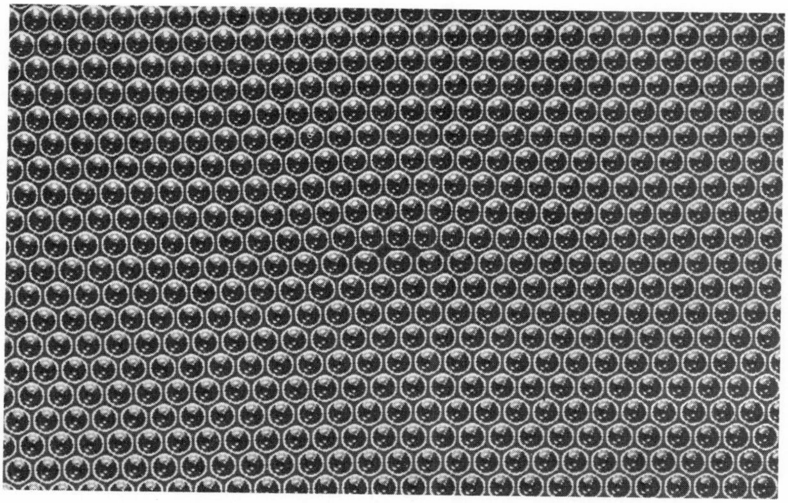

Figure 5 A dislocation in a two-dimensional bubble raft. The dislocation is most easily seen by turning the page by 30° in its plane and sighting at a low angle. (Courtesy of W. M. Lomer, after Bragg and Nye.)

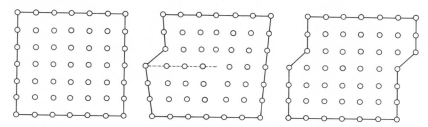

Figure 6 Motion of a dislocation under a shear tending to move the upper surface of the specimen to the right. (After Taylor.)

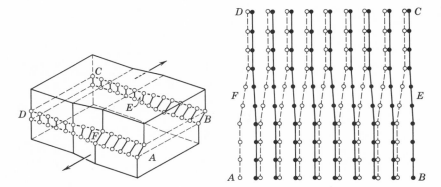

Figure 7 A screw dislocation. A part *ABEF* of the slip plane has slipped in the direction parallel to the dislocation line *EF*. A screw dislocation may be visualized as a helical arrangement of lattice planes, such that we change planes on going completely around the dislocation line. (After Cottrell.)

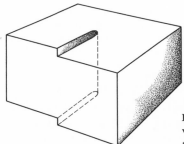

Figure 8 Another view of a screw dislocation. The broken vertical line which marks the dislocation is surrounded by strained material.

The external stress required to move a dislocation has been calculated and is quite small, probably below 10^5 dynes/cm^2, provided that the bonding forces in the crystal are not highly directional. Thus dislocations may make a crystal very plastic. Passage of a dislocation through a crystal is equivalent to a slip displacement of one part of the crystal.

The second simple type of dislocation is the **screw dislocation,** sketched in Figs. 7 and 8. A screw dislocation marks the boundary between slipped and unslipped parts of the crystal. The boundary *parallels* the slip direction, instead of lying perpendicular to it as for the edge dislocation. The screw dislocation may be thought of as produced by cutting the crystal partway through with a knife and shearing it parallel to the edge of the cut by one atom spacing. A screw dislocation transforms successive atom planes into the surface of a helix; this accounts for the name of the dislocation.

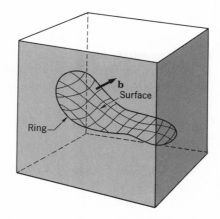

Figure 9 General method of forming a dislocation ring in a medium. The medium is represented by the rectangular block. The ring is represented by the closed curve in the interior in the block. A cut is made along the surface bounded by the curve and indicated by the contoured area. The material on one side of the cut is displaced relative to that on the other by vector distance **b**, which may be oriented arbitrarily relative to the surface. Forces will be required to effect the displacement. The medium is filled in or cut away so as to be continuous after the displacement. It is then joined in the displaced state and the applied forces are relaxed. Here **b** is the Burgers vector of the dislocation. (After Seitz.)

Burgers Vectors

Other dislocation forms may be constructed from segments of edge and screw dislocations. Burgers has shown that the most general form of a linear dislocation pattern in a crystal can be described as shown in Fig. 9. We consider any closed curve not necessarily planar within a crystal, or an open curve terminating on the surface at both ends: (a) Make a cut along any simple surface bounded by the line. (b) Displace the material on one side of this surface by a vector **b** relative to the other side; here **b** is called the **Burgers vector**. (c) In regions where **b** is not parallel to the cut surface, this relative displacement will either produce a gap or cause the two halves to overlap. In these cases we imagine that we either add material to fill the gap or subtract material to prevent overlap. (d) Rejoin the material on both sides. We leave the strain displacement intact at the time of rewelding, but afterwards we allow the medium to come to internal equilibrium. The resulting strain pattern is that of the dislocation characterized jointly by the boundary curve and the Burgers vector. The Burgers vector must be equal to a lattice vector in order that the rewelding process will maintain the crystallinity of the material.

The Burgers vector of a screw dislocation (Figs. 7 and 8) is parallel to the dislocation line; that of an edge dislocation (Figs. 3 and 4) is perpendicular to the dislocation line and lies in the slip plane.

Stress Fields of Dislocations

The stress field of a screw dislocation is particularly simple. Figure 10 shows a shell of material surrounding an axial screw dislocation. The shell of circumference $2\pi r$ has been sheared by an amount b to give a shear strain $e = b/2\pi r$. The corresponding shear stress in an elastic continuum is

$$\sigma = Ge = \frac{Gb}{2\pi r} \ . \tag{4}$$

This expression cannot hold in the region immediately around the dislocation line, as the strains here are too large for continuum or linear elasticity theory to apply.

The elastic energy of the shell is $dE_s = \frac{1}{2}Ge^2 \, dV = (Gb^2/4\pi) \, dr/r$ per unit length. The total elastic energy per unit length of a screw dislocation is found on integration to be

$$E_s = \frac{Gb^2}{4\pi} \log \frac{R}{r_0} \ , \tag{5}$$

where R and r_0 are appropriate upper and lower limits for the variable r. A reasonable value of r_0 is comparable to the magnitude b of the Burgers vector or to the lattice constant; the value of R cannot exceed the dimensions of the crystal. In many applications R should be considerably smaller than the dimensions of the crystal. The value of the ratio R/r_0 is not very important because it enters in a log term.

We now calculate the energy of an edge dislocation at the origin of the co-ordinate system (Fig. 4). We let σ_{rr} and $\sigma_{\theta\theta}$ denote the tensile stresses in the radial and circumferential directions, and we let $\sigma_{r\theta}$ denote the shear stress. In an isotropic elastic continuum σ_{rr} and $\sigma_{\theta\theta}$ are proportional to $(\sin \theta)/r$: we need a function which falls off as $1/r$ and which changes sign when y is replaced by $-y$. The shear stress $\sigma_{r\theta}$ is proportional to $(\cos \theta)/r$: considering the plane $y = 0$ we see from Fig. 4 that the shear stress is an odd function of x. On dimensional grounds the constants of proportionality in the stress are proportional to the shear modulus G and to the Burgers vector b of the displacement. The final result, which is derived in the books cited in the references, is

$$\sigma_{rr} = \sigma_{\theta\theta} = -\frac{Gb}{2\pi(1-\nu)}\frac{\sin \theta}{r} \ , \qquad \sigma_{r\theta} = \frac{Gb}{2\pi(1-\nu)}\frac{\cos \theta}{r} \ , \tag{6}$$

where ν is the Poisson ratio defined in Problem 4.1 ($\nu \approx 0.3$ for most crystals). The strain energy of a unit length of edge dislocation is

$$E_e = \frac{Gb^2}{4\pi(1-\nu)} \log \frac{R}{r_0} \ , \tag{7}$$

closely similar to the result for a screw dislocation.

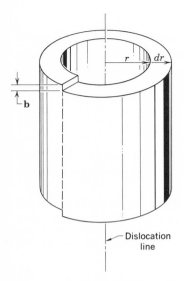

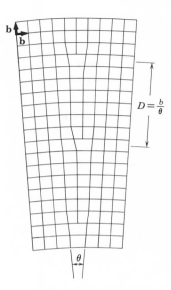

Figure 10 Shell of elastically distorted crystal surrounding screw dislocation with Burgers vector **b**.

Figure 11 Low-angle grain boundary. (After Burgers.)

We want an expression for the shear stress component σ_{xy} on planes parallel to the slip plane in Fig. 4. From the stress components σ_{rr}, $\sigma_{\theta\theta}$, and $\sigma_{r\theta}$ evaluated on the plane a distance y above the slip plane, we find

$$\sigma_{xy} = \frac{Gb}{2\pi(1-\nu)} \frac{\sin 4\theta}{4y} . \tag{8}$$

It is shown in Problem 3 that the force caused by a uniform shear stress σ is $F = b\sigma$ per unit length of dislocation. This result holds also for the force that one dislocation exerts upon another. The force that an edge dislocation at the origin exerts upon a similar one at the location (y, θ) is

$$F = b\sigma_{xy} = \frac{Gb^2}{2\pi(1-\nu)} \frac{\sin 4\theta}{4y} \tag{9}$$

per unit length.[5]

Low-angle Grain Boundaries

Burgers[6] suggested that low-angle boundaries between adjoining crystallites or crystal grains consist of arrays of dislocations. A simple example of the Burgers model of a grain boundary is shown in Fig. 11. The boundary occupies a (010) plane in a simple cubic lattice and divides two parts of the crystal that

---

[5] Strictly speaking, F is the component of force in the slip direction. There is another component of force perpendicular to the slip direction, but it is of no importance at low temperatures where the only possible motion of a dislocation is in the slip plane.

[6] J. M. Burgers, Proc. Koninkl. Ned. Akad. Wetenschap. **42**, 293 (1939); Proc. Phys. Soc. (London) **52**, 23 (1940).

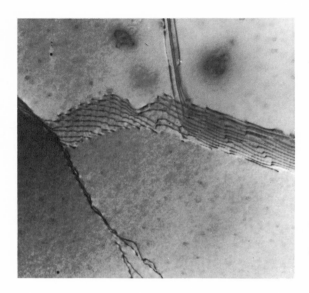

Figure 12 Electron micrograph of dislocation structures in small-angle grain boundaries in an Al–7% Mg solid solution. Magnification ×17,000. (Courtesy of R. Goodrich and G. Thomas.)

have a [001] axis in common. Such a boundary is called a pure tilt boundary: the misorientation can be described by a small rotation θ about the common [001] axis of one part of the crystal relative to the other. The tilt boundary is represented as an array of edge dislocations of spacing $D = b/\theta$, where b is the Burgers vector of the dislocations.

Experiments have substantiated this model. Figure 12 shows the distribution of dislocations along small-angle grain boundaries, as observed with an electron microscope. Further, Read and Shockley[7] derived a theory of the interfacial energy as a function of the angle of tilt, with results in excellent agreement with measurements. We note that the region of elastic distortion near a grain boundary does not extend very far into the two crystals, but is essentially confined to a slab whose thickness equals the dislocation spacing D. Each dislocation is surrounded by its own strain field and by the strain fields of dislocations above and below it. The latter strain fields nearly cancel each other, being equal in magnitude and opposite in sign, so that the strain energy near each dislocation results primarily from its own strain field. In this approximation (7) gives the elastic strain energy per unit length of dislocation in the boundary

[7] W. T. Read and W. Shockley, Phys. Rev. **78**, 275 (1950).

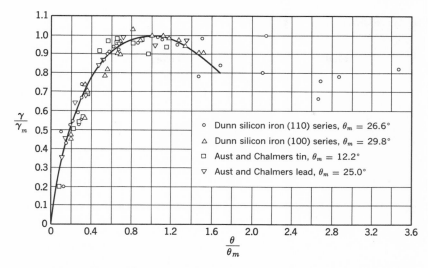

Figure 13 Comparison of theoretical curve with measurements of relative grain-boundary energy as function of the angle between grains. The theoretical curve is given by $\gamma/\gamma_m = (\theta/\theta_m)\ [1 - \log\ (\theta/\theta_m)]$. (By permission from *Dislocations in crystals*, by W. T. Read, Jr. Copyright, 1953. McGraw-Hill Book Company, Inc.)

as $[Gb^2/4\pi(1 - \nu)]\ \log\ (\alpha D/b)$, where r_0 has been set equal to b and α is a number near 1. There are $1/D = \theta/b$ dislocations per unit length of boundary, so that the grain boundary energy is

$$\gamma = \left(\frac{Gb}{4\pi(1 - \nu)}\right)\theta(-\log\theta + \log\alpha)\ . \tag{10}$$

The grain boundary energy γ is zero when θ is zero, and rises with an initial vertical slope as θ increases. With further increase in θ, γ reaches a maximum γ_m and begins to decline. If we let θ_m be the value of θ for which $\gamma = \gamma_m$, (10) reduces to

$$\frac{\gamma}{\gamma_m} = \left(\frac{\theta}{\theta_m}\right)\left(1 - \log\frac{\theta}{\theta_m}\right)\ . \tag{11}$$

This equation is compared in Fig. 13 with the experimental measurements of relative grain boundary energy as a function of θ. The agreement is remarkably

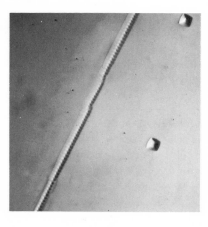

Figure 14 Dislocation etch pits in low-angle boundary on (100) face of germanium; the angle of the boundary is 27.5 sec. The boundary lies in a (011) plane; the line of the dislocations is [100]. The Burgers vector is the shortest lattice translation vector, or $|\mathbf{b}| = a/\sqrt{2} = 4.0$ Å. [After F. L. Vogel, Jr., Acta Metallurgica **3**, 245 (1955).]

good up to $\theta \geq 30°$; this is beyond the point where the dislocation model is expected to apply.

Direct verification of the Burgers model is provided by the quantitative x-ray and optical studies of low-angle boundaries in germanium crystals by Vogel[8] and co-workers. By counting etch pits along the intersection of a low-angle grain boundary with an etched germanium surface (Fig. 14), they determined the dislocation spacing D. They assumed that each etch pit marked the end of a dislocation. The angle of tilt calculated from the relation $\theta = b/D$ agrees well with the angle measured directly by means of x-rays.

The interpretation of low-angle boundaries as arrays of dislocations is further supported by the fact that pure tilt boundaries move normal to themselves on application of a suitable stress. The motion has been demonstrated in a beautiful experiment by Washburn and Parker (Fig. 15). The specimen is a bicrystal of zinc containing a 2° tilt boundary with dislocations about thirty atomic planes apart. One side of the crystal was clamped, and a force was applied at a point on the opposite side of the boundary. Motion of the boundary took place by cooperative motion of the dislocations in the array, each dislocation moving an equal distance in its own slip plane. The motion was produced by stresses of the order of magnitude of the yield stress for zinc crystals, strong evidence that ordinary deformation results from the motion of dislocations.

Grain boundaries and dislocations offer relatively little resistance to diffusion of atoms in comparison with diffusion in perfect crystals. A dislocation is an open passage for diffusion. Diffusion is known to be greater in plastically deformed material than in annealed crystals. Diffusion along grain boundaries controls the rates of some precipitation reactions in solids: the precipitation of tin from lead-tin solutions at room temperature proceeds about 10^8 times faster than expected from diffusion in an ideal lattice.

[8] F. L. Vogel, W. G. Pfann, H. E. Corey, and E. E. Thomas, Phys. Rev. **90**, 489 (1953); F. L. Vogel, Jr., Acta Met. **3**, 245 (1955).

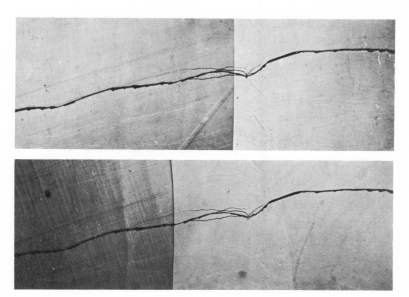

Figure 15 Motion of a small-angle grain boundary under stress. The boundary is the straight vertical line, and it is photographed under vertical illumination, thereby making evident the 2° angular change in the cleavage surface of the zinc crystal at the boundary. The irregular horizontal line is a small step in the cleavage surface which serves as a reference mark. The crystal is clamped at the left; at the right it is subject to a force normal to the plane of the page. *Top*, original position of boundary; *bottom*, moved back 0.4 mm. (Courtesy of J. Washburn and E. R. Parker.)

Dislocation Densities

The density of dislocations is the number of dislocation lines that intersect a unit area in the crystal. The density ranges from 10^2 to 10^3 dislocations/cm^2 in the best germanium and silicon crystals to 10^{11} or 10^{12} dislocations/cm^2 in heavily deformed metal crystals. The methods available for estimating dislocation densities are compared in Table 2. The book by Amelinckx is an excellent reference for all present methods of observing dislocations.

Table 2 Methods for estimating dislocation densities°

Technique	Specimen thickness	Width of image	Maximum practical density, per cm^2
Electron microscopy	$>$1000 Å	$\sim$100 Å	10^{11}–10^{12}
X-ray transmission	0.1–1.0 mm	5μ	10^4–10^5
X-ray reflection	$<$2μ (min.) — 50μ (max.)	2μ	10^6–10^7
Decoration	$\sim$10μ (depth of focus)	0.5μ	2×10^7
Etch pits	no limit	0.5μ†	4×10^8

° Based on W. G. Johnston. Prog. Ceramic Sci. **2**, 1 (1961).
† Limit of resolution of etch pits.

Figure 16 Cell structure of three-dimensional tangles of dislocations in deformed aluminum. (Courtesy of P. R. Swann.)

The actual dislocation configurations in cast or annealed (slowly cooled) crystals correspond either to a group of low-angle grain boundaries or to a three-dimensional network of dislocations arranged in cells, as shown in Fig. 16.

The dislocation density in heavily deformed crystals may be estimated from the increased internal energy that results from plastic deformation. From (5) and (7), the energy per unit length of a dislocation is about $(Gb^2/4\pi)$ log (R/r_0). We take R comparable to the dislocation spacing and $r_0 \approx b$, and so $R/r_0 \approx 10^3$. The dislocation energy then is about 5×10^{-4} erg/cm, or about 8 eV per atom plane through which the dislocation passes. The maximum energy stored in lattice distortions as a consequence of severe plastic deformation, as by twisting, filing, or compressing, has been measured thermally for several metals. If the deformation is not too great, about 10 percent of the energy expended in plastic flow is stored in the lattice. Upon continued plastic flow, however, the stored energy approaches a saturation value. The observed values of the stored energy are about 10^8 ergs/cm³. If the energy per unit length of dislocations is 5×10^{-4} erg/cm, then there must be about 10^{11} cm of dislocation lines per cubic centimeter of crystal, or about 10^{11} dislocations/cm². This is about one dislocation per square 100 atoms on a side. Such a concentration is characteristic of severely deformed metals.

A problem is posed by the presence of dislocations in cast and annealed crystals. No dislocations can be present in thermal equilibrium, because their energy is much too great in comparison with the increase in entropy that they produce. Dislocations must therefore be introduced in a nonequilibrium manner during the solidification of crystals from the melt, and the dislocations must remain even during the most careful annealing. The mechanism of their introduction during solidification is not known, although it is thought to be associated with the precipitation of lattice vacancies as the crystal cools.

Figure 17 Electron micrograph of dislocation loops formed by aggregation and collapse of vacancies in Al–5% Mg quenched from 550°C. The helical dislocations are formed by the "climb" of screw dislocations as a result of the precipitation of vacancies. Mag. ×43,000. (Courtesy of A. Eikum and G. Thomas.)

Lattice vacancies which precipitate along an existing edge dislocation will eat away a portion of the extra half-plane of atoms and cause the dislocation to climb, which means to move at right angles to the slip direction. If no dislocations are present, the crystal will become supersaturated with lattice vacancies; their precipitation in penny-shaped vacancy plates may be followed by collapse of the plates and formation of dislocation rings which grow with further vacancy precipitation, as in Fig. 17.

A second problem discussed below is posed by the very great increase in dislocation density that is caused by plastic deformation. Dislocation density measurements typically show an increase from about 10^8 to about 10^{11} dislocations/cm^2 during deformation, a 1000-fold increase. Equally striking is the fact that if a dislocation were to move completely across its slip plane, an offset of only one atom spacing would be produced, whereas offsets of as much as 100 to 1000 atom spacings are actually observed.

Dislocation Multiplication and Slip

Consider a closed circular dislocation loop of radius r surrounding a slipped area of the same radius. A dislocation loop will have to be partly edge, partly screw, and mostly intermediate in character. Because the strain energy of the dislocation loop increases in proportion to its circumference, the loop will tend to shrink. However, if a shear stress that favors slip is acting, the loop

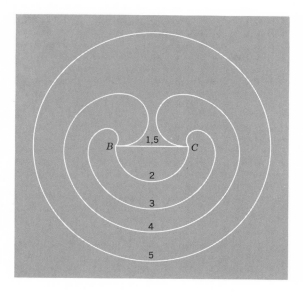

Figure 18 Frank-Read mechanism for multiplication of dislocations, showing successive stages in the generation of a dislocation loop by the segment *BC* of a dislocation line. The process can be repeated indefinitely.

will tend to expand. A dislocation segment pinned at each end (Fig. 18) is called a **Frank-Read source**,[9] and as in the figure it can lead to the generation of a large number of concentric dislocation loops on a single slip plane. This and related types of dislocation multiplication mechanisms give rise to slip and to the increased density of dislocations during plastic deformation. A beautiful example of a dislocation source is shown in Fig. 19.

STRENGTH OF ALLOYS

Pure crystals are very plastic and yield at very low stresses. There appear to be four important ways of increasing the yield strength of an alloy so that it will withstand shear stresses as high as 10^{-2} G. They are mechanical blocking of dislocation motion, pinning of dislocations by solute atoms, impeding dislocation motion by short-range order, and increasing the dislocation density so that tangling of dislocations results. All four strengthening mechanisms depend for their success upon impeding dislocation motion. A fifth mechanism, that of removing all dislocations from the crystal, may operate for certain fine hairlike crystals (whiskers) that are discussed in the section on crystal growth.

Mechanical blocking of dislocation motion can be produced most directly by introducing tiny particles of a second phase into a crystal lattice. This process is followed in the hardening of steel, where particles of iron carbide are precipitated into iron, and in hardening aluminum, where particles of Al_2Cu are precipitated. The pinning of a dislocation by impurities is shown in Fig. 20.

[9] The classical Frank-Read source is not observed as often as certain modifications: see G. Thomas, J. of Metals, Apr. 1964.

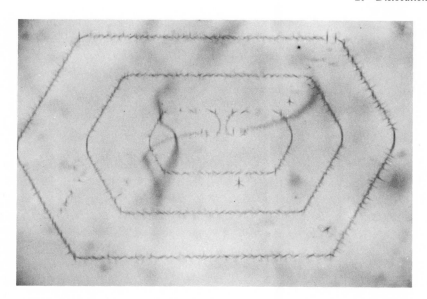

Figure 19 A Frank-Read dislocation source in silicon, decorated with copper precipitates and viewed with infrared illumination. Two complete dislocation loops are visible, and the third, innermost loop is near completion. (After W. C. Dash, *Dislocations and Mechanical Properties of Crystals*, Fisher et al., eds., Wiley, 1957, pp. 57–68.)

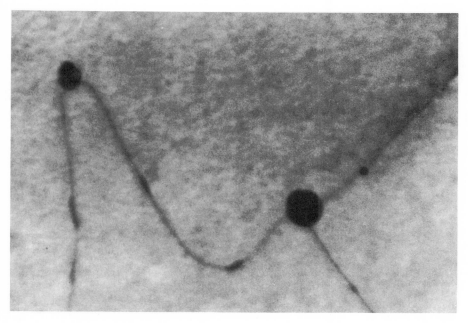

Figure 20 Dislocations pinned by impurities in magnesium oxide. (Electron micrograph courtesy of G. Thomas and J. Washburn.)

In strengthening by the addition of small particles there are two cases to be considered: either the particle can be deformed with the matrix, which requires that the particle can be traversed by the dislocation, or the particle cannot be traversed by the dislocation. If the particle cannot be cut[10] the stress necessary to force a dislocation between particles spaced L apart on a slip plane should be approximately

$$\frac{\sigma}{G} = \frac{b}{L} .$$ (12)

The smaller the spacing L, the higher is the yield stress σ. Before particles precipitate, L is large and the strength is low. Immediately after precipitation is complete and many small particles are present, L is a minimum and the strength is a maximum. If the alloy is then held at a high temperature, some particles grow at the expense of others, so that L increases and the strength drops.

The strength of dilute solid solutions is believed to result from the pinning of dislocations by solute atoms. Cottrell pointed out that the solubility of a foreign atom will be greater in the neighborhood of a dislocation than elsewhere in a crystal. For example, an atom that tends to expand the crystal will dissolve preferentially in the expanded region near an edge dislocation. A small atom will tend to dissolve preferentially in the contracted region near the dislocation—a dislocation offers both expanded and contracted regions. As a result of the affinity of solute atoms for dislocations, each dislocation will collect a cloud of associated solute atoms during cooling, at a time when the mobility of solute atoms is high. At still lower temperatures diffusion of solute atoms effectively ceases, and the solute atom cloud becomes fixed in the crystal. When a dislocation moves, leaving its solute cloud behind, the energy of the crystal must increase. The increase in energy can only be provided by an increased stress acting on the dislocation as it pulls away from the solute atom cloud, and so the presence of the cloud strengthens the crystal.

The passage of a dislocation across a slip plane in pure crystals does not alter the binding energy across the plane after the dislocation is gone. The internal energy of the crystal remains unaffected. The same is true for random solid solutions, because the solution is equally random across a slip plane after slip. Most solid solutions, however, contain short-range order, as defined in Chap. 19. Atoms of different species are not arranged at random on the lattice sites, but tend to have an excess or a deficiency of pairs of unlike atoms. Thus in ordered alloys dislocations tend to move in pairs: the second dislocation reorders the local disorder left by the first dislocation.

The strength of a crystalline material increases with plastic deformation. The phenomenon is called **work-hardening** or **strain-hardening.** The strength

[10] Hard intermetallic phases, such as refractory oxides, cannot be cut by dislocations.

is believed to increase because of the increased density of dislocations and the greater difficulty of moving a given dislocation across a slip plane that is threaded by many dislocations. The difficulty is particularly acute when one screw dislocation tries to cross another; lattice vacancies and interstitial atoms may be generated in the process. Strain-hardening frequently is employed in the strengthening of materials, but its usefulness is limited to low enough temperatures so that annealing does not occur.

The important factor in strain-hardening is not the total density of dislocations, but their arrangement. In most metals dislocations tend to form cells (Fig. 16) of dislocation-free areas of dimensions of the order of 1 micron. But unless we can get a uniform high density of dislocations we cannot strain-harden a metal to its theoretical strength, because of slip in the dislocation-free areas. A uniform density is accomplished by explosive deformation or by special thermal-mechanical treatments.[11]

Each of the mechanisms of strengthening crystals can raise the yield strength to the range of 10^{-3} G to 10^{-2} G. All mechanisms begin to break down at temperatures where diffusion can occur at an appreciable rate. When diffusion is rapid, precipitated particles dissolve; solute clouds drift along with dislocations as they glide; short-range order repairs itself behind slowly moving dislocations; and dislocation climb and annealing tend to decrease the dislocation density. The resulting time-dependent deformation is called **creep.** This irreversible motion precedes the elastic limit. The search for alloys for use at very high temperatures is a search for reduced diffusion rates, so that the four strengthening mechanisms will survive to high temperatures. But the central problem of strong alloys is not strength, but ductility, for failure is often by fracture.

DISLOCATIONS AND CRYSTAL GROWTH

It has been shown by Frank[12] and his collaborators that in some cases the presence of dislocations may be the controlling factor in crystal growth. When crystals are grown in conditions of low supersaturation, of the order of 1 percent, it has been observed that the growth rate is enormously faster than that calculated for an ideal crystal. The actual growth rate is explained by Frank in terms of the effect of dislocations on growth.

The theory of growth of ideal crystals predicts that in crystal growth from vapor a supersaturation (pressure/equilibrium vapor pressure) of the order of 10 is required to nucleate new crystals, of the order of 5 to form liquid drops, and of 1.5 to form a two-dimensional monolayer of molecules on the face of a

[11] V. F. Zackay, "Strength of steel," Sci. Amer. **209**, 71 (August, 1963).

[12] For a full review of this field see Neugebauer, ed., *Growth and perfection of crystals*, Wiley, 1959; also Chap. 7 of J. Friedel, *Dislocations*, Addison-Wesley, 1964. Long-chain polymer molecules have an unusual pattern of crystallization into flat plates; see the review by A. Keller, Physics Today **23**, 42 (May, 1970).

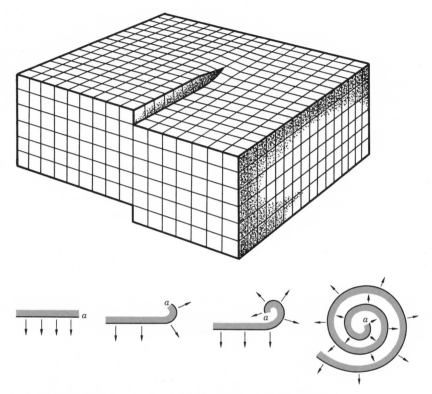

Figure 21 Development of a spiral step produced by intersection of a screw dislocation with the surface of a crystal. Each cube represents a molecule. (After F. C. Frank.)

perfect crystal. Volmer and Schultze observed growth of iodine crystals at vapor supersaturations down to less than 1 percent, where the growth rate should have been down by the factor e^{-3000} from the rate defined as the minimum observable growth. This is a substantial disagreement between observation and theory.

The large disagreement expresses the difficulty of nucleating a new monolayer on a completed surface of an ideal crystal. But if a screw dislocation is present (Fig. 21) it is never necessary to nucleate a new layer: the crystal will grow in spiral fashion at the *edge* of the discontinuity shown. (An atom can be bound to a step more strongly than to a plane.) The calculated growth rates for this mechanism are in good agreement with observation. We expect that nearly all crystals in nature grown at low supersaturation will contain dislocations, as otherwise they could not have grown.

Spiral growth patterns have been observed on a large number of crystals. A beautiful example of the growth pattern from a single screw dislocation is given in Fig. 22. Growth from several dislocations is shown in Fig. 23.

Figure 22 Phase contrast micrograph of a hexagonal spiral growth pattern on a SiC crystal. The step height is 165 Å. (Courtesy of A. R. Verma.)

Figure 23 Phase contrast micrograph of elementary growth spirals on SiC with nearly circular symmetry. In the left half of the figure, one right-handed and the other left-handed screw dislocation emerge on the crystal surface. The spiral attached to the left-handed screw dislocation is developed for about three turns when it meets the ledges originating from the right-handed screw dislocation. Thereafter, in accordance with the theory, successive closed loops are generated. These loops are circular for most part with only a slight distortion over a small portion. The step height of these circular spirals has been measured to be 15 Å. The crystal is SiC of type 6H. The right half of the figure shows a single left-handed spiral and the curve of intersection with the steps originating from the left. (Courtesy of A. R. Verma.)

If the growth rate is independent of direction of the edge in the plane of the surface, the growth pattern is an Archimedes spiral,

$$r = a\theta , \tag{13}$$

where a is a constant. The limiting minimum radius of curvature near the dislocation is determined by the supersaturation. If the radius of curvature is too small, atoms on the curved edge evaporate until the equilibrium curvature is attained. Away from the origin each part of the step acquires new atoms at a constant rate, so that $dr/dt = $ const.

Whiskers

Fine hairlike crystals or **whiskers** have been observed to grow under conditions of high supersaturation without the necessity for more than perhaps one dislocation. It may be that these crystals contain a single axial screw dislocation that aids their essentially one-dimensional growth. From the absence of dislocations we would expect these crystal whiskers to have high yield strengths, of the order of the calculated value $G/30$ discussed earlier in this chapter. A single axial screw dislocation, if present, could not cause yielding, because in bending the crystal the dislocation is not subjected to a shear stress parallel to its Burgers vector. That is, the stress is not in a direction which can cause slip. Herring and Galt[13] observed whiskers of tin of radius $\sim 10^{-4}$ cm with elastic properties near those expected from theoretically perfect crystals. They observed yield strains of the order of 10^{-2}, which correspond to shear stresses of order 10^{-2} G, about 1000 times greater than in bulk tin, thereby confirming the early estimates of the strength of perfect crystals.

Theoretical or ideal elastic properties have been observed for a number of materials. Elastic deformation of a whisker of copper at a high strain is shown in Fig. 24. Ferromagnetic domain structures on whiskers of iron were illustrated in Chapter 16; a single domain whisker of nickel is shown in Figure 25.

[13] C. Herring and J. K. Galt, Phys. Rev. **85**, 1060 (1952); W. W. Piper and W. L. Roth, Phys. Rev. **92**, 503 (1953).

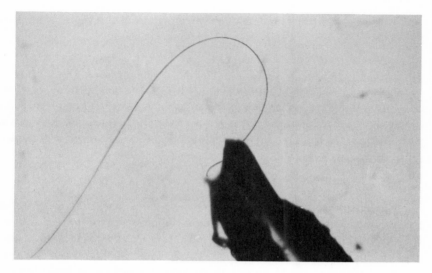

Figure 24   Copper whisker strained 1.5 percent. (Courtesy of S. S. Brenner.)

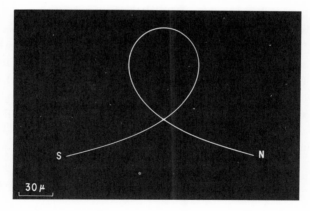

Figure 25   A nickel whisker of diameter 1000 Å bent in a loop. (Courtesy of R. W. De Blois.)

Problems

1. **Lines of closest packing.** Show that the lines of closest atomic packing are $\langle 110 \rangle$ in fcc structures and $\langle 111 \rangle$ in bcc structures.

2. **Dislocation pairs.** (a) Find a pair of dislocations equivalent to a row of lattice vacancies; (b) find a pair of dislocations equivalent to a row of interstitial atoms.

3. **Force on dislocation.** Consider a crystal in the form of a cube of side L containing an edge dislocation of Burgers vector $\mathbf{b}$. If the crystal is subjected to a shear stress σ on the upper and lower faces in the direction of slip, show by considering energy balance that the force acting on the dislocation is $F = b\sigma$ per unit length.

References

J. Friedel, *Dislocations*, Addison-Wesley, 1964. (A thorough modern discussion.)

J. Weertman and J. R. Weertman, *Elementary dislocation theory*, Macmillan, 1964. (A good introduction.)

S. Amelinckx, *Direct observation of dislocations*, Academic Press, 1964.

G. Thomas and J. Washburn, ed., *Electron microscopy and strength of crystals*, Wiley, 1963.

G. Thomas, *Transmission electron microscopy of metals*, Wiley, 1962.

C. S. Barrett and T. B. Massalski, *Structure of metals*, McGraw-Hill, 3rd ed., 1966.

J. P. Hirth and J. Lothe, *Theory of dislocations*, McGraw-Hill, 1968.

F. R. N. Nabarro, *Theory of crystal dislocations*, Oxford U. Press, 1967.

Advanced Topic* A

PROPAGATION OF ELECTROMAGNETIC WAVES
IN A PERIODIC STRUCTURE

We study the normal modes of the electromagnetic field in an infinite crystal[1] in order to obtain the dispersion relation $\omega(\mathbf{k})$ for the frequency as a function of the wavevector $\mathbf{k}$. The dispersion relation exhibits forbidden bands at values of $\mathbf{k}$ that satisfy the Bragg relation $2\mathbf{k} \cdot \mathbf{G} = G^2$. For frequencies within a forbidden band the wavevectors are not real; this means that an undamped wave cannot propagate in the crystal when the Bragg relation is satisfied. We use a powerful Fourier-analysis method.

Our basic assumption is that the polarization $\mathbf{P}(x)$, the electric dipole moment per unit volume, is linearly related to the electric field intensity $\mathbf{E}(\mathbf{x})$ by

$$\mathbf{P}(\mathbf{x}) = \chi(\mathbf{x})\mathbf{E}(\mathbf{x}) \; , \tag{1}$$

where $\chi(\mathbf{x})$ is the dielectric susceptibility (Chapter 13) of the crystal at the point $\mathbf{x}$.

In writing (1) we have assumed for convenience that the connection between $\mathbf{P}$ and $\mathbf{E}$ is *local;* that is, the polarization at $\mathbf{x}$ is determined by the electric field at $\mathbf{x}$ and not by the electric field at other points $\mathbf{x}'$. At x-ray frequencies χ is generally of the order of 10^{-4} or less. The susceptibility may be a function both of position $\mathbf{x}$ and frequency ω, but except for the frequencies near an x-ray absorption edge or an optical resonance we need not concern ourselves here with the frequency dependence of χ.

We may analyze the susceptibility in a crystal lattice in a Fourier series:

$$\chi(\mathbf{x}) = \sum_{\mathbf{G}} \chi_{\mathbf{G}} e^{i\mathbf{G}\cdot\mathbf{x}} \; , \tag{2}$$

where $\mathbf{G}$ runs over all reciprocal lattice vectors, including $\mathbf{G} = 0$. It is easily shown that $\chi_{\mathbf{G}}$ must equal $\chi^*_{-\mathbf{G}}$ if $\chi(\mathbf{x})$ is real, for then the pair of terms

$$\chi_{\mathbf{G}} e^{i\mathbf{G}\cdot\mathbf{x}} + \chi_{-\mathbf{G}} e^{-i\mathbf{G}\cdot\mathbf{x}}$$

is identically equal to its complex conjugate. The reason that the sum is only

---

* In the Advanced Topics only CGS units are used.

[1] This problem is discussed in several papers, including K. Forsterling, Ann. d. Phys. **19**, 261–289 (1934); E. Fues, Z. Physik **109**, 14–24 and 236–259 (1938).

over reciprocal lattice vectors was shown in Chapter 2, for $\chi(\mathbf{x})$ is invariant under a fundamental lattice translation.

The Fourier components $\chi_{\mathbf{G}}$ of the susceptibility are of central importance and are directly proportional to the x-ray structure factors $\mathcal{S}_{\mathbf{G}}$ of Chapter 2. We can invert (2) to obtain $\chi_{\mathbf{G}}$ in terms of $\chi(\mathbf{x})$:

$$\chi_{\mathbf{G}} = V^{-1} \int_V d^3x\, \chi(\mathbf{x}) e^{-i\mathbf{G}\cdot\mathbf{x}} \ , \tag{4}$$

where V is the volume of the crystal. To obtain this result multiply both sides of (2) by $e^{-i\mathbf{G}'\cdot\mathbf{x}}$ and integrate over the volume V:

$$\int d^3x\, \chi(\mathbf{x}) e^{-i\mathbf{G}'\cdot\mathbf{x}} = \sum_{\mathbf{G}} \chi_{\mathbf{G}} \int d^3x\, e^{i(\mathbf{G}-\mathbf{G}')\cdot\mathbf{x}} = \chi_{\mathbf{G}'} V \ , \tag{5}$$

because the integral of $\exp\left[i(\mathbf{G} - \mathbf{G}') \cdot \mathbf{x}\right]$ is V if $\mathbf{G} = \mathbf{G}'$ and zero otherwise. To see that it is zero otherwise, consider the example of a one-dimensional lattice of N cells of lattice constant a. Then $G - G' = 2\pi h/a$, where h is an integer, and

$$\int_0^{Na} dx\, e^{i2\pi hx/a} = \frac{e^{i2\pi Nh} - 1}{2\pi iha} = 0 \ , \tag{6}$$

because Nh is an integer.

Electromagnetic wave equation

The electromagnetic field in a crystal need not be invariant under a lattice translation, so that the general Fourier analysis of the electric field vector in an infinite crystal will require a Fourier integral instead of merely a sum such as (2):

$$\mathbf{E}(\mathbf{x}) = \int d^3K\, \mathbf{E}(\mathbf{K})\, e^{i\mathbf{K}\cdot\mathbf{x}} \ . \tag{7}$$

On substituting (2) and (7) in (1) we find the polarization

$$\mathbf{P}(\mathbf{x}) = \left(\sum_{\mathbf{G}} \chi_{\mathbf{G}} e^{i\mathbf{G}\cdot\mathbf{x}}\right)\left(\int d^3k\, \mathbf{E}(\mathbf{k}) e^{i\mathbf{k}\cdot\mathbf{x}}\right) = \sum_{\mathbf{G}} \chi_{\mathbf{G}} \int d^3K\, \mathbf{E}(\mathbf{K} - \mathbf{G}) e^{i\mathbf{K}\cdot\mathbf{x}} \ , \tag{8}$$

where on the right we have written $\mathbf{K} = \mathbf{k} + \mathbf{G}$ as the integration variable.

We are concerned with solutions of the electromagnetic wave equation

$$c^2 \nabla^2 \mathbf{E} = \frac{\partial^2}{\partial t^2}(\mathbf{E} + 4\pi\mathbf{P}) \ . \tag{9}$$

We look for a solution of (9) for which Fourier components $\mathbf{E}(\mathbf{K} - \mathbf{G})$ for all $\mathbf{G}$ have the identical time dependence $e^{-i\omega t}$. Such solutions are the normal modes of the electromagnetic field in the crystal. Now, from (7),

$$c^2 \nabla^2 \mathbf{E} = -c^2 \int d^3K\, K^2 \mathbf{E}(\mathbf{K}) e^{i\mathbf{K}\cdot\mathbf{x}} \tag{10}$$

and, from (7) and (8),

$$\frac{\partial^2}{\partial t^2}(\mathbf{E} + 4\pi\mathbf{P})$$

$$= -\omega^2 \int d^3K \Big\{ (1 + 4\pi\chi_0)\mathbf{E}(\mathbf{K}) + 4\pi \sum_{G \neq 0} \chi_G\mathbf{E}(\mathbf{K} - \mathbf{G}) \Big\} e^{i\mathbf{K}\cdot\mathbf{x}} . \quad (11)$$

The wave equation (9) will be satisfied if the coefficients of $e^{i\mathbf{K}\cdot\mathbf{x}}$ in (10) and (11) are equal. Thus

$$\boxed{c^2 K^2 \mathbf{E}(\mathbf{K}) = \omega^2 \sum_{\mathbf{G}} \epsilon_G \mathbf{E}(\mathbf{K} - \mathbf{G}) ,} \qquad (12)$$

where we denote $1 + 4\pi\chi_0$ by ϵ_0 and $4\pi\chi_G$ by ϵ_G. This is a set of linear homogeneous algebraic equations that depend on the structure of the crystal through the values of the Fourier coefficients ϵ_G. We now look for approximate solutions of (12) in the x-ray region where we may assume $\epsilon_0 \gg \epsilon_G$ because $\chi_G \ll 1$.

Bragg Reflection

For x-rays the wavevector $\mathbf{k}$ may be of the order of magnitude of a $\mathbf{G}$. The most interesting situation occurs when for a particular $\mathbf{G}$ it happens that

$$(\mathbf{k} - \mathbf{G})^2 = k^2 , \qquad \text{or} \qquad 2\mathbf{k}\cdot\mathbf{G} = G^2 , \qquad (13)$$

which is the condition for Bragg diffraction. When this condition is nearly satisfied both $c^2(\mathbf{k} - \mathbf{G})^2$ and c^2k^2 can be nearly equal to the same value of $\omega^2\epsilon_0$. We can show that the dominant waves in the crystal are now $\mathbf{E}(\mathbf{k})$ and $\mathbf{E}(\mathbf{k} - \mathbf{G})$, the incident wave and the Bragg-reflected wave. Equation (12) reduces to, with $\mathbf{k}$ written for $\mathbf{K}$,

$$c^2k^2\mathbf{E}(\mathbf{k}) = \omega^2[\epsilon_0\mathbf{E}(\mathbf{k}) + \epsilon_G\mathbf{E}(\mathbf{k} - \mathbf{G})] \qquad (14)$$

and, with $\mathbf{k} - \mathbf{G}$ written for $\mathbf{K}$,

$$c^2(\mathbf{k} - \mathbf{G})^2\mathbf{E}(\mathbf{k} - \mathbf{G}) = \omega^2[\epsilon_0\mathbf{E}(\mathbf{k} - \mathbf{G}) + \epsilon_{-G}\mathbf{E}(\mathbf{k})] . \qquad (15)$$

We may write ϵ_G^* for ϵ_{-G} by the argument following (2).

Equations (14) and (15) are two coupled linear equations in the field amplitudes $\mathbf{E}(\mathbf{k})$ and $\mathbf{E}(\mathbf{k} - \mathbf{G})$. They have a nontrivial solution when the determinant

$$\begin{vmatrix} c^2k^2 - \omega^2\epsilon_0 & -\omega^2\epsilon_G \\ -\omega^2\epsilon_G^* & c^2(\mathbf{k} - \mathbf{G})^2 - \omega^2\epsilon_0 \end{vmatrix} = 0 , \qquad (16)$$

or

$$\omega^4(\epsilon_0{}^2 - |\epsilon_G|^2) - \omega^2\epsilon_0 c^2[k^2 + (\mathbf{k} - \mathbf{G})^2] + c^4k^2(\mathbf{k} - \mathbf{G})^2 = 0 . \quad (17)$$

The Bragg condition is that $(\mathbf{k} - \mathbf{G})^2$ is exactly equal to k^2. Let $\mathbf{k}_0$ denote such a value of $\mathbf{k}$. Then (17) reduces to

$$\omega^4(\epsilon_0{}^2 - |\epsilon_\mathbf{G}|^2) - 2\omega^2\epsilon_0 c^2 k_0{}^2 + c^4 k_0{}^4 = 0 \ . \tag{18}$$

On solving for ω^2 we have the two roots

$$\omega_\pm{}^2 = \frac{c^2 k_0{}^2}{\epsilon_0 \pm |\epsilon_\mathbf{G}|} \ . \tag{19}$$

For real values of ω between ω_+ and ω_- the roots of (17) for k are complex, corresponding to waves damped in space. The roots are shown in Fig. 1. The frequencies between ω_+ and ω_- are said to lie in a **forbidden band**. Waves in this frequency region do not propagate in the crystal, but are reflected strongly.

This is most easily exhibited by an example in which $\mathbf{k}$ and $\mathbf{k} - \mathbf{G}$ are nearly equal in magnitude but exactly opposite in direction. In (17) we set $k = \frac{1}{2}G + \delta$, where $\delta \ll \frac{1}{2}G$. Then

$$\delta^4 - 2\delta^2\left(\frac{1}{4}G^2 + \frac{\omega^2\epsilon_0}{c^2}\right) + \left(\frac{1}{4}G^2\right)^2 - 2\left(\frac{1}{4}G^2\right)\left(\frac{\omega^2\epsilon_0}{c^2}\right)$$
$$+ \left(\frac{\omega^4}{c^4}\right)(\epsilon_0{}^2 - |\epsilon_\mathbf{G}|^2) = 0 \ . \tag{20}$$

We neglect the term in δ^4. Let us solve for δ at the arbitrary frequency $\omega^2 = (\frac{1}{2}Gc)^2/\epsilon_0$, which lies between ω_+ and ω_-. Equation (22) becomes

$$-2\delta^2\left(\frac{1}{2}G^2\right) + \left[-\left(\frac{1}{4}G^2\right)^2 + \left(\frac{\frac{1}{4}G^2}{\epsilon_0}\right)^2(\epsilon_0{}^2 - \epsilon_\mathbf{G}{}^2)\right] = 0 \tag{21}$$

or

$$4\delta^2 = -\left(\frac{1}{4}G^2\right)\left(\frac{\epsilon_\mathbf{G}}{\epsilon_0}\right)^2 \ ; \qquad \delta = \pm i\left(\frac{1}{2}G\right)\left(\frac{\epsilon_\mathbf{G}}{2\epsilon_0}\right) \ , \tag{22}$$

so that at a frequency near the middle of the forbidden band the wavevector is given by

$$k = \frac{1}{2}G\left(1 \pm \frac{i\epsilon_\mathbf{G}}{2\epsilon_0}\right) \ . \tag{23}$$

The spatial attenuation of the mode is governed essentially by $\epsilon_\mathbf{G}$, the component of the dielectric constant at the reciprocal lattice vector $\mathbf{G}$.

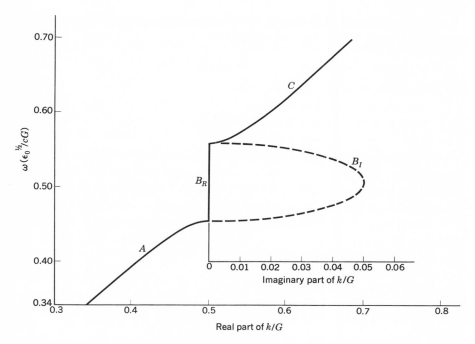

Figure 1 Frequency gap for Bragg reflection. If we substitute real frequencies in Eq. (17) and solve for the wavevector k, we find that k is real on the branches A and C of the dispersion relation. Branch A runs from zero to just below $k = \frac{1}{2}G$; branch C starts just above $k = \frac{1}{2}G$. Branch B at $k = \frac{1}{2}G$ is complex: the real part of k on branch B is always equal to $k = \frac{1}{2}G$; the imaginary part of k on branch B is plotted as B_I. The dispersion relation is plotted for the sake of visibility for an unrealistically high value of ϵ_G, namely $\epsilon_G = 0.2\epsilon_0$.

Advanced Topic B

DERIVATION OF THE VAN DER WAALS INTERACTION

We consider two identical linear harmonic oscillators 1 and 2 separated by R. Each oscillator bears charges $\pm e$ with separations x_1 and x_2, as in Fig. 1. The particles oscillate along the x axis. Let p_1 and p_2 denote the momenta $m(dx_1/dt)$ and $m(dx_2/dt)$. Then the hamiltonian of the unperturbed system is

$$\mathcal{H}_0 = \frac{1}{2m}p_1{}^2 + \tfrac{1}{2}\beta x_1{}^2 + \frac{1}{2m}p_2{}^2 + \tfrac{1}{2}\beta x_2{}^2 \; . \tag{1}$$

The uncoupled oscillators each have the resonance frequency

$$\omega_0 = \left(\frac{\beta}{m}\right)^{\frac{1}{2}} \tag{2}$$

appropriate to a simple harmonic oscillator.

Let $\mathcal{H}_1$ be the interaction energy of the two oscillators. The geometry is shown in the figure. The internuclear coordinate is R. Then

$$\mathcal{H}_1 = \frac{e^2}{R} + \frac{e^2}{R + x_1 - x_2} - \frac{e^2}{R + x_1} - \frac{e^2}{R - x_2} \; ; \tag{3a}$$

in the approximation $|x_1|, |x_2| \ll R$ we have

$$\mathcal{H}_1 \cong -\frac{2e^2 x_1 x_2}{R^3} \; . \tag{3b}$$

The total hamiltonian with the approximate form (3b) for $\mathcal{H}_1$ can be diagonalized by the normal mode transformation

$$x_s \equiv \frac{1}{\sqrt{2}}\,(x_1 + x_2) \; ; \qquad x_a \equiv \frac{1}{\sqrt{2}}\,(x_1 - x_2) \; , \tag{4}$$

or, on solving for x_1 and x_2,

$$x_1 = \frac{1}{\sqrt{2}}\,(x_s + x_a) \; ; \qquad x_2 = \frac{1}{\sqrt{2}}\,(x_s - x_a) \; . \tag{5}$$

The subscript s and a denote symmetric and antisymmetric modes of motion. Further, we have the momenta p_s, p_a associated with the two modes:

$$p_1 \equiv \frac{1}{\sqrt{2}}\,(p_s + p_a) \; ; \qquad p_2 \equiv \frac{1}{\sqrt{2}}\,(p_s - p_a) \; . \tag{6}$$

Figure 1   Coordinates of the two oscillators.

The total hamiltonian $\mathcal{H}_0 + \mathcal{H}_1$ after the transformations (5) and (6) is

$$\mathcal{H} = \left[\frac{1}{2m}p_s{}^2 + \frac{1}{2}\left(\beta - \frac{2e^2}{R^3}\right)x_s{}^2\right] + \left[\frac{1}{2m}p_a{}^2 + \frac{1}{2}\left(\beta + \frac{2e^2}{R^3}\right)x_a{}^2\right] . \quad (7)$$

The two frequencies of the coupled oscillators are found by inspection of (7) to be

$$\omega = \left[\left(\beta \pm \frac{2e^2}{R^3}\right)\Big/m\right]^{\frac{1}{2}} \cong \omega_0\left[1 \pm \frac{1}{2}\left(\frac{2e^2}{\beta R^3}\right) - \frac{1}{8}\left(\frac{2e^2}{\beta R^3}\right)^2 + \cdots\right] , \quad (8)$$

with ω_0 given by (2). In (8) we have expanded the square root.

The zero point energy of the system is $\frac{1}{2}\hbar(\omega_s + \omega_a)$; the sum is lowered because of the interaction by the amount

$$\Delta U = \tfrac{1}{2}\hbar(\Delta\omega_s + \Delta\omega_a) = -\hbar\omega_0 \cdot \frac{1}{8}\left(\frac{2e^2}{\beta R^3}\right)^2 . \quad (9)$$

This is an attractive interaction which varies as the minus sixth power of the separation of the two oscillators.

The polarizability of an oscillator is

$$\alpha \equiv \frac{\text{electric dipole moment}}{\text{electric field intensity}} = \frac{e^2}{\beta} , \quad (10)$$

by the argument of Chap. 13. Thus

$$\Delta U = -\hbar\omega_0\frac{\alpha^2}{2R^6} = -\frac{C}{R^6} , \quad (11)$$

where $C \equiv \hbar\omega_0\alpha^2$. This is the **van der Waals interaction** of Eq. (3.3). The interaction is a quantum effect, in the sense that $\Delta U \to 0$ as $\hbar \to 0$. To summarize: the zero point energy of the system is lowered by the dipole-dipole coupling $\mathcal{H}_1 = -2e^2x_1x_2/R^3$ of Eq. (3), even though the dipole moment $-ex_1$ averaged over an isolated atom is zero.

There is even an attractive van der Waals interaction between the plates of a parallel plate capacitor. The theory for this geometry is simple and elegant: it is due to Casimir.[1] The interaction of parallel plates has been measured directly by Derjaguin.[2]

Calculations of the van der Waals interaction between two H atoms are given in the text by Pauling and Wilson.[3] Retardation effects change the form of the interaction at very large distances: an account of the corrections is given by Casimir and Polder.[4]

[1] H. G. B. Casimir, Proceedings of the Amsterdam Academy of Sciences **51**, 793 (1948).

[2] B. Derjaguin, Scientific American, July 1960, pp. 47–53.

[3] L. Pauling and E. B. Wison, *Introduction to quantum mechanics,* McGraw-Hill, 1935, Sec. 47.

[4] H. B. G. Casimir and D. Polder, Physical Review **73**, 360 (1948).

Advanced Topic C

VAN HOVE SINGULARITIES IN THE
DENSITY OF ORBITALS

From the result (6.34) for the density of orbitals,

$$\mathfrak{D}(\omega) = \left(\frac{L}{2\pi}\right)^3 \int \frac{dS_\omega}{v_g} \, , \tag{1}$$

we see that there is a singularity in the integrand wherever the group velocity $v_g \equiv |\nabla_\mathbf{K} \omega|$ vanishes, that is, where there is a locally flat region of the frequency ω as a function of the wavevector $\mathbf{K}$. The points in $\mathbf{K}$ space at which this occurs are called **critical points**. A critical point may be a maximum, a minimum, or a saddle point. We consider the density of orbitals near each of these in turn. The arguments apply to dispersion relations ω versus $\mathbf{K}$ in general and are not limited to phonons; thus the arguments apply to electron energy bands (Chapter 9) and to spin wave spectra (Chapter 16). The saddle points give particularly conspicuous features in the plots of density of orbitals versus frequency (Figs. 1c and 1d).

We let $\mathbf{K}_c$ be a critical point, and we expand the frequency about this point in terms of $\mathbf{q} = \mathbf{K} - \mathbf{K}_c$. The leading terms in the expansion will have the form

$$\omega(\mathbf{q}) = \omega_c + a_1 q_1^2 + a_2 q_2^2 + a_3 q_3^2 + \cdots \, , \tag{2}$$

where q_1, q_2, q_3 are the local principal axes of a surface of constant frequency, and ω_c is the frequency at the critical point.

(*a*) *Maximum.* For simplicity we suppose that the local frequency surface is spherical:

$$\omega(q) = \omega_c - aq^2 \, . \tag{3}$$

The volume of a sphere of radius q in Fourier space is

$$\Omega = \frac{4\pi}{3} q^3 = \frac{4\pi}{3} \cdot \left(\frac{\omega_c - \omega}{a}\right)^{\frac{3}{2}} \, ,$$

and the density of orbitals near ω_c is

$$\mathfrak{D}(\omega) = \left(\frac{L}{2\pi}\right)^3 \left|\frac{d\Omega}{d\omega}\right| = \left(\frac{L}{2\pi}\right)^3 \frac{2\pi}{a^{\frac{3}{2}}} (\omega_c - \omega)^{\frac{1}{2}} \tag{4}$$

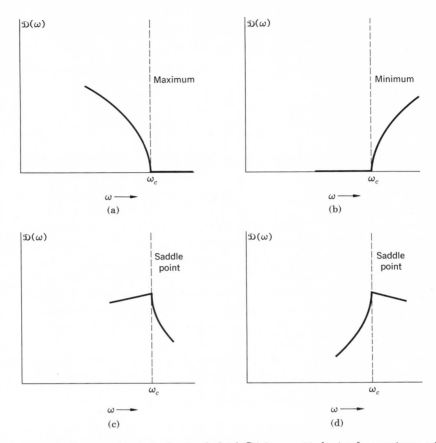

Figure 1 Discontinuities in the density of orbitals $\mathfrak{D}(\omega)$ near critical points for a maximum, minimum, and two types of saddle points. The curves plotted are only the *contributions* to the density of orbitals from the local region of the frequency surface near a critical point, and these contributions must be added to the general background of the density of orbitals from the rest of the frequency surface $\omega(\mathbf{k})$.

for $\omega < \omega_c$, and

$$\mathfrak{D}(\omega) = 0 \tag{5}$$

for $\omega > \omega_c$, as far as concerns the contribution of this region of the frequency surface. The contribution is shown in Fig. 1a.

(*b*) *Minimum.* We suppose that

$$\omega(q) = \omega_c + aq^2 . \tag{6}$$

By the above argument

$$\mathfrak{D}(\omega) = 0 , \qquad \omega < \omega_c ; \tag{7}$$

and

$$\mathfrak{D}(\omega) = \left(\frac{L}{2\pi}\right)^3 \frac{2\pi}{a^{\frac{3}{2}}} (\omega - \omega_c)^{\frac{1}{2}} , \qquad \omega > \omega_c , \qquad (8)$$

as shown in Fig. 1b.

(c) **Saddle point.** We suppose that

$$\omega(q) = \omega_c - a(q_1{}^2 + q_2{}^2 - q_3{}^2) . \qquad (9)$$

To find the density of orbitals for $\omega < \omega_c$ we make a coordinate transformation such that

$$q_1 a^{\frac{1}{2}} = (\omega_c - \omega)^{\frac{1}{2}} \cosh \xi \cos \varphi ;$$
$$q_2 a^{\frac{1}{2}} = (\omega_c - \omega)^{\frac{1}{2}} \cosh \xi \sin \varphi ; \qquad (10)$$
$$q_3 a^{\frac{1}{2}} = (\omega_c - \omega)^{\frac{1}{2}} \sinh \xi .$$

By substitution we may verify that (10) satisfies (9); the relations

$$\cosh^2\xi - \sinh^2\xi = 1 ; \qquad \cos^2\varphi + \sin^2\varphi = 1 , \qquad (11)$$

are needed.

The element of volume in Fourier space is

$$dq_1 \, dq_2 \, dq_3 = \left| J\!\left(\frac{q_1, q_2, q_3}{\omega, \xi, \varphi}\right) \right| d\omega \, d\xi \, d\varphi , \qquad (12)$$

where J is the Jacobian of the transformation:

$$J = \begin{vmatrix} \dfrac{\partial q_1}{\partial \omega} & \dfrac{\partial q_2}{\partial \omega} & \dfrac{\partial q_3}{\partial \omega} \\[2mm] \dfrac{\partial q_1}{\partial \xi} & \dfrac{\partial q_2}{\partial \xi} & \dfrac{\partial q_3}{\partial \xi} \\[2mm] \dfrac{\partial q_1}{\partial \varphi} & \dfrac{\partial q_2}{\partial \varphi} & \dfrac{\partial q_3}{\partial \varphi} \end{vmatrix} = \frac{(\omega_c - \omega)^{\frac{1}{2}}}{2a^{\frac{3}{2}}} \cosh \xi . \qquad (13)$$

The evaluation is direct and only slightly tedious.

The density of orbitals (still for $\omega < \omega_c$) is given by

$$\mathfrak{D}(\omega) = \left(\frac{L}{2\pi}\right)^3 d\omega \int J \, d\xi \int d\varphi = \left(\frac{L}{2\pi}\right)^3 d\omega \cdot \frac{(\omega_c - \omega)^{\frac{1}{2}}}{2a^{\frac{3}{2}}} \int d(\sinh \xi) \int d\varphi , \tag{14}$$

where $d\omega \int J \, d\xi \int d\varphi$ is the volume in Fourier space enclosed by the constant frequency surfaces at ω and $\omega + d\omega$. We have used $(\cosh \xi) \, d\xi = d(\sinh \xi)$. The integral over $d\varphi$ is equal to 2π. We limit the region of interest around the critical point to the interior of a fixed sphere of radius Q about $\mathbf{K}_c$; thus $\mathfrak{D}(\omega)$ gives only the contribution from the interior of this sphere.

The upper limit on $\sinh \xi$ is found from

$$Q^2 = q_1{}^2 + q_2{}^2 + q_3{}^2 = \frac{\omega_c - \omega}{a}(\cosh^2\xi + \sinh^2\xi)$$

$$= \frac{\omega_c - \omega}{a}(1 + 2\sinh^2\xi) \ , \tag{15}$$

whence

$$\int d(\sinh \xi) = \frac{1}{\sqrt{2}}\left(\frac{aQ^2}{\omega_c - \omega} - 1\right)^{\frac{1}{2}} \tag{16}$$

and

$$\mathfrak{D}(\omega) \propto (Q^2 + \omega - \omega_c)^{\frac{1}{2}} \ , \qquad \omega < \omega_c \ . \tag{17}$$

For $\omega - \omega_c \ll Q$ we may expand (17) to obtain

$$\mathfrak{D}(\omega) \propto Q - \frac{\omega_c - \omega}{2Q} \ . \tag{17a}$$

To find the density of orbitals for $\omega > \omega_c$ we use the transformation

$$q_1 a^{\frac{1}{2}} = (\omega - \omega_c)^{\frac{1}{2}} \sinh \xi \cos \varphi \ ;$$
$$q_2 a^{\frac{1}{2}} = (\omega - \omega_c)^{\frac{1}{2}} \sinh \xi \sin \varphi \ ; \tag{18}$$
$$q_3 a^{\frac{1}{2}} = (\omega - \omega_c)^{\frac{1}{2}} \cosh \xi \ .$$

The volume element is

$$dq_1 \, dq_2 \, dq_3 = \frac{(\omega - \omega_c)^{\frac{1}{2}}}{2a^{\frac{3}{2}}} \, d\omega \, d\xi \, d\varphi \ , \tag{19}$$

and the limits on $\cosh \xi$ are

$$1 \le \cosh \xi \le \frac{1}{\sqrt{2}}\left(\frac{aQ^2}{\omega_c - \omega} + 1\right)^{\frac{1}{2}} \ . \tag{20}$$

The density of orbitals

$$\mathfrak{D}(\omega) \propto \tfrac{1}{2}(Q^2 + \omega - \omega_c)^{\frac{1}{2}} - (\omega - \omega_c)^{\frac{1}{2}} \ , \qquad \omega > \omega_c \ . \tag{21}$$

The derivative of $\mathfrak{D}(\omega)$ with respect to ω is infinite at $\omega = \omega_c$, for the leading term in the derivative is

$$\frac{d\,\mathfrak{D}(\omega)}{d\omega} \propto -\frac{1}{2(\omega - \omega_c)^{\frac{1}{2}}} \ , \tag{22}$$

which approaches $-\infty$ as $\omega \to \omega_c$ from above. The behavior of $\mathfrak{D}(\omega)$ near a saddle point of the form (9) is shown in Fig. 1c; the behavior near a saddle point of the form

$$\omega(q) = \omega_c - a(q_1{}^2 - q_2{}^2 - q_3{}^2) \tag{23}$$

is shown in Fig. 1d as the reverse of Fig. 1c.

Van Hove theorem. Van Hove[1] proved by topological methods that each type of saddle point [of which (9) and (23) are special cases] will occur at least three times for any sheet of a dispersion relation ω versus **k**. The dispersion curves for phonons in aluminum (as shown in Fig. 6.12) reflect many of the expected discontinuities.

---

[1] L. Van Hove, Phys. Rev. **89**, 1189 (1953). The index of a critical point is the number of positive signs in the principal axis quadratic expansion of ω near the critical point. In three dimensions each branch of $\omega(q)$ has at least one maximum, three saddle points of each type (indices 1 and 2), and one minimum, according to a theorem in topology due to M. Morse.

Advanced Topic D

WAVEVECTOR-DEPENDENT DIELECTRIC FUNCTION OF FREE ELECTRON FERMI GAS

Important properties of an electron gas are described by the dielectric function $\epsilon(K, \omega)$ defined below. To find the dielectric function we consider the response of the electrons to an applied external electrostatic field. We start with a uniform gas of electrons of charge concentration $-n_0 e$ superposed on a background of positive charge of concentration $n_0 e$. The positive charge background is now deformed mechanically to produce a sinusoidal variation:

$$\rho^+(x) = n_0 e + \rho_K^{\text{ext}} \sin Kx \ . \tag{1}$$

The term $\rho_K^{\text{ext}} \sin Kx$ gives rise to an electrostatic field that we shall call the external field applied to the electron gas.

The electrostatic potential φ of a charge distribution is found from the Poisson equation $\nabla^2 \varphi = -4\pi\rho$. With

$$\varphi = \varphi_K^{\text{ext}} \sin Kx \ ; \qquad \rho = \rho_K^{\text{ext}} \sin Kx \ , \tag{2}$$

the Poisson equation gives the relation

$$K^2 \varphi_K^{\text{ext}} = 4\pi \rho_K^{\text{ext}} \ . \tag{3}$$

The electron gas will be deformed by the combined influences of the electrostatic potential φ_K^{ext} of the positive charge distribution and of the as yet unknown induced electrostatic potential

$$\varphi_K^{\text{ind}} \sin Kx$$

of the deformation of the electron gas itself. The electron charge density is

$$\rho^-(x) = -n_0 e + \rho_K^{\text{ind}} \sin Kx \ , \tag{4}$$

where ρ_K^{ind} is the amplitude of the charge density variation induced in the electron gas. We want to find ρ_K^{ind} in terms of ρ_K^{ext}.

The amplitude of the total electrostatic potential[1]

$$\varphi_K = \varphi_K^{\text{ext}} + \varphi_K^{\text{ind}} \tag{5}$$

of the positive and negative charge distributions is related to the total charge density variation

$$\rho_K = \rho_K^{\text{ext}} + \rho_K^{\text{ind}} \tag{6}$$

by the Poisson equation. Then, as in Eq. (3),

$$K^2 \varphi_K = 4\pi \rho_K \ . \tag{7}$$

[1] Here φ_K^{ext} is due to the positive charge background, and φ_K^{ind} is due to the electron gas.

But the electron deformation ρ_K^{ind} is related to the total static electric potential φ_K by the Thomas-Fermi equation[2] (8.21b):

$$-e[n(\mathbf{r}) - n_0] = \rho_K^{\text{ind}} \sin Kx = -\frac{3n_0 e^2}{2\epsilon_F} \varphi_K \sin Kx \ , \tag{8}$$

or

$$\varphi_K = -\frac{2\epsilon_F}{3n_0 e^2} \rho_K^{\text{ind}} \ . \tag{9}$$

We can combine (7) and (9) to obtain the ratio of the induced charge variation to the total charge variation

$$\frac{\rho_K^{\text{ind}}}{\rho_K} = -\frac{6\pi n_0 e^2}{\epsilon_F K^2} = -\frac{\lambda^2}{K^2} \ , \tag{10}$$

written in terms of the reciprocal screening length $\lambda \equiv (6\pi n_0 e^2/\epsilon_F)^{\frac{1}{2}}$.

The dielectric function $\epsilon(K, \omega)$ measures the response of the electron gas to an external electric field of wavevector K and frequency ω. The dielectric function is defined by the following relation between the amplitude ρ_K^{ext} of the applied charge density and the amplitude φ_K of the total potential:

$$\boxed{K^2 \varphi_K = \frac{4\pi}{\epsilon(K, \omega)} \rho_K^{\text{ext}} \ ,} \tag{11}$$

where both φ_K and ρ_K^{ext} have the identical frequency ω. Equation (11) is the Poisson equation (7), but with the external charge density ρ_K^{ext} written in place of the total charge density ρ_K on the right-hand side. Note that at a zero of $\epsilon(K, \omega)$ the potential φ_K may be finite even in the absence of an external charge. In this condition the system will oscillate freely, without an external driving force.

If we divide (11) by (7) we obtain an expression for $\epsilon(K, \omega)$ in terms of the charge densities:

$$\epsilon(K, \omega) = \frac{\rho_K^{\text{ext}}}{\rho_K} = 1 - \frac{\rho_K^{\text{ind}}}{\rho_K} \ . \tag{12}$$

---

[2] The use of the Thomas-Fermi equation is an approximation which is good for long wavelengths $(K \to 0)$ and zero frequency.

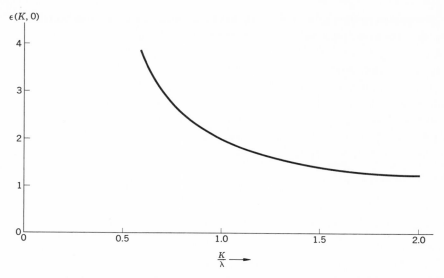

Figure 1 Static dielectric function of a free electron Fermi gas versus wavevector, in the Thomas-Fermi approximation. The horizontal axis is the wavevector times the screening length $1/\lambda$.

With use of (10) we have the result for the Thomas-Fermi dielectric function

$$\epsilon(K, 0) = 1 + \frac{\lambda^2}{K^2} \; ; \qquad \lambda^2 \equiv \frac{6\pi n_0 e^2}{\epsilon_F} \; , \tag{13}$$

as in Fig. 1. A more accurate approximation to the dielectric function is due to Lindhard.[3]

By comparison of (7) and (11) we can express the dielectric function in another way:

$$\varphi_K = \frac{\varphi_K^{\text{ext}}}{\epsilon(K, \omega)} \; , \tag{14}$$

so that the total electrostatic potential φ_K is equal to the applied potential divided by the dielectric function. If we define φ_K^{ext} as the change in potential of the positive ion cores in a metal on passage of a phonon of wavevector K, and $\epsilon(K, \omega)$ is the dielectric function of the electron gas, then φ_K is the total potential associated with the phonon, including conduction electron and ion core contributions.

---

[3] A good discussion of the Lindhard dielectric function is given by J. Ziman, *Principles of the theory of solids*, Cambridge, 1964, Chap. 5. The algebraic steps in the evaluation of Ziman's equation (5.33) are given in detail by C. Kittel, *Solid state physics* **22**, 1 (1968), Section 6.

Because $\epsilon(K, 0) \to \infty$ as $K \to 0$, the total potential φ_K approaches 0 for finite external potentials of long wavelength: the electron gas very effectively screens a long wavelength perturbation. A short wavelength perturbation is screened less effectively. The screening of a free charge is shown in Fig. 8.8a: The sharply-varying (high K) components of the Coulomb potential $1/r$ are screened less than slowly-varying (low K) components. Thus the long-range tail of the Coulomb potential is screened by the electron gas, but the steep central core of the potential is not screened. It is shown in *QTS*, Eq. (1.24), that the Fourier components of the **bare Coulomb potential** $\varphi(r) = 1/r$ are

$$\varphi_K^{\text{ext}} = \frac{4\pi}{K^2} \; . \tag{15}$$

Thus by (13) and (14) we have for the **screened Coulomb potential**

$$\varphi_K = \frac{4\pi}{\epsilon(K)K^2} = \frac{4\pi}{K^2 + \lambda^2} \; . \tag{16}$$

The difference between the bare and screened potentials is most marked for $K \ll \lambda$. In this limit the screened potential is independent of the wavevector:

$$\varphi_K \cong \frac{4\pi}{\lambda^2} = \frac{2}{3}\epsilon_F \cdot \frac{1}{n_0 e^2} \; . \tag{17}$$

We have derived two limiting expressions for the dielectric function of an electron gas:

$$\epsilon(K, 0) = 1 + \frac{\lambda^2}{K^2} \; ; \qquad \epsilon(0, \omega) = 1 - \frac{\omega_p^2}{\omega^2} \; . \tag{18}$$

We notice that $\epsilon(K, 0)$ as $K \to 0$ does not approach the same limit as $\epsilon(0, \omega)$ as $\omega \to 0$. This means that great care must be taken with the dielectric function near the origin of the ω-K plane. The full theory shows that in the region $\omega < v_F K$, where v_F is the Fermi velocity, we obtain the Thomas-Fermi dielectric function for the electron gas:

$$\epsilon(K, \omega) \cong 1 + \frac{\lambda^2}{K^2} \; , \tag{19}$$

for low K. Calculations for the general function $\epsilon(K, \omega)$ tend to be quite complicated.

We apply this result to the vibrational modes of a lattice of positive ions of mass M in a sea of electrons of mass m. The dielectric function of the positive ions alone is

$$1 - \frac{4\pi n e^2}{M\omega^2} \; , \tag{20}$$

provided their spacing is sufficiently large that the ions are independent. We now add the electron gas to the problem and consider the *total* dielectric function, lattice plus electrons:

$$\epsilon(K, \omega) = 1 - \frac{4\pi n e^2}{M\omega^2} + \frac{\lambda^2}{K^2} \, , \tag{21}$$

with the electron screening constant λ given by (13). At low K and ω we may neglect the term 1. The longitudinal modes of oscillation of the system are given by the zeros of $\epsilon(K, \omega)$, as we have argued above and in Chapter 5. At a zero of $\epsilon(K, \omega)$ we have, with $\epsilon_F \equiv \frac{1}{2} m v_F^2$,

$$\omega^2 = \frac{4\pi n^2 e^2}{M\lambda^2} K^2 = \frac{4\pi n e^2}{M} \cdot \frac{\epsilon_F}{6\pi n e^2} K^2 = \frac{m}{3M} v_F^2 K^2 \, , \tag{22}$$

or

$$\omega = vK \; ; \qquad v = \sqrt{\frac{m}{3M}} v_F \, , \tag{23}$$

which describes long wavelength acoustic phonons. The velocity is in quite good agreement with experimental values of the velocity of longitudinal waves in the alkali metals. For potassium we calculate $v = 1.8 \times 10^5$ cm/sec; the observed longitudinal wave velocity in the [100] direction is 2.2×10^5 cm/sec at 4°K. Because $v \ll v_F$ the condition $\omega < v_F K$ for the applicability of the dielectric function (19) to the electron gas is satisfied.

There is yet another type of zero of $\epsilon(K, \omega)$ for the problem of positive ions imbedded in an electron sea. For high frequencies we add to (20) the dielectric contribution $-\omega_p^2/\omega^2$ of the electron gas as given by $\epsilon(0, \omega)$ in (18), in place of (19). Then

$$\epsilon(0, \omega) = 1 - \frac{4\pi n e^2}{M\omega^2} - \frac{4\pi n e^2}{m\omega^2} \, , \tag{24}$$

and this function has a zero when

$$\omega^2 = \frac{4\pi n e^2}{\mu} \; ; \qquad \frac{1}{\mu} = \frac{1}{M} + \frac{1}{m} \, . \tag{25}$$

This is just the electron plasma frequency including the reduced mass correction for the motion of the positive ions.

Advanced Topic E

FERMI-DIRAC DISTRIBUTION

With the help of a little preparation in statistical physics the Fermi-Dirac distribution function may be derived quite easily. The Fermi-Dirac distribution gives the probability $f(\epsilon)$ that a single particle orbital of energy ϵ is occupied when the system of which the orbital is a part is in thermal equilibrium at temperature T. According to the Gibbs factor,[1] which is a generalization of the Boltzmann factor, the probability $P(N, \epsilon_l)$ that a system contains N particles and is in a state of total energy ϵ_l, is proportional to

$$\exp\left[(\mu N - \epsilon_l)/k_B T\right] , \qquad (1)$$

where μ, the chemical potential, is to be determined to make the total number of particles come out correctly, as discussed in Chapter 7.

Let us apply (1) to an orbital that can contain at most one electron. We choose the energy to be zero when the orbital is vacant, whence

$$P(0, 0) \propto e^0 = 1 . \qquad (2)$$

The energy is ϵ when the orbital contains one electron, so that

$$P(1, \epsilon) \propto e^{(\mu - \epsilon)/k_B T} . \qquad (3)$$

The probability that the orbital is occupied may be written as

$$f(\epsilon) \equiv P(1, \epsilon) = \frac{P(1, \epsilon)}{P(0, 0) + P(1, \epsilon)} = \frac{e^{(\mu - \epsilon)/k_B T}}{1 + e^{(\mu - \epsilon)/k_B T}} = \frac{1}{e^{(\epsilon - \mu)/k_B T} + 1} ,$$

$$\qquad (4)$$

the desired result.

[1] C. Kittel, *Thermal physics*, Wiley (1969).

Advanced Topic F

TIGHT BINDING APPROXIMATION FOR ELECTRONS IN METALS

It is useful to look at the formation of allowed and forbidden electron bands in another way. We start from the energy levels of the neutral separated atoms and watch the changes in the levels as the charge distributions of adjacent atoms overlap when the atoms are brought together to form the metal. We can understand the origin of the splitting of free atom energy levels into bands as the atoms are brought together by considering two hydrogen atoms, each with its electron in the $1s$ (ground) state. In Fig. 1 the wavefunctions ψ_A, ψ_B on the separated atoms are shown in (a). As the atoms are brought closer together and their wavefunctions overlap, we are led to consider the two combinations $\psi_A \pm \psi_B$. Each combination shares the electrons equally as between the two protons, but an electron in the state $\psi_A + \psi_B$ will have a somewhat lower energy than in the state $\psi_A - \psi_B$, for the following reason. In the state $\psi_A + \psi_B$ shown in (b) the electron spends part of the time in the region midway between the two protons, and in this region it is under the influence of the attractive potential of both protons at once, thereby increasing the binding energy. In the state $\psi_A - \psi_B$ shown in (c) the probability density vanishes midway between the nuclei; an extra contribution to the binding does not appear. Thus as two atoms are brought together two separated energy levels are formed for each level of the isolated atom. For N atoms, N orbitals are formed for each orbital of the isolated atom and these N orbitals will be associated with one or more bands.

As free atoms are brought together the Coulomb interaction between the atom cores and the overlapping parts of the electron distribution will split the energy levels of the combined system, spreading the levels out into bands. A state of quantum number ns of the free atom is spread out in the metal into a band of energies. Here n denotes the principal quantum number, and s indicates zero orbital angular momentum. The width of the band is proportional to the strength of the overlap interaction between neighboring atoms, each in the state ns. There will also be bands formed from $p, d, \ldots$ states ($l = 1, 2, \ldots$) of the free atoms. The $(2l + 1)$ states degenerate in the free atom will form $(2l + 1)$ bands. Each of these bands will not have in general the same energy as any other band over any substantial proportion of the range of the wavevector. Two or more bands may coincide in energy at certain values of $\mathbf{k}$ in the Brillouin zone.

The approximation which starts out from the wavefunctions of the free atoms is known as the **tight binding approximation**. A simple example of its

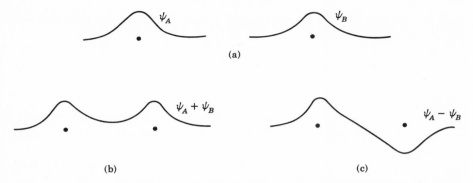

Figure 1   (a) Schematic drawing of wavefunctions of electrons on two hydrogen atoms at large separation. (b) Ground state wavefunction at closer separation. (c) Excited state wavefunction.

use appears below. The tight binding approximation is thought to be quite good for the inner electrons of atoms but it is not often a good description of the conduction electrons themselves. It is used to describe approximately the d bands of some of the transition metals and the valence bands of inert gas crystals.

Suppose that the ground state of an electron moving in the potential $U(r)$ of an isolated atom is $\phi(\mathbf{r})$ and that the energy is E_0; suppose further that ϕ is an s state. The treatment of bands arising from degenerate $(p, d, \ldots)$ atomic levels is more complicated. If the influence of one atom on another is small, we obtain an approximate wavefunction for one electron in the whole crystal by taking

$$\psi_{\mathbf{k}}(\mathbf{r}) = \sum_j C_{\mathbf{k}j}\phi(\mathbf{r} - \mathbf{r}_j) \ , \tag{1}$$

where the sum is over all lattice points. (We assume that the primitive basis contains one atom.) This function is of the Bloch form if $C_{\mathbf{k}j} = N^{-\frac{1}{2}}e^{i\mathbf{k}\cdot\mathbf{r}j}$, which gives

$$\psi_{\mathbf{k}}(\mathbf{r}) = N^{-\frac{1}{2}}\sum_j e^{i\mathbf{k}\cdot\mathbf{r}j}\phi(\mathbf{r} - \mathbf{r}_j) \ , \tag{2}$$

for a crystal of N atoms.

We prove (2) is of the Bloch form by considering the effect of a translation by a vector $\mathbf{T}$ connecting two lattice points:

$$\psi_{\mathbf{k}}(\mathbf{r} + \mathbf{T}) = N^{-\frac{1}{2}}\sum_j e^{i\mathbf{k}\cdot\mathbf{r}j}\phi(\mathbf{r} + \mathbf{T} - \mathbf{r}_j)$$

$$= e^{i\mathbf{k}\cdot\mathbf{T}}N^{-\frac{1}{2}}\sum_j e^{i\mathbf{k}\cdot(\mathbf{r}_j - \mathbf{T})}\phi[\mathbf{r} - (\mathbf{r}_j - \mathbf{T})] = e^{i\mathbf{k}\cdot\mathbf{T}}\psi_{\mathbf{k}}(\mathbf{r}) \ , \tag{3}$$

which is exactly the Bloch requirement.

We find the first-order energy by calculating the diagonal matrix elements of the hamiltonian of the crystal. We have

$$\langle \mathbf{k}|H|\mathbf{k}\rangle = N^{-1}\sum_j \sum_m e^{i\mathbf{k}\cdot(\mathbf{r}_j-\mathbf{r}_m)}\langle \varphi_m|H|\varphi_j\rangle \ , \tag{4}$$

where $\varphi_m \equiv \varphi(\mathbf{r}-\mathbf{r}_m)$. Writing $\boldsymbol{\rho}_m = \mathbf{r}_m - \mathbf{r}_j$,

$$\langle \mathbf{k}|H|\mathbf{k}\rangle = \sum_m e^{-i\mathbf{k}\cdot\rho_m}\int dV\, \varphi(\mathbf{r}-\boldsymbol{\rho}_m)H\varphi(\mathbf{r}) \ . \tag{5}$$

We now neglect all integrals in (5) except those on the same atom and those between nearest neighbors connected by $\boldsymbol{\rho}$. We write

$$\int dV\, \varphi^*(\mathbf{r})H\varphi(\mathbf{r}) = -\alpha \ ; \qquad \int dV\, \varphi^*(\mathbf{r}-\boldsymbol{\rho})H\varphi(\mathbf{r}) = -\gamma \ ; \tag{6}$$

and we get

$$\langle \mathbf{k}|H|\mathbf{k}\rangle = -\alpha - \gamma\sum_m e^{-i\mathbf{k}\cdot\rho_m} \ .$$

Thus the first-order energy is

$$\epsilon_\mathbf{k} = -\alpha - \gamma\sum_m e^{-i\mathbf{k}\cdot\rho_m} \ . \tag{7}$$

For a simple cubic lattice the nearest-neighbor atoms are at the positions

$$\boldsymbol{\rho}_m = (\pm a, 0, 0) \ ; \qquad (0, \pm a, 0) \ ; \qquad (0, 0, \pm a) \ , \tag{8}$$

so that (7) becomes

$$\epsilon_\mathbf{k} = -\alpha - 2\gamma(\cos k_x a + \cos k_y a + \cos k_z a) \ . \tag{9}$$

Thus the energies are confined to a band of width 12γ. The weaker the overlap, the narrower is the energy band. Several surfaces of constant energy for a band of the form of (9) are shown in Fig. 2.

For $ka \ll 1$,

$$\epsilon_\mathbf{k} \cong -\alpha - 6\gamma + \gamma k^2 a^2 \ . \tag{10}$$

The energy at the bottom of the band is independent of the direction of motion. The effective mass is

$$m^* = \frac{\hbar^2}{2\gamma a^2} \ . \tag{11}$$

When the overlap integral γ is small the band width is narrow and the effective mass is high.

For every orbital of an electron in the free atom there exists a band of energies in the crystal. We have considered here one orbital of the free atom and have obtained one band. The number of orbitals in the zone which corre-

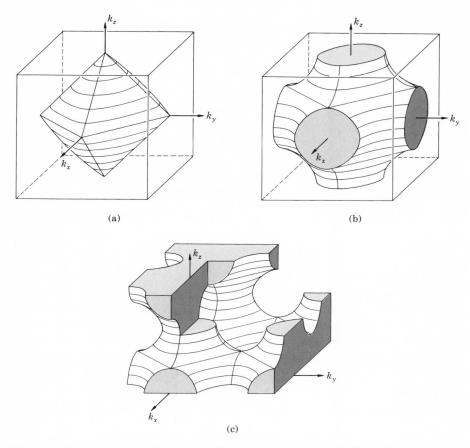

Figure 2 Constant energy surfaces in the Brillouin zone of a simple cubic lattice, for the assumed energy band

$$\epsilon_{\mathbf{k}} = -\alpha - 2\gamma(\cos k_x a + \cos k_y a + \cos k_z a).$$

(a) Constant energy surface $\epsilon = -\alpha + 2|\gamma|$. (b) Constant energy surface $\epsilon = -\alpha$. The filled volume contains one electron per primitive cell. (c) Constant energy surface $\epsilon = -\alpha$ exhibited in the periodic zone scheme. The connectivity of the orbits is more clearly shown here than in (b). Hole and electron orbits are easily found. (The sketches are from A. Sommerfeld and H. Bethe, and were made by H. Ruehle.)

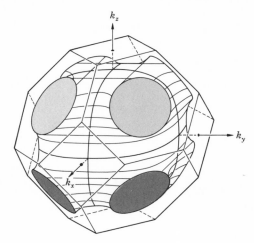

Figure 3 A constant energy surface of an fcc crystal structure, in the nearest-neighbor tight-binding approximation. The surface shown has $\epsilon = -\alpha + 2|\gamma|$.

sponds to a nondegenerate atomic level is equal to $2N$, where N is the number of atoms. We see this directly: (9) is periodic in $\mathbf{k}$, and thus only values of $\mathbf{k}$ lying within the first Brillouin zone in $\mathbf{k}$ space will define independent wavefunctions. In the simple cubic case the polyhedron is defined by $-\pi/a < k_x < \pi/a$, etc. The volume of the polyhedron is $8\pi^3/a^3$; but the number of orbitals (counting both spin orientations) per unit volume of $\mathbf{k}$ space is $V/4\pi^3$, so that the number of orbitals is $2V/a^3 = 2N$. Here V is the volume of the crystal, and $1/a^3$ is the number of atoms per unit volume.

For the bcc structure with eight nearest neighbors the same model gives

$$\epsilon_{\mathbf{k}} = -\alpha - 8\gamma \cos \tfrac{1}{2}k_x a \cos \tfrac{1}{2}k_y a \cos \tfrac{1}{2}k_z a , \tag{12}$$

as a consequence of Eq. (7). For the fcc structure with twelve nearest neighbors,

$$\epsilon_{\mathbf{k}} = -\alpha - 4\gamma(\cos \tfrac{1}{2}k_y a \cos \tfrac{1}{2}k_z a + \cos \tfrac{1}{2}k_z a \cos \tfrac{1}{2}k_x a + \cos \tfrac{1}{2}k_x a \cos \tfrac{1}{2}k_y a) . \tag{13}$$

A constant energy surface of (13) is shown in Fig. 3; notice that the surface makes contact with the hexagonal faces of the zone boundary.

Advanced Topic G

PARTICLE MOTION IN WAVEVECTOR SPACE AND IN REAL SPACE UNDER APPLIED ELECTRIC AND MAGNETIC FIELDS

A little difficulty is sometimes experienced in visualizing the collisionless motion of an electron on energy bands of various forms. It is helpful to consider explicit solutions for the situations commonly encountered. We want to describe the motion in coordinate $\mathbf{r}$ space and in wavevector $\mathbf{k}$ space, because the equations of motion involve and connect $\mathbf{r}$ and $\mathbf{k}$.

Spherical Conduction Band

We consider the motion of a wave packet (which contains one electron) on one energy band in a cubic crystal. The energy of the band is assumed to have a simple minimum at $\mathbf{k} = 0$, and near this point the energy has the form

$$\epsilon(\mathbf{k}) = \frac{\hbar^2}{2m_e} k^2 \ , \tag{1}$$

where m_e is the effective mass of an electron. The surfaces of constant energy in $\mathbf{k}$ space are spheres, hence such a band is said to be **spherical.**

No energy band in a solid has the form (1) over an entire Brillouin zone; actual energy bands are deformed by the presence of zone boundaries. For example, the band (F.9) has the form

$$\epsilon(\mathbf{k}) = 2\gamma(3 - \cos k_x a - \cos k_y a - \cos k_z a) \ , \tag{2}$$

referred to $\epsilon(0) = 0$. Here γ is a constant that depends on the overlap of atomic wavefunctions on neighboring atoms. If we expand the cosines to order $(ka)^4$ we obtain

$$\epsilon(\mathbf{k}) = \gamma[k^2 a^2 - \tfrac{1}{12}(k_x{}^4 + k_y{}^4 + k_z{}^4)a^4 + \cdots] \ , \tag{3}$$

which is spherical to within one percent for $ka < 0.1\pi$.

The motion of the wave packet is particularly simple as long as we stay within that portion of the BZ for which the spherical approximation applies. The group velocity in coordinate space is then given by

$$\frac{d\mathbf{r}}{dt} \equiv \mathbf{v} \equiv \frac{1}{\hbar} \nabla_{\mathbf{k}}\epsilon(\mathbf{k}) \cong \frac{\hbar}{m_e}\mathbf{k} \ , \tag{4}$$

where $\mathbf{r}$ is the mean position of the wave packet in real space. On integration of the motion over time we have

$$r(t) = r(0) + \frac{\hbar}{m_e} \int_0^t dt\, k(t) \quad . \tag{5}$$

In an electric field $\mathbf{E}$ the rate of change of wavevector is described by $\hbar\, d\mathbf{k}/dt = -e\mathbf{E}$. If the electric field is uniform and constant, we integrate this to obtain

$$k(t) = k(0) - \frac{e}{\hbar} Et \quad , \tag{6}$$

so that $\mathbf{k}$ increases in the direction of $\mathbf{E}$ at a uniform rate, independent of the form of the energy band. We integrate once again to obtain

$$\int_0^t dt\, k(t) = k(0)t - \frac{1}{2}\frac{e}{\hbar} Et^2 \quad . \tag{7}$$

We substitute (7) in (5) to obtain the motion of the wave packet in coordinate space, referred to the position $\mathbf{r}(0)$ and wavevector $\mathbf{k}(0)$ of the packet at $t = 0$:

$$r(t) = r(0) + \frac{\hbar k(0)}{m_e} t - \frac{1}{2}\frac{e}{m_e} Et^2 \quad . \tag{8}$$

This result is exactly that for free particle of mass m_e and charge $-e$ in the electric field $\mathbf{E}$, because for a spherical band $\hbar\mathbf{k}(0)/m_e$ is the group velocity $\mathbf{v}(0)$ at $t = 0$.

We now treat the motion in a uniform constant magnetic field $\mathbf{B}$ directed parallel to the z axis. The analog of the Lorentz force equation is

$$\hbar \frac{d\mathbf{k}}{dt} = -\frac{e}{c}\, \mathbf{v} \times \mathbf{B} \quad . \tag{9}$$

Here $\mathbf{v} \equiv \hbar^{-1} \nabla_k \epsilon$ is the velocity of the wave packet in coordinate space, as in (4) for a spherical band. We combine (4) with (9) to obtain

$$\hbar \frac{d\mathbf{k}}{dt} = -\frac{e\hbar}{m_e c}\, \mathbf{k} \times \mathbf{B} \quad . \tag{10}$$

For $\mathbf{B} = B\hat{z}$ we have, with the cyclotron frequency ω_c defined as $\omega_c \equiv eB/m_e c$, the equations of motion

$$\frac{dk_x}{dt} = -\omega_c k_y \; ; \qquad \frac{dk_y}{dt} = \omega_c k_x \; ; \qquad \frac{dk_z}{dt} = 0 \; . \qquad (11)$$

The solutions of (11) are of the form

$$k_x(t) = K \cos(\omega_c t + \varphi) \; ; \qquad k_y(t) = K \sin(\omega_c t + \varphi) \; ;$$
$$k_z = \text{constant}, \qquad (12)$$

as may be verified by substitution. Here K and φ are constants to be selected to fit the initial conditions of the motion. If initially the electron is on the Fermi surface, then K^2 will satisfy the condition

$$K^2 + k_z{}^2 = k_F{}^2 = k_x{}^2 + k_y{}^2 + k_z{}^2 \; . \qquad (13)$$

If this condition is satisfied at $t = 0$, it will be satisfied by the solutions (12) at all times, because K and k_z are constant. Thus in $\mathbf{k}$ space the particle moves at the frequency ω_c on the Fermi surface in a circle of radius K at constant k_z.

The position of the particle in real space is found by integration of $\mathbf{v} = \hbar^{-1} \nabla_{\mathbf{k}} \epsilon(\mathbf{k})$, with use of (12):

$$x(t) = x(0) + \frac{\hbar}{m_e} \int_0^t dt\, k_x = x(0) + \frac{\hbar K}{m_e \omega_c}[\sin(\omega_c t + \varphi) - \sin \varphi] \; ;$$

$$y(t) = y(0) - \frac{\hbar K}{m_e \omega_c}[\cos(\omega_c t + \varphi) - \cos \varphi] \; ; \qquad z(t) = z(0) + \frac{\hbar k_z t}{m_e} \; .$$
$$(14)$$

In real space the particle moves in a helix about the axis of the magnetic field. The radius R of the helix is

$$R = \frac{\hbar K}{m_e \omega_c} = \frac{\hbar c K}{eB} \; , \qquad (15)$$

which is equivalent to the classical relation $v_\perp = \omega_c R$, where $v_\perp$ is the velocity in real space in the plane normal to $\mathbf{B}$.

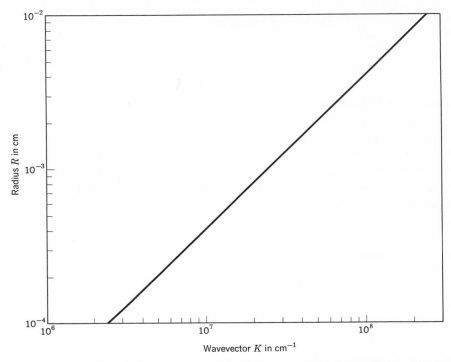

Figure 1   Electron in a magnetic field $H = 1 \times 10^4$ gauss. Radius R of orbit in real space versus radius K of orbit in Fourier space.

Notice that the radius R of the orbit in real space is proportional to the radius K of the orbit in **k** space, with $\hbar c / eB$ as the constant of proportionality. The radius R is plotted in Fig. 1 as a function of the wavevector K for a magnetic field intensity of 10 kilogauss.

Advanced Topic H

MOTT TRANSITION

According to the independent-electron picture of solids, a perfect crystal with an odd number of electrons per primitive cell is always a metal. The picture fails for many transition metal oxides, as de Boer and Verwey observed in 1937. For example, CoO is a semiconductor and not a metal, although the number of electrons is odd. In a number of papers beginning in 1949, Mott[1] put forward the hypothesis that a space lattice of hydrogen-like atoms may not necessarily form a metal, but may form an insulator (or a semiconductor). On this hypothesis a simple cubic lattice of hydrogen atoms at absolute zero will be a metal only if the lattice constant is less than some critical value a_c, where on Mott's early estimates

$$a_c \cong 4.5 a_0 \ , \tag{1}$$

with

$$a_0 = \frac{\epsilon \hbar^2}{me^2} \tag{2}$$

as the radius of the first Bohr orbit of a hydrogen atom in a medium of dielectric constant ϵ. At larger values of the lattice constant $(a > a_c)$ the crystal will be an insulator. There is now a good deal of experimental evidence in support of such a metal-insulator transition. We discuss three theoretical arguments for a transition.

We consider two hydrogen atoms at a large separation in vacuum. The energy necessary to remove an electron from one atom is the ionization energy, $E_I = me^4/2\hbar^2 = 13.60$ eV, and the energy regained if the electron is added to the second atom is the electron affinity, $E_B = 0.77$ eV, the experimental value given in Table 3.4. Thus the energy needed to form a polar state, or the **polarity energy**, is

$$E_p = E_I - E_B = 13.60 - 0.77 = 12.83 \text{ eV} \ , \tag{3}$$

for hydrogen.

We know from Advanced Topic F, which discusses the tight-binding approximation to the band structure of a simple-cubic metal, that the energy

---

[1] For surveys of this question, see N. F. Mott, Canadian J. Physics **34**, 1356 (1956); Phil. Mag. **6**, 287 (1961); the behavior of transition metal oxides is surveyed by I. G. Austin and N. F. Mott, Science **168**, 71 (1970). Among crystals that display a metal-insulator as the temperature increases are VO_2, V_2O_3, Ti_2O_3, Fe_3O_4, NiS, and NbO_2. The proceedings of a conference on the metal-insulator transition appear in Rev. Mod. Phys. **40**, 673 (1968). (The value of the overlap integral given by Mott in the first reference should apparently be multiplied by two.)

of the ground state of the conduction band is lower than that of the $1s$ state of the atom by

$$E_h = 6\gamma \; , \tag{4}$$

where γ is the overlap energy of the wavefunctions on nearest-neighbor atoms. The quantity E_h is often called the **hopping energy.** For two hydrogen atoms at separation a, the overlap energy[2] is

$$\gamma = 2\left(\frac{me^4}{2\hbar^2}\right)\left(1 + \frac{a}{a_0}\right)e^{-a/a_0} \; . \tag{5}$$

The Mott hypothesis is that the crystal will be metallic if the hopping energy is larger than the polarity energy, or if $6\gamma > E_p$. This criterion is indicated graphically in Fig. 1. If we neglect E_B in comparison with E_I in (3), then the criterion for the transition is that

$$12\left(1 + \frac{a}{a_0}\right)e^{-a/a_0} = 1 \; . \tag{6}$$

The root of this equation is

$$a_c \cong 4.1a_0 \; , \tag{7}$$

as compared with $a_c \cong 4.2a_0$ if the result (3) is used for E_p.

The polarity energy will be lower if the electron removed from an atom then remains on an adjacent atom. The polarity energy corrected for the formation of such an electron-hole pair is

$$E_p(a) = E_p - \frac{e^2}{a} = E_p - \frac{e^2}{a_0}\cdot\frac{a_0}{a} \; , \tag{8}$$

where

$$\frac{e^2}{a_0} = \frac{me^4}{\hbar^2} = 2\,\mathrm{Ry} \; . \tag{9}$$

The corrected polarity energy is also plotted in Fig. 1, and we see that this curve intercepts the hopping energy at $a \cong 4.8a_0$. In the region between $4.2a_0$ and $4.8a_0$ it is possible that an insulator may contain bound electron-hole pairs that may be mobile. Such an insulator is called an **excitonic insulator** (see the discussion of excitons in Chapter 18).

Estimates for copper give 2 eV for the polarity energy and 6 eV for the hopping energy. The ratio $E_p/E_h \cong \frac{1}{3}$ is compatible with metallic behavior.

---

[2] L. Pauling and E. B. Wilson, *Introduction to quantum mechanics*, McGraw-Hill, 1935, Eq. (42-12).

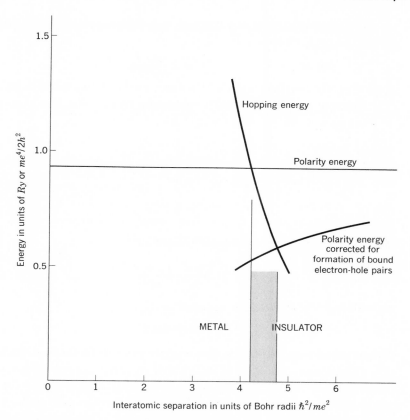

Figure 1 Dependence on interatomic separation of the hopping energy and polarity energy, for two hydrogen atoms in their ground states. The polarity energy is taken as 12.83 eV. The hopping energy is calculated for a simple cubic lattice of hydrogen atoms. The corrected polarity energy is lowered by the attractive Coulomb energy of an electron and a hole—that is, an H^- ion and a proton—at the indicated separation. Thus the shaded region may represent an insulator with mobile electron hole pairs— an excitonic insulator.

Screening of Electron-hole Pairs.

We can approach the question of a transition from another viewpoint. We start in the metallic state and expand the lattice of hydrogen-like atoms until the solid becomes an insulator. We assume that the change of state begins when a conduction electron of the metal can form a bound state with an ion. We shall see that this question involves the screening of the Coulomb interaction by other conduction electrons: at low densities a bound state can be formed, so that the metal becomes an insulator.

The screened potential energy of an electron-hole pair or of an electron-proton pair is given by Eq. (8.25) as

$$U(r) = -\frac{e^2}{r} e^{-\lambda r} ,$$ (10)

where

$$\lambda^2 = \frac{6\pi n_0 e^2}{\epsilon_F} = \frac{4me^2 n_0^{\frac{1}{3}}}{\hbar^2} \cdot \left(\frac{3}{\pi}\right)^{\frac{1}{3}} \cong \frac{4n_0^{\frac{1}{3}}}{a_0} ,$$ (11)

with $\epsilon_F = (\hbar^2/2m)(3\pi^2 n_0)^{\frac{2}{3}}$.

The potential (10) is known[3] to have a bound state for electrons in the field of a fixed positive charge e only if

$$\lambda < \frac{1}{a_0} .$$ (12)

With the use of (11) this inequality becomes

$$\frac{4n_0^{\frac{1}{3}}}{a_0} < \frac{1}{a_0{}^2} ;$$

or, with $n_0 = 1/a^3$, we have an insulator if

$$a > 4a_0 .$$ (13)

The condition derived in this way is quite close to the earlier result (8).

Knox Model of Excitonic State

Knox[4] has considered excitons in semiconductors with an indirect band gap. The energy necessary to form an exciton is $E_g - E_B$, where E_B is the binding energy of an exciton. For the hydrogenic model of an exciton (Chapter 18),

$$E_B = \frac{\mu e^4}{2\epsilon^2 \hbar^2} ,$$ (14)

where μ is the reduced mass of an electron-hole pair:

$$\frac{1}{\mu} = \frac{1}{m_e} + \frac{1}{m_h} ,$$ (15)

and ϵ is the dielectric constant. The dielectric constant and the effective masses will be insensitive to the value of the indirect gap, although they may depend strongly on the direct gap. It is plausible to suppose that under pressure the indirect gap E_g may decrease and become arbitrarily small, while E_B remains finite. When E_g becomes less than E_B, the energy necessary to create an exciton is negative, and now the normal ground state of the crystal is unstable with respect to the formation of excitons.

[3] J. M. Blatt and V. F. Weisskopf, *Theoretical nuclear physics*, Wiley, 1953, p. 55.

[4] R. S. Knox, "Theory of excitons," *Solid state physics*, Suppl. 5, Academic Press, 1963, p. 100.

Advanced Topic I

VECTOR POTENTIAL, INCLUDING FIELD MOMENTUM, GAUGE TRANSFORMATION, AND QUANTIZATION OF ORBITS

This section is included because it is hard to find the magnetic vector potential **A** discussed thoroughly in one place, and we need the vector potential in superconductivity. It may seem mysterious that the hamiltonian of a particle in a magnetic field has the form derived in (18) below:

$$\mathcal{H} = \frac{1}{2M}\left(\mathbf{p} - \frac{Q}{c}\mathbf{A}\right)^2 + Q\varphi \ , \tag{1}$$

where Q is the charge; M is the mass; **A** is the vector potential; and φ is the electrostatic potential. This expression is valid in classical mechanics and in quantum mechanics. Because the kinetic energy of a particle is not changed by a static magnetic field, it is perhaps unexpected that the vector potential of the magnetic field enters the hamiltonian. As we shall see, the key is the observation that the momentum **p** is the sum of two parts, one the kinetic momentum

$$\mathbf{p}_{\text{kin}} = M\mathbf{v} \tag{2}$$

which is familiar to us, and one the potential momentum or field momentum

$$\mathbf{p}_{\text{field}} = \frac{Q}{c}\mathbf{A} \ . \tag{3}$$

The total momentum is

$$\boxed{\mathbf{p} = \mathbf{p}_{\text{kin}} + \mathbf{p}_{\text{field}} = M\mathbf{v} + \frac{Q}{c}\mathbf{A} \ ,} \tag{4}$$

and the kinetic energy is

$$\frac{1}{2}Mv^2 = \frac{1}{2M}(Mv)^2 = \frac{1}{2M}\left(\mathbf{p} - \frac{Q}{c}\mathbf{A}\right)^2 \ . \tag{5}$$

The vector potential[1] is related to the magnetic field by

$$\mathbf{B} = \text{curl}\,\mathbf{A} \ . \tag{6}$$

---

[1] For an elementary treatment of the vector potential see E. M. Purcell, *Electricity and magnetism*, McGraw-Hill, 1965.

We assume that we work in nonmagnetic material so that **H** and **B** are treated as identical.

Lagrangian Equations of Motion

To find the hamiltonian, the prescription of classical mechanics is clear: we must first find the Lagrangian. The Lagrangian in generalized coordinates is

$$L = \frac{1}{2}M\dot{q}^2 - Q\varphi(\mathbf{q}) + \frac{Q}{c}\dot{\mathbf{q}} \cdot \mathbf{A}(\mathbf{q}) \ . \tag{7}$$

This is correct because it leads to the correct equation of motion of a charge in combined electric and magnetic fields, as we now show.

In Cartesian coordinates the Lagrange equation of motion is

$$\frac{d}{dt}\frac{\partial L}{\partial \dot{x}} - \frac{\partial L}{\partial x} = 0 \ , \tag{8}$$

and similarly for y and z. From (7) we form

$$\frac{\partial L}{\partial \dot{x}} = -Q\frac{\partial \varphi}{\partial x} + \frac{Q}{c}\left(\dot{x}\frac{\partial A_x}{\partial x} + \dot{y}\frac{\partial A_y}{\partial x} + \dot{z}\frac{\partial A_z}{\partial x}\right) \ ; \tag{9}$$

$$\frac{\partial L}{\partial \dot{x}} = M\dot{x} + \frac{Q}{c}A_x \ ; \tag{10}$$

$$\frac{d}{dt}\frac{\partial L}{\partial \dot{x}} = M\ddot{x} + \frac{Q}{c}\frac{dA_x}{dt} = M\ddot{x} + \frac{Q}{c}\left(\frac{\partial A_x}{\partial t} + \dot{x}\frac{\partial A_x}{\partial x} + \dot{y}\frac{\partial A_x}{\partial y} + \dot{z}\frac{\partial A_x}{\partial z}\right) \ . \tag{11}$$

Thus (8) becomes

$$M\ddot{x} + Q\frac{\partial \varphi}{\partial x} + \frac{Q}{c}\left[\frac{\partial A_x}{\partial t} + \dot{y}\left(\frac{\partial A_x}{\partial y} - \frac{\partial A_y}{\partial x}\right) + \dot{z}\left(\frac{\partial A_x}{\partial z} - \frac{\partial A_z}{\partial x}\right)\right] = 0 \ , \tag{12}$$

or

$$M\frac{d^2x}{dt^2} = QE_x + \frac{Q}{c}\left[\mathbf{v} \times \mathbf{B}\right]_x \ , \tag{13}$$

with

$$E_x = -\frac{\partial \varphi}{\partial x} - \frac{1}{c}\frac{\partial A_x}{\partial t} \ ; \tag{14}$$

$$\mathbf{B} = \text{curl } \mathbf{A} \ . \tag{15}$$

Equation (13) is the Lorentz force equation. This confirms that (7) is correct. We note in (14) that **E** has one contribution from the electrostatic potential φ and another from the time derivative of the magnetic vector potential **A**.

Derivation of the Hamiltonian

The momentum **p** is defined in terms of the Lagrangian as

$$\mathbf{p} \equiv \frac{\partial L}{\partial \dot{\mathbf{q}}} = M\dot{\mathbf{q}} + \frac{Q}{c}\mathbf{A} \ , \tag{16}$$

in agreement with (4). The hamiltonian $\mathcal{H}(\mathbf{p}, \mathbf{q})$ is defined by

$$\mathcal{H}(\mathbf{p}, \mathbf{q}) \equiv \mathbf{p} \cdot \dot{\mathbf{q}} - L \ , \tag{17}$$

or

$$\mathcal{H} = M\dot{q}^2 + \frac{Q}{c}\dot{\mathbf{q}} \cdot \mathbf{A} - \frac{1}{2}M\dot{q}^2 + Q\varphi - \frac{Q}{c}\dot{\mathbf{q}} \cdot \mathbf{A} = \frac{1}{2M}\left(\mathbf{p} - \frac{Q}{c}\mathbf{A}\right)^2 + Q\varphi \ , \tag{18}$$

as in (1).

Field Momentum

The momentum in the electromagnetic field that accompanies a particle moving in a magnetic field is given by the volume integral of the Poynting vector, so that

$$\mathbf{p}_{\text{field}} = \frac{1}{4\pi c} \int dV \, \mathbf{E} \times \mathbf{B} \ . \tag{19}$$

We work in the nonrelativistic approximation with $v \ll c$, where v is the velocity of the particle. At low values of v/c we consider **B** to arise from an external source alone, but **E** arises from the charge on the particle. For a charge Q at $\mathbf{r}'$,

$$\mathbf{E} = -\nabla\varphi \ ; \qquad \nabla^2\varphi = -4\pi Q\,\delta(\mathbf{r} - \mathbf{r}') \ . \tag{20}$$

Thus

$$\mathbf{p}_f = -\frac{1}{4\pi c} \int dV \, \nabla\varphi \times \text{curl } \mathbf{A} \ . \tag{21}$$

By a standard vector relation [Madelung, p. 115] we have

$$\int dV \, \nabla\varphi \times \text{curl } \mathbf{A} = -\int dV \, [\mathbf{A} \times \text{curl } (\nabla\varphi) - \mathbf{A}\,\text{div } \nabla\varphi - (\nabla\varphi)\,\text{div } \mathbf{A}] \ . \tag{22}$$

But curl $(\nabla\varphi) = 0$, and we can always choose the gauge such that div $\mathbf{A} = 0$. This is the transverse gauge.

Thus we have

$$\mathbf{p}_f = -\frac{1}{4\pi c}\int dV \, \mathbf{A}\,\nabla^2\varphi = \frac{1}{c}\int dV \, \mathbf{A}Q\,\delta(\mathbf{r} - \mathbf{r}') = \frac{Q}{c}\mathbf{A} \ . \tag{23}$$

This is the interpretation of the field contribution to the total momentum $\mathbf{p} = M\mathbf{v} + Q\mathbf{A}/c$.

GAUGE TRANSFORMATION

Suppose $\mathcal{H}\psi = \epsilon\psi$, where

$$\mathcal{H} = \frac{1}{2M}\left(\mathbf{p} - \frac{Q}{c}\mathbf{A}\right)^2 . \tag{24}$$

Let us make a gauge transformation to $\mathbf{A}'$, where

$$\mathbf{A}' = \mathbf{A} + \nabla\chi , \tag{25}$$

where χ is a scalar. Now $\mathbf{B} = \operatorname{curl} \mathbf{A} = \operatorname{curl} \mathbf{A}'$, because curl $(\nabla\chi) \equiv 0$. The Schrödinger equation becomes

$$\frac{1}{2M}\left(\mathbf{p} - \frac{Q}{c}\mathbf{A}' + \frac{Q}{c}\nabla\chi\right)^2\psi = \epsilon\psi . \tag{26}$$

What ψ' satisfies

$$\frac{1}{2M}\left(\mathbf{p} - \frac{Q}{c}\mathbf{A}'\right)^2\psi' = \epsilon\psi' , \tag{27}$$

with the same ϵ as for ψ? Equation (27) is equivalent to

$$\frac{1}{2M}\left(\mathbf{p} - \frac{Q}{c}\mathbf{A} - \frac{Q}{c}\nabla\chi\right)^2\psi' = \epsilon\psi' . \tag{28}$$

We try

$$\psi' = e^{iQ\chi/\hbar c}\psi . \tag{29}$$

Now

$$\mathbf{p}\psi' = e^{iQ\chi/\hbar c}\mathbf{p}\psi + \frac{Q}{c}(\nabla\chi)e^{iQ\chi/\hbar c}\psi ,$$

so that

$$\left(\mathbf{p} - \frac{Q}{c}\nabla\chi\right)\psi' = e^{iQ\chi/\hbar c}\mathbf{p}\psi$$

and

$$\frac{1}{2M}\left(\mathbf{p} - \frac{Q}{c}\mathbf{A} - \frac{Q}{c}\nabla\chi\right)^2\psi' = e^{iQ\chi/\hbar c}\frac{1}{2M}\left(\mathbf{p} - \frac{Q}{c}\mathbf{A}\right)^2\psi = e^{iQ\chi/\hbar c}\epsilon\psi . \tag{30}$$

Thus $\psi' = e^{iQ\chi/\hbar c}\psi$ satisfies the Schrödinger equation after the gauge transformation (25). The energy ϵ is invariant under the transformation. The gauge transformation on $\mathbf{A}$ merely changes the local phase of the wavefunction. We see that

$$\psi'^*\psi' = \psi^*\psi , \tag{31}$$

so that the charge density is invariant under a gauge transformation.

Gauge in the London Equation

Because of the equation of continuity in the flow of electric charge we require that in a superconductor

$$\text{div } \mathbf{j} = 0 \ ,$$

so that the vector potential in the London equation $\mathbf{j} = -c\mathbf{A}/4\pi\lambda_L^2$ must satisfy

$$\text{div } \mathbf{A} = 0 \ . \tag{32}$$

Further, there is no current flow through a vacuum/superconductor interface. The normal component of the current across the interface must vanish: $j_n = 0$, so that the vector potential in the London equation must satisfy

$$A_n = 0 \ . \tag{33}$$

The gauge of the vector potential in the London equation of superconductivity is to be chosen so that (32) and (33) are satisfied.

QUANTIZATION OF ORBITS IN A MAGNETIC FIELD

To treat the de Haas-van Alphen effect we assume that the orbits of a particle of charge Q in a magnetic field are quantized by the Bohr-Sommerfeld relation

$$\oint \mathbf{p} \cdot d\mathbf{r} = (n + \gamma)2\pi\hbar \ , \tag{34}$$

where n is an integer and γ is a phase correction that for free electrons has the value $\frac{1}{2}$. We write Eq. (4) as

$$\mathbf{p} = \hbar\mathbf{k} + \frac{Q}{c}\mathbf{A} \ , \tag{35}$$

where $\hbar\mathbf{k}$ is the kinetic momentum of the particle and $\mathbf{A}$ is the vector potential of the magnetic field. Then

$$\oint \mathbf{p} \cdot d\mathbf{r} = \oint \hbar\mathbf{k} \cdot d\mathbf{r} + \frac{Q}{c}\oint \mathbf{A} \cdot d\mathbf{r} \ . \tag{36}$$

The equation of motion of a particle of charge Q in a magnetic field is

$$\hbar\frac{d\mathbf{k}}{dt} = \frac{Q}{c}\frac{d\mathbf{r}}{dt} \times \mathbf{B} \ ; \tag{37}$$

this may be integrated with respect to time to give

$$\hbar\mathbf{k} = \frac{Q}{c}\mathbf{r} \times \mathbf{B} \ , \tag{38}$$

apart from an additive constant which does not contribute to the final result. Thus one of the path integrals in (36) is

$$\oint \hbar \mathbf{k} \cdot d\mathbf{r} = \frac{Q}{c} \oint \mathbf{r} \times \mathbf{B} \cdot d\mathbf{r} = -\frac{Q}{c} \mathbf{B} \cdot \oint \mathbf{r} \times d\mathbf{r} = -\frac{2Q}{c} \Phi , \quad (39)$$

where Φ is the magnetic flux contained within the orbit in real space. We have used the geometrical result that

$$\oint \mathbf{r} \times d\mathbf{r} = 2 \times (\text{area enclosed by the orbit}). \quad (40)$$

The other path integral in (36) is

$$\frac{Q}{c} \oint \mathbf{A} \cdot d\mathbf{r} = \frac{Q}{c} \int \text{curl } \mathbf{A} \cdot d\boldsymbol{\sigma} = \frac{Q}{c} \int \mathbf{B} \cdot d\boldsymbol{\sigma} = \frac{Q}{c} \Phi , \quad (41)$$

by the Stokes theorem; here $d\boldsymbol{\sigma}$ is the area element in real space. The momentum path integral

$$\oint \mathbf{p} \cdot d\mathbf{r} = -\frac{Q}{c} \Phi = (n + \gamma)2\pi\hbar , \quad (42)$$

so that the orbit of an electron is quantized in such a way that the flux through it is

$$\boxed{\Phi_n = (n + \gamma)\frac{2\pi\hbar c}{e} .} \quad (43)$$

We shall return to this in Advanced Topic J.

The value of the flux quantum may be expressed conveniently in terms of the fine structure constant $e^2/\hbar c$:

$$\frac{2\pi\hbar c}{e} = 2\pi e \cdot \frac{\hbar c}{e^2} = 2\pi e(137.04) = 4.14 \times 10^{-7} \text{ gauss-cm}^2 . \quad (44)$$

For the theory of the de Haas-van Alphen effect we need the area of the orbit in wavevector space. We have obtained in (43) the flux through the orbit in real space. By (37) we know that a line element Δr in the plane normal to $\mathbf{B}$ is related to Δk by

$$\Delta r = \frac{\hbar c}{eB} \Delta k , \quad (45)$$

so that the area S_n in $\mathbf{k}$ space is related to the area A_n of the orbit in $\mathbf{r}$ space by

$$A_n = \left(\frac{\hbar c}{eB}\right)^2 S_n . \quad (46)$$

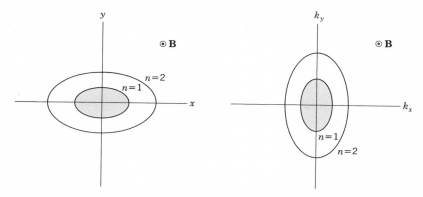

Figure 1 Orbits in coordinate space and in wavevector space for an electron in a magnetic field. The orbits are drawn for $\gamma = 0$. The flux through the inner orbit is $2\pi\hbar c/e$ in coordinate space.

It follows that

$$\Phi_n = \left(\frac{\hbar c}{e}\right)^2 \frac{1}{B} S_n = (n + \gamma)\frac{2\pi\hbar c}{e} \, , \tag{47}$$

from (43), whence the area in **k** space of an orbit will satisfy

$$\boxed{S_n = (n + \gamma)\frac{2\pi e}{\hbar c} B \, .} \tag{48}$$

This result is due to L. Onsager and to I. M. Lifshitz. Two orbits are drawn in Fig. 1.

Advanced Topic J

FLUX QUANTIZATION IN A SUPERCONDUCTING RING

We give a proof that the total magnetic flux which passes through a superconducting ring may assume only quantized values, integral multiples of the quantum of flux $2\pi\hbar c/q$, where by experiment the charge $|q| = 2e$. This result confirms the significance of electron pairs in the composition of the superconducting state. Flux quantization is a beautiful example of a long-range quantum effect, in this instance the coherence of the superconducting state extends over a ring or solenoid.

The electromagnetic field is an example of a boson field. The electric field intensity $\mathbf{E}(\mathbf{r})$ acts qualitatively as a field amplitude. The energy density may be written as, in a semiclassical approximation,

$$\frac{E^*(\mathbf{r})E(\mathbf{r})}{4\pi} \cong n(\mathbf{r})\hbar\omega \ ,$$

where $n(\mathbf{r})$ is the number of photons of frequency ω, per unit volume. We assume that the total number of photons in the volume is large in comparison with unity. Then

$$E(\mathbf{r}) \cong (4\pi\hbar\omega)^{\frac{1}{2}}n(\mathbf{r})^{\frac{1}{2}}e^{i\theta(\mathbf{r})} \ ; \qquad E^*(\mathbf{r}) = (4\pi\hbar\omega)^{\frac{1}{2}}n(\mathbf{r})^{\frac{1}{2}}e^{-i\theta(\mathbf{r})} \ ,$$

where $\theta(\mathbf{r})$ is the phase of the field. We now introduce similar particle probability amplitudes into the description of particle bosons, where a particle is an electron pair. (The analogy with photons is not exact, but it is helpful.)

The ground state of a superconductor is made up of weakly-bound electron pairs, called Cooper pairs. An electron pair will act as a boson,[1] although a single electron is a fermion. The arguments that follow apply specifically to a boson gas with a very large number of bosons in the same orbital. We then can treat the boson probability amplitude as a classical quantity, just as the electromagnetic field is used for photons. The arguments do not apply to a metal in the normal state because an electron in the normal state acts as a single unpaired fermion.

We first show that a charged boson gas obeys the London equation in the form (11.26). Let $\psi(\mathbf{r})$ be the particle probability amplitude. We suppose that the concentration

---

[1] The boson condensation temperature calculated for metallic concentrations is of the order of the Fermi temperature ($10^4 - 10^5$ K). The superconducting transition temperature is much lower and takes place when the electron pairs break up into two fermions. The model of a superconductor as composed of noninteracting bosons cannot be taken too literally, for there are about 10^6 electrons in the volume occupied by a single Cooper pair.

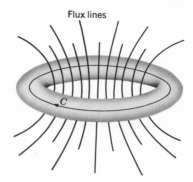

Flux lines

Figure 1   Path of integration C through the interior of a superconducting ring.

$$n = \psi^*\psi = \text{constant}. \tag{1}$$

At absolute zero n is one-half of the concentration of electrons in the conduction band, for n refers to pairs. Then we may write

$$\psi = n^{\frac{1}{2}}e^{i\theta(\mathbf{r})} \; ; \qquad \psi^* = n^{\frac{1}{2}}e^{-i\theta(\mathbf{r})} \; . \tag{2}$$

The phase $\theta(\mathbf{r})$ is important for what follows. We make the good approximation that ψ is a classical amplitude rather than a quantum field operator.

The velocity of a particle is, using (I.4),

$$\mathbf{v} = \frac{1}{m}\left(\mathbf{p} - \frac{q}{c}\mathbf{A}\right) = \frac{1}{m}\left(-i\hbar\nabla - \frac{q}{c}\mathbf{A}\right) \; . \tag{3}$$

The particle flux is given by

$$\psi^*\mathbf{v}\psi = \frac{n}{m}\left(\hbar\nabla\theta - \frac{q}{c}\mathbf{A}\right) \; , \tag{4}$$

so that the electric current density in the ring (which is a multiply-connected region) is

$$\mathbf{j} = q\psi^*\mathbf{v}\psi = \frac{nq}{m}\left(\hbar\nabla\theta - \frac{q}{c}\mathbf{A}\right) \; . \tag{5}$$

We may take the curl of both sides to obtain

$$\text{curl } \mathbf{j} = -\frac{nq^2}{mc}\mathbf{B} \; , \tag{6}$$

with use of the fact that the curl of the gradient of a scalar is identically zero. Equation (6) is one form of the London equation.

The quantization of the magnetic flux through a ring is a dramatic consequence of Eq. (5). Let us take a closed path C through the interior of the superconducting material, well away from the surface (Fig. 1). The Meissner effect tells us that $\mathbf{B}$ and $\mathbf{j}$ are zero in the interior. Now (5) is zero if

$$\hbar c\nabla\theta = q\mathbf{A} \; . \tag{7}$$

We have

$$\oint_C \nabla\theta \cdot d\mathbf{l} = \theta_2 - \theta_1 \tag{8}$$

for the change of phase on going once around the ring. The boson probability amplitude is measurable in the classical approximation, so that ψ must be single-valued and we must have

$$\theta_2 - \theta_1 = 2\pi s \ , \tag{9}$$

where s is an integer.

We also have, by the Stokes theorem,

$$\oint_C \mathbf{A} \cdot d\mathbf{l} = \int_C (\text{curl } \mathbf{A}) \cdot d\boldsymbol{\sigma} = \int_C \mathbf{B} \cdot d\boldsymbol{\sigma} = \Phi \ , \tag{10}$$

where $d\boldsymbol{\sigma}$ is an element of area on a surface bounded by the curve C, and Φ is the magnetic flux through C. From (7), (9), and (10) we have

$$2\pi\hbar c s = q\Phi \ ,$$

or

$$\Phi = \left(\frac{2\pi\hbar c}{q}\right)s \ . \tag{11}$$

Thus the flux through the ring is quantized[2] in integral multiples of $2\pi\hbar c/q$. By experiment $q = -2e$ as appropriate for electron pairs, so that the quantum of flux in a superconductor is

$$\frac{2\pi\hbar c}{2e} \cong 2.07 \times 10^{-7} \text{ gauss-cm}^2 \ . \tag{12}$$

This unit of flux is called a **fluxoid.** We remark that the simple result (11) does not hold if the flux penetrates the ring itself, as when the material of the ring is thin.

The flux through the ring is the sum of the flux Φ_{ext} from external sources and the flux Φ_{sc} from the superconducting currents which flow in the surface of the ring:

$$\Phi = \Phi_{\text{ext}} + \Phi_{\text{sc}} \ . \tag{13}$$

The flux Φ is quantized. There is normally no quantization condition on the flux from external sources, so that Φ_{sc} must adjust itself appropriately in order that Φ assume a quantized value.

---

[2] The effect was discovered by B. S. Deaver, Jr., and W. M. Fairbank, Phys. Rev. Letters **7**, 43 (1961), and R. Doll and M. Näbauer, Phys. Rev. Letters **7**, 51 (1961).

Advanced Topic K

JOSEPHSON SUPERCONDUCTOR TUNNELING EFFECTS

Quantum tunneling[1] of single electrons from a superconductor across an insulator and into a superconductor or into a normal metal was discussed in Chapter 12. The results shown there are typical of tunneling experiments unless exceptional precautions are taken in the construction of the junction, as described below.

Under suitable conditions we observe remarkable effects associated with the tunneling of superconducting electron pairs from a *superconductor* through a layer of an *insulator* into another *superconductor* (Fig. 1). The effects of pair tunneling are quite unlike single particle tunneling and include:

Dc Josephson effect. A dc current flows across the junction in the absence of any electric or magnetic field.

Ac Josephson effect. A dc voltage applied across the junction causes rf current oscillations across the junction. This effect has been utilized in a precision determination of the value of $\hbar/e$. Further, an rf voltage applied with the dc voltage can then cause a dc current across the junction.

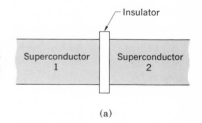

Figure 1 (a) A tunneling junction composed of two superconductors separated by a thin layer of an insulator, which may be an oxide layer of the order of 10 Å in thickness formed on one of the superconductors.

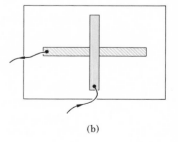

Figure 1 (b) An actual tunneling junction may be fabricated by evaporation of a rectangular strip of lead on the surface of a glass microscope slide; the strip then is allowed to oxidize, and a second strip of lead is deposited at right angles to the first strip. The resistance of the junction may be of the order of 1 ohm; the contact area of the order of 10^{-4} cm^2; the maximum Josephson current of the order of 1 mA. The earth's magnetic field has a deleterious dephasing effect across the area of the junction, so that the field may have to be screened out.

[1] All tunneling is a quantum effect: by tunneling we mean the penetration of a particle through a potential barrier—that is, through a region which is forbidden to it on classical mechanics.

Macroscopic long range quantum interference. A dc magnetic field applied through a superconducting circuit containing two junctions causes the maximum supercurrent to show interference effects as a function of magnetic field intensity. This effect can be utilized in sensitive magnetometers.

Our discussion of the Josephson junction phenomena follows the discussion of Advanced Topic J.

DC JOSEPHSON EFFECT

Let ψ_1 be the probability amplitude of electron pairs on one side of a junction, and let ψ_2 be the amplitude on the other side. For simplicity, let both superconductors be identical. For the present we suppose that they are both at zero potential.

The time-dependent Schrödinger equation $i\hbar \partial\psi/\partial t = \mathcal{H}\psi$ applied to the two amplitudes gives

$$i\hbar \frac{\partial \psi_1}{\partial t} = \hbar T \psi_2 \; ; \qquad i\hbar \frac{\partial \psi_2}{\partial t} = \hbar T \psi_1 \; , \tag{1}$$

a brilliant oversimplification of the problem. Here $\hbar T$ represents the effect of the electron-pair coupling or **transfer interaction** across the insulator; T has the dimensions of a rate or frequency. It is a measure of the leakage of ψ_1 into the region 2, and of ψ_2 into the region 1. If the insulator is very thick, T is zero and there is no pair tunneling.

Let

$$\psi_1 = n_1^{\frac{1}{2}} e^{i\theta_1} \; ; \qquad \psi_2 = n_2^{\frac{1}{2}} e^{i\theta_2} \; . \tag{2}$$

Then

$$\frac{\partial \psi_1}{\partial t} = \frac{1}{2} n_1^{-\frac{1}{2}} e^{i\theta_1} \frac{\partial n_1}{\partial t} + i\psi_1 \frac{\partial \theta_1}{\partial t} = -iT\psi_2 \; , \tag{3}$$

with use of (1) in the form $\partial\psi_1/\partial t = -iT\psi_2$. Similarly,

$$\frac{\partial \psi_2}{\partial t} = \frac{1}{2} n_2^{-\frac{1}{2}} e^{i\theta_2} \frac{\partial n_2}{\partial t} + i\psi_2 \frac{\partial \theta_2}{\partial t} = -iT\psi_1 \; . \tag{4}$$

We multiply (3) by $n_1^{\frac{1}{2}} e^{-i\theta_1}$ to obtain, with $\delta \equiv \theta_2 - \theta_1$,

$$\frac{1}{2} \frac{\partial n_1}{\partial t} + in_1 \frac{\partial \theta_1}{\partial t} = -iT(n_1 n_2)^{\frac{1}{2}} e^{i\delta} \; . \tag{5}$$

We multiply (4) by $n_2^{\frac{1}{2}} e^{-i\theta_2}$ to obtain

$$\frac{1}{2} \frac{\partial n_2}{\partial t} + in_2 \frac{\partial \theta_2}{\partial t} = -iT(n_1 n_2)^{\frac{1}{2}} e^{-i\delta} \; . \tag{6}$$

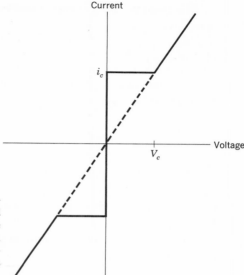

Figure 2 Current-voltage characteristic of a Josephson junction. Dc currents flow under zero applied voltage up to a critical current i_c: this is the dc Josephson effect. At voltages above V_c the junction has a finite resistance, but the current has an oscillatory component of frequency $\omega = 2eV/\hbar$: this is the ac Josephson effect.

Now equate the real and imaginary parts of (5) and similarly of (6):

$$\frac{\partial n_1}{\partial t} = 2T(n_1 n_2)^{\frac{1}{2}} \sin \delta \; ; \qquad \frac{\partial n_2}{\partial t} = -2T(n_1 n_2)^{\frac{1}{2}} \sin \delta \; ; \qquad (7)$$

$$\frac{\partial \theta_1}{\partial t} = -T\left(\frac{n_2}{n_1}\right)^{\frac{1}{2}} \cos \delta \; ; \qquad \frac{\partial \theta_2}{\partial t} = -T\left(\frac{n_1}{n_2}\right)^{\frac{1}{2}} \cos \delta \; . \qquad (8)$$

If $n_1 \cong n_2$ as for identical superconductors 1 and 2, we have from (8) that

$$\frac{\partial \theta_1}{\partial t} = \frac{\partial \theta_2}{\partial t} \; ; \qquad \frac{\partial}{\partial t}(\theta_2 - \theta_1) = 0 \; . \qquad (8a)$$

From (7) we see that

$$\frac{\partial n_2}{\partial t} = -\frac{\partial n_1}{\partial t} \; . \qquad (9)$$

The current flow from (1) to (2) is proportional to $\partial n_2/\partial t$ or, the same thing, $-\partial n_1/\partial t$. We therefore conclude from (7) that the current J of superconductor pairs across the junction depends on the phase difference δ as

$$\boxed{J = J_0 \sin \delta = J_0 \sin(\theta_2 - \theta_1) \; ,} \qquad (10)$$

where J_0 is proportional to the transfer interaction T. The current J_0 is the maximum zero-voltage current that can be passed by the junction.

With no applied voltage a dc current will flow across the junction, (Fig. 2), with a value between J_0 and $-J_0$ according to the value of the phase difference $\theta_2 - \theta_1$. This is the dc Josephson effect.

AC JOSEPHSON EFFECT

Let a voltage V be applied across the junction. We can do this because the junction is an insulator. An electron pair experiences a potential energy difference qV on passing across the junction, where $q = -2e$. We can say that a pair on one side is at potential energy $-eV$ and a pair on the other side is at eV. The equations of motion that replace (1) are

$$i\hbar \frac{\partial \psi_1}{\partial t} = \hbar T\psi_2 - eV\psi_1 \; ; \qquad i\hbar \frac{\partial \psi_2}{\partial t} = \hbar T\psi_1 + eV\psi_2 \; . \qquad (11)$$

We proceed as above to find in place of (5) the equation

$$\frac{1}{2} \frac{\partial n_1}{\partial t} + in_1 \frac{\partial \theta_1}{\partial t} = ieV \, n_1 \hbar^{-1} - iT(n_1 n_2)^{\frac{1}{2}} e^{i\delta} \; . \qquad (12)$$

This equation breaks up into the real part

$$\frac{\partial n_1}{\partial t} = 2T(n_1 n_2)^{\frac{1}{2}} \sin \delta \; , \qquad (13)$$

exactly as without the voltage V, and the imaginary part

$$\frac{\partial \theta_1}{\partial t} = \frac{eV}{\hbar} - T\left(\frac{n_2}{n_1}\right)^{\frac{1}{2}} \cos \delta \; , \qquad (14)$$

which differs from (8) by the term $eV/\hbar$.

Further, by extension of (6),

$$\frac{1}{2} \frac{\partial n_2}{\partial t} + in_2 \frac{\partial \theta_2}{\partial t} = -i \, eV \, n_2 \hbar^{-1} - iT(n_1 n_2)^{\frac{1}{2}} e^{-i\delta} \; , \qquad (15)$$

whence

$$\frac{\partial n_2}{\partial t} = -2T(n_1 n_2)^{\frac{1}{2}} \sin \delta \; ; \qquad (16)$$

$$\frac{\partial \theta_2}{\partial t} = -\frac{eV}{\hbar} - T\left(\frac{n_1}{n_2}\right)^{\frac{1}{2}} \cos \delta \; . \qquad (17)$$

From (14) and (17) with $n_1 \cong n_2$, we have

$$\frac{\partial (\theta_2 - \theta_1)}{\partial t} = \frac{\partial \delta}{\partial t} = -\frac{2eV}{\hbar} \; . \qquad (18)$$

We see by integration of (18) that with a dc voltage across the junction the relative phase of the probability amplitudes vary as

$$\delta(t) = \delta(0) - \frac{2eVt}{\hbar} \; . \qquad (19)$$

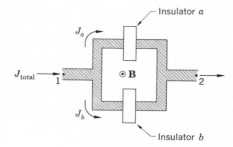

Figure 3 The arrangement for experiment on macroscopic quantum interference. A magnetic flux Φ passes through the interior of the loop.

The current is now given by (10) with (19) for the phase:

$$J = J_0 \sin\left[\delta(0) - \frac{2eVt}{\hbar}\right] .$$

(20)

The current oscillates with frequency

$$\omega = \frac{2eV}{\hbar} .$$

(21)

This is the **ac Josephson effect.** A dc voltage of 1 μV produces a frequency of 483.6 MHz.

The relation (21) says that a photon of energy $\hbar\omega = 2\,eV$ is emitted or absorbed when an electron pair crosses the barrier. By measuring the voltage and the frequency it is possible to obtain a very precise value[2] of $e/\hbar$.

MACROSCOPIC QUANTUM INTERFERENCE

We saw in Advanced Topic J that the phase difference $\theta_2 - \theta_1$ around a closed circuit which encompasses a total magnetic flux Φ is given by

$$\theta_2 - \theta_1 = \frac{2e}{\hbar c}\Phi .$$

(22)

The flux is the sum of that due to external fields and that due to currents in the circuit itself. We consider two Josephson junctions in parallel, as in Fig. 3. No voltage is applied.

Let the phase difference between points 1 and 2 taken on a path through junction a be δ_a. When taken on a path through junction b, the phase difference is δ_b. In the absence of a magnetic field these two phases must be equal. Now let the flux Φ pass through the interior of the circuit. We can accomplish this

[2] W. H. Parker, B. N. Taylor, and D. N. Langenberg, Phys. Rev. Letters **18**, 287 (1967).

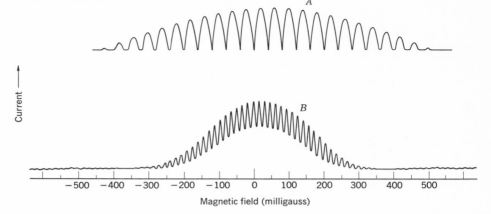

Figure 4 Experimental trace of J_{max} versus magnetic field showing interference and diffraction effects for two junctions A and B. The field periodicity is 39.5 and 16 mG for A and B, respectively. Approximate maximum currents are 1 mA (A) and 0.5 mA (B). The junction separation is 3mm and junction width 0.5 mm for both cases. The zero offset of A is due to a background magnetic field. [After R. C. Jaklevic, J. Lambe, J. E. Mercereau and A. H. Silver, Phys. Rev. **140**, A1628 (1965).]

by a straight solenoid normal to the plane of the paper and lying inside the circuit. By (22)

$$\delta_b - \delta_a = \frac{2e}{\hbar c}\,\Phi\ . \tag{23}$$

or

$$\delta_b = \delta_0 + \frac{e}{\hbar c}\,\Phi\ ; \qquad \delta_a = \delta_0 - \frac{e}{\hbar c}\,\Phi\ . \tag{24}$$

The total current is the sum of J_a and J_b. The current through each junction is of the form (10), so that

$$J_{\text{Total}} = J_0\left\{ \sin\left(\delta_0 + \frac{e}{\hbar c}\,\Phi\right) + \sin\left(\delta_0 - \frac{e}{\hbar c}\,\Phi\right)\right\} = 2(J_0 \sin\delta_0)\cos\frac{e\Phi}{\hbar c}\ . \tag{25}$$

The current varies with Φ. The magnitude of the current has maxima when

$$\frac{e\Phi}{\hbar c} = r\pi\ , \qquad r = \text{integer}. \tag{26}$$

The periodicity of the current is shown in Fig. 4. The short period variation is the interference effect of the two junctions predicted by (25) and (26). The longer variation is a diffraction effect and arises from the finite dimensions of each junction, which causes Φ to depend on the particular path we integrate over. The diffraction effect was responsible for the failure to observe pair tunneling in early experiments on single particle tunneling. The single particles are not liable to diffraction effects, but unless special care in construction and magnetic shielding is taken the contribution of the pairs is washed out by diffraction.

References

B. D. Josephson, Physics Letters **1**, 251 (1962). (Difficult.)

R. Feynman, *Lectures on physics,* Addison-Wesley, 1963, Vol. 3, Sec. 21/9. (Simplest discussion.)

J. Clarke, "The Josephson effect and $e/\hbar$," Amer. J. Phys. **38**, 1071 (1970). (Excellent elementary review.)

G. F. Zharkov, Sov. Phys. Uspekhi **9**, 198 (1966). (Good review.)

P. W. Anderson, in *Lectures on the many-body problem,* E. R. Caianiello, Ed., Academic Press, 1964; Vol. 2, pp. 113–135. (For physical insight.)

P. W. Anderson and J. M. Rowell, "Probable observation of the Josephson superconducting tunnel effect," Phys. Rev. Letters **10**, 230 (1963).

R. C. Jaklevic, J. Lambe, A. H. Silver, and J. E. Mercereau, "Quantum interference effects in Josephson tunneling," Phys. Rev. Letters **12,** 159 (1964); "Quantum interference from a static vector potential in a field-free region," Phys. Rev. Letters **12,** 274 (1964); "Macroscopic quantum interference in superconductors," Phys. Rev. **140**, A1628 (1965). (Excellent paper.)

FOR AC JOSEPHSON EFFECT:

S. Shapiro, Phys. Rev. Letters **11,** 80 (1963).

R. E. Eck, D. J. Scalapino, and B. N. Taylor, Phys. Rev. Letters **13**, 15 (1964).

D. D. Coon and M. D. Fiske, Phys. Rev. **138**, A744 (1965).

P. L. Richards, S. Shapiro, and C. C. Grimes, "Student laboratory demonstration of flux quantization and the Josephson effect in superconductors," Amer. J. Phys. **36**, 690 (1968).

Advanced Topic L

BCS THEORY OF THE SUPERCONDUCTING ENERGY GAP

The BCS theory in its original form[1] is not difficult to understand, and it appears even simpler with the help of the spin-analog method of P. W. Anderson. A careful discussion is given by *QTS*, Chap. 8. The essence of the theory of the ground state can be understood without great mathematical complications by making further simplifications in the assumptions. We first consider a mathematical problem which establishes useful background for what comes later. The problem we treat first is *not* the superconductivity problem, but the connection is close.

A Special Eigenvalue Problem

Consider an unperturbed one-particle system with an energy level spectrum such that one of the levels is R-fold degenerate and well separated in energy from all other levels. That is, we consider R independent states of the system, all having the same energy. Now let there be an additional interaction, a weak perturbation. The perturbation may split up the R states so that they occupy a certain range of energy, instead of all having the same energy.

We denote the R states associated with the degenerate level as $\varphi_1, \varphi_2, \ldots, \varphi_R$. These states satisfy the Schrödinger equation of the unperturbed problem. We may choose the zero of energy so that $H_0\varphi = 0$ for these φ. In the first approximation the new states of the system in the presence of a perturbation U may be written as linear combinations of the old states:[2]

$$\psi_j = \sum_{s=1}^{R} c_{js}\varphi_s \ . \tag{1}$$

Suppose that the ψ's formed in this way are exact solutions of the Schrödinger equation of the perturbed problem:

$$(H_0 + U)\psi_j = U\psi_j = \epsilon_j\psi_j \ . \tag{2}$$

Here $H_0\psi_j = 0$ because of the choice of the zero of energy. Using (1), we may write (2) as

$$\sum_s c_{js}U\varphi_s = \epsilon_j \sum_s c_{js}\varphi_s \ . \tag{3}$$

---

[1] J. Bardeen, L. N. Cooper, and J. R. Schrieffer, Phys. Rev. **106**, 162 (1957); **108**, 1175 (1957).

[2] This is just first-order degenerate perturbation theory in quantum mechanics.

We now multiply both sides of (3) by any φ_m^* and integrate over the volume. Using the matrix element notation $\langle m|U|s \rangle$ for the integral involving U, we have

$$\sum_s c_{js}\langle m|U|s\rangle = \epsilon_j c_{jm} \ , \tag{4}$$

where the term on the right follows from the orthogonality and normalization of the φ's.

There are R equations of the form of (4), one for each of the R possible choices of the φ_m^* by which we multiply (3). For each j we have an identical set of R independent simultaneous equations in the R unknowns c_{jm}. These equations have a nontrivial solution only if the determinant of the coefficients of the c_{jm} vanishes:

$$\begin{vmatrix} \langle 1|U|1\rangle - \epsilon \ldots \langle 1|U|R\rangle \\ \vdots \qquad\qquad \vdots \\ \langle R|U|1\rangle \ldots \langle R|U|R\rangle - \epsilon \end{vmatrix} = 0 \ . \tag{5}$$

The problem is to find the roots ϵ of this determinant; this may be a complicated numerical problem.

There is, however, one particularly simple case for which the roots can be found by informed inspection. Suppose every matrix element is equal to unity. Then (5) reduces to

$$\begin{vmatrix} 1 - \epsilon & 1 & \ldots & 1 \\ 1 & 1 - \epsilon & \ldots & 1 \\ \vdots & & & \\ 1 & 1 & \ldots & 1 - \epsilon \end{vmatrix} = 0 \ . \tag{6}$$

We can find the eigenvalues of (6) by a simple device.

A mathematical theorem states that the sum of all the eigenvalues of a hermitian matrix is equal to the sum of all the diagonal elements $\langle j|U|j\rangle$. The sum of the diagonal elements of (6) is R, so that

$$\sum_{j=1}^{R} \epsilon_j = R \ . \tag{7}$$

Another theorem states that the sum of the squares of all the eigenvalues is equal to the sum of the squares of all the elements of the matrix. Thus

$$\sum_{j=1}^{R} \epsilon_j^2 = R^2 \ . \tag{8}$$

We now assert that the determinant (6) has one root, say ϵ_1, equal to R

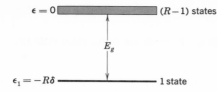

$\epsilon = 0$ ⬛⬛⬛⬛⬛⬛ $(R-1)$ states

E_g

$\epsilon_1 = -R\delta$ ————————— 1 state

Figure 1 Energy eigenvalue spectrum for R states connected by a perturbation U with equal matrix elements $\langle j|U|s\rangle = -\delta$, for all states j and s. Notice that a single state is split off with an energy gap $E_g = R\delta$.

and the other $R - 1$ roots are all equal to zero. This solution satisfies both theorems (7) and (8). We verify that there is actually one root equal to R. Form the symmetrical combination of the basis vectors φ_s:

$$\psi_1 = R^{-\frac{1}{2}} \sum_s \varphi_s \ , \tag{9}$$

for which the energy is

$$\epsilon_1 = \langle \psi_1|U|\psi_1\rangle = \frac{1}{R}\sum_{js}\langle j|U|s\rangle = \frac{1}{R}R^2 = R \ . \tag{10}$$

But this root exhausts the sum in (8), so that all other roots must be zero. Q.E.D.

If the potential U is an attractive and highly localized interaction, the matrix elements in (4) and (5) will be negative and nearly equal. Suppose that

$$\langle j|U|s\rangle = -\delta \tag{11}$$

for all states j, s. Here δ is a positive constant. Then the appropriate modification of (6), (8), and (10) gives

$$\epsilon_1 = -R\delta \ . \tag{12}$$

As shown in Fig. 1 this single-state is split off below the rest of the spectrum; there exists an energy gap $R\delta$ between the ground state and the first excited state. We see the remarkable stabilization of a single state; even though the interaction δ may be weak, $R\delta$ may be large if the degeneracy is high.

Electron Pairs and the Superconducting State

In the problem just treated the states φ_s were states of a one-particle system. Suppose that we have a system of N free electrons, initially without mutual interaction. The various states Φ of the N-particle system can be specified by giving the occupancy (between zero and one, because of the Pauli principle) of the one-electron orbitals. We may label a one-electron state as $\mathbf{k}\uparrow$; this means that the wavevector is $\mathbf{k}$ and the spin is up. It is convenient to label a state of the total N-particle system in terms of one-particle orbitals by giving the labels of the *occupied* one-particle orbitals. In the absence of interactions among the electrons each one-particle orbital is either occupied or vacant. We might write

$$\Phi_s \equiv \mathbf{k}_1\uparrow; \ \mathbf{k}_2\downarrow; \ \mathbf{k}_3\downarrow; \ \ldots; \ \mathbf{k}_N\uparrow \ , \tag{13}$$

where the subscripts on the $\mathbf{k}$'s denote specific values of the wavevector according to some arbitrary code or assignment.

We now let the electrons interact with each other by some interaction[3] between all pairs of electrons:

$$\mathcal{U} = \sum_{nm} U(\mathbf{r}_m - \mathbf{r}_n) \ . \tag{14}$$

Each term of the sum acts to scatter two electrons (say those in the p and r positions in the wavefunction Φ_s) into two vacant one-electron orbitals, not represented as occupied in Φ_s. That is, one scattering event takes the system from the N-particle state Φ_s to some other N-particle state Φ_u.

Can we possibly make the drastic assumption as we did before[4] that matrix elements of the interaction $\mathcal{U}$ are equal between all states Φ_s, Φ_u? No, this is not possible in general. In the first place, we cannot scatter at all from $\Phi_s \equiv \mathbf{k}_1\uparrow; \mathbf{k}_2\downarrow; \mathbf{k}_3\downarrow \ldots$ to a state with different total spin, such as $\Phi_v \equiv \mathbf{k}_a\downarrow; \mathbf{k}_b\downarrow; \mathbf{k}_3\downarrow \ldots$, because the interaction (14) does not contain spin operators and therefore cannot change the total spin of the system. That is, particles in $\mathbf{k}_1\uparrow$ and $\mathbf{k}_2\downarrow$ cannot be scattered to $\mathbf{k}_a\downarrow$ and $\mathbf{k}_b\downarrow$; therefore the matrix element of $\mathcal{U}$ between Φ_s and Φ_v will be zero. In the second place, matrix elements can appear with either $+$ or $-$ signs: if the particles in $\Phi_s \equiv \mathbf{k}_1\uparrow; \mathbf{k}_2\downarrow; \mathbf{k}_3\downarrow \ldots$ are scattered to $\Phi_w \equiv \mathbf{k}_c\uparrow; \mathbf{k}_d\downarrow; \mathbf{k}_3\downarrow \ldots$ with a positive matrix element, they can also be scattered to $\Phi_x \equiv \mathbf{k}_d\downarrow; \mathbf{k}_c\uparrow; \mathbf{k}_3\downarrow \ldots$. But Φ_x differs from Φ_w only in the interchange of the order of $\mathbf{k}_c\uparrow$ and $\mathbf{k}_d\downarrow$. According to the Pauli exclusion principle the interchange of the order of two states or two electrons in a wavefunction changes the sign of the wavefunction. Thus

$$\langle w|\mathcal{U}|s\rangle = -\langle x|\mathcal{U}|s\rangle; \tag{15}$$

this means that all matrix elements cannot have the same sign.

There is an ingenious way in which we can arrange for all matrix elements to have the same sign. We can find a many-body problem for which all matrix elements can be allowed to be equal. Let us agree to consider *only* those many-body states which are occupied by pairs of electrons. We can agree on a definition of a pair: the usual definition is $\mathbf{k}\uparrow; -\mathbf{k}\downarrow$. That is, if $\mathbf{k}\uparrow$ is occupied, we insist that $-\mathbf{k}\downarrow$ be occupied.[5] The many-particle states we consider are of the form

$$\Phi_A \equiv \mathbf{k}_1\uparrow; -\mathbf{k}_1\downarrow; \mathbf{k}_2\uparrow; -\mathbf{k}_2\downarrow; \ldots \tag{16}$$

These states form only a subspace of the problem, but by restricting ourselves to the subspace we get a problem we can solve. It has been shown that the

---

[3] In superconductors the important contributions to U are from the Coulomb repulsion and the indirect coupling of two electrons by the lattice distortions associated with the electron-phonon interaction. The total interaction can be attractive among electrons near the Fermi surface, particularly among those within a Debye energy $\pm\hbar\omega_D$ of the Fermi surface. It is these electrons which cooperate in (16) to form the superconducting ground state.

[4] See Eq. (11), where we were dealing with a one particle problem.

[5] We could form pairs with parallel spin, as $\mathbf{k}\uparrow; -\mathbf{k}\uparrow$, but their energy is higher because of exchange energy effects.

restriction introduces errors only of order $1/N$, where N is the number of electrons.

Within the subspace of pair states of the form of (16) it is not unreasonable to take all matrix elements of $\mathcal{U}$ as equal. Then by direct analogy with the earlier solution (12) we obtain a spectrum which has a single ground state separated by an energy gap E_g from the excited states. Our discussion has neglected the kinetic energy of the unperturbed electrons[6] so that the original states are degenerate, but BCS show that inclusion of the kinetic energy does not destroy the energy gap.

---

[6] This approximation is known as the strong-coupling approximation.

Advanced Topic M

TOPICS IN THE QUANTUM THEORY OF MAGNETISM

QUANTUM THEORY OF DIAMAGNETISM OF MONONUCLEAR SYSTEMS

From (I. 18) we know that the effect of a magnetic field is to add to the hamiltonian the terms

$$\mathcal{H}' = \frac{ie\hbar}{2mc}(\boldsymbol{\nabla} \cdot \mathbf{A} + \mathbf{A} \cdot \boldsymbol{\nabla}) + \frac{e^2}{2mc^2}A^2 \; ; \tag{1}$$

for an atomic electron these terms may usually be treated as a small perturbation. If the magnetic field is uniform and in the z direction, we may write

$$A_x = -\tfrac{1}{2}yB \; , \qquad A_y = \tfrac{1}{2}xB \; , \qquad A_z = 0 \; , \tag{2}$$

and (1) becomes

$$\mathcal{H}' = \frac{ie\hbar B}{2mc}\left(x\frac{\partial}{\partial y} - y\frac{\partial}{\partial x}\right) + \frac{e^2B^2}{8mc^2}(x^2 + y^2) \; . \tag{3}$$

The first term on the right is proportional to the orbital angular momentum component L_z if $\mathbf{r}$ is measured from the nucleus. In mononuclear systems this term gives rise only to paramagnetism. The second term gives for a spherically symmetric system a contribution

$$E' = \frac{e^2B^2}{12mc^2}\langle r^2 \rangle \; , \tag{4}$$

by first-order perturbation energy. The associated magnetic moment is diamagnetic:

$$\mu = -\frac{\partial E'}{\partial B} = -\frac{e^2\langle r^2 \rangle}{6mc^2}B \; , \tag{5}$$

in agreement with the classical result. For further details see the book by Van Vleck.

VAN VLECK TEMPERATURE-INDEPENDENT PARAMAGNETISM

We consider an atomic or molecular system which has no magnetic moment in the ground state, by which we mean that the diagonal matrix element of the magnetic moment operator μ_z is zero.

Suppose that there is a nondiagonal matrix element $\langle s|\mu_z|0\rangle$ of the magnetic moment operator, connecting the ground state 0 with the excited state s of energy $\Delta = E_s - E_0$ above the ground state. Then by standard perturbation theory the wavefunction of the ground state in a weak field ($\mu_z B \ll \Delta$) becomes

$$\psi'_0 = \psi_0 + (B/\Delta)\langle s|\mu_z|0\rangle\psi_s \ , \tag{6}$$

and the wavefunction of the excited state becomes

$$\psi'_s = \psi_s - (B/\Delta)\langle 0|\mu_z|s\rangle\psi_0 \ . \tag{7}$$

The perturbed ground state now has a moment

$$\langle 0'|\mu_z|0'\rangle \cong 2B|\langle s|\mu_z|0\rangle|^2/\Delta \ , \tag{8}$$

and the upper state has a moment

$$\langle s'|\mu_z|s'\rangle \cong -2B|\langle s|\mu_z|0\rangle|^2/\Delta \ . \tag{9}$$

There are two interesting cases to consider:

Case (a). $\Delta \ll k_B T$. The surplus population in the ground state over the excited state is approximately equal to $N\Delta/2k_B T$, so that the resultant magnetization is

$$M = \frac{2B|\langle s|\mu_z|0\rangle|^2}{\Delta} \cdot \frac{N\Delta}{2k_B T} \ , \tag{10}$$

which gives for the susceptibility

$$\chi = N|\langle s|\mu_z|0\rangle|^2/k_B T \ . \tag{11}$$

Here N is the number of molecules per unit volume. This contribution is of the usual Curie form, although the mechanism of magnetization here is by polarization of the states of the system, whereas with free spins the mechanism of magnetization is the redistribution of ions among the spin states. We note that the splitting Δ does not enter in (11).

Case (b). $\Delta \gg k_B T$. Here the population is nearly all in the ground state, so that

$$M = \frac{2NB|\langle s|\mu_z|0\rangle|^2}{\Delta} \ . \tag{12}$$

The susceptibility is

$$\chi = \frac{2N|\langle s|\mu_z|0\rangle|^2}{\Delta} \ , \tag{13}$$

independent of temperature. This type of contribution is known as Van Vleck paramagnetism.

QUENCHING OF THE ORBITAL ANGULAR
MOMENTUM BY CRYSTALLINE ELECTRIC FIELDS

We consider a single electron with orbital quantum number $L = 1$ moving about a nucleus, the whole being placed in an inhomogeneous crystalline electric field. We omit electron spin.

In a crystal of orthorhombic symmetry (Chapter 1) the charges on neighboring ions will produce an electrostatic potential V about the nucleus of the form

$$eV = Ax^2 + By^2 - (A + B)z^2 , \qquad (14)$$

where A and B are constants. This expression is the lowest degree polynomial in x, y, z which is a solution of the Laplace equation $\nabla^2 V = 0$ and compatible with the symmetry of the crystal.

In free space the ground state is three-fold degenerate, with magnetic quantum numbers $m_L = 1, 0, -1$. In a magnetic field these levels are split by energies proportional to the field B, and it is this field-proportional splitting which is responsible for the normal paramagnetic susceptibility of the ion. In the crystal the picture may be different. We take as the three wavefunctions associated with the unperturbed ground state of the ion

$$U_x = xf(r) \; ; \qquad U_y = yf(r) \; ; \qquad U_z = zf(r) \; . \qquad (15)$$

These wavefunctions are orthogonal, and we assume that they are normalized. Each of the U's can be shown to have the property

$$\mathcal{L}^2 U_i = L(L + 1) U_i = 2 U_i , \qquad (16)$$

where $\mathcal{L}^2$ is the operator for the square of the orbital angular momentum, in units of $\hbar$. The result (16) confirms that the selected wavefunctions are in fact p functions, having $L = 1$.

We observe now that the U's are diagonal with respect to the perturbation, as by symmetry the nondiagonal elements vanish:

$$\langle U_x | eV | U_y \rangle = \langle U_x | eV | U_z \rangle = \langle U_y | eV | U_z \rangle = 0 . \qquad (17)$$

Consider for example

$$\langle U_x | eV | U_y \rangle = \int xy |f(r)|^2 \{ Ax^2 + By^2 - (A + B)z^2 \} \, dx \, dy \, dz \; ; \quad (18)$$

the integrand is an odd function of x (and also of y) and therefore the integral must be zero. The energy levels are then given by the diagonal matrix elements:

$$\langle U_x | eV | U_x \rangle = \int |f(r)|^2 \{ Ax^4 + By^2x^2 - (A + B)z^2x^2 \} \, dx \, dy \, dz$$
$$= A(I_1 - I_2) , \qquad (19)$$

where

$$I_1 = \int |f(r)|^2 x^4 \, dx \, dy \, dz \; ; \qquad I_2 = \int |f(r)|^2 x^2 y^2 \, dx \, dy \, dz . \quad (20)$$

In addition,

$$\langle U_y | eV | U_y \rangle = B(I_1 - I_2) \;\; ; \qquad \langle U_z | eV | U_z \rangle = -(A + B)(I_1 - I_2) \;\; . \quad (21)$$

The three eigenstates in the crystal field are p functions with their angular lobes directed along each of the x, y, z axes, respectively.

The orbital moment of each of the levels is zero, because

$$\langle U_x | L_z | U_x \rangle = \langle U_y | L_z | U_y \rangle = \langle U_z | L_z | U_z \rangle = 0 \;\; .$$

This effect is known as **quenching**. The level still has a definite total angular momentum, since $\mathcal{L}^2$ is diagonal and gives $L = 1$, but the spatial components of the angular momentum are not constants of the motion and their time average is zero in the first approximation. Therefore the components of the orbital magnetic moment also vanish in the same approximation. The role of the crystal field in the quenching process is to split the originally degenerate levels into nonmagnetic levels separated by energies $\gg \mu H$, so that the magnetic field is a small perturbation in comparison with the crystal field.

At a lattice site of cubic symmetry there is no term in the potential of the form (14), that is, quadratic in the electron coordinates. Now the ground state of an ion with one p electron (or with one hole in a p shell) will be triply degenerate. However, the energy of the ion will be lowered if the ion displaces itself with respect to the surroundings, thereby creating a noncubic potential such as (14). Such a spontaneous displacement is known as a **Jahn-Teller effect** and is often large and important, particularly with the Mn^{3+} and Cu^{2+} ions[1] and with holes in alkali and silver halides.

---

[1] See L. Orgel, *Introduction to transition metal chemistry*, Wiley, 1960; extensive references are given by M. D. Sturge, Phys. Rev. **140**, A880 (1965).

Author Index

Subject Index